HANDBOOK OF ADVANCED MATERIALS

HANDBOOK OF ADVANCED MATERIALS

ENABLING NEW DESIGNS

Editor-in-chief

James K. Wessel

Wessel & Associates
Oak Ridge, Tennessee

A JOHN WILEY & SONS, INC., PUBLICATION

Published by John Wiley & Sons, Inc., Hoboken, New Jersey.
Published simultaneously in Canada.

Library of Congress Cataloging-in-Publication Data

Wessel, James K.
Handbook of advanced materials: enabling new designs / James K. Wessel.
p. cm.
"Wiley-Interscience publication."
ISBN 0-471-45475-3 (cloth)
1. Materials—Handbooks, manuals, etc. I. Title.
TA403.4.W48 2004
620.1′1—dc22

2004004219

Printed in the United States of America.

10 9 8 7 6 5 4 3 2 1

CONTRIBUTORS

D. C. Agarwal Krupp-VDM Technologies Corp., 11210 Steeplecrest Drive, Suite 120, Houston, Texas 77065-4439

Robert Akid Materials Research Institute, Sheffield Hallam University, Howard Street, Sheffield, United Kingdom S1 1WB

Chris Deemer Energy Technology Division, Argonne National Laboratory, 9700 South Cass Avenue, Bldg. 212ET, Argonne, Illinois 60439

William A. Ellingson Energy Technology Division, Argonne National Laboratory, 9700 South Cass Avenue, Bldg. 212ET, Argonne, Illinois 60439

F. H. (Sam) Froes Institute for Materials and Advanced Processes (IMAP), University of Idaho, Moscow, Idaho 83844

G. N. Haidemenopoulos Department of Mechanical and Industrial Engineering, University of Thessaly, Volos, Greece

Michael G. Jenkins Department of Mechanical Engineering, University of Detroit Mercy, 4001 W. McNichols Rd., Detroit, Michigan 48219

J. Randolph Kissell The TGB Partnership, 1325 Farmview Road, Hillsborough, North Carolina 27278

John U. Knickerbocker IBM, Inc., 208 Creamery Road, Hopewell Junction, New York 12533

Sarah H. Knickerbocker IBM, Inc., 208 Creamery Road, Hopewell Junction, New York 12533

Syros G. Pantelakis Department of Mechanical Engineering and Aeronautics, University of Patras, Patras, Greece

Theodore P. Philippidis University of Patras, Department of Mechanical Engineering and Aeronautics, Patras, Greece

Ivar E. Reimanis Metallurgical and Materials Engineering Department, Colorado School of Mines, Golden, Colorado 80401

David W. Richerson Department of Materials Science and Engineering, University of Utah, Salt Lake City, Utah 84112

Chris A. Rudopoulos Materials and Engineering Research Institute, Sheffield Hallam University, City Campus, Sheffield, United Kingdom S1 1WB

John Shaffer Oak Ridge National Laboratory, P.O. Box 2009, Oak Ridge, Tennessee 37831

Vinod Sikka Oak Ridge National Laboratory, 1 Bethel Valley Road, Oak Ridge, Tennessee 37831

Anastasios P. Vassilopoulos University of Patras, Department of Mechanical Engineering and Aeronautics, Patras, Greece

James K. Wessel Wessel & Associates, 127 Westview Lane, Oak Ridge, Tennessee 37830

Eric Whitney Pennsylvania State University, State College, Pennsylvania 16804

CONTENTS

PREFACE

The use of improved materials enables engineers to design new and better products and processes. Benefits include increased sales of improved products and, where new materials are used in manufacturing, reduced plant cost. Society benefits through the use of improved products that use these new materials.

Sophisticated new materials save lives (artificial hearts, shatterproof glass, bulletproof vests), conserve energy (lightweight cars) and expand human horizons (aircraft, spacecraft, computers through the World Wide Web). In the twenty-first century a new generation of materials promises to again reshape our world and solve some of the planet's most pressing problems. Although there is a tremendous array of materials, this book focuses on so-called advanced materials, especially those offering the latest advancements in properties. They are materials of construction with exceptional properties enabling improvement in the engineering components or final products in which they are used. They are also the latest in revolutionary materials and the latest improvement in more traditional advanced materials.

As a designer of "hardware," you may be tempted to assume that the best material for your use is the one you have been using. If so, you will find that this book includes many common materials of construction that have seen recent improvements. For the more adventuresome, we include revolutionary materials whose use may result in great benefit, enabling unique and cost-effective product design.

This handbook presents the most recently introduced advanced materials in an effort to inform you as soon as possible of materials that may improve your product or process. Each chapter describes material characteristics from which materials can be tentatively selected for further exploration. Additional information is available from the references, engineering societies, and trade associations. Examples include The Composite Fabricators Association, The United States Advanced Ceramic Association, ASM International, The American Society of Mechanical Engineers, The Aluminum Association, The American Iron & Steel Institute, The Steel Manufacturers Association, International Titanium Association, and others. All are available through their websites.

This book's purpose is not to provide all the data you need to select materials. Each chapter describes an individual class of materials. Most include corrosion-resistant data plus a separate chapter on this important property. The book's purpose is to narrow your material selection. For your final decision, work with the material supplier as a partner, sharing your problem's parameters. Material suppliers have broad experience that will benefit your material selection. Treat them as a joint problem solver rather than a vendor. Be open to a design change that will realize the benefits of using a new material. Always test materials before use.

Some of the materials presented have revolutionary performance compared to the existing materials that you are using. Others are improvements over existing materials, but, unlike revolutionary materials, they are more familiar, with abundant engineering data, and some similarity to your existing material. Revolutionary materials, like continuous fiber ceramic composites (CFCCs), offer a breakthrough in performance in extreme environments like superior resistance to high temperature, corrosion, and wear. Others, including CFCCs, are also stronger and lighter weight.

Some of the materials presented are high priced, reflecting their high performance. They are used where the result economically benefits the provider and the user. Life-cycle costing will reveal if this is true for your application.

Designing a product involves selecting a material, shape, and manufacturing process. Finding an optimal combination of these to maximize performance and minimize cost is essential for innovation in engineering design and education.

Psychologists tell us that 5% of designers are willing to try something new and 80% will follow if the 5% are successful. Be one of the 5%. The use of new materials can save money, reduce downtime, reduce maintenance, increase operating temperature, increase efficiency, lower emissions, and reduce life-cycle costs.

James K. Wessel
Oak Ridge, TN

CHAPTER 1

Polymer Composites

JOHN SHAFFER*

Oak Ridge National Laboratory, P.O. Box 2009, Oak Ridge, Tennessee 37831

THEODORE P. PHILIPPIDIS and ANASTASIOS P. VASSILOPOULOS†

University of Patras, Department of Mechanical Engineering and Aeronautics, Patras, Greece

*Sections 1.1–1.7
†Sections 1.8–1.13

Handbook of Advanced Materials Edited by James K. Wessel
ISBN 0-471-45475-3

1.A POLYMER COMPOSITES

1.1 DESCRIPTION

1.1.1 Scope

Polymer composites can cover a broad range of material combinations. For this chapter, we will consider those combinations that are between the stages of those still being invented and those in wide use. We will also restrict our consideration to those combinations that are intended for structural application. Many, if not most, of the basic concepts and principles of use will be applicable across the total range of materials developed. The specific characteristics of the materials discussed or used as examples will be of those that are advanced in the sense that their full use potential has not yet been realized. For that reason, a great deal of attention will be given to those material combinations that incorporate continuous carbon or graphite fibers as a reinforcing material in a high-performance polymer matrix. Unlike many metals, polymer composite formulas are often proprietary to their suppliers. Contact the supplier to determine the best polymer composite for your application. Suppliers can be identified by contacting the Composite Fabricators Association at www.cfa-hq.org. They are located at 1010 North Glebe Road, Suite 450, Arlington, VA 22201, telephone 703-525-0511.

1.1.2 History and Future Developments

Modern polymer composites can trace their origins back to the 1950s when researchers at Wright-Patterson Air Force Base in Ohio began to investigate the properties of plastics that had within them embedded glass fibers. The motivation for these investigations was the search for materials that would meet the ever-increasing demands for higher performance aircraft. Lighter, stronger, and stiffer were the guiding principles. In conjunction with companies such as Owens-Corning Fiberglas and Union Carbide, a high-performance composite of continuous S-Glass and epoxy was developed. This composite found applications in such places as the Poseidon missile casing and ballistic armor. It is still an important material today.

In the 1960s, fibers composed of oriented carbon or graphite began to be developed. The fibers were of low density and higher stiffness than glass fiber. As the demands of agencies such as the Air Force and National Aeronautics and Space Administration (NASA) grew for higher stiffness materials than metal or glass fiber composites, these carbon/graphite fibers and their composites became the materials of choice. Today, many consider *advanced composites* to be those reinforced with carbon or graphite fiber. In actuality, glass-fiber-reinforced composites continue to find new, advanced uses. The design, manufacturing, testing, and performance measuring methods for polymer composites containing any fiber were developed during the time when glass-reinforced composites were finding expanded usage.

The history of glass and carbon-fiber-reinforced composite development is documented by several authors. It is not the intent here to review that history beyond the simple introduction given above. It needs to be pointed out, however, that the composites developed as a result of the search for stiffer, lighter, stronger has had some fortunate side effects in other areas. The new materials also gave the designers more choices of materials for their electrical, thermal, and corrosion needs. These nonstructural properties will be further explored later in the chapter.

The future of polymer composite development is mixed. The decade of the 1990s has seen a slowdown in the drive for improvements led by aerospace. Companies that competed with each other in the need to produce ever more advanced products have seen the market drastically change. Performance used to be the differentiating factor. In today's world, performance with affordability or value is the key. The industry is looking for new customers in application areas that were not even imagined when advanced polymer composites were developed. Golf clubs, tennis rackets, hockey sticks, softball bats, pole vault poles, canoes, fishing poles, and the like are but the tip of the iceberg for new applications. Automobile, truck cab and trailer, railroad car, and ship applications are under active development. The success of these applications will depend upon designers embracing these materials in their work.

As inventors and applications engineers begin to be comfortable with the type and nature of these advanced materials, application areas will expand and costs will come down. It is hoped that this chapter will give to the designer the basic knowledge and understanding of how these material work, how they are made, and, most importantly, how they can open design imagination.

1.1.3 Definition

Stating a simple definition of a composite is a deceptively complex task. It gets even more difficult if the definition is intended to convey the multitude of options available. Here are a few examples:

1. Made up of distinct parts or elements
2. A macroscopic combination of two or more distinct materials, having a recognizable interface between them
3. Two or more materials judiciously combined, usually with the intent of achieving better results than can be obtained by using individual materials by themselves
4. High-strength fiber—primarily continuous, oriented carbon, aramid, or glass rather than randomly distributed chopped fibers or whiskers—in a binding matrix that enhances stiffness, chemical and hydroscopic resistance, and processability properties

Each of these definitions is equally correct. They express an increasing degree of complexity to the product being defined. They also imply the ability (or difficulty) to define a material simultaneously with its application. *Engineered materials*, as they are often called, now require the designer to consider materials other than those available to him in the "handbook." The material he will use is now his to define, as he needs. This material will be made from parts and elements put together in a manner chosen to best fulfill the need. The possibilities are immense; the solutions only limited by imagination.

1.2 CONSTITUENT MATERIALS AND PROPERTIES

The materials that make up the parts of a composite are usually referred to as the constituents. For a polymer composite, the two basic parts are the polymer matrix, or resin system, and the fiber reinforcement. In the next section, the options available for each of these two parts will be presented along with some specialized intermediate forms of product that form the starting point in the design of a structure made from a polymer composite.

1.2.1 Fibers

Polymer composites have developed into important structural materials due to the wide variety of reinforcing fibers that are available. Glass and carbon fibers are by far the most common types and are produced by a number of manufacturers worldwide. Other fiber materials such as aramid, quartz, boron, ceramic, or polyethylene are also available and provide unique properties. For applications in advanced polymer composites, the most common form of the fiber is continuous tow (carbon) or roving (glass). In this form, continuous filaments have been gathered as untwisted bundles and packaged in spool form. Typically, these packages weigh between 2 and 20 lb and are supplied on 11 by 3-in. cores. This product is the basic element for further processing (either directly or via intermediate forms) into a polymer composite structure.

Carbon fibers were first commercially produced from a regenerated cellulose fiber (rayon). Because of high production costs and environmental concerns, rayon-based carbon fiber is not widely used today. The majority of carbon fiber available today is made from an acrylic precursor fiber (polyacrylonitrile, or PAN) and is the most commonly used structural fiber. Fibers made from petroleum or coal tar pitch are also available and, because of their high modulus and unique thermal properties, find uses in thermal management applications. PAN-based carbon fibers are available from a number of sources. Tables 1.1, 1.2, and 1.3 present typical properties of carbon fiber products. The tables are grouped by tensile modulus grade; low or standard (33–35 Msi), intermediate (40–50 Msi), and high (>50 Msi).

Today, new fiber developments are producing material with heavier tow count and lower costs. These materials are usually of the low modulus type and will find applications in high-volume applications such as automotive, construction, and infrastructure.

TABLE 1.1 Low Modulus (<275 GPA) Carbon Fibers

Supplier	Trade Name	Designation	Tensile Modulus (GPa)	Tensile Strength (GPa)	Elongation (%)	Density (g/cm^3)
Toray	Torayca	T300	230	3.53	1.5	1.76
		T300J	230	4.21	1.8	1.78
		T400H	250	4.41	1.8	1.80
		T700S	230	4.90	2.1	1.80
BP Amoco	Thornel	T300	231	3.75	1.4	1.76
		T300C	231	3.75	1.4	1.76
		T650/35	255	4.28	1.7	1.77
Hexcel		AS4	228	4.07	1.8	1.79
		AS4C	231	4.15	1.8	1.78
		AS4D	241	4.28	1.8	1.79
SGL Carbon	Sigrafil C	C10	180–240	2.00	1.0	1.75
		C25	215–240	2.50	1.05–1.40	1.78
		C30	220–240	3.00	1.25–1.60	1.78
Grafil		34–700	234	4.48	1.9	1.80
		34–600	200	4.00	1.7	1.79
Zoltek	Panex	33 (45K)	228	3.79	1.5	1.80
Toho Rayon	Besfight	G30–400	235	3.80	1.6	1.76
		G30–500	235	3.92	1.7	1.76
		G30–700	240	4.81	2.0	1.76
Fortafil		F3(C)50K	227	3.80	1.7	1.80
Nippon	Granoc	XN-20	200	2.73		
		HT	230	4.80		

TABLE 1.2 Intermediate Modulus Carbon Fibers

Supplier	Trade Name	Designation	Tensile Modulus (GPa)	Tensile Strength (GPa)	Elongation (%)	Density (g/cm^3)
Toray	Torayca	T800H	294	5.49	1.9	1.81
		T100G	294	6.27	2.2	1.80
		M35J	343	4.70	1.4	1.75
		M30	294	3.92	1.3	1.80
Hexcel		IM7	276	5.45	2.0	1.78
		IM8	303	5.73	1.9	1.79
		IM9	276	6.00	2.2	1.79
Grafil	Pyrofil	MS40	345	4.83	1.3	1.77
		MR50	296	5.52	1.9	1.80
Toho Rayon	Besfight	G40-600	295	4.51	1.5	1.74
		G40-800	285	5.79	2.0	1.80
		G50-500	345	2.94	0.9	1.79

TABLE 1.3 High Modulus Carbon Fibers

Supplier	Trade Name	Designation	Tensile Modulus (GPa)	Tensile Strength (GPa)	Elongation (%)	Density (g/cm^3)
Toray	Torayca	M40J	377	4.41	1.2	1.77
		M50J (6K)	475	4.12	0.8	1.88
		M60J (6K)	588	3.92	0.7	1.94
BP Amoco	Thornel	P55S (4K)	379	1.90	0.5	2.00
		P75S (2K)	517	2.10	0.4	2.00
Hexcel		UHM	440	3.73	.08	1.87
Grafil	Pyrofil	HS40	455	4.41	1.0	1.85
		HR40	393	4.83	1.2	1.82
Toho Rayon	Besfight	G55-700	380	4.90	1.2	1.79
		G80-600	540	3.82	0.7	1.92
		G100-300	650	3.33	0.5	1.97
Nippon	Granoc	HM	377	4.40		
		XN60	600	3.50		
		YS95A	920	3.53		

Other types of fibers are used in polymer composites and impart special properties. Table 1.4 lists many of these along with typical properties and uses. See Chapter 3 for a more thorough description of these fibers.

1.2.2 Resins

Polymer composites get their name from the type of matrix or binder used to hold the fibers together to make a solid material of designed properties. The most important function of the polymer matrix is to allow the fibers to share the loads. This requires that the matrix be more flexible than the fiber and be attached in

TABLE 1.4 Miscellaneous Fibers

Fiber Type	Manufacturer	Trade Name	Tensile Modulus (GPa)	Tensile Strength (GPa)	Density (g/cm^3)	Uses
PBO	Toyobo	Zylon AS	180	5.8	1.54	Ballistic protection, sailcloth,
		Zylon HM	270	5.8	1.56	High-temperature filters
Boron	Textron		400	3.6	2.57	Bicycle frames, skis, aircraft repairs
Quartz	Quartz Products	Quartzel	78	3.6	2.2	Radomes, heat shields, high-temperature applications
Ceramic	Nippon Carbon	Nicalon	193	2.9	2.55	High-temperature applications
Aramid	DuPont	Kevlar	55–143	2.3–3.4	1.44–1.47	Armor, ballistic protection
Polyethylene	Allied-Signal	Spectra	86–103	2.1–2.4	0.97	Chemical resistance, impact properties

some manner to the fiber. While the method used to manufacture the composite (to be discussed later) can have a large influence on the effectiveness of the loading transfer, reinforcing fibers are usually sold with a *sizing*, or coating, on them specifically designed to promote chemical bonding between the matrix and the fiber surface.

The matrix also serves as a coating or protector for the fibers and must therefore be chosen not only for its ability to work with the fiber as the load transfer medium but also for its environmental performance. Polymer matrices can be divided into two general classifications: thermoset and thermoplastic. As their names imply, heat is used during processing. A thermoset material is generally processed as a liquid and crosslinked, or cured, through the application of heat to form a nonreversible chemical structure. In contrast, a thermoplastic is melted, formed and then cooled in a reversible process wherein the materials are not crosslinked. There are even materials, such as the polyimides, that exhibit characteristics of both types.

The field of polymer chemistry is very broad. Many excellent reference books exist that detail the molecular structure, processing, and performance of polymers. In this section, only property information on the most common types of polymers used as composite matrices will be presented.

Thermoset matrix materials include epoxies, polyesters, bismaleimides, polyimides, cyanate esters, and phenolics. Epoxies are by far the most common matrix material for advanced polymer composites. Table 1.5 lists major types of matrix materials available, their physical properties, and service limits.

The curing of a thermoset material usually requires the use of a hardener or catalyst in order to promote the crosslinking process. Three types of materials are common for crosslinking epoxies: amines, anhydrides, and Lewis acids. Each

TABLE 1.5 Matrices for Advanced Polymer Composites

Resin Family	Typical Cure Temperature (°F)	Maximum Service Temperature (°F)	Typical Tensile Properties		
			Strength (ksi)	Modulus (ksi)	Elongation (%)
Epoxy	350	350	8–13	375–500	3–7
	350	300			
Phenolic	300	300	1.0–1.6	75–150	
Bismaleimide	375	450	11.9	620	
Cyanate	180	350	12.7	470	

type of curing agent will modify the physical properties of the polymer and can change the processing methods. Matrix suppliers will assist in the choice of materials and processes for a given application.

Thermoplastic matrix materials differ from thermoset in that they are not crosslinked materials that require hardeners and heat. Thermoplastic materials are solids that are formed to shape by heat and pressure. When combined with a fibrous reinforcement, the composite is pressed or molded into the final desired shape. The differences in manufacturing methods can sometimes result in savings of time and equipment cost. Property differences in the final product exist and are usually the determining factor in the selection of the resin type. Many types of thermoplastic matrix materials exist. Conventional materials such as polyesters, polystyrene, nylon, and the like are not usually thought of as advanced thermoplastic matrices even though they are widely used in automotive, medical, and other commercial applications.

1.2.3 Prepregs

Composites are manufactured by combining fibers and resin in a mold or on a form that defines the final product shape. This can be done in one step by a wet lay-up method or through the use of an intermediate product known as a *prepreg.* A prepreg is a product form in which the reinforcing fibers are preimpregnated with the polymer matrix resin and partially cured to form a sheet or tapelike material. Many fiber and resin suppliers also supply prepregs. Other companies are just prepreg suppliers. The prepreg allows precise control over the relative proportions of resin and fiber in the composite and allows fiber orientation to be controlled. The development of this intermediate product form has had a large impact on the expanding use of polymer composite structures. While there are generic types of prepregs available, almost any fiber–resin combination is possible. The reinforcements can be contained in the prepreg as parallel fibers (unidirectional), woven fabrics of textile types, nonwoven cloth, or braids. The choice of a particular product form is closely related to the manufacturing process to be used and to the complexity of the final product.

1.3 DESIGN OPTIONS

The preferential incorporation of reinforcing fibers into a polymer matrix opens up the design options available. At one end of the spectrum there are unidirectional fiber-reinforced materials that maximize the use of the available strength and stiffness of the fiber and produce a product that is highly directional in its properties. At the other end are random or multidirectional fiber materials whose properties approach isotropy.

1.3.1 Final Products

Today, advanced polymer composites find their greatest use in the aerospace sector where they were initially developed. Stealth aircraft such as the F-177 and the B-2 are only possible because of the unique properties of advanced polymer composites such as high strength and light weight. From helicopter blades to rocket motor casings to ballistic armor, these materials have fueled a revolution in new product applications. Initially, many projects attempted to replace a metal part with composite parts by direct substitution. This did not often work well. The unique properties of composites could not be incorporated in a part substitution and the resultant product frequently was more expensive than the original. Fortunately, as time passed, designers became more familiar with composite design methodologies and designed new products with composites in mind from the concept stage. The following section outlines the design approach.

1.3.2 Introduction to Methodology

The design of structures with advanced polymer composites proceeds through the application of classical lamination theory. Individual laminae, or plies, are stacked with the fibers oriented in various directions to build a laminate with the desired properties. Designers are used to working with materials such as plastics and metals that are described as homogeneous and isotropic. That is, the materials properties are not dependent upon the position or orientation in the material. For these classes of materials in a plane stress state, the relationship between stress and strain is described through the elastic constants Young's modulus E, and Poisson's ratio, ν.

However, a laminated composite material cannot usually be accurately described this simply. Homogeneous orthotropic, homogeneous anisotropic, heterogeneous orthotropic, and heterogeneous anisotropic are additional descriptions that may be required to accurately analyze a laminated material. Fortunately, this complexity is not often required and, with the advent of modern software, is even manageable on desktop computers. For the balance of this section, we will make the assumption that a composite exhibits homogenous orthotropic behavior. We will also consider a special ply configuration that approaches isotropic behavior—quasi-isotropic.

1.3.3 Laminae

As indicated previously, a homogenous, isotropic material requires two independent constants to describe its stress–strain behavior. A homogeneous, orthotropic composite material has three perpendicular planes of material property. If the axes are chosen to coincide with the reinforcing filament direction, then this set is called the principal lamina direction. Dimensionally, these laminae are physically thin compared to their length and width. Although the thickness stresses are small and as applied in a laminated structure, a state of plane stress or plane strain is assumed. This leads to the need for four independent elastic constants in order to describe the stress–strain response: E_{11} and E_{22}, Young's modulus; G_{12}, shear modulus; and ν_{12}, major Poisson's ratio. This description of a lamina, or ply, is most common in the design of a laminated composite structure. Testing of ply materials is most oriented toward establishing these constants.

1.3.4 Laminates

When multiple layers of lamina are combined and act structurally as a single layer, a laminated composite is created. To analyze a laminated composite structure, the designer must know the properties of each layer and how the reinforcing fibers are oriented with respect to one another, that is the stacking sequence. For example, a laminate consisting of 16 individual layers may have the fibers oriented in the following fashion:

Two layers with fibers at 0°
Two layers with fibers at 90°
One layer with fibers at +45°
Three layers with fibers at −45°
Three layers with fibers at −45°
One layer with fibers at +45°
Two layers with fibers at 90°
Two layers with fibers at 0°

This description is quite lengthy and shorthand methods have been developed to present the information:

$[0_2/90_2/45/-45_3/-45_3/45/90_2/0_2]_T$ or
$[0_2/90_2/45/-45_6/45/90_2/0_2]_T$ or
$[0_2/90_2/45/-45_3]_S$

Each of these methods describes the laminate. In the first method, each of the orientations is given along with the number of layers indicated by the subscript. The [ ]'s and the subscript T indicate that this is a description of the total laminate. The second method simply combines the two −45° layer groups into one. The third description, however, recognizes an important property of this particular lay-up sequence. It is symmetrical about the centerline of the laminate. Only one-half of the stacking sequence is explicitly listed, and the subscript T is replace by S to indicate the symmetry. While lamination theory can accurately analyze any stacking sequence, the condition of midplane symmetry is an important one for the designer of polymer composite structures. Nonsymmetrical lamina lay-up can result in out-of-plane bending and twisting under mechanical or thermal stress that must be considered.

In addition to midplane symmetry, there is one other design concept that is usually followed by designers. That is, the stacking sequence is usually "balanced." This means that there are an equal number of plies at angles of $+\theta$ and $-\theta$. Construction that follows this convention will avoid the shear coupling that is present in a single orthotropic lamina.

Lamination theory describes the stress–strain response of stacked orthotropic lamina. This behavior can be used to analyze the strength of the laminate if the assumption is made that the basic strength criteria for the lamina remain valid in the laminate. Under this assumption, a strength analysis proceeds by determining the individual ply stresses and/or strains in the laminate and comparing them to the allowable for the ply. Failure is often deemed to have occurred when one of the plies exceeds an allowable stress–strain limit. This *first ply failure* does not necessarily lead to complete failure of the laminate, as the failed ply may transfer some or all of the load it carried to another ply in the laminate and not exceed an allowable at that location. Procedures are available to analyze *ply-by-ply failure* sequences but are usually used as part of a failure analysis process rather than a design study.

A final word on composite laminate and ply failure. Since a fiber-reinforced lamina is modeled most frequently as an orthotropic material, the use of a failure criteria such as the maximum principal strain criteria used with isotropic materials is not applicable. A maximum strain criteria for an orthotropic material requires that the strains developed under load be referred to the lamina principal axes and evaluated against the tensile and compressive allowable for the lamina. This leads to the need for five failure strains; tensile and compressive limits in the fiber direction, tensile and compressive limits in the transverse to the fiber direction, and an in-plane shear limit. Other failure criteria, such as the Tsai–Wu criterion, are developed as yield surfaces that depend upon the interaction between the lamina principal direction and shear yield strengths. Commercial computer software for analyzing laminated composite structures is available. These packages can be customized to allow input of new materials, modified failure limits, and failure analysis methods.

1.4 TESTING/ANALYSIS

1.4.1 Mechanical Properties

The mechanical properties of polymer matrix composite materials depend upon the type of fiber and resin used, the relative percentages of each, the laminate lay-up, and the method of manufacture. The properties presented in this chapter are focused upon the fiber and resin materials that make up the composite. These are the product forms most often purchased by a user who combines them into a laminate. In this section, the methods used to determine the constituent and laminate properties commonly used in selecting materials will be reviewed.

The fiber properties presented in the previous tables are typical of what a potential buyer will encounter. The tensile strength, tensile modulus, and elongation are usually determined by the impregnated strand test. Over the past few years, industry standard test methods have been developed for determining these properties. The properties are, thus, reasonably comparable between manufactures in a general sense. The test is also useful for quality control purposes.

Fiber density is an important property. Often it is specific strength or modulus that controls the applicability, especially in weight and stiffness critical areas. There are also industry standards that can be used to measure this property.

The important properties of the polymer matrix resins used in advanced composites are both chemical and mechanical. For uncured thermoset resins, the important properties are related to the processing method to be employed. Viscosity, gel time, cure temperature, and the like all must be considered in order to properly process and cure the composite. The test methods used are common in the polymer manufacturing business and can be found in many references.

One of the most important properties of a cured thermoset resin is the glass transition temperature (T_g). This parameter is both a measure of the completeness of the cure and an indication of the maximum service temperature of the composite. Differential scanning calorimetry (DSC) and dynamic mechanical analysis (DMA) are two common techniques. DSC measures the amount of heat given off (or absorbed) in a resin sample as the temperature is increased. When the T_g no longer changes the resin is completely cured. With the DMA technique the response of the resin to mechanical stress is monitored with respect to temperature. The temperature at which a significant change to the elastic moduli is observed is the T_g. Since these methods measure two different parameters, they can give two different estimates of T_g. Care should be taken when reviewing supplier data as the method used is not always indicated.

As prepregs are an intermediate product from that combines fiber and resin in a specific ratio and partially processes the resin, it is important to know that the ratio and the "b-staged" resin are properly prepared. The fiber–resin ratio is measured by the aerial weight, or in grams per square meter, of fiber. Since this property is chosen by the application requirements, it is not a handbook type of quantity. A typical value for this parameter will place the fiber fraction at ~60% by volume.

In the cured laminate, the calculation of the relative amounts of fiber and resin is an important measure of the quality and proper processing history of the material. ASTM methods are available for determining these ratios and for determining the void content. Void content can have a detrimental effect on the properties of the composite and is usually limited to 1 or 2% of the material's volume. The technique involves burning (in the case of glass fiber) or chemically digesting (in the case of carbon fiber) the resin matrix. The relative weights (W) of fiber and resin, when combined with the densities (D) of the composite (C), fiber (F) and resin (R) will yield the void content:

$$V_v = 100[1 - D_C/W_C(W_R/D_R + W_F/D_F)]$$

The tensile properties of an advanced polymer composite material are usually measured with a flat coupon. ASTM D3039 is one test method standard that can be used. The test is applicable to unidirectional and oriented laminates. It differs in purpose from the impregnated strand test previously discussed. The structural fiber–resin ratios are more closely represented in the coupon test, and the results are more applicable to the actual planned use. The influences of the matrix and fiber-to-matrix interface are more evident. Testing at elevated temperatures and after exposure to other environmental conditions often use this specimen. ASTM methods also are available to govern the procedures used.

Compressive properties of polymer–matrix composites are difficult to measure. The ASTM provides a recommended method but many users develop their own. Again, the purpose of compressive testing is often to evaluate the performance of a fiber–resin combination to various service environments. Numerous tests for shear properties have been developed. A shear test is often used to measure the effectiveness the fiber–resin interface. The ASTM, again, provides methods to follow. A simple test such as ASTM 2344, apparent interlaminar shear strength, is often used for quality control and comparative purposes. ASTM D3518 is a procedure for measuring shear strength and modulus design data.

1.5 NONDESTRUCTIVE TESTS (QUALITY ASSURANCE)

Nondestructive tests of polymer matrix composites have received a great deal of attention. The difficulty and expense of performing destructive tests on actual structures have spurred the search for testing techniques that verify performance (quality assurance) without destroying the product. While no standard nondestructive tests for product quality exists, the use of ultrasonic techniques have become quite sophisticated. The ability to detect delaminations, inclusions, and voids on complicated geometries has made the test routine easier in many programs. Similarly, the use of infrared thermography to detect flaws or damage has developed recently.

1.6 ENVIRONMENTAL PERFORMANCE

1.6.1 Temperature

The service or operating temperature of a polymer matrix composite is probably the most important parameter considered in choosing the chemical nature of the matrix. In Table 1.5, the glass transition temperature, T_g, is an indication of the maximum service environment. The operating temperature is kept below the T_g. Polymer matrix composites are limited to 260°–316°C (500–600°F) applications.

Above these temperatures, metal or ceramic matrices are required. Testing for temperature effects is usually done by performing several of the mechanical tests previously described at elevated temperature. In general, tests that stress the matrix, such as shear and compression, will show the greatest effect. Temperature effects are generally reversible provided that the temperature exposure has not been high enough to cause physical damage to the matrix.

1.6.2 Moisture Exposure

Moisture tends to "plasticize" or soften the matrix. As with temperature effects, the composite properties are measured after exposure to water for varying times and at varying temperatures. Moisture effects, like elevated temperature effects, are generally reversible.

1.7 FABRICATION

1.7.1 Methods and Processes

1.7.1.1 Overview

The manufacture of a composite structure requires that the constituent fiber and resin be combined in a specified ratio, with the fibers in a chosen orientation and heated to cure or form the final product. The details of how this process is accomplished will ultimately determine the properties of the composite structure. Many of the techniques used have evolved from processing knowledge for plastic molding. Indeed, in the automotive sector, the composites manufacturing methods used most frequently are termed *liquid molding* and are similar to the resin transfer molding process used in the aerospace sector. The principal difference is the speed requirements for the product. And therein lies the challenge for modern advanced composites. The tolerable cost of manufacturing is dependent upon the end use. Low-volume application areas, such as aircraft or space, typically utilize the more expensive methods, and high-volume areas, such as automotive or infrastructure, require that costs be low. The processing methods that will be outlined in this section will follow manufacturing evolution from manual, labor-intensive methods to highly automated and rapid methods.

1.7.1.2 Processes

Manual Lay-up The simplest technique used to make a composite structure is the manual lay-up method. Fibers are laid on a form and liquid resin is added and distributed throughout the fibers by hand rolling. After the desired thickness is attained, the product is allowed to cure, either at room temperature or in an oven. This method is time consuming and produces composites of low quality. Much effort has been undertaken in the industry to improve the manual lay-up method. The development of prepreg materials was a significant advancement. Better control of the fiber–resin ratio and simpler lay-ups, combined with autoclave curing, produced better parts. Figure 3.15 shows the Filament winding technique used for composites.

Automated Tape Laying New machines have been developed that aid in the lay down of prepreg. These tape-laying machines are programmed to follow the contours of the mold, laying down prepreg tape in prescribed orientations and applying heat and pressure automatically. The head can follow reasonably gentle contours and, with some models, can automatically add or drop tape layer. The lay-up usually still requires vacuum bagging and autoclave curing.

Filament Winding The filament winding process can be a very cost-effective method for producing a composite part. As its name implies, the method consists of wrapping fibers around a mandrel in layers until the desired thickness is reached. A winding machine allows the fiber orientation to be varied thereby allowing the composite part to develop the design property profile. Matrix curing is most often done in an oven, although autoclave curing is occasionally used.

Resin Transfer Molding In resin transfer molding (RTM), a mold is filled with reinforcement and injected with resin. Cure takes place in the mold and the composite takes the shape of the mold. There are variations on this basic technique depending upon how and when the fiber and reinforcement are combined and cured. Reaction injection molding (RIM), structural reaction injection molding (SRIM), vacuum-assisted resin transfer molding (VARTM), and resin film infusion (RFI) are types that have been developed, usually first for a specific part need.

Pultrusion Pultrusion is the process where bundles of resin-impregnated fibers are cured by pulling them through a heated die. The addition of glass or carbon fiber to the pulling process yields a product that maximizes strength and stiffness in the pulling direction. When combined with part rotation and overwrapping techniques, pultrusion can produce a wide variety of structural composite shapes.

1.7.1.3 Tools

Advanced composites are formed on tools. The preceding process illustrations contain tooling adapted for the composite forming method used. Pressure and cure/forming temperatures are primary drivers for the design and materials chosen. Production quantity is also an important factor in tooling selection. Composites, themselves, are often used as tooling materials. As the cost of raw materials comes down, manufacturing costs, tooling, and speed became the barriers to the introduction of an advanced composite part into a high-volume application.

Machining The machining of polymer composites differs from both the machining of metals and plastics and requires consideration of techniques used in both.

TABLE 1.6 Adhesive Bonding vs. Mechanical Fastening

Property/Performance	Adhesive Bonding	Mechanical Fastening
Stress concentration/delamination	×	
Peel strength		×
Bearing Strength	×	
Ease of construction		×
Environmental performance		×
Disassembly		×
Cost	×	

Composites are usually made near net shape. They usually require trimming, sanding, painting, drilling, grinding, and the like. Composites are weak in the directions transverse to the fibers and are subject to delaminating. Generally, the same types of tools that are used for metalworking can be used. Tooling companies sell special tools designed for composites with specific kinds of reinforcement. Carbon tends to be brittle and Kevlar tough. Tools tipped with carbide or impregnated with diamond flakes are common. Cooling may be necessary to prevent overheating and damaging the matrix material.

Assembly/Joining Adhesive bonding is the most common method used for joining polymer composites. The adhesives used can be one-part or two-part adhesives and cure at room temperature or elevated temperature. The materials are similar to those used for matrix materials and chosen with many of the same considerations in mind. Surface preparation is extremely important to the quality of the bond as is the choice cure cycle. Mechanical fastening uses methods similar to metal joining, that is, rivets, bolts, pins, and the like. Care must be used as a hole will reduce the strength of the composite and increase the potential for delamination. Often, reinforcing pads, doublers, must be used. Fastener materials, especially in carbon composites, can cause galvanic corrosion. Hence, nickel, nonmetal, and titanium are commonly used. Table 1.6 lists some of the property and performance considerations in the choice of assembly method.

1.B FATIGUE OF GLASS-FIBER-REINFORCED PLASTICS UNDER COMPLEX STRESS STATES

1.8 INTRODUCTION

Design allowables of general applicability for fatigue-critical composite structures cannot be easily established. Different material systems, that is, type of reinforcement and matrix, lamination sequence, load cases definition, and geometry of structural component usually result in case-specific situations treated more or less as such. The reason is that aforementioned parameters affect differently a multitude of failure mechanisms, for example, fiber breaks, matrix cracking, debonding, delaminations and the like, that are propagating in a different way and rate. Therefore, what has been observed in the past during the development of a structural composite application is an initial phase with intensive experimental efforts to produce large databases on fatigue strength of specific material systems and a subsequent assessment period in which design allowables, fit to purpose, are extracted. Safety levels are set by design standards and are mainly based on empirical partial safety factor approaches.

Fatigue behavior of carbon-fiber-reinforced epoxies (CFRP) has been extensively investigated the last 25 years due to the concentrated effort in developing composite structural components for aeronautical applications. Most aspects of fatigue-related engineering problems, that is, life prediction, property degradation, joints design and the like, were confronted leading to the adoption of design allowables and large amount of published data, for example [1–5]. Yet, damage tolerance issues have not been treated efficiently [6] due to many reasons, the main one being the lack of definition of a generalized damage metric, for example, such as the crack length in metals, that could be of use with different lay-ups and material configurations [7]. In addition, the effect of variable amplitude loading on remaining life and fatigue under complex stress states have only received limited attention.

Structural response to cyclic loads of glass-fiber-reinforced plastics (GFRP) extensively used in a number of mechanical engineering applications such as leisure boats, transportation cars, and the like, has not been investigated at any significant extent until 15 years ago. Due to the amazing growth of wind energy industry, especially in Europe, much effort was spent the last decade in establishing fatigue design allowables of GRP (glass-reinforced polyester), in particular, laminated composites for wind turbine rotor blades. Lots of experimental data were produced characterizing fatigue strength of matrix systems such as polyester, epoxies, and vinylester reinforced by continuous glass fibers in the form of woven or stitched fabrics and unidirectional roving [8–17]. The effect of both constant and variable amplitude, that is, spectral, loading conditions was investigated.

However, limited experimental data and design guidelines are available of the complex stress state effect, produced either by multiaxial or off-axis loading, on fatigue behavior of GFRP laminates. Existing studies [18–22] point out the strong dependency of fatigue response on load direction, as a result of material anisotropy and indicate the need to continue research on this topic including effects of spectral and nonproportional loading.

Experimental results are presented herein from a comprehensive program consisting of static and fatigue tests on straight edge coupons cut at various on- and off-axis directions from a GRP multidirectional (MD) laminate of $[0/(\pm 45)_2/0]_T$ lay-up. Fatigue behavior of off-axis loaded laminates, that is, complex state of stress in material principal directions, is investigated in depth for several off-axis orientations. This includes derivation of signal–noise (S–N) curves at various R ratios ($R = \sigma_{min}/\sigma_{max}$), statistical evaluation of fatigue strength results and determination of design allowables at specific reliability levels. Constant life diagrams are extracted for the various off-axis directions and are compared with existing data from similar material systems.

Several investigators have been concerned in the past with the multiaxiality of fatigue stresses. Hashin and Rotem [23] first, proposed a fatigue strength criterion for fiber-reinforced plastic (FRP) materials, based on the observed failure modes. For unidirectional materials two distinct failure modes exist, fiber and matrix dominated, respectively, whereas for laminated composites a third mode was introduced to cope with delaminations [24]. To use the criterion, experimental determination of three S–N curves is assumed, that is, axial loading in the fiber direction, transversely to it and shear loading in the principal material directions. Application of the criterion is limited to materials for which failure modes can be separated, that is, it cannot be used for woven or stitched fabrics. Fawaz and Ellyin [25] proposed a multiaxial fatigue strength criterion that needs less experimental data as input, that is, only one S–N curve and the static strength properties. Other authors have also attempted to modify existing static failure criteria to cope with cyclic loads [18, 19, 26, 27].

A quadratic failure polynomial criterion, introduced in [20] to predict fatigue strength under complex stress states, is shown to forecast satisfactorily material response under off-axis and multiaxial loading for all the cases of stress ratio R considered in this study.

Besides strength prediction and fatigue behavior under off-axis loading, stiffness reduction measurements were performed as well. By continuously monitoring force-displacement loops, longitudinal Young's modulus is derived as a function of the number of cycles. Its variation, depending on the applied stress ratio and off-axis load orientation, is modeled by a simple empirical equation [28], which is shown to fit satisfactorily the experimental data. It is observed in general [21, 22] that the higher the cyclic stress range, the lower the stiffness reduction with increasing number of cycles, and this is particularly true for alternating load, $R = -1$ Furthermore, a systematic statistical analysis for all stress ratios, R, and off-axis orientations proved that irrespective of stress amplitude level, modulus degradation data are fitted satisfactorily by standard statistical distributions.

Stiffness degradation measurements for various R values were used to define fatigue design curves corresponding to specific modulus degradation and not to failure. In that case, test points in the S–N plane denote that under cyclic stress, σ, a predetermined stiffness reduction is reached after N cycles. The corresponding, stiffness-controlled, fatigue design curves, denoted as Sc–N, can serve better the requirements of design and full-scale testing of structural components made of FRP materials. For example, in wind turbine rotor blade testing [29], functional failure is said to correspond to irreversible stiffness reduction of up to 10% and therefore, Sc–N based fatigue design of the blade must be used instead to comply with eventual certification requirements.

For the GRP material database presented herein, Sc–N curves were determined and compared to fatigue strength allowables [22]. It was shown that these two families of curves can be correlated, and, therefore, it was possible to derive design allowables corresponding to predetermined levels of stiffness degradation and survival probability. Most interesting is the fact that by considering only half of the data used, that is, virtual 50% test cost reduction, Sc–N curves could be still accurately defined, pointing out the way for a potential testing time reduction for other composite material systems as well.

1.9 THEORETICAL CONSIDERATIONS

1.9.1 Fatigue Strength

In predicting fatigue life of structural components made of, for example, composites, there are at least two alternative methodologies that could be used depending on the *damage tolerant* or *safe-life* design concept adopted for the specific structural part. In the former case, it is assumed to consider a damage metric such as crack length, delaminated area, residual stiffness, or residual strength, and by means of a criterion correlate this metric to fatigue life. In safe-life design situations, cyclic stress or strain amplitude are directly associated to operational life through S–N or ε–N curves. Under complex stress states, multiaxial limit state functions are introduced that are usually generalizations of static failure theories to take into account factors relevant to the fatigue life of the structure, that is, number of cycles, stress ratio, and loading frequency. Due to the fact that damage-tolerant fatigue design of composite structures is still in its infancy, and much more research is needed to establish reliable methodologies of general applicability, most of the industrial applications with this type of materials are safe-life parts.

One of the first attempts for generalizing a multiaxial static failure theory to account for fatigue, was made by Hashin and Rotem [23]. They presented a fatigue strength criterion based on the different damage modes developing upon failure. For unidirectional materials there are two such modes, mode I, or fiber failure mode, and mode II, or else matrix failure mode. The discrimination between these two modes is based on the off-axis angle of the reinforcement with respect to the loading direction. The critical angle, as shown in [23], is given by:

$$\tan\theta_c = \frac{\tau^s}{\sigma_A^s}\frac{f_\tau(R, N, \nu)}{f'(R, N, \nu)}, \tag{1.1}$$

where τ^s and σ_A^s stand for the static shear and longitudinal (axial) strength, respectively, while functions $f_\tau(R, N, \nu)$, $f'(R, N, \nu)$ are the fatigue functions of the material along the same directions. The S–N curves of the material are given as the product of the static strength along any direction and the corresponding fatigue function. In the above equation $R = \sigma_{\min}/\sigma_{\max}$, N is the number of cycles and ν the loading frequency.

If the reinforcement forms an angle less than θ_c, with respect to the loading direction, then mode I is the prevailing mode of failure, else mode II is the one that leads to fatigue failure. Thus, the failure criterion has two forms:

$$\sigma_A = \sigma_A^u$$
$$\left(\frac{\sigma_T}{\sigma_T^u}\right)^2 + \left(\frac{\tau}{\tau^u}\right)^2 = 1, \tag{1.2}$$

where superscript u denotes fatigue failure stress, or else the S–N curve of the material in the corresponding direction, and subscript T stands for transverse to the fiber direction. It can be shown that any off-axis fatigue function (failure mode II) can be given as a function of f_τ, f_T, τ^s, σ_T^s and the angle θ [23]:

$$f''(R, N, n) = f_\tau \sqrt{\frac{1 + (\tau^s/\sigma_T^s)\tan^2\theta}{1 + [(\tau^s/\sigma_T^s)(f_\tau/f_T)]^2 \tan^2\theta}} \tag{1.3}$$

Equation (1.3) can be used for the calculation of any off-axis fatigue function but also to calculate fatigue functions f_τ, f_T from two different off-axis, experimentally obtained, fatigue functions. For the application of this criterion three S–N curves need to be defined experimentally, along with the static strengths of the material.

For multidirectional laminates [24], the situation is far more complicated. As each lamina is under a different stress field, failure may occur at a ply after a certain amount of load cycling while the other plies could be still intact. These differences, along with inherent inhomogeneity, produce interlaminar stresses, capable to cause successive failure, probably with different damage mechanisms. In order to take into account these stresses, another failure mode, interlaminar, is established and the set of equations (1.2) is supplemented by:

$$\left(\frac{\sigma_d^c}{\sigma_d^u}\right)^2 + \left(\frac{\tau_d^c}{\tau_d^u}\right)^2 = 1, \tag{1.4}$$

where superscript c denotes cyclic stress and subscript d delamination failure mode, respectively.

The Hashin and Rotem [23] failure criterion can predict fatigue behavior of a unidirectional (UD) or multidirectional (MD) laminate subjected to uniaxial or multiaxial cyclic loads provided that the discrimination between the failure modes exhibited during fatigue failure is possible.

Fawaz and Ellyin [25] proposed a fatigue strength criterion suitable for UD and MD materials under multiaxial cyclic loading. The criterion has attractive features as it needs only one experimentally obtained S–N curve and some static strengths. The multiaxiality is entered through any acceptable static failure criterion, and the predicted S–N curve is given by:

$$S(\alpha_1, \alpha_2, \theta, R, N) = h(\alpha_1, \alpha_2, \theta)[g(R)m_r \log(N) + b_r], \tag{1.5}$$

as a function of a reference S–N curve, known by experiment, given by:

$$S_r = m_r \log(N) + b_r. \tag{1.6}$$

In the above two equations, subscript r denotes reference direction and α_1 is the first biaxial ratio, $\alpha_1 = \sigma_y/\sigma_x$, while α_2 is the second biaxial ratio, $\alpha_2 = \tau_{xy}/\sigma_x$; x and y refer to a global coordinate system rotated at an angle θ from the principal material system and R is the cyclic stress ratio defined, as usual, by $R = \sigma_{\min}/\sigma_{\max}$. Functions h and g are dimensionless and are defined by:

$$h(\alpha_1, \alpha_2, \theta) = \frac{\sigma_x(\alpha_1, \alpha_2, \theta)}{X_r}, \tag{1.7}$$

$$g(R) = \frac{\sigma_{\max}(1 - R)}{\sigma_{\max_r} - \sigma_{\min_r}}, \tag{1.8}$$

where $\sigma_x(\alpha_1, \alpha_2, \theta)$ is the static strength in the x direction and X_r is the static strength in the reference direction.

As can be seen from Eq. (1.8), function g is introduced to account for different stress ratios R. When stress ratio for the reference S–N curve (R_r) is the same as the stress ratio of the S–N curve under prediction, then $g = 1$, while for $R = 1$, $g = 0$.

Recently, Jen and Lee [26, 27] modified Tsai–Hill failure criterion to cope with cyclic loading, and as shown in [26], fatigue life prediction of AS4 carbon/PEEK APC-2 laminates at various stress ratios was quite successful. The failure functions read:

$$\left(\frac{\sigma_1}{\overline{\sigma}_1}\right)^2 + \left(\frac{\sigma_2}{\overline{\sigma}_2}\right)^2 - \frac{\sigma_1\sigma_2}{\overline{\sigma}_1^2} + \left(\frac{\sigma_6}{\overline{\sigma}_6}\right)^2 - 1 = 0, \tag{1.9}$$

where $\sigma_i = \sigma_i(N, R, \nu)$, $i = 1, 2, 6$ are the applied cyclic stresses and $\overline{\sigma}_i = \overline{\sigma}_i(N, R, \nu)$ denote the respective fatigue strengths in principal material coordinates, being functions of cycle number N, stress ratio R, and frequency ν.

Depending on the loading, tension–tension (T–T), compression–compression (C–C), or T–C, $\overline{\sigma}_i$ in Eq. (9) are derived from experiments under similar loading conditions.

A modification of failure tensor polynomial [30] to account for fatigue loading, henceforth denoted by FTPF, was used in [20] to predict fatigue strength under multiaxial stress. The failure tensor polynomial for orthotropic media expressed in material principal coordinate system, under plane stress, is given by:

$$F_{11}\sigma_1^2 + F_{22}\sigma_2^2 + 2F_{12}\sigma_1\sigma_2 + F_1\sigma_1 + F_2\sigma_2 + F_{66}\sigma_6^2 - 1 \leq 0, \tag{1.10}$$

with the components of the failure tensors given by:

$$\begin{aligned} F_{11} &= \frac{1}{XX'}, \qquad F_{22} = \frac{1}{YY'}, \qquad F_{66} = \frac{1}{S^2}, \\ F_1 &= \frac{1}{X} - \frac{1}{X'}, \qquad F_2 = \frac{1}{Y} - \frac{1}{Y'}, \end{aligned} \tag{1.11}$$

where X, X' stand for tension and compression strengths along direction 1 of the material symmetry coordinate system, Y, Y' are the corresponding values for the transverse direction, while S is the shear strength. Failure tensor polynomial criterion with the form of Eq. (1.10) is valid for orthotropic materials or materials of higher symmetry. The choice of the off-diagonal term of failure matrix F_{ij}, F_{12} was shown to lead to completely different failure theories [31]. Nevertheless, for simplicity, the form of F_{12} used in this study is given by [32]:

$$F_{12} = -\tfrac{1}{2}\sqrt{F_{11}F_{22}}. \tag{1.12}$$

Failure tensor polynomial in fatigue (FTPF) assumes the same functional form as Eq. (1.10):

$$F_{ij}\sigma_i\sigma_j + F_i\sigma_i - 1 \leq 0, \qquad i, j = 1, 2, 6. \tag{1.13}$$

However, the components of failure tensors F_{ij}, F_i are functions of the number of cycles N, the stress ratio R, and the frequency ν, of the loading:

$$F_{ij} = F_{ij}(N, R, \nu), \qquad F_i = F_i(N, R, \nu). \tag{1.14}$$

Experimental evidence gained so far for any type of continuous fiber-reinforced polymers, at least, strongly suggests the form of functional dependence of failure tensor components shown in relation (1.14). This implies an increased complexity of experimental strength characterization with respect to static loading since it is not sufficient anymore to discriminate just between tension or compression, but also between the same type of loading, for example, tension, at different R values or loading frequency ν. Therefore, fatigue strength of an orthotropic material, in-plane stressed, is characterized by three S–N curves [20]:

$$\begin{aligned} X(N, R, \nu) &= A_X + B_X \log N, \\ Y(N, R, \nu) &= A_Y + B_Y \log N, \\ S(N, R, \nu) &= A_S + B_S \log N. \end{aligned} \tag{1.15}$$

Failure tensor components are defined by:

$$\begin{aligned} &F_{11} = \frac{1}{X^2(N, R, \nu)}, \qquad F_{22} = \frac{1}{Y^2(N, R, \nu)}, \qquad F_{66} = \frac{1}{S^2(N, R, \nu)}, \\ &F_{12} = -\frac{1}{2X(N, R, \nu)Y(N, R, \nu)}, \qquad F_1 = F_2 = F_6 = 0. \end{aligned} \tag{1.16}$$

Strength for certain values of R and ν is predicted using the equation:

$$\frac{\sigma_1^2}{X^2(N)} + \frac{\sigma_2^2}{Y^2(N)} - \frac{\sigma_1\sigma_2}{X(N)Y(N)} + \frac{\sigma_6^2}{S^2(N)} - 1 = 0, \tag{1.17}$$

provided the three S–N curves $X(N)$, $Y(N)$, and $S(N)$ are derived for the same loading conditions, R, ν. In case the complex stress field is produced by multiaxial loading of nonproportional characteristics, resulting in different R_i ratios and ν_i values for each stress component, σ_i, then, the corresponding $X(N)$, $Y(N)$, and $S(N)$ strengths, to be used in Eq. (1.17), must be known experimentally for the same R_i, ν_i conditions.

The experimental characterization of $X(N)$, $Y(N)$ is performed through uniaxial testing of straight-edge flat coupons cut along the respective principal material direction. For $S(N)$, it was proposed in [20] to use the value of half the fatigue strength of a flat coupon cut off-axis at 45° and loaded uniaxially. This choice yielded satisfactory results for alternating loads, $R = -1$, but its performance was less effective for other loading types. Therefore, it is proposed to determine $S(N)$ by fitting Eq. (1.17) to the experimental S–N data derived from uniaxial testing at a suitable off-axis orientation.

1.9.2 Stiffness Reduction During Fatigue Life

Monitoring and evaluation of stiffness changes during operation can give useful information on the integrity of a composite structure. Prediction of gradual decrease of elastic moduli due to the cyclic loading is essential, inter alia, for design purposes. Many investigators, even from the early years of composite material applications, were considering stiffness as a suitable damage metric and used stiffness degradation to account for damage accumulation in the material under consideration, for example, [3, 33–38]. However, to the author's knowledge, no fatigue theory based on stiffness degradation has gained wide acceptance among scientists nor has inspired confidence to designers. One could mention several reasons, but the truth is that prediction of stiffness degradation during fatigue life is not a simple matter. There is a number of parameters that influence the variation of stiffness, such as material system, loading conditions, cyclic stress level, stress multiaxiality, and the like. The knowledge of interaction rules of all these factors is essential for the formulation of a viable theory for prediction of stiffness degradation. Therefore, for design purposes, simple as possible models predicting stiffness reduction should be used, having parameters that can be reliably defined through standard fatigue tests under specific loading conditions.

For GFRP composites and especially those material systems and stacking sequences used in the fabrication of wind turbine rotor blades, stiffness degradation measurements were intensively carried out the last decade and reported in the literature [8–12, 21, 22, 28]. The resulting trend from all the experimental studies on stacking sequences of practical interest, that is, usually combinations of (0°) and (±45°) fabrics, is that after the very few first cycles there is an abrupt stiffness reduction, followed by a long period of slow linear degradation and finally a steepest stiffness variation is observed prior to final failure.

Based on the experimental evidence, an empirical model for the description of stiffness changes and the prediction of stiffness-controlled design curves was introduced in [28] and further validated for different material systems and loading conditions in [21, 22]. A brief outline of the model, which is further exploited in the present study for predicting stiffness degradation and derive corresponding design allowables, is given below.

The degree of damage in a polymer matrix composite coupon can be evaluated by measuring stiffness degradation, E_N/E_1, where E_1 denotes the Young's modulus of the material measured at the first cycle, different in general from the static value E_0, and E_N the corresponding one at the Nth cycle. It is assumed that stiffness degradation can be expressed by [28]:

$$\frac{E_N}{E_1} = 1 - K\left(\frac{\sigma_a}{E_0}\right)^c N. \tag{1.18}$$

Material constants, K and c, in Eq. (1.18) are determined by curve fitting the respective experimental data for E_N/E_1, depending on the number of stress cycles, N, and the level of applied cyclic stress amplitude, σ_a. Recasting Eq. (1.18) in the following form:

$$\frac{1 - (E_N/E_1)}{N} = K\left[\frac{\sigma_a}{E_0}\right]^c, \tag{1.19}$$

allows easy determination of model constants. Notice that these model characteristics depend strongly on applied stress ratio R and stress multiaxiality. Relation (1.18) also establishes a stiffness-based design criterion since for a predetermined level of E_N/E_1, for example, p, one can solve for σ_a to obtain an alternative form of design curve, henceforth denoted by Sc–N, corresponding not to material failure but to a specific stiffness degradation percentage (1 - p)%. The Sc–N curves for any specific stiffness degradation level, E_N/E_1, can be easily calculated by means of the following equation:

$$\sigma_a = E_0\left[\frac{1 - (E_N/E_1)}{KN}\right]^{1/c}. \tag{1.20}$$

1.9.3 Statistical Evaluation of Fatigue Strength Data

Strength data from each set of on- and off-axis fatigue tests were subjected to statistical analysis to determine characteristic values. The methodology used [39, 40] is briefly discussed below. The form of the S–N equation is assumed to be given by:

$$\sigma_a = \sigma_0 N^{-1/k}, \tag{1.21}$$

where N is the number of cycles to failure, σ_a denotes the stress amplitude level, and k, σ_0 are material constants.

Irrespective of stress level, the probability of survival after N cycles is assumed to be given by a two-parameter Weibull distribution:

$$P_S(N) = \exp\left[-\left(\frac{N}{\overline{N}}\right)^{\alpha_f}\right]. \tag{1.22}$$

Calculation of constants σ_0 and k of Eq. (1.21) is performed as follows:

A two-parameter Weibull distribution is fitted to the data of the ith stress level:

$$P_S(N_i) = \exp\left[-\left(\frac{N_i}{\overline{N}_i}\right)^{\alpha_{fi}}\right], \qquad i = 1, \ldots, m. \tag{1.23}$$

The parameters α_{fi} and $\overline{N_i}$ of each Weibull distribution are determined by the following maximum-likelihood estimators:

$$\frac{\sum_{j=1}^{r_i} N_{ij}^{\hat{\alpha}_{fi}} \ln N_{ij} + (n_i - r_i) N_{S_i}^{\hat{\alpha}_{fi}} \ln N_{S_i}}{\sum_{j=1}^{r_i} N_{ij}^{\hat{\alpha}_{fi}} + (n_i - r_i) N_{S_i}^{\hat{\alpha}_{fi}}} - \frac{1}{r_i} \sum_{j=1}^{r_i} \ln N_{ij} - \frac{1}{\hat{\alpha}_{fi}} = 0, \tag{1.24}$$

and

$$\widehat{\overline{N}}_i = \left\{ \frac{1}{r_i} \left[\sum_{j=1}^{r_i} N_{ij}^{\hat{\alpha}_{fi}} + (n_i - r_i) N_{S_i}^{\hat{\alpha}_{fi}} \right] \right\}^{1/\hat{\alpha}_{fi}}. \tag{1.25}$$

In the above equations, n_i is the total number of coupons tested under the ith stress level, σ_{a_i}, r_i is the number of failed coupons under that stress level, and N_{S_i} is the number of cycles after which the test was stopped.

The values of the number of cycles to failure for every coupon at each stress level are subsequently normalized by the corresponding estimated characteristic number, $\hat{\overline{N}}_i$. Thus, the following normalized data set is formed:

$$X(X_{i1}, X_{i2}, \ldots, X_{in_i}), \qquad i = 1, 2, \ldots, m \tag{1.26}$$

where

$$X_{ij} = \frac{N_{ij}}{\hat{\overline{N}}_i}. \tag{1.27}$$

It is assumed that this set of data also follows a two-parameter Weibull distribution:

$$P_S(X) = \exp\left[-\left(\frac{X}{X_0}\right)^{\alpha_f}\right]. \tag{1.28}$$

The parameters of the distribution of Eq. (1.28) are estimated by:

$$\frac{\sum_{i=1}^{m} \sum_{j=1}^{r_1} X_{ij}^{\hat{\alpha}_f} \ln X_{ij} + \sum_{i=1}^{m} (n_i - r_i) Y_i^{\hat{\alpha}_f} \ln Y_i}{\sum_{i=1}^{m} \sum_{j=1}^{r_i} X_{ij}^{\hat{\alpha}_f} + \sum_{i=1}^{m} (n_i - r_i) Y_i^{\hat{\alpha}_f}} - \frac{1}{r_T} \sum_{i=1}^{m} \sum_{j=1}^{r_i} \ln X_{ij} - \frac{1}{\hat{\alpha}_f} = 0, \tag{1.29}$$

$$\widehat{X}_0 = \left\{ \frac{1}{F} \left[\sum_{i=1}^{m} \sum_{j=1}^{r_i} X_{ij}^{\hat{\alpha}_f} + \sum_{i=1}^{m} (n_i - r_i) Y_i^{\hat{\alpha}_f} \right] \right\}^{1/\hat{\alpha}_f}. \tag{1.30}$$

The following notation was used:

$$Y_i = \frac{N_{S_i}}{\hat{\overline{N}}_i}, \quad r_T = \sum_{i=1}^{m} r_i,$$

where r_T is the total number of failed coupons.

The value of $\hat{X}_0$ has to be unity for a perfect fit. If $\hat{X}_0$ takes any value other than unity, the characteristic number of cycles for each stress level can be adjusted to produce $\hat{X}_0 = 1$. In particular:

$$\overline{N}_{0i} = \hat{X}_0 \hat{\overline{N}}_i. \tag{1.31}$$

The slope of the S–N curve, $1/k$, and the y intercept, σ_0, can be determined by fitting $\log \sigma_{\alpha_i}$ versus $\log \overline{N}_{0i}$ to a straight line. With σ_0, k, and α_f already determined, the S–N curve at any specified level of reliability can be calculated by:

$$\sigma_a = \sigma_0 \{[-\ln P_S(N)]^{1/\alpha_f k}\} N^{-1/k}. \tag{1.32}$$

1.10 EXPERIMENTAL PROCEDURE

1.10.1 Material and test Coupons

A comprehensive experimental program was realized consisting of static and fatigue tests of straight-edge coupons cut from a multidirectional laminate. The stacking sequence of the plate consists of four layers, 2 × UD, unidirectional lamina of 100% aligned warp fibers, with a weight of 700 g/m^2 and 2 × stitched, ±45°, of 450 g/m^2, 225 g/m^2 in each off-axis angle. The material used was E-glass/polyester, E-glass from AHLSTROM GLASSFIBRE, while the polyester resin was CHEMPOL 80 THIX by INTERCHEM. This resin is a thixotropic unsaturated polyester and was mixed with 0.4% cobalt naphthenate solution (6% Co), accelerator, and 1.5% methyl ethyl ketone peroxide (MEKP, 50% solution), catalyst. Rectangular plates were fabricated by hand lay-up technique and cured at room temperature. Considering as 0° direction that of the UD layer fibers, the lay-up can be encoded as $[0/(\pm 45)_2/0]_T$. Coupons were cut, by a diamond saw wheel, at 0°, on-axis, and 15°, 30°, 45°, 60°, 75°, and 90° off-axis orientations. All data from cyclic loading were used to characterize anisotropic mechanical properties of the material, for the verification of theoretical predictions from FTPF strength criterion and for the study of stiffness variation during life.

The coupons were prepared according to ASTM 3039–76 standard, and aluminum tabs were glued at their ends. Coupon edges were trimmed with sandpaper. The coupons were 250 mm long and had a width of 25 mm. Their nominal thickness was 2.6 mm. The length of the tabs, with a thickness of 2 mm, was 45 mm leaving a gage length of 160 mm.

Static and fatigue tests were performed. The number of coupons tested, 307 in total, was partitioned as follows: 50 coupons for static tests to provide baseline data, both in tension and compression, while 257 coupons were tested under uniaxial cyclic stress for the determination of 17 S–N curves at various off-axis directions and loading conditions.

1.10.2 Test Program and Results

1.10.2.1 Static and Fatigue Strength

Static tests were performed in tension and compression on an MTS machine of 250 kN capacity under displacement control at a speed of 1 mm/min. The coupons used for static compression tests had a gage length of 30 mm to avoid buckling. Fatigue tests, of sinusoidal constant amplitude waveform, were also carried out on the same MTS machine. In total, 17 S–N curves were determined experimentally, under 4 different stress ratios, namely, $R = 10$ (C–C), $R = -1$ (T–C), $R = 0.1$, and $R = 0.5$ (T–T). The frequency was kept constant at 10 Hz for all the tests, which were continued until coupon ultimate failure or 10^6 cycles, whichever occurred first. In particular, for the on-axis coupons, 0°, under reversed loading, $R = -1$, tests were continued for up to 5×10^6 cycles. For stress ratios comprising compression, the antibuckling jig of Fig. 1.1 was used. Its geometry and operational characteristics are in essence those described in [41].

Uniaxial tests on coupons cut off-axis from principal material directions were performed to induce complex stress states in the principal coordinate system (PCS). Denoting by σ_i, $i = 1, 2, 6$, the in-plane stress tensor components in the

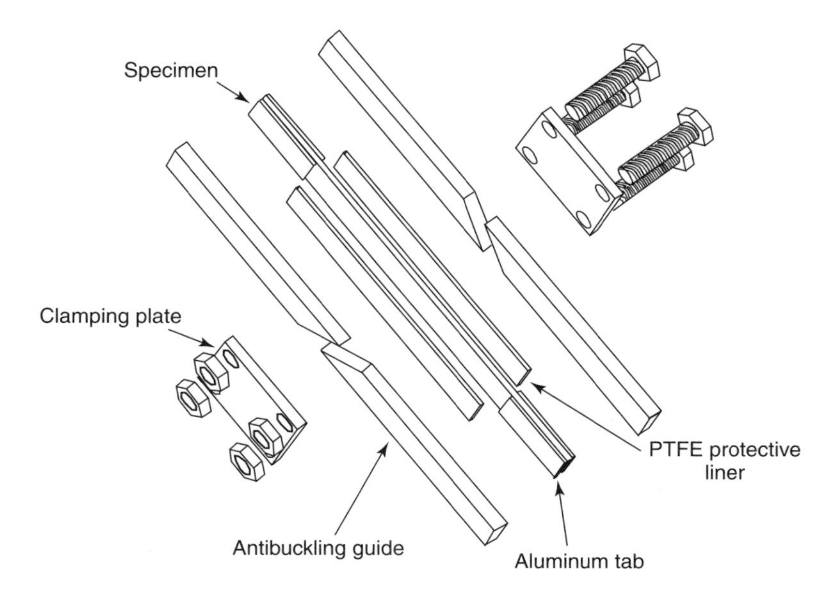

FIGURE 1.1 Sketch of antibuckling device.

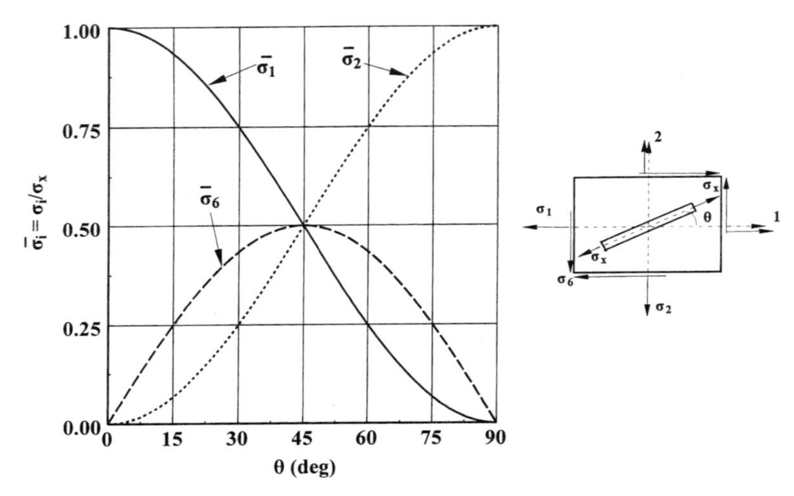

FIGURE 1.2 Complex stress state in principal material system of off-axis loaded coupon.

PCS of the multidirectional laminate, (see Fig. 1.2), and by σ_x, the applied normal stress at an off-axis angle θ, the following transformation relations are valid:

$$\sigma_1 = \sigma_x \cos^2\theta,$$
$$\sigma_2 = \sigma_x \sin^2\theta,$$
$$\sigma_6 = \sigma_x \sin\theta\cos\theta, \tag{1.33}$$

The biaxiality ratios σ_2/σ_1 and σ_6/σ_1 as a function of θ take values that are proportional to $\tan^2\theta$ and $\tan\theta$, respectively.

With respect to material properties that need to be determined experimentally in order to use the FTPF criterion, Eq. (1.17), it is clear from Fig. 1.2 as well, that $X(N)$ and $Y(N)$ are the S–N curves determined from on-axis and 90° off-axis coupon tests. And $S(N)$ is determined by fitting Eq. (1.17) to the experimental fatigue strength data from any off-axis orientation. By substituting relations (1.33) into Eq. (1.17) and solving for $S(N)$ one has:

$$S(N) = \left(\frac{\sin^2\theta\cos^2\theta}{\dfrac{1}{\sigma_x{}^2} + \dfrac{\sin^2\theta\cos^2\theta}{X(N)Y(N)} - \dfrac{\cos^4\theta}{X^2(N)} - \dfrac{\sin^4\theta}{Y^2(N)}} \right)^{1/2}. \tag{1.34}$$

For every experimental point corresponding to a coupon loaded off-axis under stress amplitude σ_x and failed at N cycles, one derives by means of Eq. (1.34) the corresponding $S(N)$ value.

Static strength results from both tension and compression plotted as a function of the off-axis angle θ are presented in Fig. 1.3. Theoretical predictions, solid

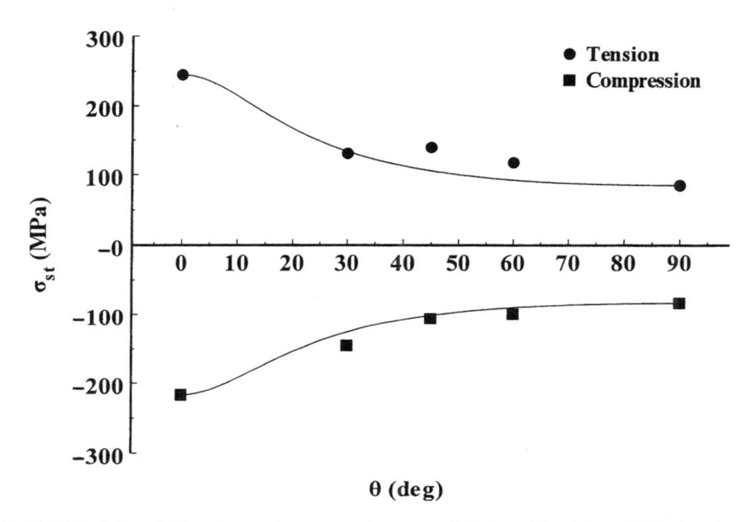

FIGURE 1.3 Off-axis static strength, σ_{st}, of $[0/(\pm45)_2/0]_T$ GRP laminate.

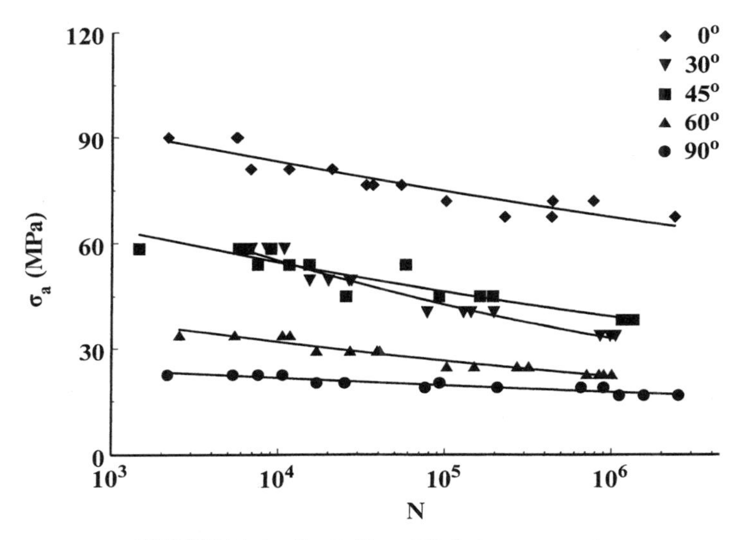

FIGURE 1.4 C–C ($R = 10$) fatigue strength.

line, according to the FTPF criterion are also shown along with data points. To derive the calculated strengths, the multidirectional laminate is considered as homogeneous orthotropic medium. Details on failure predictions under static loading are given in [20].

The S–N curves for $R = 10, -1, 0.1$, and 0.5 are presented in Figs. 1.4–1.7, respectively, where the coordinates of data points correspond to stress amplitude, σ_a, and number of cycles to failure, N. Detailed results on life cycles for every

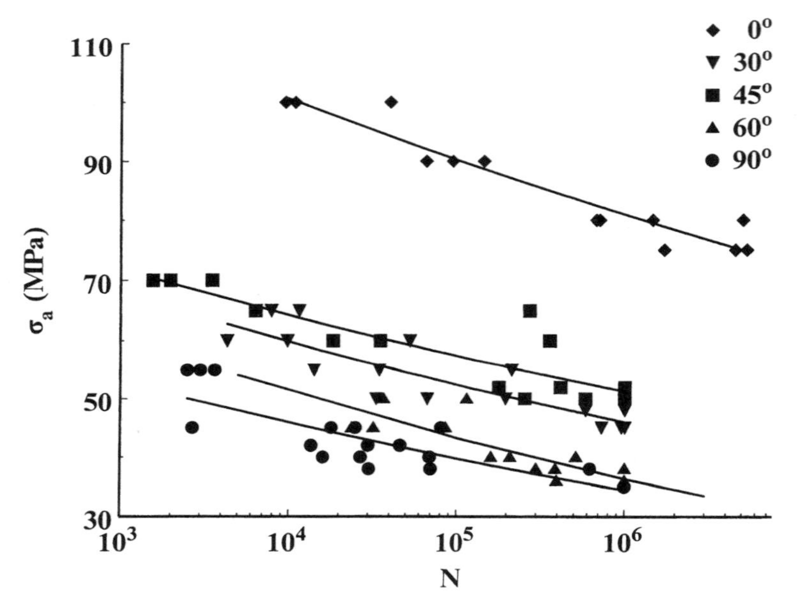

FIGURE 1.5 T–C ($R = -1$) fatigue strength.

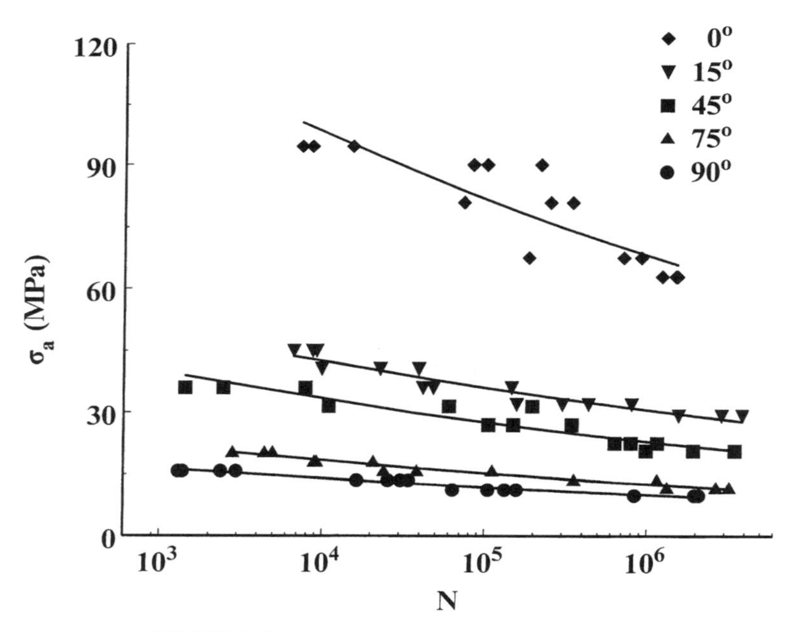

FIGURE 1.6 T–T ($R = 0.1$) fatigue strength.

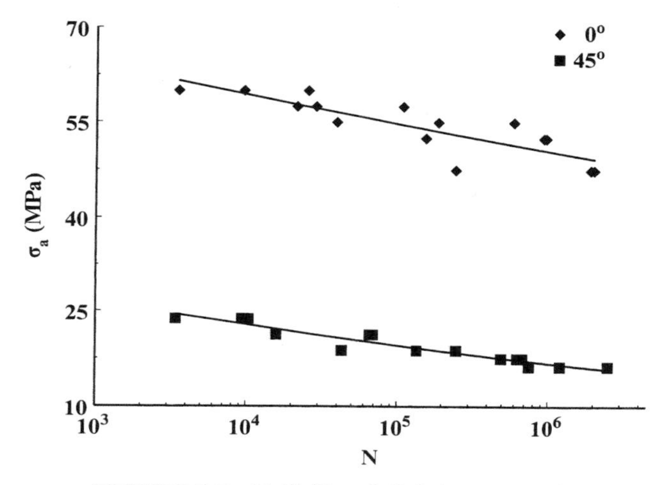

FIGURE 1.7 T–T ($R = 0.5$) fatigue strength.

coupon tested under various stress ratios R and off-axis angle θ are displayed in Tables 1.7–1.10. Linear regression curves, shown as solid lines, in Figs. 1.4–1.7 are of the form $\sigma_a = \alpha N^{-1/b}$.

As seen from these figures, fatigue strength of this specific GRP laminate is higher in tensile loading ($R = 0.1$) than in the respective compressive one ($R = 10$), only for the case of on-axis coupons, 0°, and $N < 10^6$. The opposite is suggested by the experimental data, that is, compressive strength is higher than tensile fatigue strength for any other off-axis loading configuration. It is furthermore observed that the material is more fatigue sensitive in tension than in compression as indicated from the slope of the respective S–N curves.

1.10.2.2 Stiffness Degradation

For all the coupons tested in fatigue, hysteresis loops were monitored continuously by recording load and cross-head displacement signals. Stiffness changes with respect to the number of cycles were studied in terms of dynamic modulus, E_N, determined by linear curve fitting of data samples at every stress–strain hysteresis loop. A dimensionless measure of stiffness degradation is given by the ratio E_N/E_1, E_1 being the modulus at the first load cycle, greater in general from the static Young's modulus, E_0, due to the higher strain rate of deformation.

According to the simple model of Section 1.9.2, the variation of the ratio E_N/E_1 with respect to cycle number, N, is linear, its slope depending on the cyclic stress amplitude. This was postulated for the main central life period of a coupon, excluding initiation and final failure phases. By means of Eq. (1.19), model parameters were derived by fitting the experimental data for the various stress ratios, R, and off-axis angle, θ, values; see Table 1.11.

TABLE 1.7 Number of Cycles to Failure from C–C ($R = 10$) Tests for Various Stress Ranges and Off-axis Loading Orientations

σ_a (MPa)	σ_{max} (MPa)	0°	30°	45°	60°	90°
90	200	5500, 2161, 5607				
81	180	20,776, 11,400, 6728				
76.5	170	53,626, 33,052, 36,350				
72	160	770,046, 437,115, 100,184				
65.5	150	2,357,018, 431,315, 225,912				
58.5	130		6819, 5713, 8500, 10,727	5716, 8972, 1453, 6519		
54	120			7465, 15,257, 11,500, 57,500		
49.5	110		27,173, 26,292, 19,888, 15,329			
45	100			91,597, 25,317, 192,288, 161,427		
40.5	90		128,527, 195,000, 77,433, 142,397			
38.25	85			1,145,696, 1,338,602, 1,221,080		
33.75	75		850,000, 976,497, 1,050,000		11,675, 5442, 10,545, 2540	
29.25	65				16,943, 38,911, 26,841, 40,316	
24.75	55				150,000, 270,000, 317,000, 102,412	
22.5	50				710,316, 840,316, 896,316, 1,000,000[a]	5316, 2158, 7567, 10,595
20.25	45					93,315, 17,042, 92,141, 25,006
18.9	42					75,724, 896,052, 206,244, 658,432
16.65	37					1,111,693, 2,505,659, 1,554,429

[a]Test stopped without coupon failure.

TABLE 1.8 Number of Cycles of Failure from T–C ($R = -1$) Tests for Various Stress Ranges and Off-axis Loading Orientations

σ_a (MPa)	σ_{max} (MPa)	0°	30°	45°	60°	90°
100	100	10,700, 39,637, 9350				
90	90	64,871, 93,498, 143,896				
80	80	670,275, 702,056, 1,446,527, 5,000,000[a]				
75	75	4,500,000[a], 1,700,786, 5,269,524				
70	70			1557, 3500, 1972		
65	65		7820, 11,407	6317, 270,633		
60	60		4298, 52,316, 9749	18,375, 357,155, 35,012		
55	55		14,098, 34,538, 212,856			3641, 2510, 3000
52	52			179,000, 415,000, 1,000,000[a]		
50	50		195,710, 33,149, 66,559	255,000, 580,995, 1,000,000	36,774, 115,000	
48	48		581,997, 1,000,000			
45	45		956,933, 726,537, 1,000,000[a]		85,700, 32,000, 23,769	2700, 18,000, 80,556, 25,000
42	42					29,850, 13,580, 46,120
40	40				209,560, 512,000, 161,000	26,916, 69,134, 16,027
38	38				388,000, 298,317, 1,000,000[a]	70,030, 30,206, 620,305
36	36				395,000, 1,000,000[a], 1,007,000	
35	35					1,000,000, 1,000,000[a]

[a] Test stopped without coupon failure.

TABLE 1.9 Number of Cycles to Failure from T–T ($R = 0.1$) Tests for Various Stress Ranges and Off-axis Loading Orientations

σ_a (MPa)	σ_{max} (MPa)	0°	15°	45°	75°	90°
94.5	210	8400, 7284, 15,000				
90	200	82,000, 214,300, 99,783				
81	180	337,760, 245,995, 72,100				
67.5	150	898,645, 701,093, 182,123				
63	140	1,204,333, 1,464,000, 1,500,000				
45	100		9125, 6610, 8654			
40.5	90		22,613, 9834, 38,891			
36	80		41,330, 144,730, 47,894	7819, 2450, 1420		
31.5	70			10,935, 59,821, 195,116		
31.95	71		298,233, 155,864, 432,283, 798,467			
29.25	65		3,828,947, 2,852,760, 1,549,331			
27	60			149,100, 105,000, 342,000		
22.5	50			793,000, 1,150,000, 632,089		
20.7	46			985,000, 1,915,000, 3,462,000		
20.25	45				4419, 2815, 4965	
18	40				9226, 20,715, 8924	
15.75	35				111,152, 38,237, 24,111	1296, 2954, 2370, 1370
13.5	30				1,149,039, 354,521, 354,109	30,441, 25,581, 16,440, 34,181
11.7	26				1,325,554, 2,654,235, 3,200,000	
11.25	25					133,043, 104,913, 156,936, 63,520
9.9	22					839,958, 1,955,673, 2,075,673

TABLE 1.10 Number of Cycles to Failure from T–T ($R = 0.5$) Tests for Various Stress Ranges and Off-axis Loading Orientations

σ_a (MPa)	σ_{max} (MPa)	0°	45°
60	240	25,210, 9500, 3500	
57.5	230	107,441, 28,461, 21,333	
55	220	38,970, 586,000, 185,694	
52.5	210	923,020, 956,833, 154,000	
47.5	190	1,996,000, 243,500, 1,900,000	
23.75	95		3411, 9370, 10,393
21.25	85		65,459, 15,858, 68,947
18.75	75		42,900, 135,872, 249,194
17.5	70		630,000, 493,345, 678,643
16.25	65		1,208,000, 750,000, 2,500,000[a]

[a]Test stopped without coupon failure.

TABLE 1.11 Stiffness Reduction Model Parameters

	$R = 10$		$R = -1$		$R = 0.1$		$R = 0.5$	
θ (deg)	K	c	K	c	K	c	K	c
0	8.74×10^{31}	16.91	2.66×10^{33}	17.1	7.20×10^{12}	8.962	2.12×10^{33}	19.91
15					2.26×10^{26}	13.65		
30	4.20×10^{13}	8.78	2.57×10^{24}	12.04				
45	1.43×10^{23}	13.2	2.01×10^{22}	11.76	4.18×10^{14}	10.53	9.30×10^{20}	13.9
60	1.07×10^{19}	10.49	2.05×10^{17}	9.871				
75					1.25×10^{17}	1.01E + 01		
90	3.94×10^{51}	22.69			8.81×10^{16}	10.39		

Recalling Eq. (1.18), one has:

$$1 - \frac{E_N}{E_1} = K\left(\frac{\sigma_a}{E_0}\right)^c N. \tag{1.35}$$

The left-hand side of the above equation can be thought of as a measure of damage, ranging from 0 for a virgin specimen and approaching asymptotically 1 for a damaged one to failure. Using this form of presentation, that is, Eq. (1.35), experimental data from some of the cases studied are shown in Figs. 1.8–1.12, where logarithms base 10 of the quantities of interest are displayed. Experimental points are fitted by regression lines and, as shown, the degree of fit supports, in general, the assumptions of the linear model, that is, Eq. (1.18).

Data on stiffness changes, collected during fatigue of each set of coupons, of a certain R value and off-axis angle θ, were fitted by a number of probability distributions [22]. The scope of the analysis was to examine the stochastic behavior of stiffness degradation and statistically test the validity of acquired data. In Table 1.12, estimated parameters by maximum likelihood are given for

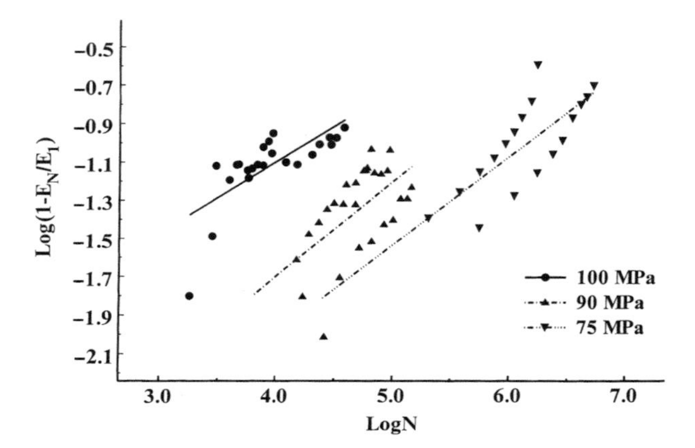

FIGURE 1.8 Stiffness degradation data. $R = -1, 0°$ on-axis.

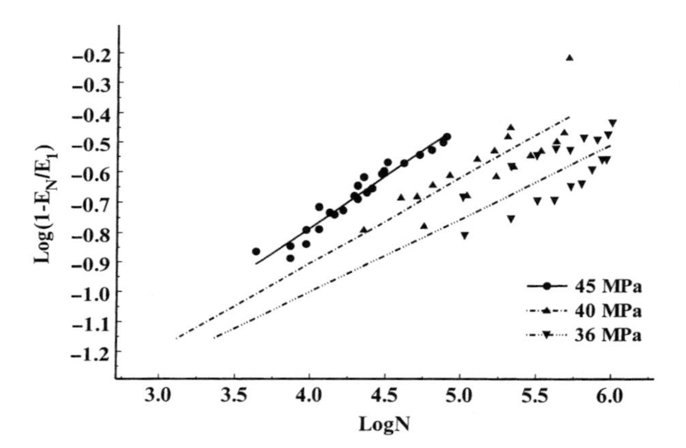

FIGURE 1.9 Stiffness degradation data. $R = -1, 60°$ off-axis.

each set of data for five distributions:

$$
\begin{aligned}
&\text{Normal:} && F(x;\mu,\sigma) = \int_{-\infty}^{x} \frac{1}{\sigma\sqrt{2\pi}} \exp\left[-\frac{(z-\mu)^2}{2\sigma^2}\right] dz,\\
&\text{Lognormal:} && F(x;\mu,\sigma) = \int_{-\infty}^{x} \frac{1}{\sigma z\sqrt{2\pi}} \exp\left[-\frac{(\ln z-\mu)^2}{2\sigma^2}\right] dz,\\
&\text{Weibull:} && F(x;\mu,\sigma) = 1-\exp\left[-\left(\frac{x}{\sigma}\right)^n\right], \qquad x\geq 0, \qquad (1.36)\\
&\text{Largest element:} && F(x;\mu,\sigma) = \exp\left[-\exp\left(-\frac{x-\mu}{\sigma}\right)\right],\\
&\text{Smallest element:} && F(x;\mu,\sigma) = 1-\exp\left[-\exp\left(\frac{x-\mu}{\sigma}\right)\right].
\end{aligned}
$$

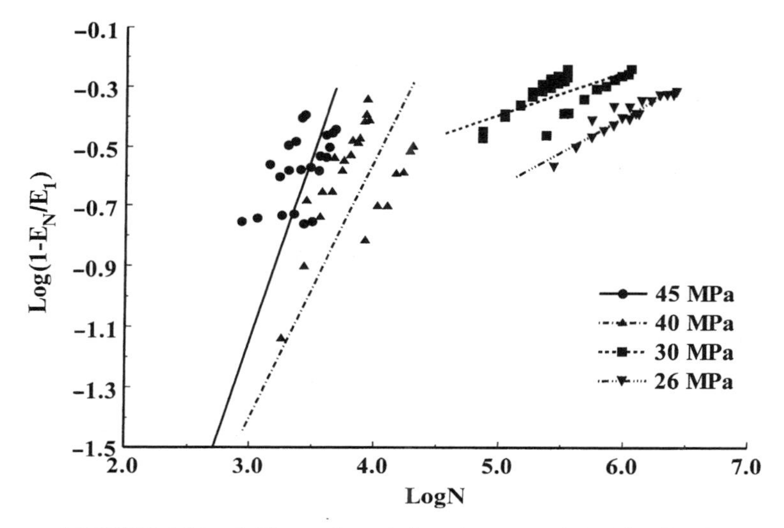

FIGURE 1.10 Stiffness degradation data. $R = 0.1$, 75° off-axis.

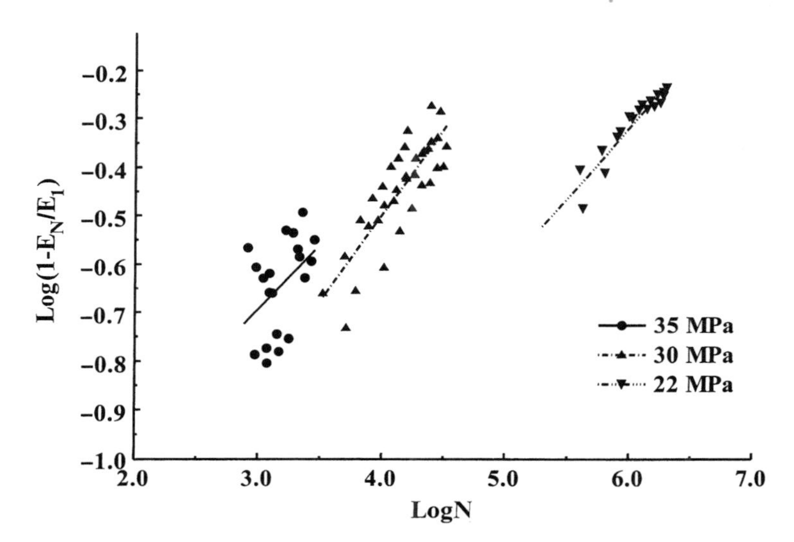

FIGURE 1.11 Stiffness degradation data. $R = 0.1$, 90° off-axis.

Kolmogorof–Smirnof (KS) goodness-of-fit tests were performed for each one of the hypotheses. KS test results, that is, the D_N statistic and its probability, $P(D_N)$, for all the aforementioned distributions are given in Table 1.13. Values of $P(D_N)$, greater or equal to 0.05 correspond to goodness of fit at a significance level of 5% or higher. Calculations for the KS test were performed using the method described in Press et al. [42].

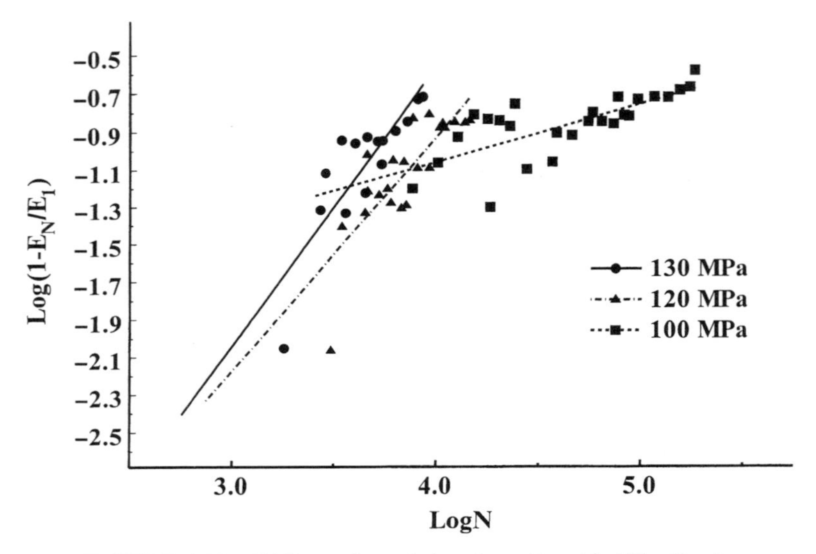

FIGURE 1.12 Stiffness degradation data. $R = 10, 45^\circ$ off-axis.

As seen from the results of Table 1.13, stiffness degradation data, at a specific R value and off-axis angle θ, can be modeled by a single statistical distribution for the whole range of stress levels considered for an S–N curve determination. Notice that for the cases studied herein, the number of experimental points taken into account was at least 80.

An example of favorable and unfavorable comparison between experimental data and theoretical distributions is given in Figs 1.13 and 1.14, respectively, for different material configurations. The upper and lower 95% confidence interval bounds correspond to the less satisfactory distribution, which for the case of Fig. 1.13 is the two-parameter Weibull distribution, whereas for Fig. 1.14 is the lognormal distribution. In this latter case, since the experimental sampling distribution intersects the 95% confidence bounds, the respective null hypothesis, that is, the lognormal distribution, is not accepted at the significance level of 5% [43].

At noted from the results of Table 1.13, the best performing distribution is the two-parameter Weibull distribution, which succeeds in 14 of 16 treated cases to fit the data at a significance level greater than 5%. The statistical distributions are sorted, in this table, according to their fitting capability and, therefore, the second best performing function is the normal. It must be emphasized that in all cases treated except that of $R = 10$ at $\theta = 90^\circ$, stiffness degradation data are satisfactorily fitted by a single statistical distribution, at a significance level of 5% or higher, irrespective of stress level in the same S–N curve.

TABLE 1.12 Stiffness Degradation: Parameters for Various Statistical Distributions

	Degrees	0				15	30		45				60		75	90	
	R Ratio	10	−1	0.1	0.5	0.1	10	−1	10	−1	0.1	0.5	10	−1	0.1	10	0.1
Distribution																	
Weibull	σ	0.926	0.941	0.734	0.870	0.897	0.951	0.849	0.932	0.833	0.917	0.889	0.945	0.799	0.685	0.968	0.692
	η	18.79	23.99	5.915	11.63	13.29	26.84	8.868	16.74	8.530	15.14	17.02	23.27	8.970	5.042	27.82	4.904
Normal	μ	0.902	0.919	0.684	0.834	0.862	0.932	0.805	0.904	0.788	0.884	0.866	0.945	0.761	0.632	0.946	0.638
	σ	0.050	0.050	0.119	0.085	0.083	0.040	0.104	0.059	0.104	0.079	0.046	0.057	0.080	0.129	0.070	0.129
Smallest	μ	0.925	0.942	0.738	0.872	0.900	0.950	0.852	0.931	0.835	0.919	0.886	0.945	0.797	0.690	0.978	0.697
	σ	0.039	0.039	0.094	0.067	0.065	0.031	0.082	0.046	0.082	0.617	0.036	0.036	0.063	0.101	0.055	0.101
Lognormal	μ	−0.104	−0.086	−0.396	−0.188	−0.154	−0.072	−0.226	−0.103	−0.248	−0.128	−0.146	−0.081	−0.278	−0.479	−0.059	−0.469
	σ	0.056	0.056	0.178	0.108	0.104	0.043	0.136	0.066	0.137	0.094	0.053	0.051	0.102	0.201	0.088	0.195
Largest	μ	0.880	0.897	0.630	0.795	0.824	0.914	0.758	0.877	0.741	0.848	0.845	0.903	0.725	0.574	0.914	0.580
	σ	0.039	0.039	0.094	0.067	0.065	0.031	0.082	0.046	0.082	0.617	0.036	0.036	0.063	0.101	0.055	0.101

TABLE 1.13 KS Test Results for Distributions of Table 1.12

	Degrees	0				15	30		45				60		75	90	
	R Ratio	10	−1	0.1	0.5	0.1	10	−1	10	−1	0.1	0.5	10	−1	0.1	10	0.1
Distribution																	
Weibull	D_N	0.113	0.093	0.091	0.103	0.065	0.090	0.069	0.099	0.075	0.108	0.160	0.085	0.108	0.082	0.154	0.107
	$P(D_N)$	0.250	0.480	0.325	0.130	0.554	0.311	0.883	0.357	0.769	0.121	0.001	0.327	0.181	0.387	0.040	0.139
Normal	D_N	0.088	0.141	0.064	0.115	0.110	0.110	0.048	0.099	0.072	0.159	0.143	0.166	0.087	0.077	0.253	0.104
	$P(D_N)$	0.549	0.076	0.769	0.069	0.055	0.125	0.997	0.346	0.769	0.005	0.005	0.002	0.419	0.455	0.000	0.160
Smallest	D_N	0.100	0.103	0.126	0.102	0.049	0.112	0.104	0.113	0.113	0.096	0.213	0.101	0.157	0.144	0.237	0.174
	$P(D_N)$	0.386	0.343	0.063	0.143	0.872	0.115	0.425	0.206	0.234	0.214	0.000	0.158	0.013	0.012	0.002	0.018
Lognormal	D_N	0.099	0.151	0.069	0.140	0.122	0.113	0.067	0.104	0.091	0.172	0.133	0.078	0.067	0.048	0.274	0.068
	$P(D_N)$	0.399	0.047	0.679	0.013	0.024	0.109	0.907	0.290	0.486	0.002	0.010	0.438	0.755	0.943	0.000	0.659
Largest	D_N	0.158	0.202	0.098	0.185	0.178	0.153	0.104	0.159	0.136	0.226	0.075	0.129	0.061	0.064	0.322	0.047
	$P(D_N)$	0.035	0.002	0.244	0.000	0.000	0.010	0.420	0.023	0.090	0.000	0.369	0.031	0.844	0.692	0.000	0.959

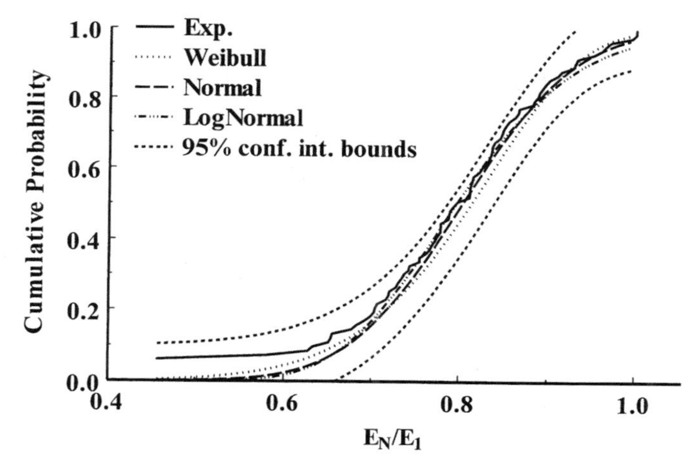

FIGURE 1.13 Comparison of experimental and theoretical cumulative distributions of stiffness degradation. $R = -1$, 30° off-axis.

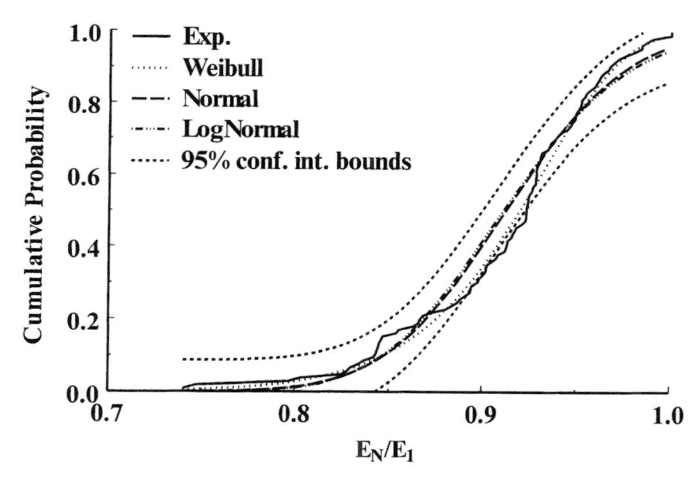

FIGURE 1.14 Comparison of experimental and theoretical cumulative distributions of stiffness degradation. $R = -1$, 0° on-axis.

1.11 DISCUSSION

1.11.1 Survival Probability, Constant Life Diagrams

Experimental data on number of cycles to failure, shown in Tables 1.7–1.10, were processed following the statistical procedure exposed in Section 1.9.3 to derive design allowables at a given reliability level. Fatigue strength curves were then defined at a survival probability of 95% using Eq. (1.32) and parameter values of the statistical model from Table 1.14. The results are shown in

TABLE 1.14 Fatigue Strength Model Parameters

R / Direction	10			−1			0.1			0.5		
	a_f	σ_0	1/k	a_f	σ_0	1/k	a_f	σ_0	1/k	a_f	σ_0	1/k
0°	1.750	130.4	0.04723	1.548	163.4	0.04996	2.668	224.7	0.08606	1.379	93.38	0.04451
15°							2.326	85.95	0.07391			
30°	4.797	160.0	0.1138	1.367	121.3	0.06977						
45°	1.776	122.1	0.08294	1.012	113.6	0.05589	1.669	81.09	0.09089	2.296	45.25	0.07107
60°	3.071	72.95	0.08638	1.725	114.8	0.08094						
75°							2.196	39.57	0.08128			
90°	1.734	35.33	0.05006	1.389	95.96	0.07463	3.789	27.41	0.07182			

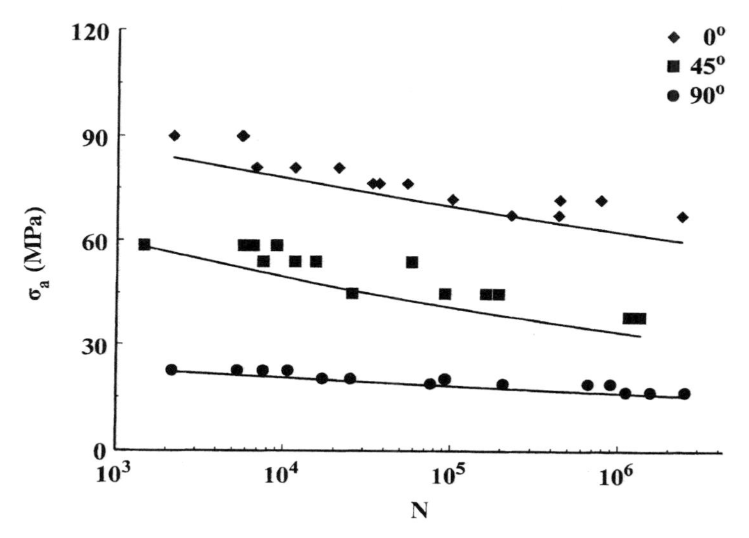

FIGURE 1.15 S–N curves of 95% survival probability under C–C loading. $\theta = 0°$, 45°, 90°.

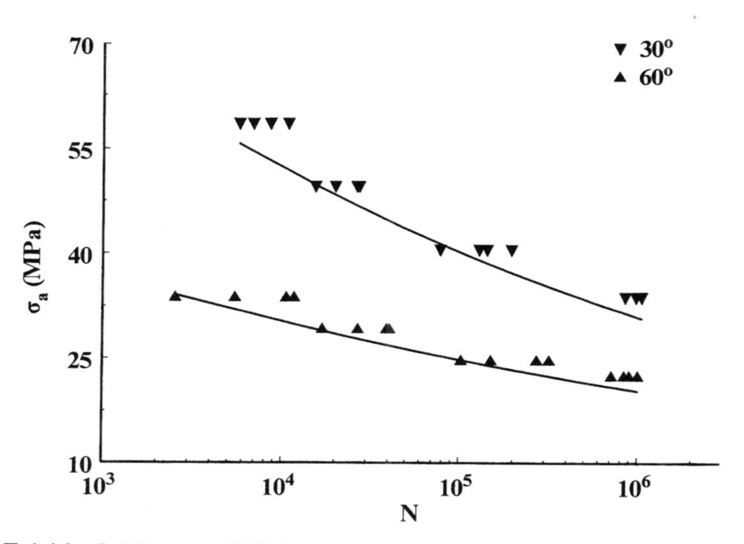

FIGURE 1.16 S–N curves of 95% survival probability under C–C loading. $\theta = 30°, 60°$.

Figs. 1.15–1.21, where experimental data are plotted along with theoretical predictions for comparison. It is clearly seen for all the cases treated, that is, 17 S–N curve series of tests at various R ratios and θ values, that the statistical procedure implemented performs very well and thus allowables derived that way can be reliably used in design.

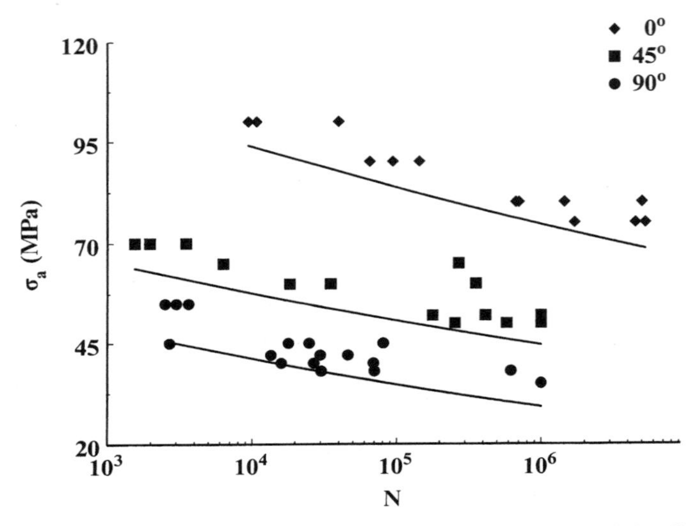

FIGURE 1.17 S–N curves of 95% survival probability under T–C loading. $\theta = 0°$, 45°, 90°.

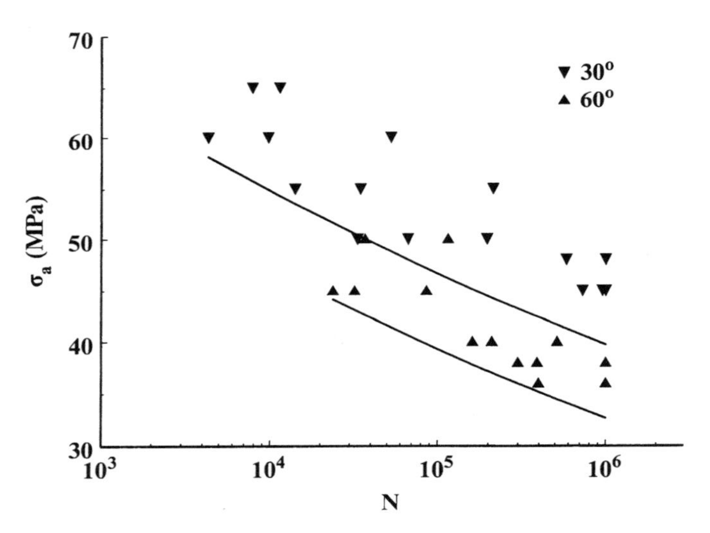

FIGURE 1.18 S–N curves of 95% survival probability under T–C loading. $\theta = 30°$, 60°.

The effect of stress ratio R, on fatigue strength, depending also on off-axis loading orientation, was discussed in Section 1.10.2.1. The same trends are exhibited by 95% reliability S–N curves, and this can be shown by plotting in the same graph curves with different R values. The fatigue strength dependence on stress ratio for on-axis loaded coupons is illustrated in Fig. 1.22 while respective

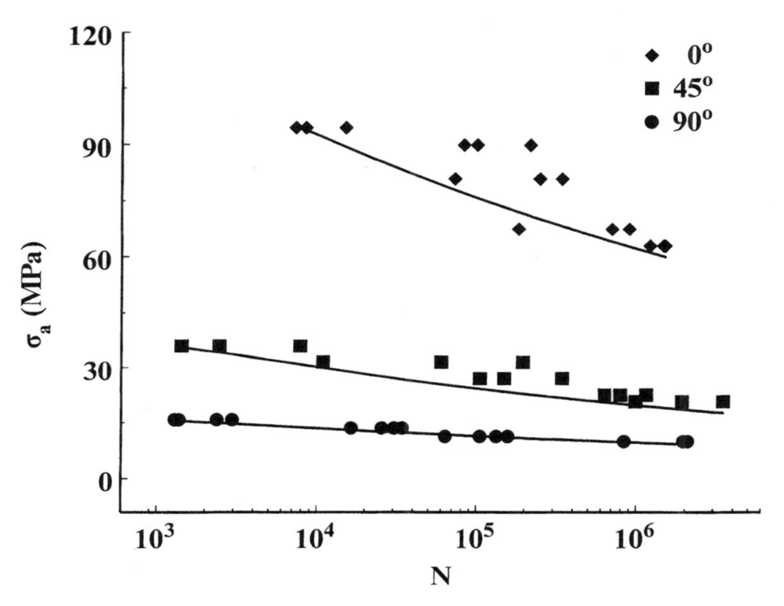

FIGURE 1.19 S–N curves of 95% survival probability under T–T ($R = 0.1$) loading. $\theta = 0°, 45°, 90°$.

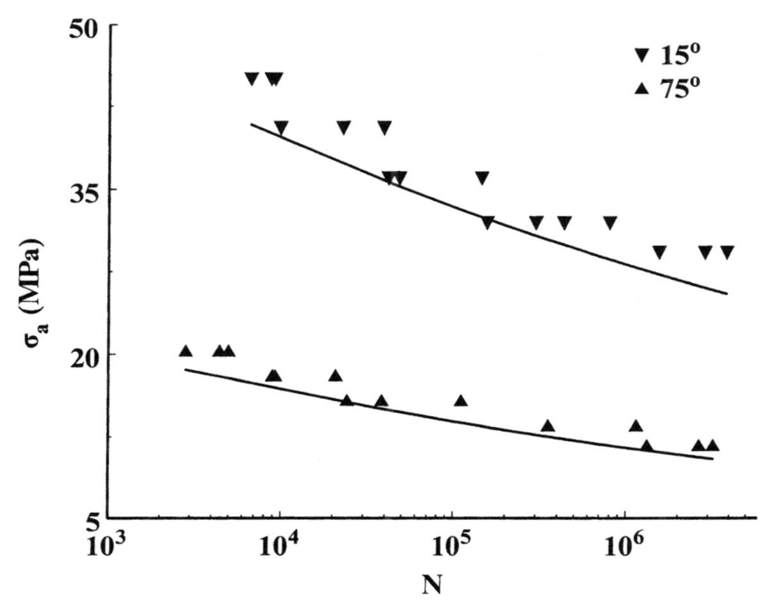

FIGURE 1.20 S–N curves of 95% survival probability under T–T ($R = 0.1$) loading. $\theta = 15°, 75°$.

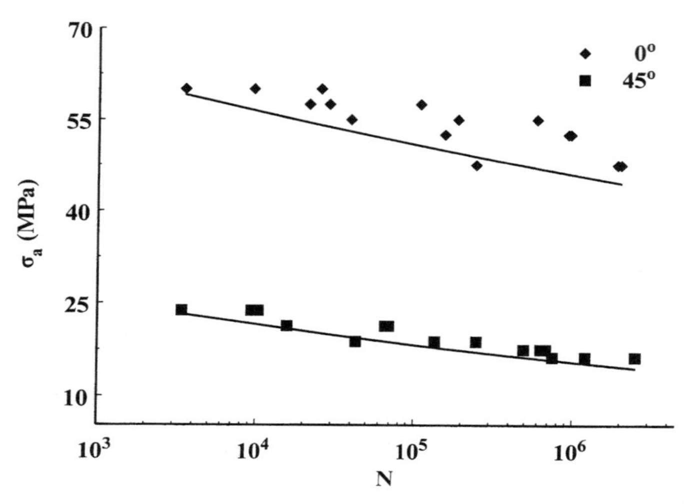

FIGURE 1.21 S–N curves of 95% survival probability under T–T ($R = 0.5$) loading. $\theta = 0°, 45°$.

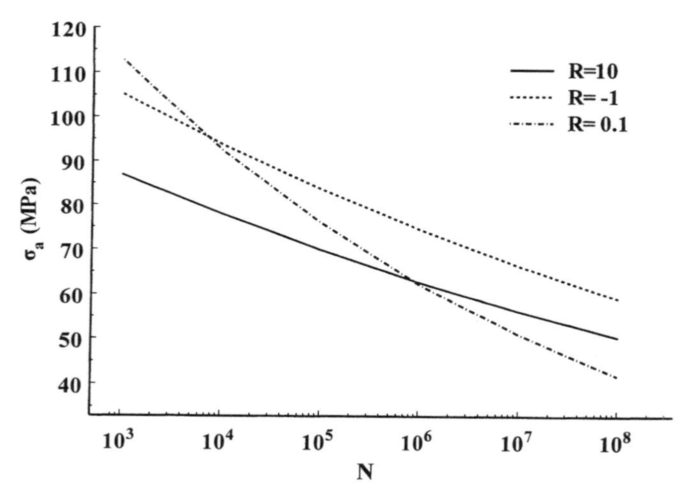

FIGURE 1.22 Effect of stress ratio R on 95% survival probability S–N curves. On-axis loading.

results from 45° off-axis tests are shown in Fig. 1.23. It is therefore verified that for on-axis loading the GRP laminate investigated is weaker to compressive stress ranges when $N < 10^6$, while for high-cycle fatigue it can withstand lower tensile stress ranges than compressive ones. In the contrary, for off-axis loading, compressive stress ranges withstood by material coupons were almost double the respective tensile ones; see Fig. 1.23.

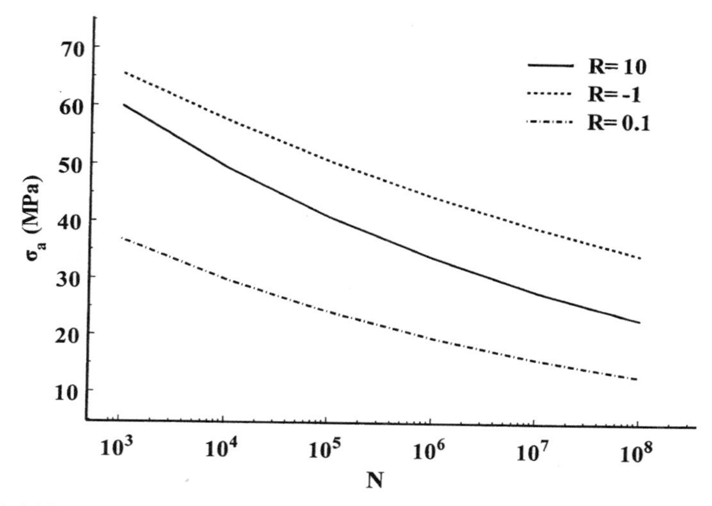

FIGURE 1.23 Effect of stress ratio R on 95% survival probability S–N curves. Off-axis, 45°, loading.

To comply with the needs of designing a composite structure against variable amplitude fatigue, that is, spectrum loading, strength data are projected in the plane having as coordinates stress amplitude and mean stress, $\sigma_a - \sigma_m$. Radial lines emanating from the origin are expressed by:

$$\sigma_a = \left(\frac{1-R}{1+R}\right)\sigma_m, \tag{1.37}$$

and thus, every such curve of the family corresponds to a constant R value. Therefore, points along these lines are points of the S–N curve for that particular stress ratio R. Constant life diagrams are formed by joining points of consecutive radial lines, all corresponding to a certain value of cycles, N. Some useful characteristics of $(\sigma_a - \sigma_m)$ plane are shown in Fig. 1.24. As shown, the positive half-plane is divided into three sectors, the central one being of double surface area. The tension–tension sector is bounded by the radial lines $R = 1$ and $R = 0$, the former corresponds to static loading and the latter to tensile cycling with $\sigma_{\min} = 0$. The S–N curves, that is, radial lines belonging to this sector have positive R values smaller than unity. Similar type comments for the other sectors can be derived from the annotations shown in Fig. 1.24. It is interesting also to mention that to every radial line with $0 < R < 1$, that is, in the T–T sector, corresponds its symmetric with respect to the σ_a axis, which lies in the C–C sector and whose R value is the inverse of the tensile one, for example, $R = 0.1$ and $R = 10$.

For the experimental data presented in Tables 1.7–1.10 and using the model parameters shown in Table 1.14, it is possible to form constant life diagrams, for each off-axis loading configuration, corresponding to any specific survival

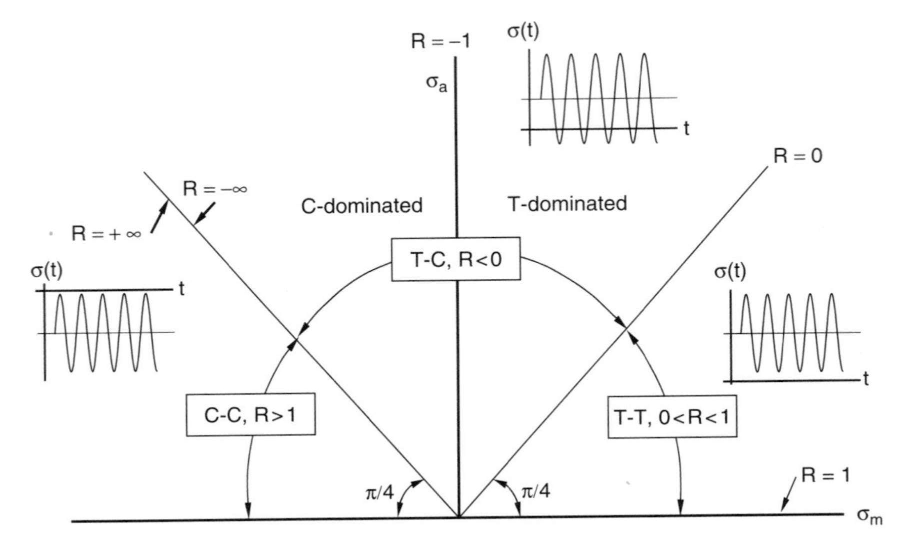

FIGURE 1.24 $\sigma_a - \sigma_m$-plane notation for constant life diagrams.

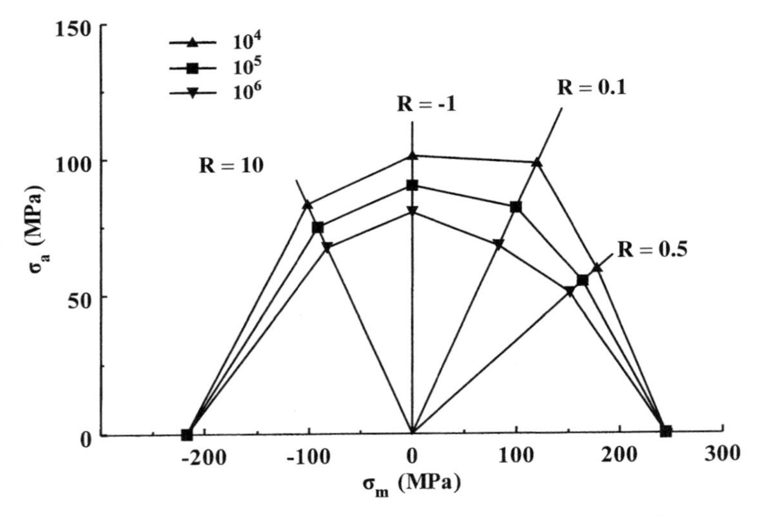

FIGURE 1.25 Constant life diagram for $[0/(\pm 45)_2/0]_T$ laminate at 0°. UTS = 244.84 MPa, UCS = −216.68 MPa.

probability. Such an exercise was performed for a 50% reliability level, and the results were shown in Figs. 1.25–1.27 for $\theta = 0°$, 45°, and 90°, respectively.

For on-axis loading (Fig. 1.25) constant life curves are closer to a Gerber-like prediction rather than a Goodman straight line. Since several design codes, for composite structures, suggest the use of Goodman criterion to account for variable

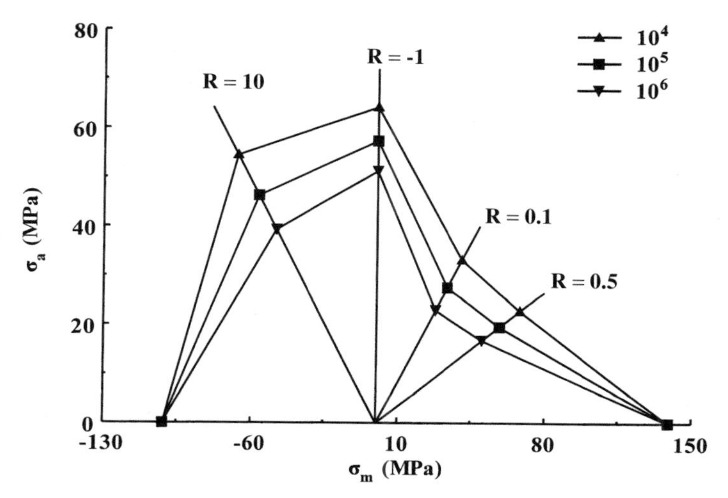

FIGURE 1.26 Constant life diagram for $[0/(\pm 45)_2/0]_T$ laminate at 45°. UTS = 139.1 MPa, UCS = −101.59 MPa.

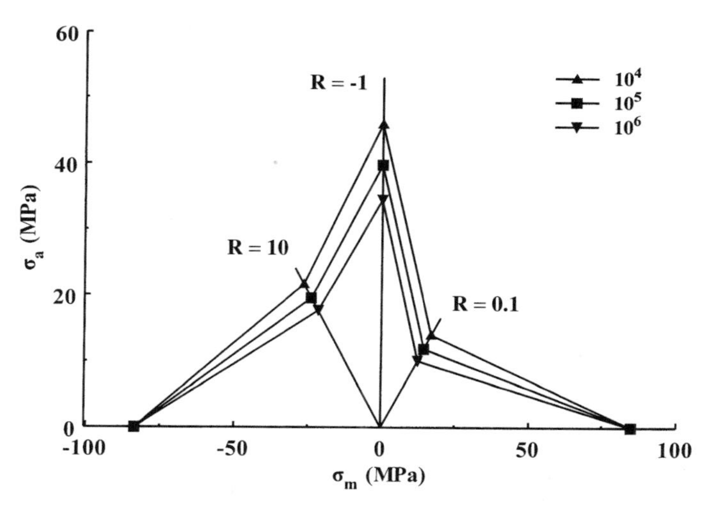

FIGURE 1.27 Constant life diagram for $[0/(\pm 45)_2/0]_T$ laminate at 90°. UTS = 84.94 MPa, UCS = 83.64 MPa.

amplitude cyclic loading, this would lead to conservative design decisions, along with Palmgren–Miner damage accumulation rule. Another interesting aspect of this graph, already mentioned, is that the material proved to be stronger in tension, than in compression, for small number of cycles, while the opposite holds true with an increasing number of cycles. The same trend was reported for a similar material and stacking sequence by other researchers [14], and their results were

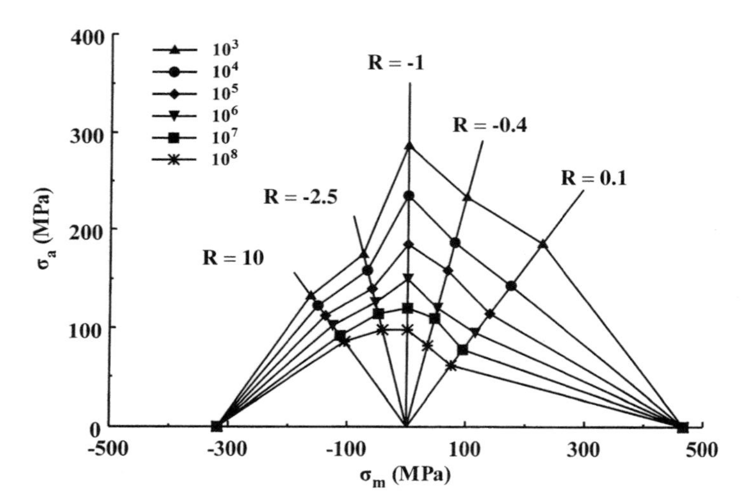

FIGURE 1.28 Constant life diagram for GRP laminate composed of (0/ ± 45) fabrics. UTS = 467 MPa, UCS = −318 MPa. Data from [14].

reproduced in Fig. 1.28 for comparison. It is worthwhile noting in this figure that at the high-cycle range ($N > 10^7$), even the Goodman straight line is an optimistic approach to the real $\sigma_a - \sigma_m$ relation.

Off-axis loading reveals the anisotropic nature of the GRP laminate investigated since, as observed in Figs. 1.26 and 1.27, for $\theta = 45°$ and $90°$, respectively, the fatigue response of the material differs significantly from what was discussed already. What is common in these figures is the higher fatigue strength in compressive rather than in tensile stress ranges, and the poor performance of Goodman law in describing the relation between mean stress and cyclic amplitude.

1.11.2 Fatigue Strength Prediction

Efficient and reliable prediction of fatigue life of any structural component under complex stress states is of paramount importance in design. Such a task can be carried out by means of the FTPF criterion, discussed in Section 1.9.1, which for plane stress conditions is expressed by Eq. (1.17). For the formulation of the criterion in the principal material directions of a laminate possessing similar strength symmetries as the one investigated herein, the S–N curves along the two orthogonal symmetry directions as well as the respective shear fatigue strength must be known. Determination of the latter, always under the same R value loading, is performed using the methodology proposed in Section 1.10.2.1, Eq. (1.34).

Comparison of FTPF prediction with experimental data from various material systems as well as with theoretical predictions from other strength criteria can be found in [20]. For the experimental off-axis data presented in Tables 1.7–1.9 under $R = 10, -1$ and 0.1, respectively, calculations following the aforementioned

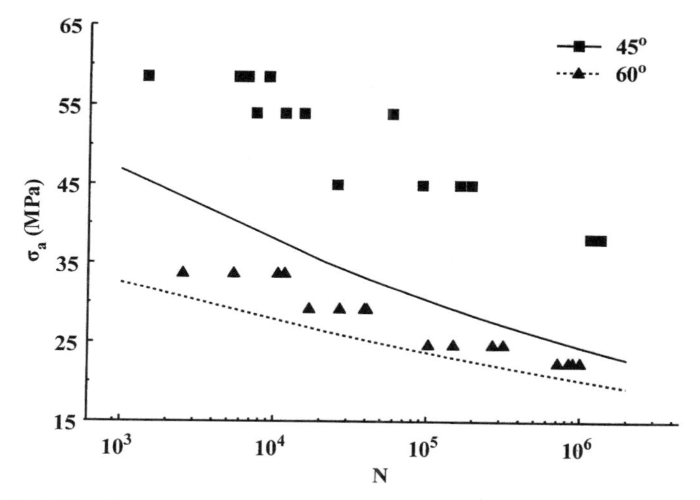

FIGURE 1.29 Comparison of experimental data and FTPF predictions ($R = 10$).

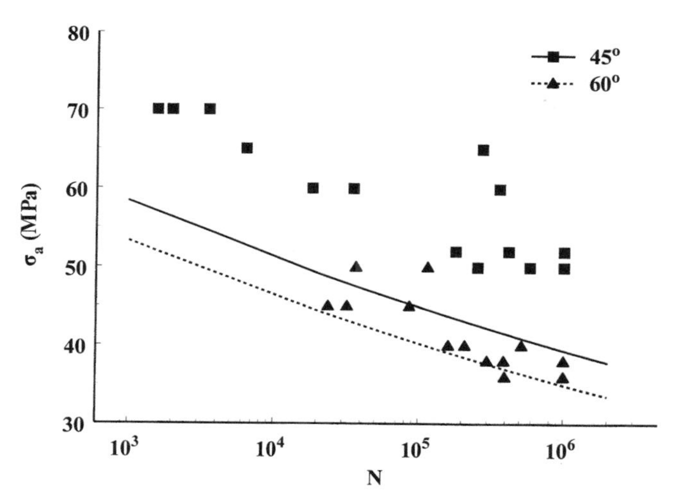

FIGURE 1.30 Comparison of experimental data and FTPF predictions ($R = -1$).

methodology were performed, and the results are presented in Figs. 1.29–1.34. Shear strength S–N formulation is derived for all cases by fitting the experimental data of smaller off-axis angles, that is, 30° for $R = 10, -1$ and 15° for $R = 0.1$. Reliable predictions, that is, conservative, of the criterion are produced that way for the other off-axis directions.

The results of Figs. 1.29–1.31 were derived by solving Eq. (1.34) for σ_x. The expressions used for $X(N)$, $Y(N)$, and $S(N)$ are given in Table 1.15 and

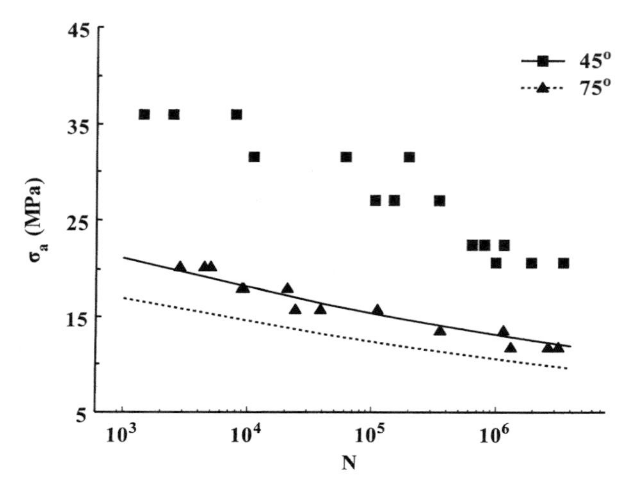

FIGURE 1.31 Comparison of experimental data and FTPF predictions ($R = 0.1$).

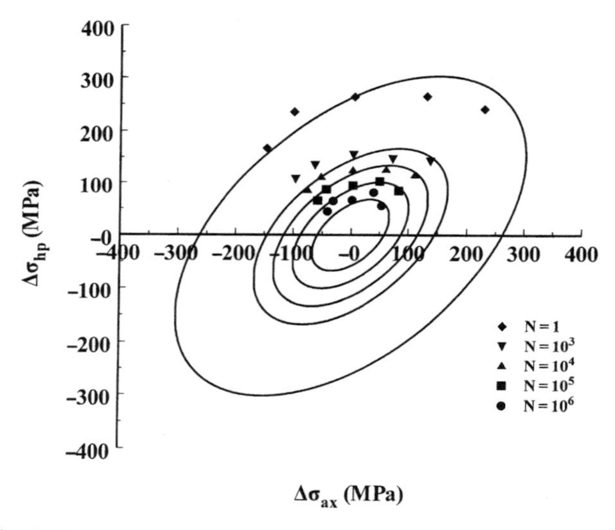

FIGURE 1.32 FTPF predictions vs. experimental data from woven GRP cylindrical specimens biaxially loaded at 0° [20].

correspond to the median survival probability approximately. If a higher reliability level is required, the procedure for the determination of $S(N)$ has to be repeated by using values for $X(N)$, $Y(N)$, and off-axis test results to be fitted by Eq. (1.34), corresponding at that survival probability.

As concluded from Figs. 1.29–1.31 the predictions of the FTPF criterion for off-axis orientations such as 60° or 75° are good and always on the safe side. The same is valid also for 45°, but the predictions are too conservative. However,

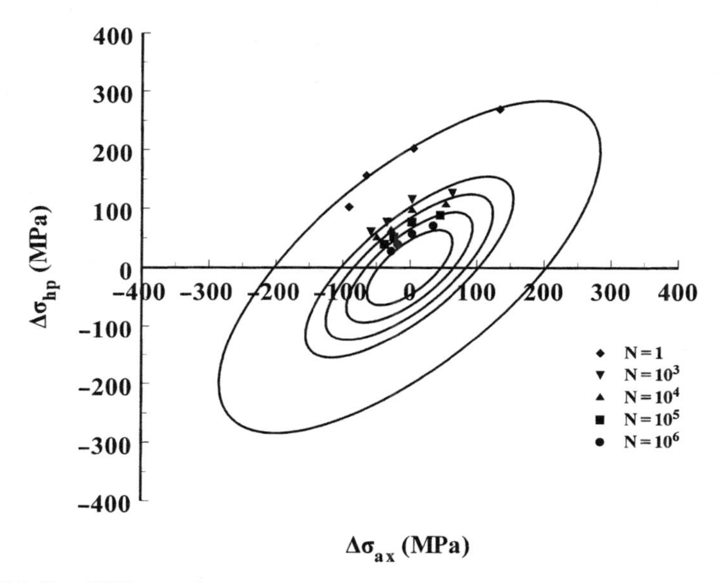

FIGURE 1.33 FTPF predictions vs. experimental data from woven GRP cylindrical specimens biaxially loaded at 45° [20].

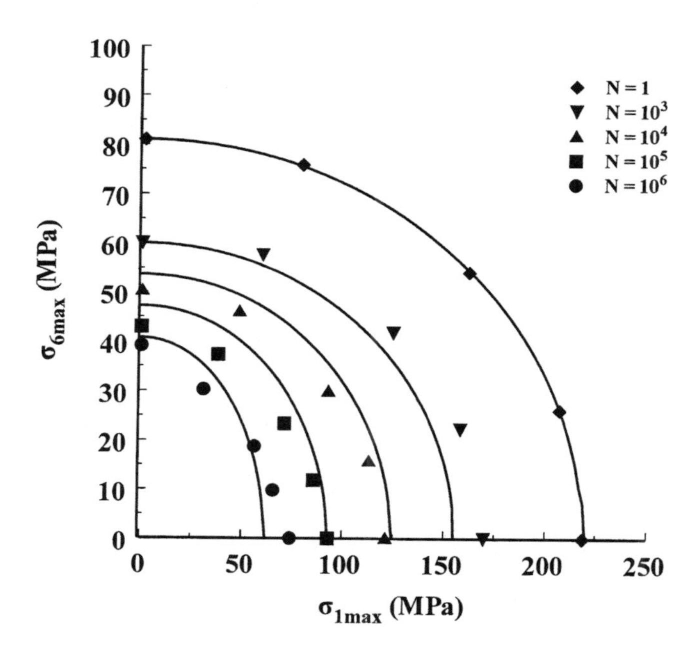

FIGURE 1.34 FTPF predictions vs. experimental data from woven GRP cylindrical specimens loaded under combined tension–torsion [20].

TABLE 1.15 Experimental Fatigue Strength Equation along Principal Material Directions of $[0/(\pm 45)_2/0]_T$ GRP Laminate (in MPa)

	$R = 10$	$R = -1$	$R = 0.1$
$X(N)$	$\sigma_a = 125.6N^{-0.04507}$	$\sigma_a = 155.3N^{-0.04724}$	$\sigma_a = 204N^{-0.07957}$
$Y(N)$	$\sigma_a = 32.66N^{-0.04459}$	$\sigma_a = 81.68N^{-0.06249}$	$\sigma_a = 26.86N^{-0.07108}$
$S(N)$	$\sigma_a = 102.3N^{-0.13580}$	$\sigma_a = 48.55N^{-0.05860}$	$\sigma_a = 21.44N^{-0.06940}$

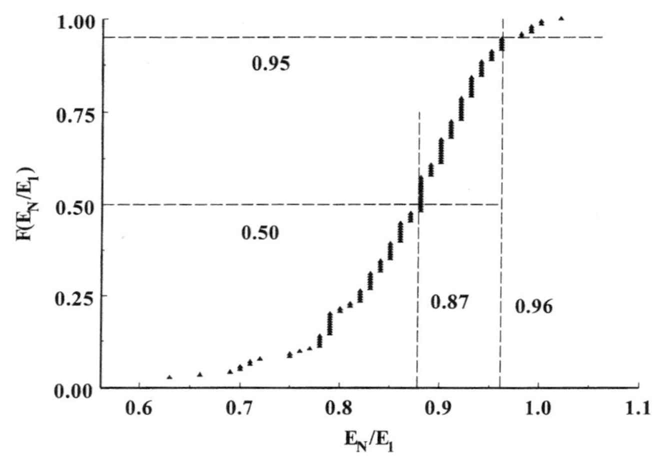

FIGURE 1.35 Sampling distribution of stiffness degradation data. $R = 0.1$, 15° off-axis.

this was observed for static strength results also (see Fig. 1.3) and it is due to the presence of fiber bundles along ±45°, that is, the GRP laminate investigated is not in essence a homogeneous orthotropic medium. For unidirectional glass epoxy laminates tested off-axis, it was shown in [20] that predictions of fatigue strength by the FTPF criterion were corroborated satisfactorily by the experimental data for the entire range of off-axis directions. Then, it is logical to conclude that the quadratic version of the failure tensor polynomial is adequate for design calculations, where one needs safe and reliable predictions, but if higher accuracy is needed, from the material characterization point of view, higher order tensor formulation [44] could be necessary.

Besides uniaxial loading test cases, the FTPF criterion was shown to predict satisfactorily fatigue strength under multiaxial cyclic loads as well [20]. Theoretical predictions were compared to experimental data of Owen and Griffiths [18] on woven glass polyester cylinders cycled under biaxial hoop, σ_{hp}, and axial, σ_{ax}, stresses. Two separate loading cases were reported at 0° and 45° with respect to the fiber's direction. Suitable experimental data were also found in a study by Fujii and Lin [19], from an experimental program consisting of fatigue tests under tension–torsion loading on cylindrical specimens made of woven glass/polyester.

The stress ratio R considered was equal to 0, that is, $\sigma_{min} = 0$, while the test frequency was limited to 2 Hz.

Predicted failure locii by the FTPF criterion plotted against experimental data from [18] are shown in Figs. 1.32 and 1.33 for $1-10^6$ cycles. Failure locii for the cylindrical specimens loaded at 0° with respect to the fibers direction are shown in Fig. 1.32, while corresponding locii for the specimens loaded at 45° off-axis are shown in Fig. 1.33. In both figures, $\Delta\sigma_{ax}$ and $\Delta\sigma_{hp}$ denote ranges of axial and hoop stress, respectively.

The applicability of FTPF criterion in reliably predicting fatigue strength under multiaxial loading is further demonstrated in Fig. 1.34, where predicted fatigue failure locii for $1-10^6$ cycles are shown along with experimental data from [19].

It is clearly shown in both cases examined that predictions made by the FTPF criterion are very close to, and are corroborated well by, the experimental data from multiaxial cyclic loads.

1.11.3 Stiffness Controlled, Sc–N, Fatigue Design Curves

Based on stiffness degradation data, already discussed in Section 1.10.2.2, stiffness-controlled Sc–N curves, corresponding to specific E_N/E_1 values, were calculated by means of Eq. (1.20). Fatigue strength curves were also defined at predetermined survival probability levels based on the parameters of the statistical model from Table 1.14 and were plotted in Figs. 1.15–1.21. Comparing these two sets of fatigue design curves, it was concluded that they could be correlated as follows. To any survival probability level, $P_s(N)$, corresponds a unique stiffness degradation value, E_N/E_1, which can be determined from the cumulative distribution function, $F(E_N/E_1)$, of the respective data. It is this value of E_N/E_1, for which $F(E_N/E_1) = P_s(N)$; see Fig. 1.35. Observing the two different curves derived as stated above, it was concluded that they are similar for all cases considered in this work, with the Sc–N being slightly more conservative in general. Therefore, one can use in design an Sc–N curve bearing information on both issues: survival probability and residual stiffness.

The derivation procedure of an Sc–N curve is schematically demonstrated in Figs. 1.35 and 1.36 for the data of 15° off-axis coupons under $R = 0.1$. In Fig. 1.36 both design curves, for 50 and 95% survival probability, are plotted together along with experimental failure data. It is observed indeed that Sc–N and S–N curves from each set lie very close, and that the former type of design curve is slightly more conservative. Using as design allowable the Sc–N at $E_N/E_1 = 0.96$, as seen from Fig. 1.35, a 95% reliability level is at least guaranteed while stiffness reduction will be less than 5%. Similar comments are also valid for Figs. 1.37–1.40, where results are shown for coupons cut at different off-axis angles and tested under different R ratios. It has to be mentioned that this good correlation between stiffness-based and reliability S–N curves is the rule followed by all other types of coupon, tested under different loading conditions. In Table 1.16, S–N curve equations are given for 95% reliability level for all data sets used in this study and compared to the corresponding stiffness-based Sc–N curve equations.

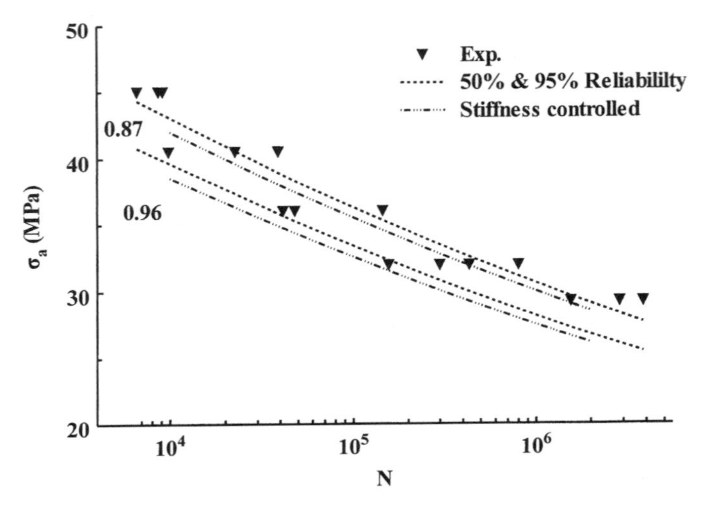

FIGURE 1.36 Sc–N vs. S–N curves. $R = 0.1$, 15° off-axis.

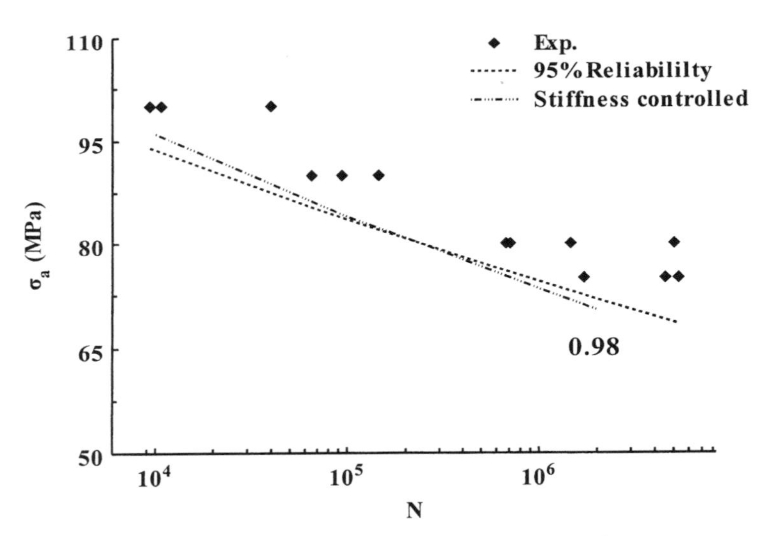

FIGURE 1.37 Sc–N vs. S–N curves. $R = -1$, 0° on-axis.

These results strongly recommend that despite the observed discrepancies, which are not significant in most of the cases, stiffness-based Sc–N curves be used instead of reliability S–N curves in design. Curves of the former type refer to two design parameters, reliability and stiffness degradation level. Thus, they can be used in design to cover requirements of design codes and regulations. In addition, Sc–N curves can be determined much faster, as stiffness degradation trends are readily captured with only a small number of coupons tested.

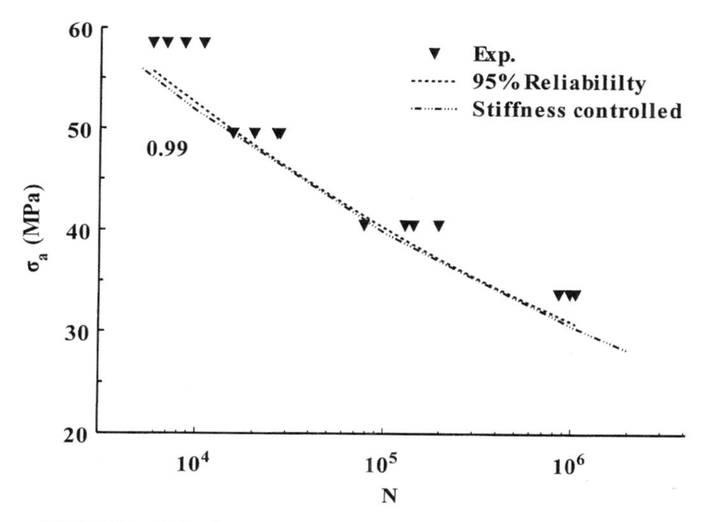

FIGURE 1.38 Sc–N vs. S–N curves. $R = 10$, 30° off-axis.

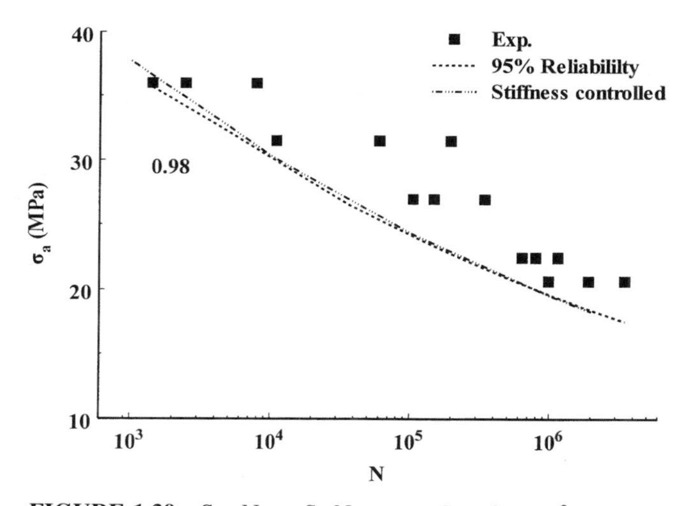

FIGURE 1.39 Sc–N vs. S–N curves. $R = 0.1$, 45° off-axis.

To demonstrate this, the procedure for the determination of stiffness-based Sc–N curves was repeated by using half of the coupons. A fraction of 50% of the coupons from each set was randomly selected and the calculations were repeated. The Sc–N curves, determined that way, were then compared to the original ones. Probability cumulative distributions were almost identical in most of the cases studied, for example, see Fig. 1.41. Thus, Sc–N curves were similar to those determined by using the full data set as shown, for example, in Fig. 1.42 for 30° off-axis coupons, tested under alternating stress, $R = -1$.

TABLE 1.16 Stiffness-Controlled and S–N Curves of 95% Survival Probability[a]

Direction \ R	10		−1		0.1		0.5	
	Sc–N	S–N	Sc–N	S–N	Sc–N	S–N	Sc–N	S–N
0°	$137.3N^{-0.0591}$ (0.97)	$120.4N^{-0.0472}$	$164.6N^{-0.0585}$ (0.98)	$148.5N^{-0.0499}$	$263.4N^{-0.1116}$ (0.85)	$204.2N^{-0.0861}$	$91.6N^{-0.0502}$ (0.95)	$84.9N^{-0.0445}$
15°					$75.7N^{-0.0733}$ (0.96)	$78.2N^{-0.0739}$		
30°	$147.5N^{-0.1139}$ (0.99)	$149.1N^{-0.1138}$	$113.2N^{-0.0807}$ (0.96)	$104.2N^{-0.0698}$				
45°	$97.6N^{-0.0758}$ (0.98)	$106.3N^{-0.0829}$	$133.4N^{-0.0850}$ (0.95)	$96.4N^{-0.0559}$	$72.7N^{-0.0950}$ (0.98)	$69.0N^{-0.0909}$	$41.7N^{-0.0719}$ (0.93)	$41.3N^{-0.0711}$
60°	$69.4N^{-0.0953}$ (0.98)	$67.1N^{-0.0864}$	$121.7N^{-0.0986}$ (0.89)	$99.9N^{-0.0809}$				
75°					$42.5N^{-0.0995}$ (0.83)	$35.5N^{-0.0813}$		
90°	$30.2N^{-0.0441}$ (0.99)	$32.4N^{-0.0501}$		$81.8N^{-0.0746}$	$31.8N^{-0.0963}$ (0.84)	$25.9N^{-0.0718}$		

[a] Numbers in parentheses indicate the respective E_N/E_1 value.

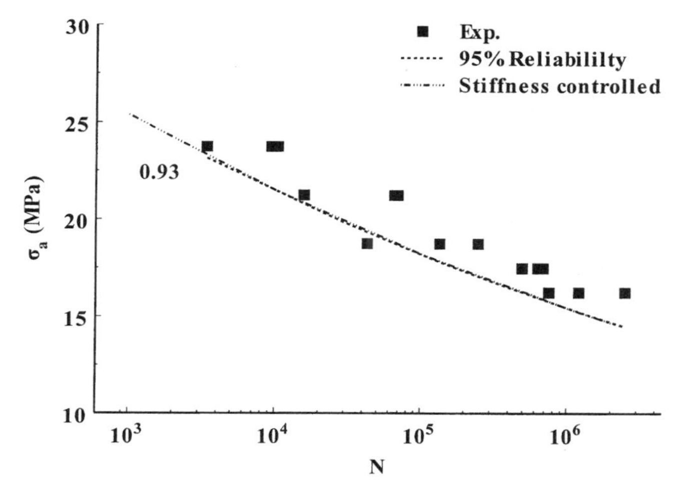

FIGURE 1.40 Sc–N vs. S–N curves. $R = 0.5$, 45° off-axis.

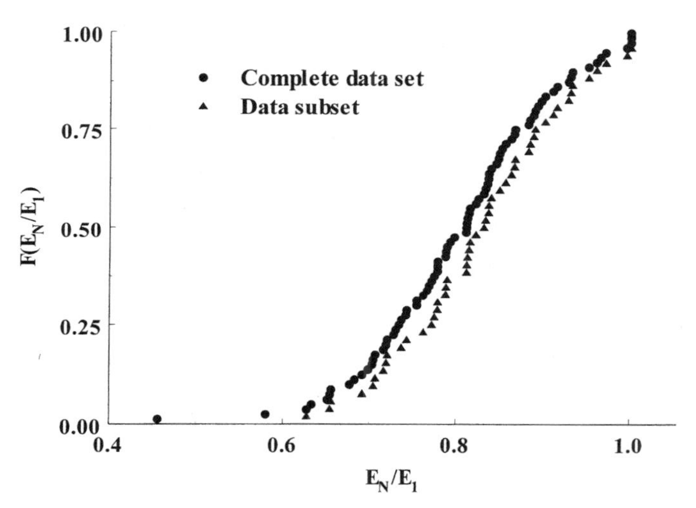

FIGURE 1.41 Sampling distributions of complete and half of data set. $R = -1$, 30° off-axis.

1.12 CONCLUSIONS

Fatigue performance of glass fiber-reinforced plastics under complex stress states was considered in this study. Prediction of operational life of structures made of said materials is feasible and can be based on measurements of fatigue strength and stiffness degradation.

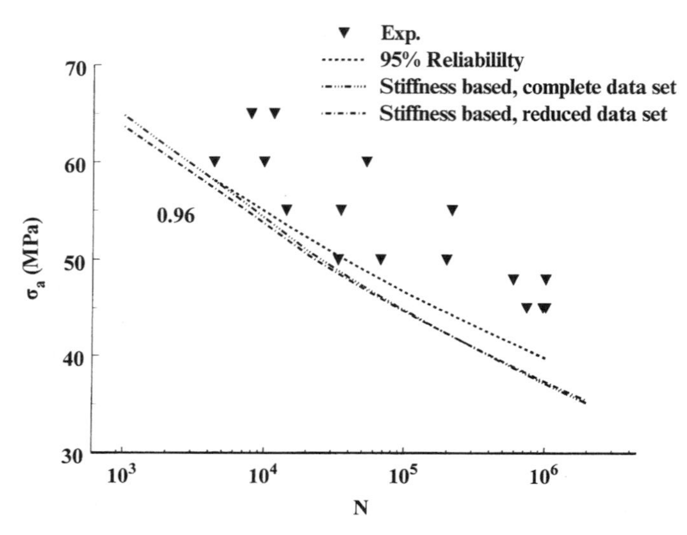

FIGURE 1.42 Comparison of Sc–N curves determined using all and half of experimental stiffness degradation data.

Quadratic failure tensor polynomial criterion, FTPF, forecasts satisfactorily material response under off-axis and multiaxial loading for all cases of stress ratio R considered in this study. Application of FTPF assumes the experimental determination of three S–N curves along principal material directions. Efficiency of the criterion is proved through comparison of theoretical predictions and experimental fatigue strength data. Fatigue strength variation with off-axis angles is similar to static strength variation, irrespective of stress ratio R. Off-axis experimental strengths, both in static and fatigue loading, are well predicted in most of the cases by failure tensor polynomial criterion.

For on-axis loaded coupons, 0°, it is shown that fatigue strength is higher in tensile loading ($R = 0.1$) than in compressive ($R = 10$) for $N < 10^6$. This is not the case, however, for all the other sets of coupons tested, where it was clearly demonstrated that compressive fatigue strength is higher than tensile strength for any off-axis loading orientation.

Constant life diagrams, composed for the median survival probability point out that for on-axis loading constant life curves are closer to a parabolic curve than to a Goodman straight line. This suggests that the use of the Goodman relation, along with the Palmgren–Miner rule, could lead to conservative design decisions. On the other hand, however, high cycle fatigue results, especially in the T–T sector of the $\sigma_a - \sigma_m$ plane and off-axis strength results, in general, suggest that a Goodman-like relationship between σ_a and σ_m is very optimistic.

During fatigue life the stiffness of a structural element is decreased. Observed stiffness degradation is correlated to the damage accumulated in the material. Herein, coupons cut at several off-axis angles from a multidirectional laminate $[0/(\pm45)_2/0]_T$, were subjected to cyclic loading and their stiffness changes were investigated. Longitudinal Young's modulus is defined as the average slope of the stress–strain loop and is a function of the number of cycles, N. Its variation, depending on the applied stress ratio and off-axis load orientation, is modeled by a simple empirical equation that produces acceptable fits of the experimental data.

Records of stiffness reduction for various R values were used to define fatigue design curves corresponding to specific modulus degradation and not to failure. The corresponding, stiffness-controlled, fatigue design curves, Sc–N, can serve better the requirements of design and full-scale testing of structural components made of FRP materials.

It was shown that Sc–N and S–N curves of 95% reliability can be correlated and, therefore, it was possible to define design allowables corresponding to predetermined levels of stiffness degradation and survival probability.

ACKNOWLEDGMENTS Part of this study was supported by the Greek General Secretariat of Research and Technology under contract EPET II #573 and by the Center for Renewable Energy Sources (CRES). Composite plates were prepared by rotor blade manufacturers Geobiologiki S.A. The authors gratefully acknowledge their assistance.

REFERENCES

1. K. L. Reifsnider and K. N. Lauraitis (Eds.), *Fatigue of Filamentary Composite Materials*, ASTM STP 636, American Society for Testing and Materials, Philadelphia, 1977.
2. K. N. Lauraitis (Ed.), *Fatigue of Fibrous Composite Materials*, ASTM STP 723, American Society for Testing and Materials, Philadelphia, 1981.
3. R. Talreja, *Fatigue of Composite Materials*, Technomic, Lancaster, 1987.
4. P. A. Lagace (Ed.), *Composite Material: Fatigue and Fracture*, 2nd ed. 5 ASTM STP 1012, American Society for Testing and Materials, Philadelphia, 1989.
5. K. L. Reifsnider (Ed.), *Fatigue of Composite Materials*, Vol. 4, Composite Materials Series, Elsevier, New York, 1991.
6. T. K. O'Brien, "Towards a Damage Tolerance Philosophy for Composite Materials and Structures," in S. P. Garbo (Ed.), *Composite Materials: Testing and Design*, 9th ed., ASTM STP 1059, American Society for Testing and Materials, Philadelphia, 1990.
7. S. G. Pantelakis, T. P. Philippidis, and T. B. Kermanidis, "Damage Accumulation in Thermoplastic Laminates Subjected To Reversed Cyclic Loading," in S. A. Paipetis and A. G. Yioutsos (Eds.), *High Technology Composites in Modern Applications*, University of Patras, Patras, Greece, 1995, pp. 156–164.
8. C. W. Kensche (Ed.), European Commission, DG XII, 1996.
9. R. M. Mayer, *Design of Composite Structures Against Fatigue: Applications to Wind Turbine Blades*, Mechanical Engineering Publications, Suffolk, 1996.
10. B. J. de Smet and P. W. Bach, *Database Fact: Fatigue of Composites for Wind Turbines*, ECN-C-94-045, 1994.

11. J. F. Mandell and D. D. Samborsky, DOE/MSU Composite Material Fatigue Database: Test Methods, Material and Analysis, Technical Report SAND97-3002, Sandia Laboratories, 1997.
12. D. R. V. van Delft, H. D. Rink, P. A. Joosse, and P. W. Bach, "Fatigue Behaviour of Fibreglass Wind Turbine Blade Material at the Very High Cycle Range," European Wind Energy Conference Proceedings, Vol. 1, Thessaloniki, Greece, 1994, pp. 379–384.
13. A. T. Echtermeyer, "Fatigue of Glass Reinforced Composites Described by One Standard Fatigue Lifetime Curve," European Wind Energy Conference Proceedings, Vol. 1, Thessaloniki, Greece, 1994, pp. 391–396.
14. P. A. Joosse, D. R. V. van Delft, and P. W. Bach, "Fatigue Design Curves Compared to Test Data of Fibreglass Blade Material," European Wind Energy Conference Proceedings, Vol. 3, Thessaloniki, Greece, 1994, pp. 720–726.
15. C. W. Kensche, "Lifetime of Gl-Ep Rotor Blade Material under Impact and Moisture," 3rd Symposium on Wind Turbine Fatigue Proceedings, Petten, Holland: IEA, April 21–22, 1994, p. 137–143.
16. D. R. V. van Delft, G. D. de Winkel, and P. A. Joosse, "Fatigue Behaviour of Fiberglass Wind Turbine Blade Material under Variable loading," 4th Symposium on Wind Turbine Fatigue Proceedings, Stuttgart, Germany: IEA, February 1–2, 1996, pp. 75–80.
17. C. W. Kensche, "Which Slope for Gl-EP Fatigue Curve?" 4th Symposium on Wind Turbine Fatigue Proceedings, Stuttgart, Germany: IEA, February 1–2, 1996, pp. 81–85.
18. M. J. Owen and J. R. Griffiths, *J. Mat. Sci.*, **13**, 1521–1537 (1978).
19. T. Fujii and F. Lin, *J. Comp. Mat.*, **29**(5), 573–590 (1995).
20. T. P. Philippidis and A. P. Vassilopoulos, *J. Comp. Mat.*, **33**(17), 1578–1599 (1999).
21. T. P. Philippidis and A. P. Vassilopoulos, *Int. J. Fat.*, **21**, 253–262 (1999).
22. T. P. Philippidis and A. P. Vassilopoulos, "Fatigue Design Allowables of GRP Laminates Based on Stiffness Degradation Measurements," *Comp. Sci. Tech.*, **60**, 2819–2828 (2000).
23. Z. Hashin and A. Rotem, *J. Comp. Mat.*, **7**, 448–464 (1973).
24. A. Rotem, *AIAA J.* **17**(3), 271–277 (1979).
25. Z. Fawaz and F. Ellyin, *J. Comp. Mat.*, **28**(15), 1432–1451 (1979).
26. M.-H. R. Jen and C.-H. Lee, *Int. J. Fat.*, **20**(9), 605–615 (1998).
27. M.-H. R. Jen and C.-H. Lee, *Int. J. Fat.*, **20**(9), 617–629 (1998).
28. S. I. Andersen, P. Brondsted, and H. Lilholt, "Fatigue of Polymeric Composites for Wingblades and the Establishment of Stiffness-controlled Fatigue Diagrams," Proceedings of 1996 European Union Wind Energy Conference. Göteborg, Sweden, May 20–24, 1996, pp. 950–953.
29. IEC-TC88-WG8 test guideline, "Full-scale Structural Testing of Rotor Blades for WTGS's," IEC 61400-23, 1998.
30. S. W. Tsai and E. M. Wu, *J. Comp. Mat.*, **5**, 58–80 (1971).
31. P. S. Theocaris and T. P. Philippidis, *Comp. Sci. Tech.*, **40**, 181–191 (1991).
32. S. W. Tsai and H. T. Hahn, *Introduction to Composite Materials*, Technomic, Lancaster, 1980.
33. H. T. Hahn and R. Y. Kim *J. Comp. Mat.*, **10**, 156–180 (1976).

34. A. L. Highsmith and K. L. Reifsnider, "Stiffness Reduction Mechanisms in Composite Laminates," in K. L. Reifsnider (Ed.), *Damage in Composite Materials*, ASTM STP 775, American Society for Testing and Materials, Philadelphia, 1982, pp. 103–117.
35. K. L. Reifsnider, K. Schulte, and J. C. Duke "Long-Term Fatigue Behaviour of Composite Materials," in T. K. O'Brien (Ed.), *Long-Term Behaviour of Composites*, ASTM STP 813, American Society for Testing and Materials, Philadelphia, 1983, pp. 136–159.
36. W. Hwang and K. S. Han *J. Comp. Mat.*, **20** 154–165 (1986).
37. B. Liu and L. B. Lessard, *Comp. Sci. Tech.*, **51** 43–51 (1994).
38. H. A. Whitworth *Composite Structures*, **40**(2), 95–101 (1998).
39. J. M. Whitney, I. M. Daniel, and R. B. Pipes, *Experimental Mechanics of Fiber Reinforced Composite Materials*, Prentice-Hall, 1984.
40. J. M. Whitney, "Fatigue Characterisation of Composite Materials," in K. N. Lauraitis (Ed.), *Fatigue of Fibrous Composite Materials*, ASTM STP 723, American Society for Testing and Materials, Philadelphia, 1981, pp. 133–151.
41. Din 65 586, Fatigue Strength Behaviour of Fiber Composites under One Stage Loading, Vorlage, April 1992.
42. W. H. Press, S. A. Teukolsky, W. T. Vetterling, and B. P. Flannery, *Numerical Recipes in Fortran. The Art of Scientific Computing*, 2nd ed., Cambridge University Press, Cambridge, MA, 1994.
43. F. J. Massey, *J. Am. Stat. Assoc.*, **46**, 68–78 (1951).
44. E. M. Wu and J. K. Scheublein, "Laminate Strength—A Direct Characterization Procedure," in *Composite Materials: Testing and Design*, ASTM STP 546, American Society for Testing and Materials, Philadelphia, 1974, pp. 188–206.

CHAPTER 2

Advanced Ceramic Materials

DAVID W. RICHERSON

Department of Materials Science and Engineering, University of Utah, Salt Lake City, Utah 84112

2.1 DEFINITION OF ADVANCED CERAMICS

We live in the "age of engineered materials." The properties of materials are either selected or developed to meet the needs of specific applications. Ceramics are a very broad class of materials with a wide range of properties [1]. Some advanced ceramics have special optical, electrical, or magnetic properties. Others have special mechanical or thermal properties. The focus of this chapter and the

Handbook of Advanced Materials Edited by James K. Wessel
ISBN 0-471-45475-3

next chapter will be on ceramics for structural applications where the mechanical and thermal properties are especially important.

Structural applications have varied material requirements. Some require high hardness to provide wear resistance, especially industrial applications that involve sliding, rolling, and fluid or particulate flow. Others require high strength to resist mechanical stresses or thermal stresses. Examples include bearings, cutting tools, and heat engine components. Others require high-temperature stability, corrosion resistance, or thermal shock resistance.

Several categories of advanced ceramics are discussed in this chapter: (1) monolithic (noncomposite) polycrystalline ceramics, (2) self-reinforced ceramics with composite microstructures, and (3) particle-reinforced or whisker-reinforced ceramic matrix composites. The subsequent chapter addresses continuous fiber-reinforced ceramic matrix composites.

Dramatic advances have occurred in ceramics technology in recent years [2, 3]. New and improved ceramics are now available that have much higher strength and toughness than prior ceramics. New design methods have been developed—especially through the use of finite-element codes for thermal and stress analysis—that are leading to substantial improvements in reliability and reduction in risk. This chapter describes some of the advances in key ceramic materials and reviews some of the success stories of applying these ceramics to challenging applications.

2.2 GENERAL IMPROVEMENTS IN MECHANICAL PROPERTIES

Major progress has been accomplished in the past 20–30 years to increase the capability of ceramics for thermal, wear, corrosion, and structural applications. In particular, the strength and toughness have been dramatically improved to the degree that ceramics are now available that can compete with metals in applications previously thought impossible for ceramics. Figure 2.1 illustrates the level of increases in the key structural characteristics of strength, toughness, and Weibull modulus. [Weibull modulus is the slope of the log–log plot of probability of failure versus fracture stress (strength) for test bars prepared from a block of the material.]

Strength is a measurement of the resistance to formation of a crack or structural damage in the material when a load is applied. Toughness is a measurement of the resistance of the material to propagation of a crack or extension of damage to the point of failure. The Weibull modulus is a measurement of the uniformity in strength. The lower the Weibull modulus, the higher the likelihood that the material will fail at a stress substantially below the average strength. Thus, high Weibull modulus means better material reliability and greater ease in designing with the material.

Most ceramics in the 1960s had strength well below 345 MPa (50,000 psi). Now aluminum oxide, silicon carbide, silicon nitride, and toughened zirconia are available with strength above 690 MPa (100,000 psi). Strength at elevated

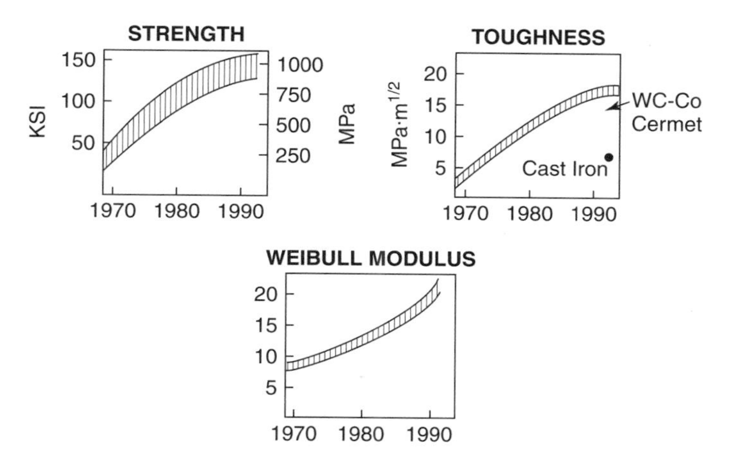

FIGURE 2.1 Strength, toughness, and uniformity of ceramic materials have been dramatically increased since 1970 [2].

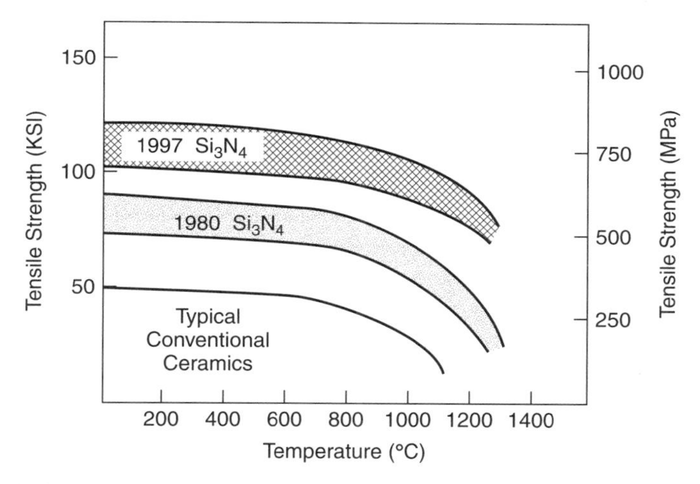

FIGURE 2.2 Silicon nitride ceramics strength improvements [2].

temperatures also has been improved, as shown in Figure 2.2 for silicon nitride materials. SiC ceramics retain strength to even higher temperatures.

One of the most significant advances in ceramics during the past 20 years has been to increase fracture toughness. Increased fracture toughness is important to industry because it reduces risk of fracture during installation and service, a risk that has always been a concern with glass and traditional ceramics. Figure 2.3 compares the toughness of some of the new ceramic materials with typical ceramics and other key engineering materials. Glass has fracture toughness of about 1 $MPa \cdot m^{1/2}$, and conventional ceramics range from about 2–3 $MPa \cdot m^{1/2}$. Steel

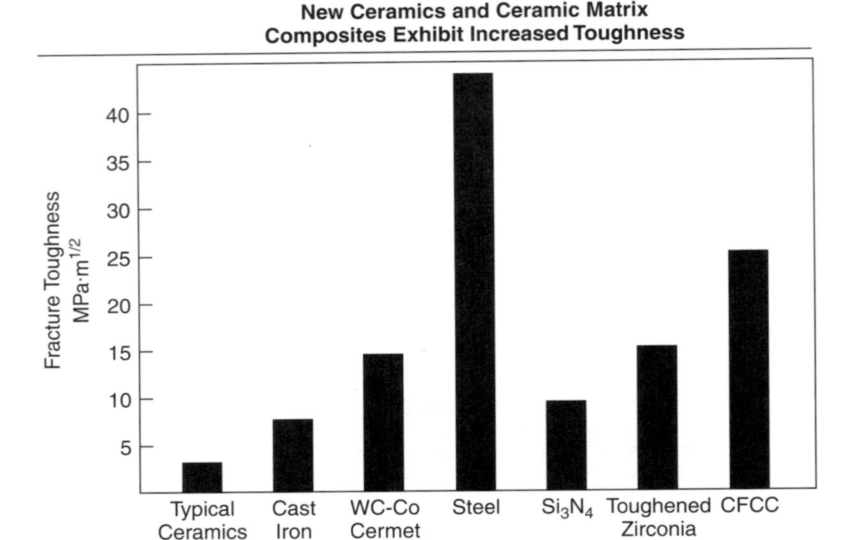

FIGURE 2.3 Fracture toughness of new ceramics compared to other materials [2].

is around 40 MPa $\cdot$ m$^{1/2}$. Some silicon nitride materials now approach 10 MPa $\cdot$ m$^{1/2}$, which is tougher than cast-iron. Some transformation-toughened zirconia materials have toughness around 15 MPa $\cdot$ m$^{1/2}$, which is higher than that of many tungsten carbide–cobalt (WC–Co) cermets. These new tougher ceramics have demonstrated dramatically improved resistance to impact, contact stress, and handling damage and are providing high reliability and durability that users are accustomed to receiving with metals and WC–Co. New continuous fiber-reinforced ceramic composites (CFCCs) are under development that provide further improvements in fracture toughness, as discussed in the next chapter.

2.3 ALUMINUM OXIDE

Aluminum oxide (generally referred to as alumina) is the same composition as sapphire (Al_2O_3), which accounts for its high hardness and durability. Alumina ceramic is produced by compacting alumina powder into a shape and firing the powder at high temperature to allow it to densify (sinter) into a solid, polycrystalline, nonporous part. Alumina is the most mature high-technology ceramic in terms of quantity produced and variety of industrial uses. Approximately 5 million metric tons were produced in 1995 for wear, chemical, electrical, medical, and other applications. Table 2.1 lists some of the applications of alumina.

Alumina is used in these applications because of its excellent combination of properties, including high hardness and wear resistance, chemical resistance, smooth surface, reasonable strength, and moderate thermal conductivity. Table 2.2

TABLE 2.1 Production Applications of Aluminum Oxide Ceramics

Seal rings	Chute and conveyor liners
Rotary and gate valves	Nozzle, pipe, and tubing liners
Pump parts and liners	Wear pads for any application
Papermaking foils, suction box covers, palm guides, liners	
Cyclone liners	Spark plug insulators
Liners in coal-handling systems	Sodium vapor lamp arc tubes
Wire drawing tooling	Thermocouple protection tubes
Thread guides in the chemicals and textile industries	Radomes
Chemical laboratory ware	Grinding wheels, abrasives, polishes
Molten metal filters, crucibles	Glass tank linings
Mill liners and grinding media	Cutting tool inserts
	Heat exchange media
	Medical applications

TABLE 2.2 Comparison of Typical Properties of Aluminum Oxide and Other Advanced Ceramic

	Coors AD-96 Alumina	Ceradyne Ceralloy 147-E1Si_3N_4	Carborundum Hexoloy SA Silicon Carbide	Coors ZTA	Coors TZP	Coors YTZP
Flexural Strength (MPa)	358	700	380	450	620	900
Elastic Modulus (GPa)	300	310	410	360	200	200
Fracture Toughness ($MPa \cdot m^{1/2}$)	4–5	6	4.6	5–6	11	13
Thermal Conductivity (W/mK)	24.7	26	125	27	2.2	2.2
Coefficient of Thermal Expansion ($10^{-6}/°C$)	8.2	3.1	4.0	8.2	10.1	10.3
Density (g/cm^3)	3.75	3.25	3.1	4.0	6.02	6.02

Source: Company data sheets from Coors Ceramics, Ceradyne Inc., and Saint-Gobain Advanced Ceramics Corporation. ZTZ = zirconia-toughened alumina, TTZ = transformation-toughened zirconia, and YTZP = yttria tetragonal zirconia polycrystal.

lists some properties of commercial grades of alumina compared to several other advanced ceramics. New laboratory grades of alumina have even better properties. For example, one fine-grained (0.56 μm) sintered alumina being evaluated for cutting tool inserts has average bending strength of 650–700 MPa and Vickers hardness (with 10 kgf load) of 22.1 ± 0.9 GPa [4]. This alumina has substantially out-performed cubic boron nitride and Al_2O_3–TiC for cutting cast-iron and hardened steel by turning on a lathe at 300 m/min.

Alumina is presently the lowest cost high-performance ceramic because of the large quantity produced. The alumina suppliers have an enormous design and experience database so that they should be consulted when seeking an alternate material for increased corrosion resistance, wear resistance, dimensional stability, decreased friction, and higher temperature use.

2.4 SILICON NITRIDE

Silicon nitride refers to a family of ceramics whose primary constituent is Si_3N_4. The ceramics in this family have a favorable combination of properties that includes high strength over a broad temperature range, high hardness, moderate thermal conductivity, low coefficient of thermal expansion, moderately high elastic modulus, and unusually high fracture toughness for a ceramic [5, 6]. This combination of properties leads to excellent thermal shock resistance, ability to withstand high structural loads at high temperatures, and superior wear resistance.

Silicon nitride has been under development continuously in the United States since the late 1960s. Initial efforts were directed toward development of components for gas turbine engines, but this turned out to be a very difficult challenge. Although extensive testing has been conducted, silicon nitride has not yet reached a significant level of production for turbine engines [7]. However, silicon nitride ceramics have reached large-scale production for cutting tools, bearings, turbocharger rotors, diesel cam follower rollers, and diesel prechambers and have reached moderate levels of production for other applications such as thermocouple protection tubes, grit-blast nozzle liners, wire-forming rolls and guides, papermaking dewatering foil segments, check valve balls, downhole oil well parts, aluminum die-casting tooling, and a variety of custom wear parts [8]. Figures 2.4 and 2.5 illustrate some silicon nitride parts.

Although the initial driver for silicon nitride development was gas turbine engine components, the first major application was cutting tool inserts [9]. Cutting hard metals such as cast-iron, tool steels, and superalloys results in high temperature at the tool–workpiece interface. Tool failure was usually caused by a combination of wear and high-temperature corrosion. WC–Co, the traditional workhorse for metals machining, wears/corrodes rapidly if the temperature gets too high, so the cutting speed must be limited to around 120 m/min (~400 surface feet per minute) and sometimes even down to around 25 m/min. Silicon nitride is much more resistant that WC–Co to temperature and chemical corrosion. Cutting speeds higher than 1520 m/min have been demonstrated with silicon nitride at a depth of cut of 5 mm and feed rate of 0.4 mm per revolution. Such a rapid rate of metal removal heats the silicon nitride cutting edge to around 1100°C and imposes extreme conditions of thermal shock, impact, contact stress, and erosion/corrosion. This gives an indication of the severe conditions that silicon nitride materials can survive.

The use of silicon nitride cutting tool inserts has had a dramatic effect on manufacturing output [10]. For example, face milling of gray cast-iron gear-case housings with silicon nitride inserts doubled the cutting speed, increased tool life from one part to six parts per edge, and reduced the average cost of inserts by

FIGURE 2.4 Silicon nitride parts including blast nozzle liners, wire-forming rolls and guides, papermaking dewatering foil segments, check-valve balls, downhole oil well parts, custom wear parts, and centrifugal dewatering screen and scraper blade for potash and coal dewatering. (Photo courtesy of Ceradyne, Inc., Costa Mesa, CA).

FIGURE 2.5 Experimental silicon nitride gas turbine engine components. (Photos courtesy of Honeywell Engines, Systems, and Services, Phoenix, AZ).

50%. Outside grinding of diesel truck cylinder liners increased the number of parts machined per tool index from around 130 to 1200 and totally eliminated a prior problem with insert breakage. As a result, tool life was increased to achieve 9600 cylinders per cutter load of inserts compared to 450. The decreased downtime alone increased the output per shift by 25%.

A more recent application for silicon nitride that is having major impact on many industries is bearings [5, 11]. Silicon nitride was first demonstrated as a superior bearing material in 1972 [12] but did not reach production until nearly 1990 because of challenges in reducing the cost. Since 1990 the cost has been reduced substantially as production volume has increased. Although silicon nitride bearings are still 2–5 times more expensive than the best bearing steel, their superior performance and life have resulted in rapid escalation in their use. About 15–20 million silicon nitride bearing balls were being produced in the United States by 1996, and the number has increased dramatically each year since. Table 2.3 lists some applications for silicon nitride bearings.

One of the most important applications of silicon nitride bearings is in machine tool spindles [5]. Because of their light weight (60% lighter than steel), silicon nitride bearings can be operated at much higher speed than metal bearings without generating a critical level of centrifugal stress. Because of their low thermal expansion (one-fifth that of steel) and high elastic modulus, the silicon nitride bearings can operate to much closer tolerances than metal bearings, which enables machines with higher precision and lower vibration. Because of their high hardness and smoother surface, the silicon nitride bearings run smoother and wear at about one-seventh the rate of the best metal bearings. All of these factors together result in 3–10 times the life of metal bearings, up to 80% higher speed capability, about 80% lower friction, higher operating temperature, and 15–20% reduction in energy consumption.

In addition to cutting tool inserts, bearings and check valves, silicon nitride is being vigorously evaluated for diesel and auto engine valves, valve guides, stator vanes and rotors for turbines, a variety of wear parts, forging dies for aluminum, and many other potential products. As additional production applications are achieved and current production levels increase, it is anticipated that the cost of silicon nitride will be significantly reduced, which will remove the primary barrier that has limited broad use of advanced silicon nitride materials.

TABLE 2.3 Some Applications of Silicon Nitride Bearings

Machine tool spindles	Gas turbine engines	Pumps
High-speed hand grinders	High-speed compressors	Gas meters
Food-processing equipment	High-speed train motors	Check-valve balls
CAT scanners	Air-driven power tools	Chemical-processing equipment
Spectroscopes	Gyroscopes	Galvanizing lines
Photo copier roll bearings	Optical-kinematic mounts	High-speed dental drills
Medical centrifuges	Racing cars	In-line skates
Aircraft anti-icing valves	Semiconductor processing equipment	Textile equipment
Gearboxes	Actuators	Radar
Helicopter pitch blades	Aircraft wing flap ball screws	Butterfly valves
Shuttle liquid oxygen pumps	Shuttle main engine	Instruments

The key message from the above examples is that the silicon nitride family is a new generation of ceramics that are much more durable and resistant to brittle fracture than many engineers may realize and may be viable options to consider. The key properties that distinguish silicon nitride from traditional ceramics are the high toughness, thermal shock resistance, and both chemical and structural stability at high temperature.

2.5 SILICON CARBIDE

Another ceramic that is well established in the marketplace is silicon carbide (SiC). Silicon carbide has many of the same applications as aluminum oxide and silicon nitride. It is more expensive than alumina and has lower toughness than silicon nitride, so it is not the optimum material for all corrosion or wear applications. But where it can be used, it normally provides superior wear resistance and long life. Table 2.4 identifies some of the production applications of silicon carbide.

Silicon carbide also is important for tooling in the semiconductor industry, for laser mirrors, as a substrate for wear-resistant diamond coatings, as an abrasive and grinding wheel, as heating elements and igniters, as an additive for reinforcement of metals, and for numerous refractories applications.

Like silicon nitride, silicon carbide is a family of materials each with its special characteristics. Most of the silicon carbide materials have very high hardness (harder than alumina and silicon nitride) and thus have superior wear resistance. Most have unusually high thermal conductivity for a ceramic, low thermal expansion compared to metals, and very high temperature capability. Some actually increase in strength at elevated temperature, such as sintered silicon carbide from Saint-Gobain Advanced Ceramics Corporation that has room temperature flexure strength slightly above 413 MPa (60,000 psi) and that increases in strength to around 580 MPa (80,000 psi) at 1800°C.

Relatively pure SiC also has excellent resistance to corrosion in the presence of hot acids and bases. In one series of tests reported by Saint-Gobain Advanced Ceramics Corporation in one of its product brochures, dense SiC was immersed in different acids and bases for 125–300 h. For 98% sulfuric acid at 100°C, the SiC lost only 1.8 mg/cm^2/year compared to >1000 for tungsten carbide with 6% cobalt and 65 for 99% pure alumina. For 50% NaOH at 100°C, the SiC

TABLE 2.4 Production Applications of Silicon Carbide

Seals	High-temperature liners, refractories
Thrust bearings	Heat exchanger tubes
Valves	Thermocouple protection tubes
Pump parts	Links for high-temperature belt furnace
Cyclone liners	Bearings in magnetic drive pumps
Radiant burners	Grit blast nozzle liners

FIGURE 2.6 Silicon carbide seal and pump parts. (Photo courtesy of Saint-Gobain Advanced Ceramics Corporation, Niagara Falls, NY).

lost only 2.5 $mg/cm^2/year$ compared to 5 for WC–Co and 75 for alumina. The SiC exhibited even less weight loss (>0.2 $mg/cm^2/year$) for exposures in highly concentrated hot nitric and phosphoric acids and room temperature HCl and HF. Because of the high corrosion resistance combined with high wear resistance, SiC is important for seals and pump components, as illustrated in Figure 2.6.

2.6 TRANSFORMATION-TOUGHENED ZIRCONIA

Transformation-toughened zirconium oxide (TTZ) is another family of important high-strength, high-toughness ceramics that have been developed during the last 20–25 years [13, 14]. TTZ materials have fracture toughness values ranging from about 6–15 $MPa \cdot m^{1/2}$, compared to conventional ceramics with fracture toughness of about 2–3 MPa · $m^{1/2}$. The mechanism of toughening in TTZ materials involves a volume increase due to a polymorphic transformation that is triggered when an applied stress causes a crack to form in the TTZ [15]. The volume increase only occurs for material adjacent to the crack and presses against the crack to keep it from propagating through the TTZ. Some forms of steel have a similar mechanism of toughening, so TTZ has sometimes been called *ceramic steel.*

Figure 2.7 shows the microstructure of one type of TTZ called *partially stabilized zirconia* (PSZ). It consists of lenticular-shaped precipitates of the tetragonal form of zirconia distributed throughout larger grains of the cubic phase of zirconia. The tetragonal grains are the ones that transform adjacent to a crack. Another TTZ is made up completely of tiny grains of the tetragonal phase and is referred to as *tetragonal zirconia polycrystal* (TZP). Both types are mentioned because they each have different properties, and one may be preferable for a specific application.

Transformation toughening was a breakthrough in achieving high-strength, high-toughness ceramic materials. For the first time in history a ceramic material was now available with an internal mechanism for actually inhibiting crack propagation. A crack in a normal ceramic travels all the way through the ceramic with little inhibition, resulting in immediate fracture. TTZ has fracture toughness (resistance to crack propagation) three to six times higher than normal zirconia

FIGURE 2.7 Microstructure of PSZ type of transformation-toughened zirconia. (Photo courtesy of Professor Arthur Heuer, Case Western Reserve University, Cleveland, OH).

TABLE 2.5 Successful Applications of Transformation-Toughened Zirconia Ceramics

Tooling for making aluminum cans	Knife and scissor blades
Wire-drawing capstans, pulleys, rolls, guides, and some dies	Cutting tool inserts
Dies for hot extrusion of metals	Hip replacements
Golf cleats, putters, drivers	Buttons

and most other ceramics. It is tougher than cast-iron and comparable in toughness to some compositions of WC–Co cermet.

Table 2.5 lists some of the applications where TTZ has been successful. TTZ ceramics typically cost around four times as much as steel and two times as much as WC–Co for a part such as an extrusion die. In spite of the higher cost, though, TTZ often can provide sufficient increased life to justify its use on a life-cycle cost basis. The suppliers can provide information on life-cycle cost for existing applications and can probably estimate for similar applications.

2.7 OTHER MONOLITHIC ADVANCED CERAMICS

The ceramic materials discussed so far each are being used successfully in a wide range of applications. Many other monolithic ceramics have proven themselves in niche applications. For example, cordierite (a magnesium aluminosilicate that has low thermal expansion) has been used as a honeycomb structure catalyst substrate in automotive catalytic converter pollution control devices. Cordierite-based catalytic converters have saved us from over 1.5 billion tons of air pollution since they were introduced in the mid-1970s.

ZrB_2 and some other diborides have demonstrated very high-temperature capability and are being evaluated for rocket nozzle liners and for the leading edges of hypersonic vehicles [16]. AlN has been developed with very high thermal conductivity and is beginning to find applications as tooling in the manufacture of integrated circuits. Boron carbide (B_4C) has incredible hardness and has been used successfully as armor for personnel and military vehicles and also for wear resistance applications such as liners for sand blast nozzles [17].

2.8 SELF-REINFORCED CERAMIC COMPOSITES

A composite is a mixture of materials engineered with the intention of obtaining the best characteristic of each material. In the case of ceramics, composite microstructures can result in an increase in fracture toughness that can enhance durability and reliability. Several general approaches have been developed in recent years: self-reinforcement, addition of a ductile metal phase, addition of a dispersion of particles or whiskers, and addition of a network of continuous fibers [18, 19]. WC–Co cermets are examples of addition of a ductile metal but will not be discussed. Continuous fiber reinforcement is discussed in the next chapter. The following sections discuss self-reinforcement and addition of particles and whiskers.

The simplest and generally most cost-effective method of forming a ceramic composite microstructure is self-reinforcement. It is also often referred to as *in situ* reinforcement or toughening. That is because the composite microstructure is achieved in place during the sintering (densification) of the material by control of chemistry and temperature, rather than by mixing in a second phase prior to sintering. Self-reinforcement has been obtained in several ways: (1) forming a multiphase microstructure where one phase acts as the matrix and another acts as a reinforcement, (2) heat-treating to cause a phase to precipitate or crystallize into the matrix phase, and (3) growth of elongated intertwined grains.

2.8.1 Multiphase Microstructure

An example that illustrates a ceramic composite with a multiphase microstructure achieved *in situ* during sintering is shown in Figure 2.8. This ceramic composite

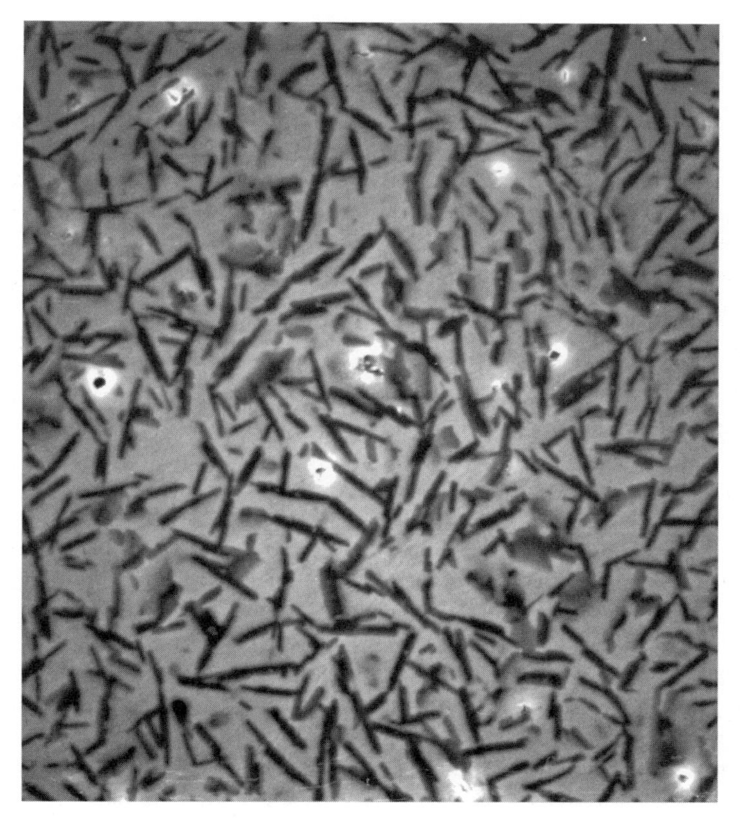

FIGURE 2.8 Microstructure showing aluminate platelets formed *in situ* during sintering. (Photograph courtesy of Raymond Cutler, Ceramatec Inc., Salt Lake City, UT).

consists of CeO_2-doped transformation-toughened ZrO_2 containing an interlacing network of platelets of strontium aluminate [20]. Two wt % $SrZrO_3$ and 30 vol % Al_2O_3 powders were mixed with a coprecipitated powder of ZrO_2-12 mol % CeO_2 and sintered at 1500°C. Thin platelets of strontium aluminate about 0.5 μm wide and 5–10 μm long formed throughout the zirconia matrix during sintering. The resulting strength (in four-point bending) was 726 MPa, and the fracture toughness was 11.2 MPa · $m^{1/2}$. The mechanisms of toughening were a mixture of transformation toughening and crack deflection.

A crack passing through a monolithic ceramic cuts across grains, goes around grains, or follows along natural crystallographic cleavage planes. The crack's path is relatively short, so that the total energy needed to drive the crack through is generally small. The resulting fracture toughness is typically 1–2 MPa · $m^{1/2}$ for a single-crystal ceramic and 2–3.5 MPa · $m^{1/2}$ for a polycrystalline ceramic. The aluminate platelets deflect the crack, forcing it to follow a longer and more tortuous path through the ceramic. This increases the amount of energy required to drive the crack through the material, resulting in higher fracture toughness.

2.8.2 Formation of a Precipitate or Dispersion of Crystals During Heat Treating

Transformation toughening in PSZ is a good example of formation of a reinforcing phase through the careful control of temperature (and chemistry). As mentioned before, toughening in PSZ involves a dispersion of nanoscale (typically under 500 nm) lenticular tetragonal zirconia in larger grains of cubic zirconia. The nanoscale reinforcement develops during the fabrication process. Zirconia powder mixed with MgO or CaO powder is compacted into the desired shape and sintered at a high enough temperature that the material densifies and that the zirconia and oxide additive form a solid solution with a cubic zirconia structure. The temperature is then reduced such that the thermodynamically stable phases are a mixture of cubic zirconia and tetragonal zirconia. Tiny nuclei of tetragonal zirconia begin to precipitate in the cubic zirconia grains. By controlling the temperature and time of heat treatment, the precipitates are allowed to grow to an optimum size ranging from about 100 to 300 nm.

The fracture toughness of PSZ is typically 6–10 MPa · $m^{1/2}$, although some values have been reported exceeding 15 MPa · $m^{1/2}$. The toughening mechanism is referred to as *crack shielding*. The compressive stress produced due to the volume increase as the tetragonal precipitates transform to monoclinic zirconia shields the tip of the crack from tensile stress.

Another interesting example of manipulation of microstructure during heat-treating is Macor, a material developed in the early 1970s by Corning. In this case the *in situ* reinforcement is achieved by crystallization during the fabrication process [21]. First, a composition nominally 47.2% SiO_2, 16.7% Al_2O_3, 8.5% B_2O_3, 9.5% K_2O, 14.5% MgO, and 6.3% F is melted and cast as slabs or cylinders of glass. The glass is then heat treated to form tiny nuclei of magnesium fluorophlogopite mica crystals. Further heat treatment grows these crystals to 5–10 μm diameter flakes that form a "house-of-cards" structure throughout the glass matrix, resulting in a very high degree of toughening by crack deflection. The composite is not particularly strong (~60–102 MPa), but it is very resistant to fracture and is soft enough (roughly between Teflon and brass) that it can be machined with conventional metallic drill bits and cutters. Macor and similar glass–ceramic composites have been used extensively for glass-sealed electrical feedthroughs, face seals, positioning and heat-treating fixtures, dental repairs, and many other applications.

2.8.3 Microstructure Containing Elongated, Intertwined Grains

Flat platelets in the microstructure, such as mica crystals and aluminate crystals, cause a crack to deflect only in a single plane. Elongated rod-shaped grains in the microstructure force a crack to deflect in more than one plane to get around the grain. This requires more energy, so highly elongated grains have the potential to achieve higher toughness than platelets. Faber and Evans [22] predicted and verified experimentally that a dispersion of disk-shaped particles or grains can increase toughness by a factor of 3 and rod-shaped ones by a factor of 4.

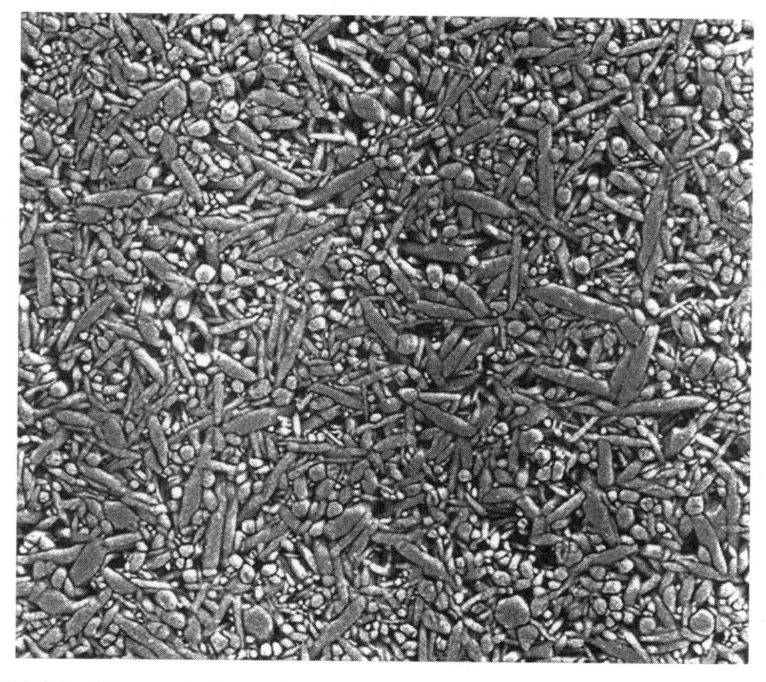

FIGURE 2.9 Elongated, intertwined microstructure of high-strength, high-toughness AS-800 silicon nitride material manufactured by Honeywell Engines, Systems, and Services, Phoenix, AZ. (Photo courtesy of George Graves, University of Dayton Research Institute, Dayton, OH).

Microstructures with elongated grains have been achieved for some silicon nitride materials. These are referred to as *self-reinforced silicon nitride* [23] and have been reported to have fracture toughness values ranging from about 6–14 $MPa \cdot m^{1/2}$ [24–26]. Alpha phase Si_3N_4 powder is blended with MgO, Y_2O_3, $Al_2O_3 + Y_2O_3$, or other oxide sintering aids. At high-temperature, the oxide reacts with a thin layer of SiO_2 that coats each Si_3N_4 particle to form a liquid-phase. The α-Si_3N_4 particles dissolve and recrystallize as elongated β-Si_3N_4 grains. By control of chemical composition, temperature, and time at temperature, an intertwined structure such as that shown in Figure 2.9 results. The high toughness of this type of intertwined structure has been a significant factor in the success of silicon nitride in surviving applications that prior ceramics have not been able to survive.

2.9 PARTICLE-REINFORCED CERAMIC COMPOSITES

The options for achieving a composite microstructure by self-reinforcement are limited. Another approach is to add particles of a second ceramic into a ceramic matrix during the fabrication process. This opens up many additional options for making ceramic matrix composites.

Most powders that are added to a ceramic for toughening are prepared by crushing and grinding or by chemical processes. These powders are typically equiaxed (roughly spherical), like a grain of sand, and between 0.5 and 40 μm in diameter. The composite is prepared by mixing the reinforcing powder with the matrix ceramic powder and compacting the powders into the desired shape by a conventional ceramic fabrication processes such as pressing. The compact is then placed in a high-temperature furnace and sintered the same way that the matrix would be sintered if the reinforcing powder had not been present. This conventional sintering works for small to moderate volume fraction of reinforcing particles, generally up to about 15–20%. For larger volume fraction of particles, hot pressing or postsintering hot isostatic pressing may be required if a pore-free composite is desired. Both are more costly than conventional sintering.

An important ceramic matrix composite with equiaxed particle reinforcement is Al_2O_3 with a dispersion of nominally 30–35% titanium carbide (TiC) particles. This material was initially developed as a cutting tool insert. Alumina without reinforcement had been used intermittently as a cutting tool since the 1920s, but only with limited success. The alumina–TiC had higher hardness and slightly higher toughness and could cut a wider range of alloys including hardened steel, chilled cast-iron, and cast-iron with an abrasive surface scale. It had improved reliability and could even survive interrupted cuts.

An important early success of alumina–TiC cutting tool inserts was in the steel industry [10]. Large steel rolls (typically over 4 m long and 75 cm in diameter) in steel rolling mills require frequent refurbishing. This refurbishing was previously done using an expensive ceramic grinding wheel and required 14–18 h per roll. Use of alumina–TiC cutting tool inserts reduced the refurbishing time to 5 h per roll [10] and became standard practice.

Alumina–TiC also has become important in the computer industry as the substrate material for read–write heads. Its attributes for this application are light weight, high stiffness, and ability to be machined chip-free to a precision smooth surface.

Other particles have been added to alumina in efforts to increase toughness. A 10 vol% of 30-μm-diameter flat plates of Ba-mica was reported to increase the toughness to 8.6 $MPa \cdot m^{1/2}$ [27]. A 5 vol% of titanium diboride particles was reported to result in toughness of 6.5 $MPa \cdot m^{1/2}$ [28]. Addition of dispersions of particles to SiC and Si_3N_4 also have resulted in increase in toughness. Examples are listed in Table 2.6 and the strength and toughness values compared with *in situ* reinforced silicon nitride, particle-reinforced alumina, whisker-reinforced ceramics, and a couple of ductile metal-reinforced ceramics.

The particulate-reinforced ceramics listed in Table 2.6 resulted in increased toughness primarily due to the mechanism of crack deflection. Addition of transformation-toughened zirconia particles to other ceramics can result in toughness increase by the mechanism of crack shielding. Toughening occurs if the particles are small (usually under 0.5 μm), if the host ceramic is strong enough to prevent the particles from transforming during cooling for the sintering temperature, and if there is no chemical interaction between the materials. Alumina with 15–20% addition of transformation-toughened zirconia particles has been reported to have toughness between 6.5 and 15 $MPa \cdot m^{1/2}$ and flexure strength between 480 and 1200 MPa [38]. These values of toughness and strength are comparable to values reported for pure transformation-toughened zirconia, and the transformation-toughened alumina (TTA) has higher hardness and is thus

TABLE 2.6 Comparison of Strength and Toughness for Various Ceramic Matrix Composites

Material	Flexural Strength (MPa)	Fracture Toughness ($MPa \cdot m^{1/2}$)	Reference
Alumina with 30 wt% TiC particles	638	4.5	29
Si_3N_4 with 30 vol% 8-μm SiC particles	885	4.9	30
SiC with 16 vol% TiB_2 particles	478	6.8–8.9	31
Alumina with 30 vol% 30-μm Ba-mica	—	8.6	27
Alumina with 30 vol% SiC whiskers	660	8.6	32
Si_3N_4 with 30 vol% 0.5-μm SiC whiskers	970	6.4	33
Si_3N_4 with 30 vol% 5-μm SiC whiskers	450	10.5	34
Si_3N_4 with 30 vol% BN-coated Si_3N_4 whiskers	428	9.2	35
$MoSi_2$ with 20 vol% SiC whiskers	310	8.2	36
In situ reinforced silicon nitride	785	8.2	24
In situ reinforced silicon nitride	900	9.7	25
In situ reinforced silicon nitride	550	10.6	26
ZrC-ZrB_2 with 24.2 vol% Zr metal	880	~20	37
ZrC-ZrB_2 with 2.5 vol% Zr metal	870	~11	37

Source: Adapted from Ref. 19.

more resistant to some forms of wear. The TTA also is lighter in weight and the raw materials are lower in cost than for TTZ. However, the TTA is more notch sensitive and also tends to chip during grinding [39].

2.10 WHISKER-REINFORCED CERAMIC MATRIX COMPOSITES

Ceramic whiskers are usually single crystals that have grown preferentially along a specific crystal axis under vapor or liquid–vapor conditions. The whiskers typically range in size from 0.5 to 10 μm in diameter and a few microns to a few centimeters in length and can have very high strength. Some silicon carbide whiskers have been reported with a strength of 21,000 MPa and a Young's modulus of 840 GPa [40].

Whiskers are more difficult than particles to disperse uniformly in a ceramic matrix. Furthermore, the matrix is more difficult to densify to a pore-free condition; the whiskers form an infrastructure that inhibits shrinkage of the ceramic during sintering. For whisker volume fraction greater than about 10%, either hot pressing or liquid-phase sintering are generally required. Table 2.6 included some examples of whisker-reinforced ceramic matrix composites.

The most extensive development has been conducted on the addition of SiC whiskers to aluminum oxide [41, 42]. Figure 2.10 illustrates improvement in strength with different volume percent of SiC whiskers. The whiskers increase the high-temperature strength as well as the room-temperature strength and also improve the creep resistance, the stress rupture life, the thermal shock resistance, and the Weibull modulus. A data sheet from Advanced Composite Materials Corp., Greer, South Carolina, reported that alumina with SiC whiskers survived quenching from 900°C into room temperature water, while unreinforced alumina

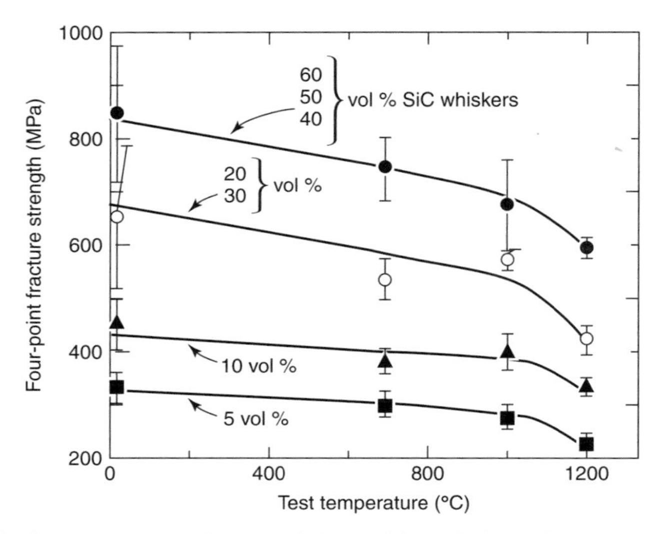

FIGURE 2.10 Improvement in strength by addition of SiC whiskers to alumina [32].

showed a severe drop in strength after a quench from only 200°C. Rhodes and co-workers [43] reported an increase in Weibull modulus to 22.5 for alumina containing 25 vol% SiC whiskers, compared to Weibull modulus of only 4.6 for the baseline alumina with no whiskers. High modulus means low scatter in strength data and generally high reliability. For example, the baseline alumina strength test bars showed large variation in strength, ranging from 300 to 650 MPa, while the alumina with 25% SiC whiskers varied only from about 580 to 700 MPa.

As was the case with silicon nitride and alumina–TiC, the first major industrial application of alumina–SiC_w composites was for cutting tool inserts. The increased toughness allowed these cutting tools to survive better than monolithic alumina, and the excellent high-temperature strength and chemical corrosion resistance allowed longer life and higher rate of cutting than cemented carbide tools. An interesting example is the use of alumina–SiC_w inserts for turning and threading an aircraft landing gear lever arm made of hardened (56–58 HRC) 300 M alloy steel. The turning operation was reduced from 12.5 to 1.5 h and the threading operation was reduced from 75 to 10 min [10].

Alumina–SiC_w composites have also been used successfully in other industrial applications including wire drawing dies, extrusion dies, tooling for making aluminum cans, valve seats, seals, and plungers for chemical pumps.

SiC whiskers have been added as reinforcement to other ceramic materials. Buljan and co-workers [33] prepared by hot pressing silicon nitride matrix samples that contained 30 vol% SiC whiskers. They reported strength/toughness of 970 MPa/6.4 $MPa \cdot m^{1/2}$ at room temperature, 820 MPa/7.5 $MPa \cdot m^{1/2}$ at 1000°C, and 590 MPa/7.7 $MPa \cdot m^{1/2}$ at 1200°C. These values are not significantly different from those that can be achieved with self-reinforced silicon nitride at substantially lower cost.

Other researchers have explored the use of SiC whiskers to increase the low-temperature toughness and the temperature capability of molybdenum disilicide ($MoSi_2$). $MoSi_2$ is brittle and has low toughness ($\sim 5.3\ MPa \cdot m^{1/2}$) at temperatures up to about 1000°C. Around 1000°C, $MoSi_2$ goes through a brittle/ductile transition. At 1200°C $MoSi_2$ has a yield strength of 139 MPa that drops to 19 MPa at 1400°C and 8 MPa at 1500°C. Addition of 20 vol% SiC whiskers increases the room temperature toughness to 6.8 $MPa \cdot m^{1/2}$, 1200°C yield strength to 386 MPa, the 1400°C yield strength to 125 MPa, and the 1500°C yield strength to 70 MPa [36, 44]. The $MoSi_2$–SiC_w composite has potential to increase the life and durability of industrial heating elements and also critical parts in glass melting furnaces.

2.11 NOVEL CERAMIC MATRIX COMPOSITE FABRICATION APPROACHES

2.11.1 Reaction Formed with a Ductile Metal Reinforcement Phase [19]

One reactive method, often referred to as *directed metal oxidation*, involves reaction of molten aluminum with the oxygen in the air [45]. Normally, when aluminum metal is exposed to oxygen, a thin crust of aluminum oxide forms on the molten metal surface and acts as a barrier to further oxidation. However, if Mg and Si are added to the Al, the molten alloy wets the edges of the alumina crystals that nucleate on the surface of the Al such that the metal is wicked to the surface continually as the alumina is formed. An intertwined network of metal plus ceramic results. If the reaction temperature is high, around 1150°C, the ceramic–metal composite is mostly ceramic and has fracture toughness of only 4.7 MPa $\cdot$ m$^{1/2}$. If the reaction temperature is lower, around 900°C, the percentage of metal increases and the fracture toughness also increases to about 7.8 MPa $\cdot$ m$^{1/2}$. In both cases, the material has very high wear resistance and higher temperature capability than aluminum.

Other composite structures can be achieved by this directed metal oxidation process [45, 46]. If a bed or layer of particles or fibers is placed adjacent to the molten aluminum, the aluminum–alumina will grow right through to produce a particulate or fiber-reinforced ceramic–metal composite. This approach has been used with hard silicon carbide particles to achieve very wear-resistant parts that have performed well in many industrial applications.

Another interesting material that has been prepared by a reactive process combines toughening by platelet-shaped grains plus a ductile metal phase [37]. A porous perform of compacted boron carbide powder is placed in a graphite mold with zirconium metal. When heated to 1850–2000°C in an inert atmosphere, the Zr metal melts, infiltrates the preform, and reacts with the boron carbide to form a mixture of zirconium diboride platelets and zirconium carbide grains. The quantity of Zr can be varied either to be completely consumed by the reaction or to be retained as a residual metal phase to over 30% by volume. A sample with only 1% Zr had fracture toughness of 11 MPa $\cdot$ m$^{1/2}$, while a sample with about 30% Zr had fracture toughness over 20 MPa $\cdot$ m$^{1/2}$.

2.11.2 Fibrous Monolith Composites

Even though the fracture toughness of *in situ*, particulate, and whisker-reinforced ceramics is improved compared to monolithic ceramics, these ceramic matrix composites still fracture in a brittle mode. As is discussed in the next chapter, addition of continuous (long) fibers can result in nonbrittle fracture modes. However, continuous fibers are currently very expensive, and the resulting composites generally are not cost competitive for most industrial applications. The fibrous monolith was a concept introduced in 1988 by Coblenz [47], envisioned to produce a composite structure comparable to the use of continuous fibers, but starting with inexpensive powders.

The concept of a fibrous monolith can best be understood by describing the steps in fabrication and the resulting microstructure. A silicon nitride/boron nitride material is used as the example [48]. The first step is to extrude a viscous mixture of silicon nitride powder (plus sintering aids) and a polymer binder through a small orifice to form a long filament. The filament is then coated with a surface layer of boron nitride. Coated filament also has been achieved in a single coextrusion step. Strands of the BN-coated filament are stacked in the desired orientation (usually unidirectional) in a die and warm pressed at a temperature where the polymer deforms, typically 100–150°C. The filaments deform into flattened hexagonal "cells" that extend through the complete length of the sample and are separated from each other by the BN. The binder is burned off and the sample or part is hot pressed at about 1750°C to densify the silicon nitride. The BN is a nonreactive layer that prevents the cells from bonding to each other during hot pressing, so the silicon nitride retains a pseudofiber form. The material fractures similarly to wood with a high degree of crack deflection, debonding, and cell pull-out. Flexural strength is typically in the range 500–700 MPa, elastic modulus 270–280 GPa, and work of fracture 7000–10, 000 J/m^2 [49].

Fibrous monoliths have been fabricated from a variety of other cell/cell boundary combinations including SiC/BN, SiC/C, ZrB_2/BN, alumina/aluminum titanate, alumina/metals, and even a novel arrangement of diamond and WC–Co. [48–50]. The fibrous monolith that appears closest to industrial application is the diamond/WC–Co material. It has been constructed into inserts for drill bits for rock drilling and has performed very well in laboratory and field tests [50]. In the most extensive field test, a hammer bit with fibrous monolith inserts cut through 2500 ft of hard, silicified sandstone in search of a natural-gas deposit.

2.12 SUMMARY

Broad progress has been achieved during the past 30–40 years to improve the properties and reliability of ceramic materials for structural applications. Strength, Weibull modulus, hardness, fracture toughness, and resistance to thermal shock, high-temperature creep, and environmental attack all have been improved dramatically. Aluminum oxide has continued to be a workhorse for corrosion and wear resistance applications. Silicon carbide, silicon nitride, and transformation-toughened zirconia have all emerged as viable structural materials and are rapidly growing in importance.

Improved understanding of the relationships of properties, microstructure, and processing have contributed to the improvement of monolithic ceramics and also encouraged the development of ceramic matrix composites. Carefully engineered microstructures are now available to optimize individual ceramic materials for specific needs such as high hardness, creep resistance, and high toughness. Ceramic-based materials are now more than ever important for an engineer to consider as alternatives when seeking the optimum material for a specific application.

REFERENCES

1. D. W. Richerson, *Modern Ceramic Engineering: Properties, Processing, and Use in Design*, 2nd ed., Marcel Dekker, New York, 1992.

2. D. W. Freitag and D. W. Richerson, *Opportunities for Advanced Ceramics to Meet the Needs of the Industries of the Future*, Oak Ridge National Laboratory, Report DOE/ORO 2076, Oak Ridge, TN, Nov. 1998.
3. J. B. Wachtman, Jr. (Ed.), *Structural Ceramics, Treatise on Materials Science and Technology*, Vol. 29, Academic, San Diego, 1989.
4. A. Krell, P. Blank, L.-M. Berger, and V. Richter, *Am. Ceram. Soc. Bull.*, **78**(12), 65–73 (1999).
5. Riley, F. L. (Ed.), *Progress in Nitrogen Ceramics*, Martinus Nijhoff, The Hague, 1983.
6. D. A. Bonnell and T. Y. Tien, (Eds.), *Materials Sci. Forum, Vol. 47, Preparation and Properties of Silicon Nitride Based Materials*, Trans Tech, Zurich, 1989.
7. D. W. Richerson, "Evolution of Ceramics for Turbines," *Mechanical Engineering Magazine*, September, 1997.
8. D. W. Richerson, *The Magic of Ceramics*, American Ceramic Society, Westerville, OH, 2000.
9. B. North, *Int. J. High Tech. Ceramics*, **3**, 113–127 (1987).
10. J. H. Adams, B. Anschuetz, and G. Whitfield, "Ceramic Cutting Tools," in *Engineered Materials Handbook, Vol. 4, Ceramics and Glasses*, ASM International, Metals Park, OH, 1991, p. 966.
11. R. N. Katz, "Ceramic Materials for Rolling Element Bearing Applications," in S. Jahanmir (Ed.), *Friction and Wear of Ceramics*, Marcel Dekker, New York, 1994, p. 313.
12. H. R. Baumgartner, "Evaluation of Roller Bearings Containing Hot Pressed Silicon Nitride Rolling Elements," in J. J. Burke, A. E. Gorum, and R. N. Katz (Eds.), *Ceramics for High Performance Applications*, Brook Hill, Chestnut Hill, MA, 1973, pp. 713–727.
13. A. H. Heuer and L. W. Hobbs (Eds.), *Advances in Ceramics, Vol. 3, Science and Technology of Zirconia*, American Ceramic Society, Westerville, OH, 1981.
14. N. Claussen, M. Ruhle, and A. H. Heuer (Eds.), *Advances in Ceramics, Vol. 11, Science and Technology of Zirconia II*, American Ceramic Society, Westerville, OH, 1984.
15. D. B. Marshall, M. C. Shaw, R. H. Dauskardt, R. O. Ritchie, M. Readey, and A. H. Heuer, *J. Am. Ceram. Soc.*, **73**, 2659–2666 (1990).
16. R. A. Cutler, "Engineering Properties of Borides," in *Engineered Materials Handbook, Vol. 4, Ceramics and Glass*, ASM Int., Metals Park, OH, 1991, pp. 787–803.
17. P. T. B. Shaffer, "Engineering Properties of Carbides," in *Engineered Materials Handbook, Vol. 4, Ceramics and Glass*, ASM Int., Metals Park, OH, 1991, pp. 804–811.
18. D. W. Richerson, Chapter 19, "Ceramic Matrix Composites," in P. K. Mallick (Ed.), *Composites Engineering Handbook*, Marcel Dekker, New York, 1997, pp. 983–1038.
19. D. W. Richerson, "Industrial Applications of Ceramic Matrix Composites," in M. G. Bader, K. K. Kedward, and Y. Sawada (Eds.), *Comprehensive Composite Materials*, Vol. 6, Elsevier Science, Oxford, UK, 2000.
20. R. A. Cutler, R. J. Mayhew, K. M. Prettyman, and A. V. Virkar, *J. Am. Ceram. Soc.*, **74**(1), 179–186 (1991).
21. A. E. McHale, "Engineering Properties of Glass-Ceramics," in *Engineered Materials Handbook, Vol. 4, Ceramics and Glasses*, ASM Int., Metals Park, OH, 1991, pp. 870–878.

22. K. T. Faber and A. G. Evans, *Acta. Metall.*, **31**(4), 565–584 (1983).
23. F. F. Lange, *J. Am. Ceram. Soc.*, **56**(10), 518 (1973).
24. E. Tani, S. Umebayashi, K. Kishi, K. Kobayashi, and M. Nishijima, *Am. Ceram. Soc. Bull.*, **65**(9), 1311–1315 (1986).
25. K. Matsuhiro and T. Takahashi, *Ceram. Eng. Sci. Proc.*, **10**(7–8), 807–816 (1989).
26. C. W. Li and J. Yamanis, *Ceram. Eng. Sci. Proc.*, **10**(7–8), 632–645 (1989).
27. J. W. McCauley, *Ceram. Eng. Sci. Proc.*, **2**(7–8), 649 (1981).
28. J. Liu and P. D. Ownby, *J. Am. Ceram. Soc.*, **74**(1), 241–243 (1991).
29. R. A. Cutler, A. C. Hurford, and A. V. Virkar, "Pressureless-sintered Alumina-TiC Composites," in B. K. Sarin (Ed.), *Hard Materials, Vol. 3*, Elsevier, New York, 1988, pp. 183–192.
30. S. T. Buljan, J. G. Baldoni, J. Neil, and G. Zilberstein, *Dispersoid-Toughened Silicon Nitride Composites*, Final Report, ORNL/Sub/85-22011/1, Sept., 1988.
31. C. H. McMurtry, W. D. G. Boecker, S. G. Seshadri, J. S. Zanghi, and J. E. Garnier, *Am. Ceram. Soc. Bull.*, **66**(2), 325–329 (1987).
32. T. N. Tiegs and P. F. Becher, "Alumina-SiC Whisker Composites," in *Proceedings of the 23rd Auto. Tech. Devt. Contractors Coord. Meeting*, SAE P-165, Society of Automotive Engineers, Warrendale, PA, 1985.
33. S. T. Buljan, J. G. Baldoni, and M. L. Huckabee, *Am. Ceram. Soc. Bull.*, **66**(2), 347–352 (1987).
34. P. D. Shalek, J. J. Petrovic, G. F. Hurley, and F. D. Gac, *Am. Ceram Soc. Bull.*, **65**(2), 351–352 (1986).
35. L. J. Neergaard and J. Homeny, *Ceram. Eng. Sci. Proc.*, **10**(9–10), 1049–1062 (1989).
36. D. H. Carter, J. J. Petrovic, R. E. Honnell, and W. S. Gibbs, *Ceram. Eng. Sci. Proc.*, **10**(9–10), 1121–1129 (1989).
37. T. D. Claar, W. B. Johnson, C. A. Andersson, and G. H. Schiroky, *Ceram. Eng. Sci. Proc.*, **10**(7–8), 599–609 (1989).
38. N. Claussen, "Transformation-Toughened Ceramics," in H. Krockel et al. (Eds.), *Ceramics in Advanced Energy Technologies*, Reidel, Dordrecht, 1984, pp. 51–86.
39. J. D. Sibold, "Wear Applications," in *Engineered Materials Handbook, Vol. 4, Ceramics and Glasses*, ASM Int., Metals Park, OH, 1991, pp. 973–977.
40. R. A. Lowden, ORNL Rept. TM-11039, March 1989.
41. G. C. Wei and P. F. Becher, *Am. Ceram. Soc. Bull.*, **64**(2), 298–304 (1985).
42. T. N. Tiegs and P. F. Becher, *Am. Ceram. Soc. Bull.*, **66**(2), 339–342 (1987).
43. J. F. Rhodes, H. M. Rootare, C. A. Springs, and J. E. Peters, "Al_2O_3-SiC Whisker Composites," presented at the 88th Annual Meeting of the American Ceramic Society, Chicago, IL, April 28, 1986.
44. J. J. Petrovic and R. E. Honnell, *Ceram. Eng. Sci. Proc.*, **11**(7–8), 733–744 (1990).
45. M. S. Newkirk, H. D. Lesher, D. R. White, C. R. Kennedy, A. W. Urquhart, and T. D. Claar, *Ceram. Eng. Sci. Proc.*, **8**(7–8), 879–885 (1987).
46. A. S. Fareed, B. Sonuparlak, C. T. Lee, A. J. Fortini, and G. H. Schiroky, *Ceram. Eng. Sci. Proc.*, **11**(7–8), 782–794 (1990).
47. W. S. Coblenz, Fibrous Monolithic Ceramic and Method for Production, U.S. Patent 4,772,524, Sept. 20, 1988.

48. D. Kovar, B. H. King, R. W. Trice, and J. W. Halloran, *J. Am. Ceram. Soc.*, **80**(10), 2471–2487 (1997).

49. D. Popovic, G. Danko, K. Stuffle, B. H. King, and J. W. Halloran, *Ceram. Eng. Sci. Proc.*, **17**(3), 278–286 (1996).

50. T. Mulligan, Z. Fang, J. A. Sue, G. E. Hilmas, and M. J. Rigali, "Commercialization of Fibrous Monolith Processed Drill Bit Inserts," presented at the 23rd Annual Conference on Composites and Advanced Materials, Cocoa Beach, Florida, January 1999.

CHAPTER 3

Continuous Fiber Ceramic Composites

JAMES K. WESSEL

Wessel & Associates, 127 Westview Lane, Oak Ridge, Tennessee 37830

3.1 INTRODUCTION

As shown in the previous chapter, ceramics are finding use where temperatures exceed the capability of other materials, especially metals. Even so, they are

Handbook of Advanced Materials Edited by James K. Wessel
ISBN 0-471-45475-3

not selected for many applications because of the brittleness of these monolithic ceramics. In the search for improved toughness, material scientists conceived the idea of reinforcing ceramics with continuous strands of high-temperature ceramic fiber, analogous to continuous fiberglass-reinforced plastics. Embedded continuous ceramic fibers reinforce the ceramic matrix by deflecting and bridging fractures.

These continuous fiber-reinforced ceramic composite (CFCC) materials offer the advantages of ceramics: resistance to heat, erosion, and corrosion—while adding toughness and thermal shock resistance. The result is a lightweight, hard, tough, high-temperature, thermal shock, erosion, and corrosion-resistant structural material. These materials are used where designers seek less downtime, reduced maintenance, lower operating costs, increased operating temperature, increased efficiency, lower emissions, and reduced life-cycle costs (see Table 3.1). Designers are evaluating and using them in applications in major industries.

Monolithic ceramics, although strong in tension, tend to fracture suddenly with total loss of strength. Conversely, when the yield strength of CFCC is exceeded, failure occurs "gracefully," with the material able to continue to bear load. This feature reduces the risk of catastrophic failure and encourages designers to use CFCC materials for this and other benefits (see Fig. 3.1).

All CFCC materials are composed of a ceramic fiber, a fiber–matrix interface coating and a ceramic matrix, arranged to form a continuously reinforced material. The fiber is converted to useful form by using conventional textile-forming techniques: single-fiber filaments can be grouped into a tow, woven into fabrics, cut, sewn, laminated, and tooled to form a net-shape preform for subsequent processing. Other forming processes include winding the coated fiber filaments onto a mandrel to form tubes, cylinders, and related shapes. This formed fiber shape, or preform, is infiltrated with a ceramic matrix by various techniques and converted to a ceramic by the application of heat and pressure.

The fibers provide toughness by arresting cracks, bridging cracks, and by a phenomenon known as fiber "pull-out."

For a crack to grow, energy must be expended. When the crack comes to a fiber, it must divert around that fiber. This consumes more energy than linear growth and the crack will stop. If the crack is propagated by sufficient energy to pass around the fiber, the fiber can bridge the crack and hold the composite together. Finally, if the forces are sufficient to fail the composite, the fiber must be pulled out of the composite. This pull-out requires additional energy and, as the fibers continue to carry the load, a noncatastrophic, load-bearing failure mode

TABLE 3.1 CFCC Characteristics

Characteristic	Advantage
Resists corrosion	Survives hostile environments
Resists high temperatures	Use temperatures to 2200°F
Fiber reinforced	Survives cyclic loading
Near-net-shape fabrication	Lowers life-cycle cost

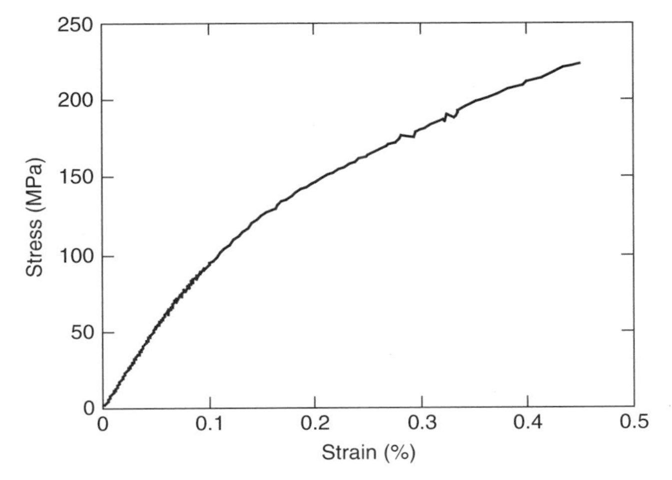

FIGURE 3.1 Continuous fiber reinforcement changes the shape of typical ceramic stress–strain curve. Yields occurs "gracefully." This slow yield eliminates sudden failure.

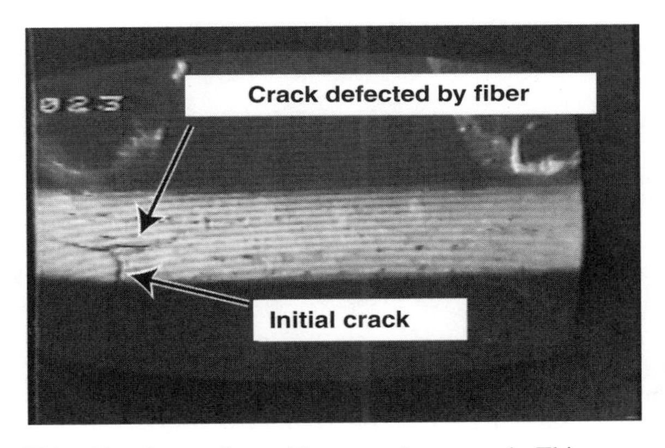

FIGURE 3.2 This video frame shows fiber stopping a crack. This consumes energy and maintains CFCC integrity.

or metal-like behavior results. Therefore, the shape of the CFCC stress–strain curve beyond the elastic limit is determined by the ability of the ceramic fiber to "slip" through the matrix. This slip is facilitated by appropriate fiber coatings (see Figs. 3.2 and 3.3).

Typical properties of CFCCs are shown in Table 3.2. They are presented as a range since CFCCs are a family of products composed of various ceramic fibers, coatings, ceramic matrices, and made by various processes by several manufacturers. This offers the advantage of selecting from a range of CFCC materials or customizing a formula to meet your specific requirements. Future compositions may have properties outside this range.

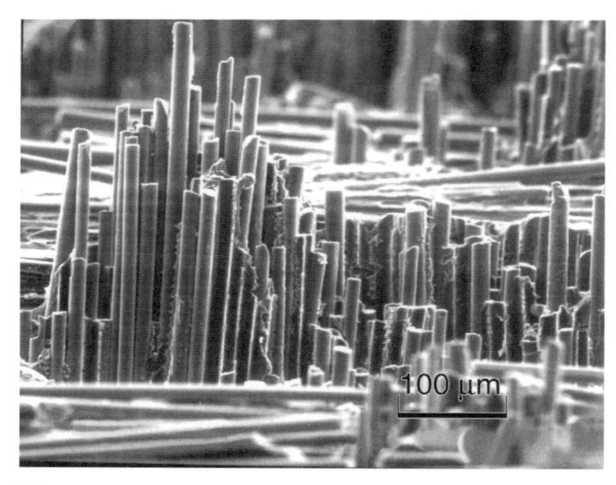

FIGURE 3.3 When overstressed, fibers pull away from matrix. Fiber–matrix interface coating allows fiber to slip within matrix, taking any load onto fiber, lending strength to composite.

TABLE 3.2 CFCC Properties at Room Temperature

Property	Range of Values	
Density	2.1–3.1 g/cm^3	0.076–0.112 lb/in.^{3}
Open porosity	0–20%	
Tensile properties		
Strength	207–400 MPa	31–60 ksi
Modulus	90–250 GPa	13–36 Msi
Strain to failure	0.4–0.8%	
Flexural properties		
Strength	200–480 MPa	29–70 ksi
Modulus	83–240 GPa	12–35 Msi
Compressive strength	450–1100 MPa	65–159 ksi
Shear strength	28–68 MPa	4.0–9.8 ksi
Room temperature thermal conductivity	1–40 W/m°C	

A major reason for choosing ceramics is their high-temperature performance. In Fig. 3.4, the specific strength of CFCCs is compared to other high-temperature materials. The various types of CFCCs and their processes are given in Section 3.3.

Continuous fiber reinforcement is used as monofilament or multifilament tows. A composite using fiber tows costs less because it is easier to process into complex shapes. Some of the more common fibers include oxides (alumina and mullite) and nonoxides (silicon carbide and silicon nitride). Where application temperatures are below 1100°C (2012°F) or the exposure time is limited, the oxide fiber mullite is most widely used because of its lower cost. Silicon carbide is favored where engineers desire a stronger, harder, stiffer composite with superior thermal stability.

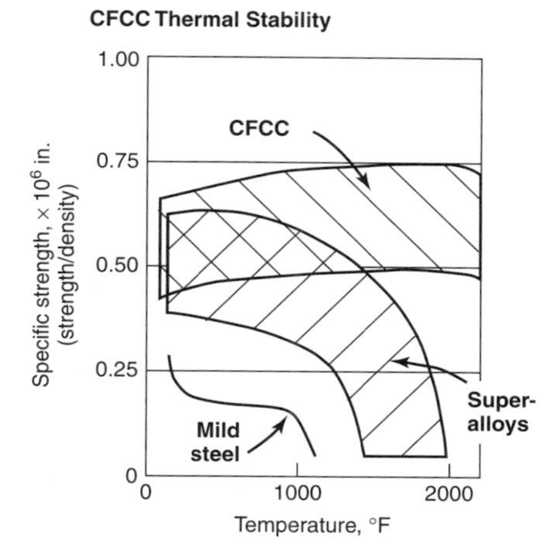

FIGURE 3.4 CFCCs operate beyond temperature range of best metals.

Ceramic matrices used in CFCCs are either metal oxides or nonoxides. Oxides are desired for their inherent oxidative stability. Oxide matrices are alumina, silica, mullite, barium aluminosilicate, lithium aluminosilicate, and calcium aluminosilicate. Alumina and mullite are the most frequently used oxides because of their thermal stability, chemical resistance, and their compatibility with various fiber reinforcements. Although the oxide matrices have a longer history, nonoxide CFCCs are stronger, harder, stiffer, and uniquely resist certain environments. Common nonoxides include silicon carbide and silicon nitride.

Composites of fiber and matrix can be mixed and matched, that is, oxide matrix reinforced with nonoxide fibers, and visa versa, as well as oxide fibers with oxide matrix and nonoxide fibers with nonoxide matrix. The most widely used reinforcement is silicon carbide fiber because of its compatibility with a wide range of oxide and nonoxide matrices. Table 3.3 shows the variety of CFCC materials.

TABLE 3.3 CFCC Materials and Processes

Process	Matrix	Fiber	Manufacturer
Chemical vapor infiltration (CVI)	SiC	SiC	Honeywell Advanced Composites
Direct metal oxidation (DMO)	Al_2O_3	SiC, Al_2O_3	Honeywell Advanced Composites
Polymer impregnation pyrolysis	SiC	SiC, SiOC, SiNC	COI Ceramics
Melt infiltration	SiC, Si	SiC	General Electric
Reaction bonding	Nitride-bonded SiC	SiC	Textron Systems
Sol–gel	Mullite, Al_2O_3	SiC, Al_2O_3	McDermott, COI Ceramics

3.2 APPLICATIONS

The many applications of CFCCs illustrate the variety of their shapes and sizes (Fig. 3.5). Cylinders or tubes are widely used in industry. They are usually made of metals, either wrought or centrifugally cast. They commonly fail by corrosion from products of combustion, or erosion, or creep when operated at high temperatures, and are relatively heavy. They are subject to fouling, especially when catalyzed by metal-alloyed constituents required for high-temperature use. Ceramics have advantages over metals: higher temperature capability, lighter weight (requiring less support structure), corrosion resistance (permitting reduced wall thickness), and reduced fouling, especially with silicon carbide.

3.2.1 Heater Tubes

Process streams are often heated by immersion heaters. Engineers are evaluating CFCC materials in immersion heaters to melt aluminum. Aluminum is normally melted in reverberatory furnaces or in furnaces with radiant burners. These methods have several limitations: (1) efficiency is limited, resulting in at least 60% of the heat going up the stack; (2) vapor-phase reactions in the space between the tubes, furnace ceiling, and melt result in the formation of oxide scale that contaminates and lessens the quality of the aluminum; and (3) heating is nonuniform. All affect aluminum quality and cost.

An alternative method is gas-fired immersion tube heaters, with monolithic silicon carbide ceramic tubes immersed directly into the aluminum. These tubes have demonstrated increased efficiency, more uniform heating, and substantially reduced contamination. However, they are susceptible to thermal shock.

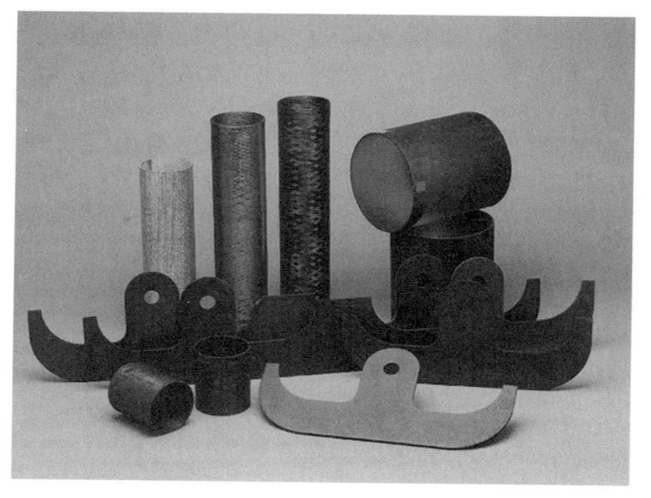

FIGURE 3.5 CFCCs can be made in wide variety of shapes and sizes.

Engineers are evaluating tough CFCC materials because they do not have these limitations. Also, CFCC materials are not wetted by molten aluminum and are not chemically attacked by either the aluminum on the outside of the tube nor the combustion gases or heating element on the inside (Fig. 3.6).

Textron Systems Corporation, Lowell, Massachusetts, is using a computer-controlled 5-axis winding machine with a seamless mandrel to fabricate and test these immersion tubes. Textron is teamed with Deltamation (a furnace designer), F. W. Shafer (a furnace manufacturer), Doehler-Jarvis (an aluminum caster), and an automotive manufacturer to evaluate their tubes. These tubes have survived 1752 h operating in a production aluminum caster. This exposure included the normal practice of cycling through 30 h of melting at 870°C (1600°F) with 15 min of pour time. The CFCC survived the thermal shock of 2 cold starts and the 50 tube withdrawals during the pours. They continue operating toward a goal of 3000 h. Success would realize benefits of reduced downtime, increased product yield, improved quality, increased efficiency, reduced energy consumption, reduced emissions, and lower operating costs. Energy savings accrue due to an increased heat transfer improvement of 40% and increased product yield. Fifty trillion Btu/year would be saved if CFCC tubes were used to melt all U.S. aluminum. Emissions are reduced because less fuel is consumed and recuperation is enabled.

FIGURE 3.6 Immersion tubes contain the heating element. Tubes are inserted into the pot of metal and heated, melting the metal. Thermal shock and stress are severe, especially at the melt line.

3.2.2 Gas Turbine Engine Applications

The use of CFCC components in gas turbine engines increases their efficiency, resulting in fuel savings, reduced emissions, reduced downtime, and other benefits. A turbine is a rotary engine that uses a continuous stream of fluid to turn a shaft that drives machinery. A gas turbine engine uses gas as fuel. This engine consists of a rotary bladed shaft passing through a compressor, combustor, and exhaust sections. Air is compressed, mixed with fuel, ignited in the combustor, and then exhausted through rotary blades that spin, driving the upstream compressor as well as downstream machinery. Fuel can be natural-gas, kerosene, or gas rendered from coal. The hot exhaust gases can be used to power pumps, other equipment, electrical generators or to generate steam for industrial processes or both, a so-called cogeneration system.

Natural-gas-fired turbines are slated to provide 80% of new electrical power capacity in the United States. Of the 200 planned power plants, 96% will use natural-gas fuel, most fueling gas turbines.

Turbine manufacturers are interested in reducing downtime and emissions and improving engine efficiency. Engine shutdowns are the bane of utility operators. The resulting severe thermal shocks damage these large expensive engines. This is the primary reason these engines are limited to lower temperature operation resulting in lower efficiency.

Turbine engine efficiency, as with all heat engines, is determined by operating temperature. The higher the temperature, the higher the efficiency. A turbine engine efficiency increase of 0.4% results in fuel savings of $460,000/ year for a 160-megawatt (MW) engine. A 0.5% additional airflow through the combustor (as a result of reduced cooling to the shroud), at base load conditions, could reduce NO_x emission levels 10–25%. A 1.25% reduction in pressure drop, as a result of less cooling, could lead to a $370,000 fuel savings per year per engine. One turbine engine is designed with CFCC components with an efficiency increase of 15% because its use allows for near-stoichiometric fuel combustion for increased power without the cooling air requirement penalties associated with metallic structures. Replacing steel with CFCC combustor liners, shrouds, and interstage seals enables this increase in efficiency (Fig. 3.7).

Combustors have inner and outer liners. The inner liner faces the flaming gases. It is a cylinder, perforated to pass fuel that combines with compressed air and ignited in the combustor. Cooling air flows between the inner and outer liners to preserve the inner liner. Diverting air for cooling also reduces efficiency. The more thermally stable CFCC combustor liners lower and potentially eliminate the need for cooling. At a given NO_x level, metal liners showed higher CO levels than the CFCC liner. This is attributable to the quenching effect of cooling air. With no cooling air, the CFCC combustor produced NO_x levels below 10 ppm (15% oxygen) with low CO emissions. Metallic liners were limited to NO_x emissions near 20 ppm.

The very high-temperature gases from the combustor pass through the first-stage turbine stator into the first-stage rotor. Concentric to the outer diameter of the rotor blades is a ring of stationary components called shrouds. Shrouds are

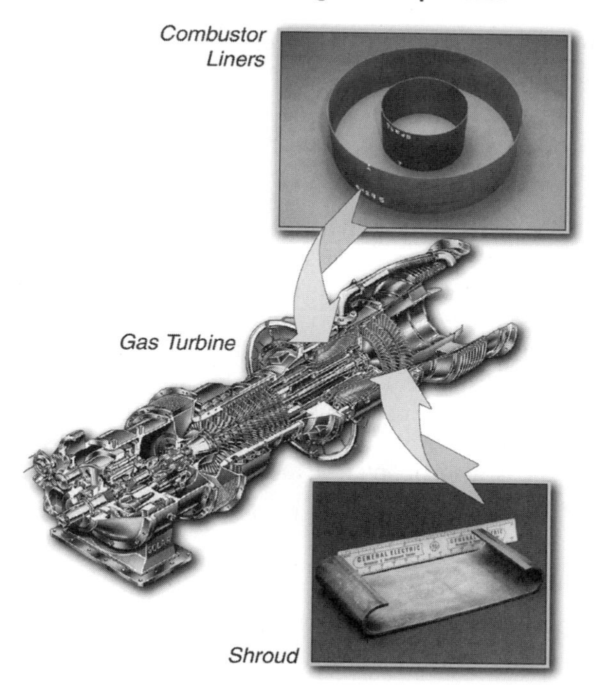

FIGURE 3.7 CFCC thermal stability enables higher operating temperatures resulting in increased efficiency.

a series of open-top, curved walled boxes attached to the engine inner case and concentric to the outer diameter of the rotating blades. They seal between the inner engine case and the end of the rotating blade. The clearance between the ring of shrouds and the rotor blade is minimized to reduce exhaust gas leakage around the end of the blade. The shroud of a 160-MW engine is 2.44 m in diameter, consisting of 96 rectangular segments 7.6 cm wide, 15.2 cm long and 1.3 cm high. Shroud temperatures can reach 1290°C (2354°F) if uncooled. They are presently made of metallic super alloys and require about 1.2% of the compressor output for cooling. CFCC shrouds enable higher temperature operation, reduce the amount of cooling air required by 80%, resulting in a higher efficiency turbine operation, reducing emissions by 10–25%.

CFCCs successfully performed during 100 shutdowns after steady-state operation at 1115°C (2120°F). To create an excessive thermal shock condition, high airflow rates were maintained after the fuel was turned off resulting in dramatic temperature reductions, high thermal stresses, and complex mechanical stresses in the CFCC shroud. CFCCs achieved the primary goal of demonstrating risk reduction to the engine operator. CFCC combustors, compared to metal combustors reduce NO_x by 47–60% and CO emissions by 33–60%.

Downstream, another type of interstage seal is used. Since the metal rotating blade and the metal seal thermally expand as the engine warms-up, it cannot be designed to end-seal exhaust gases at all operating temperatures. An abradable seal is placed around the inner case surface. The rotating blade expands into and cuts a path in this material, forming a perfect seal, preventing exhaust gas leakage and increasing efficiency. CFCCs possess the appropriate physical properties and heat resistance to perform satisfactorily as this interstage seal. This seal will improve efficiency, resulting in fuel savings of 0.5%.

Malden Mills, a Polartec™ textile mill in Lawrence, Massachuesetts, has a Solar Centaur 50S gas turbine outfitted with CFCC components. The turbine generates steam, electricity and heat. It uses 25–40% less fuel than todays coal-fired plants and emits 40% less carbon dioxide, a greenhouse gas. The CFCC turbine has successfully operated for 16,000 h and continues to perform.

CFCC thermal stability, thermal shock resistance, strength and oxidation resistance is enabling gas turbine engines with higher operating temperatures, increased efficiency, reduced downtime, maintenance, emissions and operating costs. CFCC light weight is also of interest to airborne turbine users where 30% of turbine weight would be eliminated.

3.2.3 Hot-Gas Filters

McDermott Corporation, Lynchburg, Virginia, engineers use this same process technology to produce CFCC flanged closed-end porous tubes. These tubes perform as filters used to remove solids from gases. The solids may be either the desired product or a contaminant or a catalyst to be recovered and recycled. Their removal protects downstream equipment from erosion. Filters must withstand chemical corrodants, high-speed hard-particle impingement, long-term strength retention, pulse stress, vibration, fatigue, temperature, pressure, and high mechanical and thermal stresses. They must maintain low pressure drops and high flow rates to sustain production rates. Filter materials must not contaminate the product stream (Table 3.4).

Metals and monolithic ceramics have been the materials of choice. Metals tend to corrode and have temperature limitations. They also require cooling the gas stream prior to filtering, thereby decreasing efficiency, increasing costs and complexity by requiring gas dilution air scrubbers or heat exchangers. Monolithic

TABLE 3.4 CFCC Candle Filter Material Typical Properties

Construction: Nextel™ Ceramic Fibers in a silicon carbide matrix
Continuous use temperature: 1204°C (2200°F)
Maximum short-term temperature: 1315°C (2400°F)
Coefficient of thermal expansion: 4.6×10^{-6}/°C (2.5×10^{-6}/°F)

ceramics are limited due to susceptibility to thermal stresses, mechanical shock, and damage during installation.

One specific application is filtering coal ash from coal gas. Coal gasification plants generate electricity from gas created by heating coal. The gas is fed, as fuel, to a turbine engine. The engine turns an electric generator.

Coal gasification plants are of interest since they offer certain advantages over traditional coal or natural-gas-fired utility plants. Gasification plants emit less carbon dioxide and oxides of nitrogen and cost less to operate.

The turbines require particle-free gas fuel for safety, cost, and meeting clean air regulations. Coal ash is filtered from the coal gas stream through the use of candle filters. These filters are porous, hollow tubes 1.5 m (59 in.) long. They are ganged into arrays as shown in Fig. 3.8.

Hot coal gas is pumped onto and through the filter leaving the coal ash on the outside. The clean gas passes through the open end of the tube and into the turbine. The coal ash accumulates until the system is occasionally back-pulsed, every 15 min at 90 psi gage, to drop the ash into collectors. The ash buildup can become so great that it bridges between tubes and creates a surprisingly high mechanical stress. The back-pulse is also a major physical shock on the filters.

Power plants in Karhula, Finland, and Wilsonville, Alabama, are evaluating CFCC hot-gas filtration systems. The Karhula plant has operated CFCC filters successfully for 580 h. The Wilsonville plant is operated by Southern Company Services. Its CFCC filters continue to operate successfully beyond 3000 h at 850°C (1800°F). The CFCC filters are resisting the corrosive coal ash, high temperatures, and both thermal and physical shocks.

CFCCs – Tough, Lightweight Gas Filters

- **Survives at Elevated Temperatures**
- **Stops Damage Downstream**
- **Reduces Downtime**
- **Lowers Cost**
- **Resists Corrosion**
- **Reduces Emission**

Filter Assembly

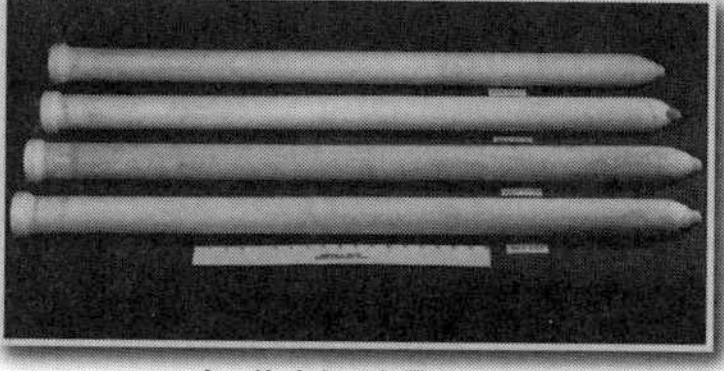

Individual Filters

FIGURE 3.8 Individual filters on left are ganged into assembly shown on right. Porosity is built into CFCC cylinder to filter fine solid particles from coal-derived gas prior to burning in turbine engine. Engine energy is converted to electricity. These filters are also useful for filtering other solids from other gas streams.

Within other gaseous streams there is often a need to remove particulates before further processing. Conventional filtration materials and techniques either are inefficient or insufficiently rugged to survive in the process environment. Energy efficiency and throughput can be accomplished with CFCC filters that withstand both corrosive and high-temperature environments and mechanical and thermal stress. The proposed filtration concept cleans itself via back-pulsing, with apparent indefinite life.

A candle filter system offers two advantages over electrostatic systems that are coupled with liquid scrubbers. It eliminates the need to cool the stream saving energy, capital, and maintenance. Second, the filtered stream is available for heat recovery.

McDermott Corporation CFCC filters have a density of 0.8 g/cm^3. The combined effect of lower weight and thinner walls enables a smaller, simpler supporting structure. CFCC hot-gas filters offer thermal stability, strength, resistance to thermal shock, resistance to fatigue, corrosion, erosion, and general inertness. Anticipated benefits include reduced downtime, increased throughput, reduced energy consumption, longer life, reduced emissions, increased product yield, increased efficiency, and reduced cost.

3.2.4 Heat Exchangers

High-pressure heat exchanger efficiency can be significantly increased and downtime reduced with CFCCs. Heat exchangers made of tough CFCCs survive thermal shock, operate at higher temperatures longer, and resist fouling erosion and corrosion. Where reactions are conducted in the exchanger, higher operating temperatures lead to faster reactions, less residence time, and improved efficiency. Many processes use heat exchangers to capture heat from exhaust streams to preheat inlet streams. In one example, in a CFCC natural-gas preheater, compared to metals, the overall efficiency improved from 35% to a new efficiency of 47%. The metal heat exchanger was limited to 816°C (1500°F) or less. It required cooling the gas stream and reheating it to 1260°C (2300°F) downstream. The use of CFCCs eliminated cooling the gas upstream of the exchanger and the reheating step. This saved 33% of the thermal loading. Reduced fuel consumption reduced cost and lowered emissions (Fig. 3.9).

In another example a heat exchanger preheats a stream of combustables prior to incineration. This facility incinerates a wide variety of waste, both solid and liquid, except polychlorinated biphenyls (PCBs) and dioxin, from 80 locations. The flue gas typically contains HCl, water vapor, oxides of carbon, sulfur, and nitrogen. The ash is comprised of oxides of aluminum, calcium, iron, sodium, potassium, and silicon along with small amounts of heavy metals. A combination of solid and liquid waste was burned at a rate of 1360–1810 kg/h (3000–4000 lb/h). CFCC heat exchanger tubes were exposed to inlet air and flue gas temperatures of 425–980°C (800–1800°F). After 6 months operation the strength of CFCC heat exchanger tubes did not change. This test was the first successful demonstration of a high-temperature CFCC heat exchanger in a highly corrosive environment under actual industrial conditions.

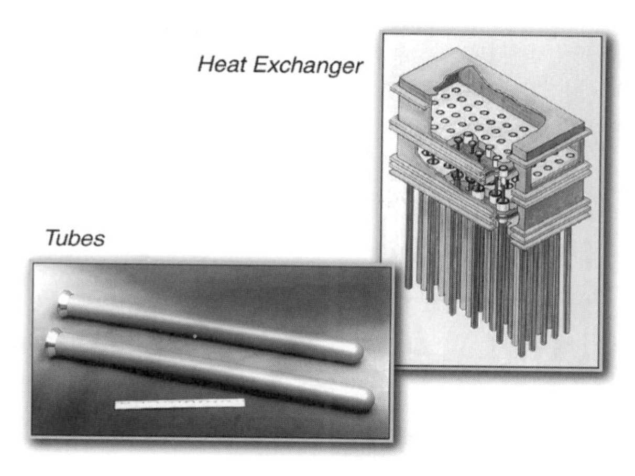

FIGURE 3.9 Individual tubes on left are assembled into heat exchanger on right. Thermal stability and nonbrittle nature of CFCCs make them candidate components of high-temperature or corrosive environment heat exchangers.

Since CFCCs withstand higher temperatures than the previously used metal, the downstream incinerator burns the incoming stream more completely with less noxious emissions, reduced energy consumption, reduced operating cost, and reduced landfill.

High-pressure heat exchangers are used as the reaction vessel in a new process to form ethylene (Fig. 3.10). This new method will dramatically improve this process. The thermal stability and corrosion resistance of a CFCC heat exchanger will improve reformer efficiency. This is particularly important because ethylene production requires more energy than any other organic chemical process. Steam cracking, the process in place for 40 years, was optimized long ago. The new process, called *reforming*, will improve efficiency and reduce energy consumption. Materials of construction must withstand methanol, hydrogen, and ammonia. As an intermediate step, CFCCs are being evaluated to improve the conventional steam cracking process that is used today to form ethylene and other hydrocarbons.

In conventional steam cracking systems, the feedstock is mixed with steam and passed at high-temperature and pressure through metal tubes in a direct-fired furnace heat exchanger. The process is constrained by the metal alloys used. By replacing those alloys with CFCC, higher temperature and pressure can be achieved that will significantly improve ethylene yields. Boosting process temperatures to 980°C (1800°F) from the current maximum of 900°C (1650°F) will increase the yield from 27 to 37%, an increase of 36%.

Continuous fiber-reinforced ceramic composites resist high-temperature corrosive reforming by-products: methanol, hydrogen, and ammonia. Coking, a process-retarding carbon deposition catalyzed by the metals normally used, is a problem. Steam, normally mixed with the feedstock, is added, in part, to reduce coking. The use of CFCCs minimizes coking and is expected to allow the process

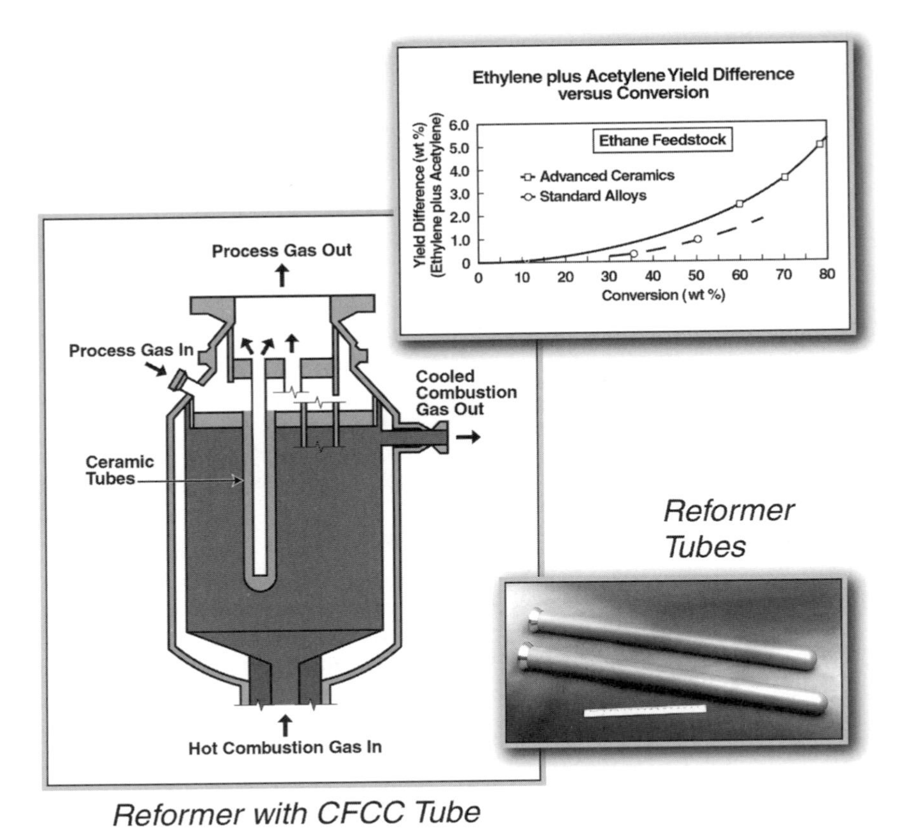

FIGURE 3.10 CFCC thermal stability and toughness improve reformer yield.

to run 50% longer before downtime for maintenance. Run length is expected to increase from 60 to 90 days in the case of ethane as the feedstock. Steam use is also reduced.

Overall, the combination of increased yield, greater run lengths, reduced feedstock, and steam and energy use is expected to increase ethylene production capacity by 10%. Similar results are expected for reformers making cleaner gasoline. Participants in this endeavor include Stone & Webster Engineering Corporation and the CFCC suppliers.

3.2.5 Pump Components

The resistance of CFCCs to corrosion and their near transparency to electrical eddy currents make them an attractive candidate containment shell for canned and magnetic driven pumps. Canned motor pumps, widely used in chemical processing, move hazardous liquids where a leak cannot be tolerated. The pumps are used for acids, alkali salts, alcohols, aromatics, monomers, polymers, hydrocarbons, halogenides, and other chemicals. These liquids are often at high-temperature.

The outer case of the pump contains coils of electrical wires analogous to the stator windings of an electrical motor. The inner, rotating, portion of the pump contains coils of electrical wire like the rotor of an electrical motor. When an electrical potential is applied to the stator, the rotor spins, driving the pump impeller attached to it. A containment shell separates the stator from the rotor, is the guide/housing for the rotor, and seals the pumped liquid away from the pump driving mechanism. Shell materials need high hoop strength, corrosion resistance, and low electrical conductivity.

Metallic containment shells conduct electricity, causing a substantial loss of power. CFCC shells are not electrically conductive so they eliminate any eddy current and drag and thus reduce the electrical energy required to operate the pump as well as reduce heat transfer to the liquid. Dow Corning Corporation engineers teamed with Sundstrand Corporation to evaluate CFCCs as canned pump shells handling hazardous liquids in processes up to a maximum temperature of 450°C (840°F). The technology can be extended to magnetic pumps.

The thermal stability, toughness, corrosion resistance, and unique electrical properties of CFCCs and their use in canned pumps will result in reduced downtime, increased throughput, reduced energy use, and reduced operating costs. This application also demonstrates the ability to fabricate CFCCs into thin-walled structures.

3.2.6 Separator Components

Other applications requiring excellent toughness and erosion resistance include internal components of cyclones that separate solids.

Engineers in a major U.S. city evaluated CFCCs vortex finders in a waste separating cyclone (Fig. 3.11). The cyclone separates combustible from noncombustable trash. A downstream incinerator converts the combustable trash to steam for plant use and generation of electricity. Trash, mixed with sand to facilitate separation, is fed at high-speed into the heated cyclone. The cyclone has a centrally located hollow tube, called a vortex finder. Sand and trash rapidly wear the metal superalloy vortex finder. Panels of CFCC are being evaluated to replace the metal because of their superior erosion and high-temperature resistance. They also resist the corrosive atmosphere that contains various acids, chlorides, sulfides, trash, and sand.

City engineers, teamed with Honeywell and Foster Wheeler engineers, tested CFCCs in this rigorous application. CFCCs retained 100% of their original properties after 2500 h of operation. This included a 300-h startup followed by 2200 h of operation at 900°C (1650°F). The erosion rate of the 4×8 ft CFCC panels is $\frac{1}{13}$ that of the superalloy metal panels.

The reduced maintenance and downtime saves money as well as enabling increased operating temperatures, more complete combustion, and less landfilling

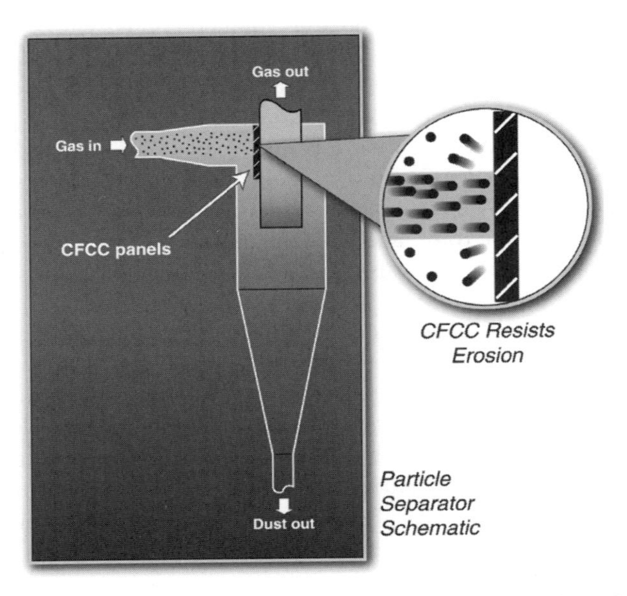

FIGURE 3.11 Erosion resistance of CFCCs improve durability of this cyclone separator.

of trash. Maintenance is particularly costly because the cyclone is located 60 ft above ground, requiring scaffolding, adding to downtime. The lighter panels of CFCC ease handling and installation. CFCCs enable reduced maintenance, a higher operating temperature, increased efficiency, reduced energy use, and produce less pollution. Higher operating temperatures also allow hotter air to reach the heat exchanger, increasing steam and energy production.

Other solid–solid separation processes also need erosion-resistant parts. Using industries include chemical processing, power generation, and others. The successful use of CFCCs in municipal waste incinerators make them candidates for applications in these other industries.

3.2.7 Infrared Burners

Another application of tubular CFCC materials is in gas-fired infrared burners. These burners are used to dry paper, paint, textiles, and cure plastics. They are used instead of steam boxes, black-body infrared burners, and convective and electrical infrared heaters because they are more reliable, versatile, flexible, exhibit less heat loss, and are more efficient. Their heat can be applied exactly where it is needed. They have been made of metal, which has limited thermal stability, spectral control, and efficiency.

Continuous fiber-reinforced ceramic composite infrared heaters offer the advantages of ceramics plus the ability to tailor the spectral emission, through the addition of rare earths, to match that of the heated material. Typical burners of this type emit radiation over the full spectrum, although only a narrow spectrum of energy is required. By modifying the CFCC formula with the addition of rare earths, the spectral emission is controlled, matching water absorption wavelengths, thereby optimizing heat transfer. Selective emittance burners lower fuel consumption, result in faster heat-up and cool-down cycles, and increase the throughput of dried product, resulting in improved operating efficiency. CFCC thermal stability and thermal shock resistance improves durability, provides additional operating efficiency, and reduces life-cycle cost.

3.2.8 Radiant Burners

Continuous fiber-reinforced ceramic composites are easily formed into flat sheets and cut into shapes. Radiant burner screens are an example. Radiant burners are among the most efficient with very low NO_x formation. One design has a manifold through which gaseous fuel is fed and ignited. Efficiency is increased by adding a mesh screen directly in front of the manifold (Fig. 3.12). The screen protects the burner from damage and sustains emissivity. It also acts as a reverberatory screen because it reflects the burners heat back onto it, completing

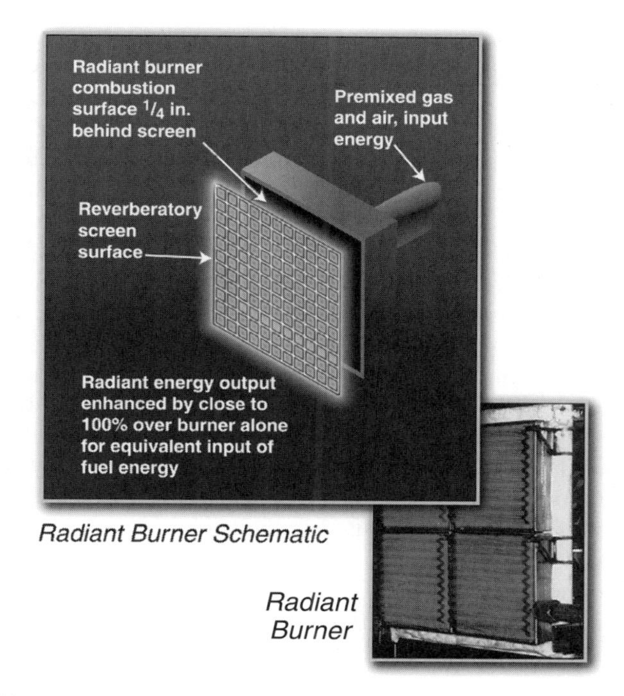

FIGURE 3.12 Radiant burners and their screens may be flat as shown here or cylindrical. These screens protect the burner from rough handling and falling refractory and increase burner efficiency. CFCC screens operate at higher temperatures enabling maximum efficiency.

combustion and spreading the flame uniformly across its surface. The screen nearly doubles the radiated heat. Radiant burners use natural-gas to provide uniform high heat transfer efficiently with low emissions. They are used in residential warm air furnaces, commercial deep-fat fryers, dryers for the plastics industry, commercial greenhouses, and residential and commercial water heaters. Other industrial applications include fire-tube and fluid-tube boilers, process heaters, paper dryers, and high-efficiency volatile organic compound incineration. Other processes that could benefit from this method include metal treating, organic chemical oxidizing, plastic curing, and drying paint and paper and other materials.

Honeywell, Alzeta, Visteon, and an automotive manufacturer teamed to evaluate CFCC reverberatory screen radiant burners in a glass treating facility. The CFCC has survived the thermal fatigue of 1000 h of operation at 1100°–1200°C (2000°–2200°F) plus 15,000 thermal cycles, and 32,000 on–off cycles with no sign of deterioration. They experienced higher throughput, faster processing, lower fuel consumption, greater furnace temperature control, reduced emissions, and lower energy costs. Fuel consumption was reduced by 33%. Heat flux from the burner doubled. Overall performance increased by 35%. Retrofit of the previous process with the CFCC reverberated radiant burner would result in less downtime, lower maintenance, increased efficiency, reduced energy consumption, longer life, and lower life-cycle costs.

Alzeta Corporation manufactures several types of radiant burners. One type is used to incinerate perfluorocarbons from the semiconductor industry. A second is used in a high-efficiency boiler system. CFCCs enable an operating efficiency of 88% versus a previous 84%. This saves 40 billion Btu/year. Each unit emits 10 ppm less NO_x than previous ones. A third CFCC unit is used in residential water heaters. It operates at 87% versus 80–85% for burners with metal screens and produces less than 20 ppm NO_x. The CFCC burner saves 30 billion Btu/year.

A fourth type of CFCC screened burner is the Pyrocore radiant burner product line. CFCCs enable Alzeta to offer a longer warranty on this burner. Pyrocore burners are generally used in boilers in virtually every industry. It reduces the emission of global warming gases by 90–98% from each burner.

Continuous fiber-reinforced ceramic composites enable higher temperature operation, increased efficiency, true of many industrial processes. Higher temperature operation results in more complete combustion, reduced fuel consumption, and lower emissions. CFCCs screens are tough, durable, thermally stable, dimensionally stable, corrosion-resistant, and thermally shock resistant. They can be retrofitted onto existing burners or used in their original manufacture. If 10% of radiant burners used CFCC screens, 50 billion cubic feet of natural-gas would be saved per year and NO_x emissions reduced by 35,000 tons. Additional savings would be realized through elimination of downstream NO_x control equipment and higher product yield as a result of more uniform heat transfer.

3.2.9 Pipe Hangers

Refineries heat crude oil by pumping it through pipes suspended in a natural-gas-fired furnace. Stainless steel hangers have traditionally been used to support

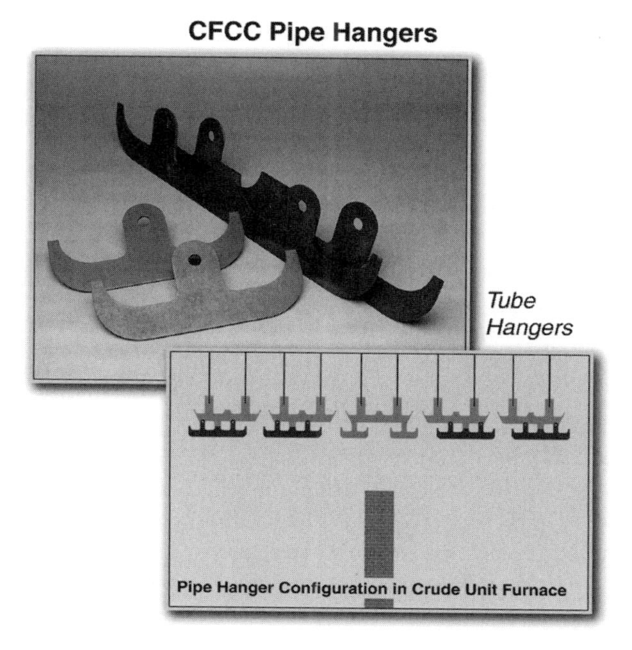

FIGURE 3.13 Tube (or pipe) hangers support flammable-fluid-filled pipes in refinery furnaces. Operating at higher temperatures, CFCC hangers improve safety and reduce downtime. Hangers supporting twin tubes are 18 in. wide and $\frac{1}{4}$ in. thick, resulting in less weight and lower stress on fasteners.

these long horizontal tubes (Fig. 3.13). The hangers are exposed to near 1100°C (2000°F) in order to heat the oil to 815°C (1500°F). These temperatures are too close to steels limit and the hangers sag, oxidize, and become brittle. Sometimes they fail before 3 years, the required service time. Hanger failure results in burst pipes, fire, and explosion. CFCC pipe hangers enable higher furnace temperatures, resulting in increased crude throughput and higher production rates for the refinery.

The crude furnace consists of a firebox, which is refractory lined, steel or alloy tubing through which the crude oil is transported and heated, and metal tube hangers that support the tubes. A typical firebox environment contains nitrogen, carbon dioxide, and carbon monoxide along with smaller amounts of oxygen, water vapor, and sulfur dioxide. The furnace temperature is limited by the susceptibility of the metal tubes to thermal degradation.

After 8500 h exposure in a major refinery, a silicon carbide/silicon carbide CFCC had a flexural strength of 20 ksi, well above the required 6 ksi. The data indicates that they will perform the required 3 years, the time of their next evaluation.

The use of CFCC pipe hangers gives a 150°C (300°F) safety margin and a hanger weight of 3 lb versus 30 lb for steel, enabling a safer, reduced stress application.

CFCC pipe hangers offer thermal stability, dimensional stability, corrosion resistance, fatigue resistance, and light weight compared to steel. The use of CFCC pipe hangers will result in longer life, improved efficiency, safety, and reduced downtime and maintenance. In this application, safety is an added benefit.

3.2.10 Furnace Fans

Continuous fiber-reinforced ceramic composites can also be made into complex shapes. An example is fan blades. The thermal stability and other properties of CFCCs make them attractive for high-temperature fans such as those used in heat-treating furnaces.

Metals and other materials are heat treated in furnaces to assure uniform and desired properties. This requires uniform heating produced by locating circulation fans to eliminate thermal "dead spots" in the furnace. Historically, these fans were made of superalloy steel. Performance has been limited due to the limited thermal stability of the steel as well as fatigue and creep, which lead to imbalance and failure. Fan imbalance requires lower fan speed, excessive maintenance, downtime, limited furnace loads, and increased operating costs. To preserve metal fans, furnace atmosphere is cooled before the fan and reheated upon exit, adding to the complexity and cost of the furnace.

The use of CFCC fans enables an increase of furnace temperature to 1200°C (2200°F) from the previous <1040°C (<1900°F). The creep resistance of CFCCs permits fan speeds above 1100 rpm, increasing circulation, resulting in faster heat-up rates, reduced load turnaround time, and uniform heat-treated product at a higher yield. CFCC fans reduce fuel consumption by 3%. In addition, CFCC fans are much lighter. They require only one person to install them (Fig. 3.14).

FIGURE 3.14 CFCC furnace fans reduce weight, requiring only one-man installation and consuming less operating energy.

COI Ceramics Corporation, San Diego, California, provides fans to furnace manufacturer Surface Combustion Corporation to place in its heat-treating furnaces. After 6 months use, the CFCC retains 75% of its original strength. Tests indicate that CFCC fans have twice the life of metal ones.

3.2.11 Diesel Engine Components

The development of low heat rejection diesel engine technology, for both stationary and mobile power plants, will require new in-cylinder materials capable of withstanding the higher temperatures produced from insulating the combustion chamber and nearby components. One of those applications involves the evaluation of CFCCs as a self-lubricating valve guide.

Tests demonstrate continuous carbon fiber-reinforced silicon nitride provides a low-friction surface, resulting in decreased wear of the valve stem. Since they require no oil for lubrication, CFCC valve guides will save fuel, increase engine efficiency 30%, increase reliability, and reduce particulate emissions by 25%.

Current cast-iron valve guides operate for 20,000 h. Commercial liquid lubricants are fed into the guide-valve stem interface. At temperatures above 300°C (572°F), the cast-iron has insufficient yield strength to support valve stem side loads. The resulting deformation allows the valve to contact the cylinder liner and not seat properly in the valve seat, decreasing its performance and life. This problem is intensified in new designs that operate at higher temperatures. At 500°C (932°F), lubricants cannot withstand the severe thermo-oxidative environments without deposit formation and wear. These deposits cause valves to stick and form particles that accelerate wear. The CFCC valve guide will be used initially in stationary diesel engines and applications may extend to internal combustion engines in general.

The top piston ring in a diesel engine is exposed to very severe thermal, mechanical, and chemical conditions. Each cycle of the engine exposes the top of the piston and the first piston ring to a burst of high-pressure and high-temperature combustion gases. They force the piston to move through the cylinder, resulting in high mechanical stresses imposed upon the ring at both the interface with the piston and the cylinder liner. The life of the rings is limited by wear and cyclic fatigue.

Lubricant must be supplied to minimize these mechanical stresses. Lubricant leaks past the ring, burns, and exits as pollution. Ceramic rings have the potential for lower wear, greater resistance to cyclic fatigue, and higher temperature capability. They have less tendency to distort, resulting in less lubricant leakage. Their thermal stability will permit more complete combustion and less pollution.

3.2.12 Thermophotovoltaic Burner/Emitter

Tremendous progress has been accomplished in increasing the efficiency of photovoltaic devices. Further progress can be achieved by linking high-efficiency photovoltaic cells with a nonsolar energy source such as a high-temperature burner/emitter. Thermophotovoltaic (TPV) power systems convert the energy radiated from an incandescent source directly to electricity. This method of generating electricity is similar to solar electric systems except for the source of radiant energy. In TPV, the energy source is a man-made emitter. The emitter is heated to the desired temperature to optimize the energy conversion. The heater can be heated by many sources including fuel combustion or chemical or nuclear reactions. Photocells convert the thermal energy to electricity. TPVs are lightweight/portable, mechanically simple, efficient, and quiet. They are attractive for residential, light industrial, appliance, and recreational power supply/cogeneration applications. Viable applications include (1) off-grid remote power, (2) self-propelled appliances such as furnaces and water heaters, (3) small power generators for recreational vehicles and boats, (4) back-up power for critical loads such as communications, and (5) portable generators such as battery chargers for remote commercial and military applications.

The higher the emitter temperature, the higher the efficiency. The thermal stability and toughness of CFCCs make them prime candidates for the burner/emitter portion of TPV devices. One CFCC design uses a porous construction similar to a hot-gas filter. The porous construction maximizes on–off response time, increases resistance to thermal shock, and allows silent, low emission, surface combustion. To achieve reasonable efficiency, the CFCC composition is selected to emit radiation matching the band gap of the photovoltaic cells. The result is a quiet, remote-capable, multifuel, lower emissions generator with improved efficiency. Studies suggest an efficiency of 10–15% is possible, which is competitive with current small (under 1 kW) heat engines for some applications.

3.2.13 Flame Stabilizer Ring

Over 55,000 MW of electricity is produced by steam generators using low NO_x burners fueled with pulverized coal. The burner requires a flame-stabilizing ring at the end of the fuel nozzle. The ring anchors the flame to the end of the nozzle and promotes stability and rapid fuel devolatilization. The flame stabilizer ring is exposed to corrosive gases at 1100–1200°C (2000–2200°F). Metal rings degrade faster than the other burner components. CFCCs have the potential to meet the life goals and reduce downtime and maintenance.

Tests show that oxide–oxide CFCCs are the most chemically stable ring material in this application. Aluminum oxide fibers are woven into cloth, cut to the desired shape, and stacked in layers to the desired thickness. This fiber preform is infiltrated with a liquid sol matrix and then converted to aluminum oxide ceramic by heating.

3.3 MATERIALS AND PROCESSES

3.3.1 Ceramic Fibers

The most commonly used fibers in CFCCs are alumina, mullite, and silicon carbide. Silicon carbide fiber is produced by Nippon Carbon, COI Ceramics, Textron, and Ube. COI Ceramics distributes Nippon Carbon's fiber tradename Nicalon. It also produce its own silicon carbide fiber under the tradename Sylramic. Textron manufactures silicon carbide fiber under the tradename SCS. Ube sells silicon carbide fiber under the tradename Tyranno. 3M Corporation produces mullite fibers and alumina fibers under the tradename Nextel.

Nicalon ceramic fiber is a silicon carbide-type fiber manufactured by a polymer pyrolysis process. The fiber is homogeneously composed of ultrafine beta-SiC crystallites and an amorphous mixture of silicon, carbon, and oxygen. It is produced in several grades including those named Ceramic, HVR, LVR, and Carbon-Coated Ceramic. The Ceramic grade is Nippon Carbon's standard product, offering optimum mechanical properties and performance at elevated temperatures. HVR is a low-dielectric fiber that sacrifices some strength in order to achieve low-volume resistivity. The LVR fiber has a low-volume resistivity (high dielectric), once again balancing electrical properties with mechanical strength. The Carbon-Coated Ceramic grade fiber is uniformly coated with pyrolytic carbon to a nominal thickness of 1 nm. It is commonly used to reinforce CFCCs because the carbon coating allows the fiber to slip with the matrix. It is supplied with polyvinyl sizing that, if not desired, is removed by hot water rinsing. Nicalon fiber is available as continuous fiber tow or woven in 5, 8, and 12 harness satin weaves (Tables 3.5 and 3.6).

COI Ceramics Sylramic fibers are 10 μm in diameter (fibers or fibers derived from a polymer composed of silicon, carbon, oxygen, and nitrogen).

TABLE 3.5 Typical Properties of Nicalon™ Fibers

		Grade			
Property	Unit	Ceramic	HVR	LVR	Carbon-Coated CG
Density	g/cm^3	2.55	2.35	2.5	2.55
Tensile strength	GPa	3.0	2.8	2.8	2.8
Tensile modulus	MPa	210	180	200	200
Volume resistivity	$\Omega \cdot$ cm	10^3-10^4	10^6	0.5–5	0.8
Dielectric constant		5	4.5	7	8
Loss factor		0.06	0.02	2	5
Thermal conductivity	W/m · K	2.97			
Coefficient of thermal expansion	10^{-6}/K	3.2			
Specific heat	kJ/kg · K	0.72			

TABLE 3.6 Percent Room Temperature Tensile Strength Retention of Nicalon™ Fiber after Thermal Exposure

Exposure:	75 h/1000°C (1830°F)	300 h/1000°C (1830°F)	75 h/1200°C (2190°F)
Argon atmosphere:	100	–	–
Air:	81	70	41

TABLE 3.7 Textron Fibers Properties

	Fiber		
Property	SCS 6	SCS 9A	SCS ULTRA
Diameter	5.6 mils (140 μm)	3.2 mils (80 μm)	5.6 mils (140 μm)
Density	0.11 lb/in.^{3} (3 g/cm^3)	0.1 lb/in.^{3} (2.8 g/cm^3)	
Tensile at room temperature	500 ksi (3450 MPa)	500 ksi (3450 MPa)	900 + ksi (6210 MPa)
Tensile			
2192°F (1200°C)	500 ksi (3450 MPa)		
2800°F (°C)	125 ksi (862 MPa)		
Modulus at room temperature	58 Msi (400 GPa)	44.5 Msi (307 GPa)	60 Msi (414 GPa)
CTEx10-6 at room temperature	2.3/°C	4.3/°C	

Textron manufactures silicon carbide-containing fibers derived from carbon. SCS-6 is a round fiber, measuring 5.6 mils in diameter. A smaller fiber, named SCS-9A, is 3.2 mils in diameter, more easily bent, and suitable for parts with small radii. This round fiber is 50/50 Si/C. Textron's newest and strongest fiber is SCS-ULTRA, a round fiber measuring 5.6 mils diameter (Table 3.7).

3M Corporation provides metal oxide fibers as its Nextel™ alumina–boria–silica fibers. Oxide-based CFCCs are appropriate for oxidizing high-temperature environments.

All fibers are coated to produce slippage of fibers in the matrix and to protect the fiber during composite manufacture and corrosive attack in use. All of these fibers may be woven in 2 or 3 dimensions.

3.3.2 Composites

Textron Systems Corporation makes tubular CFCCs products by gas-phase reaction, combining elements of ceramic slip casting, filament winding, and gas-phase nitridation bonding. The result is either a nitride-bonded silicon carbide or nitride-bonded silicon nitride ceramic depending upon the materials and process.

Textron tubular products are formed by drawing silicon carbide monofilaments (or yarn) through an aqueous-based slurry of silicon carbide containing a binder (polymer), silicon powder, and silicon carbide particulates. The coated filaments are wound onto a drum or segmented mandrel at $\pm 30°$ and dried. After the binder is removed, the silicon powder is converted to silicon nitride by nitriding/heating in a gas, typically containing nitrogen, ammonia, and hydrogen. This converts the silicon to silicon nitride, creating a matrix of silicon nitride to bond the silicon carbide powders and fibers together into a strong composite.

The green silicon carbide preform is densified by submerging it in a liquid silicon carbide precursor and heating by induction to a temperature above the decomposition temperature of the precursor. The extreme heat transfer environment imposed on the preform causes a steep thermal gradient to develop through

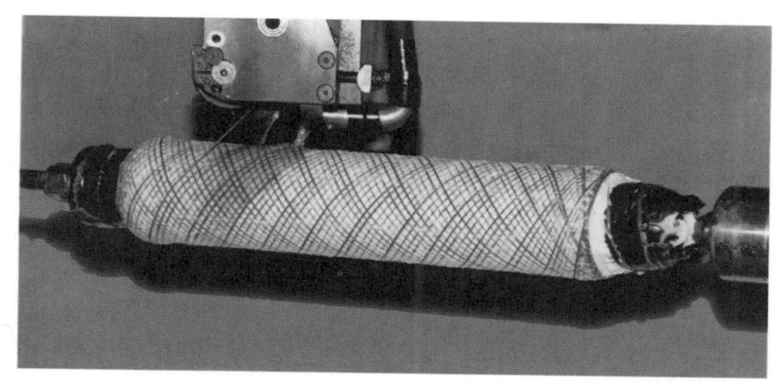

FIGURE 3.15 Cylindrical shapes are usually formed by winding ceramic fibers onto mandrel and then impregnating with ceramic matrix.

the thickness. Densification occurs by deposition of silicon carbide into open porosity of the green form, beginning at the inside of the composite and moving to the external surface through control of the thermal gradient. The final cylinder is completely net shape.

For most tubes, Textron uses its large SCS-6 silicon carbide fiber. For small cylinders, such as a 1- to 4-in.-diameter combustor liner of a small missile turbine engine, a smaller diameter fiber, such as its SCS-9 is required to bend around the small radius.

McDermott makes tubes by winding onto mandrels and infiltrating with aluminum oxide precursor sol (Fig. 3.15). The sol is converted into aluminum oxide by heating. It also uses an alternate process of laying-up cloth onto round mandrels, hardening, and removing from the mandrel.

Honeywell Advanced Composites Corporation, Newark, Delaware, makes CFCCs of silicon carbide and oxide fibers, silicon carbide and aluminum oxide matrices in various combinations depending upon the desired CFCC properties. It manufactures CFCCs with three processes: chemical vapor infiltration (CVI), melt infiltration, and directed oxidation.

At the start of each process, tooling holds the shape of a preform of fiber made of silicon carbide or carbon or metal oxide. The CVI process deposits a carbon or boron nitride (from boron chloride, ammonia, and hydrogen) coating onto the fiber. The coating is selected based upon the operating environment of the finished product. Once the coating is applied, the preform is rigid and free-standing in its desired geometry. The preform is infiltrated with methyltrichlorosilane and hydrogen vapors that react to form a silicon carbide matrix between the coated fibers. This is an isobaric, isothermal infiltration process conducted near 1000°C (1652°F) under reduced pressure. This process results in parts of various shapes and sizes including flat plates, cylinders, and more complex parts. Tables 3.8 and 3.9 list typical properties of Honeywell's Enhanced CFCC containing silicon carbide fiber in a plain weave and five harness satin fabric, respectively. Note that properties are retained at high temperatures.

Honeywell's DIMOX directed metal oxidation (DMO) process involves the growth of an oxide matrix through a preform of silicon carbide or oxide fibers.

TABLE 3.8 Properties of Honeywell Enhanced CFCC with Plain Weave Nicalon™ Fibers

Property	Units	Nominal Value at Temperature	
		23°F	2012°F
Density	lb/in.^{3}	0.08	
Tensile			
Strength	ksi	31.1	30.8
Elongation	%	0.41	0.5
Modulus	Msi	20.3	16.6
Compressive			
Strength	ksi	72.8	
Contraction	%	0.35	
Modulus	Msi	21.0	
Interlaminar			
Shear strength	ksi	4.4	
In-plane			
Fracture toughness	ksi · in.$^{1/2}$	14.8	

TABLE 3.9 Properties of Honeywell Enhanced CFCC with 5 Harness Satin Weave Nicalon™ Fibers

Property	Units	Nominal Value at Temperature	
		73°F	1562°F
Density	lb/in.^{3}	0.08	
Tensile			
Strength	ksi	34.3	38.5
Elongation	%	0.47	0.63
Modulus	Msi	18.1	17.3
Compressive			
Strength	ksi	83.7	
Contraction	%	0.43	
Modulus	Msi	20.4	
Flexural			
Strength	ksi	61.1	
Interlaminar			
Shear strength	ksi	5.5	

The fiber preform first undergoes the CVI process in which it is coated with a dual layer of boron nitride and silicon carbide. The treated fiber preform is then placed in contact with molten aluminum metal in the presence of air at elevated temperatures. The aluminum oxidizes and forms a matrix rich in aluminum oxide. The coatings protect the fibers from the molten aluminum, and the silicon carbide wets to facilitate infiltration of the aluminum. Residual aluminum, present in the matrix as microscopic interconnected channels, is removed from the CFCC. For cylindrical parts the matrix is grown through the preform by reaction of aluminum metal with air, resulting in an aluminum oxide matrix. The process is capable of manufacturing parts 4 ft in diameter and 7 ft high.

McDermott Technologies Corporation has the only oxide/oxide-based CFCC process. It is fabricating CFCCs from a powder slurry and sol–gel (a liquid converted to solids) impregnation technique. Metal oxide fibers are wound onto a mandrel or woven into cloth or preforms; the residual voids are filled with reactive, fine particles of alumina, mullite, and/or yttrium alumina garnet (YAG) matrices. Densification occurs at 1100°C (2012°F). Fiber coatings include carbon as a fugitive interface. This fugitive interface is achieved by depositing a carbon or Scheelite ($CaWO_4$) coating onto the fiber during composite processing. It is oxidized to create fiber–matrix slip. The "fugitive" approach creates a void along the fiber. This void deflects cracks, leaving continuous fibers to bear the load.

McDermott also makes porous CFCC filter tubes. Any one of three methods can be used. One method involves winding high-strength aluminum oxide ceramic fibers (3M Corporation's Nextel™ 610) onto a mandrel and filling a portion of the space between fibers with aluminum oxide particles utilizing a liquid-to-solid sol–gel multiple infiltration process. The second method involves fabrication of an open network skeleton of high-strength silicon carbide ceramic fibers rigidized by chemical vapor deposition, then filling with a porous oxide matrix by vacuum casting. The third method substitutes a silicon carbide matrix for the aluminum oxide matrix in the first method. In all of these methods the final step is the application of heat and pressure to form an oxide–oxide CFCC. All methods result in a CFCC filter of the appropriate porosity, and meeting the filtration and strength specifications of this application.

COI Ceramics Corporation produces CFCC materials by a polymer infiltration and pyrolysis (PIP) process. This is a versatile way to fabricate large, complex-shaped structures. The process uses low-temperature forming and molding steps typically used in plastic matrix composites. A preform, composed of silicon carbide fibers or fibers derived from a polymer composed of silicon, carbon, oxygen, and nitrogen, is impregnated with a polymer matrix and cured by conventional methods. The composite is pyrolyzed to temperatures beyond 980°C (1800°F) to convert the preceramic matrix polymer to a ceramic. Subsequent impregnation and pyrolysis steps are carried out to achieve the desired density. Both the initial shaping and fabrication of the composite are carried out at low temperature.

Continuous fiber-reinforced ceramic composites fabricated by the PIP process can consist of various fiber, interface coating, and matrix chemistries. Fiber architecture preforms can include filament windings, braids, or two- and three-dimensional weaves. An important aspect of this PIP process is its adaptability to polymer matrix processing equipment. Aside from reducing initial capital investment by using existing equipment, PIP works well with various preforming techniques such as hand lay-up, filament winding, braiding, reaction transfer molding, three-dimensional weaving and conventional cure, such as autoclaving. COI Ceramics makes flat plates, cylinders and complex shaped parts of CFCCs.

COI Ceramics fabricates cylinders by two methods. One method involves winding a silicon carbide fiber onto a cylindrical mandrel followed by liquid infiltration and heat. The second method is similar to the first except than the fiber is, woven into a cloth and then wrapped onto the mandrel.

COIs PIP CFCCs are available in two classes: the Sylramic 100 series—a carbon-coated Nicalon™ fiber in an amorphous SiOC matrix for maximum use temperature <450°C (842°F) in oxidizing environments and up to 1100°C (2012°F) in inert environments—and Sylramic™ 200 and 300 series of proprietary coated Nicalon™ fiber in an amorphous SiNC matrix for use up to 1200–1250°C (2192°–2282°F) in an oxidizing environment.

COI Ceramics also produce a CFCC based on a sol–gel derived alumino-silicate matrix that can be combined with a variety of commercially available fiber reinforcements such as 3M Corporation's Nextel™ fibers. This latter fiber provides the highest temperature resistance and creep resistance. The baseline oxide–oxide system relies on controlled matrix porosity for toughness, eliminating the need for fiber coatings.

Tables 3.10 and 3.11 contain data generated on a COI Ceramic Nextel™ 720 reinforced alumino-silicate CFCC at 982°C (1800°F) and 1093°C (2000°F).

The residual strengths after fatigue are equal or greater than the unexposed composite. The creep rupture tests are just as encouraging, with residual stress after 100 h at 150 MPa (21.6 ksi) equal to 30.3 ksi (Table 3.12).

COI Ceramics also manufactures SiC/SiC CFCC composites with properties as shown in Table 3.13.

TABLE 3.10 COI Ceramics Oxide CFCC Properties with Various 3M Oxide Fabrics

	Nextel™ 312	Nextel™ 550	Nextel™ 720	Nextel™ 610
Composite density (g/cm^3)	2.3	2.41	2.6	2.83
Nominal fiber volume (%)	48	36	48	51
RT tensile strength (ksi)	18.1	21.4	28.3	53.1
RT tensile modulus (Msi)	4.5	5.8	11.6	18
Coef of thermal exp (ppm/°F)	2.7	3.0	3.5	4.4
Dielectric constant (X-band)	4.4	4.8	5.6	5.8

TABLE 3.11 Typical Properties of Nextel™ 720 Reinforced Alumino-Silicate CFCC

	Temperature	Strength (ksi)	Modulus (Msi)	Strain (%)
Tension	RT[a]	28.3	11.6	0.3
Tension (after 100 h/1800°F)	RT	27.2	11.9	0.29
Tension	1800°F	24	10.6	0.27
Tension (after 100 h/1800°F)	1800°F	25.5	11.4	0.28
Flexure	RT	31.4	14.3	0.22
Compression	RT	26.9	11.6	0.22
In-plane shear	RT	4.5	2.0	0.49
Interlaminar shear	RT	1.7		

[a] RT = room temperature.

TABLE 3.12 Residual Tensile Strengths after Fatigue and Creep Loadings in Air

	Test Temperature (°F)	Stress Level	Loading Conditions	Residual Tensile Strength (ksi)
Fatigue	1800	18.1	100,000 cycles	28.0
	1800	21.6	100,000 cycles	30.6
	1800	23.2	100,000 cycles	28.3
	2000	21.6	100,000 cycles	24.4
Creep rupture	1800	21.6	100 h	30.3

TABLE 3.13 COI Ceramics SiC/SiC CFCC Physical Properties

Property	Unit	Nicaloceram™	Hi-Nicaloceram™	Hi-Nicaloceram S™
Tensile strength	MPa	110	240	330
Tensile modulus	GPa	60	80	110
Flexural modulus	MPa	110	400	550
Density	g/cm^3	2.0	2.2	2.3

The melt infiltration technology of General Electric Corporation, Schenectady, New York, produces CFCCs composed of continuous silicon carbide fibers in a matrix of silicon carbide and silicon. The composite is made by a silicon melt infiltration process in which densification takes place in a matter of minutes. A boron-nitride-based coating is applied to the fiber. It provides fiber pull-out and protects the fiber from the molten silicon during the infiltration step. The coated fibers are pulled through a liquid mixture containing polymers and fillers and wound onto a drum to produce unidirectional plates.

The composite preform plates are loaded into a vacuum furnace while in contact with silicon. Initial heating of the preform to 500–600°C (932°–1112°F) is done slowly to allow burnout of any binders, leaving a body of 35–40% porosity. Heated above 1410°C (2570°F), the silicon melts and infiltrates the porous preform by capillary action. No external pressure is required. During infiltration the silicon reacts with the any free carbon in the preform (incorporated as a particulate in the matrix slurry or from pyrolysis of a binder constituent) to form silicon carbide. Any residual porosity is filled with silicon. The overall infiltration process is near net shape, with less than 0.5% change in preform dimensions.

The plates are cut into tapes, stacked, and pressed by die pressing, compression molding, vacuum bagging, or autoclaving to form a laminated composite. This process is also used to form silicon carbide/alumina CFCCs and alumina/alumina CFCCs when aluminum is used in place of silicon. An aluminum oxide matrix is grown through a silicon carbide fiber preform. The fiber preform is first coated with a boron nitride/silicon carbide dual layer. The boron nitride is derived from boron chloride, ammonia, and hydrogen. The silicon carbide is derived from methyltrichlorosilane. Both coatings are deposited employing CVI. These coatings protect fibers from the molten aluminum, and the silicon carbide coating wets the fiber to facilitate infiltration of the aluminum. When molten aluminum metal is placed in contact with the silicon carbide fiber preform, it oxidizes in the presence of air at elevated temperatures and forms a matrix that is rich in aluminum oxide. Residual aluminum, present in the matrix as microscopic interconnected channels, is removed from the CFCC.

Amercom, Chatsworth, California, pioneered a CFCC process employing liquid infiltration of preforms with preceramic polymers, and phenolic resin, as a way of rigidizing the preform in the first-stage of CVI. This eliminates the need for graphite tooling to hold and configure the fiber-reinforced preform during the initial rigidization and densification steps. Instead, aluminum or other metal tooling can be used repeatedly and requires heat up to only 204°C (400°F). The batch-processing capacity of a reactor for the early CVI processing stages are increased by 300% or more. This technology is owned by COI Ceramics, Incorporated.

A process of Allied Signal Ceramic Components, Torrance, California, illustrates the versatility of ceramic composite chemistry. The process results in a carbon fiber-reinforced silicon nitride ceramic composite. Cold isostatic pressing is used to form the composite. Glass encapsulation and hot isostatic pressing are employed to densify it. Silicon nitride provides excellent mechanical properties and resists corrosion. The carbon fiber provides self-lubrication and toughness.

3.4 CORROSION RESISTANCE DATA

The excellent corrosion resistance of silicon carbide and aluminum oxide leads to similar expectations of CFCCs based on these precursors. Table 3.14 data illustrates that expectation.

Although CFCCs resist corrosion by a wide variety of agents, the addition of a fiber and a fiber–matrix interface does introduce opportunity for corrosive paths. Therefore, CFCC corrosion resistance cannot be taken for granted but must be evaluated for every application. This section gives the corrosion data

TABLE 3.14 Corrosion Resistance of Some Nonfiber-Reinforced Silicon Carbide Ceramics

		Weight Loss ($mg/cm^2/yr$)	
Test Environment[a] Reagent (wt %)	Temperature (°C)	Sintered SiC (no free Si)	Aluminum Oxide (99%)
98% H_2SO_4	100	1.8	65
50% NaOH	100	2.5	75
53% HF	25	<0.2	20
85% H_3PO_4	100	<0.2	>1000
70% HNO_3	100	<0.2	7
45% KOH	100	<0.2	60
25% HCl	70	<0.2	72
57% HNO_3	25	<0.2	16

[a] 125–300 h, submerged, stirred.

that has been generated on CFCCs except for some that is included in the earlier application section. In most cases, corrosion data was generated with a particular application in mind. Application-oriented data includes simulation of filters, heat exchangers, steam cracker materials, chemical pump housings, and gas turbine combustor liners.

A chemical company is evaluating CFCC filters to remove silicon from a stream of moist hydrogen chloride and another stream containing various chlorosilanes. Success would eliminate the need for additional, very large, bag houses. The facility would substantially benefit both in process stream efficiency and effluent cleanup from improved high-temperature filtration technology. CFCCs have withstood over 1000 h in each of these extremely corrosive streams, at 290°C (550°F) and 1040°C (1900°F), respectively.

CFCC filter samples consisting of rings, cut from candle filters (wall thickness approximately $\frac{1}{2}$ in.), measuring 2.5 cm in height and 7.5 cm in diameter were placed in each gas stream. Moist hydrogen chloride gas passed over each sample at a rate of 9.6 cm^3/min plus 887 cm^3/min of mixed gas of nitrogen, oxygen, carbon dioxide, and water vapor, in an operating temperature of 1040°C, simulating the operating environment. After 1000 h exposure, the burst strength of the CFCC rings were tested at room temperature and compared to pretest strength. The CFCC retained 65% of its original strength in this extremely corrosive environment (Fig. 3.16).

In the other filter test, chlorosilanes were passed over CFCC rings at a rate of 500 cm^3/min of methyl chloride plus 2.7 mL/min of dimethyldichlorosilane.

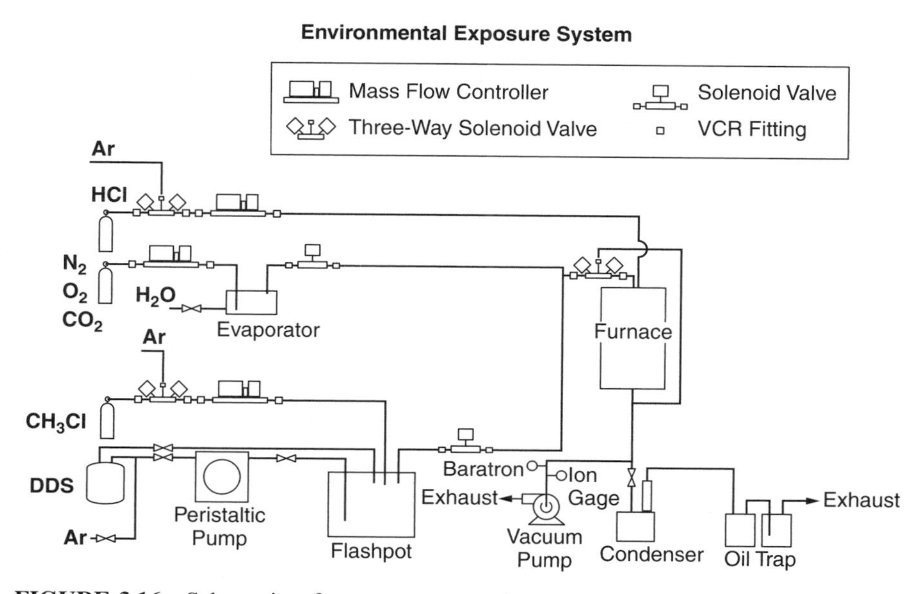

FIGURE 3.16 Schematic of apparatus to evaluate CFCC filters in chemical stream. Apparatus was designed and is operated by Oak Ridge National Laboratory.

After 1000 h in this 290°C (554°F) chlorosilane stream the silicon carbide-based CFCC had retained 85% of its original strength.

Corrosion tests were conducted to see if oxide-based CFCCs would survive as a material for constructing high-temperature heat exchangers. One heat exchanger, used to recover heat immediately downstream from an afterburner-equipped waste incinerator, needed materials that would withstand not only the heat but the corrosive exhaust gases from the incinerator. The incinerator feed contained heavy metals, alkali metals, and transition metals. Immediately downstream from the incinerator is an afterburner. CFCC samples were held in the middle of the stream of hot, corrosive gases immediately behind the afterburner. Fifteen test samples were mounted on three fixtures, five per fixture. Each sample was 15 cm in length. During the first 58 days of testing, the CFCC samples were exposed to 925–980°C (1690–1800°F) for a total of 560 h. The balance of the time was used in 37 cycles between this operating temperature and 370–450°C (700°–842°F), mimicking normal daily shutdown. Four cycles of longer duration also occurred. An additional 13 days exposure at 980–1040°C (1800–1900°F) completed the testing. Twenty additional cycles occurred during the final 13 days.

After exposure, the CFCC samples were cut into C-rings and their strength measured at room temperature and compared to pretest strength. The alumina and zirconia matrix CFCC specimens retained 73 and 75% of their strength, respectively.

Continuous fiber-reinforced ceramic composites have also been exposed to coal gas and ash to estimate their performance in a coal-fired hot-air heat exchanger. The exchanger tubes must perform at 1150–1260°C (2100–2300°F) during exposure to coal ash and flue gas. The ash contains high levels of alkali condensed out of the flue gas. A CFCC comprised of a silicon carbide matrix reinforced with silicon carbide fiber, produced by CVI, was chosen due to its excellent retention of physical properties in oxidizing atmospheres. Since it is susceptible to attack by alkali, previous testing predicted that this CFCC would perform satisfactorily if alkali is scrubbed from the flue gas and temperatures do not exceed 1150°C (2100°F). The current metal capability would be defined as suitable for an environment of low alkali content at 982°C (1800°F).

Foster Wheeler Corporation engineers inserted a horizontal probe into a direct-fired coal-burning boiler at Gallatin station, operated by the Tennessee Valley Authority (TVA). Placed onto the probes were four samples of CFCC tube measuring 2 in. inner diameter × 3 in. long × $\frac{1}{8}$ in. thick. The environment included high levels of alkali and a temperature of 1750°F. After 8000 h of exposure, the CFCC sample were removed, inspected, and tested. The specimen had thick layers of coal ash on them, which was easily removed. There was no evidence of corrosion when analyzed both visually and by metallographic techniques. There was no measurable thinning of the CFCC specimens. There was no loss of physical strength.

Another application-oriented corrosion evaluation of CFCCs concerned their evaluation in steam crackers used to crack ethylene. Several CFCCs were evaluated in conditions simulating a steam cracker environment. The flexure bar-shaped test

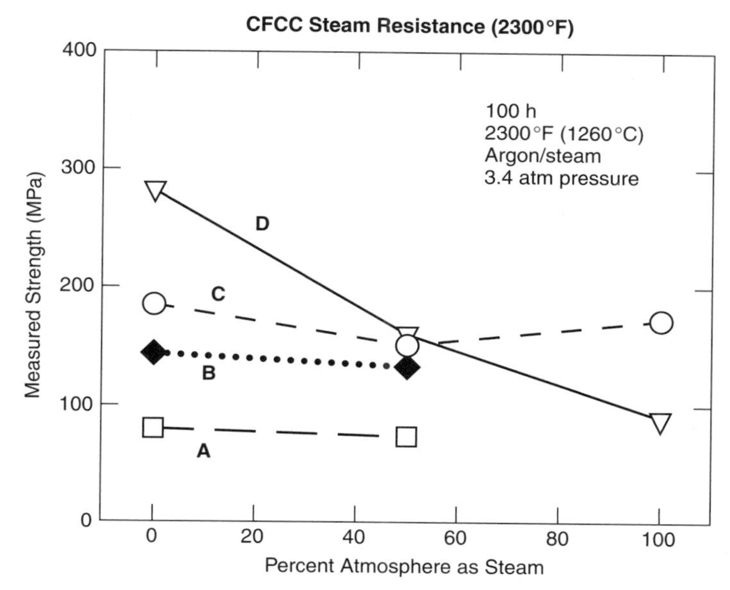

FIGURE 3.17 (A) Silicon carbide fibers reinforcing an alumina matrix, (B) silicon carbide fibers reinforcing a silicon carbide matrix, (C) unidentified fibers reinforcing a silicon nitride bonded silicon nitride matrix, and (D) other fibers reinforcing a proprietary matrix. Materials available from Honeywell, Inc.

specimens measuring 3 × 4 × 50 mm and tensile bars 12 cm long with a gage area of 8 × 40 mm were exposed for 100 h at 1260°C (2300°F) in an argon–steam atmosphere at a total pressure of 3.4 atm (50 psia) and various concentrations of steam, conditions that exceed those experienced in steam cracking. Argon was used instead of hydrocarbons since earlier studies showed no corrosion by hydrocarbons. Strength tests were conducted at room temperature.

Figure 3.17 shows the excellent retention of CFCC physical strength after this severe exposure to steam at high-temperature and pressure. Additional experiments are required because some specimen were too narrow to accommodate the fiber weave, resulting in lower strength due to configuration rather than the harsh environment.

The pump housing application described earlier required examining the corrosion resistance of CFCCs to commonly pumped chemicals. Figure 3.18 presents this data. Flexural bars measuring 0.25 × 3 × 0.1 in. were immersed in each chemical for 500 h. Temperatures were chosen to match those expected in the pump. After exposure, the bars were dried and flexure strength measured. This relatively simple test and the obvious corrosion resistance of CFCCs encouraged more specific and complicated testing.

Since an initial focus of the CFCC development was for gas turbine combustor liners, a lab-scale high-temperature and pressure apparatus was designed to simulate that application. Exposure conditions included a pressure of 150 psia,

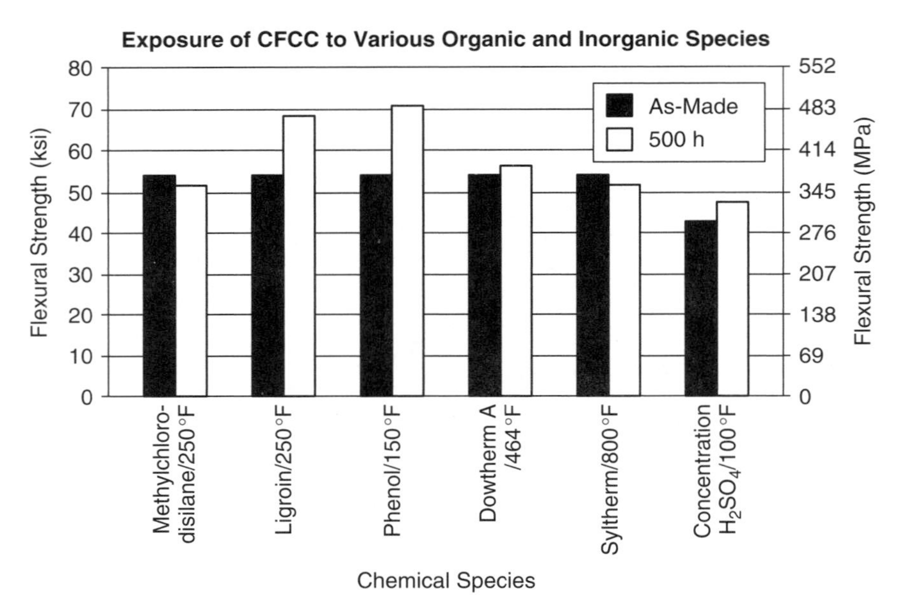

FIGURE 3.18 Resistance to commonly pumped chemicals. This CFCC was made by polymer infiltration and pyrolysis (PIP) process.

temperature of 1204°C (2200°F), and 15% steam atmosphere. Test specimen were evaluated at 500-h intervals, and the results compared to specimen exposed in actual engines operated on typical duty cycles.

Tensile and flexural CFCC specimens were exposed to the pressurized atmosphere mentioned above. Test specimens were evaluated every 500 h.

The CFCC evaluated was a CVI silicon carbide matrix reinforced with silicon carbide fiber, with a pyrocarbon interface coating and a silicon carbide seal coat. When specimen thickness was measured at 500 and 100 h, it was determined that surface recession was occurring at a steady rate of 45 μm/500 h, identical to the actual engine experience operating on a typical duty cycle (Fig. 3.19). See the gas turbine engine description in Section 3.2.2 of this chapter for additional actual engine gas turbine performance data.

Additional simulated gas turbine engine data had been generated earlier. Instead of recession rate, this data focused on strength retention of CFCC combustion liners. In these tests, the CFCC consisted of a silicon-carbide-reinforced silicon carbide/silicon matrix composite formed by melt infiltration. The fiber–matrix interface coating was boron nitride based applied by CVD.

Rectangular tensile bars measuring 150 × 12 × 2.5 mm with a gage length of 2.5 mm were exposed in a induction-heated tube furnace to a temperature of 1200°C (2192°F), 90% water and 10% oxygen atmosphere (water vapor pressure of 0.9 atms) with a gas velocity of 0.04 m/s. The test was conducted for 500 h, with the temperature cycled to room temperature every 2 h. The test specimen

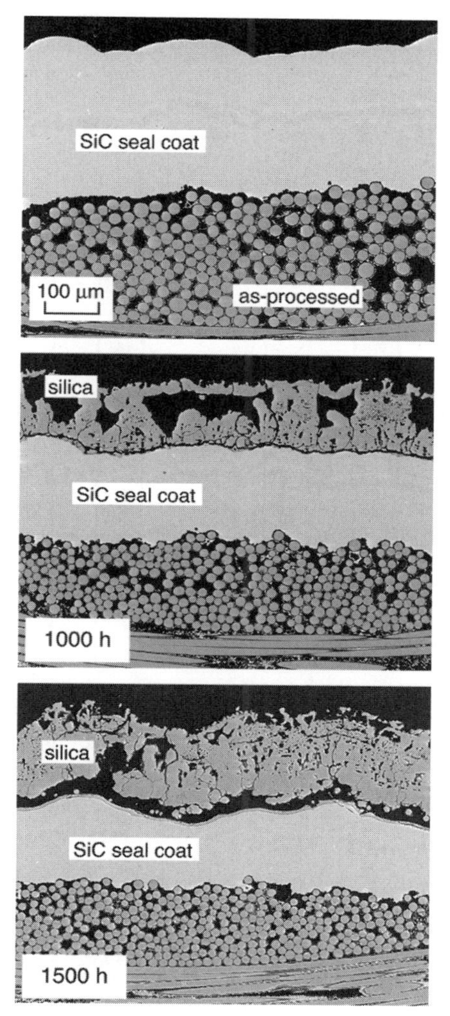

FIGURE 3.19 Seal coat of silicon carbide protects CFCC and prolongs its life. Note low erosion rate under these harsh conditions.

bars were pulled at a strain rate of 0.0002 m/s. The results are given in Fig. 3.20. Note the excellent retention of strength.

In a second test, both apparatus and conditions were changed to simulate other duty cycles. In this second test, tensile bars were exposed for 100 h in the exhaust section of a high-pressure combustion rig. The exposure conditions replicated an actual engine combustion atmosphere using natural-gas fuel at an equivalency ratio of 0.32, a gas temperature of 980°C (1800°F), a gas velocity of 85 m/s, and a pressure of 10 bar. During the 100 h exposure, the temperature was cycled to room temperature six times.

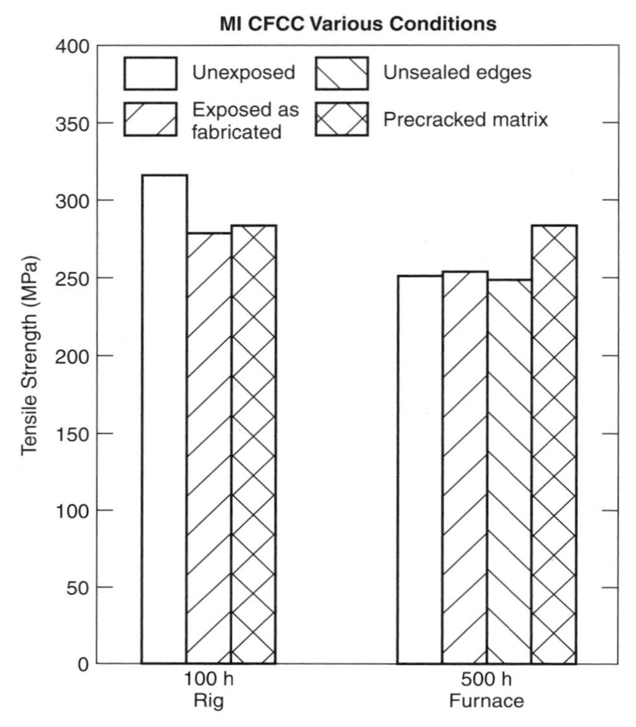

FIGURE 3.20 Note that this melt infiltration CFCC resists environment even precracked or with unsealed edges exposed.

The CFCCs demonstrated good thermal stability and the combustion exposure was very promising.

3.5 U.S. ADVANCED CERAMICS ASSOCIATION (USACA)

An excellent resource for additional CFCC information is the U.S. Advanced Ceramic Association (USACA). The people and companies producing CFCCs are represented by this trade association. Members include all companies with production capability in North America, even if their headquarters are elsewhere. Throughout its history, USACA has spent time educating policymakers in the executive and legislative branches of the federal government on the advantages and applications of CFCCs and advocating industry positions of interest.

In December 1998, USACA published a report entitled, *Opportunities for Advanced Ceramics to Meet the Needs of the Industries of the Future*. It reports the actual and potential applications of advanced ceramics, including CFCCs, in the chemical, forest products, steel, glass, aluminum, and metalcasting industries. The publication number for this report is DOE/ORO 2076 and is available from the U.S. Department of Energy, `www.oit.doe.gov/catalog/`.

3.6 SOURCES FOR ADDITIONAL INFORMATION

Honeywell Advanced Composites, Incorporated, PO Box 9559, Newark, DE, 19714-9559, www.honeywell-aci.com, Phil Craig, telephone (302) 456–6577, Fax (302) 456–6480, email phil.craig@honeywell.com

Textron Systems, Incorporated, Two Industrial Avenue, Lowell, MA, 01851, Ray Suplinskas, telephone (987) 454–5600, Fax (978) 454–5632, email rsuplins@systems.textron.com

McDermott Technologies, Incorporated, Lynchburg Research Center-MC 76, PO Box 11165, Lynchburg, VA 24504, Richard Goettler, telephone (804) 522–6418, Fax (804) 522–6980, email richard.w.goettler@mcdermott.com

General Electric, Incorporated, Corporate Research and Development, Building K1-RM 3B4, 1 Research Circle, Niskayuna, NY 12309, Krishna Luthra, telephone (518) 387–6348, Fax (518) 387–7563, email luthra@crd.ge.com

COI Ceramics, Incorporated, 9617 Distribution Avenue, San Diego, CA 92121, Andy Szweda, telephone (858) 621–7463, Fax (858) 621–7451, email aszweda@coiceramics.com

United States Advanced Ceramics Association, Suite 300, 1800 M Street NW, Washington, DC 20036-5802, www.advancedceramics.org, telephone (202) 293–6253, Fax (202) 223–5537, email usaca@ttcorp.com

U.S. Department of Energy, Office of Industrial Technology, EE-23, Forrestal Building, 1000 Independence Avenue SW, Washington, DC 20585, www.oit.doe.gov/cfcc, Sara Dillich, telephone (202) 586–7925, email sara.dillich@ee.doe.gov

Oak Ridge National Laboratory, 1 Bethel Valley Road, Oak Ridge, TN 37831, www.ms.ornl.gov/programs/energyeff/cfcc/, Peter Angelini, telephone (865) 574–4565, Fax (865) 576–4963, email angelinip@ornl.gov

3.7 COLLABORATIVE PROGRAM

Continuous fiber-reinforced ceramic composites were developed in a collaborative program that combined the experience and facilities of industry with the expertise and specialized talents available at universities and national laboratories. CFCC producers worked with users to determine appropriate formulas and processes for industrial applications. Universities and national laboratories had a supporting role, conducting the most basic studies such as composite design, material characterization, test method development, and investigation of performance-related phenomena.

The U.S. Department of Energy (DOE) Office of Industrial Technology (OIT), in communication with industry, initiated the CFCC program in 1992 as a 10-year collaborative effort between industry, national laboratories, academia, and government. The goal of the program is to advance processing methods for reliable and cost-effective ceramic composites to a point where industry will assume full risk of development and commercialization. The CFCC materials under development support the OIT Industries of the Future program, including chemical, steel, aluminum, metalcasting, forest products, glass, agriculture, and mining industries. Together, these industries consume 80% of the total U.S. manufacturing energy use. Industries that implement CFCC components in their applications will realize substantial energy, economic, and environmental benefits, including lower maintenance, higher efficiency, and decreased operating costs. Additional benefits accrue from optimization of process operating conditions, reduced downtime, and increased useful lifetimes.

Ten teams were established, headed by individual material suppliers, that included component manufacturers, end users, national laboratories, and universities.

Seventeen national laboratories and universities supported this activity with fundamental research, materials characterization, test methods, environmental exposure and other data, computer design codes, nondestructive inspection techniques, standards development, and life prediction techniques. Their basic research enabled fundamental understanding of CFCC chemistries and processing conditions. They are leading the creation of standards and aiding industry in developing procedures for determining material performance relationships. This activity was managed by Oak Ridge National Laboratory.

Prior to the program, industry had only a concept of how to best reinforce ceramics. Some material suppliers had made small hand lay-ups that demonstrated a promising reinforcement. Today, the teams incorporate CFCCs into commercial products and customers are realizing the benefits envisioned at the start of the program. Numerous other applications are evaluating CFCCs. The thousands of hours of successful performance means that many of these other applications will soon go commercial.

3.8 CONCLUSION

Advanced monolithic ceramics are used throughout industry, demonstrating superior performance compared to conventional materials. As user confidence grows—and as energy savings, increased productivity, and reduced maintenance are realized—the need is emerging for advanced ceramics having improved toughness. Continuous fiber ceramic composites are viewed as the ultimate solution with many applications rapidly becoming commercially viable.

REFERENCES

1. D. Haught, M. Smith, and J. Wessel, *CFCCs—Enablers of Higher Temperature Processes*, National Engineering Meeting, Chicago, IL, 1999.
2. J. Wessel, "Breaking Tradition with Ceramic Composites," *Chem. Engr. Mag.*, **103**(10), October (1996).
3. D. Freitag and D. Richerson, *Opportunities for Advanced Ceramics to Meet the Needs of the Industries of the Future*, United States Advanced Ceramic Association, United

States Department of Energy, Office of Industrial Technologies, Washington, DC, 1998, pp. 2–14, 17, 19.

4. U.S. Department of Energy, Office of Industrial Technology, *Continuous Fiber Ceramic Composite Program Update-Executive Summary*, DOE, Washington, DC, 1999.
5. U.S. Department of Energy, *CFCC Program Plan*, DOE, Washington, DC, 1997.
6. U.S. Department of Energy, *CFCC Program Plan Update*, DOE, Washington, DC, 2000.
7. D. Freitag, "Progress and Opportunities in the Development and Application of Advanced Ceramics," 14th International Conference on Ultra-High Temperature Materials, Ube City, Japan, 2000.
8. U.S. Department of Energy, *CFCC Accomplishments and Program Completion Roadmaps*, DOE, Washington, DC, 2000.
9. U.S. Department of Energy, *CFCC Project Fact Sheets*, DOE, Washington, DC, 1999.
10. U.S. Department of Energy, *CFCC Working Group Proceedings*, DOE/OR-2047, Oak Ridge, TN, 1996.
11. U.S. Department of Energy, *Petroleum Refining-High Pressure Heat Exchanger Systems for Chemical Production*, DOE, Washington, DC, 1999.
12. U.S. Department of Energy, *Oak Ridge Natl. Lab. Rev.*, **33**(1), Oak Ridge, TN, 2000, p. 16.
13. D. Hindman and C. Billis, "Performance of an Advanced Heat Exchanger Using Ceramic Composite Tubes in an Hazardous Waste Incinerator," in *Advanced Heat/Mass Transfer and Energy Efficiency*, American Society of Mechanical Engineers, HTD-Vol. 320, PIP-Vol. 1, San Francisco, 1995.
14. 3M Corporation, Type 203 Ceramic Composite Filters XN-529 Product Data Sheet, Minneapolis, MN, 2001.
15. U.S. Department of Energy, Enabling Technologies, No. 454-567/80093, DOE, Washington, DC, 1999.
16. U.S. Department of Energy, CFCC Program, Washington, DC, 1996.
17. U.S. Department of Energy, CFCC Program, Washington, DC, 1997.
18. V. McConnell, *High Performance Composites, Fail Safe Ceramics*, March–April, 25–31 (1996).
19. U.S. Department of Energy, Office of Industrial Technologies, Company One Pagers, Washington, DC, 1997.
20. A. Szweda, *Applications for Advanced Ceramics*, U.S. Advanced Ceramics Association, Office of Industrial Technologies Conference, Washington, DC, 1997.
21. P. Craig, memo to M. Karnitz and D. Haught, April 23, 1998, p. 3.
22. McDermott Technologies, Inc., Continuous Fiber Ceramic Composite Data Sheet, Alliance, OH, 1997.
23. R. Goettler, J. Keeley, and R. Wagner, *Ceramic Composite Technical Development for Industrial Applications at Babcock and Wilcox*, American Society of Mechanical Engineers, Composite and Functionally Graded Materials, MD-Vol. 80, Book No. HO 1115–1997.
24. McDermott Technologies, Inc., TPV System Data Sheet, Alliance, OH, 1998.
25. R. Wagner, *Ceramic Composite Hot-Gas Filter Development*, McDermott Technologies, Lynchburg, VA, 1999.

26. Textron Corporation, Ceramic Matrix Composite Data Sheet, CERIDAT 8, Lowell, MA, 1998.
27. Textron Corporation, SCS-6 Silicon Fiber Data Sheet, SCS6DAT7, 1998.
28. Textron Corporation, SCS-Ultra Silicon Carbide Fiber Data Sheet, 1998.
29. Textron Corporation, SCS-9A Silicon Carbide Fiber Data Sheet, 1998.
30. U.S. Department of Energy, *CFCC Newsletter*, No. 11, Oak Ridge, TN, 1999, p. 4.
31. U.S. Department of Energy, *CFCC Newsletter*, No. 9, 1997, pp. 4, 9.
32. U.S. Department of Energy, *CFCC Newsletter*, No. 8, 1997, pp. 1, 2, 4, 6, 7, 18.
33. U.S. Department of Energy, *CFCC Newsletter*, No. 7, 1996, pp. 6, 7, 10–12.
34. U.S. Department of Energy, *CFCC Newsletter*, No. 6, 1995, pp. 3, 7, 9.
35. U.S. Department of Energy, CFCC Newsletter, No. 5, 1994, pp. 6, 9.
36. U.S. Department of Energy, CFCC Newsletter, No. 4, 1993, p. 7.
37. U.S. Department of Energy, CFCC Newsletter, No. 3, 1993, p. 7.
38. U.S. Department of Energy, CFCC Newsletter, No. 2, 1993, pp. 1, 2, 5, 7.
39. S. Richlen, Continuous Fiber Ceramic Composites, *ASTM Standardization News*, September 1990, pp. 64, 68.
40. COI Ceramics, Inc., Nicalon Data Sheets, San Diego, CA, 2001.
41. COI Ceramics, Inc., CMC Data Sheets, 2001.
42. Honeywell Advanced Composites, Inc., Enhanced Ceramic Grade Nicalon CFCC Data Sheets, Newark, DE, 2001.

CHAPTER 4

Low-Temperature Co-Fired Ceramic Chip Carriers

JOHN U. KNICKERBOCKER and SARAH H. KNICKERBOCKER

IBM, Inc., 208 Creamery Road, Hopewell Junction, NY 12533

4.1 INTRODUCTION AND BACKGROUND

The 1990s brought revolutionary new technologies to computing, and one such new technology has been that of multilayer ceramic interconnecting substrates.

Handbook of Advanced Materials Edited by James K. Wessel
ISBN 0-471-45475-3

We have witnessed many high-performance products transition from aluminum oxide dielectric with molybdenum or tungsten conductor materials to much higher performance glass–ceramic dielectric with copper conductor integrated wiring. In particular, cordierite glass–ceramics have enabled greater processing speeds due to their lower dielectric constants and copper's much higher electrical conductivity than traditional molybdenum, and tungsten has further aided signal speeds. And, new processing technologies have allowed greater dimensional control of the finished product, which in turn has permitted tighter tolerances and more advanced design ground rules. Additionally, the exceptional reliability of this technology has been demonstrated across many different form factors and applications.

This chapter reviews glass–ceramic/copper multilayer interconnect substrate technology. It reviews the fundamental materials properties as well as the key processing parameters. This chapter also reviews some future directions and challenges for this emerging technology in the new millennium.

The very first glass–ceramic/copper multilayer interconnect substrates were introduced by IBM in 1990. These interconnect substrates were as large as 127×127 mm and possessed as many as 70 distinct layers. The copper interconnect wiring was done with 90-μm pitch. The first applications for these interconnect substrates were for mainframe computers as multichip modules (MCMs). In this case, as many as 121 devices were flip-chip or C4 (controlled collapse chip connection) joined to the substrate, as well as hundreds of C4 decoupling capacitors. While these first interconnect packages were used for mainframe computers, subsequent applications have been found for single-chip, chip-scale-sized modules, and their use in high-speed and high-frequency applications has continued to grow with passing years. How did these new interconnect materials come about and why were they chosen? We hope to answer these questions and others like them in this chapter.

4.2 MARKET APPLICATIONS

Ceramic chip carriers are used to bridge between silicon die and organic boards for electronic systems within a specified market. The package or chip carrier generally provides the electrical, mechanical, and sometimes the thermal support base for semiconductor dies at a desired cost. Ceramic chip carriers are utilized in a wide array of products including high-performance applications, cost–performance applications, commodity applications, hand-held applications, automotive applications, and memory applications. Low-temperature co-firing ceramic applications include high performance such as mainframes and servers, cost–performance such as workstations, commodity such as high-performance games, and wireless applications for hand-held applications and use in automotive applications such as global positioning systems (GPS). Pin grid array (PGA) and land grid array (LGA) are the ceramic chip carriers of choice for high-performance applications that require a demountable capability. For portable, lower cost solutions, ball grid array (BGA) and column grid array (CGA) solutions are often used, or when demounting is not a requirement.

4.3 MATERIALS

4.3.1 Materials Characteristics

4.3.1.1 Interconnect Substrate Materials Selection

Cost is a critical factor in material selection. Beyond cost, often the single most important property used in the selection of new dielectric substrate material is that of dielectric constant. The propagation delay of a signal within a computer is inversely proportional to the square root of the substrate material dielectric constant. The lower the dielectric constant of the insulator, the better the performance. Since traditionally aluminum oxide with a dielectric constant of approximately 9.5 to 10 has been used for many years, any new material, if it was going to perform better, should have a dielectric constant well below 10. High-performance silicon chips are frequently joined to the multilayer interconnect substrates through flip-chip or C4 technology, and for large die sizes a close thermal expansion match between the chip and carrier is desirable. Silicon chips can generate many watts of heat in use; so the heating and cooling associated with a computer being turned on and off could be a source of significant fatigue if the thermal expansion mismatch is large, although underfills have much improved fatigue life between die and chip carriers for a variety of module solutions, including both ceramic and especially organic chip carriers where large thermal expansion mismatch is typically large. Another important property of a substrate dielectric material is the processing compatibility between the metal and the ceramic. Can they be processed together to form strong, dense bodies with the desired properties? Molybdenum and tungsten metals have been commonly used with aluminum oxide ceramic because it is necessary to reach temperatures in excess of 1500°C in order to coalesce the Al_2O_3 particles during firing. At this temperature many metals are molten. While we would like the metal powder particles to sinter into a dense solid conductor, if the metal melts, its properties and the ceramic substrate itself could be altered significantly. Historically, if a highly conductive metal such as copper, gold, or silver are to be used, then the ceramic material's firing temperature must be considerably lower than that of aluminum oxide. Because copper is an attractive conductor material in terms of cost and resistance to electromigration, the firing temperature of the composite should ideally be kept below 1083°C since copper melts at this temperature. Some recent mixtures of a refractory metal such as molybdenum or tungsten mixed with copper have been shown to cofire with alumina-based dielectrics when sintered at reduced temperatures even above the melting point of copper and may provide a means to improve the conductors conductivity but still have a lower propagation speed. One final set of material characteristics that should not be ignored for a chip carrier is the resulting composite mechanical properties. Aluminum oxide is a very robust material. Compared to Al_2O_3, many materials are much weaker, and if this is not taken into account, many problems can result. Although glass–ceramics generally have a lower modulus compared to Al_2O_3, they are much stronger than most glass-only compositions. So, even though glass–ceramics are less resistant to cracking or fracture than Al_2O_3, they are just as good or significantly better than many other possible low dielectric constant ceramic candidates. Listed in Table 4.1 is a summary of the important materials properties for choosing a dielectric substrate material and a comparison between traditional Al_2O_3 and Mo/W and that of glass–ceramic and copper.

In Table 4.2 a comparison between alumina and Mo/W and glass–ceramics and Cu shows the superiority of glass–ceramics for high-performance applications over alumina ceramic. Many materials were considered before glass–ceramics were selected. When using the properties as shown in Table 4.1 as the selection criteria, some of the candidate materials considered included Si_3N_4, AlN, BeO, mullite, borosilicate glass, and silica. Each one of these

TABLE 4.1 Materials and Properties Data Comparison

	Materials							
Characteristic	Al_2O_3	Glass –Ceramic	Mullite	BeO	Si_3N_4	SiO_2	AlN	BoroSilicate Glass
Coefficient of thermal expansion (CTE) (ppm)	6.5	2.4–11.5	4.2	6.8	2.3	3–5	3.3	3–6
Modulus (MPa)	360	150–300	300	300	320	140	280	120–160
K'	9.5–10	4.2–8.8	6.4	6.8	7	3.8	8.5	4–6
Loss factor	0.001	0.001–0.0005	0.001	0.001	0.001	0.0001	0.0005	0.005
Co-fire conductor	W, Mo	Cu, Au, Ag	W, Mo	W, Mo	W	W	W	Cu
Co-fire temperature (°C)	1600	<1000	1200+	2000	1600	1500	1700+	<1000

TABLE 4.2 Important Interconnect Materials Comparison Between Traditional Aluminum Oxide and Mo/W and Glass–Ceramic/Cu

Property	Al_2O_3 and Mo/W	Glass–Ceramic and Copper
K ceramic (dielectric constant)	10	5
Coefficient of thermal expansion (CTE) ($Si = 3 \times 10^{-6}/°C$)	$7 \times 10^{-6}/°C$	$3 \times 10^{-6}/°C$
Firing temperature	1600°C	1000°C
Strength	50,000 psi	30,000 psi
Metal electrical resistivity 10^{-6} Ω cm	5.2–5.5	1.7

materials, while potentially superior to alumina, were deemed less desirable than glass–ceramic. Many had significantly higher dielectric constants. The materials that had lower dielectric constants, silica and borosilicate glass, were known to be much weaker in strength than glass–ceramics by as much as 70%. And, importantly, silica is viscous to 1500°C, making it impossible to co-fire with nonnoble metals. And, while the thermal expansion coefficient of these materials is very low, the thermal expansion coefficient of glass–ceramic can be tailored to a near match with silicon. Silica and borosilicate glass cannot. For these reasons, glass–ceramics were deemed the best choice for a new interconnect packaging material.

Within the category of glass–ceramics lie a whole host of materials. Some of those considered were cordierite ($2MgO{\cdot}2Al_2O_3{\cdot}5SiO_2$), beta-spodumene ($Li_2O{\cdot}Al_2O_3{\cdot}4SiO_2$), celsian ($BaO{\cdot}Al_2O_3{\cdot}2SiO_2$), and anorthite ($CaO{\cdot}Al_2O_3{\cdot}2SiO_2$). Cordierite was chosen as the best material due to the unacceptably high thermal expansion coefficients, when sintering additives were used, of celsian and anorthite. Beta-spodumene has a very low thermal expansion coefficient, but its dielectric constant was found to be unacceptably high at approximately 9.

4.3.1.2 Cordierite Composition Optimization

Although cordierite was singled out as the best next-generation interconnect substrate material, because of its superior properties, stoichiometric cordierite by itself would not be acceptable. In order for a new interconnect substrate material to be fabricated using many of the traditional multilayer ceramic processing technologies such as green-sheet casting, punching, metal paste screening, lamination and sintering, it is required that a powder of this new material possess certain sintering characteristics. In particular, the powder of ceramic and metal must coalesce and sinter to near theoretic density. Later in this chapter, processing will be discussed in more detail. Since stoichiometric cordierite powder does not sinter to anywhere near theoretical density, sintering additives were needed. At IBM many additives were tried to improve sintering including Li_2O, Na_2O, B_2O_3, P_2O_5, CaO, and Fe_2O_3. The effects on sintering of a similar glass–ceramic, spodumene have been published. After making many cordierite-based compositions and studying their sintering characteristics using a dilatometer, it was determined that B_2O_3 and P_2O_5 showed the most promise. Although Li_2O did significantly increase the sintered density, it also greatly decreased the sintering temperature. For reasons that will be described in more detail later in this chapter, this was not desirable. While the other additives such as Na_2O and Fe_2O_3 were found to improve sintering, the sintered bodies still possessed significant amounts of porosity. When determining the proper sintering additives and the quantity to use, other characteristics besides density were used. It is also important to ensure that the thermal expansion coefficient be as close to Si as possible. And although a high-density is important, it is equally important that the sintering time and temperature be compatible with the metal conductor powder sintering. Since these ceramic powders will be tape cast into green sheets and eventually the green-sheet organics will need to be removed through pyrolysis, it is necessary that the sintering not occur before this process can be completed. For these reasons B_2O_3 and P_2O_5 were found to be the best candidates. In order to more fully understand the important role of these additives, compositions were prepared both with and without these additives, and the final properties of the ceramics were measured. Also, the three major components in cordierite were adjusted to achieve better sintering properties as well as a closer thermal expansion coefficient match to silicon. Stoichiometric cordierite has a thermal expansion coefficient that is much lower than Si, that is, 0.5×10^{-6} vs. $3 \times 10^{-6}/°C$. By decreasing the alumina content and increasing the MgO content, the expansion coefficient increased. It was also found that by increasing the MgO content, the degree of sintering increased and more dense bodies were formed.

By adjusting the quantities of the 3 major components and by adding small quantities of B_2O_3 and P_2O_5 an optimized glass-ceramic composition was achieved.

4.3.2 Dielectric and Conductor Compatibility

Since glass–ceramic and copper powders are being fired together in the same body at the same time, it is extremely important that they are compatible. Copper also has the added requirement that it be fired in A reducing atmosphere. At the same time that the atmosphere is reducing to copper, the binder and plasticizer in the ceramic green sheet and the organics in the metal inks or pastes must be removed. If residual carbonaceous material is left behind after firing, as little as 400 ppm can significantly increase the effective dielectric constant of the substrate. In fact, as little as 800 ppm can cause the dielectric constant to increase from 5 to as much as 1000. Another challenge when co-firing glass–ceramic and copper powder simultaneously is to closely control the dimensions and integrity of the very fine lines and spaces. One very serious problem can arise if the copper powder sinters or coalesces at a much lower or higher temperature than the glass–ceramic. Most fine grained pure copper powders sinter quickly and completely by 400°C. Due to carbon removal requirements in the green sheet and metal paste, we would like the copper to sinter at temperatures as high as 800°C, which is also approaching the sintering temperature of the cordierite glass–ceramic.

These challenges can be solved by several process and material enhancements. The sintering process will be discussed in detail in the processing section. The copper chemical composition could be altered to delay the sintering to a higher temperature, but the additives must not increase the electrical resistivity of the sintered conductor. Several approaches were found to work. Copper particles could be coated with an organic barrier that prevented diffusion between the particles and thereby inhibiting the driving force for sintering. The organic barrier is removed at the desired temperature either through thermal energy or by using an atmosphere that will oxidize the organic barrier. Examples of these materials include polyvinyl alcohol, polyvinyl formate, polyvinyl butyral, acrylonitriles, epoxies, and many others.

Another approach to delaying the sintering of the copper powder until higher temperatures is to intersperse copper particles with metal or metal oxides. The particles act as grain boundary inhibitors. Some examples of these materials include chromium, molybdenum, aluminum, gold, nickel, and palladium. The resulting resistivity for copper that contained a sufficient quantity of oxide material, which sintered at a high-temperature, was measured to be less than twice the resistivity of pure copper.

4.4 LOW-TEMPERATURE MULTILAYER CERAMIC (MLC) PROCESSING

The MLC process creates multiple layers of dielectric and conductor powders prior to sintering and then uses co-firing to create a three-dimensional network of conductive wiring and dielectric. The co-fired multilayer ceramic chip carrier process lends itself to a wide array of high-density area array chip carriers. The ability of MLC chip carriers to provide high-density interconnection in each layer in addition to high-density *X* and *Y* wiring permits a wide range of chip input/output (I/O) and package I/O with various package sizes to be processed through the same tool set. Similarly, MLC lends itself to support a variety of voltages through the addition of one or more fully metallized planes. Figure 4.1 shows an overview of the process flow used for multilayer ceramics from raw materials through final test and inspection.

4.4.1 Raw Materials Preparation and Casting

The multilayer ceramic process begins with ceramic and/or glass powders with organic binders and solvents to create a ceramic slurry. The ceramic slurry is formulated to create a system capable of tape casting into thin sheets. The ceramic slurry typically is comprised of ceramic powder and glass additive or a crystallizable glass powder, organic binder, plasticizer, and solvents. The mixture is cast into thin ceramic green tape, which may range in thickness from approximately 0.025 to 0.750 mm depending on the tape thickness required. Figure 4.2 shows a schematic representation for a slurry being tape cast using a doctor blade. Once cast, the green or unfired tape is cut into standardized green sheets of a common *X*-*Y* size. Several sizes are currently used in production, 150 mm^2 to 250 mm^2 are most common although many manufacturers utilize rectangular sizes where one dimension can be 200–300 mm. Advantages of larger sizes are reduced costs; however, ability to maintain dimensional control over the entire green sheet during subsequent processes can impact yields. In either case, green sheets should have good dimensional stability and be free of defects such as voids, pin holes, impurities, contamination, and nonhomogeneities.

The thick film pastes contain metal powders dispersed in an organic vehicle plus additives to enhance shelf life and screening properties. Pastes for low-temperature co-firing ceramic (LTCC) applications generally contain silver, silver

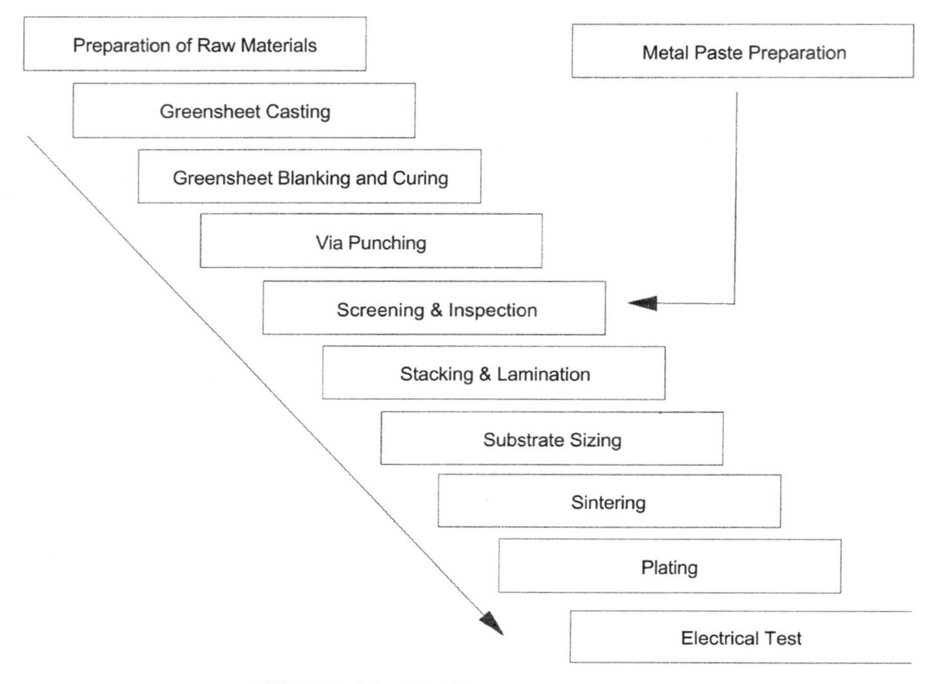

FIGURE 4.1 Multilayer ceramic process.

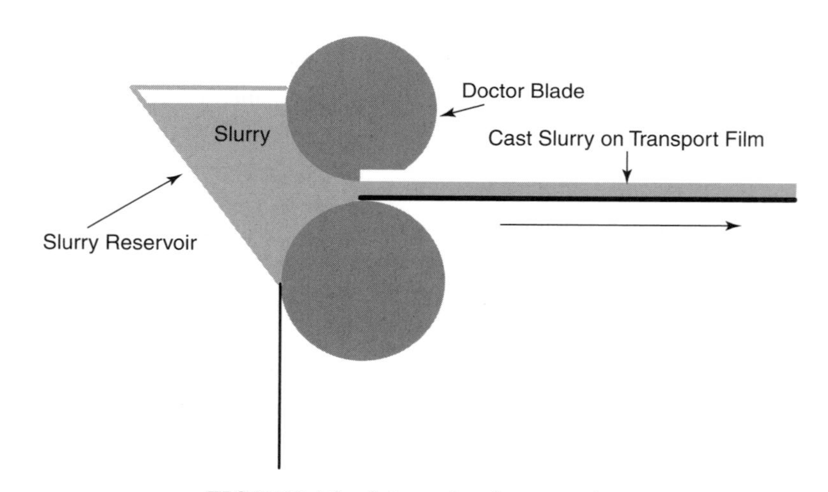

FIGURE 4.2 Schematic of tape casting.

palladium, or copper powder plus organic resins and solvents. The paste constituents are weighted, mixed with a large blender, and then processed through a three-roll mill whose shearing action disperses the powder particles into a thick film paste with suitable rheological properties for screen printing *X-Y* patterns and filling via's in green sheets for *Z* connections. The paste must be compatible when co-fired with ceramic to provide the proper onset, rate, and volume shrinkage compared to the ceramic. Sintered thick-film metallizations must provide adequate surface pad adhesion characteristics to ensure mechanical integrity for package interconnection to die and board I/O pads.

4.4.2 Green Sheet Punching, Metallization, and Inspection

In the next step a series of vertical holes, termed *vias*, are punched into the green sheet prior to metallization with the thick film paste. A mechanical punch is normally used to create the vias with diameters of 0.05–0.25 mm (see Fig. 4.3). For via diameters of 0.075 mm or smaller, laser via formation, E-beam formation, or photo-defined vias can be utilized. Aspect ratios of green sheet thickness to via diameter of 3:1 to 1 · 5:1 are commonly used. Mechanical punching can provide low cost and flexibility of pattern formation, but the process must be optimized for very high via counts and densities (>80, 000 vias per layer for 150 × 150 mm green sheet size) to minimize mechanical stresses that lead to via location error due to green sheet distortion imposed by mechanical punching. Decreasing green sheet thickness and increasing punch diameter to die bushing clearance tends to reduce via positional error, because thinner sheets have less mechanically imposed stress when forming vias for a given via diameter. Green sheets with over 78,000 vias (0.01 mm via diameter) and an active area of 150 × 150 mm have been reported.

As shown in Fig. 4.4, the thick film via fill and metal printing of surface patterns ties together the punched ceramic green sheet and thick film paste in the powder (presintering) process often referred to as screen printing or green sheet personalization. In this process either vias and *X-Y* metallizations are achieved sequentially with paste drying after each operation or vias and *X-Y* pattern can be created simultaneously by extrusion printing the conductor paste. A variety of mask patterns are employed to create the circuitry for MLC chip carriers,

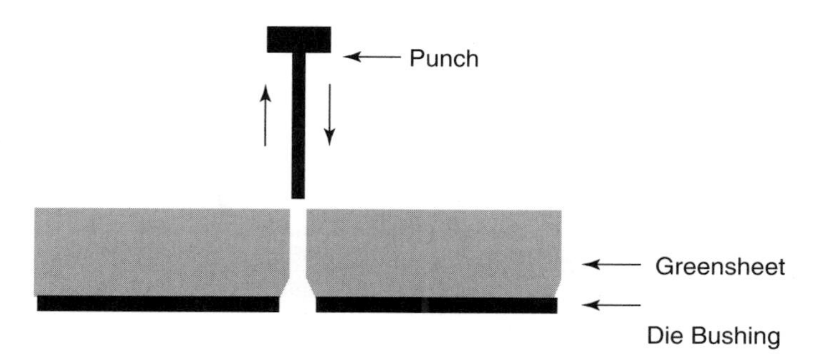

FIGURE 4.3 Schematic of punching.

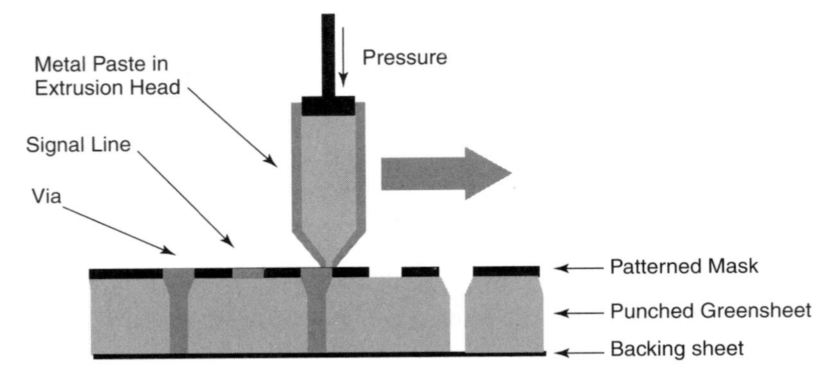

FIGURE 4.4 Multilayer ceramic green sheets are patterned using metal paste extrusion.

including top surface through vias, *X* and *Y* signal patterns, voltage, power, and ground planes and bottom surface I/O pads. In screening thick film paste, a patterned mesh or metal mask is utilized for via hole filling and for surface pattern definition. Line widths typically ranging from 0.050 mm to over 0.250 mm can be screen printed. Solid or mesh power planes can also be screen printed as illustrated in Fig. 4.5.

Green sheet inspection is routinely conducted to ensure via fill and patterns meet specifications for complex substrates. This operation employs optical comparisons to ensure line connectivity, absence of shorts, and via fill. Inspection is most commonly used for complex packages with high layer counts, greater than 10–15 layers, or for fine line patterns and line widths of less than 70 μm. Figure 4.6 shows a number of metallized green sheets. Inspection can also be helpful when introducing new part numbers into manufacturing to aide in yield learning. Inspections can be skipped in many volume applications of products, and is required only for reasonable yield for complex substrates.

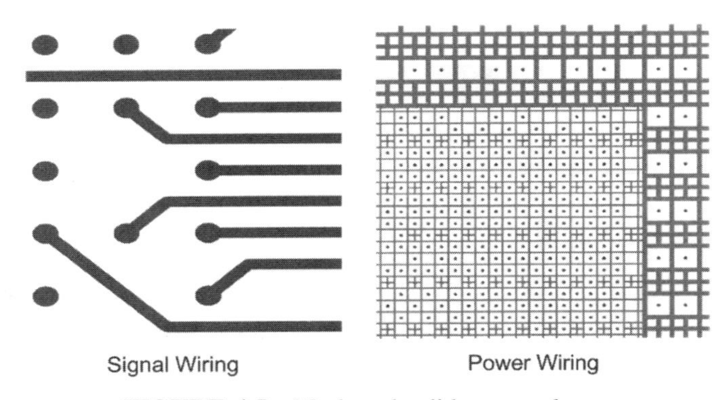

FIGURE 4.5 Mesh and solid power planes.

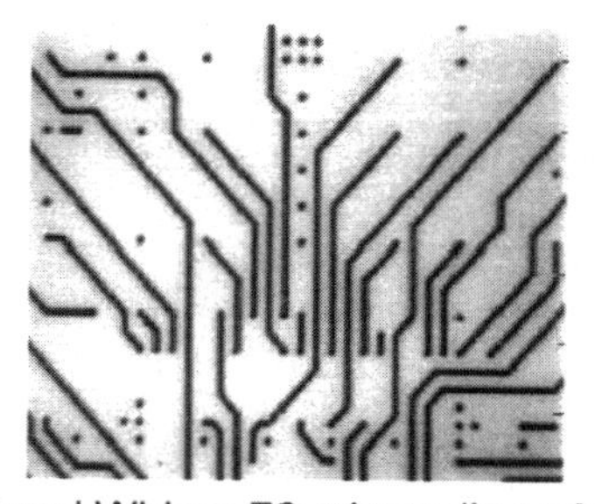

Signal Wiring: 70 micron line width

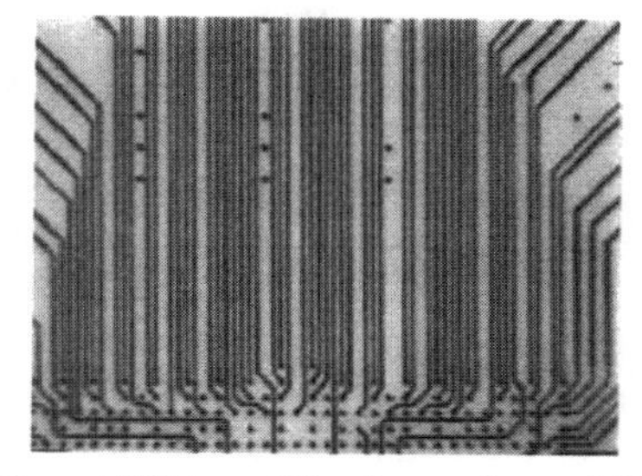

Signal Wiring: 35 micron line width

FIGURE 4.6 Patterned green sheets.

4.4.3 Stacking, Lamination, and Green Sizing

A variety of green sheets sequentially stacked and laminated using pressure and temperature are required to create a laminate consisting of a three-dimensional network of conductor and dielectric materials. The ceramic powders are held together by an organic binder previously used in green sheet fabrication and conductive paste fabrication or by an added adhesive used between layers during stacking.

Alignment of each layer is critical during stacking to properly align the many *Z* wiring connections to create via stacks, interconnect lines between layers using vias, and avoid shorts or opens between layers. Alignment can be achieved by using registration holes in green sheets previously used for punching, screening, and inspection, by locating the holes over stacking pins. Alternatively, each layer can be optically aligned and stacked. A green laminate typically consisting of multiple chip carriers is singulated into individual carriers by green saw sizing prior to sintering. Similarly, packages are green machined to obtain corner chamfers or rounded edges.

4.4.4 Sintering

During sintering, the organic components used in "green" processing MLC laminates are removed through decomposition or pyrolysis during the initial stages of heating and up to temperatures below about 600°C. Choice of binders and plasticizers and the development of the proper sintering profile are critical in order to provide a minimum of residuals from this process, which can be removed subsequently in the sintering process. In the pyrolysis section of the sintering process one must avoid rapid rise in temperature, which leads to the highest decomposition of the organics so as to not delaminate or cause other defects in the multilayer component structure. Next, residues containing carbon are removed by oxidation. For LTCC, carriers containing silver or silver–palladium conductors, carbon oxidation can be performed in an oxygen atmosphere such as in air. For copper metallization, a controlled atmosphere is utilized to oxidize the carbon without oxidizing copper, using temperatures between 600–800°C. Here the rate of reaction must be adequate to complete the removal of carbon within a reasonable time period (to a low level so as to not impact the resistance of the dielectric or impede the controlled sintering of the metal or dielectric materials) but must also avoid oxidation of the metal powders in the structure. Once the organics are removed, the dielectric and conductor particles undergo densification to create an insulating dielectric chip carrier with conductive metal wiring. Densification for LTCC packages occurs below about 1000°C. Compatible shrinkage of both conductor and dielectric must be developed to ensure good dimensional control and mechanical and electrical integrity. It is important to control the onset, rate, and total volume shrinkage of the various materials in the multilayer chip carrier. Factors that can influence dimensional stability include all operations from raw materials, through "green" processing operations, up to and including sintering time, temperature, atmosphere, and sinter fixturing. The key to repeatability producing high-quality ceramic chip carriers lies in the manufacturing process, which must be controlled to consistently operate within the materials and process specifications. These controls are in addition to an engineered compatible conductor and dielectric system optimally matched for shrinkage onset, rate, and volume. Figure 4.7 shows a schematic of typical process steps for LTCC sintering.

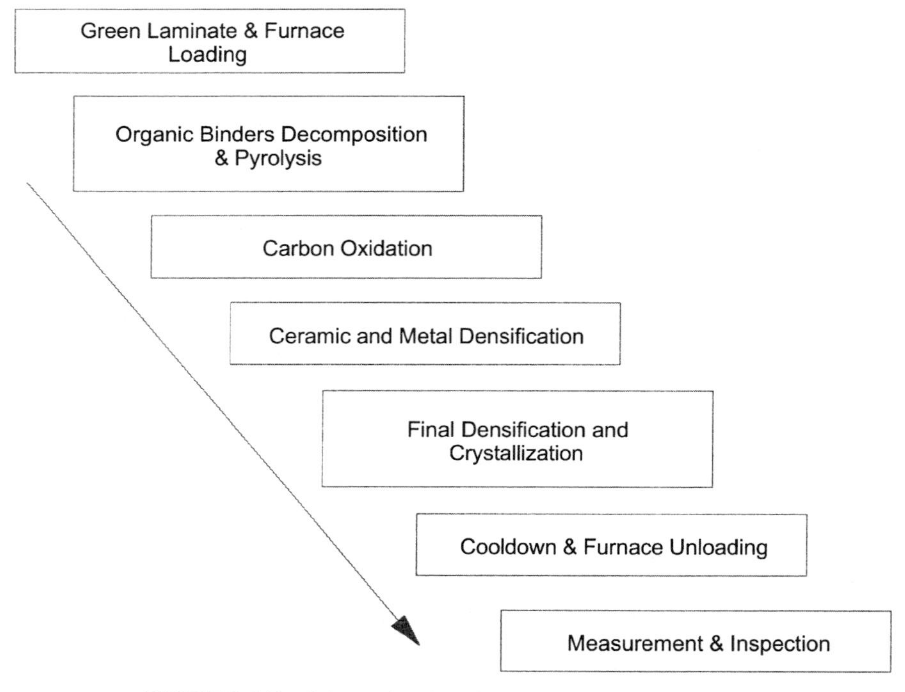

FIGURE 4.7 Schematic of typical LTCC sintering process.

4.4.5 Postsinter Finish Metallization

Subsequent to sintering, ceramic chip carriers can undergo precision machining to provide surfaces sufficiently flat and smooth to accommodate package applications utilizing one or more layers of advanced thin-film wiring or where an advanced sealing process requires precision tolerances such as a flange or module encapsulation. Lapping and polish techniques have been developed that can provide a surface with nominal flatness of about 1 micron per inch and a surface finish of 300 Å average roughness (R_{av}). Thin-film materials, process, and attributes have been described elsewhere [56].

The LTCC chip carriers that do not require thin-film processing are finished with nickel and gold plating on the surface features to enhance solder and braze wetting characteristics and provide a barrier to other metallizations used in MLC fabrication. Finish metallizations can be electroplated or electroless plated. A relatively thick nickel layer (0.0005–0.004 mm) is utilized to support rework capability (i.e., multiple solder reflows). Gold thickness is dependent on application and type of interconnection. For example, thin gold ($\ll$ 0.001 mm) is used on flip-chip die attach pads whereas for wirebond and pin grid array pads, heavy gold (0.001–0.002 mm) is required. For some LTCC carriers, alternate finishing metallizations can be used, and in some cases none are required depending on surface co-fire metals used during MLC processing. For example, silver–palladium based alloys are Pb–Sn solder wettable, and for copper or copper–nickel alloys, which have been co-fired, a gold finish is often the only postfire metal deposition required.

4.4.6 Test and Inspection

Electrical test can include direct electrical shorts and opens tests. Tests can also include a functional test to support high frequency, assess simultaneous switching noise, line resistance, propagation delay, and others. Chip carriers can also be tested for defects and can be repaired with a variety of techniques reported elsewhere [21]. Chip carriers that meet electrical and customer specifications are finally inspected for surface defects prior to shipping or subsequent module processing operations.

4.5 MECHANICAL AND THERMAL PROPERTIES

Table 4.3 shows a comparison of high-performance chip carriers used in Hitachi and IBM mainframes and servers. Table 4.4 shows feature sizes common for production LTCC chip carriers. Table 4.5 shows characteristics of glass plus ceramic LTCC compared to crystallizable glass–ceramic LTCC products.

TABLE 4.3 High-Performance Chip Carriers Comparison for High-Performance Data Processing

	Hitachi	Hitachi	IBM	IBM
Application	M-880	MP5800	Enterprise-G4	Enterprise-G5
Technology	HTCC	LTCC	HTCC/TF	LTCC/TF
Ceramic/conductor	Mullite/W	GlassCeram/Cu	Alumina/Mo	Glass
Ceram/Cu				
Carrier Size (mm)	106 × 106	115 × 115 and 136 × 136	127 × 127	127 × 127
K (effective)	5.9	5.3	9.5	5
Time of flight (ps/mm)			12	7.8
SCM/MCM	MCM	MCM	MCM	MCM
Chips/MCM	36–41	25 and 36	30	29
Number of layers	44	50	68	75
Plane pairs	9	13	19	17
Total wiring (m)	200	400 and 600	486	445
Chip I/O pitch (μm)			225	225
Internal wiring pitch (μm)		300	225	225
Linewidth (μm)		60	80	80
Board I/O	2521	3593 and 5105	3526	4224
Thin film				
K			3.2	3.2
Time of flight (ps/mm)			6.7	6.7
Number of layers			4	6
Number of plane pairs			0(Redist. Only)	1
Total wiring (m)			65	212
Internal wiring pitch (μm)			N/A	45

Source: Koybayashi et al. (in press).

TABLE 4.4 Feature Sizes in Chip Carriers Using LTCC Multilayer Ceramics

	Multilayer Ceramic
Feature	LTCC
Chip I/O pitch	150–225 μm
Ceramic carrier	
Via diameter	50–150 μm
Via pitch	150–225 μm
Line width	50–125 μm
Line pitch	75–125 μm
Number layers	2–100
Board I/O pitch	
PGA	2.5 mm
BGA	1.27–1.0 mm
CGA	1.27–1.0 mm
LGA	1.27–1.0 mm

TABLE 4.5 Electrical Performance of LTCC Ceramic Chip Carriers

	Glass–Ceramic	Glass Plus Ceramic
Coefficient of thermal expansion (CTE) (ppm)	2.4–8.0	3.0–11.5
Strength (MPa)	170–230	150–300
K' (dielectric constant)	4.9–5.6+	4.2–8.8
Thermal conductivity (W/m °K)	0.001–0.0001	
Metallurgy	Ag, Ag/Pd, Au, Cu	Ag, Ag/Pd, Au, Cu

4.6 INTEGRAL PASSIVES

4.6.1 Advantages and Limitations of Integral Passives

Discrete components continue to offer low cost and value in area array interconnection solutions. An increasing number of applications can take advantage of solutions where miniaturization and/or functionality require *integral passive technology* or integrated technology solutions.

Integrated and integral passive components are defined by the Passive Components Technology Working Group of the National Electronics Manufacturing Initiative (NEMI). Although some definitions may change, those given by NEMI generally apply to all chip carriers and package solutions [49, 50]

Integral passive or integrated packages can use thick- or thin-film processes and can be fabricated using ceramic technology, thin film, laminate technologies or within silicon die. Utilizing buried integral passives permits increased space on boards or packages that can not be integral. Most approaches to date have utilized thick-film paste printed on layers. More recent developments include sputtered or evaporated metals on dielectric films.

4.6.1.1 Resistors

Resistors have been integral to ceramic packaging since the 1960s where palladium–silver–palladium oxide–glass thick films were screen printed on chip carriers. Today, resistors support small or preferably no change of properties with time. Resistor properties are defined as positive or negative temperature coefficient of resistance (TCR), which relates to changes in resistance with temperature. Resistors ranging from a few ohms to thousands of ohms are available using ruthenium oxides and lead borosilicates resistor pastes as examples. In addition, a wide range of compositions are used, depending upon the temperature coefficient of resistance desired. Increase control of tolerances in advanced applications are continuing to be sought. Laser trimming of resistors after sintering is utilized to fine tune the desired resistor values.

Thin-film resistors have also been widely used on top of blank ceramic carriers, but the major focus in recent years has been on the integration of resistors into multilayer, multichip thin-film modules. Materials such as copper–nickel alloys, tantalum–nitride, tantalum–silicide, ruthenium–dioxide, and titanium–tungsten have been vacuum deposited on ceramic and/or polymer layers to integrate both within and on the surface of a package. Most of these applications are in the development or prototyping stage.

4.6.1.2 Capacitors

Integral capacitors have been used in multilayer ceramics where thin dielectric tape layers (sometimes with increased dielectric constant for increased capacitance) are incorporated during the building of a chip carrier. Cavity wire bond packages and area array flip-chip packages used from one to about five thin dielectric layers to provide capacitance near or directly under a die. High temperature co-fired ceramic (HTCC) with integral capacitors uses dielectric thickness down to $\sim$20 μm fired and a dielectric constant of 9–14 for alumina ceramic. Integral capacitors in LTCC can provide higher capacitance values where dielectric constants of 8–40 have been utilized with similar dielectric thickness compared to HTCC. Potential integral capacitor solutions for LTCC are being developed with dielectric constants of over 200.

Thin-film capacitor development is also primarily driven by the need to provide high decoupling capacitance coupled with a low-impedance path as close as possible to the chip. These factors become more and more important as chip frequencies increase and chip voltages decrease. Several materials and techniques are being investigated to produce thin, high-dielectric-constant, defect-free films compatible with multilayer copper/polyimide thin-film structures. Materials such as barium titanate, aluminum oxide, and tantalum pentoxide either deposited as the oxide or subsequently anodized have been used to achieve capacitance values as high as 70 pF/cm^2.

This work continues at the exploratory and protype stage. As the performance of discrete capacitors continues to improve, the performance requirements for economic use of integral capacitors also continue to increase.

4.6.1.3 Other Applications

As discussed in the introduction, communications is becoming an increasingly important area for ceramic applications. A wireless radio product utilizes hundreds of discrete components. Increasing functionality along with manufacturing cost and miniaturization requirements are causing increased focus on low-temperature co-fired ceramic technology with integral passives. Motorola has reported on the development of multilayer ceramic integrated circuits (MCIC) where low-temperature co-fire ceramic (LTCC) could be used to integrate radio frequency functions for wireless communications application.

4.7 PRODUCTS APPLICATIONS FUTURE TRENDS

The future trends for area array ceramic include all aspects of advances in packaging with application focus on:

1. Higher performance server applications, high I/O single chip module (SCM) and complex MCMs for example
2. Communications for both wired and wireless applications, internet switches for example, and integrated function wireless digital–radio frequency mixed signal packages for hand-held applications requiring low power
3. High function interposer capacitors that provide capacitance directly to a chip
4. Base support for virtual chips or system on a package solutions either with or without thin-film wiring
5. High-performance and high-reliability applications such as space, automotive, telecommunications, military, and critical application specific integrated circuit (ASIC) requirements with high I/O and large die size
6. High-density probes and space transformers for electrical test probes and wafer probes

The future trends for area array ceramic include:

1. Automated design for advanced wiring and increasing numbers of nets
2. Automated design and modeling for function including integrated passives
3. Increased I/O and I/O density to a chip and from the package to the board
4. Higher wiring density in the chip carrier
5. Capability to tailor impedance for multiple circuits
6. Capability to support multiple voltages (3 plus)
7. High reliability surface mount and land grid array for connection to boards without reduction in thermal cycle performance, cost impact, or electrical impact
8. Support to each market segment where ceramic is optimized for cost and performance, using advances in materials and process technology to meet customer applications

4.8 SUMMARY

Low-temperature co-fire ceramics have many attributes that meet the needs for applications ranging from simple, high-frequency, hand-held communications needs to complex switches and multichip modules that support high-performance servers and supercomputers. The advantages of the LTCC materials such as low dielectric constant and high conductivity copper conductivity provide low-cost solutions for many customers needs. LTCC processing begins with raw materials, including dielectric powders and conductive metal powders, and continues through multilayer ceramic processing, resulting in a three-dimensional array of conductive wires with a dielectric of desired properties. LTCC also can be fabricated with integrated passives and has extendability in the wiring and interconnection ground rules compared to those used today. LTCC will continue to be used in a variety of applications where it is best at meeting the cost and performance objectives sought by the customer.

REFERENCES

1. Sarah Knickerbocker, Michelle Tuzzolo, and Samuel Lawhorne, *J. Am. Ceram. Soc.*, **72**(10), 1873–1879 (1989).
2. Amanda Kumar, Sarah Knickerbocker, and Rao Tummala, "IEEE Trans Comp. Hybrids and Man Tech.," 1992 Proceedings, 42nd Electronic Components and Technology Conf., p. 678–681.
3. *National Technology Roadmap for Semiconductors*, Semiconductor Industry Association, San Jose, CA, 1997.
4. R. R. Tummala and E. Rymaszewski (Eds.), *Microelectronics Packaging Handbook*, Van Nostrand Rienhold, New York, 1989.
5. A. J. Blodgett, *Sci. Am.*, **249**, 86 (1983).
6. P. E. Garrou and I. Turlik (Eds.), *Multichip Modules Technology Handbook*, McGraw-Hill, New York, 1998.
7. K. Yokouchi, N. Kamehara, and K. Niwa, *Proceed. ISHM*, **1**, 183 (1991).
8. M. Yamada, M. Nishiyama, T. Tokaichi, and M. Okano, *Proceed, ECTC*, **1**, 745 (1992).
9. Y. Shimada, Y. Koybayashi, K. Kata, M. Kurano, and H. Takamizawa, *Proceed ECTC*, **1**, 76 (1990).
10. S. K. Ray, H. Quinnones, S. Iruvanti, E. Atwood, and L. Walls, "Ceramic Column Grid Array (CCGA) Module for A High Performance Work Station Application," Proc. 47th Electronic Components and Technology Conference, May 18–21, 1997, San Jose, CA, pp. 319–324.
11. A. Shaikh, *Advancing Microelectronics*, P. Garrou (Ed.), Special AlN edition, 18 (1994).
12. R. Master, personal communications, 1998.
13. J. Tetar, *Electronics News*, **44**(2102), 16–18 (1998).
14. S. Knickerbocker, personal communications on SiGe technology solutions, 1999.
15. D. L. Wilcox, R. F. Huang, and D. Anderson, *Proceed. IMAPS*, Proc. 1997 International Symposium on Microelectronics, SPIE, **3235**, 17–23 (1997).
16. J. P. Cazenave and T. Suess, *Proceed. ISHM*, **2105**, 483 (1993).
17. J. Knickerbocker, G. Leung, W. Miller, S. Young, S. Sands, and R. Indyk, *IBM J. Res. Develop.*, **35**(3), 330 (1991).
18. D. Bendz, R. Gedney, and J. Rasile, *IBM J. Res. Develop.*, **26**, 278 (1982).
19. M. Williams, *IEEE 40th ECTC Proc.*, 408 (1990).
20. T. Watari and H. Murano, *IEEE Trans. Components, Hybrids, Manuf. Technol.*, **CHMT-8**, 462 (1985).
21. G. Katopis, W. Becker, and H. Stoller, "First Level Package Design Considerations for the IBM S/390 G5 Server," IEEE 7th Topical Meeting on Electrical Performance of Electronic Packaging, 1998, pp. 15–16.
22. Franz Bechtold, "Innovative Solutions for Multi-chip Modules and Microsystems on LTCC," 11th European Microelectronics Conference, 1997, pp. 508–515.

23. K. Selvaraj, V. Ramaswamy, and A. V. Ramaswamy, *Asian J. Phys.*, **6**(1–2), 132–137 (1997).
24. J. Joly, K. Kurzweil, and D. Lambert, "MCMs for Computers and Telecom in CHIP-PAC Programme," Multichip Modules with Integrated Sensors, Proc. of the NATO Adv. Res. Workshop, 1996, pp. 63–73.
25. P. W. McMillan, *Glass-Ceramics*, 2nd ed., Academic, New York, 1979.
26. Daniel Amey and Samuel Horowitz, *Proc. 3rd Int. Sym. Adv. Packaging Mat. Processes, Prop. Interfaces*, 158–161 (1997).
27. Rao Tummala and Eugene Rymaszewski, *Microelectronic Packaging Handbook*, Van Nostrand, New York, 1989, pp. 501–511.
28. H. Kanda, R. Mason, C. Okabe, J. Smith, and R. Velasquez, *Proc. Int. Sym. Microelectronics (ISHM)*, **2649**, 47–52 (1995).
29. S. Ladd, J. Mandry, D. Amey, S. Horowitz, J. Page, and D. Holmes, "Design Trade-offs of MCMC vs Ball Grid Array on Printed Wiring Board," *Proc. SPIE, Int. Soc. Optical Eng.*, **2794**, 33–38 (1996).
30. P. C. Donahue, B. E. Taylor, D. I. Amey, R. R. Draudt, M. A. Smith, S. J. Horowitz, and J. R. Larry, "A New Low Loss Lead Free LTCC System for Wireless and RF Applications," Proc. Int. Conf. Multichip Modules and High Density Packaging, 1998, pp. 196–199.
31. Michael O'Neill, *Proc. NEPCON West*, **2**, 769–776 (1998).
32. Daniel Amey, Samuel Horowitz, and Roupen Keusseyan, "High Frequency Electrical Characterization of Electronic Packaging Materials: Environmental and Process Considerations," Proc. 4th Int. Sym. Adv. Packaging Materials Processes, Prop. Interfaces, 1998, pp. 123–128.
33. P. Pruna, R. D. Gardner, D. L. Hankey, and S. P. Turvey, Microwave Characterization of Low Temperature Co-fired Ceramic," Proc. 4th Int. Sym. Adv. Packaging Mat. Proc., Prop. Interfaces, 1998, pp. 134–137.
34. Peter Barnwell, *Advanced Packaging*, **7**(1), 34–35 (1998).
35. Peter Hardin, George Melvin, and Mike Nealon, "The ES/9000 Glass Ceramic Thermal Conduction Module-Design for Manufacturability," 11th IEEE/CHMT Int. Electronics Manf. Tech. Sym., 1991, pp. 351–355.
36. Michael Richtarsic and Jack Thornton, "Characterization and Optimization of LTCC for High Density Large Area MCM's," Proc. Int. Conf. Multichip Modules and High Density Packaging, 1998, pp. 92–97.
37. C. L. Haertling and D. J. Smith, *Intern. J. Microcircuits and Electronic Packaging*, **18**(2), 169–178 (1995).
38. A. L. Kovacs and D. F. Elwell, "Integrated Brazed LTCC Packages," Proc. Int. Conf. Multichip Modules, 1994, pp. 591–596.
39. S. Horowitz, D. Amey, and R. Draudt, *Printed Circuit Fabrication*, **21**(6), 34–36 (1998).
40. W. S. Hackenberger, T. R. Shrout, J. P. Dougherty, and R. F. Speyer, *Inter. SAMPE Electronics Conf.*, **7**, 643–650 (1994).
41. E. Jung, R. Aschenbrenner, E. Busse, and H. Reichl, *Proc. 1st Electronic Tech. Conf., EPTC*, 238–243 (1997).

42. Y. W. Yau, M. Sarfaraz, and N. Sandhu, "Novel Manufacturing for Microelectronics Packaging Fabrication," Proc. 43rd. Electronic Comp. Tech. Conf., 1993, pp. 163–165.
43. D. Partlow, J. Gipprich, A. Bailey, A. Piloto, and K. Zaki, *Int. J. Microcircuits Electronic Packaging*, **19**(2), 155–161 (1996).
44. O. Samela and P. Ikalainen, "Ceramic Packaging Technologies for Microwave Applications," Proc. 1997 Wireless Comm. Conf., 1997, pp. 162–164.
45. R. Bauer, M. Luniak, L. Rebenklau, K. J. Wolter, and W. Sauer, "Realization of LTCC-Multilayer with Special Cavity Applications," Proc. 1997 International Symposium Microelectronics, SPIE, **3235**, 659–664 (1997).
46. W. S. Hackenberger, T. R. Shrout, J. P. Dougherty, and R. F. Speyer, *Proc. SPIE-Int. Soc. Opt. Eng.*, **2105**, 215–220 (1993).
47. M. Massiot, E. Perchais, J. W. Cicognani, and E. Polzer, "LTCC-An Economical and Industrial Interconnect and Packaging Solution for MCM's," 9th European Hybrid Microelectronic Conf. Proc., 1993, pp. 150–157.
48. D. Schroeder and L. Rexing, *Proc. SPIE-Inter. Soc. Optical Eng.*, **1986**, 313–319 (1993).
49. R. Brown and A. Shapiro, *Proc. SPIE-Inter. Soc. Opt. Eng.*, **1986**, 287–294 (1993).
50. S. Gallo, A. Mones, J. Hormadalay, J. Cicognani, R. Brown, M. Massiot, and E. Perchais, *Proc. SPIE-Int. Soc. Opt. Eng.*, **1986**, 178–186 (1993).
51. T.–D. Ni, J. DeMarco, D. Sturzebecher, and M. Cummings, *1996 IEEE MTT-S, Inter. Microwave Symp. Digest*, **3**, 1627–1630 (1996).
52. J. Muller, H. Thust, and D. Schwanke, *Int. J. Microcircuits Electronic Packaging*, **19**(4), 475–482 (1996).
53. E. Palmer and M. Newton, *Int. J. Microcircuits Electronic Packaging*, **16**(4), 279–284 (1993).
54. R. Tummala and C. P. Wong, "Materials in Next Generation of Packaging," Proc. 3rd Int. Symp. Adv. Packaging Materials Processes, Properties and Interfaces, 1997, pp. 1–3.
55. T. Goodman and Y. Murakami, *IEEE Trans. Comp. Packaging, Manuf. Tech.*, **18**(1), 168–173 (1995).
56. T. Redmond, C. Prasad, and G. Walker, "Polyimide Copper Thin Film Redistribution on Glass Ceramic/Copper Multilevel Substrates," Proc. Electronics Components Conf., 1991, pp. 689–692.
57. M. Rytivaara, "Buried Passive Elements Manufactured in LTCC," Pkg. and Int. at Microwave and mm-Wave Freq., IEE Seminar, 2000.
58. R. Poddar and M. A. Brooke, "Accurate High Speed Empirically Based Predictive Modeling of Deeply Embedded Gridded Parallel Plate Capacitors Fabricated in a Multilayer LTCC Process," *IEEE Trans. Comp. Pkg. Mfg. Tech.*, **22**, pp. 26–31, February 1999.

CHAPTER 5

Intermetallics

JAMES K. WESSEL

Wessel & Associates, 127 Westview Lane, Oak Ridge, Tennessee 37830

VINOD SIKKA

Oak Ridge National Laboratory, 1 Bethel Valley Road, Oak Ridge, Tennessee 37831

5.1 INTRODUCTION

Intermetallics are a broad class of metals resulting from the combination of various elements including nickel aluminide, titanium aluminide, niobium aluminide, iron aluminide, iron silicide, and various other silicides. Each has a unique set of properties. Titanium aluminide is valued for light weight (lower than nickel-based superalloys), oxidation resistance, and stiffness. Niobium aluminide is light weight and, with a melting point of 2060°C (3740°F), operates at higher temperatures than nickel-based superalloys but has low fracture toughness and poor oxidation resistance at elevated temperatures.

Various silicides have been commercially available for many years, particularly molybdenum disilicide, used in heating elements. Iron silicide (FeSi) is sold under

Handbook of Advanced Materials Edited by James K. Wessel
ISBN 0-471-45475-3

the trade name Hastalloy D and used in high-temperature castings. Another iron silicide (Fe_3Si) is available as Duriron™. Other silicides are used for their oxidation resistance. All have attractive melting temperatures with some reaching 2400°C (4352°F). They are also used as coatings to protect other materials such as niobium aluminide from oxidation.

The nickel aluminide composition Ni_3Al has been known for years as an intermetallic material that, due to its ordered crystal structure and high melting temperature, is strong, hard, and thermally stable. It is particularly attractive because it combines lower density (25% less than superalloys) and resistance to wear, deformation, fatigue, oxidation, carburization, and coking. Particularly attractive is the unusual characteristic of increasing strength with increasing temperature. Despite such attractive properties, Nickel aluminide did not find wide use because it was too brittle to fabricate and too expensive. In 1982, Oak Ridge National Laboratory scientists, led by C. T. Liu, discovered a way to make this desirable material ductile. The result is a material (IC-221M) lighter and five times stronger than stainless steel that becomes stronger as the temperature approaches 800°C (1472°F), as shown in Fig. 5.1.

The research, development, and commercialization of Ni_3Al included a lower cost, safer process developed by a team of Oak Ridge National Laboratory scientists led by Vinod Sikka. All of this effort combined to produce a useful structural material of value in many industrial structural applications ranging from furnace furniture and steel processing rollers to making dies for forming beverage containers.

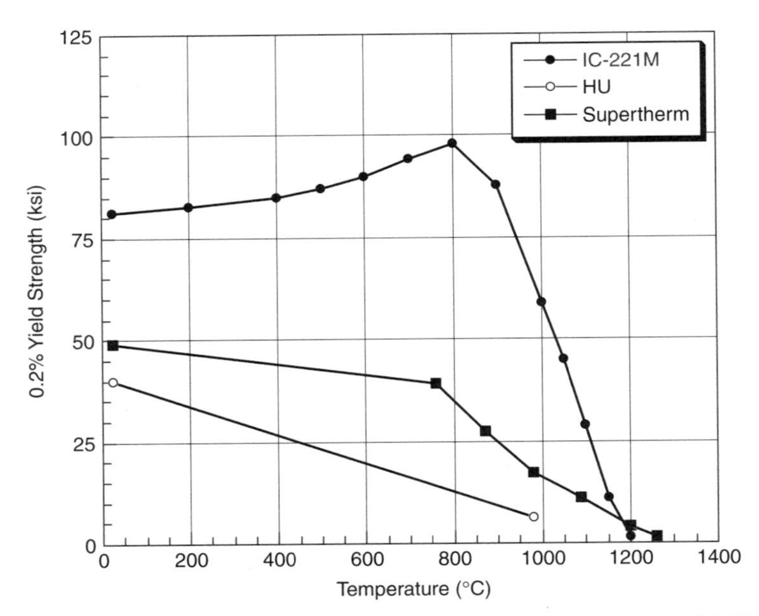

FIGURE 5.1 Tensile properties as a function of temperature. (Courtesy of ASM, International.)

The Ni_3Al-based alloy is produced by Alloy Engineering and Casting Company (Champaign, Illinois), United Defense Corporation (Anniston, Alabama), BiMac Corporation (Dayton, Ohio), Sandusky International Corporation (Sandusky, Ohio), and Alcon Industries (Cleveland, Ohio). Ni_3Al powder is available from Ametek Corporation (Eighty Four, Pennsylvania).

5.2 PHYSICAL PROPERTIES

The physical properties of the Ni_3Al alloy designated IC-221M are given in Table 5.1 and Figs 5.2 and 5.3. Data highlights include tensile yield strength and

TABLE 5.1 Design Data for Ni_3AL-Based Alloy IC-221M

	Temperature (°C)							
Property	Room	200	400	600	800	900	1000	1100
Density (g/cm^3)	7.86	—	—	—	—	—	—	—
Hardness (R_C)	30	—	—	—	—	—	—	—
Microhardness (dph)	260	270	280	290	280	230	120	—
Modulus (GPa)	200	190	174	160	148	139	126	114
Mean coeff. of thermal expansion (10^{-6}/°C)	12.77[a]	13.08[b]	13.72[b]	14.33[b]	15.17[b]	15.78[b]	16.57[b]	—
Thermal conductivity (W/m · K)	11.9	13.9	16.7	20.3	25.2	27.5	30.2	—
0.2% Tensile yield strength (MPa)	555	570	590	610	680	600	400	200
Ultimate tensile strength (MPa)	770	800	850	850	820	675	500	200
Total tensile elongation (%)	14	14	17	18	5	5	7	10
10^2 h rupture strength (MPa)	—	—	—	—	—	—	—	—
10^3 h rupture strength (MPa)	—	—	—	—	—	—	—	—
10^3 h rupture strength (MPa)	—	—	—	—	—	—	—	—
Charpy impact toughness (J)	40	40	40	35	15	10	—	—
Fatigue 10^6 cycle life (MPa)	—	—	—	630[c]	—	—	—	—
Fatigue 10^7 cycle life (MPa)	—	—	—	550[c]	—	—	—	—

[a] Room temperature to 100°C (212°F).

[b] Room temperature to specified temperature.

[c] Data at 650°C (1202°F) for investment-cast test bars.

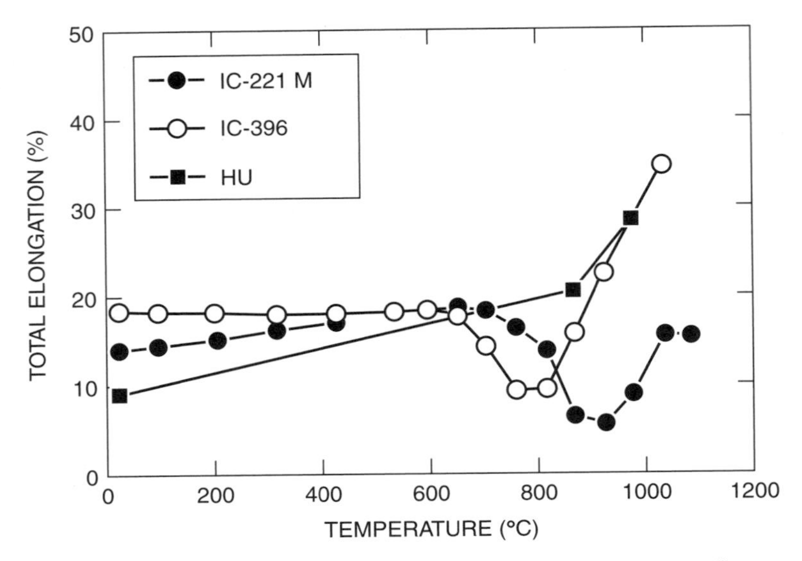

FIGURE 5.2 Elongation vs. temperature comparison with superalloys.

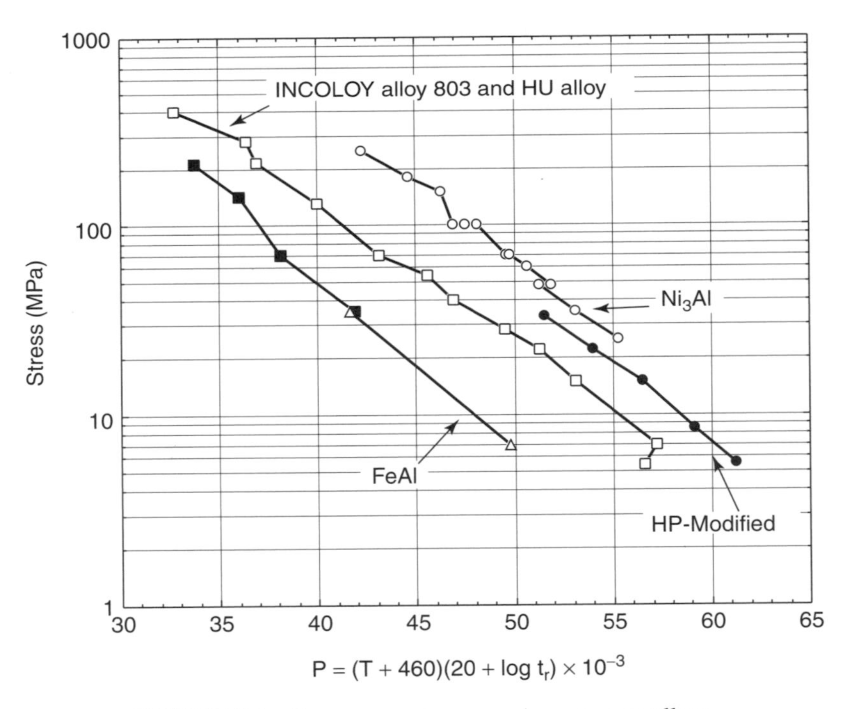

FIGURE 5.3 Creep property comparison to superalloys.

modulus at 1100°C (2012°F) (Table 5.1), creep strength (Fig. 5.3), and properties superior to other high-temperature alloys (Figs, 5.1–5.3 and Table 5.1). The high melting temperature of 1400°C (2552°F) enables the high-temperature properties of Ni_3Al.

Ni_3Al possesses good compressive yield strength at 650–1100°C (1202–2012°F) and superior fatigue resistance compared to many nickel-based superalloys.

Its hardness results in excellent wear resistance even at temperatures above 600°C (1112°F).

5.3 CORROSION RESISTANCE

A couple of attributes contribute to the corrosion resistance of IC-221M. The aluminum at the surface oxidizes to protect the bulk of the alloy from corroding substances, including oxygen. The aluminum content of the bulk alloy is set at a level to maximize its corrosion resistance to other materials, such as sulfuric acid.

Carbon has a low affinity for aluminum, giving the Ni_3Al alloy excellent resistance to a carburizing atmosphere and coking. This resistance is maintained to 1100°C (2012°F).

The resistance of IC-221M to these conditions is illustrated in Fig. 5.4.

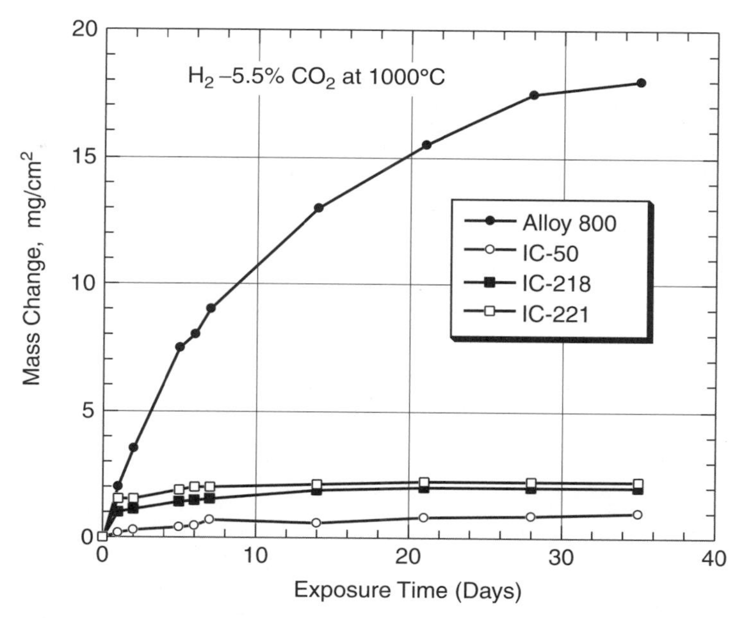

FIGURE 5.4 Reducing carburizing environment for Ni_3Al vs. nickel-based alloys.

5.4 APPLICATIONS

Ni_3Al alloy IC-221M is being used in furnace furniture, steel transfer rollers, and in a number of other high-temperature applications.

Heat treaters use tables, trays, posts, fixtures, and other supports, commonly called "furnace furniture," to hold treated parts in the furnace (Figs. 5.5 and 5.6).

FIGURE 5.5 Pusher carburizing furnace trays.

FIGURE 5.6 Batch carburizing furnace trays.

Delphi Automotive Systems Corporation uses furnace furniture at its Saginaw (Michigan) Steering Gear plant to hold gears for power steering systems while hardening them in a high-temperature carbon atmosphere. At the end of this carburizing process, the gears and their supporting furniture are quenched from 900°C to room temperature. The same furniture is subjected to this cycle with every load of gears processed. This repeated thermal cycling and the carburizing atmosphere makes this a severe challenge to any furnace material. Delphi engineers are replacing chromium alloy furniture with nickel aluminide. The nickel aluminide forms a thin film of aluminum oxide on its surface, preventing carbon from diffusing into the body of the metal. HP, a chromium nickel alloy, furniture lasts 6 months. The nickel aluminide furniture is still in operation after 39 months.

The superior strength of Ni_3Al permits heavier tray loading while decreasing the tray size. The use of nickel aluminide has resulted in furniture replacement savings, higher operating temperature, a 10% throughput increase, reduced cost, reduced energy use, and less waste. The efficiency gain allows Delphi to postpone plans to build a new furnace.

A major U.S. steel maker heat treats steel plates to soften them and then pass them over transfer rollers to provide a smooth surface finish and move them through the process. Due to the high process temperatures, conventional rollers sag, jostle the plates, develop oxide particles, and blister, scratching the plates. Nickel aluminide rollers withstand the heat and are three times stronger than the conventional rollers. The nickel aluminide rollers are only inspected once per year versus the conventional practice of shutting down every 6 weeks, inspecting rollers, grinding out particles, and replacing sagging rollers. The use of nickel aluminide rollers reduces downtime, the number of spare rollers and the cost of repairing rollers, resulting in savings for the plant. The rollers are centrifugally cast of nickel aluminide alloy, IC-221M, manufactured by Sandusky International Corporation, product number 184687. Cast rings were welded together at Oak Ridge National Laboratory for the initial tests. Engineers used IC-221LA filler metal provided by Stoody Company (Bowling Green, Kentucky) (See Figs. 5.7 and 5.8). Preapplication testing included weld tensile strength, fatigue (360 cycles, 871°C (1600°F) to room temperature), hardness versus time, microstructure examination, and oxidation behavior.

Nickel aluminide is also being evaluated as statically cast triunions for rollers in austenitizing and hydrogen annealing furnaces.

Ni_3Al is being evaluated as thick-walled tubes and pipe because iron oxide does not stick to it. Ni_3Al has less creep and better resistance to carbon and oxygen than conventional materials such as cast stainless steels (HU and HP modified).

United Defense is supplying nickel aluminide rails to Rapid Technologies Corporation (Newnan, Georgia) for its walking-beam furnaces. Its heat-treating furnace moves steel bars rapidly through a high-temperature zone. The concept requires rapid heating, resulting in less natural-gas and cooling water. Without nickel aluminide beams, these savings and their product would not be possible.

FIGURE 5.7 Furnace roll for hydrogen annealing furnace.

FIGURE 5.8 Blistering of conventional furnace roll.

Ni_3Al is being evaluated as statically cast die blocks for the hot forging process to increase life because of its yield strength at 850°C (1562°F) and its resistance to oxidation, which is better than cast stainless steels. Ni_3Al is being evaluated as cast hot pressing dies for permanent magnet material. It has excellent chemical compatibility and high-temperature yield strength compared to IN-718.

Cast-shaped rods of Ni_3Al are used as industrial furnace heating elements to increase life by increased oxidation resistance and sagging. They replace FeCrAl alloys.

Ni_3Al powder is used as a binder for tungsten and chromium carbide as tool and die materials. It improves wear resistance and aqueous corrosion resistance in certain acid solutions. It replaces expensive cobalt currently used as the binder material.

Ni_3Al is being used in circuit boards as a metallic core that absorbs more heat than a monolithic Al_2O_3 board. This enables the board to handle higher power devices.

Other applications include tube hangers, ethylene cracker furnace tubes, gas filters, radiant burner tubes (Fig. 5.9) (see Chapter 3 for descriptions of these four applications), glass processing equipment, furnace belt links (Fig. 5.10), container dies, binder for tool and die material, auto belt tooling, furnace heater (and other heater elements), mufflers, return bends, firing legshot forging dies, hot pressing dies, trays, mufflers, brake components (Fig. 5.11), boiler tubes, catalytic converter substrate, salt bath containers, sulfuric acid containers, mixers, and other parts in corrosive solutions.

In 1995, 50,000 lb of Ni_3Al were made.

5.5 SPECIFICATIONS

Ni_3Al-based alloy is the first intermetallic alloy to have an approved ASTM specification, A1002-99, Standard Specification for Castings, Nickel-Aluminum Ordered Alloy (Table 5.2).

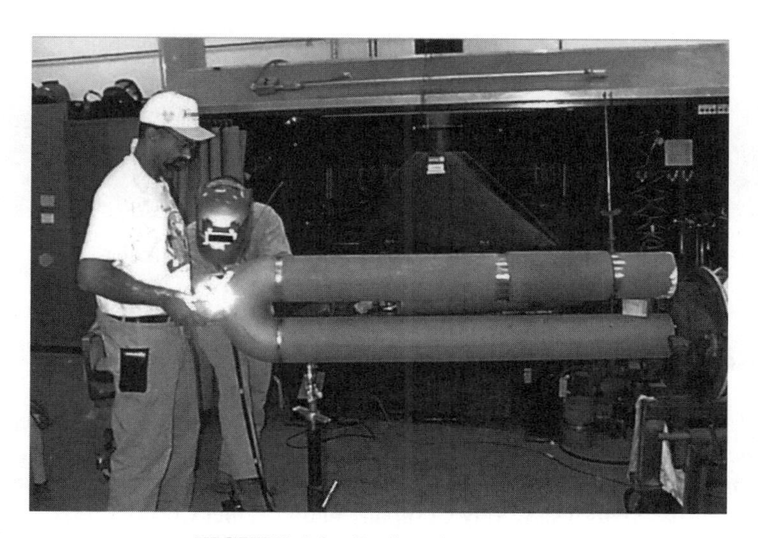

FIGURE 5.9 Radiant burner tube.

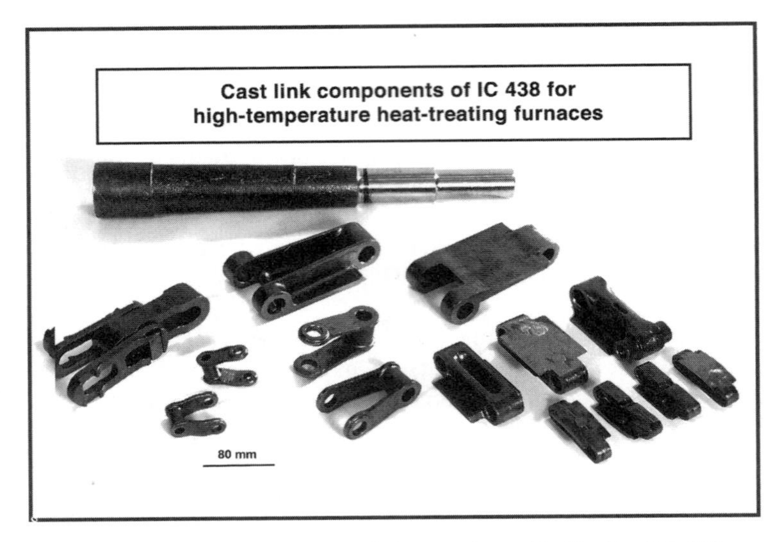

FIGURE 5.10 Belt furnace links cast from Ni_3Al alloy IC-438.

FIGURE 5.11 Truck brake stabilizer forged using Ni_3Al-based die.

5.6 WELDING

IC-221LA is the recommended weld wire for cast repair of Ni_3Al-based alloys in applications to 800°C (1472°F). Above 800°C (1472°F), use IC-221 W. This latter wire is also useful for weld overlay deposits on steels, stainless steel, and other nickel-based alloys. Stoody Company and Polymet Corporation (Cincinnati, Ohio) produce weld wire (Fig. 5.12).

TABLE 5.2 Chemical Requirements Specified under A1002-99 for Castings of Ni_3Al-Based Alloy

Element	Composition (wt %)
C[a]	0.08
S[b]	0.02
Al	7.3–8.3
Cr	7.5–8.5
Mo	1.20–1.70
Zr	1.60–2.10
B	0.003–0.012
Si[b]	0.20
Fe[a]	1.00
Ni	[c]

[a] Maximum.
[b] Maximum. For welding applications, the sulfur shall be 0.003% by weight or less and silicon shall be 0.05% by weight or less.
[c] Balance.

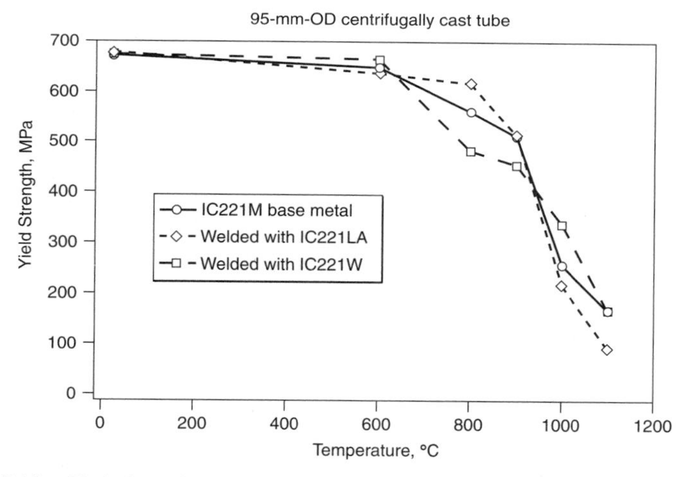

FIGURE 5.12 Variation of yield strength with temperature for IC122M tube weldments.

5.7 SUPPLIERS

Several manufacturers supply Ni_3Al and weld wire. Static castings are provided by Alcon Industries Incorporated (Cleveland, Ohio), Alloy Engineering and Casting Company (Champaign, Illinois), and United Defense LP (Anniston, Alabama). The latter two companies also supply centrifugal castings as does Sandusky International (Sandusky, Ohio).

Ametek (Eighty Four, Pennsylvania) supplies Ni_3Al powder. Stoody and Polymet supply weld wire.

5.8 MATERIALS AND PROCESSES

A new process was invented to assure that the aluminum was safely contained during the manufacture of Ni_3Al. In this Exomelt™ process, the exothermic heat from the reaction of aluminum with nickel melts the alloying elements that give this Ni_3Al its desirable ductility and strength (Fig. 5.13).

The crucible loading sequence is critical. Nickel is placed on the crucible floor. It is topped with alloying elements, then another layer of nickel sandwiched between two layers of aluminum. The crucible is heated to 800°C (1472°F) to melt the aluminum. It reacts with the nickel to form NiAl. The resulting heat release melts the NiAl (>1639°C) (2980°F) and forms droplets that drip onto the remaining aluminum. Maintaining the heat results in exposure of the bottom nickel layer and the formation of Ni_3Al.

The important feature of the Exomelt™ process (Fig. 5.14) is the layering described above. It uses the exothermic heat to drive the process, reducing energy input and resulting in an economical process, saving one-third to one-half of the energy consumed by the conventional process. The rapid heating minimizes

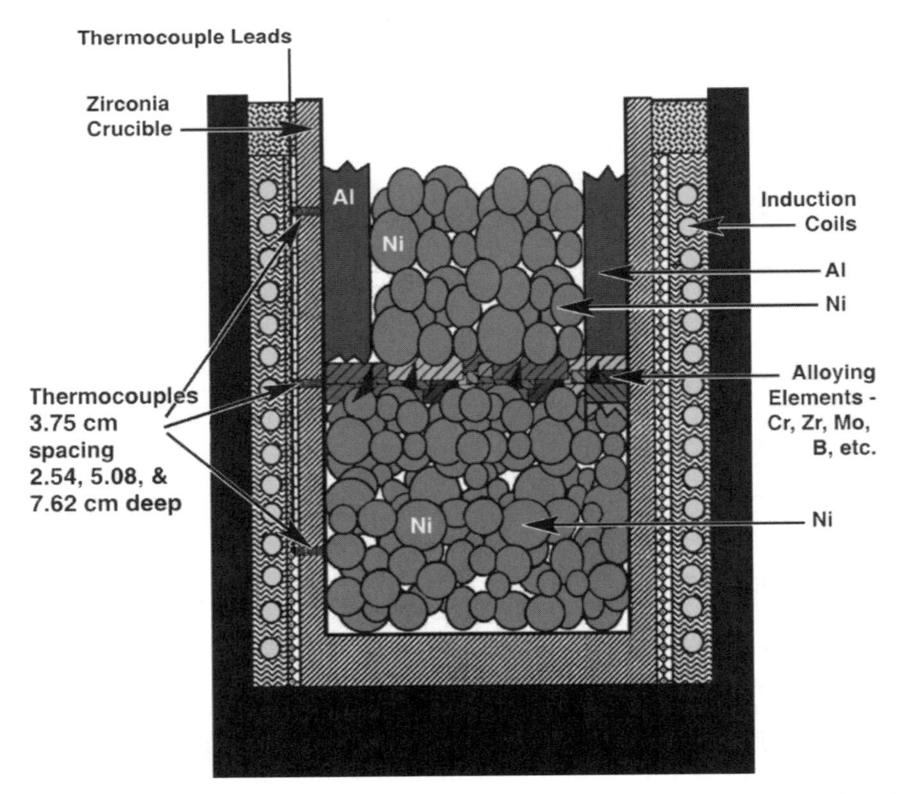

FIGURE 5.13 Furnace schematic of the Exomelt process. (Courtesy of John Wiley & Sons, Inc.)

FIGURE 5.14 Exomelt process loading sequence.

oxidation of the alloying elements, increases furnace life by minimizing time at high-temperature, and safely protects heating elements from attack by aluminum.

Vacuum melting in the Exomelt process is helped because all alloying elements can be loaded at the start of the process. Ni_3Al is produced in heats of 150–3000 lb.

The Exomelt™ process reduces process time by 50% (Fig. 5.15). It is reproducible and extends furnace crucible life. Temperature is controllable. It is inheritantly safe to operate, and on-stream commercially, producing more than 100,000 lb in 1998. It utilizes vacuum melting by loading aluminum at the start of

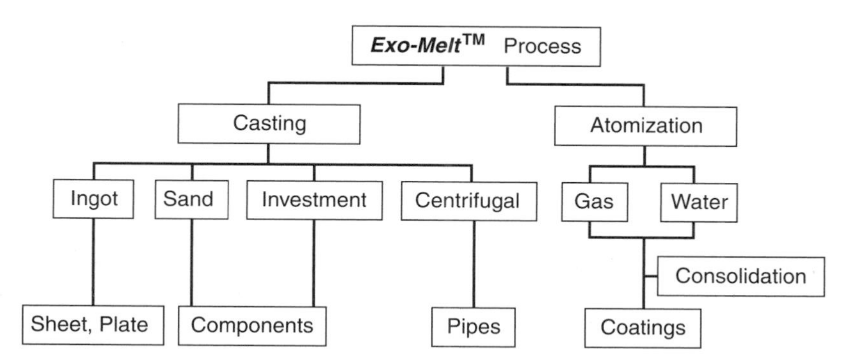

FIGURE 5.15 Exomelt process decision tree.

the process. Remelting can be used. Totally, savings add up to 50% over earlier processes.

5.9 U.S. DEPARTMENT OF ENERGY PARTICIPATION

The research of Ni_3Al was funded by the U.S. Department of Energy and conducted by C.T. Liu of Oak Ridge National Laboratory.

A team of Oak Ridge National Laboratory scientists, led by Vinod Sikka, partnered with Phillip Morris engineers to develop the award-winning Exo-Melt™ process. They invented the Exo-Melt™ process, a safe process that earned an R&D award from *R&D* magazine. Funding was a $21 million investment over 20 years.

The U.S. Department of Energy, Office of Industrial Technologies, and the Delphi Automotive Systems Corporation provided the funding to jointly develop the Ni_3Al furnace fixtures.

5.10 SUMMARY

Innovative product and process development has resulted in an economical Ni_3Al alloy of unique and beneficial properties compared to other high-temperature alloys. Several manufacturers provide Ni_3Al in the form of furnace furniture, steel process rollers, and other shapes for many other applications.

5.11 FOR MORE INFORMATION

Contact Vinod Sikka (865) 574-5112, `sikkavk@ornl.gov`, telephone: 865-574-5112. See the Oak Ridge National Laboratory website: `www.ornl.gov`

Contact these suppliers: Alloy Engineering and Casting Company (Champaign, Illinois), Alcon Industries Incorporated (Cleveland, Ohio), Sandusky International Corporation (Sandusky, Ohio), United Defense LP (Anniston, Alabama), Stoody Division of Thermadyne Holdings Corporation (Bowling Green, Kentucky), and BiMac Corporation (Dayton, Ohio).

REFERENCES

1. C. Krause, *Oak Ridge National Lab. Rev.* **28**(4), (1995).
2. K. Natesan and P. F. Tortorelli, "High-Temperature Corrosion and Applications of Nickel and Iron Aluminides in Coal-Conversion Power Systems," in S. C. Deevi, V. K. Sikka, P. J. Maziasz, and R. W. Cahn (Eds.), *Proc. Int'l. Symp. on Nickel and Iron Aluminides: Processing, Properties, and Applications*, ASM International, Materials Park, OH, 1997, pp. 265–280.

3. G. Welsch and P. D. Desai (Eds.), *Oxidation and Corrosion of Intermetallic Alloys*. Metals Information Analysis Center, Purdue University, West Lafayette, IN, 1996.

4. M. P. Brady, B. A. Pint, P. F. Tortorelli, I. G. Wright, and J. Hanrahan, Jr., "High-Temperature Oxidation and Corrosion of Intermetallics," in M. Schütze (Ed.), *Materials Science and Technology: A Comprehensive Treatment, Vol. 19 B. Corrosion and Environmental Degradation of Materials*. Wiley-VCH, Weinheim, Germany, 1999, Chapter 6.

5. V. Sikka, S. Deevi, and M. Santella, ASM Materials Solution Conference and Exposition, St. Louis, Missouri, 2000.

6. Delphi Adopts Ni_3Al for Heat Treat Fixtures, *Adv. Mater. Proc.*, **159**(6), 9–11 (2000).

CHAPTER 6

Metal Matrix Composites

CHRIS A. RODOPOULOS

Materials and Engineering Research Institute, Sheffield Hallam University, City Campus, Sheffield, United Kingdom S1 1WB

JAMES K. WESSEL

Wessel & Associates, 127 Westview Lane, Oak Ridge, Tennessee 37830

6.1 INTRODUCTION

Just as plastics and ceramics can be composited, so can metals. One of the newest class of commercially available materials are metal matrix composites (MMCs). These composites are analogous to organic (plastic) and ceramic composites because they are also composed of a reinforcement, usually a fiber, and a matrix. In the case of MMCs, as indicated by the name, the matrix is one or more

Handbook of Advanced Materials Edited by James K. Wessel
ISBN 0-471-45475-3

TABLE 6.1 Range of Common Metal Matrix Composites

Density, kg/m^3	2000–5500
Tensile strength, room temperature, MPa	400–1000
Bend strength, MPa	600–1300
Compression strength, MPa	500–1000
Elastic modulus, GPa	120–300

TABLE 6.2 Properties of Carbon Fiber-Reinforced Aluminum Matrix MMC

Tensile strength, MPa	825
Bend strength, MPa	1300
Elastic modulus, GPa	200

metals. Metal matrices include aluminum, magnesium, nickel, titanium, copper, and other metals. The fiber composition also varies, but the most common are ceramic, usually carbon or silicon carbide. Although they can be reinforced with particulates, MMCs are typically reinforced with fibers in continuous strands or chopped or whiskerlike. Because the matrix and fiber selections are many and varied, a wide range of properties result from the many possible formulas. Table 6.1 gives the range of properties due to the range of compositions.

A popular MMC is carbon-fiber-reinforced aluminum. Table 6.2 gives the properties of one of these compositions.

Composite composition can be further tailored to provide other properties required by specific applications. For instance, weldability can be altered. If the MMCs need to be welded to titanium, titanium is added to the MMC composition to facilitate weldability.

Metal matrix composites can operate at temperatures up to 800°C (1470°F), between that of plastic matrix composites and continuous fiber ceramic composites. At this temperature range and lower, MMCs are used where low weight, stiffness, toughness, fatigue resistance, erosion resistance, high thermal conductivity, and dimensional stability are desired. This has led to applications in aerospace, electronic, automotive, and other industrial applications.

6.2 APPLICATIONS

Like many advanced materials, MMCs were initially used in the aerospace industry where a high value is placed on weight savings and high performance. The high-strength density of MMCs coupled with their stiffness makes them useful in aircraft fins, helicopter rotor components, and exit vanes in turbine engines.

Engineers select MMCs for spacecraft struts and booms because weight saving is particularly important.

Like plastic composites, sporting goods markets find value in materials used in aerospace applications. MMC golf club shafts and heads, bicycle frames, and other bike components are used for both their stiffness and weight savings.

The automotive industry is finding that stiffness, wear resistance, fatigue resistance, and weight savings are important MMC features. Engines utilize MMC pistons, cylinder liners, brake rotor, and drums value MMC wear resistance, high thermal conductivity, and dimensional stability. Driveshafts and other applications take advantage of MMC stiffness.

Engineering data is available in Military Handbook 17 and at ASM International, Materials Park, Ohio 44073-0002, telephone (440) 338–5151 ext. 5663, Fax (440) 338–4634, website at `www.asminternational.org`

6.3 PROCESSING

Metal matrix composites are formed by methods similar to other composites. MMCs can be formed by continuous casting (drawn through a die to form rods and beams), plasma spraying (tapes), powder metallurgy, and liquid infiltration. Sheets and plates are formed by rolling into a die and extruding.

Continuous casting is very efficient. It also exposes the composite to less time at temperature, so a wider selection of metal matrices can be used.

In all of these processes, the matrix is heated above its melting point to facilitate forming and also wetting of the fiber. The "wet" perform is placed in a vacuum or pressurized to maximize fiber wetting, then solidified under heat and pressure in an autoclave. Variations in the subsequent heat-treating steps also affects the final properties of the MMC, giving the designer a variety of properties from which to choose. Final steps include machining techniques common to metals processing.

Like ceramic composites, the fiber requires a ceramic coating to protect it from matrix attack and a transition metal coating to enhance wetting of the fiber by the matrix. In MMCs, unlike continuous fiber ceramic composites, bonding between the fiber and matrix improves the composites strength.

6.4 DAMAGE TOLERANCE

6.4.1 Introduction

Even though the first unidirectional metal matrix composites (the other two types, namely whisker reinforced and particle reinforced, do not exhibit any remarkable fatigue resistance superiority when compared to monolithic materials, and therefore the fatigue research has somehow isolated them) date back to the mid-1950s, many engineers still refer to them as "exotic" materials. Back in the 1950s, as a solution to many engineering problems, a material was created that could withstand extreme mechanical conditions (these are defined in cases where ambient conditions, i.e. heat, can deteriorate the mechanical properties of the material). At that time, the space program was receiving a lot of governmental funding. Metallurgists and material's research engineers thought that a universal solution to many so-called extreme engineering problems could be the "artificial" creation of a material that could embed the mechanical characteristics of *one* material (or phase or matrix) with the physical characteristics of a *second* material (fiber). This distinction between mechanical and physical characteristics can be appreciated if we consider that a ductile material exhibits a higher fatigue resistance (mechanical characteristic) than a brittle (material, while a brittle material is showing better performance to resist, for example, heat (physical characteristic). Considering the potentials emanating from such theoretical assumptions, it is not difficult to understand why unidirectional metal matrix composites (uMMCs) has achieved such a degree of research interest during the last 40 years.

Undoubtedly, the same potentials characterize all composite material, however, the true benefit from the use of uMMCs is their superior fatigue resistance. When engineers conducted the first-cyclic loading tests, they noted that as the crack was advancing, the crack propagation rate was dropping. According to Paris and Erdogan [1] this is impossible because the growth of a crack under cyclic loading is in direct connection to the stress intensity factor (SIF):

$$\frac{da}{dN} = C\ \Delta K^m, \tag{6.1}$$

where *da/dN* stands for the change in crack length per loading cycle; *C*, *m* are empirical constants (depending on microstructure, load ratio environment, etc.); ΔK is the SIF range defined as:

$$\Delta K = Y\ \Delta\sigma\sqrt{\pi a}, \tag{6.2}$$

where a is the crack length, Y is a geometric correction factor, and $\Delta\sigma$ is the stress range. The only similar behavior known to date was the closure effect provided by the extra plasticity positioned close to the crack tip due to residual plastic stretch at crack flanks (the phenomenon is known as plasticity—induced crack closure, or as Elber's theory [2]). It was therefore evident that such behavior is controlled by the conditions of contact behind the crack tip. The first hard evidence of this new phenomenon, published in 1973 by Avenston and Kelly [3], where *fiber bridging* (FB) was quoted as responsible for this quasi-static toughening improvement. The basic idea around FB is that if the fiber strength is high enough to sustain the intense stress field ahead of the crack tip without failure, then with further propagation of the crack, the fiber is inserted into the crack. As a result of the continuous opening and closing of the crack flanks, the bridged fiber slides against the crack flanks producing a reaction stress known as bridging stress, σ_1. Since only the "openings" of the crack are really responsible for crack growth in a tensile fatigue crack, the reaction stress σ_1 has an opposite direction

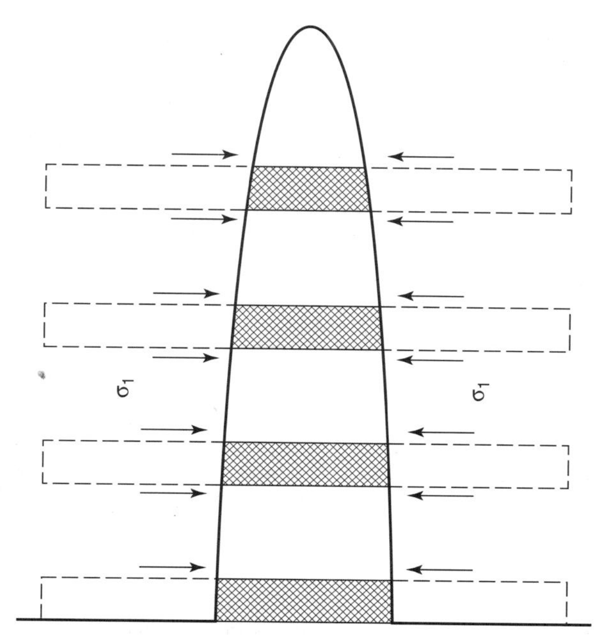

FIGURE 6.1 Fiber bridging induced closure.

to the direction of the applied stress and therefore is considered as closure stress. The bridging process is schematically shown in Fig. 6.1.

In 1995 the theoretical background around fatigue crack growth in uMMCs was impacted by the introduction of the so-called *fiber constrain effect* (FCE) [4]. Experiments conducted in a 32% SCS-6/Ti-15-3$[0]_8$ under tensile fatigue loading showed a complete different crack growth behavior when the crack growth was monitored on a 100-cycle base than on a typical 1000-cycle base [5]. This thorough examination revealed that when a crack is approaching the fiber, the crack growth rate experiences some kind of growth retardation. This retardation, which is not visible in the 1000-cycle base monitoring, was found to become more considerable as the crack was advancing closer to the fiber. However, at some distance from the fiber, the crack growth rate was found to accelerate rapidly to a value close to that attained before the onset of the retardation. This "ladder"-type (Fig. 6.2) growth behavior was attributed to the fact that when the crack approaches a fiber (or fiber row in a two-dimensional definition), the high stiffness fiber (the stiffness of fiber phase is up to five times higher than the matrix in the most known uMMCs) exerts some kind of constrain to the freely propagation of crack tip plasticity [4]. The term *freely* should not be misunderstood; the matrix yield stress definitely dictates the propagation of plasticity. However, it is taken to have a constant effect (without microstructural changes).

To better understand the idea of FCE, let us recall some basic principles of fracture mechanics. In any crack system, the stress ahead of the crack tip is assumed

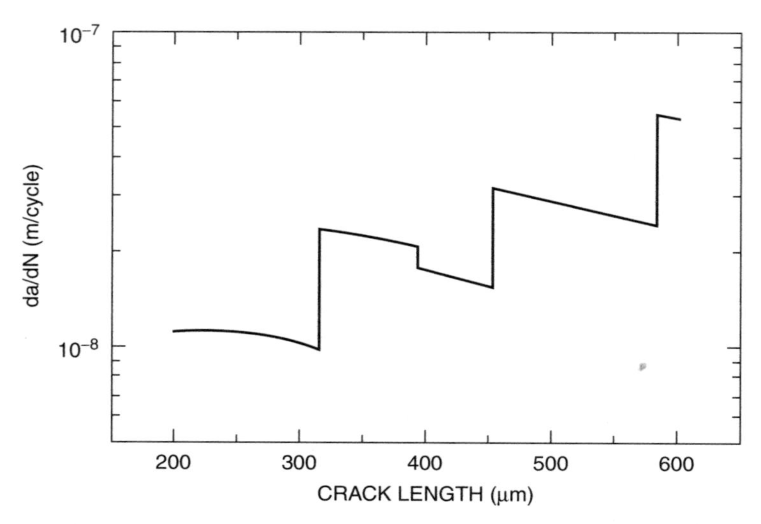

FIGURE 6.2 Typical "ladder"-type growth behavior due to FCE.

to follow a relaxation pattern with three distinct points, as shown in Fig. 6.3*a*. At point $A(r = 0)$ the stress is assumed infinite (maximum); at point $B(r = r_p)$ the stress is equal to yield stress of the material; and at point $C(r \to \max)$ the stress is assumed equal to the far-field stress (applied stress). The distance r_p, also known as plastic zone (first recognized by Paris and Erdogan [1]), represents the area where the von Mises equivalent stress exceeds the yield stress of the material [6]. If now the crack is allowed to propagate an amount Δa, a similar displacement will occur to the plastic zone. It should be noted that further extension of the crack will result in a proportional increase of r_p.

However, by positioning a brittle fiber at the end of the plastic zone, Fig. 6.3*b*, the plastic zone is not able to attain similar displacement (Δa), since the "yield stress/fracture strength" of the fiber is much higher than that of the matrix material. Under physical meaning definition, the above indicates a flow transition (at point B) of the matrix strain corresponding to σ_y into $\sigma_y/E_m E_f$, where E_f, E_m is the stiffness of the fiber and the matrix, respectively.

In 1995 de los Rios et al. [5] published a work where the continuous blockage of the plastic zone displacement was assumed to produce a tensile stress concentration (FCE) around the constraining fiber. In the same work, the authors supported the idea that the FCE will continue to proportionally increase with crack length until the FCE is able to achieve: (a) interfacial debonding or (b) failure of the fiber. When one of the two damage conditions is achieved, the plastic zone is allowed to propagate, and hence the crack growth rate recovers back to a primeval value. Furthermore, experimental work on Ti-based uMMCs [7] has revealed that the crack tip plasticity constrain provided by the intact (undebonded or unbroken) fiber is so severe that when the crack tip plasticity is allowed to propagate it actually jumps to the next fiber. Whether interfacial debonding or

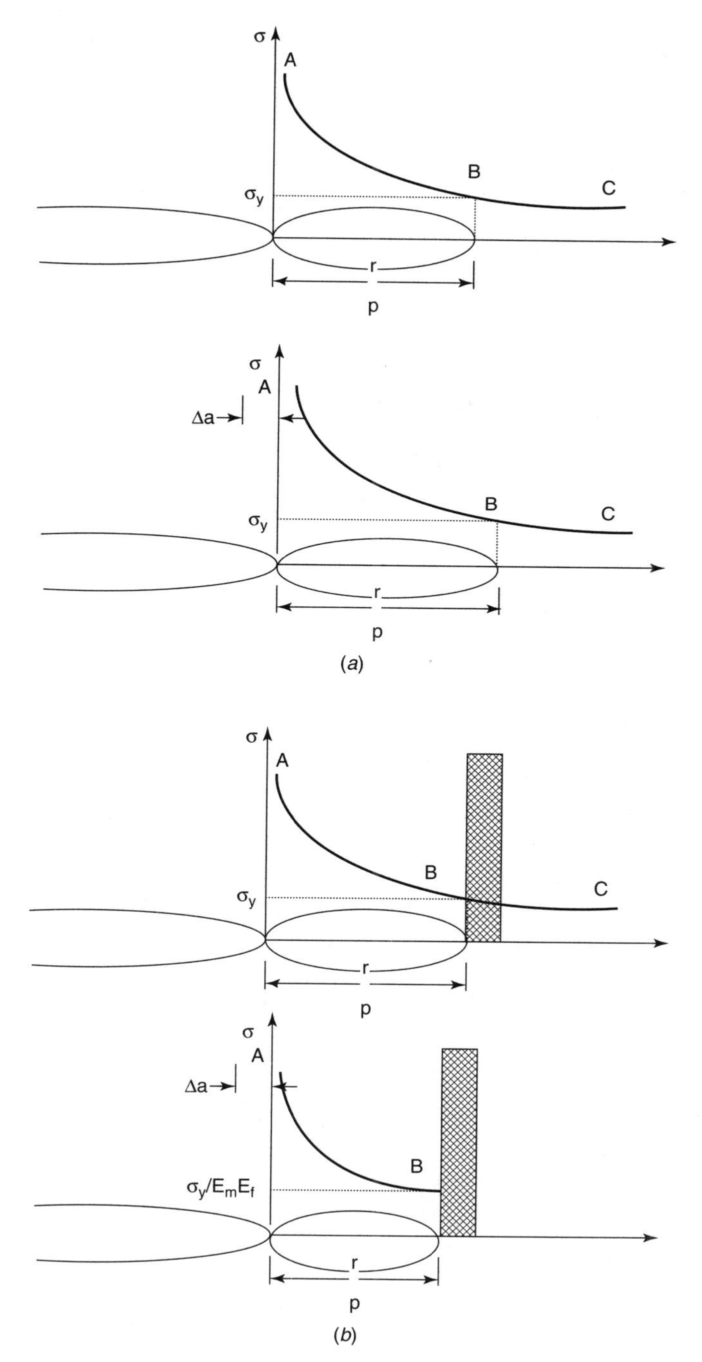

FIGURE 6.3 (*a*) Crack extension of Δa results into similar propagation of the plastic zone. (*b*) Effect of FCE on crack propagation.

fiber failure will be achieved first is indicated by the quality of the interface. Generally "weak" interfaces, which proclaim the existence of a low value of interfacial shear strength (a parameter that defines the amount of stress that can be transferred from the matrix to the fiber and vice versa), tend to produce interfacial debonding, while "strong" interfaces tend to produce fiber failure. The distinction between weak and strong interface can be found in any updated composite materials handbook. It should be noted that the first statement regarding crack tip shielding by stiffer elastic boundary was made by Ritchie [8].

Between 1996 and 1999, published experimental and theoretical results [9–11] indicated that the basic damage mechanisms (FB, FCE, and yield stress) dictating fatigue damage in uMMCs are subjected to some kind of deterioration with loading history. For example, FB, FCE, and yield stress were reported to fade away with crack length. This *time-dependent behavior* (TDB), as we will see later, defines the boundaries between reliable and nonreliable damage-tolerant design.

To include a detailed analysis of damage-tolerant design in uMMCs within the limits of a handbook is an impractical task. This is due to the massive theoretical background hidden behind each damage mechanism. However, a damage-tolerant design analysis without the slightest dash of a supportive background is hopeless. Therefore, the following section discusses essential concepts regarding each damage mechanism.

6.4.2 Fatigue Damage in uMMCs

6.4.2.1 Role of the Interface

There are several reported works regarding the influence of fiber–matrix interface on the fatigue performance and failure modes of matrix growing cracks in uMMCs [12, 13]. However, the complex chemical, physical, and mechanical interfacial phenomena, observed especially in the ductile titanium matrix interfaces [14], have constrained the classification of the role of the interface on fatigue into two limiting types considering their resistance to debonding.

The first type considers that the interface is weak in comparison to the matrix yield strength [13], and so the stresses near the tip of the matrix crack as it approaches the fiber could cause the fiber to debond from the matrix (Fig. 6.4). This mechanism was first proposed by Cook and Gordon [15] (also known as the Cook–Gordon effect), who suggested that the matrix crack deviates along the interface. During debonding, the stress and deformation fields developed at crack tips located at or near the bimaterial interface induce a mixed-mode fracture character, $K = K_{\mathrm{I}} + K_{\mathrm{II}}$, ($K_{\mathrm{I}}$, K_{II} represents the strength of stress singularities in tensile and in-plane shear loading, respectively), which in turn reduces the driving force at the vicinity of the crack tip and consequently increases the fracture toughness of the material [16–18].

It was reported that the duration of the debonding propagation is increased in cases where high residual stresses prevail [19]. Clearly, high residual stresses reduce the microstructural resistance at the interface. Matrix cracks may revert back to mode I matrix cracking ($K_{\mathrm{II}} = 0$) when the stress field ahead of the

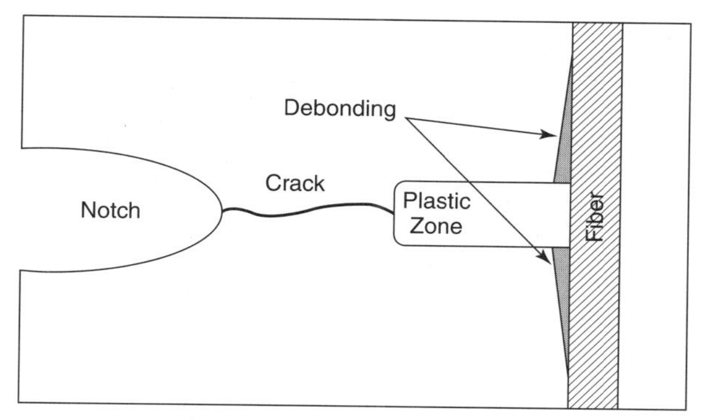

FIGURE 6.4 Weak interfaces allow interfacial debonding (bimaterial propagation) under high crack tip stresses. Bimaterial propagation is achieved, when K_{Π} is equal to the fracture toughness of the interface. Interfacial defects will increase the stress concentration at the interface, and hence they will promote debonding.

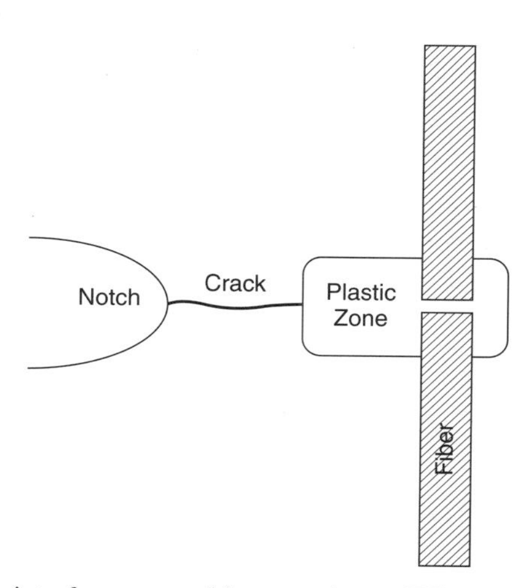

FIGURE 6.5 Strong interfaces can achieve maximum FCE. However, due to fiber failure, they minimize any FB.

interface crack is less than the strength of the interface [20, 21]. Moreover, He and Hutchinson [22] argued that the debonding process is promoted by the existence of microstructural defects in the reaction zone. High values of fiber volume fraction were reported to enhance the possibility of interfacial debonding since the statistical number of weak interfaces ahead of the crack tip is increased [23]. Additionally, fiber orientations between 45° and 90° were found to promote the deflecting crack tip behavior [24].

When the fibers are well bonded to the matrix (strong interface), debonding may not occur as the crack approaches the fibers, but instead fiber failure in a brittle manner will take place [25] (Fig. 6.5). This type of damage behavior is explained since for high values of the interfacial shear strength the FCE could become higher than the ultimate tensile strength of the composite. Even though this type of damage reduces the fracture toughness and the fatigue resistance of the material [26], numerical work has shown that strong interfaces enhance the performance of the transverse mechanical properties [27, 28]. The multiple fiber fractures observed in this type of material is a consequence of transferring to the matrix the loss of stiffness after fiber failure, which in turn is directly transferred back to the broken fiber, which may fail again [7].

Several experimental techniques have been proposed in the last few years to quantify the fiber–matrix interface. These include fiber push-out, fiber pull-out, and fragmentation testing. A fiber push-out (also known as push-through) test has been considered as the most straightforward of these techniques designed to obtain a single debonding and frictional shear stress measurements [29]. The basic configuration of the push-out test consists of a microhardness testing machine with a Vickers indenter, controlled by a tensile test machine usually equipped with a 100-N compression load cell (especially for SCS-6 monofilaments tests) [30]. Indenters with 0.1-mm diameter and 0.1-mm tip length usually made of tungsten carbide, diamond, or high-strength steel, are used. Axial displacements of the indenter are measured by a piezotranslator [31]. Generally, the push-out test has the advantage that a direct shear stress is applied to the interface, while repeated loading of the interface can provide data for strength degradation due to fatigue. However, repeated push-out test results should be treated with conservatism since they represent loading conditions extremely lower than those close to crack tip. A typical push-out test set-up is shown in Fig. 6.6.

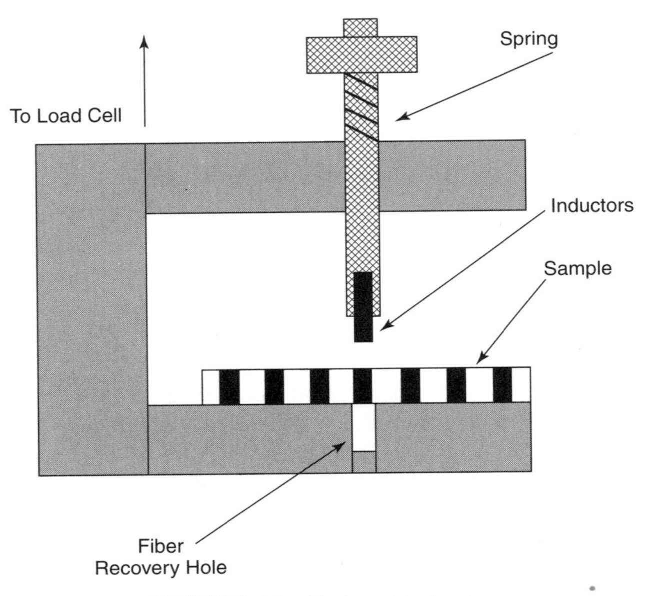

FIGURE 6.6 Push-out test setup.

The interfacial shear strength τ is calculated directly from the experimental load–displacement data, considering a simple force balance at the interface:

$$\tau = \left(\frac{\sigma_f d}{4L}\right), \tag{6.3}$$

where σ_f is the axial stress at the fiber at the time of debonding, d is the fiber diameter, and L is the embedded fiber length or specimen thickness. In Fig. 6.7 fiber push-out load–fiber displacement curve of a SCS-6/Ti-15-3 composite is shown. Push-out tests conducted on virgin SCS-6/titanium interfaces showed that the sliding resistance declines with fiber–matrix displacement after debonding. This behavior has been attributed to wear process and plastic deformation of the matrix [31]. Additionally, reductions in sliding resistance have also been reported for interfaces subjected to cycle loading. In the latter case fiber coating fragmentation, asperity wear, and relief of residual stresses through local plastic deformation of the matrix due to asperity mismatch were suggested as the main reasons for such behavior [31].

Fiber pull-out tests, on the other hand, are mainly used to determine the influence of a matrix crack on debonding and sliding. During loading, a crack is initiated in the matrix at the circumferential notch. Pull-out tests are generally utilized in cases where the effect of local stress field, fiber failure, and fiber pull-out on the debonding and the sliding process are in question [32, 33]. The machine setup usually consists of a universal tensile machine while the axial displacements are measured using once more a piezotranslator. For the measurement of the interface strength, the shear-lag analysis [34] is usually used.

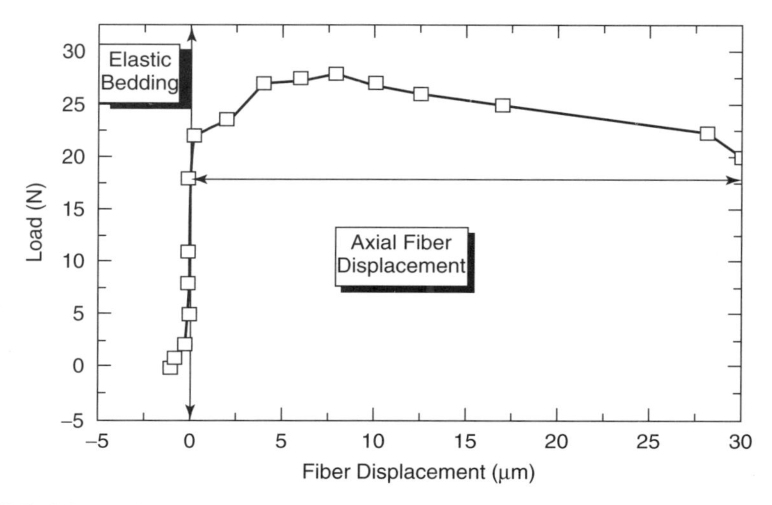

FIGURE 6.7 Typical load–displacement data from fiber push-out test. (Figure reproduced from [31]).

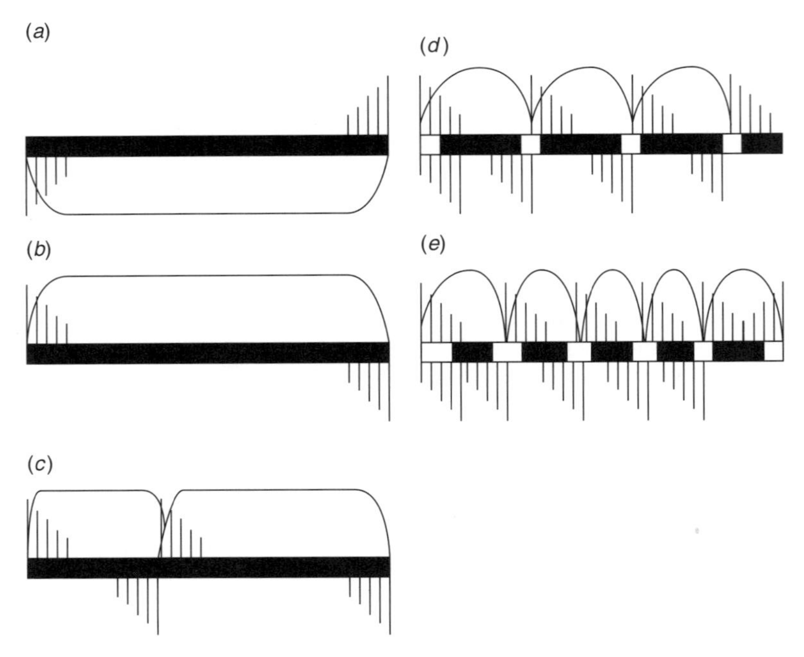

FIGURE 6.8 Five steps of fragmentation process. Specimen is loaded in tension along the fiber direction. Clouds of points represent local strength of fiber and white areas on the black fiber stands for debonded lengths: (*a*) Residual thermal stresses, (*b*) elastic loading, (*c*) first fiber fracture, (*d*) extension of debonded, and (*e*) saturation. (Reproduced from [33]).

The fragmentation procedure involves heavy straining of the matrix in which a single fiber is embedded. The test consists of five observation steps leading to the evaluation of bond strength and sliding distance [30]. The first step (Fig. 6.8*a*) corresponds to axial compressive stress (residual thermal stress) resulting from cool-down from the fabrication process. By increasing progressively the strain on the matrix, the fiber is subjected to an elastic uniform loading (Fig. 6.8*b*). As the strain continues to increase, the fiber breaks at a critical defect and the fragment ends are first stressed elastically until the corresponding shear stress reaches the interface strength, (Fig. 6.8*c*). After further straining the fiber breaks in more segments. The segment ends are now stressed only by friction stresses (Fig. 6.8*d*) up to the final step when the displacement becomes constant (Fig. 6.8*e*). The maximum shear stress is then calculated using the Kelly–Tyson equation [35]:

$$\tau = \frac{d\sigma_{f(l_c)}}{2l_c}, \tag{6.4}$$

where l_c is the so-called critical or ineffective fiber length (defines the minimum fiber length able to support maximum load), d is the fiber diameter, and $\sigma_{f(lc)}$

is the fiber strength for the particular segment length as calculated by Weibull analysis. The advantages of this method are the ability to correlate fiber strength characteristics and residual thermal stresses with debonding and sliding. However, the complex behavior due to several fiber breaks raised strong opposition for the reliability of this method [36].

Analysis of the stress field in the vicinity of the fiber end, for example, is at a fiber break (where a fiber length l is perfectly bonded to a matrix of lower modulus, the interface is thin and no load is transferred through the fiber ends), revealed that load is transferred between the fiber and the matrix by shear stresses. The pioneer work of Cox [*shear-lag analysis (SLA)*] [34] showed that the shear stress is a maximum at the fiber ends and declines to a minimum at the center of the fiber in a nonlinear manner. Since the fiber ends do not carry the full load, the average fiber strength is less than the strength of a continuous fiber of the same length subjected to the same loading conditions. Thus the tensile stress on the fiber is zero at the ends, which attains the maximum value at the fiber center, (Fig. 6.9). Extensive review on the SLA can be found in [36–38].

When the shear stress τ exceeds the interfacial shear strength of the interface, interfacial debonding takes place over a specific fiber length given by [23],

$$l_d = \frac{\sigma_d d}{4\tau_s}, \tag{6.5}$$

where σ_d is the tensile stress at the fiber required to debond fiber length of l_d and τ_s is the interfacial shear strength. Equation 6.5, however, has been reported to

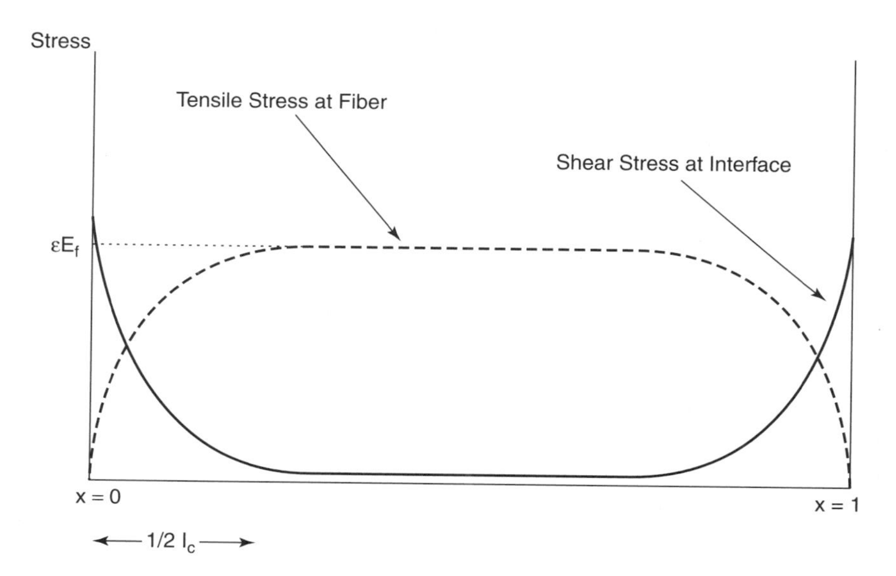

FIGURE 6.9 Schematic representation of variation of tensile stress at fiber and shear stress at interface according to SLA.

represent an overestimation of the debond length since the debond energy of the interface has been considered low, and thus the effect of the elastic shear stress outside the debond length has been consequently taken as negligible [39, 40]. An alternative solution, where the debond length was found to be smaller but closer to experimental observations, has been reported by Chiang et al. [41]. In their work strain compatibility between fiber and matrix was assumed to apply outside the debond length.

After debonding, the damage pattern of the interface is dictated by frictional sliding at the fiber–matrix interface during progressive debonding and fiber pull-out. Several models, developed in the last 5 years [42, 33], quantify the sliding resistance with two parameters: (a) the coefficient of friction μ and (b) the interfacial clamping stress σ_c, which acts perpendicular to the interface and is caused by thermal expansion mismatch between the fiber and the matrix or roughness interaction between fiber and matrix. The sliding resistance is then described as:

$$\tau = \mu(\sigma_c - \sigma_\nu), \tag{6.6}$$

where σ_ν is the stress relief due to Poisson's contraction of the fiber, which has been reported to be proportional to the fiber axial stress [32]:

$$\sigma_\nu = \kappa\sigma_f, \tag{6.7}$$

where

$$\kappa = \frac{E_m \nu_f}{E_f(1 + \nu_m) + E_m(1 - \nu_f)}$$

E_f, E_m, and ν_f, ν_m are the Young's modulus and the coefficient of Poisson's contraction of the fiber and the matrix, respectively.

Push-out tests conducted on virgin SCS-6/titanium interfaces showed that the sliding resistance declines with fiber–matrix displacement after debonding. This behavior has been attributed to wear process and plastic deformation of the matrix [43]. Additionally, reductions in sliding resistance have also been reported for interfaces subjected to cycle loading. In the latter case fiber coating fragmentation, asperity wear, and relief of residual stresses through local plastic deformation of the matrix due to asperity mismatch were suggested to be the main reasons for such behavior [31].

Typical values of the interfacial shear strength have been quoted between 100 MPa [44] and 360 MPa [45], for the as-fabricated SCS-6/Ti-6-4 and between 90 MPa [46] and 125 MPa [44] for the as-fabricated SCS-6/Ti-15-3. Differences between published data are possibly related to experimental techniques. On the other hand, values for frictional resistance were found to be 81 MPa for the as-fabricated SCS-6/Ti-15-3 [44] and 88 MPa for the as-fabricated SCS-6/Ti-6-4 [44]. Heat treatment has also been observed to influence the shear stress values, especially in composites with metastable β-matrices such as Ti-15-3. This was attributed to volume changes of the α-phase precipitate and the corresponding changes of the interfacial clamping stress [29]. On the contrary, no significant changes were reported for the SCS-6/Ti-6-4. Interfacial shear strength has been reported to increase with specimen thickness up to a value of approximately four to five times the fiber diameter and found to be independent of specimen thickness thereafter [46].

6.4.2.2 Fatigue Damage — Notched Specimens

The propagation of fatigue cracks from local stress concentrations could be extremely complex due to differences in the initial value of composite strength and the corresponding differences in the fatigue damage pattern. In general, the presence of notches leads to a reduction of strength due to (a) reductions in the effective cross-sectional area of the component and (b) stress concentrations at the notch tip. In brittle monolithic materials, the ultimate tensile strength, σ_{uts}, of a panel containing a circular hole may be reduced up to a value of onethird of the strength of the unnotched panel. Alternatively, in notched ductile materials (especially in the form of thin sheets), Bilby et al. [47] argued that tensile fracture is achieved by the formation of a narrow zone of intense plasticity ahead of the notch and the nucleation of a crack within the plastic zone. In the same work it was suggested that if plastic flow and microcracking processes take place, then the stress concentrations due to the notch are altered, and the material is characterized by the term *notch insensitive*. In the case of unidirectional MMCs two different cases have been identified to control notch sensitivity [48]. If the fiber–matrix interface is weak, the stress concentration effect leads to large-scale fiber debonding, and so elastic relaxation due to unloading of the fibers causes the notch to open and to remove stress concentrations. In this case the composite is considered notch insensitive. On the other hand if the interface is strong, then stress relaxation can only be achieved by matrix cracking and fiber failure. In this case the composite is considered notch sensitive, and the degree of sensitivity increases with bond strength and fiber volume fraction. Published experimental work conducted in 32% SCS-6/Ti-6-4 (strong interface) [49] revealed that the presence of a hole or a notch could lead to a reduction of about 50% of the strength of the unnotched specimen.

Using single-edge notch specimens, several workers have attempted to distinguish the various damage patterns that may develop in Ti-based uMMCs [23, 50, 51]. In as-fabricated SCS-6/Ti-6-4 single mode I matrix cracks were observed to grow in the first 100–1000 cycles following three different propagation patterns as schematically shown in Fig. 6.10. When the applied stress or the stress intensity factor is sufficiently high ($\approx$75% of the σ_{uts}), then cracks were observed to propagate in a catastrophic mode I manner (pattern A), with increasing crack growth rates, as in monolithic materials [23]. In this condition, the majority of bridging fibers were observed to fail close to crack tip [52]. By reducing the stress level ($\approx$55% of the σ_{uts}), the fibers did not fail immediately, and thus friction stresses would develop as a result of fiber bridging. These friction (closing) stresses would cause deceleration in crack propagation rate (pattern B). In this situation, some of the bridging fibers, especially those close to the notch tip or crack mouth, might fail and the load carried by the broken fibers should be redistributed

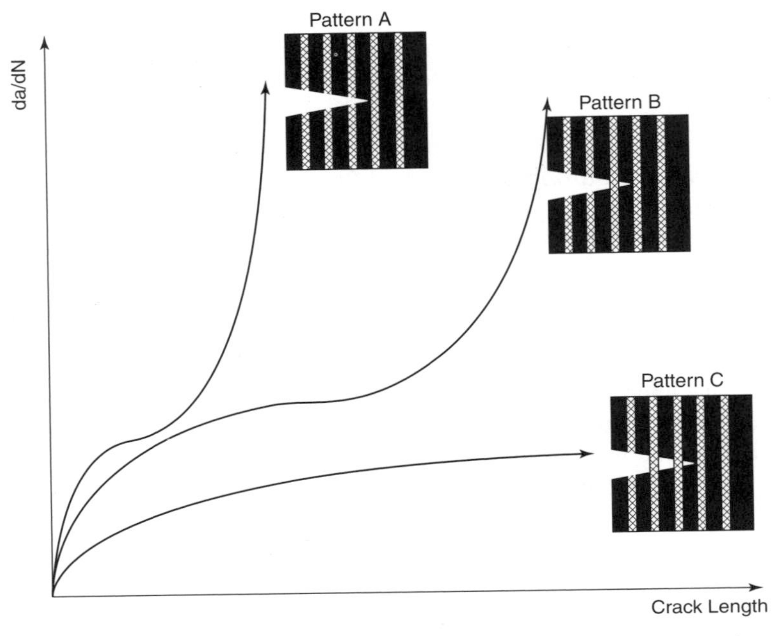

FIGURE 6.10 Fatigue damage behavior of notched uMMCs in relation to the applied stress.

leading to more fiber failure and causing a rapid increase in the propagation rate. Within this pattern two different modes of crack growth were observed [51]. If after the incidence of the first fiber failure no further fibers bridge the crack, the crack propagation rate would become unstable. Otherwise stable crack propagation would be expected. The boundary between stable and unstable propagation was found to be particularly sensitive to stress ratio $R(R = \sigma_{min}/\sigma_{max})$. Tests at high positive ratio ($R = 0.5$) in notched SCS-6/Ti-6-4 three-point bend specimens [51] showed an increased tendency to unstable propagation in relation to those conducted at lower ratios ($R = 0.1$). Additionally, pattern B behavior was reported to became more frequent with increases in notch length, at a particular stress level [53]. This behavior was attributed to the increase on the unbridged length, which causes the stress sustained by the bridged fibers in the crack wake to increase, and thus fiber failure is achieved at lower stress levels. Elevated temperatures and low frequencies were also reported to promote this behavior [54, 55] (the understanding of such behavior is still under investigation). Fiber pull-out within this pattern has been reported to be not significant [51]. On further reduction of the applied stress ($\approx$30% of the σ_{uts}), bridging fibers remained largely intact along the crack, and crack growth rates decrease as the bridging zone increases. Cracks either attained a steady-state growth condition or arrested due to reduction in the matrix stress intensity factor (pattern C) [52].

From tests conducted on SCS-6/Ti-15-3 in the as-fabricated state, mode I cracks were observed to initiate in a manner similar to that discussed above. At high stresses (pattern A), extensive bridging fiber failure was observed after small amounts of crack extension [56]. After subsequent crack growth, fiber failure ahead of the crack tip was observed to govern the damage pattern [53]. At lower stresses or shorter notch lengths, fewer fiber failures were noted [57]. In tests conducted on center-notched tensile specimens, crack growth was observed initially to decrease while steady-state propagation was observed when the crack was long compared to the initial notch [58]. This sudden decrease was explained by Harmon and Saff [59], by suggesting that this behavior may derive from the extensive debonding of the fiber next to the notch tip and the corresponding stress relaxation. However in fatigue tests conducted on single edge-notched specimens at positive mean stresses [52], secondary cracks were found to initiate from the original crack or notch tip and propagate along the fiber–matrix interface. After these deflected cracks had propagated parallel to the fiber direction for a certain distance, the cracks were deflected and propagated perpendicular to the fiber direction. Similar multiple matrix cracking was observed about circular holes [60], where mode II secondary cracks nucleated at locations of maximum shear stress [59]. Controversial opinions have been reported on the effect of heat treatment on SCS-6/Ti-15-3. In one work [52], it was reported that heat treatment decreases the crack propagation rate. In another work it was found that extensive thermal exposure could lead to catastrophic failure even if low stress levels were applied [54]. For thermal exposed composites, the stress ratio was found to have a small effect at low and mean stresses [54].

6.4.2.3 Fatigue Damage — Unnotched Specimens

W. S. Johnson is considered a pioneer in the field of fatigue damage in uMMCs. From the early days of his Ph.D. thesis in 1979 [61] and up to 1988 [62], he suggested that the main cause of fatigue failure in any unidirectional composite system is the damage accommodated by the fiber (failure) and the corresponding loss of stiffness. Experimental observations in alumina-fiber-reinforced aluminum composites, conducted by the U.S. Air Force Materials Laboratory [63], convinced researchers that the most significant damage mechanism is extensive fiber damage, including multiple fractures of individual fibers. They found that even at the late stages of the fatigue life of a specimen, the fatigue resistance of a composite was still superior to that of an unreinforced matrix, since a sufficient number of broken fibers were still able to carry load quite effectively. The so-called *matrix-dominated damage* is based on the belief that the matrix material requires less strain than the fibers to initiate damage. With the introduction of SiC-reinforced Ti–matrix composites, this failure mechanism was more than ever verified. It was suggested in [64] that the superior endurance of the SCS-6/Ti-15-3 system is mainly controlled by the high fatigue limit of the SCS-6 fiber (approximately 1300 MPa [65]), as compared to an average strength of 3800 [66] or 4500 MPa [27]).

However, based on further observations, Johnson [57, 67] suggested that the fatigue failure of a SCS-6/Ti-15-3 and SCS-6/Ti-6-4 uMMCs occurs in a *self-similar damage* manner. This was supported by the similar endurance limit strains of the fiber and the matrix, and scanning electron microscopy (SEM) examination of fatigue fracture surfaces, which revealed low-level or negligible fiber pull-out (Fig. 6.11). Minimum fiber pull-out signifies fiber failure close to the crack plane. Furthermore, since the fatigue limit (approximately 600 MPa for the SCS-6/Ti-6-4 [68] and 650–700 MPa for the SCS-6/Ti-15-3 [69]) is significantly lower than the yield stress, the matrix may nucleate fatigue cracks without global yielding [70]. Johnson argued that since the strain at matrix fatigue limit is close to the fiber failure strain, matrix and fibers should fail simultaneously. Self-similar damage was also reported for boron/titanium uMMCs [71].

In 1991, subsequent studies of fatigue crack growth in SCS-6/Ti-15-3 and SCS-6/Ti-6-4 composites [72] indicated that fatigue failure does not occur in a self-similar damage manner because cracks were found to be bridged by intact fibers. In the same work, cracks were found to initiate from several different fabrication and manufacturing defects (broken fibers at edges, touching fibers, voids at the interface, etc.).

Even though extensive research has been conducted on crack initiation and growth of unnotched Ti-alloy-based MMCs [64, 68, 73, 74], most of the workers have agreed that the issue is quite confused since it involves the understanding of three basic parameters that could act individually as well as simultaneously. The first observation concerns the ratio of the applied strain to the time-dependent

FIGURE 6.11 SEM micrograph showing minimum fiber pull-out due to fiber failure close to crack plane for 32% SCS-6/Ti-15-3 $[0]_8$ tested at $\sigma_{max} = 600$ MPa, $R = 0.1$. The light gray area represents plasticity passage. (Photo taken from [69]).

fracture strain of the interface especially at medium stress levels. The second parameter is the tendency of the matrix material to initiate secondary cracks, particularly at the center of the specimen. The last parameter is the ability of the MMC to arrest these secondary cracks by constraining the crack micro-plasticity, FCE, or by producing adequate FB, and stress relaxation during bridging and debonding, respectively.

Furthermore, from experimental observations conducted on SCS-6/Ti-15-3 and SCS-6/Ti-6-4 smooth specimens (40% SCS-6/Ti-15-3 $[0]_6$, 35% SCS-6/Ti-6-4 $[0]_6$, 32% SCS-6/Ti-15-3 $[0]_8$, and 32% SCS-6/Ti-6-4 $[0]_8$ [52, 68, 72]), the fatigue damage behavior of both materials was classified into three distinct regimes depending on the maximum applied stress level.

At high peak applied stresses (about 80% of the quoted tensile strength), the fracture surface of the two types of SCS-6/Ti-15-3 materials were found to exhibit a flat morphology, that is, most of the fiber breakages were found close to the crack plane (no fiber pull-out). Fatigue damage was observed close to the fiber–matrix interfaces with a random distribution throughout the specimen (changes in the reaction layer thickness could develop stress concentrations [68]). Small number of secondary cracks were observed, which suggests that fatigue failure was mainly controlled by fiber breakage accumulation. Similar observations have been quoted for the two types of SCS-6/Ti-6-4 composites.

At medium stresses (40–80% of the tensile strength), the fracture surface of the SCS-6/Ti-15-3 was reported as irregular (similar to a tensile fracture surface [72]) and composed of several relatively flat fatigue cracking regions that extend from the specimen surface [72]. Significant fiber pull-out, with random distribution was also detected (Fig. 6.12). Cracks were found to initiate from broken fibers and

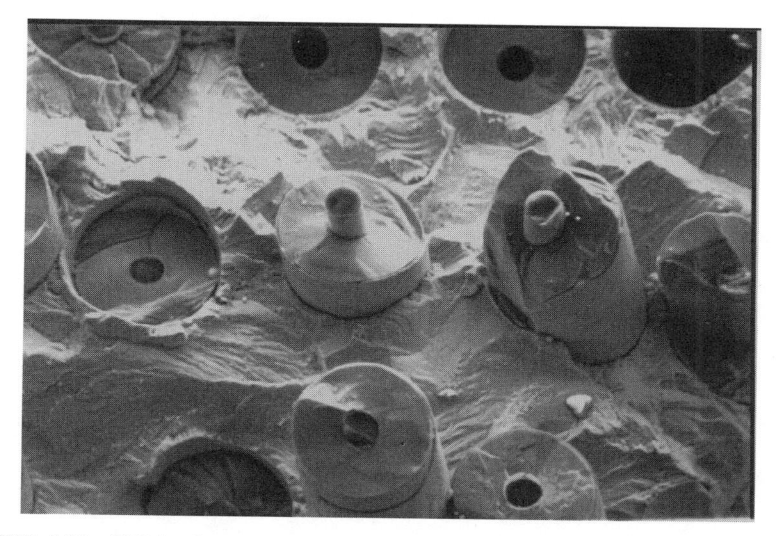

FIGURE 6.12 SEM micrograph showing fiber pull-out for 32% SCS-6/Ti-15-3 $[0]_8$ tested at $\sigma_{max} = 960$ MPa, $R = 0.1$. (Photo taken from [69]).

interfaces, especially at the machined edges, while a large number of secondary cracks was detected at the specimen center [73, 68]. For the SCS-6/Ti-6-4 both reference sources confirmed a similar fracture surface to that of the SCS-6/Ti-15-3. However, a small number of secondary cracks at the specimen center were detected for the 32% SCS-6/Ti-6-4 [68] while no secondary cracking was found for the 35% SCS-6/Ti-6-4 [72]. This disagreement was attributed to differences on the reaction zone thickness (thicknesses of 1.7 and 2.43 μm were quoted for the SCS-6/Ti-6-4 and SCS-6/Ti-15-3, respectively, in the as-fabricated condition [72]) and the interfacial shear strength [68]. Clearly, lower interfacial shear strength increases the number of broken interfaces and thus the probability of secondary cracking. Values of 124 MPa for the SCS-6/Ti-15-3 and 156 MPa for the SCS-6/Ti-6-4 were quoted, respectively [44].

At lower stresses, when the applied strain level to the composite is lower than the fracture strain of the interface, the fatigue damage pattern of the 35% SCS-6/Ti-15-3 was reported to be limited by matrix crack initiation at the specimen edges as a result of broken fibers due to machining while no secondary matrix cracking from broken interfaces was observed [72]. Also, metallographic inspection of the specimens revealed that after 10^6 cycles the fibers were still intact and bridged the cracked matrix. The same fatigue damage pattern was reported for the 40% SCS-6/Ti-6-4 [72]. However, in tests conducted on 32% SCS-6/Ti-15-3 at 600 MPa, cracks were found to grow from the reaction layer in the same manner as at higher stresses [68]. In the same work, cracks in the 32% SCS-6/Ti-6-4 were observed to grow not only at edges but also from "warts" on the fibers (see Fig. 6.13). Fiber warts were not observed in the SCS-6/Ti-15-3.

Cycling was found to degrade the tensile properties of both composites. After 10^6 cycles fatigue testing, the elastic modulus and tensile strength of the

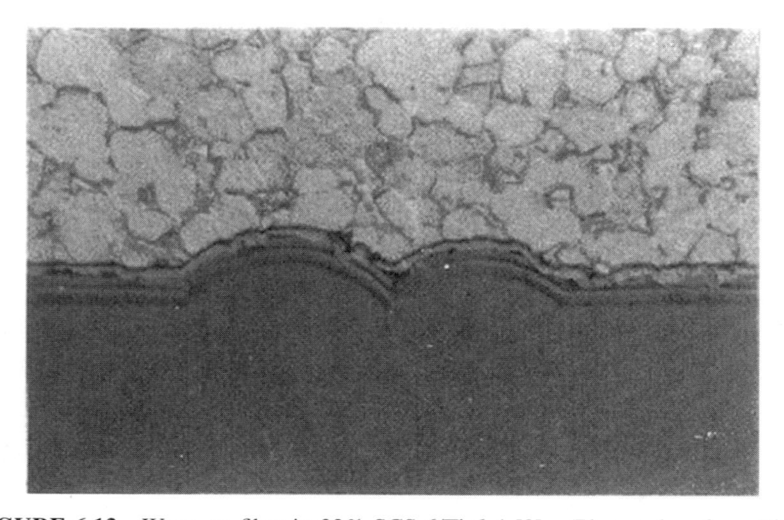

FIGURE 6.13 Warts on fiber in 32% SCS-6/Ti-6-4 $[0]_8$. (Photo taken from [68]).

35% SCS-6/Ti-15-3 were measured as 130 GPa and 1103 MPa, respectively, compared to initial values of 210 GPa and 1572 MPa [72]. For the 40% SCS-6/Ti-6-4 a similar degradation of the tensile strength was reported (postfatigue value of 1034 MPa as compared to an initial value of 1572 MPa [72]). However, cycling was found to produce lower degradation of the elastic modulus for the SCS-6/Ti-6-4 composite (initial value of 213 GPa; fatigued value 193 GPa [72]). Considering that for the same load history both materials have approximately accumulated similar crack length, differences on the degradation of the elastic modulus can only be explained by differences in the number of bridged fibers that failed during cycling.

It should be noted that the above findings do not represent a universal picture of the fatigue behavior of uMMCs for a number of reasons. The most critical are:

1. Most of the tests were conducted on strip or dogbone specimens. These specimens, even after careful polishing at the edges to minimize the effect of coarse finish, are vulnerable to additional crack initiation and therefore may underestimate the true fatigue life of the material when compared to traditional circular section specimens [68].
2. There is a minimum amount of data about the effect of load ratio [75, 76].
3. Control mode (strain-controlled or load-controlled), especially at different stress ratios, shows in most of the cases an unpredictable behavior [75].
4. The material's behavior under tension-compression loading cannot be fully appreciated since their typical thickness is about 2.5 mm and therefore are unable to withstand significant compression loads [75, 77].

6.4.2.4 Micromechanisms of Fatigue Crack Growth

A vital step to an efficient and safe damage design is the understanding of the fatigue characteristics of uMMCs. Most of the fatigue fracture micromechanisms, especially in Ti-based MMCs, include fiber failure ahead of the crack tip [74, 78], the propagation of the matrix crack under conditions of crack deflection at fibers [8], matrix crack shielding by the fibers [7], crack bridging by unbroken fibers [79–81], and fiber pull-out produced by broken fibers [82].

Fiber Failure Ahead of Crack Tip The issue of broken fibers ahead of the crack tip in uMMCs has a significant engineering interest since it incorporates fatigue damage with the residual strength of the material. In general fiber failure ahead of the crack tip has to be initially distinguished into two categories: (a) fiber failure prior to debonding and (b) fiber failure after debonding. To understand the idea of fiber failure prior to debonding, let us define two possible damage scenarios. The first scenario is that the normal stress sustained by a fiber close to the crack tip is significantly increased by a single overload. In this case, depending on the amount of the overload and the ability of the material to redistribute stress through the interfacial shear stress, a number of successive fibers close to crack tip may fail while others will experience severe debonding [69]. Both events will provide a significant loss of strength. The second scenario is based on the presence of fiber warts. Generally fiber warts play the role of stress raisers at the fiber without promoting interfacial debonding as in the case of broken interfaces [68]. On the other hand, postdebonding fiber failure is likely to happen due to stress concentration at the transition point between debonded and intact interface. It should be noted that such an event is likely to happen at long crack lengths where the operational life of the material is almost consumed [10]. This event may exhibit notable fiber pull-out. All the above are graphically shown in Fig. 6.14.

In general fiber fracture ahead of the crack tip could take place when the local stresses exceed one of the following three failure criteria [8]:

1. The normal stress σ_{yy} acting along the fiber exceeds the fracture strength of the fiber. This case is of particular interest when the fiber matrix interface is very strong and thus fiber failure is to be expected instead of interfacial debonding. This is not the case for the SCS-6/Ti composite systems in the as-fabricated condition (since the fiber attains a high value of Weibull modulus and no stress concentrations are developed due to changes on the reaction layer thickness). However, this mechanism may occur when the material is highly heat treated [72]. The stress σ_{yy} is usually computed using elastic crack tip stress fields:

$$\sigma_{yy} = \frac{K}{\sqrt{2\pi r}}. \tag{6.8}$$

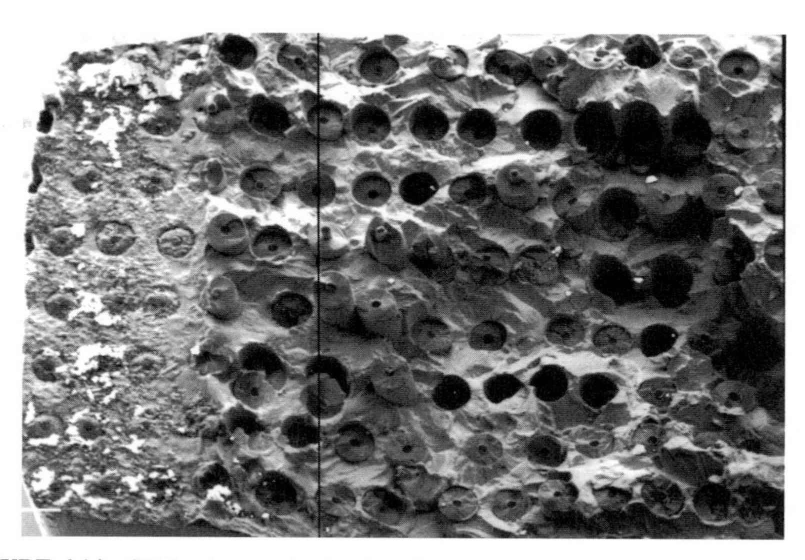

FIGURE 6.14 SEM micrograph showing fiber pull-out for 32% SCS-6/Ti-15-3 $[0]_8$ tested at $\sigma_{\max} = 800$ MPa, $R = 0.1$. The black line represents final crack length. (Photo taken from [69]).

2. The normal stress σ_{yy} exceeds the axial debonding strength of the fiber–matrix interface. In this case, the load carried by the interface is transferred to the fiber before debonding is achieved, and so fiber failure could take place.
3. Interface shear debonding occurs when the interface shear stress τ_{xy} exceeds the interfacial shear strength. In this case, the shear load sustained by the interface is transferred to the fiber. The change of the stress carried by the fiber due to debonding is given as [83]:

$$\sigma_{yy} = \frac{K}{\sqrt{2\pi r}} + 4\tau \frac{l_d}{D}, \tag{6.9}$$

where τ is the average shear strength of the interface, l_d is the debond length, and D is the fiber diameter.

It should be noted that Eqs. 6.8 and 6.9 can be effectively used only in cases where small-scale yielding conditions prevail.

Crack Deflection of Main Crack to Fiber–Matrix Interface The joining of the main matrix crack with a preexisting interface crack leads to a kinked crack deflected to propagate along the fiber–matrix interface. Crack deflection generally reduces the mode I crack driving force, since the crack path is now deviated from the direction of maximum tensile stress [8]. The mixed-mode solution proposed by Cotterell and Rice [84] for a simply kinked crack subjected to a nominal K_I stress intensity factor in a monolithic material can also be used in the case of MMCs [8]. Clearly, the corresponding stress intensity factor of a mixed-mode crack is computed in terms of the individual local K tensors using the expression

$$K_{\text{tip}} = (k_1^2 + k_2^2)^{1/2}, \tag{6.10}$$

where k_1, k_2 represent the local mode I and mode II stress intensities, respectively, of a crack with a branch forming an angle θ with the original direction of the crack. The values of k_1, k_2 are

$$\begin{aligned} k_1 &= \cos^3\left(\frac{\theta}{2}\right) K_\text{I}, \\ k_2 &= \sin\left(\frac{\theta}{2}\right) \cos^2\left(\frac{\theta}{2}\right) K_\text{I}. \end{aligned} \tag{6.11}$$

Deflection of the main crack in the direction of the interface was reported, particularly for short crack lengths in the SCS-6/Ti-15-3 composite [64, 68].

Matrix Crack Shielding by Fibers Several analyses [8, 85] have revealed that when a crack approaches a fiber or an elastic boundary stiffer than the medium where the crack propagates, the crack tip opening displacement (at a given stress intensity) decreases as the crack tip comes closer to a fiber manifested by a larger stress transferred from the matrix to the fiber (FCE). This is due to the constraint effect provided by the fibers on crack tip plasticity while the fiber–matrix interface is unbroken. Detailed analysis is provided later.

Crack Bridging by Unbroken Fibers In fiber bridging modeling there are two basic groups of models. Those based on a direct connection between FB and crack propagation energy and those that are treating FB as a distinct phenomenon.

The first work to consider the steady-state cracking of matrix cracks in a ceramic matrix composite during monotonic loading was published by Aveston et al. in 1971 [86]. The model, also known as the ACK model, was based on the determination of the matrix stress intensity factor, K_m, through a strain energy balance before and after cracking. Using shear lag assumptions, and full crack bridging, they determined that for conditions of steady-state cracking, the stress intensity factor is independent of the crack length and is controlled only by the transition zone between the crack tip and the onset of steady-state cracking. In terms of the model the matrix is considered load free since the fibers support full load. Additionally, in cases where the crack is partially bridged, the contribution of the unbridged crack portion to the stress intensity factor is negligible.

In 1986 Budiansky et al. [87] suggested a new energy balance approach, similar to the ACK analysis, to describe crack growth. In terms of the model, the fibers are initially bonded, while debonding may be achieved by the passage of the crack. Both the ACK and the Budiansky model are referred to as steady-state fiber bridging (SSFB) models [88].

Another class of models was introduced in 1985 and 1987 by Marshall et al. [81] and McCartney [89], respectively. These models combine continuum fracture mechanics principles and micromechanics analysis to determine stress intensity factor solutions for an arbitrary size matrix crack, subjected to monotonic loading. According to these models (also known as generalized fiber-bridging models, GFB [88]) the friction stresses developed by the intact fibers within the matrix crack wake are idealized by an unknown uniform closure pressure. The evaluation of the closure pressure in the GFB models is obtained by combining crack opening displacement solutions from continuum fracture mechanics and from micromechanics analysis. Even though the models differ from each other in the methodology followed to relate those two issues, identical steady-state solutions (as derived from the SSFB models [86, 87]) are used as boundaries to characterize K_m. The formulation of the GFB models as applied to fatigue loading was developed by McMeeking and Evans [90].

According to the GFB models, the restraining effect of the fiber causes a reduction in both the crack surface displacements and the crack tip stresses. Based on the Marshall, Cox, and Evans analysis (also known as MCE) [81], the composite stress intensity factor is defined by superposition of the normal stress intensity factor due to the remote stress on an unbridged crack and that due to the friction stresses due to fiber bridging. Using micromechanics analysis, the friction stresses were idealized as continuous, but with varying distributed crack flank pressure. The friction or bridging stress in relation to the fiber stress is given by

$$p(x) = \sigma_f(x)V_f, \tag{6.12}$$

where $p(x)$ is the crack flank pressure, $\sigma_f(x)$ is the fiber stress at a given distance x from the crack mouth, and V_f is the fiber volume fraction. Equation 6.12 could be considered valid only in cases where at least one fiber is positioned within the crack.

The friction stress can be related to the stress intensity factor, K_{tip}, through a modified Sneddon–Lowengrub equation [91]. The Sneddon–Lowengrub equation describes in a convenient form cracks in infinite bodies, loaded by arbitrary crack flank pressure distributions. In the case of a straight crack embedded in an infinite medium, K_{tip} is written as [88]:

$$K_{\text{tip}} = \sqrt{\frac{4a}{\pi}} \int_0^1 \left\{ \frac{\sigma_\infty - p(X)}{\sqrt{1 - X^2}} \right\} dX, \tag{6.13}$$

where a represents the crack length, $p(X)$ is the friction stress at X, X is the normalized distance along the crack length defined as $X = x/a$, and σ_∞ is the remote stress.

To evaluate Eq. 6.13, friction stresses are related to the crack opening displacement (COD) $p(x) \propto \sqrt{u(x)}$. Such direct relation is based on the assumption that there is strain compatibility between the fiber and the matrix in the slip regime while outside the slip regime the effect of the shear stress is negligible (Fig. 6.15).

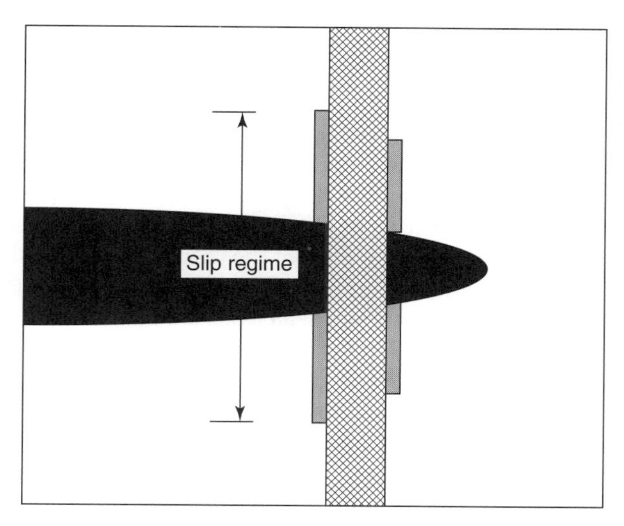

FIGURE 6.15 Schematic representation of slip regime. Note that length of slip regime is not constant, but it increases as crack propagates further and bridging fibers remain intact.

According to MCE the closure pressure is given by:

$$p(x) = 2\left[\frac{u(x)\tau V_f^2 E_f E_c}{R(1 - V_f)E_m}\right]^{1/2}, \tag{6.14}$$

where $u(x)$ is the COD at x, τ is the shear resistance of the interface, R is the fiber radius, and E_f, E_c, E_m represent the elastic moduli of the fiber, composite and matrix, respectively.

As an improved solution to the MCE closure pressure, the shear lag model was further modified by McCartney [89] in order to make the model energetically consistent. The MCE closure pressure was calculated as:

$$p(x) = 2\left[\frac{u(x)\tau V_f^2 E_f E_c^2}{R(1 - V_f)^2 E_m^2}\right]^{1/2}. \tag{6.15}$$

Moreover, later numerical comparisons between the two models have revealed that the crack opening displacement profile obtained by Eq. 6.15 is identical to that obtained by Eq. 6.14 only if the shear resistance is reduced by a factor of 3.2 [79]. Additionally, the lack of a standard method for obtaining the correct shear resistance (it was mentioned before that different methods could produce great variations) urged Kantzos [92] to suggest an alternative solution [known as the fiber pressure model (FPM)] for the determination of the closure pressure. The closure pressure in FPM is assumed to be equal to the stress carried by the fibers in the bridged region averaged out over the total bridged area $(a - a_0)$. The problem of the shear stress parameter was overcome by suggesting

$$p(x) = \sigma_\infty\left\{\frac{w}{w - a_0} + \frac{6\, w a_0[0.5(w - a_0) - (x - a_0)]}{(w - a_0)^3}\right\}, \tag{6.16}$$

where w is the specimen width, a_0 and a are the initial notch length and total crack length, and x is the distance to the bridged area measured from the free surface.

Even though most of the models described above do capture the essential features of fatigue damage, that is, matrix cracking and fiber–matrix debonding, there are several limitations to the ability of the models to predict the crack driving force: (a) the friction stresses are idealized as a continuum closure pressure; (b) the one-dimensional micromechanical analysis used to relate COD and closure pressure is based on the assumption that the fiber in the wake of the matrix crack is far from the crack tip, and so any crack tip effect is negligible [88]; (c) the complex and time-dependent conditions ahead of the crack tip are not modeled (i.e., fiber failure, extensive debonding, and the corresponding differences in composite fatigue resistance); (d) the models claimed to be applicable to situations in which cracks are long compared to the distance between two successive fibers (interfiber spacing) and partially bridged [81]; and finally (e) high crack tip plasticity coupled with crack tip–fiber interactions are difficult to reconcile with linear elastic fracture mechanics (LEFM), since matrix small-scale yielding and homogeneous continuum mechanics principles are violated in composites [88].

To overcome similar problems, de los Rios et al. [4] suggested that the fatigue crack growth in uMMCs should be addressed in terms of the crack tip opening displacement (CTOD). The model, originally developed by Navarro and de los Rios [93], implements the representation of the crack and its plastic zone by means of dislocations subjected to an applied stress σ, as first developed by Bilby et al. [47] in 1963 for monolithic materials. In its original form, the model considers infinitesimal dislocations distributed within two regions or zones, one for the crack itself and the other for the plastic zone. In 1995 [4], the model was extended further by considering a third zone to represent cases where the plastic zone is blocked by grain boundaries (this is common for short cracks). Such an approach was argued to be more realistic in physical terms since in the two-zone system an infinite stress level is sustained by the grain boundary. In terms of the three-zone system, also known as three-zone micromechanical model (TZMM), the plastic zone size (slip band ahead of the crack tip) is blocked by the grain boundary and remains blocked until the stress in the third zone, that is, the grain boundary, attains the required critical level for dislocations to cross this zone. Adapting the same system for a uMMC, the three zones of the crack system are: (a) the crack, (b) the plastic zone, and (c) the plastic constrained zone at the fiber (Fig. 6.16).

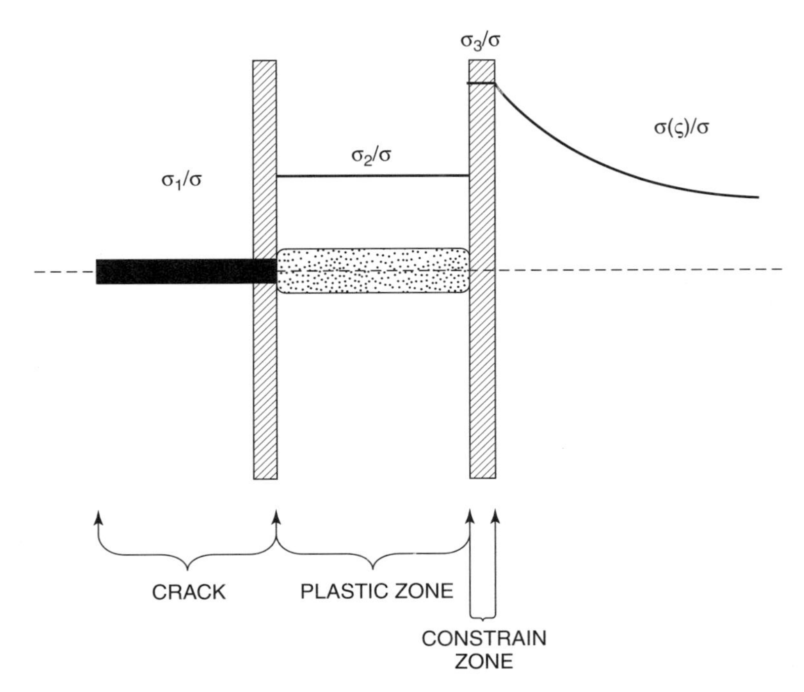

FIGURE 6.16 Three-zone system according to de los Rios et al. [4].

The advantages of such representation are: (a) the model considers plastic displacements throughout the crack system and so the near-tip plasticity effects are included; (b) the model can be effectively applied for short and long cracks; (c) the model accounts also for high crack tip plasticity (large-scale yielding), since it is based on elastoplastic principles; and (d) the friction stresses in the crack wake are not idealized as a continuum closure pressure but as point loads (can apply also for minimal degrees of anisotropy).

In terms of the model (Fig. 6.17), the fiber diameter is represented by d, the interfiber spacing by D, and the crack length by $2a$. Considering only the positive half of the crack system, the crack tip is at a, the front of the fiber at $iD/2 - d/2$, and the back of the fiber at $iD/2 + d/2 = c$. In terms of dimensionless coordinate system $x/c = \zeta$, the crack tip is at n_1, the fiber front at n_2, and the fiber back is at 1. The factor i represents the number of half fiber spacings in the crack system, that is, $i = 1, 3, 5, \ldots$

In the case where the crack is subjected to an applied stress σ in mode I, the stresses in each zone are: (a) σ_1 in the crack zone (friction stress due to fiber

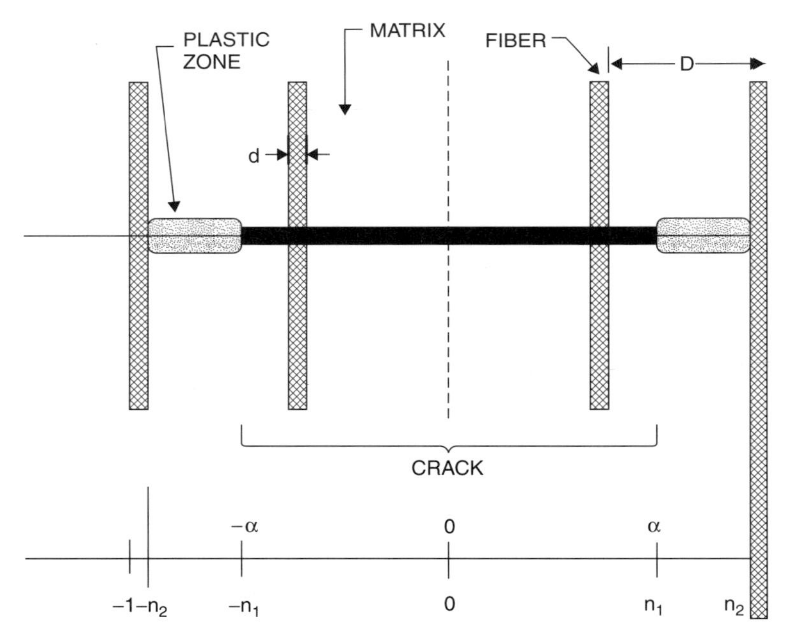

FIGURE 6.17 Schematic representation of crack system. Crack length is denoted by $2a$, the fiber diameter by d and the fiber spacing by D. Considering only the positive coordinates side, crack tip is positioned at a, the plastic zone extends to the next fiber ahead of the crack tip at $iD/2 - d/2$ and the fiber plastic constrained zone at $iD/2 + d/2 = c$, $i = 1, 3, 5\ldots$ Dimensionless coordinate ζ describes position throughout, in particular, $\zeta = n_1$ at the crack tip, $\zeta = n_2$ at the plastic zone, and $\zeta = 1$ at the end of fiber.

bridging), (b) σ_2 is the flow response in the plastic zone (resistance to plastic deformation), and (c) σ_3 in the fiber zone ahead of the plastic zone (constraint effect provided by the fiber to matrix microplasticity). A typical stress distribution along the crack system is shown in Fig. 6.16. The stress σ_1 is considered zero for a nonbridged crack and nonzero for a bridged crack. The plastic zone is assumed to be blocked by the fibers, which impedes any matrix plastic displacement at the fibers. The effect of this constraint is the development of stress σ_3 in the matrix between the fibers of a row, which on achieving a critical value, resolved along the fiber–matrix interface, will cause debonding.

The solution of the equilibrium equation of all the forces, internal and external, acting in the three-zone system, was obtained by Navarro and de los Rios [85, 93] and gives the expressions for the COD $\equiv \phi$ over the entire crack system, and for the stress σ_3. These are as follows:

$$\begin{aligned}
\phi = \mathrm{COD} = \frac{bc}{\pi^2 A}\Bigg\{&(\sigma_2-\sigma_1)\left[(\zeta_b-n_1)\cosh^{-1}\left(\left|\frac{1-n_1\zeta_b}{n_1-\zeta_b}\right|\right)\right.\\
&\left.-(\zeta_b+n_1)\cosh^{-1}\left(\left|\frac{1+n_1\zeta_b}{n_1+\zeta_b}\right|\right)\right]-(\sigma_2-\sigma_1)\left[(\zeta_a-n_1)\cosh^{-1}\left(\left|\frac{1-n_1\zeta_a}{n_1-\zeta_a}\right|\right)\right.\\
&\left.-(\zeta_b+n_1)\cosh^{-1}\left(\left|\frac{1+n_1\zeta_a}{n_1+\zeta_a}\right|\right)\right]+(\sigma_3-\sigma_2)\left[(\zeta_b-n_2)\cosh^{-1}\left(\left|\frac{1-n_2\zeta_b}{n_2-\zeta_b}\right|\right)\right.\\
&\left.-(\zeta_b+n_2)\cosh^{-1}\left(\left|\frac{1+n_2\zeta_b}{n_2+\zeta_b}\right|\right)\right]-(\sigma_3-\sigma_2)\left[(\zeta_a-n_2)\cosh^{-1}\left(\left|\frac{1-n_2\zeta_a}{n_2-\zeta_a}\right|\right)\right.\\
&\left.\left.-(\zeta_a+n_2)\cosh^{-1}\left(\left|\frac{1+n_2\zeta_a}{n_2+\zeta_a}\right|\right)\right]\right\},
\end{aligned} \tag{6.17}$$

$$\sigma_3 = \frac{1}{\cos^{-1} n_2}\left\{(\sigma_2-\sigma_1)\sin^{-1} n_1 - \sigma_2 \sin^{-1} n_2 + \frac{\pi}{2}\sigma\right\}, \tag{6.18}$$

where b is the Burgers vector, $A = Gb/2\pi$ for screw dislocations, or $A = Gb/2\pi(1-\nu_m)$ for edge dislocations, G, ν_m are the shear modulus and the Poisson's ratio of the matrix, respectively, and σ is the applied stress.

If crack growth is considered to be a function of the CTOD, ϕ_t, through a Paris-type relationship, then Eq. 6.17 determines da/dN when $\zeta_a = n_1$, $\zeta_b = 1$ [n_1 represents a dimensionless measurement of crack length, i.e., $n_1 = a/(iD/2 + d/2)$]. In addition, Eq. 6.18 establishes the condition for crack propagation across the fiber row when the axial stress σ_3 acting around the fiber is equal to the stress required for debonding, σ_{3d}. Clearly, $\sigma_3 = \sigma_{3d}$ acknowledges the condition when the clamping stress provided by the fibers to plastic displacements within the plastic zone are removed, since no or minimum interfacial shear stress is acting along the debond length.

Assuming that debonding is not a continuous process (propagation of a bimaterial interface crack), then the stress at the fiber zone required to debond a particular fiber length could be written as:

$$\sigma_{3d} = \frac{\sigma_d E_c}{E_f}, \tag{6.19}$$

where σ_d is the tensile stress at the fiber required to debond a particular embedded fiber length, E_c and E_f are the elastic moduli of the composite and the fiber, respectively.

Equation 6.19 is obtained by considering a simple force balance in a fiber push-out test and strain compatibility between the fiber and the composite in the fiber zone. If the interfacial shear strength is taken as constant along the fiber–matrix interface and the shear-lag analysis is utilized, the stress applied at the fiber to cause debonding is obtained as:

$$\sigma_d = \frac{4\tau l}{d}, \tag{6.20}$$

where l is the embedded fiber length, d is the fiber diameter, and τ is the interfacial shear strength. Published results in the literature indicate values for interfacial shear strength of SCS-6/Ti-15-3 and SCS-6/Ti-6-4 in the range of 124–148 MPa and 138–156 MPa [31, 94]. Furthermore, experiments have shown that the interfacial shear strength increases for longer embedded fiber length or thicker specimens and gradually approaches a constant value for fiber lengths approximately 4–5 times the fiber diameter ($d = 140$ μm for the Textron SCS-6 [95]) [46]. Nevertheless, Eq. 6.19 is limited by the composite ultimate tensile strength for debonding to be achieved before fiber failure. Such a premise can be considered as the distinction between a strong and a weak interface bond.

To determine the flow response of the composite is a difficult task. This is due to uncertainties raised by the presence and contribution of the fiber within the plastic zone [69]. At short crack lengths, close to the interfiber spacing, it is rational to assume that the crack tip plastic zone is fiber free and hence the flow response of the uMMC is matrix flow dependent. However, as the crack length increases, the possibility of having a number of fibers within the plastic zone is strong and rational. A better understanding of the way fibers are entering the plastic zone is provided by the following steps, recognized by the TZMM. The condition for crack propagation is achieved as follows: with the crack tip positioned between two fibers as shown in Fig. 6.17, the level of the stress σ_3 is given by Eq. 6.18. On further crack growth the level of σ_3 increases due to the increase in n_1^i until σ_3 attains the value for debonding, derived by Eq. 6.19. This is happening at a critical crack length defined by n_1^{ic}. The value of n_1^{ic} and, therefore, the crack tip position at the critical point, is obtained by substituting Eq. 6.19 into Eq. 6.18 and solving for n_1^i. At this point the crack tip plasticity constraining effect of the fiber is overcome since the plasticity is now allowed to pass around the fiber and become constrained once again by the next fiber. This behavior, which is in direct agreement with work published by Schulte and Minoshima [7], justifies that the major factor controlling the fatigue resistance of the material, especially at short crack lengths, is the plasticity constraining effect (PCE). The above are schematically shown in Fig. 6.18.

Undoubtedly, an accurate application of Eq. 6.18 requires a sound determination of the composite flow stress. If we considered that the crack is long

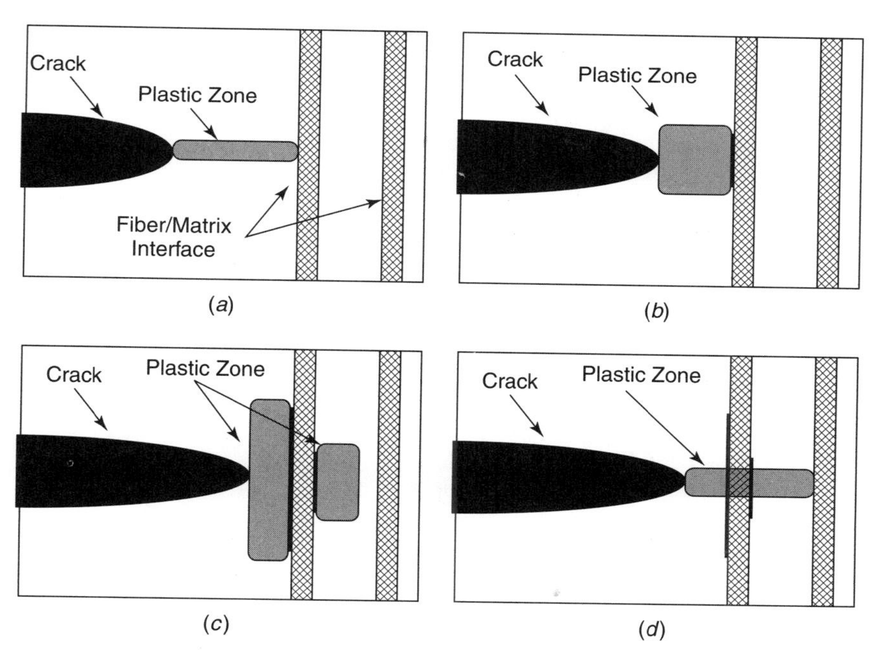

FIGURE 6.18 Conditions for crack propagation. At (*a*) the crack tip plastic zone contacts the high stiffness fiber. As the crack propagates further, the plastic zone is squeezed between the crack tip and the intact fiber (*b*). As a result a tensile stress is starting to build up around the fiber (constrain effect). When constrain effect has become sufficient to initiate debonding (*c*), the plastic flow propagates round the fiber and the constrain effect relaxes (*d*).

enough to contain fibers, then the flow response ahead of the crack tip would be controlled by the matrix yield stress and by the high stiffness phase. Such collaboration is justifiable by examining stress–strain curves of the constituent materials, (Fig. 6.19).

The fact that similar behavior to that shown in Fig. 6.19 has been observed in most uMMCs convinced many workers to accept that an isostrain condition between the fiber and the matrix within the plastic zone is somehow justifiable [48, 96, 97]. Based on the above, the flow response of the uMMC can be written as [96]:

$$\sigma_2 = \frac{\sigma^t_{ym}}{E_m} E_f V_f + \sigma^t_{ym}(1 - V_f). \tag{6.21}$$

Assuming that debonding is always achieved before fiber failure, the crack propagates through the matrix without breaking the fibers. Subsequently, intact fibers located behind the crack tip slide in relation to the matrix, producing friction (bridging) stresses σ_1, which reduce the crack driving force and improve the fatigue resistance of the MMC. The friction stress at each bridged fiber row

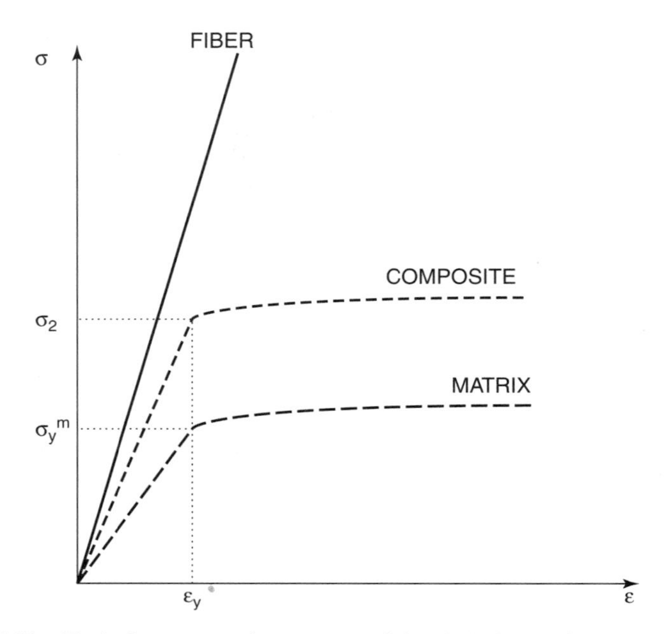

FIGURE 6.19 Typical stress–strain response of the uMMC constituent material. Yielding response of matrix also defines yielding response of composite.

in the crack zone is calculated as follows [48]:

$$\sigma_1 = \frac{\text{COD}\ G}{(l_s + \text{COD})N_N}, \tag{6.22}$$

where N_N is the number of fibers per row, l_s is the sliding distance, and G represents the matrix shear modulus. Equation 6.22 is obtained considering displacement (strain) compatibility between the fiber and the matrix at the interface (Fig. 6.20). Additionally, Eq. 6.22 acknowledges that all the fibers in the same row are subjected to an equal strain.

An expression for the sliding distance, l_s, is given in [98]

$$l_s = 2\sqrt{\frac{\text{COD}\ V_m E_m E_f d}{4\tau E_c}}, \tag{6.23}$$

where E_c is the Young's modulus of the composite (calculated by the rule of mixtures), V_m is the volume fraction of the matrix, and τ is the interfacial shear stress.

Since the COD depends on the value of σ_1, a numerical iterative method is required for the calculation of σ_1. Initially a $\sigma_1 = 0$ value is adopted, and then an interim value of σ_1 is calculated. The iterations are repeated until there is

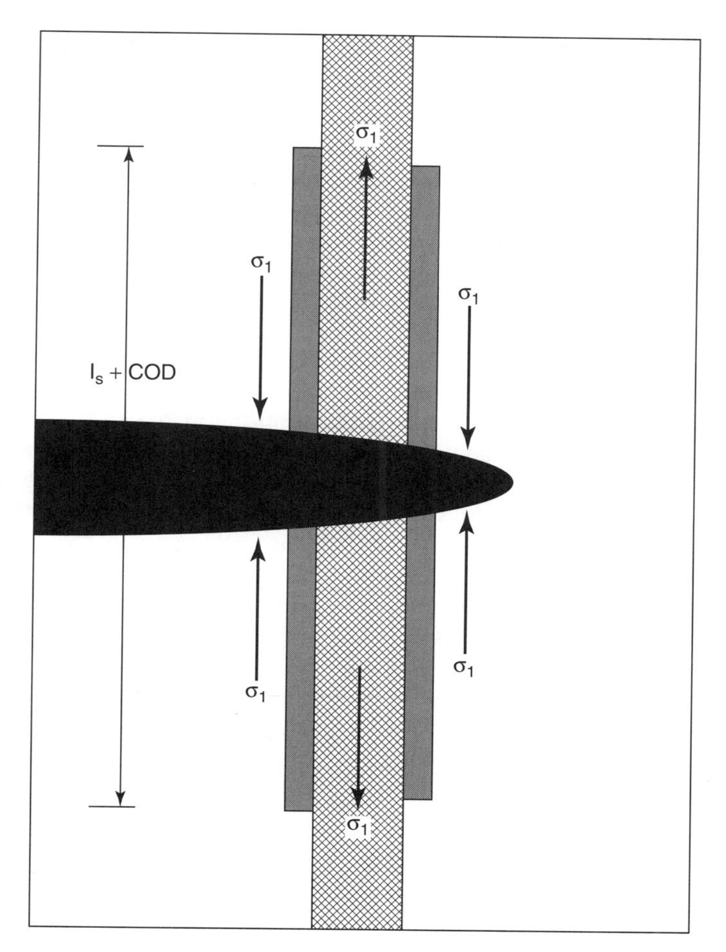

FIGURE 6.20 Schematic of strain compatibility setup at bridging fiber.

no difference between successive values of σ_1. Once the friction stresses, acting at the interface, have been determined, the stress at each fiber row, σ_f, can be evaluated [99]:

$$\sigma_f = \frac{4\sigma_1 l_s E_c}{d V_m E_m}. \tag{6.24}$$

Equation 6.24 assumes that all the fibers in the same row exhibit the same sliding distance. Since the friction stresses depend on COD, a numerical iterative method between 6.22 and 6.24 is required for the calculation of σ_f. Moreover, Eq. (6.24) is controlled by the strength integrity of the fiber, and therefore an additional equation should be included in the iteration. In general, the average strength of fibers of a given strength distribution and a particular gage length and fiber diameter is calculated by the Weibull function for average strength and is written as [100, 101]:

$$\sigma_{\mathrm{fr}} = \sigma_0 \left(\frac{L}{d}\right)^{-1/m} \Gamma\left(\frac{m+1}{m}\right), \tag{6.25}$$

where m is the Weibull modulus, L is the gage length, σ_{fr} is the average fiber strength, σ_0 is the normalizing factor, and Γ is a tabulated gamma function. A typical behavior of the most commonly used Textron SCS-6 fiber is given in Fig. 6.21.

The length L over which possible fiber failure should be expected is the sliding distance given by Eq. 6.23. In this region, shear tractions are developed at the fiber ends, allowing stress to be transferred from the matrix to the fibers. Such behavior, increases the probability of failure within the sliding distance, while outside of this region fiber strength can be considered invulnerable. Additionally, the matrix plastic displacement, COD, at the fiber, should also be considered in the sliding distance calculations. Even though the COD region is characterized by the absence of interfacial shear stress and consequently defines a different statistical environment, for simplicity reasons, it is assumed that the interfacial shear stress τ is also acting along the COD (COD very small compared to l_s). Thus, the average strength of the debonded fibers in the sliding region can be evaluated by considering a gage length equal to a sliding distance, $L = l_s + \mathrm{COD}_{\mathrm{cr}}$ ($\mathrm{COD}_{\mathrm{cr}}$ defines the critical displacement at the time of fiber failure). The evaluation of the $\mathrm{COD}_{\mathrm{cr}}$ in respect to typical strength data provided by the manufacturer is

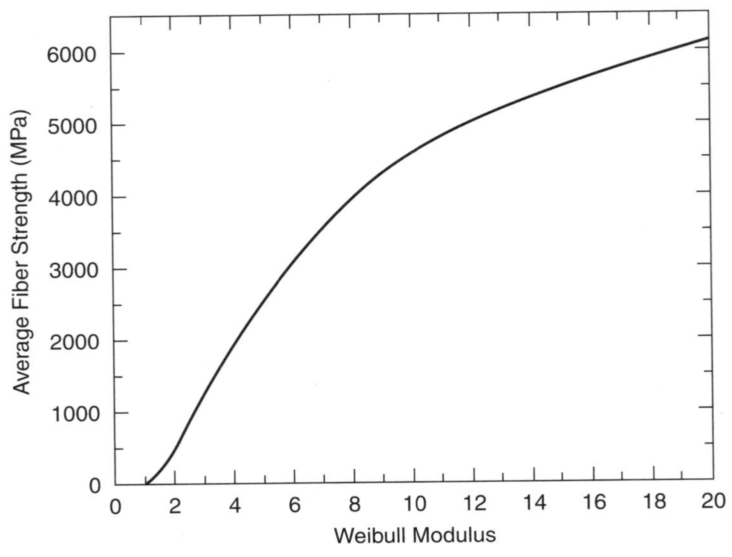

FIGURE 6.21 Average strength of SCS-6 fiber for several values of Weibull modulus.

achieved by employing the Masson and Bourgain [101] estimation of the Weibull probability:

$$\sigma_{2L_2} = \sigma_{1L_1} \left(\frac{L_1}{L_2} \right)^{1/m}, \tag{6.26}$$

where σ_{1L_1}, σ_{2L_2} are the average strengths at a given failure probability for gage lengths L_1 and L_2 respectively. Assuming that $\sigma_f = \sigma_{fr}$, where σ_{fr} is the fiber fracture strength, and applying an iterative method between Eqs. 6.22– 6.26, the critical COD necessary to fracture the fibers, is derived as a function of the Weibull modulus m (Fig. 6.22).

The above methodology makes clear that: (a) the interfacial shear strength and the debonding process, consequently, is responsible for the existence of bridging fibers; (b) the effectiveness of fiber bridging is controlled by the matrix material (G) and the strength of the fiber, and (c) the bridging life of each fiber row decreases for longer crack lengths (COD $\rightarrow \infty$).

6.4.3 Damage-Tolerant Design

Accurate knowledge of the size of a crack in a structure and its propagation rate as a function of the applied stress–strain field are the major ingredients for a fatigue damage-tolerant approach (FDTA). FDTA has been originally developed by the

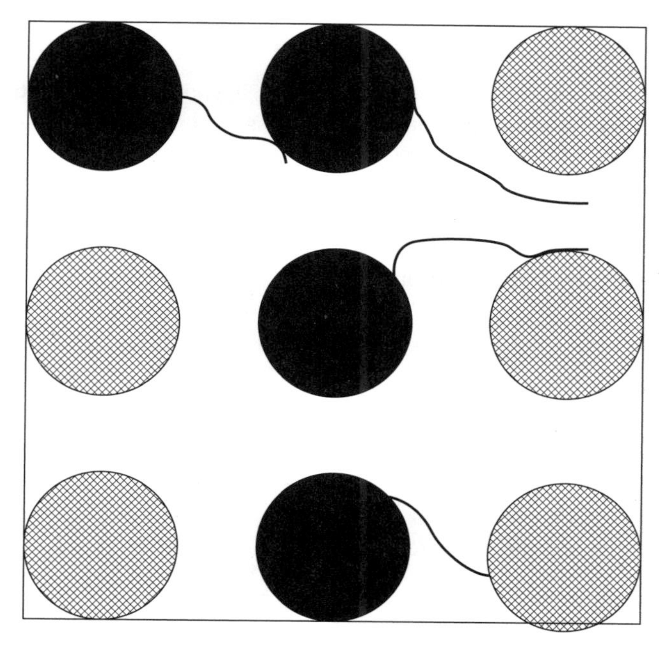

FIGURE 6.22 Schematic representation of radial cracks emanating from fiber breaks due to residual stresses.

airframe industry to ensure local failure within fail-safe boundaries previously established through failure analysis techniques. In this sense, linear elastic fracture mechanics (LEFM) is used to relate crack size with nondestructive and fracture surface observations in order to estimate load history and typical service life. Thus, the application of fracture mechanics methods as part of damage-tolerant assessment involves the consideration of four parameters: (a) initial cracks ($a_{\rm in}$); (b) threshold stress intensity factor for crack propagation ($K_{\rm th}$); (c) the final crack length ($a_{\rm f}$); and (d) the crack growth rate (da/dN). Typical FDTA techniques can be found elsewhere [102, 103].

6.4.3.1 Initial Cracks

Effect of Residual Stresses One of the main factors affecting the performance of uMMCs is the development of residual stresses during manufacturing. These stresses, which arise from the difference in the thermal expansion coefficients (CTE) between the metallic matrix and the fiber, may be responsible for the presence of radial cracks observed in the as-received condition [104, 105] (Fig. 6.22).

In most uMMCs, the matrix CTE, α_m, exceeds that of the fiber, α_f, thus, after cooling from the processing temperature, the matrix is subjected to an axial tensile stress σ_m^r. Large matrix tensile stresses may induce premature crack initiation, especially in the presence of external loading [68]. The residual thermal stresses in the matrix can be evaluated as [104]:

$$\sigma_m^r = \frac{V_f E_m E_f \ \Delta\alpha \ \Delta T}{E_c}, \qquad \Delta\alpha = \alpha_m - \alpha_f, \tag{6.27}$$

where ΔT is the effective temperature range. It should be noted that residual stresses developed during the fabrication process at absolute temperatures greater than half of the melting point of the matrix, are not considered as a result of the matrix stress relaxation due to creep or viscoplastic flow [62]. Furthermore, when the matrix contracts more than the fiber $\Delta a > 0$, as in the case of a single long fiber embedded in an infinite matrix, the matrix is placed in hoop tension, which can be simulated by Lamé distribution [106]:

$$\sigma_m^\theta = \frac{1}{2}\sigma_m^r \left(\frac{R}{x}\right)^2, \tag{6.28}$$

where R is the fiber radius and x is the distance from the fiber center. Equation 6.28 proclaims that the maximum stress due to the relaxation of the residual stresses is situated at the broken fiber radius, $x = R$. Furthermore, Eq. 6.28 can be expressed in terms of stress intensity factor by including the geometric features of broken fibers at the edge and the projected length of an initial crack:

$$K = Y\sigma_m^\theta \sqrt{\pi a_{\rm in}}. \tag{6.29}$$

Generally, maximum matrix cracking as a result of residual stress relaxation is expected in the case of broken fibers (due to machining). To encounter the effect of such geometric discontinuities, elastic stress concentration formulas are usually used. Considering a semicircular corner crack configuration to represent the geometry of a broken fiber, Eq. 6.28 can be written as [107]:

$$\sigma_m^\theta = \frac{1}{\pi}\sigma_m^r\left(\frac{R}{x}\right)^2 \tag{6.30a}$$

or in terms of stress intensity factor as:

$$K_m^r = \frac{1}{\pi}\sigma_m^r\left(\frac{R}{a_{\text{in}}}\right)^2\sqrt{\pi a_{\text{in}}}, \tag{6.30b}$$

where $x = a_{\text{in}}$ represents the statistical average of the projected area of defects on a plane normal to the maximum σ_m^r. To evaluate a_{in}, the right part of Eq. 6.30 is set to be equal to the matrix fracture toughness, K_c^m. Typical results of such methodology can be found in Fig. 6.23.

Even though the size of the radial cracks is small compared to the size of a broken fiber, at short crack lengths such cracks could make a difference in terms of crack arrest, as shown later [10, 11]. It should be noted that the above represents a typical practice for the evaluation of radial cracks. For a more thorough evaluation residual stress interactions, with neighboring broken or not fibers, should also be considered.

Effect of Manufacturing Defects Many workers [68, 108] have observed that manufacturing defects in uMMCs are characterized by varying sizes and types.

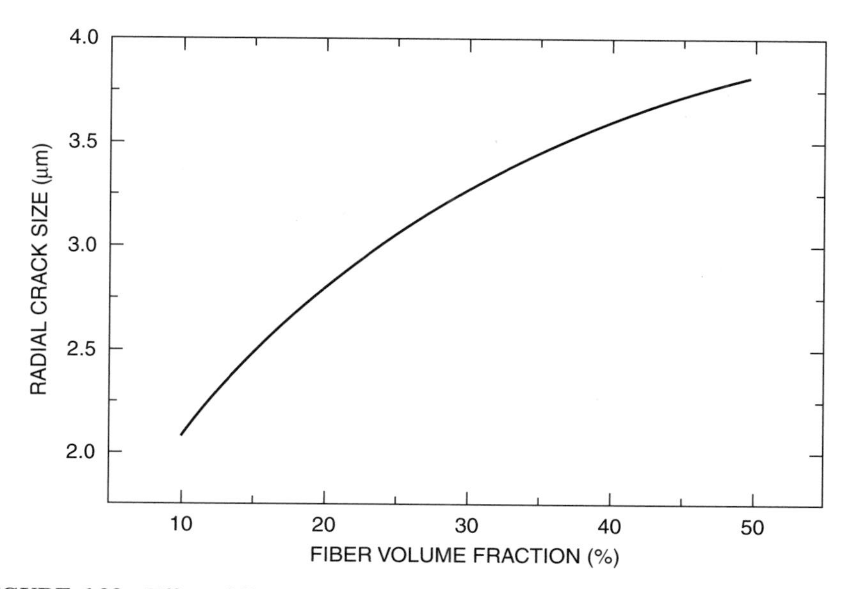

FIGURE 6.23 Effect of fiber volume fraction on radial cracks for SCS-6/Ti-15-3 uMMC.

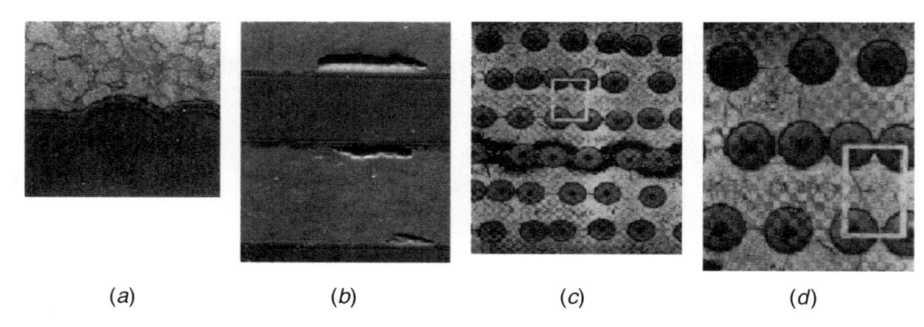

(*a*) (*b*) (*c*) (*d*)

FIGURE 6.24 Typical sources of initial crack like defects for SCS-6/Ti-15-3 uMMC: (*a*) Fiber warts, (*b*) voids at interface, (*c*) broken fibers at edges, (*d*) touching fibers. (Photos taken from [68]).

From a structural point of view, those favorable to initiate fatigue cracks are: (a) broken fibers at the edges (damaged during machining), (b) voids in the interface, (c) broken reaction layers, (d) warts on the fibers, and (e) fiber touching (Fig. 6.24).

In general, manufacturing defects in uMMCs are divided into two categories, the surface defects and the through-thickness defects. Undoubtedly, surface defects (broken fibers and touching fibers) can be easily identified by microscopic inspection after minimum surface preparation. The number of broken fibers per surface unit (the surface unit consists of a rectangle with fiber centers at the corners) can provide useful information regarding the uniformity and concentration of such defects, (Fig. 6.24*d*). On the other hand, to reveal through-thickness defects (fiber warts, voids, etc.) careful surface etching until the fiber layer should be employed. The quantification of such defects, due to the lack of a specific recommendation, is subjected to personal assessment. In [69], however, it was suggested that the number of voids per critical fiber length could provide some severity indication of these defects.

Undoubtedly, surface defects are in direct connection to the fatigue endurance of the material. This is because broken fibers at edges, touching fibers, and most importantly their combinations could host significant stress concentrations. Such concentrations are represented by circular, semicircular, or quarter-elliptical shape notches [109, 110].

6.4.3.2 Threshold Stress Intensity Factor

In ductile monolithic materials the transition of a short crack or cracklike defect into a catastrophic fatigue crack is related to a boundary condition known as: (a) fatigue limit or (b) threshold stress intensity factor, K_{th}, [111, 112]. In brief, the K_{th} boundary represents crack tip strain conditions able to create plasticity damage at some certain distance ahead of the crack tip.

In uMMCs, the evaluation of a similar boundary condition is a puzzling task due to the number of parameters involved in the fatigue damage process. During the early stages of research, many workers supported the idea that crack arrest in

MMCs can be defined in a way similar to crack arrest in monolithic material [54, 58]. In detail, they assumed that when the crack growth rate is approximately 10^{-8} mm/cycle and no crack progression is detected for at least 10^7 cycles, then conditions of crack arrest prevail. Undoubtedly, such empirical approach is not able to provide numerical solutions and consequently information for design. In 1996, de los Rios et al. [96] published a work where crack arrest (threshold) is achieved when the crack strain conditions cannot overcome the FCE and therefore cannot propagate plasticity. In other words, if the crack cannot develop the required shear stresses at the interface to initiate debonding, crack arrest conditions should be assumed. These hypothetical conditions for crack arrest are shown in Fig. 6.25.

In terms of mathematical modeling, crack arrest is achieved when two boundary conditions are met: (a) the crack contacts the fiber (negligible plasticity) and (b) the shear stress at the interface is still lower than the interfacial shear strength. According to Fig. 6.17, the above boundary conditions can be written as:

$$n_1 = n_2 \approx 1,$$

which proclaims that the crack tip plasticity is minimum, and

$$\sigma_3 \leq \sigma_{3d},$$

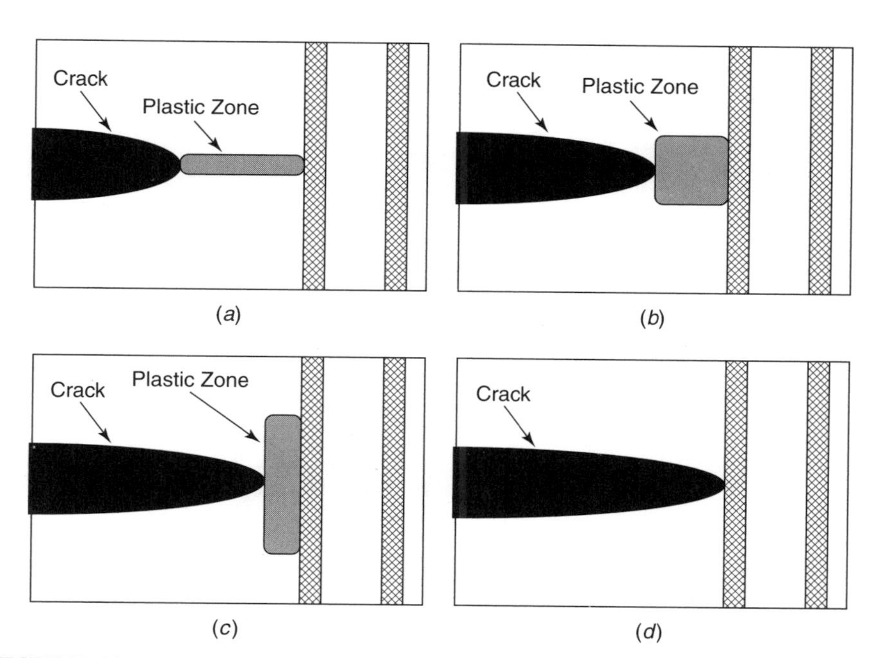

FIGURE 6.25 Conditions for crack arrest. (*a*) FCE starts instance plastic zone contacts the fiber. (*b*) and (*c*) Further propagation of crack against fiber results into plastic zone condensation and higher shear stresses at interface. (*d*) Crack arrest.

which proclaims that the developed shear stress at the interface is lower than the interfacial shear strength. Since $n_1 = n_2 \approx 1$, then $\sin^{-1} n_1 \approx \sin^{-1} n_2 \approx 1$ and $\cos^{-1} n_2 \approx \sqrt{2}\sqrt{d/(a+d+a_{\text{in}})}$. Using these approximations into Eq. 6.18, the maximum allowed applied stress, which would still lead to crack arrest of a particular crack length, a, yields

$$\sigma_{\text{arr}} = \frac{2\sqrt{2}}{\pi}\sigma_{3d}\sqrt{\frac{d}{a+d+a_{in}}} + \sigma_1, \tag{6.31}$$

where σ_{3d} is given by Eq. 6.19 and a_{in} represents defect or notch size.

Equation 6.31, shown in Fig. 6.26, represents a theoretical Kitagawa–Takahashi (KT) curve for the uMMCs. However, in contrast to the monolithic materials where the true fatigue limit is the highest stress level, which is unable to transform a fatigue flaw into a fatigue crack, in uMMCs the true fatigue limit corresponds to the inability of an already established fatigue crack to propagate beyond one or more fiber rows. The accuracy of Eq. 6.31, considering $\sigma_1 = 0$, can be seen in Fig. 6.27.

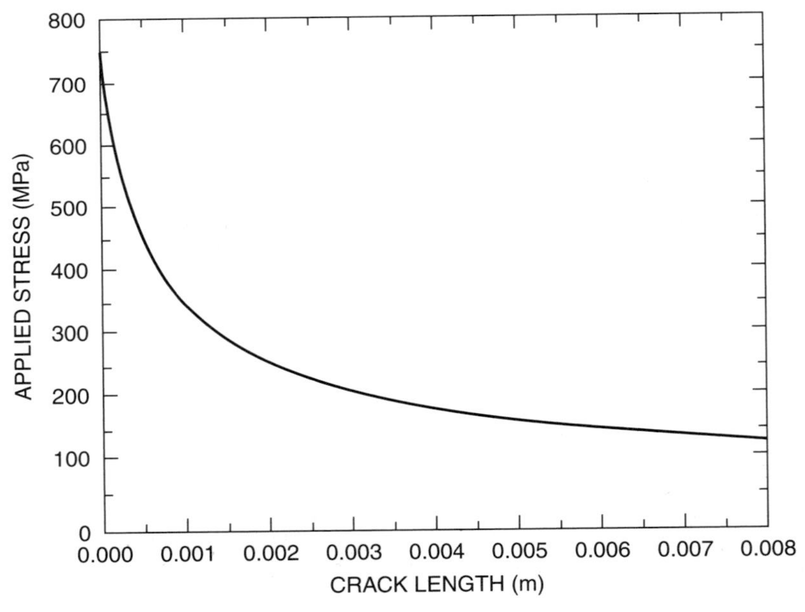

FIGURE 6.26 Crack arrest curve for 32% SCS-6/Ti-15-3 uMMC. For calculations values of $\sigma_{3d} = 1173$ MPa, $a_{\text{in}} = 140$ μm and $d = 140$ μm were used. Negligible bridging stress was considered. Such simplification is reinforced from fact that small cracks are expected to be arrested by first or second fiber row. Consequently, contribution of bridging stress is expected to be negligible or minimum.

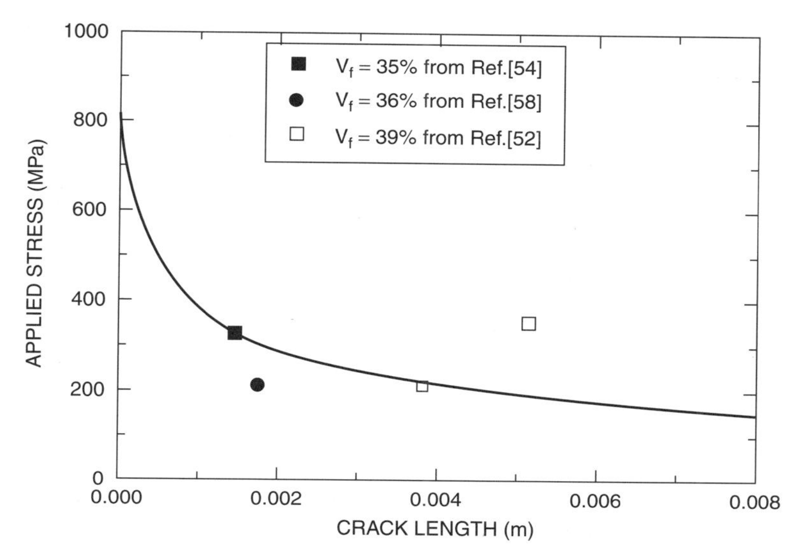

FIGURE 6.27 Comparison between experimental and predicted crack arrest for SCS-6/Ti-15-3. Curve has been created by considering $V_f = 39\%$.

Equation 6.31 in terms of stress intensity factor can be written as:

$$K_{\text{th}} = \frac{2\sqrt{2}}{\pi}\sigma_{3d}\sqrt{\frac{\pi d a}{a + d + a_{\text{in}}}}, \tag{6.32}$$

where K_{th} is the threshold stress intensity factor for MMCs. In Eq. 6.32 the effect of the closure stress σ_1 is disregarded. In Fig. 6.28 the effect of different values of fiber volume fraction on K_{th} is presented.

From Fig. 6.28 it is clear that, for cracks of approximately 1 mm of length, the value of K_{th} asymptotically tends to a constant value. This distinction could represent a transition between a short and a long crack. At this stage we have to make clear that optimization of the crack arrest capacity of the uMMC through $\sigma_{3d}(V_f)$ is limited by the ultimate tensile strength of the material and the desired transverse properties.

6.4.3.3 Final Crack Length

It is still debatable whether uMMCs fail due to conditions of fracture toughness or conditions of general yielding. This is basically because the contribution of the toughening mechanisms in a fiber-reinforced composite, namely interfacial debonding, fiber bridging, and fiber pull-out, decreases with loading cycles [9, 11]. Additionally, the possibility for an accurate analytical determination of the fracture toughness is getting even thinner if we take into account that

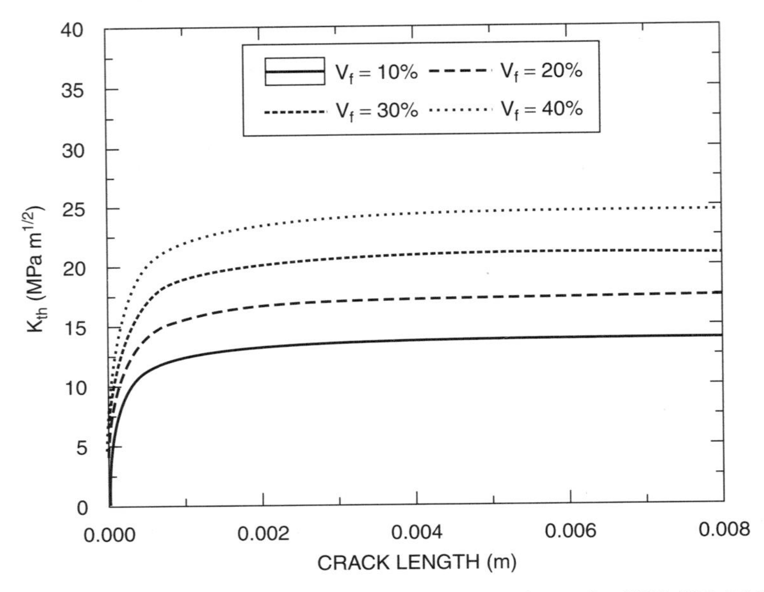

FIGURE 6.28 Predicted threshold stress intensity factor for SCS-6/Ti-15-3.

long cracks can cause failure of the fibers within the plastic zone. Fiber failure at lengths smaller than the critical will result into an immediate reduction of the flow resistance of the material. In [11] it was proposed that the fatigue failure of uMMCs is following a hybrid pattern, constituted by both failure conditions. In the same work it was suggested that as the crack propagates further the significance of the FCE becomes less vital. In other words, there is a particular crack length that could claim the simultaneous debonding of two successive fiber rows [113]. The so-called FCE degradation could deteriorate further to include the simultaneous debonding of a larger number of fiber rows. If a large number of fibers ahead of the crack tip is debonded, the FCE becomes minimum and therefore the crack tip plasticity could follow an unconditional propagation. The term *unconditional* denotes spread of plasticity controlled only by the material's flow resistance. Maximum spread of plasticity and consequently maximum crack tip opening displacement (CTOD $\rightarrow$ max), will result into fast propagation of the crack and the consequent failure of the fibers ahead of the crack tip.

Considering that at the time of crack instability the crack system is still under equilibrium of the stresses, Eq. 6.18 can be used. In terms of the TZMM maximum plasticity is modeled by assuming $n_2 \rightarrow 1$. This boundary condition yields

$$\sin^{-1} n_2 = \frac{\pi}{2} \tag{6.33a}$$

and

$$\cos^{-1} n_2 \approx \sqrt{2}\sqrt{\frac{d}{a+d+a_{in}}}. \tag{6.33b}$$

Using the above approximations, Eq. 6.18 yields

$$(\sigma_2 - \sigma_1)\sin^{-1} n_1 - \frac{\pi}{2}\sigma_2 - \sigma_{3d}\sqrt{2}\sqrt{\frac{d}{a+d+a_{in}}} + \frac{\pi}{2}\sigma = 0. \tag{6.34}$$

Solving for n_1, Eq. 6.34 becomes

$$n_1 = \frac{\dfrac{\pi}{2}\sigma_2 + \sigma_{3d}\sqrt{2}\sqrt{\dfrac{d}{a+d+a_{in}}} - \dfrac{\pi}{2}\sigma}{\sigma_2 - \sigma_1}. \tag{6.35}$$

Since $iD/2 + d/2 \to \infty$, then $n_1 \to 0$ and thus the stress for crack instability for a given crack length is given by rearranging Eq. 6.35:

$$\sigma_{\text{ins}} = \frac{2\sqrt{2}}{\pi}\sigma_{3d}\sqrt{\frac{d}{a+d+a_{\text{in}}}} + \sigma_2, \tag{6.36}$$

Or, by employing Eq. 6.21:

$$\sigma_{\text{ins}} = \frac{2\sqrt{2}}{\pi}\sigma_{3d}\sqrt{\frac{d}{a+d+a_{in}}} + \left[\frac{\sigma_{ym}^t}{E_m}E_f V_f + \sigma_{ym}^t(1 - V_f)\right]. \tag{6.37}$$

Realistically, σ_2 in Eq. 6.36 represents a time-dependent parameter. This is due to the fact that the flow resistance of the material ahead of the crack tip could take significantly high and significantly low values during crack propagation. In fact, when the plastic zone is small, the contribution of the fibers is negligible and therefore $\sigma_2 = \sigma_{ym}^t$. Larger plastic zones, on the other hand, especially those corresponding to cracks above the crack arrest curve (Fig. 6.26), the contribution of the fibers is significant and Eq. 6.21 should be employed. Such representation of the flow resistance will remain the same until the crack is long enough to claim fiber failure within the plastic zone. In this case the effective V_f will start to decrease. The decrease will continue, probably following an asymptotic descent, until the fiber reaches a length smaller than its critical value. At this stage, the V_f^{eff} should be taken as negligible and the flow resistance equal to $\sigma_{ym}^t(1 - V_f)$. Equation 6.36 also makes clear that if all fibers ahead of the crack tip are fractured, then $\sigma_{3d} \to 0$, and instability will occur due to $\sigma_{\text{ins}} = \sigma_{ym}(1 - V_f)$, which signifies conditions of general yielding [47]. Figure 6.29 shows a typical crack instability curve.

6.4.3.4 *Crack Growth Rate*

Undoubtedly, there is still uncertainty regarding the accurate modeling of fatigue crack growth. This is basically because uMMCs behave in a most complicated manner as a result of their time-dependent fracture toughening mechanisms.

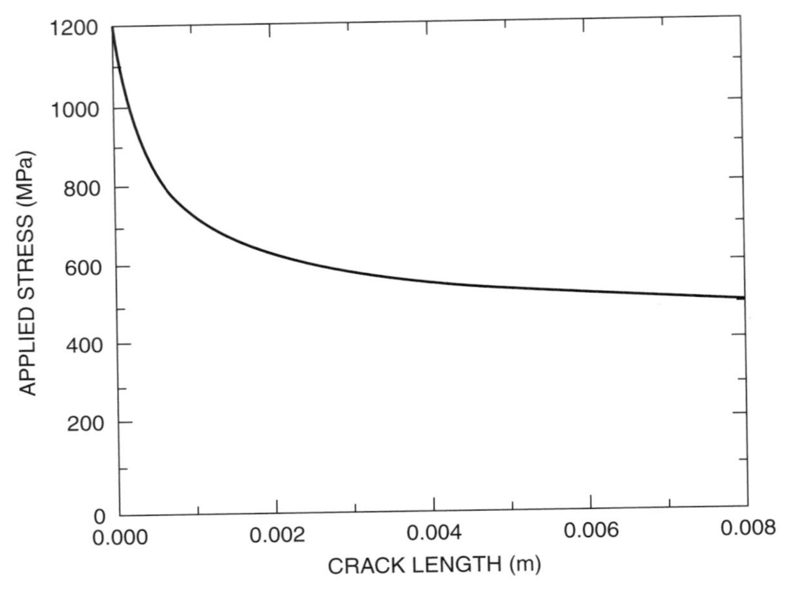

FIGURE 6.29 Crack instability curve for 32% SCS-6/Ti-15-3. Value of $\sigma^{t}_{ym} = 556$ MPa was used.

These complications urged engineers to suggest that accurate modeling using the models presented in Section 6.4.2, is possible only within specific limits that could secure the steady behavior of the toughening mechanisms [69, 96]. Such limitation reflects the incapacity of current fatigue modeling methodologies (LEFM, EPFM) to cope with: (a) fiber failure ahead of the crack tip (degradation of FB), (b) degradation of FCE, and (c) degradation of flow resistance.

In 1996, de los Rios et al. [96] published a work where the TZMM was quoted to operate extremely accurately until a specific crack length. The so-called onset of unsteady crack growth has been previously observed from fatigue experiments conducted in plane and notched SCS-6/Ti-15-3 uMMC specimens [51, 69, 92, 114]. Those experiments showed a distinct change in the slope of the fatigue crack growth rate (Fig. 6.30). The fact that this change takes place well before the final failure of the specimen urged the researchers to assume that such behavior is dominated by fiber failure in the crack wake and not ahead of the crack tip. Ibbotson et al. [51] concluded that this change may mark the initiation of an unstable crack growth. Similar, hypothesis has been upheld by de los Rios et al. [96].

In [11], it was suggested that negligible closure stress and substantial crack length could signify a hypothetical lower bound for fatigue failure (or a typical bound of the operational life of the material). In [10] it was reported that the onset of FB degradation can be predicted by assuming that the COD close to crack tip is equal to the COD_{cr} (Fig. 6.31) and employing a modified version of

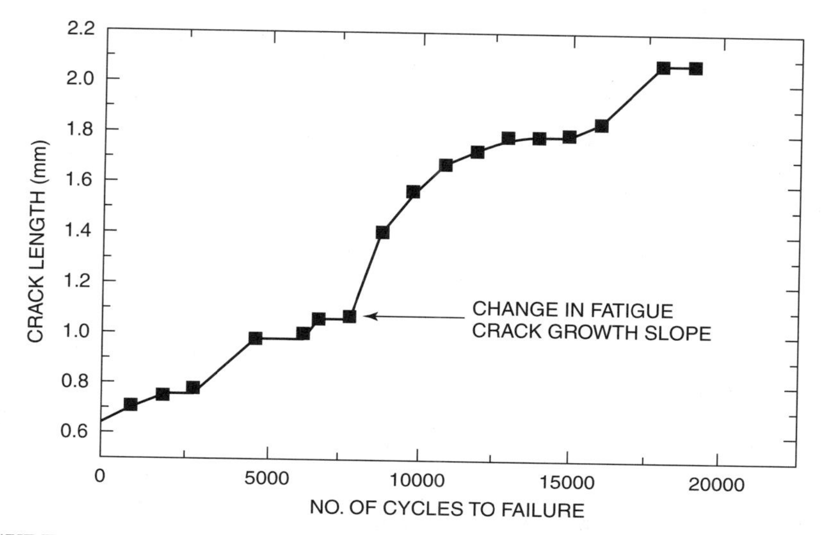

FIGURE 6.30 Crack length vs. loading cycles of 32% SCS-6/Ti-15-3 loaded at $\sigma_{max} =$ 800 MPa ($R = 0.01$) with SEN of 650 μm.

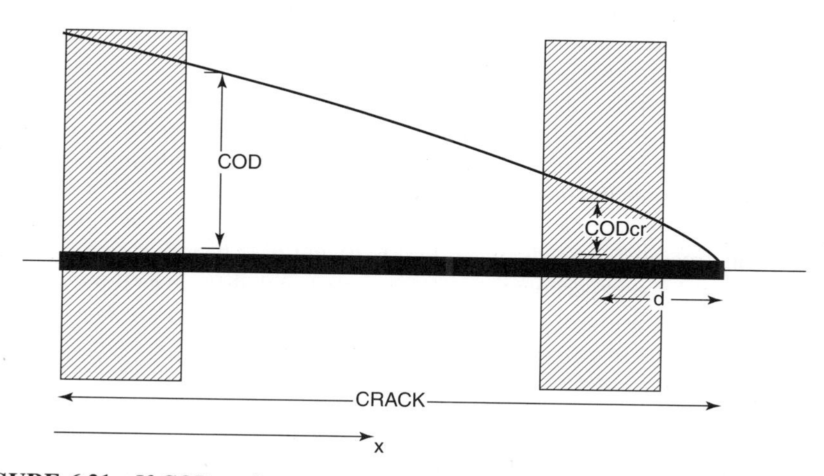

FIGURE 6.31 If COD at distance $x = a - d$ is equal to COD_{cr}, failure of all bridging fibers could be assumed.

the crack-opening solution given by Tada et al. [115]:

$$\mathrm{COD} = \frac{4\sigma}{E_c}[a^2 - x^2]^{1/2} F\left(\frac{x}{a}\right), \tag{6.38}$$

where

$$F\left(\frac{x}{a}\right) = 1.454 - 0.727\left(\frac{x}{a}\right) + 0.618\left(\frac{x}{a}\right)^2 - 0.224\left(\frac{x}{a}\right)^3.$$

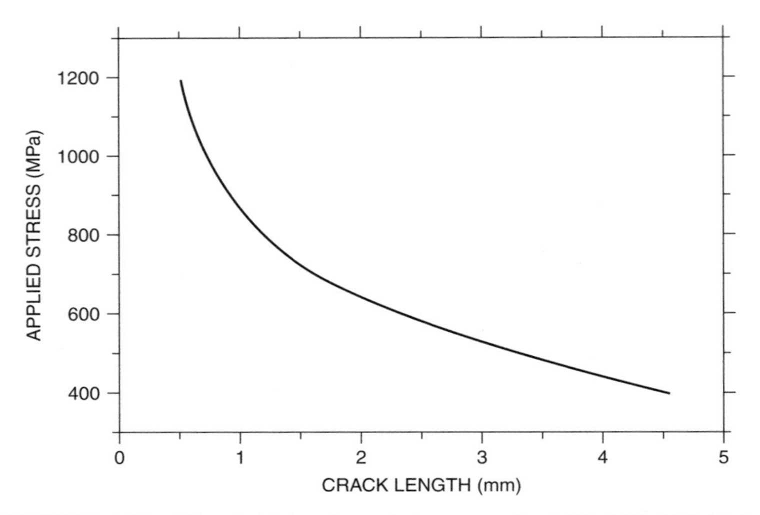

FIGURE 6.32 Fiber-bridging degradation curve for 32% SCS-6/Ti-15-3.

If the COD close to crack tip is equal to the COD_{cr}, it is then rational to assume that all the fibers within the crack wake have failed. A predicted lower bound for fatigue failure is presented in Fig. 6.32.

Figure 6.32 provides the so-called unsteady crack growth behavior of the material. Within the limits of the curve the material is expected to behave in a constant manner since the toughening mechanisms are exhibiting limited and therefore predictable fluctuations. Above the curve, values of FB will start to appear significantly degraded pinpointing the onset of unsteady crack growth. The phenomenal great difference between Fig. 6.29 and 6.32 becomes less significant if we considered that experiments have shown that such difference represents only 15–20% of the total life of the material at most stress levels [69]. It is therefore justifiable, in terms of safety, to accept values of final crack length defined by the FB degradation.

REFERENCES

1. P. C. Paris and F. Erdogan, *J. Basic Eng.*, **85**, 528–534 (1963).
2. W. Elber, *Eng. Fracture Mech.*, **2**, 37–45 (1970).
3. J. Aveston and A. Kelly, *J. Mat. Sci.*, **8**, 352–362 (1973).
4. E. R. de los Rios, C. A. Rodopoulos, and J. R. Yates, "Micro-mechanical Crack Growth Modelling of Fibre-Reinforced Composites," in H. S. Found (Ed.), *Experimental Techniques and Design in Composite Materials*, Sheffield University Press, Sheffield, UK, 1995, pp. 304–320.
5. E. R. de los Rios, C. A. Rodopoulos, and J. R. Yates, "The Effect of Fibre, Matrix Mechanical Properties on the Fatigue Crack Propagation in a Fibre-reinforced

Titanium Matrix Composite," in S. A. Paipetis and A. G. Youtsos (Eds.), *High Technology Composites in Modem Applications*, University of Patras, Applied Mechanics Laboratory, Patras, Greece, 1995, pp. 316–327.

6. G. R. Irwin, "Plastic Zone Near a Crack and Fracture Toughness," in *Proceedings of the Seventh Sagamore Ordnance Materials Conference*, Vol. IV, Syracuse University, New York, 1960, pp. 63–78.
7. K. Schulte and K. Minoshima, *Composites*, **24**(3), 197–208 (1993).
8. R. O. Ritchie, *Mat. Sci. Eng.*, **A103**, 15–28 (1988).
9. E. R. de los Rios, C. A. Rodopoulos, and J. R. Yates, "Fatigue Damage Design Principles of Fibre-reinforced Titanium Matrix Composites," in R. Cook and P. Poole (Eds.), *ICAF'97*, EMAS, Edinburgh, 1997, pp. 1051–1060.
10. C. A. Rodopoulos, "Fatigue Damage Map for Metal Matrix Composites—A Useful Tool for Design Against Fatigue," in S. A. Paipetis and E. E. Gdoutos (Eds.), *1st Hellenic Conference on Composite Materials*, Xanthi, 1997, pp. 545–569.
11. E. R. de los Rios, C. A. Rodopoulos, and J. R. Yates, *Int. J. Fatigue*, **19**(5), 379–387 (1997).
12. M. D. Senmeier and P. K. Wrigth, "The Effect of Fiber Bridging on Fatigue Crack Growth in Titanium Matrix Composites," in M. N. Gungor and P. K. Liaw (Eds.), Proceedings TMS Fall Meeting, Indiana, *Fundamental Relationships Between Microstructure and Mechanical Properties of Metal Matrix Composites*.
13. M. Y. He and A. G. Evans, *Acta Metall. Mat.*, **39**(7), 1587–1593 (1991).
14. W. D. Brewer and J. Unman, Interface Control and Mechanical Property Improvements in Silicon Carbide Titanium Composites, NASA Technical Paper 2066, 1982.
15. J. Cook and J. E. Gordon, *Proc. Roy. Soc. London*, **A282**, 508–520 (1964).
16. S. S. Wang and I. Choi, *J. Appl. Mech.*, **50**, 169–178 (1983).
17. J. R. Rice and G. C. Sih, *Trans. ASME*, **June**, 418–423 (1965).
18. M. Comninou, *J. Appl. Mech.*, **December**, 631–636 (1977).
19. H. F. Wang, W. W. Gerberich, and C. J. Skowronek, *Acta Metall. Mat.*, **41**(8), 2425–2432 (1993).
20. P. Ehrburger and J. B. Donnet, *Phil. Trans. Roy. Soc. London*, **A294**, 495–505 (1980).
21. P. W. Erickson and E. P. Plueddemann, *Historical Background of the Interface in Composite Materials*, Vol. 6, E. P. Plueddemann (Ed.), Academic Press, New York, 1974, pp. 1–29.
22. M. Y. He and J. W. Hutchinson, *J. Appl. Mech.*, **56**, 270–278 (1991).
23. B. N. Cox and D. B. Marshall, *Fatigue and Fracture of Eng. Mat. Struct.*, **14**(8), 847–861 (1991).
24. J. P. Lucas, *Eng. Fracture Mech.*, **42**(3), 543–561 (1992).
25. M. F. Kanninen and C. H. Popelar, *Advanced Fracture Mechanics*, Oxford University Press, New York, 1985.
26. S. M. Jeng, J.-M. Yang, and C. J. Yang, *Mat. Sci. Eng.*, **A138**, 181–190 (1991).
27. S. Jansson, H. E. Déve, and A. G. Evans, *Metall. Trans.*, **22A**, 2975–2984 (1991).
28. Y. S. Lee, M. N. Gungor, and P. K. Liaw, *J. Composite Mat.*, **25**, 536–556 (1991).
29. K. M. Fox, M. Strangwood, and P. Bowen, *Composites*, **25**(7), 684–691 (1994).
30. J.-P. Favre, A. Vassel, and C. Laclau, *Composites*, **25**(7), 482–487 (1994).

31. T. J. Mackin, P. D. Warren, and A. G. Evans, *Acta Metall. Mat.*, **4**, 1251–1257 (1992).
32. M. Kuntz, K.-H. Schlapschi, B. Meier, and G. Grathwohl, *Composites*, **25**(7), 476–481 (1994).
33. R. J. Kerans and T. A. Parthasarathay, *J. Am. Ceramic Soc.*, **74**(7), 1585–1596 (1991).
34. H. L. Cox, *Br. J. Appl. Phys.*, **3**, 72–79 (1952).
35. A. Kelly and W. R. Tyson, *J. Mech. Phys. Solids*, **14**, 177–186 (1966).
36. G. S. Holister and C. Thomas, *Fibre Reinforced Materials*, Elsevier, London, 1966.
37. D. Hull, *An Introduction to Composite Materials*, Cambridge Solid State Science Series, Cambridge University Press, Cambridge, UK, 1981.
38. R. F. Gibson, *Principles of Composite Materials Mechanics*, McGraw-Hill, New York, 1994.
39. M. D. Thouless and A. G. Evans, *Acta Metall. Mat.*, **36**, 517–522 (1988).
40. D. B. Marshall and B. N. Cox, *Mech. Mat.*, **7**, 127–133 (1988).
41. Y.-C. Chiang, A. S. D. Wang, and T.-W. Chou, *J. Mech. Phys. Solids*, **41**, 1137–1154 (1993).
42. C.-H. Shueh, *Acta Metall. Mat.*, **38**(3), 403–409 (1990).
43. P. D. Warren, T. J. Mackin, and A. G. Evans, *Acta Metall. Mat.*, **40**, 1243–1249 (1992).
44. S. M. Jeng, P. Alassoeur, and J. M. Yang, *Mat. Sci. Engn.*, **A138**, 155–167 (1991).
45. P. K. Wright, R. Nimmer, G. Smith, M. Sensmeier, and M. Brun, "The Influence of the Interface on Mechanical Behavior of Ti-6Al-4V/SCS-6 Composites," in R. Y. Lin, R. J. Arsenault, G. P. Martins and S. G. Fishman (Eds.), *Interfaces in Metal-Matrix Composites*, The Minerals, Metals and Materials Society, 1989, pp. 595–581.
46. C. J. Yang, S. M. Jeng, and J.-M. Yang, *Scripta Metall. Mat.*, **24**, 469–474 (1990).
47. B. A. Bilby, A. H. Cottrell, and K. H. Swinden, *Proc. Roy. Soc. London*, **A272**, 304–314 (1963).
48. D. Hull, *An Introduction to Composite Materials*, 1st ed., Cambridge University Press, Cambridge, UK, 1981.
49. S. J. Connell, F. W. Zok, Z. Z. Du, and Z. Suo, *Acta Metall. Mat.*, **42**(10), 3451–3461 (1994).
50. W. S. Johnson, *Composites*, **24**(3), 187–196 (1993).
51. A. R. Ibbotson, C. J. Beevers, and P. Bowen, *Scripta Metall. Mat.*, **25**, 1781–1786 (1991).
52. S. M. Jeng, P. Alassoeur, and J. M. Yang, *Mat. Sci. Engn.*, **A154**, 11–19 (1992).
53. D. P. Walls, G. Bao, and F. Zok, *Acta Metall. Mat.*, **41**, 2061–2071 (1993).
54. P. Bowen, "Characterisation of Crack Growth from an Unbridged Defect in Continuous Fibre Reinforced Titanium Metal Matrix Composites," in N. D. R. Goddard and P. Bowen (Eds.), *Test Techniques in Metal Matrix Composites II*, ERA Technology, London, 1992, pp. 107–126.
55. R. A. Naik, W. D. Pullock, and W. S. Johnson, *J. Mat. Sci.*, **26**, 2913–2920 (1991).
56. D. Walls, G. Bao, and F. Zok, *Scripta Metall. Mat.*, **25**, 911–916 (1991).
57. W. S. Johnson, R. A. Naik, and W. D. Pollock, "Fatigue Damage Growth Mechanisms in Continuous Fiber Reinforced Titanium Matrix Composites," in H. Kitagawa

and T. Tanaka (Eds.), *Fatigue 90*, Engineering Materials Advisory Service, Tokyo, 1990, pp. 841–850.

58. L. J. Ghosn, J. Telesman, and P. Kantzos, "Fatigue Crack Growth in Unidirectional Metal Matrix Composite," in H. Kitagawa and T. Tanaka (Eds.), *Fatigue 90*, Engineering Materials Advisory Service, Tokyo, 1990, pp. 893–898.
59. D. M. Harmon and C. R. Saff, "Damage Initiation and Growth of Fiber Reinforced Metal Matrix Composites," in W. S. Johnson (Ed.), *Metal Matrix Composites: Testing, Analysis and Failure Modes*, ASTM STP 1032, 1989, pp. 237–245.
60. D. M. Harmon, C. R. Saff, and D. L. Greaves, "Strength Predictions for Metal Matrix Composites," in W. S. Johnson (Ed.), *Metal Matrix Composites: Testing, Analysis and Failure Modes*, ASTM STP 1032, 1989, pp. 222–236.
61. W. S. Johnson, Characterization of Fatigue Damage Mechanisms in Continuous Fiber Reinforced Metal Matrix Composites, Ph.D. Thesis, Duke University, 1979.
62. W. S. Johnson, Fatigue Testing and Damage Development in Continuous Fiber Reinforced Metal Matrix Composites, NASA Technical Memorandum, 100628, 1988.
63. G. D. Menke and I. J. Toth, The Time-Dependent Mechanical Behavior of Metal Matrix Composites, AFML-TR-102, U. S. Air Force Materials Laboratory, 1971.
64. W. S. Johnson, S. J. Lubowinski, and A. L. Highsmith, "Mechanical Characterization of Unnotched SCS6/Ti-15-3 Metal Matrix Composites at Room Temperature," in J. M. Kennedy, H. H. Moeller and W. S. Johnson (Eds.), *Thermal and Mechanical Behaviour of Ceramic and Metal Matrix Composites at Room Temperatures*, ASTM STP 1080, 1990, pp. 175–191.
65. G. S. Brady and H. R. Clauser, *Materials Handbook*, 13th ed., McGraw-Hill, New York, 1991.
66. W. A. Curtin, *Composites*, **24**, 98–102 (1993).
67. W. S. Johnson, "Fatigue of Metal Matrix Composites," in K. L. Reifsnider (Ed.), *Fatigue of Composite Materials*, Elsevier Science, New York, 1990, pp. 199–229.
68. I. Greaves, The Growth of Naturally Initiating Fatigue Cracks in Titanium-Silicon Carbide MMCs, Ph.D. Thesis, University of Sheffield, 1994.
69. C. A. Rodopoulos, Fatigue Studies under Constant and Variable Amplitude Loading in MMCs, Ph.D. Thesis, University of Sheffield, 1996.
70. W. S. Johnson, Fatigue Damage Accumulation in Various Metal Matrix Composites, NASA Technical Memorandum, 89116, 1987.
71. C. R. Saff, Durability of Continuous Fiber Reinforced Metal Matrix Composites, AFWAL-TR, Air Force Wright Aeronautical Laboratories, 1987.
72. S. M. Jeng, P. Alassoeur, and J.-M. Yang, *Mat. Sci. Engn.*, **A148**, 67–77 (1991).
73. R. T. Bhatt and H. H. Grimes, "Fatigue Behaviour of Silicon-Carbide Reinforced Titanium Composites," in *Fatigue of Fibrous Composite Materials*, ASTM STP 723, 1979, pp. 274–290.
74. S. S. Yau and G. Mayer, *Mat. Sci. Engn.*, **82**, 45–57 (1986).
75. B. Lerch and G. Halford, *Mat. Sci. Engn.*, **A200**, 47–54 (1995).
76. B. P. Sanders, S. Mall, and S. C. Jackson, *Int. J. Fatigue*, **21**, 121–134 (1999).
77. D. L. Krabbel, B. P. Sanders, and S. Mall, *Composites Sci. Tech.*, **57**, 99–117 (1997).
78. G. C. Sih, "Microstructure and Damage Dependence of Advanced Composite Material Behavior," in G. C. Sih, G. F. Smith, I. H. Marshall, and J. J. Wu (Eds.),

Workshop on Composite Materials Response: Constitutive Relations and Damage Mechanisms, Elsevier Applied Science, England, 1987, pp. 1–23.

79. L. J. Ghosn, P. Kantzos, and J. Telesman, *Int. J. Fracture*, **54**, 345–357 (1992).
80. B. N. Cox, *Acta Metall. Mat.*, **39**(6), 1189–1201 (1991).
81. D. B. Marshall, B. N. Cox, and A. G. Evans, *Acta Metall. Mat.*, **33**(11), 2013–2021 (1985).
82. J. Llorca and M. Elices, *Int. J. Fracture*, **54**, 251–267 (1992).
83. J. K. Wells and P. W. R. Beaumont, *J. Mat. Sci.*, **17**, 397–405 (1982).
84. B. Cotterell and J. R. Rice, *Int. J. Fracture*, **16**, 155–169 (1980).
85. A. Navarro and E. R. de los Rios, *Proc. Roy. Soc. London*, **A437**, 375–390 (1992).
86. J. Aveston, G. A. Cooper, and A. Kelly, "Single and Multiple Fracture," in *Proceed. of a Conference on the Properties of Fibre Composite*, National Physical Laboratory, IPC Science and Technology Press, London, 1971, pp. 15–26.
87. B. Budiansky, J. W. Hutchinson, and A. G. Evans, *J. Mech. Phys. Solids*, **34**(2), 167–189 (1986).
88. J. Bacuckas, Jr., and W. S. Johnson, Application of Fiber Bridging Models to Fatigue Crack Growth in Unidirectional Titanium Matrix Composites, NASA Technical Memorandum, 107588, 1992.
89. L. N. McCartney, *Proc. Roy. Soc. London*, **A409**, 329–350 (1987).
90. R. M. McMeeking and A. G. Evans, *Mech. Mat.*, **9**, 217–227 (1990).
91. I. N. Sneddon and M. Lowengrub, *Crack Problems in the Classical Theory of Elasticity*, Wiley, New York, 1969.
92. P. Kantzos, Fatigue Crack Growth in Ti-based Metal Matrix Composites, M.S. Thesis, Pennsylvania State University, 1991.
93. A. Navarro and E. R. de los Rios, *Phil. Mag.*, **A57**, 43–50 (1988).
94. J. M. Yang, S. M. Jeng, and C. J. Yang, *Mat. Sci. Eng.*, **A138**, 155–167 (1991).
95. D. Walls, G. Bao, and F. Zok, *Acta Metall. Mat.*, **25**, 911–916 (1991).
96. E. R. de los Rios, C. A. Rodopoulos, and J. R. Yates, *Fatigue and Fracture of Eng. Mat. Struct.*, **19**(5), 539–550 (1996).
97. T. W. Clyne and P. J. Withers, *An Introduction to Metal Matrix Composites*, Cambridge University Press, Cambridge, UK, 1993.
98. D. L. Davidson, *Metall. Trans.*, **23A**, 865–879 (1992).
99. K. S. Chan, *Acta Metall. Mat.*, **41**(3), 761–768 (1993).
100. D. M. Kotchick, R. C. Hink, and R. E. Tressler, *J. Composite Mat.*, **9**, 327–336 (1975).
101. J. J. Masson and E. Bourgain, *Int. J. Fracture*, **55**, 303–319 (1992).
102. SAE Fatigue Design and Evaluation Committee, Fatigue Design Handbook, Engineering Society for Advancing Mobility Land Sea Air and Space, 1988.
103. C. C. Osgood, *Fatigue Design*, Wiley Interscience, New York, 1970.
104. A. Bartlett and A. G. Evans, *Acta Metall. Mat.*, **39**(7), 1579–1585 (1991).
105. T. C. Lu, J. Yang, Z. Suo, A. G. Evans, R. Hecht, and R. Mehrabian, *Acta Metall. Mat.*, **39**(8), 1883–1890 (1991).
106. G. Lamé, *Lèçons sur la Theorie de L'Elasticité*, Gauthier-Villars, Paris, 1852.

107. R. W. Hertzberg, *Deformation and Fracture Mechanics of Engineering Materials*, 3rd ed., Willey, New York, 1989.

108. B. A. Lerch, D. R. Hull, and T. A. Leonhardt, As-Received Microstructure of a SiC/Ti-15-3 Composite, NASA Technical Memorandum 100938, 1988.

109. A. C. Pickard, The Application of 3-Dimensional Finite Element Methods to Fracture Mechanics and Fatigue Life Prediction, Engineering Materials Advisory Services, 1988.

110. J. C. Newman, Jr., and I. S. Raju, Stress-Intensity Factor Equations for Cracks in Three-Dimensional Finite Bodies, NASA Technical Memorandum 83200, 1981.

111. F. A. McClintock, "On the Plasticity of the Growth of Fatigue Cracks," in D. C. Drucker, and J. J. Gilman (Eds.), *Fracture of Solids*, Vol. 20, Wiley, New York, 1963, pp. 65–102.

112. S. Suresh, *Fatigue of Materials*, Cambridge Solid State Science Series, Cambridge University Press, London, 1991.

113. E. R. de los Rios, C. A. Rodopoulos, and J. R. Yates, *Fatigúe and Fracture of Eng. Mat. Struct.*, **21**, 1503–1511 (1998).

114. L. Ghosn, P. Kantzos, and J. Telesman, Modeling of Crack Bridging in a Unidirectional Metal Matrix Composite, NASA Technical Memorandum 1044355, 1990.

115. H. Tada, P. C. Paris, and G. R. Irwin, *The Stress Analysis of Crack Handbook*, 2nd ed., Del Research, St. Louis, MO, 1985.

CHAPTER 7

Nickel and Nickel Alloys

D. C. AGARWAL

Krupp VDM Technologies Corp., 11210 Steeplecrest Drive, Suite 120, Houston, Texas, 77065

Handbook of Advanced Materials Edited by James K. Wessel
ISBN 0-471-45475-3

7.1 INTRODUCTION

To fully appreciate the corrosion challenges of the new millennium, one has to learn from the innovations made in the nickel alloy metallurgy of the past century. In the twenty first century, as was the case in the last century, within the various industries, after carbon steel, the 300 series stainless steels will continue to be the "most widely used tonnage" material. Other corrosion mitigation technologies such as electrochemical protection, nonmetallics, coatings, and paints and use of inhibitor technology will also play a major role. The materials of construction for these modern chemical process, petrochemical industries, and other industries not only have to resist uniform corrosion caused by various corrodents but must also have sufficient localized corrosion and stress corrosion cracking resistance as well. These industries have to cope with both the technical and commercial challenges of rigid environmental regulations, the need to increase production efficiency by utilizing higher temperatures and pressures, and more corrosive catalysts, and at the same time possess the necessary versatility to handle varied feedstock and upset conditions. Even though nickel as an element was discovered about 250 years ago, the first major nickel alloy introduced to the industry, about 100 years ago, was a Ni–Cu alloy 400. This alloy is still being widely used in a variety of industries and will continue to be used in this current century. Over the past 100 years, specially in the last 50 years, improvements in alloy

TABLE 7.1 Chronology of Historical Development of Some Austenitic Corrosion-Resistant Alloys

Decade	Alloys
Pre-1950s	300 SS, 200, 400, 600, alloys B & C
1950s	20Cb, 800, 825, alloy F, alloy X
1960s	300L series SS, 20Cb3, 904L, Al-6X alloy 700, 625, G, C-276
1970s	317LM, 254SMo, 28, G-3, C-4, B-2
1980s	A16XN, N06030, 22, 59, 1925hMo, 31
1990s	Controlled chemistry alloy B-2, B-3, B-4, B-10, 686, 2000, 33

metallurgy, melting technology, and thermomechanical processing, along with a better fundamental understanding of the role of various alloying elements, has led to new nickel alloys. These have not only extended the range of usefulness of existing alloys by overcoming their limitations but are reliable and cost-effective and have opened new areas of applications. This chapter briefly describes the various nickel alloy systems developed and in use during the last 100 years with comments as to what the future holds for the newer alloys developed in the last 20 years and the competition faced by these alloys in the new millennium. Table 7.1 gives the chronology of various aqueous corrosion alloys developed in the pre-1950 and post-1950 era. Prior to the 1950s the alloy choices available to material engineers for combating corrosion were very limited. The latter half of this century saw a phenomenal growth in the development of new nickel alloys, including the high-performance Ni–Cr–Mo C family alloys. As is evident from this listing of austenitic alloys, today's corrosion/material engineers have a much wider selection of alloys to meet their specific needs.

Some of the alloys are very recent, developed after the 1980s, whereas some date clearly back to the beginning of the twentieth century. New alloys and refinements of old ones are continually being developed. Typical composition of some of the common wrought nickel alloys of various alloy systems are given at the beginning of individual alloy sections, as described later.

Different from the aqueous corrosion alloys are a class of alloys known as "superalloys," which are intended for elevated temperature service, usually based on periodic table Group VIIIA elements, where relatively severe mechanical stressing is encountered and where high surface stability to various high-temperature modes of degradation is needed. The superalloys are divided into three classes: nickel base, cobalt base, and iron base, and these are utilized at a higher proportion of their actual melting point than any other class of commercial metallurgical materials. This chapter will not delve into the superalloys but direct the reader to many excellent books and articles in the open literature [1–5].

7.2 NICKEL AND NICKEL ALLOY SYSTEMS

Nickel and nickel alloys have useful resistance to a wide variety of corrosive environments, typically encountered in various industrial processes such as in chemical processing, petrochemical processing, aerospace engineering, power generation and energy conversion, thermal processing and heat treatment industry, oil and gas production, pollution control and waste processing, marine engineering, pulp and paper industry, agrichemicals, industrial and domestic heating, the electronics and telecommunication industries, and others. In many instances the corrosive conditions are too severe to be handled by other commercially available materials including stainless and super stainless steels. Nickel by itself is a very versatile corrosion-resistant metal, finding many useful applications in industry. More importantly, its metallurgical compatibility over a considerable composition range with a number of other metals as alloying elements has become the basis for many binary, ternary, and other complex nickel base alloy systems, having very unique and specific corrosion-resistant and high-temperature resistant properties to handle the modern day corrosive environments. These alloys are more expensive than the standard 300 series stainless steels due to their higher alloy content and hence are only used when stainless steels are not suitable or when product purity cannot be compromised and safety considerations became very important. The subject of corrosion is highly complex and is dependent on the chemical composition, the microstructural features of the alloy, the various reactions occurring at the alloy/environment interface, and the chemical nature of the environment.

This chapter will mainly concentrate on the aqueous corrosion alloy systems, with a brief description of high-temperature alloys. The intent of this chapter is not to go into the theoretical discussion and analysis of corrosion science but to present the major nickel alloy systems, their major characteristics, the effects of alloying elements, and, most importantly, the strengths, weaknesses, and application of these alloy systems in the industry. A few words on fabrication are also included because an improper fabrication may destroy the corrosion resistance of an otherwise perfectly good nickel alloy.

7.3 AQUEOUS CORROSION

There are various forms of corrosion that can be classified into different categories. Although these differ in nature, they are also interrelated. These are:

- Uniform or general corrosion
- Galvanic corrosion
- Localized corrosion (pitting, crevice, or underdeposit corrosion)
- Inter granular corrosion
- Environmentally assisted cracking (hydrogen embrittlement, chloride stress corrosion cracking, sulfide stress cracking, corrosion fatigue, liquid metal embrittlement)
- Selective leaching
- Erosion corrosion

A detailed description of these are adequately covered in the open literature [6–8].

TABLE 7.2 Alloying Elements in Aqueous Corrosion-Resistant Nickel Alloys

Alloying Elements	Main Feature	Other Benefits and Corrosion Resistance Improvements
Ni	Provides matrix for metallurgical compatibility to various alloying elements. Improves thermal stability and fabricability	Enhances corrosion in mild reducing media. Alkali media improves chloride SCC
Cr	Provides resistance to oxidizing media.	Enhances localized corrosion resistance.
Mo	Provides resistance to reducing media	Enhances localized corrosion resistance and chloride SCC. Provides solid solution strengthening.
W	Behaves similar to Mo but less effective. Detrimental to thermal stability.	Provides solid solution strengthening.
N	Austenitic stabilizer—economical substitute for nickel.	Enhances localized corrosion resistance, thermal stability, and mechanical properties.
Cu	Improves resistance to seawater.	Enhances resistance to H_2SO_4 and HF containing acid environments.

7.3.1 Alloying Effects

The effects of various alloying elements on nickel matrix for wet corrosion and high-temperature corrosion are presented in Tables 7.2 and 7.3. Some of the alloying elements are common to both but impart different property characteristics. Some may be undesirable for wet corrosion alloys but beneficial for high-temperature corrosion alloys and vice versa.

There are many good textbooks and articles available in the open literature and to get a deeper understanding and details of specific alloy systems, the reader is strongly encouraged to consult these. A reference list is provided at the end of this chapter [6–8].

7.3.2 Aqueous Corrosion Modes

Alloys can be basically divided into various binary and ternary alloy systems with very specific properties and applications as shown in the following list:

Alloy Systems	Some Major Alloys in These Systems
Ni	Commercially pure nickel, alloy 200/201
Ni–Cu alloys	Alloy 400, K-500
Ni–Mo alloys	Alloy B, B-2, B-3, B-4, B-10
Ni–Si alloys	Cast Ni-Si alloys, alloy SX, Lewmet, alloy D-205

TABLE 7.3 Alloying Effects in High-Temperature Nickel Alloys

Cr	Helps in oxidation resistance provided temperature does not exceed 950°C for long periods. Volatility of Cr_2O_3. Also not good in fluorine at high temperatures: Improves sulfidation resistance and high Cr beneficial to oil ash corrosion and attack by molten glass Decreases carbon diffusion; helps carburization resistance Detrimental to nitriding resistance. Increases high-temperature strength
Si	Improves oxidation resistance, nitriding, sulfidation, and carburizing resistance. Synergistically acts with Cr to improve scale resilience. Detrimental to nonoxidizing chlorination resistance.
Mo	Improves high-temperature strength, good in reducing chlorination. Improves creep strength, bad for oxidation at higher temperatures.
Ni	Improves carburization, nitriding, and chlorination resistance; bad for sulfidation resistance.
W	Behaves similarly to Mo.
C	Improves strength; helps nitridation resistance; beneficial to carburization and metal dusting resistance; oxidation resistance adversely effected
Y and RE	Improves adherence and spalling of oxide layer. Helps with sulfidation resistance.
Al	Independently and synergistically with Cr improves oxidation resistance. Helps with sulfidizing resistance; bad for nitriding resistance.
Ti	Bad for nitriding resistance.
Nb	Increases short-term creep strength; may be beneficial in carburizing resistance; detrimental to nitriding resistance.
Mn	Slight positive effect on high-temperature strength and creep; bad for oxidation resistance; increase solubility of N_2.
Co	Reduces rate of sulfur diffusion; hence, helps with sulfidation resistance; improves solid solution strength.

Ni–Fe alloys	Invar
Ni–Cr–Fe alloys	Alloys 600, 601, 800, 800H, 800HT, 690
Ni–Cr–Fe–Mo–Cu alloys	20, 28, 825, G, G-3, G-30, 31, 33, 1925hMo
Ni–Cr–Mo alloys	625, C-276, C-4, C-22, 686, C-2000, 59, Mat 21
High-temperature Alloys	602CA, 603GT, 2100GT, 45TM, 230, 625, X HR160, 718, 617, 690, 214, Nimonic series, Udimet series, others

7.4 ALLOY SYSTEMS

7.4.1 Commercially Pure Nickel

Alloy/UNS#	Ni	Cu	Fe	Mn	C
200/N02200	99.0 min	0.25	0.40	0.35	0.15
201/N02201	99.0 min	0.25	0.40	0.35	0.02

The two main alloys, commercially pure alloy 200 and alloy 201, have useful resistance at low to moderate temperatures to corrosion by dilute unaerated solution of the common nonoxidizing mineral acid such as HCl, H_2SO_4, or H_3PO_4. The reason for its good behavior is the fact that the standard reduction potential of nickel is more noble than that of iron and less noble than copper. Because of nickel's high overpotential for hydrogen evolution, there is no easy discharge of hydrogen from any of the common nonoxidizing acids and a supply of oxygen is necessary for rapid corrosion to occur. Hence in the presence of oxidizing species such as ferric or cupric ions, nitrates, peroxides, or oxygen, nickel can corrode rapidly. Nickel's outstanding corrosion resistance to alkalies has led to its successful use as caustic evaporator tubes. At boiling temperatures and concentration of up to 50% NaOH, the corrosion rate is less than 0.005 mm/year. The iso-corrosion diagram for nickel 200 and 201 in sodium hydroxide clearly shows its superiority and usefulness even at higher concentrations and temperatures [9]. However, when nickel is to be utilized at temperatures above 316°C (600°F) in these applications, the low-carbon version (alloy 201) is recommended to guard against the phenomenon of graphitization occurring at the grain boundaries, which leads to possible loss of ductility causing embrittlement.

Nickel is very resistant to chloride stress corrosion cracking resistance but may be susceptible to caustic cracking in aerated solution in severely stressed conditions. Use of Ni–Cr–Fe such as alloy 600 may be more resistant under such conditions. Nickel has a high resistance to corrosion by most natural freshwaters and rapidly flowing seawater. However, under stagnant or crevice conditions, severe pitting attack may occur. While nickel's corrosion resistance to oxidizing acids such as nitric acid, is poor, it is sufficiently resistant to most nonaerated organic acids and organic compounds. Nickel is not attacked by anhydrous ammonia or very dilute ammonium hydroxide solution (<2%). Higher concentrations cause rapid attack due to formation of a soluble ($Ni-NH_4$) complex corrosion product.

Nickel's good resistance to halogenic environments at elevated temperatures such as in chlorination or fluorination reactions has been utilized in many modern-day chemical processes, largely due to the fact that the nickel–halide films formed on the nickel surface have relatively low vapor pressures and high melting points.

Nickel has been successful in production of high purity caustic in 50–75% concentration range, petrochemical industry, chemical process industry, handling of food and food industry, and production of synthetic fibers. Other useful applications are due to its magnetic and magnetostrictive properties, high thermal and electrical conductivities, and low vapor pressure.

7.4.2 Ni–Cu Alloys

Alloy/UNS #	Ni + Co	Cu	Fe	Mn	C	Al, Ti
400/N04400	63.0 min	31	2.5	2.0	0.30	—
K500/N05500	63.0 min	30	2.0	1.5	0.18	2.8, 0.6

The two main alloys in this system are Monel 400 or alloy 400 and its age–hardenable version, alloy K-500. Alloy 400 was developed at the beginning of the twentieth century and, even after approximately 100 years, continues to be used in the modern-day chemical, petrochemical, marine, refineries, and many other industries. Alloy 400 containing about 30–33% copper in a nickel matrix has many similar characteristics of commercially pure nickel, while improving upon many others. Addition of some iron significantly improves the resistance to cavitation and erosion in condenser tube applications. The main uses of alloy 400 are under conditions of high flow velocity and erosion as in propeller shafts, propellers, pump-impeller blades, casings, condenser tubes, and heat exchanger tubes. Corrosion rate in moving seawater is generally less than 0.025 mm/year. The alloy can pit in stagnant seawater, however, the rate of attack is considerably less than in commercially pure alloy 200. Due to its high nickel content (approx. 65%) the alloy is generally immune to chloride stress corrosion cracking.

The general corrosion resistance of alloy 400 in nonoxidizing mineral acids is better compared to nickel. However, it suffers from the same weakness of exhibiting very poor corrosion resistance to oxidizing media such as nitric acid, ferric chloride, cupric chloride, wet chlorine, chromic acid, sulfur dioxide, or ammonia.

In unaerated dilute hydrochloric and sulfuric acid solution the alloy has useful resistance up to concentrations of 15% at room temperature and up to 2% at somewhat higher temperature, not exceeding 50°C. Due to this specific characteristic, alloy 400 is also used in processes where chlorinated solvents may form hydrochloric acid due to hydrolysis, which would cause failure in standard stainless steel.

Alloy 400 possesses good corrosion resistance at ambient temperatures to all HF concentration in the absence of air. Aerated solutions and higher temperature increase the corrosion rate. The alloy is susceptible to stress corrosion cracking in moist aerated hydrofluoric or hydrofluorosilic acid vapor. This can be minimized by deaeration of the environments or by stress relieving anneal of the component in question.

Neutral and alkaline salt solutions such as chloride, carbonates, sulfates and acetates have only minor effect even at high concentrations and temperatures up to boiling. Hence the alloy has found wide use in plants for crystallization of salts from saturated brine.

Alloy K-500, the age-hardenable alloy, which contains aluminum and titanium, combines the excellent corrosion resistance features of alloy 400 with the added benefits of increased strength, hardens, and maintaining its strength up to 600°C. The alloy has low magnetic permeability and is nonmagnetic to −134°C. Some of the typical applications of alloy K-500 are for pumpshafts, impellers, medical blades and scrapers, oil well drill collars, and other completion tools, electronic components, springs and valve trains. This alloy is primarily used in marine and oil and gas industrial applications. In contrast alloy 400 is more versatile, finding many uses in roofs, gutters, and architectural parts on a number of institutional buildings, tubes of boiler feedwater heaters, seawater applications (sheathing, others), HF alkylation process, production and handling of HF acid, and in refining of uranium, distillation, condensation units, and overhead condenser pipes in refineries and petrochemical industries, and many others.

7.4.3 Ni–Fe Alloys

Alloy	Ni	Cr/Co	Mn	Si	C	Fe	Others
Alloy 36, Invar	36	0.2/0.5	0.35	0.2	0.03	Bal	—
Magnifer 7904, Hymu 80	80	—	0.50	0.3	0.02	Bal	Mo 5

The nickel alloys containing 36–80% nickel are generally used due to their special physical properties, such as low coefficient of thermal expansion and/or magnetic properties.

Higher nickel alloys containing 76–80% nickel with some iron and some molybdenum have the highest magnetic permeability and are used as inductive components in transformers, circuit breakers, low-frequency transducers, relay parts, and screens Alloys with 36% nickel, known as Invar, has extremely low expansion characteristics. Due to its applications in cryogenic environments, this alloy has undergone extensive corrosion testing. The nickel–iron alloys have moderately good resistance to a variety of industrial environments, but are primarily used for their physical characteristics as opposed to corrosion-resistant characteristics.

7.4.4 Ni–Si Alloys

Alloy	Ni	Co	Mo	Cu	Cr	Si	Fe
Lewmet 66	Bal	6	0.2	3	31	3	16
Alloy D-205	Bal	—	2.5	2	20	5	6

Cast Ni–Si alloys typically containing 8–10% silicon were developed for handling hot or boiling sulfuric acid of most concentrations and have also been used to resist strong nitric acid above 50% concentration along with nitric–sulfuric acid mixtures. A few wrought Ni–Si alloys have also been developed in this century, such as alloy SX, Lewmet grades, and alloy D-205.

7.4.5 Ni–Mo Alloys (B Family Alloys)

Alloy/UNS #	Ni	Mo	Fe	Cr	C
B/N10001	Bal	28	5	0.5	0.03
B-2/N10665	Bal	28	1.8	0.7	0.005
B-3/N10675	Bal	28	1.5	1.5	0.005

B-4/N10629	Bal	28	3	1.2	0.005
B-10/N10624	Bal	24	6	8.0	0.005

Alloy B, the original alloy in the Ni–Mo family, developed in the 1920s, suffered from heat affected zone (HAZ) corrosion in nonoxidizing acids (i.e., acetic, formic, and hydrochloric) due to its higher carbon content. In the decade of the 1960s, improved argon–oxygen decarburization (AOD) melting technology led to development of alloy B-2. This alloy solved the HAZ corrosion problem but suffered from poor fabricability. Recent developments of controlled chemistry alloy B-2, alloy B-3, and Nimofer 6629—Alloy B-4, UNS N10629—solved both these problems by eliminating/reducing the formation of detrimental intermetallic phases with further improvement in corrosion resistance behavior. Greater details on fundamental behavior and understanding of Ni–Mo alloy systems are available in the open literature [10, 11]. Alloy B-2, B-3, and B-4 are recommended for service in handling all concentrations of HCl in the temperature range of 70–100°C handling of wet HCl gas. It has excellent resistance to pure H_2SO_4 up to boiling point in concentrations below 60%. One weakness of the alloy is its lack of chromium and hence its very poor corrosion resistance in the presence of oxidizing species. Alloy B-2 has been successfully used in the production of acetic acid, pharmaceuticals, alkylation of ethyl benzene, styrene, cumene, organic sulfonation reactions, melamine, herbicides, and many other products. Alloy B-4, the improved version of alloy B-2, is being tested and has already found applications in production of resins encountering hydrochloric acid due to the presence of aluminum chloride in the temperature range of 120–150°C. In one chemical company in Spain, alloy B-4 was tested and specified for use in production of pesticides, where severe corrosive conditions exist due to presence of hydrochloric acid. The "C" family alloys were totally inadequate under these conditions. Alloy B-4 has solved both the fabricability problems, encountered with alloy B-2 and the susceptibility to stress corrosion cracking in many corrosive environments.

A very recent development in the Ni–Mo family has been the introduction of alloy B-10 (Nimofer 6224). One of the major weaknesses of the B, B-2, B-3, and B-4 alloys was their inability to handle the presence of oxidizing species in the corrosive media. Unacceptable and very high corrosion rates resulted. Under such conditions, the C family alloys with their higher chromium contents, such as alloy C-276 or alloy 59, could easily handle the oxidizing species but lacked sufficient molybdenum to counteract against the highly acidic hydrochloric or sulfuric acid reducing conditions. Alloy B-10 was an intermediate alloy between the C and B family, where the molybdenum level was significantly higher than the C family but somewhat lower than the B family. Also, the chromium and iron levels were increased to 8 and 6%, respectively, to counteract against the presence of the oxidizing corrosive species, which may be present in the environment. This alloy has already found successful uses in very specific crevice corrosion conditions caused in waste incinerators. Many other field tests are under way with this alloy [12].

7.4.6 Ni–Cr–Fe Alloys

Alloy/UNS #	Ni	Cr	Fe	Al	Si	Others
600/N06600	Bal	16	9	—	0.3	—
601/N06601	Bal	23	14	1.4	0.3	—
800H/N08810	32	21	Bal	0.4	0.5	Ti
690/N06690	Bal	30	10	—	0.3	—
602CA/N06025	Bal	25	9.5	2.24	—	Y,Zr,Ti
45TM/N06045	Bal	27	23	—	2.7	RE

Of the many commercial Ni–Cr–Fe alloys, the major ones are alloy 600, 601, alloy 800 variations, and alloy 690. Alloy 602CA provides an overall improvement over alloy 600/601, whereas alloy 45TM is an improvement over alloy 800H in resisting sulfidation, oxidation, metal dusting, and carburization type attack.

7.4.6.1 Alloys 600, 601, 690, and 602CA

Alloy 600, due to its high nickel content, has excellent resistance to halogens at elevated temperatures and has been used in processes involving chlorination. It has good oxidation and chloride stress corrosion cracking resistance. In production of titanium dioxide by chloride routes natural titanium oxide (illmenite or rutile) and hot chlorine gases reacted to produce titanium tetrachloride. Alloy 600 has been successfully used in this process due to its excellent resistance to corrosion by hot chlorine gas. This alloy has found wide usage in the furnace and heat-treating field due to its excellent resistance to oxidation and scaling at 980°C. The alloy also has found considerable use in handling water environments, where stainless steels have failed by cracking. It has been used in a number of nuclear reactors including steam generator boiling and primary water piping systems. Alloy 690 in recent years has substituted alloy 600 in nuclear applications due to its superior stress corrosion cracking resistance. Alloy 600 has also been used in preheaters and turbine condensers with maximum service temperatures around 450°C. However, the low chromium content of alloy 600 prevented its use in applications that required extended exposure to high temperatures and requiring superior creep properties. This limitation was addressed by increasing the chromium content in alloy 601. Even with this modification, alloy 601 had some limitations. The need to extend the temperature range to 1200°C and still maintain good strength with improved resistance to environmental degradation led to the development of a new alloy, and alloy 602CA, which is described in the high-temperature section.

7.4.6.2 Alloys 800, 800H, 800HT, and 45TM

Alloy 800 (20% Cr, 32% Ni, and 46% Fe as balance) is used primarily for its oxidation resistance and strength at elevated temperatures. One of its major benefits is that the alloy does not form embrittling sigma phase even after long time exposure between 650 and 870°C. This fact coupled with its high creep and stress rupture strength have led to many applications in the petrochemical industry such as in the production of styrene (steam-heated rectors). The alloy exhibits good resistance to carburization and sulfidation and thus has been used in coal gasification in components such as heat exchangers, process piping, carburizing fixtures, and retorts. Two major applications are electric range heating element sheathing and extruded tubing for ethylene and steam methane reformer furnaces.

In aqueous corrosion service, alloy 800 is generally not widely used since its corrosion resistance is somewhere between type 304SS and 316SS. Even though alloy 800H has been used in coal gasification due to its good carburization and sulfidation resistance, it undergoes accelerated attack in some processes. To enhance resistance to these modes of degradation, a higher chromium, silicon-containing alloy was developed, known as alloy 45TM (UNS N06045). This is a chromia/silica forming alloy with excellent resistance to high-temperature attack in coal gasification and thermal waste incinerators, and in refineries and petrochemical industries, where sulfidation has been a major problem. Details on this alloy are available in the open literature [4, 17].

7.4.7 Ni–Cr–Fe–Mo–Cu Alloys

Alloy/UNS#	Ni	Cr	Mo	Cu	Fe	Others
825/N08825	Bal	22	3	2	31	Ti
G/N06007	Bal	22	6.5	2	20	Cb + Ta
G-3/N08825	Bal	22	7	2	20	Cb + Ta
G30/N06030	Bal	29	5	1.5	15	Cb + Ta
20/N08020	35	20	3	3.5	Bal	Cb
28/N08028	31	27	3.5	1	36	—
31/N08031	31	27	6.5	1.2	32	N
33/R20033	31	33	1.6	1.2	32	N
1925hMo/N08926	25	21	6.5	0.9	Bal	N

Fortification of Ni–Cr–Fe alloys with molybdenum and copper has resulted in a series of alloys with improved resistance to corrosion by hot reducing acid such as sulfuric, phosphoric, and hydrofluoric acid and acids containing oxidizing species. By maintaining copper content to about 2% or less and chromium content from 20 to 33% and molybdenum levels from 1.5 to 7.0% and replacing some of the nickel with iron to reduce cost, a group of alloys are produced that have useful corrosion resistance in a wide variety of both oxidizing and reducing acids (except hydrochloric), organic compounds and to acid, neutral, and alkaline salt solutions.

7.4.7.1 Alloy 825

Alloy 825 is developed from alloy 800 with the addition of molybdenum (3%), copper (2%), and titanium (0.9%) for providing improved aqueous corrosion resistance in a wide variety of corrosive media. Its high nickel content of about 42% provides excellent resistance to chloride-ion stress corrosion cracking, although not immune to cracking. When tested in boiling magnesium chloride solutions, this alloy has been an upgrade to the 300 series stainless steels, when localized corrosion and stress corrosion cracking has been a problem. The high nickel in conjunction with the molybdenum and copper provides good resistance to reducing environments such as those containing sulfuric and phosphoric acids. Laboratory test results and service experience have confirmed the useful resistance of alloy 825 in boiling solutions of sulfuric acid up to 40% by weight and at all concentrations up to a maximum temperature of 66°C. In the presence of oxidizing species other than oxidizing chlorides, which may form HCl by hydrolysis, the corrosion resistance in sulfuric acid is usually improved. Hence the alloy is suitable for use in mixtures containing nitric acid, cupric acid, and ferric sulfates. In pure phosphoric acid the alloy is resistant at concentrations and temperatures up to and including boiling 85% acid. The alloy's high chromium content confers resistance to a variety of oxidizing media such a nitric acid, nitrates, and oxidizing salts. The titanium addition with an appropriate heat treatment serves to stabilize the alloy against sensitization to intergranular attack.

Some typical applications include various components used in sulfuric acid pickling of steel and copper, components in petroleum-refineries and petrochemicals (tanks, agitators, valves, pumps), equipment used in production of ammonium sulfate, pollution control equipment, oil and gas recovery, acid production, nuclear fuel reprocessing and handling of radioactive waste, and phosphoric acid production (evaporators, cylinders, heat exchangers, equipment for handling fluorsilicic acid solution, and many others). Alloy 825 is a versatile alloy handling a wide variety of corrosive media but has begun to be gradually replaced in the industry by other alloys due to their superior localized corrosion resistance, such as the G family alloy and the 6% moly superaustenitic stainless steels, such as alloy 1925hMo (N08926) and alloy 31 (N08031).

7.4.7.2 G Family — G, G-3, and G-30

Alloy G was a development from alloy F, an alloy of similar composition, but with addition of about 2% copper. This addition of copper significantly improved the corrosion resistance in both sulfuric and phosphoric acid environments. Alloy G, developed in the 1960s, had excellent corrosion resistance in the as-welded condition and could handle the corrosive effects of both oxidizing and reducing agents. The alloy exhibited resistance to mixed acids, fluorsilicic acid, sulfate compounds, concentrated nitric acid, flue gases of coal-fired power plants, and hydrofluoric acid. Due to its higher nickel and molybdenum content over alloy 825, the alloy is essentially immune to chloride stress corrosion cracking and has significantly superior localized corrosion resistance. This alloy has been widely used in industries similar to those using alloy 825 as mentioned before, with the added advantage of improved corrosion resistance. However, this alloy is now obsolete and has been replaced by alloy G-3.

Alloy G-3 is an improved version of alloy G, having similar excellent corrosion resistance characteristics, but greater resistance to HAZ attack and with better weldability. Due to its lower carbon content, the alloy offers slower kinetics of carbide precipitation and its slightly higher molybdenum content provides for superior localized corrosion resistance. Alloy G-3 has replaced alloy G in almost all the industrial applications to date and alloy 825 in many applications, where better localized corrosion resistance is needed.

Alloy G-30 is a modification of the G-3 alloy with significantly increased chromium content and a lower molybdenum content. The alloy shows excellent resistance in commercial phosphoric acids as well as many complex and mixed acid environments of nitric/hydrochloric and nitric/hydrofluoric acids. The alloy has good resistance in sulfuric acid also. Some typical applications of alloy G-30 have been in phosphoric acid service, mixed acid service, nuclear fuel reprocessing, components in pickling operations, petrochemicals, agrichemicals manufacture (fertilizers, insecticides, pesticides, herbicides), and mining industries.

7.4.7.3 Standard 6% Mo Alloys and Advanced 6% Mo Alloy 31

Standard 6Mo alloys, such as Cronifer 1925hMo, 254SMO, Inco 25-6Mo, and Al-6XN were basically derived from alloy 904L metallurgy by increasing the molybdenum content by approximately 2% and fortifying it with nitrogen as a cost-effective substitute for nickel for metallurgical balance and improved thermal stability. The addition of molybdenum and nitrogen provided the added benefits of improved mechanical properties and resistance to localized corrosion. These alloys are readily weldable with an overalloyed filler metal such as alloy 625, C-276, or 59 to compensate for the segregation of molybdenum in the interdendritic regions of the weld. They have been extensively used in offshore and marine, pulp and paper, flue gas desulfurization (FGD), chemical process industry for both organic and inorganic compounds, and a variety of other applications. The 6Mo family alloys successfully bridge the performance gap between standard stainless steels and the high-performance nickel-based alloys in a cost-effective manner.

A higher chromium/higher nickel version of standard 6Mo alloys is the new advanced 6Mo alloy Nicrofer 3127hMo—alloy 31—UNS N08031. Its greatly improved corrosion resistance compared with conventional 6Mo family alloys and alloy 28 is achieved via increased Cr (27%) and Mo (6.5%) contents and fortification with nitrogen (0.2%). Alloy 31's corrosion behavior, achieved with only about half the nickel content of alloy 625, makes it a very cost-effective alternative in many applications. The pitting potential of this alloy as determined in artificial seawater makes it a suitable alloy for heat exchangers using seawater or brackish water as cooling media. Its corrosion resistance in sulfuric acid in medium concentration ranges is superior even to that of alloy C-276 and alloy 20 (Table 7.4). Its localized corrosion resistance is superior to many alloys including alloy 625. (Table 7.5). Figure 7.1 shows this alloy to maintain its corrosion pitting potential as a function of temperature in artificial seawater. This property has led to its use in tube-and-shell heat exchangers using seawater or brackish water as a cooling medium.

However, in view of the specific active/passive characteristics of alloy 31 in sulfuric acid environments, one must be extremely careful, when specifying this material for sulfuric acid use at 80% concentration and temperatures above 80°C, because these conditions will render the material active. Alloy 31 has been

TABLE 7.4 Corrosion Resistance in Sulfuric Acid [Corrosion Rate (mpy)]

	60°C			80°C			100°C		
% H_2SO_4	Alloy 20	Alloy C-276	Alloy 31	Alloy 20	Alloy C-276	Alloy 31	Alloy 20	Alloy C-276	Alloy 31
20	<5	<1	<0.1	10	4	<0.1	>25	7	0.3
40	<5	<2	<0.1	10	3	<0.2	>25	10	0.6
60	>5	<2	<0.1	11	4	0.4	>50	11	1
80	5	<1	0.2	18	15	0.8	>50	240	240

TABLE 7.5 Critical Pitting and Crevice Corrosion Temperature per ASTM G-48 (10% $FeCl_3$)

Alloy	Critical Pitting Corrosion Temp. (°C)	Critical Crevice Corrosion Temp. (°C)	PRE[a]
316	15	<0	24
904L	45	25	37
20	15	<10	29
825	30	<5	32
G-3	70	40	45
1925hMo	70	40	48
625	77.5	57.5	52
33	85	40	50
31	85[b]	65	54

[a] PRE: Pitting resistance equivalent = %Cr + 3.3(%Mo) + 30N.
[b] Above 85°C, the 10% $FeCl_3$ solution chemically breaks down.

extensively used in the most varied applications including the pulp-and-paper industry, phosphoric acid environments, copper smelters, sulfuric acid production, pollution control, wastewater treatment in uranium mining, sulfuric acid evaporators, leaching of copper ores, pressure leaching of nickel from nickel lateritic ores, flue gas desulfurization systems of coal-fired power plants, viscose rayon production, fine chemicals production, and many others [18–20].

7.4.7.4 Alloy 20

The first version of alloy 20 was introduced in 1951 for sulfuric acid applications. A few years later, a columbium-stablized version was developed as 20Cb3, which allowed its weldments to be used in the as-welded condition without the need for postweld heat treatment. Further modifications led to the modern version alloy 20Cb3 with increased nickel content. This alloy is used in many applications due to its good resistance in sulfuric acid media and resistance to stress corrosion cracking. It is used in the manufacture of synthetic rubber, high octane gasolene, solvents, explosives, plastics, synthetic fibers, chemicals and pharmaceuticals, in the food processing industry, and many others. However, due to its lower molybdenum content, the localized corrosion resistance of this alloy

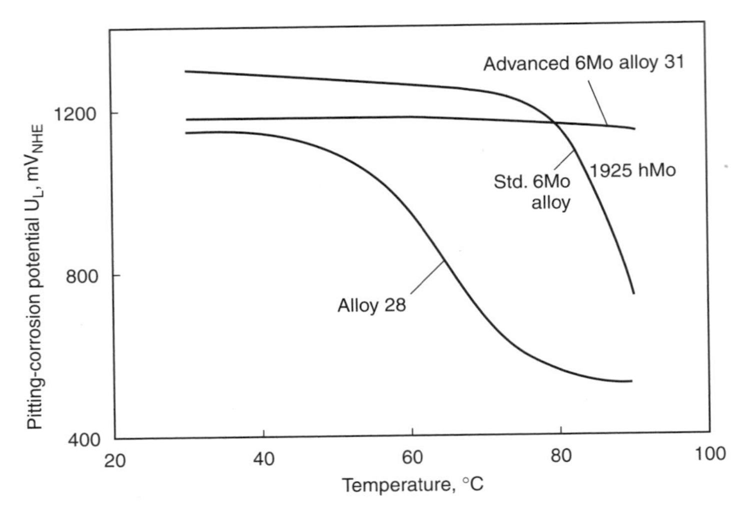

FIGURE 7.1 Corrosion pitting potential of alloy 31 vs. other alloys in artificial seawater.

is inadequate in certain applications and is being replaced by the standard and advanced 6Mo alloys.

7.4.7.5 Alloy 33

Alloy 33, a recent innovation, is a chromium-based austenitic wrought superstainless steel (33 Cr, 32 Fe, 31 Ni, 1.6 Mo, 0.6 Cu, 0.4 N). This alloy has excellent resistance to both acidic and alkaline corrosive media, mixed HNO_3/HF acids, hot sulfuric acid, localized corrosion, and stress corrosion cracking. Due to its high nitrogen content, this alloy has excellent mechanical properties. Its high pitting resistance equivalent (PRE) makes it a very cost-effective alloy in comparison to some stainless steels and some lower alloys.

In comparison to other high chromium alloys such as alloy G-30, 690, and 28, alloy 33 shows excellent corrosion resistance behavior. This alloy on a cost—performance basis has the potential of being an excellent alternative to many alloys currently in use such as 825, 904L, 20, 28, 6Mo alloys, G-3, G-30, and in some cases even alloy 625. The alloy has been successfully used in high-temperature concentrated sulfuric acid and is undergoing tests in a wide variety of diverse industries and environments. Details on the development and properties of this alloy are presented in the open literature [21, 22].

7.4.8 High-Performance Ni–Cr–Mo Alloys

Alloy/UNS#	Ni	Cr	Mo	W/Cu/Ta	Fe	Others
C/N10002	Bal	16	16	4/—/—	6	C-0.03
625/N06625	Bal	22	9	—/—/—	2	Cb 3.5
C-276/N10276	Bal	16	16	4/—/—	6	C-0.005

C-4/N06445	Bal	16	16	—/—/—	2	C-0.005
22/N06022	Bal	21	13	3/—/—	3	C-0.005
59/N06059	Bal	23	16	—/—/—	<1	C-0.005
686/N06686	Bal	21	16	4/—/—	2	C-0.005
2000/N06200	Bal	23	16	—/1.6/—	2	C-0.005
Mat21/N06210	Bal	19	19	—/—/1.8	1	C-0.005

The C family of alloys, the original being Hastelloy alloy C (1930s) was an innovative optimization of Ni–Cr alloys having good resistance to oxidizing corrosive media and Ni–Mo alloys with superior resistance to reducing corrosive media. This combination resulted in the most versatile corrosion-resistant alloy in the "Ni–Cr–Mo" alloy family with exceptional corrosion-resistance in a wide variety of severe corrosive environments typically encountered in Chemical Process Industry (CPI) and other industries. The alloy also exhibited excellent resistance to pitting and crevice corrosion attack in low pH, high chloride oxidizing environments and had total immunity to chloride stress corrosion cracking. These properties allowed this alloy to serve the industrial needs for many years although it had some limitations. The decade of the 1960s (alloy C-276), 1970s (alloy C-4), 1980s (alloy C-22 and 622), and 1990s (alloy 59, alloy 686, alloy C-2000, and MAT 21) saw newer alloy developments with improvements in corrosion resistance, which not only overcame the limitations of alloy C, but further expanded the horizons of applications as the needs of CPI became more critical, severe, and demanding.

Today the original alloy C of the 1930s is practically obsolete except for some usage in the form of castings. The chronology of the various corrosion-resistant Ni–Cr–Mo alloy developments during the twentieth century with special emphasis on the last 40 years of evolution in the C family of Ni–Cr–Mo alloys and their applications is presented below.

Prior to the 1950s the alloy choices available to material engineers for combatting corrosion were very limited. The latter half of the last century saw a phenomenal growth in the development of new alloys including the high-performance C family alloys. Table 7.1 gives a brief listing of some alloys developed during the pre-1950s period and the last five decades. As is evident from this listing of austenitic alloys, today's corrosion/material engineers have a much wider selection of alloys to meet their specific needs. The next few sections describe in detail the historical development of the C family of alloys, their corrosion resistance characteristics, both uniform corrosion and localized corrosion resistance, their thermal stability behavior, and the many industrial applications, where only the alloys of this family have provided reliable, safe, and cost-effective performance.

7.5 HISTORICAL DEVELOPMENT CHRONOLOGY OF C ALLOYS

7.5.1 Alloy C (1930s to 1965)

The element nickel has some unique electrochemical properties of its own, thus making unalloyed nickel a suitable choice in certain applications, but more important is its metallurgical compatibility with a number of other important alloying elements such as chromium, molybdenum, tungsten, copper, and iron. This compatibility and optimization between Ni–Cr and Ni–Mo alloys led to the first alloy of the C family, Hastelloy alloy C in the 1930s. The development of this alloy has been well described by McCurdy in 1939 [23].

This alloy was the most versatile corrosion-resistant alloy available in the 1930s through mid-1960s to handle the needs of the chemical process industry. However, the alloy had a few severe drawbacks. When used in the as-welded condition, alloy C was often susceptible to serious intergranular corrosion attack in HAZ in many oxidizing, low-pH, halide-containing environments. This meant that for many applications, vessels fabricated from alloy C had to be solution heat treated to remove the detrimental weld HAZ precipitates. This put a serious limitation on the alloy's usefulness. The CPI during the late 1940s and 1950s was constantly coming up with new processes that needed an alloy without these limitations of "solution heat-treating" after welding. Also in severe oxidizing media, this alloy did not have enough chromium to maintain useful passive behavior, thus exhibiting high uniform corrosion rates.

7.5.2 Alloy 625 (1960s to Present)

One of the "severe oxidizing media" limitation of alloy C was overcome by increasing the chromium content from 16 to 22% in alloy 625, an alloy developed in the late 1950s and commercialized in the 1960s. However, the molybdenum content was reduced to 9% and columbium was added for stabilization against intergranular attack, which permitted the use of this alloy in the as-welded condition without the need for a solution anneal as was the case with alloy C. The increased chromium improved corrosion resistance in a number of strongly oxidizing corrosive media, while maintaining adequate resistance to many reducing corrosive media. This alloy had a good balance of corrosion-resistant properties but was not as versatile in "reducing acid media" as alloy C, due to the lower molybdenum level in alloy 625. Also its localized corrosion resistance was significantly lower than alloy C.

This alloy is resistant to corrosion and pitting in seawater and has useful resistance to wet chlorine, hypochlorites, and oxidizing chlorides at ambient temperatures. It is resistant to various concentrations of hydrofluoric acid, even in the aerated condition, and to such acid mixtures as nitric-hydrofluoric, sulfuric-hydrofluoric, and phosphoric-hydrofluoric acids under most conditions encountered in industrial practice. Alloy 625 has good strength and resistance to scaling in air up to 980°C and many uses exist in high-temperature applications.

An alloy with lower nickel content, similar chromium and molybdenum content, alloy X (47Ni, 22Cr, 18.5Fe, 9Mo, 0,10C, 0.6 W) was developed having good high-temperature strength and resistance to oxidation and scaling up to ~1100°C. This alloy is mostly used for high-temperature applications in the heat treatment industry and for flying and land-based gas turbine components.

Another alloy, alloy N (69.5Ni, 7Cr, 16.5Mo, 5Fe) was developed for use with molten fluoride salts at 815°C in nuclear applications. Chromium was reduced to 7% to prevent intergranular attack and mass transfer of chromium in this environment.

7.5.3 Alloy C-276 (1965 to Present)

To overcome one of the above serious limitations, the chemical composition of alloy C was modified by a German company, BASF, which basically consisted of reducing both the carbon and silicon levels in this alloy by more than 10-fold to very low levels of typically 50 ppm carbon and 400 ppm silicon. This was only possible due to the invention of a new melting technology, known as the AOD–process and VOD (vacuum oxygen decarburization) process. This low-carbon and silicon content alloy came to be known as alloy C-276, which then was produced in the United States under a license from BASF Company, which was awarded a U.S. patent. The corrosion resistance of both these alloys was essentially similar in many corrosive environments, but without the detrimental effects of continuous grain boundary precipitates in the weld HAZ of alloy C-276. Thus alloy C-276 could be used in most applications in the as-welded condition without suffering severe intergranular attack. The corrosion behavior of both alloy C and alloy C-276 has been adequately covered in the open literature via numerous publications [6, 24]. The grain boundary precipitation kinetics and the time-temperature-transformation (T-T-T) diagram for these alloys are also well documented [25–27]. The industrial applications of alloy C-276 in the process industries are very extensive, diverse, and versatile due to its having excellent resistance in both oxidizing and reducing media even with halogen ion contamination. However, there were certain process conditions, where even alloy C-276 with its low carbon and low silicon was not adequately thermally stable in regard to precipitation of both carbides and intermetallic phases, thus being susceptible to corrosive attack. Within the broad scope of chemical processing, examples exist where serious intergranular corrosion of a sensitized (precipitated) microstructure has occurred. To overcome this, a modification of alloy C-276 was developed in the 1970s, called alloy C-4.

7.5.4 Alloy C-4 (1970s to Present)

In addition to the 10-fold decrease in carbon and silicon of alloy C, alloy C-4 had three other major modifications, that is, omission of tungsten from its basic chemical composition, reduction in iron level, and addition of some titanium. The above changes resulted in significant improvement in the precipitation kinetics of intermetallic phases, when exposed in the sensitizing range of 550–1090°C for extended periods of time, virtually eliminating the intermetallic and grain boundary precipitation of the "mu" phase, which has a $(Ni, Fe, Co)_3(W, Mo, Cr)_2$ type structure and various other phases. These phases are detrimental to ductility, toughness, and corrosion resistance.

The general corrosion resistance of alloy C-276 and alloy C-4 was essentially the same in many corrosive environments, except that in strongly reducing media like hydrochloric acid, alloy C-276 was better; but in highly oxidizing media, the opposite was true, that is, alloy C-4 was better. Alloy C-4 offers good corrosion resistance to a wide variety of media including organic acids and acid chloride solutions. Details of alloy C-4 development are documented in the open literature [6, 28, 29]. This alloy has found greater acceptance in European countries in contrast to alloy C-276, which is more widely used and accepted in the United States.

7.5.5 Alloy C-22 (1982 to Present)

The expiration of the alloy C-276 patent in the United States in 1982 saw the introduction of a newer development in the C family, alloy C-22. This alloy claimed that the mu-phase control in alloy C-4, which was accomplished by controlling the "electron vacancy" number by omitting tungsten and reducing iron, was done at the expense of reduced corrosion resistance to oxidizing chloride solutions, where tungsten is a beneficial element. In addition, both alloys C-276 and C-4 suffered high corrosion rates in oxidizing, nonhalide solutions, due to their relatively low chromium levels of 16%. Hence, the claim was, that there existed a need for an alloy with higher chromium levels for oxidizing environments with an optimized balance of Cr, Mo, and W, thus yielding an alloy with superior corrosion properties and good thermal stability. This led to the alloy C-22 composition (Table 7.2) with approximately 21% Cr, 13% Mo, 3% W, and 3% Fe with balance nickel. Even though the corrosion resistance of this alloy was superior to alloys C-276 and C-4 in highly oxidizing environments, slightly better pitting corrosion resistance in "green death" solution, its behavior in highly reducing environments and in severe localized crevice corrosion conditions was still inferior to the 16% molybdenum containing alloy C-276. Details on the development of alloy C-22 have been described elsewhere [30–32]. Research efforts during the 1980s at Krupp VDM led to the most advanced alloy development within the Ni–Cr–Mo family, alloy 59 [33, 34], which overcame the shortcomings of both alloys C-22 and C-276. It also provided solutions to the most severe and critical corrosion problems of the CPI, petrochemical, pollution control, and other industries.

7.5.6 Alloy 59 (1990 to Present)

As is evident from the composition of the various alloys of the C family, alloy 59 has the highest chromium plus molybdenum content with the lowest iron content of typically less than 1%. It is one of the highest nickel-containing alloy of this family and is the purest form of a "true" Ni–Cr–Mo alloy without the addition of any other alloying elements, such as tungsten, copper, or titanium. This "purity" and balance of alloy 59 in the ternary Ni–Cr–Mo system, is mainly responsible for the alloy's superior thermal stability behavior. The electron vacancy number, which used to be an important parameter in the "Phacomp calculations" for alloy development and for prediction of occurrence of various phases was later superseded by a recent and more precise "new Phacomp" methodology, proposed by

TABLE 7.6 3M Study—Hazardous Waste Incineration Scrubber Corrosion Data (1991 h)[a]

Alloy	MPY[b]	Remarks
59	1.1	Clean
686	5.4	Clean
C-22	6.7	Clean
31	7.1	Clean
622	12.1	Weld attack
C-276	35.1	Clean
625	58.6	Rough
825	117	Pitting

[a] Reference 34.
[b] To convert to mm/y multiply by 0.025.

Morinaga et al. [35], also lends support to this phenomenon of superior thermal stability of alloy 59. This is proven and discussed later.

7.5.7 Alloy 686 (1993 to Present)

This is another recent development in the C family of Ni–Cr–Mo–W alloys, which is very similar in composition to alloy C-276, but where the chromium level has been increased from 16 to 21%, while maintaining the Mo and W at similar levels. This alloy is very overalloyed with the combined Cr, Mo, and W content of around 41%. To maintain its single-phase austenitic structure, this alloy has to be solution annealed at a very high-temperature of around 1220°C followed by very rapid cooling to prevent precipitation of intermetallic phases. Its thermal behavior, as discussed later, is significantly inferior to alloy 59 and its performance in field tests in a hazardous waste incinerator at 3M Co., St. Paul, Minnesota, showed five times higher corrosion rate than alloy 59 [34]. Table 7.6 presents this data comparing alloy 59 with alloy 686 and other alloys. Alloy 59 exhibited the best performance.

7.5.8 Alloy C-2000 (1995 to Present)

This is the most recent introduction in the C alloy family in which basically 1.6% copper has been added to the alloy 59 composition. However, as shown later, addition of copper has resulted in significantly lower thermal stability behavior in comparison to alloy 59.

7.6 CORROSION RESISTANCE CHARACTERISTICS OF C ALLOYS

7.6.1 Uniform Corrosion Resistance

Table 7.7 gives the uniform corrosion resistance data of these alloys in various boiling corrosive environments. The media are both oxidizing and reducing in

TABLE 7.7 Comparison of Some Ni–Cr–Mo Alloys in Various Boiling Corrosive Environments

	Uniform Corrosion Rate (mpy)[a]				
Media	C-276	C-22	686	C-2000	59
ASTM 28A	240	36	103	27	24
ASTM 28B	55	7	10	4	4
Green Death	26	4	8	—	5
10% HNO_3	19	2	—	—	2
65% HNO_3	750	52	231	—	40
10% H_2SO_4	23	18	—	—	8
50% H_2SO_4	240	308	—	—	176
1.5% HCl	11	14	5	2	3
10% HCl	239	392	—	176	179
10% H_2SO + 1% HCl	87	354	—	—	70
10% H_2SO_4 + 1% HCl[b]	41	92	—	—	3

[a] To convert to mm/y multiply by 0.0254.
[b] 90°C.

TABLE 7.8 Critical Pitting and Crevice Corrosion Temperature per ASTM G-48 (10% $FeCl_3$)

Alloy	Cr	Mo	PRE[a]	CPT (°C)	CCT (°C)
C-22	21	13	65	>85[b]	58
C-276	16	16	69	>85	>85
686	21	16	74	>85	>85
59	23	16	76	>85	>85

[a] PRE = %Cr + 3.3(%Mo) + 30N.
[b] Above 85° the 10% $FeCl_3$ solution chemically breaks down.

nature and normally used for comparing the relative performance of alloys. As is evident from this data, alloy 59's overall performance is better than any other C family alloy.

7.6.2 Localized Corrosion Resistance

Localized corrosion has caused more failures in the CPI than any other single corrosion phenomena. Chromium, molybdenum, nitrogen, and to a lesser extent tungsten contribute significantly to enhancing pitting and crevice corrosion resistance of nickel base alloys. Table 7.8 shows the critical pitting and crevice corrosion behavior of these alloys per American Society for Testing and Materials (ASTM) G-48 (10% ferric chloride) test method. As is evident the lower molybdenum containing alloy C-22 had significantly lower critical corrosion

TABLE 7.9 Localized Corrosion Resistance in Green Death Solution ($11.5\% \ H_2SO_4 + 1.2\% \ HCl + 1\% \ FeCl_3 + 1\% \ CuCl_2$)

Alloy	PRE	CPT (°C)	CCT (°C)	Crevice Depth at 105°C (mm)
C-22 (13% Mo)	65	120	105	0.35
C-276 (16% Mo)	59	110	105	0.035
686 (16% Mo)	74	>120[a]	110	—
C-2000 (16% Mo)	76	110	100	—
59 (16% Mo)	76	>120[a]	110	0.025

[a] Above 120°C, the green death solution chemically breaks down.

TABLE 7.10A Thermal Stability per ASTM G28A after Sensitization at 1600°F (871°C)

	Corrosion Rate (mpy)[a]				
Sensitization	C-276	C-22	686	C-2000	59
Annealed	240	36	103	27	24
1 h	>500[b]	>500[b]	872[b]	116[b]	40[c]
3 h	>500[b]	>500[b]	>1000[b]	178[b]	51[c]

[a] To convert to mm/y multiply by 0.0254.
[b] Alloys C-276, C-22, C-2000, and 686. Heavy pitting attack with grains falling due to deep intergranular attack.
[c] Alloy 59. No pitting attack.

temperature. This fact is again proven by data in Table 7.9 where the maximum depth of attack at 105°C in "green death" solution was 10 times greater for alloy C-22 than for alloy C-276. In this test alloy 59 had the lowest depth of crevice attack, showing the beneficial effects of its highest PRE number.

7.6.3 Thermal Stability

The superior thermal stability characteristics of alloy 59 are shown in Tables 7.10A and 7.10B. The data clearly demonstrates the detrimental effects of tungsten and copper on the thermal stability behavior of various alloys of the C family. Figure 7.2 shows the attack on the tungsten-containing Ni–Cr–Mo alloy C22 after aging at 1600°F for one hour. This phenomenon was also observed on other tungsten-containing Ni–Cr–Mo alloys, C-276, and alloy 686 along with the copper-containing alloy C2000.

In the real world, during welding of heavy-walled vessels and/or hot forming of heavy-walled materials, this phenomenon takes on an added importance in maintaining the superior corrosion resistance behavior. Other corrosion resistance data and information on physical metallurgy, fabricability, and weldability of alloy 59 have been adequately covered elsewhere [36–38].

TABLE 7.10B Thermal Stability per ASTM G28B after Sensitization at 1600°F (871°C)

	Corrosion Rate (mpy)[a]				
Sensitization	C-276	C-22	686	C-2000	59
Annealed	55	7	10	4	4
1 h	>500[b]	339[b]	17[b]	>500[b]	4[c]
3 h	>500[b]	313[b]	85[b]	>500[b]	4[c]

[a] To convert to mm/y multiply by 0.0254.
[b] Alloys C-276, C-22, C-2000, and 686. Heavy pitting attack with grains falling due to deep intergranular attack.
[c] Alloy 59. No pitting attack.

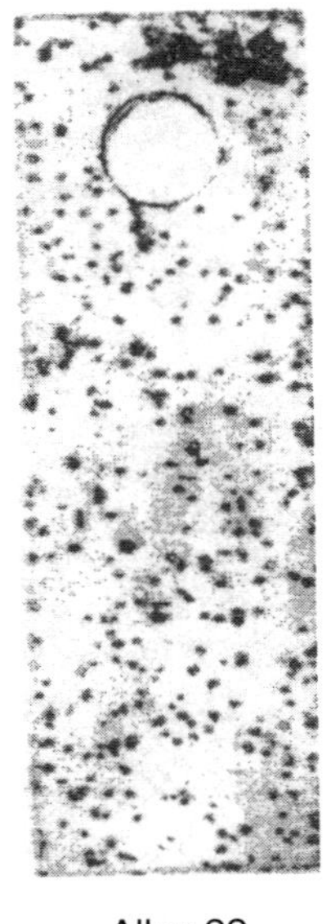

Alloy 22 Alloy 59

FIGURE 7.2 Influence of thermal stability on corrosion of alloy 22 and alloy 59 after aging at 1600°F and tested in ASTM G-28B test solution (deep intergranular attack on alloy C22. No attack on alloy 59).

7.7 APPLICATIONS OF C ALLOYS

The C family of alloys has found widespread application in chemical and petrochemical industries producing various chlorinated, fluorinated, and other organic chemicals, agrichemicals, and pharmaceutic industries producing various biocides, pollution control (FGD of coal-fired power plants, wastewater treatment, incinerator scrubbers), pulp and paper, oil and gas (sour gas production), marine, and many others.

TABLE 7.11 Major Industries Using Alloy C-276/C-4

I.	Petroleum
	Petroleum refining
	Oils/greases
	Natural-gas processing
II.	Petrochemical
	Plastic
	Synthetic organic fibers
	Organic intermediates
	Organic chemicals—chlorinated/fluorinated hydrocarbons
	Synthetic rubber
III.	CPI—chemical process industries
	Fine chemicals
	Inorganic chemicals
	Soaps/detergents
	Paints
	Fertilizer—agrichemicals—herbicides/pesticides
	Adhesives
	Industrial gases
IV.	Pollution control
	FGD
	Waste water treatment
	Incineration
	Hazardous waste
	Nuclear fuel reprocessing
V.	Pulp and paper
VI.	Marine/seawater
VII.	Pharmaceuticals
VIII.	Sour gas/oil and gas production
IX.	Mining/metallurgical

As mentioned earlier, the original alloy C is now obsolete, except for use in some castings. In the last 35 years, approximately 60,000 tons of alloy C-276 and C-4 have been used in a variety of industries, some of which are listed in Table 7.11.

Due to its higher chromium content, Alloy C-22 did improve upon the weaknesses of alloy C-276 in highly oxidizing media. The introduction of alloy 59 in 1990 and the industry realizing the benefits of alloy C-276 over alloy C-22 in highly reducing environments, and availability of alloy C-276 from multiple sources as opposed to a single source for alloy C-22, led to a resurgence in alloy C-276 usage.

Today alloy C-22, an alloy of the eighties, has been superseded by alloys of the nineties, that is, alloy 59, alloy 686, and alloy C-2000. However, there has been no major industrial applications of either alloy C-2000 or alloy 686 to date. In contrast, alloy 59 with its first commercial introduction in 1990 has already found a wide number of applications, and these continue to increase as the "corrosion universe" realizes its superior corrosion resistance, excellent fabricability, weldability, and thermal stability behavior. This alloy is covered under all appropriate ASTM, AWS, and NACE MRO-175 specifications and is also covered in ASME Boiler and Pressure Vessel Code up to 1400°F and has a code case for SCIII applications to 427°C (800°F). This alloy is also covered in the various international specifications.

7.8 SOME SPECIFIC APPLICATIONS OF ALLOY 59

A major chemical company producing chlorinated and fluorinated chemicals had to replace the reactor pressure vessel made out of alloy C-276 every 12–14 months, due to excessive corrosion. The process employed various hydrocarbons, ammonium fluoride, sulfuric acid, and a proprietary catalyst in which one atom of chlorine was replaced with one atom of fluorine. The presence of fluorides ruled out use of tantalum, titanium, and glass-lined vessels. Switching to a Ni–Mo alloy B-2 prolonged the life by only 20–25%. This was also unacceptable. Extensive tests made with alloy 59 and other alloys indicated that with alloy 59 the life of this ASME code reactor vessel could be increased by 300–400%. A vessel was built with alloy 59 in 1994. After 27 months of service, a minor repair of the "thermowell" weld had to be performed. It is expected that the life of this vessel will even surpass the original expectations. Since then two more vessels of alloy 59 have been ordered by this same company and are giving excellent service. Other companies in Europe have also selected alloy 59 in the production of chlorinated, fluorinated, and fine chemicals. Figure 7.3 shows one of the reactor vessel constructed out of alloy 59.

FIGURE 7.3 An 8000-gal alloy 59 pressure vessel for production of fine chemicals at Degussa-Hulls, Europe. (Fabricator: Apparatebau GmbH, Essen, Germany).

Alloy 59 was selected by a European company for hydrofluoric acid production after a field test in a rotary kiln, where this alloy gave superior performance to alloy 686, alloy C-2000, and alloy C-22.

The corrosive conditions in scrubbers of coal-fired power plants (FGD systems) and waste incinerators, both municipal and hazardous waste, have been so severe that only alloys of the Ni–Cr–Mo have given reliable performance. The presence of condensates with chloride levels over 100,000 ppm, and fluorides of over 10,000 ppm, very low pH (below 1), sulfuric acid, hydrochloric acid, hydrofluoric acid, various salts, and other contaminants create a situation where lower alloys have failed in a few days to a few weeks. Many thousands of tons of alloy 59 have been used in recent years in these systems in Europe and other parts of the world, giving satisfactory performance [39, 40]. Very recently alloy 59 (over 80 tons) was selected by Arizona Public Service for its FGD scrubber project for both the chimney and the outlet duct. This alloy was chosen over other C family alloys because of its superior corrosion-resistant properties and proven case histories. Another major specification of this alloy has been in the Syncrude Project in Canada, where oil is extracted from the tar sands. The FGD scrubber system will utilize ammonia solution to produce ammonium sulfate, which then will be used as an agrichemical. Over 700 tons of alloy 59 will be used in this project.

During manufacture and synthesis of acrylates and methacrylates, the process reaction at 130°C is carried under oxidizing conditions in the presence of acids, fatty alcohols, and paratoluene sulfonic acid. The previous material of construction, alloy 400 had failed rapidly with corrosion rates approaching 0.75 mm/yr. A test program with various alloys including 904L, 28, G-3, 625, C-276, 31, and 59 showed alloy 59 to be totally free from localized attack with corrosion rate of less than 0.025 mm/yr. Alloy 59 was selected and has operated without any problems for the last 5 years.

In citric acid production, a 6% Mo alloy failed rapidly. The reaction was treating with calcium citrate with concentrated H_2SO_4 around 96°C. A test program with alloy 59 led to its selection, and since then four reactors have been built. The first one, installed in 1990, continues to operate without any problems. In another citric acid plant, plate heat exchangers of alloy 20 were failing rapidly. Testing with various alloys also led to alloy 59 selection. These alloy 59 plate heat exchangers are giving very reliable service.

In a copper plant, the SO_2-rich gas from the flash furnace is scrubbed with a solution of dilute 5% contaminated H_2SO_4 at a temperature of 45–60°C. The acid produced has a concentration of typically around 50–55% H_2SO_4 and a temperature of about 75°C. The chloride and fluoride contents of this acid are both high, at about 7000 ppm. Previous materials of construction (alloy 20 and rubber lined carbon steel) had failed very rapidly. Tests were carried out using alloy 59, alloy 31 and other alloys. Corrosion rates for both alloys 59 and 31 were below 0.025 mm/yr with no localized corrosion. Following these tests, alloy 31 was purchased for the scrubber internals handling the produced acid and alloy 59 for the induced draft fans. These have been in successful operation for the last 6 years with no detectable corrosion. Since then more fans of alloys 59 and 31 have been placed in service.

In a weld overlay of burner bases, where hydrogen and chlorine are burnt to produce hydrochloric acid, a two-layer electroslag alloy 59 weld overlay performed significantly better than all previously used materials including alloy C-22. In another weld overlay application with alloy 59, superheater tubes in a waste incineration plant, extended the life by significantly reducing unusually high fireside surface wastage.

In a plant, plate heat exchangers handling acetic acid derivative effluents were failing rapidly. Corrosion testing at 100°C with alloy C-276 and alloy 59 gave corrosion rates of 0.4 mm/yr for alloy C-276 vs. 0.04 mm/yr for alloy 59, a 10-fold improvement. Hence alloy 59 was selected. The media consisted of sulfates, acetic acid, phosphates, and chlorides with a pH of 1.

Gold sponge is deposited from an electrolyte of dilute HCl containing impure gold. The deposited spongy gold cathodes are washed in water to remove the HCl and then dried in an oven at 150°C, where the evaporation of remaining dilute HCl electrolyte creates very severe corrosive conditions. After extensive testing, alloy 59 was selected for this application and has been performing well since 1990.

Details of some other applications are described elsewhere [41]. There are many more applications of alloy 59, too numerous to list here, but the above gives a flavor of the diversity, versatility, and usefulness of this alloy in a wide range of industries and applications.

7.9 HIGH-TEMPERATURE ALLOYS

The need for high-temperature materials is encountered in a wide variety of modern industries such as in aerospace, metallurgical, chemical, petrochemical, glass manufacture, heat treatment, waste incinerators, heat recovery, advanced energy conversion systems, and others. Depending on the condition of chemical makeup and temperatures, a variety of aggressive corrosive environments are produced, which could be either sulfidizing, carburizing, halogenizing, nitriding, reducing, and oxidizing in nature or a combination thereof. All high-temperature alloys have certain limitations and the optimum choice is often a compromise between the mechanical property requirement constraints at maximum temperature of operation and environmental degradation constraints imposed due to the corrosive species present.

Alloys designed to resist high-temperature corrosion have existed since the beginning of the twentieth century. Generally high-temperature metal degradation occurs at temperatures above 1000°F (540°C), but there are few cases where it can also occur at somewhat lower temperatures. Carbon steel, a very useful and the workhorse material of construction in many industries, is attacked by H_2S above 500°F (260°C), by oxygen or air above 1000°F (540°C) and by nitrogen above 1800°F (980°C). Chromium and molybdenum containing low-alloy steels significantly extend the range of usefulness of carbon steel. However, the severity of the processes as encountered in modern-day industries (chemical, petrochemical, refineries) and the new technologies of thermal destruction of hazardous and municipal waste, fluidized-bed combustion, coal gasification and chemical from coal processes, and the use of "dirty feedstock," such as heavy oil and high-sulfur coal, coupled with demands for higher efficiency and tougher environmental regulations, have necessitated the use of higher alloy systems of iron base, nickel base, and cobalt base alloys. Today alloy systems have not only to provide reliable and safe performance in a cost-effective manner but must have sufficient versatility to resist changing corrosive conditions due to starting feedstock changes.

Optimal material selection for high-temperature applications requires a thorough understanding of the mechanical requirements at the temperature of operation including upset conditions and mechanical degradation due to high-temperature corrosive attack. These specific property requirements are:

Mechanical	Corrosion Resistance
High-temperature Strength	Oxidation
Stress rupture strength	Carburization and metal dusting
Creep strength	Nitridation
Fatigue	Sulfidation
Thermal stability	Halogenation
Thermal shock	Molten salt corrosion
Toughness	Liquid metal corrosion
Others	Ash/salt deposit corrosion

These requirements will vary and be different for various industries such as heat treatment, aerospace, power generation, metallurgical processing, petrochemical and refineries, heat recovery, waste incineration, and others.

In nickel base alloys, the major alloying elements for imparting specific property or a combination of properties are tabulated in Table 7.3. These alloying elements can be classified as follows:

- Protective scale formers: Cr as chromia, Al as alumina, and Si as silica
- Solid solution strengtheners: Mo, W, Nb, Ti, Cr, Co
- Age hardening strengtheners: Al + Ti, Al, Ti, Nb, Ta
- Carbide strengtheners: Cr, Mo, W, Ti, Zr, Ta, Nb
- Improved scale adherence (spallation resistance): Rare Earths (La, Ce), Y, Hf, Zr, Ta

Most high-temperature alloys have sufficient amounts either of chromium with the addition of either aluminum or silicon to form the protective oxide scales for resisting high-temperature corrosion. Table 7.12 gives the typical chemical composition of several common high-temperature alloys in commercial use today. Optimization of the various alloying elements led to a new alloy for service

TABLE 7.12 Metallurgical Optimization of Alloy 602CA Nominal Chemistry Comparison to Other High-Temperature Alloys

Alloy	Fe	Ni	Cr	Si	C	Others
309	Bal	13	25	0.5	0.15	—
310	Bal	20	25	0.5	0.08	—
253	Bal	11	21	1.7	—	N, Ce
DS	Bal	36	18	2.2	0.06	—
800/800H	Bal	31	20	0.4	0.08	Ti, Al-0.4
120	Bal	38	25	0.6	0.06	Nb-0.7
45TM	23	Bal	27	2.7	0.08	RE
600	9	Bal	16	—	0.07	—
601	14	Bal	23	—	0.06	Al-1.4
602CA	**9.5**	**Bal**	**25**	—	**0.18**	**Y, Zr, Ti, Al-2.2**
230	1.5	Bal	22	0.4	0.10	W-14, Mo-1.2
214	2.5	Bal	16	0.10	0.03	Al-4.5, Y
X	18	Bal	22	—	0.10	W, Co, Mo-9
625	3	Bal	22	—	0.03	Cb, Mo-9
617	1.5	Bal	22	—	0.06	Co, Mo-9, Al-1.2

temperatures up to 1200°C in various industries. This alloy, known as alloy 602CA (UNS N06025), employs the beneficial effects of high chromium, high aluminum, high carbon, and microalloying with titanium, zirconium, and yttrium. Developed in the early 1990s, the alloy has found numerous applications in various industries as mentioned above. The typical chemical composition of the alloy in weight percent is given below:

Ni	Cr	Fe	Al	C	Ti	Zr	Y
Bal	25	9.5	2.2	0.18	0.15	0.06	0.08

This alloy is covered in ASTM and other international specifications. ASME code case 2359 has been approved for SC VIII, Div. 1, and SC I (steam service only) up to 1650°F. AWS coverage for weld filler metal in A5.11 and A5.14 is under progress. The major properties of interest in this alloy are:

- Excellent oxidation resistance up to 1200°C, superior to other wrought nickel base alloys currently available in the market
- Good high-temperature strength (stress rupture and stress to produce 1% creep at temperatures up to 1200°C), superior to most other Ni base alloys over 1000°C
- Excellent carburization resistance
- Excellent metal dusting resistance

Alloy 602CA employs the beneficial effects of high chromium, high aluminum, high carbon, and microalloying with titanium, zirconium, and yttrium in

FIGURE 7.4 Microstructure of annealed alloy 602CA (×500).

a nickel matrix. The relatively high carbon content of approximately 0.18–0.2% in conjunction with 25% chromium ensures the precipitation of bulky homogeneously distributed carbides, typically 5–10 in size. Transmission and scanning electron microscopy suggest these bulky carbides to be of $M_{23}C_6$-type primary precipitates. Microalloying with titanium and zirconium allows the formation of finely distributed carbides and carbonitrides (Fig. 7.4). Solution annealing even up to 1230°C does not lead to complete dissolution of these stable carbides, and thus the alloy resists grain growth and maintains relatively high creep strength due to a combination of solid solution hardening and carbide strengthening. This phenomenon of grain growth resistance is responsible for maintaining good ductility, a high creep strength up to 1200°C, and superior low-cycle fatigue strength. Table 7.12A shows the grain growth data for various high-temperature alloys where alloy 602CA had very little grain growth even after approximately 1000 h of exposure at 2050°F (1121°C). Hence repair and reconditioning of exposed parts can easily be achieved with alloy 602CA. The presence of approximately 2.2% aluminum in this alloy allows the formation of a continuous homogenous self-repairing Al_2O_3 sublayer beneath the Cr_2O_3 layer, which synergistically imparts excellent oxidation as well as carburization and metal dusting resistance: "Reactive elements" like yttrium significantly increase the adhesion and spallation resistance of the oxide layers, thereby further enhancing the high-temperature corrosion-resistant properties. Also, because

TABLE 7.12A Effects of High-Temperature Exposure on Grain Growth for Various Alloys Exposed at 2050°F (1121°C)

	Average ASTM Grain Size Number						
Exposure (h)	**602CA**	601	601GC	600	353MA	330	333
None (as annealed)	**7**	5	5.5	8	6	7	4
184	**7**	1	3.5	0	2.5	3	2.5
510	**6.5**	0	3	00	2	2	2
990	**6.5**	00	2.5	00	1.5	1.5	1

of its relatively low aluminum content, this alloy does not embrittle due to gamma prime formation, as is the case with higher aluminum-containing nickel alloys. This alloy is available in two conditions, the most common being Nicrofer 6025HT (alloy 602CA), which is solution annealed at 1220°C with typical grain size greater than 70 μm and is used in those applications where both high-temperature corrosion resistance and good stress rupture and creep properties are required. In special cases, where only the high-temperature corrosion resistance is needed, this alloy is supplied in annealed condition of 1180°C with grain size less than 70 μm.

7.9.1 High-Temperature Mechanical Properties

The mechanical properties of interest in designing high-temperature components are *time-independent properties*, for example, short-term tensile (typically below 600°C) and *time-dependent properties* (typically above 600°C), such as stress rupture and creep strength, and thermal stability, that is, maintenance of reasonable impact toughness after long aging. Table 7.13 lists some of the mechanical properties from recent production heats. Comparison with other high-temperature alloys is provided elsewhere: Recent work on determining 10,000 h average rupture strength of alloys 330, 230, and 602CA have shown that alloy 230 has higher

TABLE 7.13 High-Temperature Mechanical Properties of Alloy 602CA

	Room Temp. 25°C(77°F)	600°C (1112°F)	800°C (1471°F)	1000°C (1832°F)	1100°C (2012°F)	1200°C (2192°F)
	Typical Short-Term Tensile Properties					
UTS (ksi)	105	89	45	15.5	12	5
0.2% YS (ksi)	51	38	35	13	9	4.5

Temperature °C(°F)	$R_m/10^5$ h[a]	$R_m/10^4$ h[b]	$R_p1.0/10^5$ h[c]	$R_p1.0/10^4$ h[d]
	100,000 h and 10,000 h Creep Strength (ksi)			
650°C (1202)	20.3	31.2	17.4	26.8
700°C (1292)	14.5	22.5	12.3	19.1
800°C (1471)	2.90	6.10	2.40	4.60
900°C (1652)	1.40	2.60	1.10	1.90
950°C (1743)	1.00	1.90	0.80	1.30
1000°C (1832)	0.70	1.30	0.50	0.84
1050°C (1922)	0.45	0.90	0.28	0.52
1100°C (2012)	0.30	0.64	0.15	0.32
1150°C (2102)	0.20	0.44	0.06	0.15
1200°C (2192)	—	0.43	—	0.14

[a] $R_m/10^5$ h = Stress rupture in 100,000 h.
[b] $R_m/10^4$ h = Stress rupture in 10,000 h.
[c] $R_p.1.0/10^5$ h = 1% creep in 100,000 h.
[d] $R_p1.0/10^4$ h = 1% creep in 10,000 h.

TABLE 7.14 Impact Strength of Alloy 602CA in Joules after Aging at Various Temperatures up to 8000 h

Annealed Condition	Typical Value 78–84 J		
Exposure Temperature and Condition	1000 h	4000 h	8000 h
500°C exposure	53	35	30
10% cold worked (CW) + aged	28	26	22
CW + aged + annealed	76	77	78
640°C exposure	54	32	30
10% cold worked + aged	33	25	27
CW + aged + annealed	77	77	85
740°C exposure	55	30	27
10% cold worked + aged	40	29	25
CW + aged + annealed	79	79	76
850°C exposure	73	62	58
10% cold worked + aged	73	70	68
CW + aged + annealed	76	84	80

creep rupture strength values in comparison to alloy 602CA below 1800°F but exhibits lower values above 1800°F [42, 43]. Table 7.14 lists the impact strength after aging at various temperatures up to 8000 h. It is evident that alloy 602CA possesses adequate toughness properties for most industrial applications.

7.9.2 High-Temperature Corrosion Resistance

Oxidation It is well known that elements having greater thermodynamic affinity for oxygen tend for form passive barriers in alloy systems, thus providing the required resistance. Chromium, aluminum, and silicon are the three major elements that account for these passive barriers. The usefulness of protective chromia Cr_2O_3 is limited to around 950°C due to the formation of volatile chromium oxide (CrO_3). The higher thermodynamic stability of the alumina sublayer, at even very low partial pressures of oxygen, improves the alloy 602CA oxidation resistance in cyclic tests. Rare-earth elements further reduce the cracking, fissuring, and spalling of the protective oxide.

Table 7.15 presents the laboratory test data on cyclic oxidation testing (24-h cycles—1.5-h heat up, 16 h hold at temperature, and furnace cool down, for test temperatures up to 1100°C, and cooling in air for temperatures higher than 1100°C) for periods up to 1200 h. As is evident, alloy 602CA gave superior performance when compared to many other iron, nickel, and cobalt base alloys. Metallographic examination of alloy 602CA showed a continuous alumina sublayer without any selective internal oxidation by comparison to alloy 601 (Fig 7.5). Further tests conducted on alloy 602CA and alloy 601 for 3150 h

TABLE 7.15 Cyclic Oxidation Data—1200 h, 24-h Cycles

	Weight Change in mg/m^2 h				
Alloy	750°C	850°C	1000°C	1100°C	1200°C
602CA	**+0.4**	**+3**	**+12**	**+7**	**−310**
X	+1	+8	+5	−5	—
800H	+7	+8	−24	−162	—
625	+1	+6	−100	−1410	—
601	+1	+10	+7	−24	−820
188	+1	+4	+7	−302	—
617	+4	+12	+19	−19	—

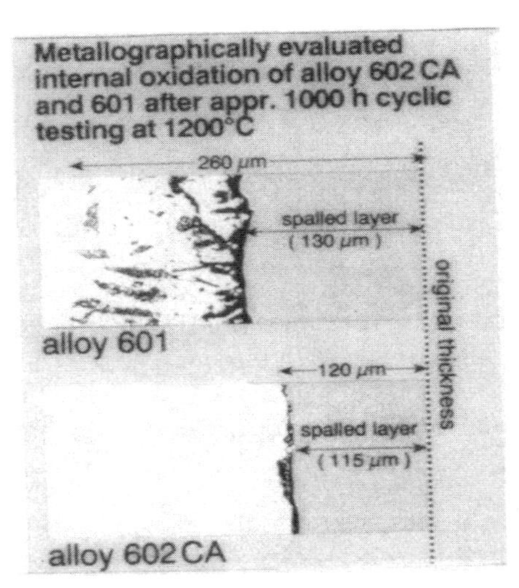

FIGURE 7.5 Internal oxidation attack on alloy 601 after cyclic testing at 1200°C in air for 1000 h (alloy 602CA—no internal attack).

at a lower temperature of 2100°F (1148°C) again showed excessive internal oxidation with alloy 601. In contrast alloy 602CA had no internal attack but only a thin surface oxide scale. This is especially beneficial in applications that utilize thin sheets such as in radiant tubes. No internal oxidation means the entire wall thickness is sound metal and the alloy retains most of its original properties. The higher thermodynamic stability and more than 5 orders of magnitude lower dissociation pressure of alumina are the primary reasons for formation of the protective alumina layers. Another series of cyclic oxidation test at 2100°F (1148°C) for 3000 h (cycle time of 160 h) measured the weight loss as well as total penetration in mils by metallographic examination. These results are shown in Table 7.16.

TABLE 7.16 Weight Gain and Internal Penetration of Various Alloys after 3000 h Exposure in Air at 2100°F (1148°C)

Alloy	Weight Gain (mg/cm^2)	Max Internal Penetration (mils)
330	55	—
333	34	15.4
446 Stainless	>530	79.6
353MA	37	16.8
617	30	6.3
230	23	5.0
602CA	**18**	**1.41**
214	9.3	2.8
HR120	206	46.3
800HT	294	54.4

7.9.3 Carburization/Metal Dusting

Besides oxygen attack, high-temperature alloys are frequently subjected to attack by carbon. Gaseous environments generated by many high-temperature industrial processes, particularly in the petrochemical/refinery industries, in the conversion of fossil fuels and in certain heat treatment operations, frequently contain gases with carbon activities of up to 1. In other cases, such as in ammonia or methanol synthesis, carbon activities can be much higher than 1. The degradation of metallic systems in carburizing environments can take two forms, namely carburization and metal dusting (sometimes referred to as catastrophic carburization). Due to the very low solubility of carbon in nickel, materials with high nickel content are considered beneficial for imparting carburization resistance. Alloys high in chromium, aluminum, and silicon may form protective oxide layers, which prevents the ingress of carbonaceous corrosive species thus providing improved resistance. However, if alternating exposure to carburizing and oxidizing environments is experienced, the precipitated carbides are converted to oxides, and the liberated CO widens the grain boundaries thus loosening the oxide layer, thereby causing accelerated deterioration.

The higher nickel plus chromium coupled with high aluminum content of alloy 602CA results in lowest weight gain in the temperature range tested as shown in Table 7.17. The reason for improved carburization behavior is due to the formation of an alumina sublayer rather than via the nickel content alone as exhibited by the oxidation data in Table 7.15 at 1200°C for alloy 602CA and alloy 601 and Table 7.16 data at 2100°F (1148°C). A recent study by Brill and Agarwal [44] examines the carburization behavior of various nickel and iron base alloys in the temperature range of 550–1200°C. The results from this study also confirm the excellent carburization resistance of alloy 602CA. Another study by Brill [45] shows the mathematical relationship between the effects of the alloying elements in an alloy with the carburization resistance. Brill introduced a constant

TABLE 7.17 Cyclic Carburization Behavior in CH_4/H_2 Environment (A_c = 0.8) in Temperature Range 750–1000°C

	Weight Change (mg/m^2 h)		
Alloy	750°C	850°C	1000°C
310	2	130	305
800H	4	143	339
625	4	105	204
617	2	50	64
X	2	93	204
601	2	69	152
602CA	**0**	**44**	**58**

K_B that showed the positive effect of Ni + Co, Mo, Si, Mn, Al, and C. The higher the value of this constant, the better was the performance of the alloy in carburizing environments.

In a recent study on metal dusting behavior of nine nickel base alloys and four Fe–Ni–Cr alloys tested in a carburizing H_2–CO–H_2O gas with a carbon activity $a_C >> 1$ at 650°C, alloy 602CA was one of the most resistant material. Table 7.18 gives the tabular data for the various materials tested. One very important point to note is that these results were obtained on unstressed

TABLE 7.18 Total Exposure Times and Final Wastage Rates after Exposure in Carburizing CO-H_2-H_2O Gas at 650°C

Alloy	Surface Condition	Total Exposure Time (h)	Final Metal Wastage Rate (mg/cm^2 h)
800H	Ground	95	0.21
HK-40	—	190	0.04
HP-40	—	190	0.038
DS	Ground	1,988	4.3×10^{-3}
600H	Ground	5,000	0.033
601	Black	6,697	7.3×10^{-3}
601	Polished	1,988	4.9×10^{-3}
601	Ground	10,000	5.8×10^{-4}
C-4	Ground	10,000	1.1×10^{-3}
214	Ground	9,665[a]	1.2×10^{-3}
160	Ground	9,665[a]	6.3×10^{-4}
45TM	Black	10,000	1.0×10^{-5}
602CA	**Black**	**10,000**	**1.1×10^{-5}**
617[b]	Ground	7,000[a]	3.7×10^{-6}
690	Ground	10,000	2.0×10^{-6}

[a] The total exposure time of these specimens was less than 10,000 h because they were inserted later than the other alloys.

[b] Alloy 617 showed evidence of metal dusting after 7,000 h.

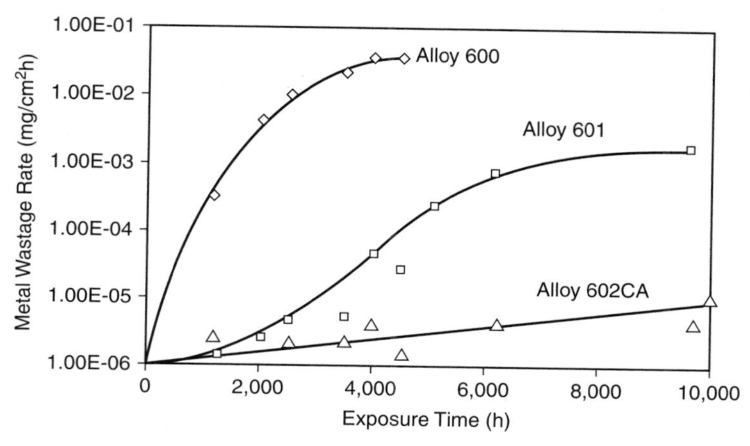

FIGURE 7.6 Metal wastage of nickel base alloys 602CA, 600, and 601 due to metal dusting after exposure in strongly carburizing $CO - H_2 - H_2$ gas at 650°C.

coupons. In the real world the components exposed to metal dusting environments are stressed and hence have certain amount of strain. Alloy 602CA, even with 1% strain, maintained its passive oxide layers, thus preventing any accelerated attack, whereas in alloy 690 the passive layer is damaged leading to accelerated metal wastage. Figure 7.6 shows the comparison between alloys 600, 601, and 602CA.

The combination of excellent high-temperature strength at temperatures greater than 1000°C and the excellent oxidation resistance up to 1200°C with good carburization/metal dusting resistance led to the selection and good performance of alloy 602CA in several diverse applications, as described later.

7.10 FABRICABILITY/WELDABILITY OF ALLOY 602CA

Welding of alloy 602CA follows the same general rules established for welding other highly alloyed nickel base materials, where cleanliness is very important and critical. Heat input should be kept low with interpass temperatures not exceeding 150°C, preferably 120°C. The use of gas tungsten arc welding (GTAW) process and matching filler metal is recommended. For shielded metal arc welding matching electrodes are available. Submerged arc welding with a GTAW top layer has also been successfully used. Preheating is not required. The shielding gas for GTAW is argon +2% nitrogen, and its use is very critical in preventing any hot cracking. For gas metal arc welding (GMAW), the shielding gas is argon with additions of helium, nitrogen, and carbon dioxide (argon +5% nitrogen +5% helium +0.05% carbon dioxide). Details on welding parameters, hot working, cold working, heat treatment, descaling, and machining are presented elsewhere [42, 43].

7.11 APPLICATIONS OF ALLOY 602CA

Due to the unique combination of the above mentioned properties, alloy 602CA has been extensively used in the following applications. More potential applications continue to develop via test programs being conducted in many industries and will be reported in future as they materialize.

- *Heat Treatment Industry*: Furnace rolls, bell furnaces, bright annealing furnaces, accessories and transport hooks for enameling furnaces, transport rollers for ceramic kilns, wire conveyor belts, anchor pins for refractories, tubes for bright annealing wires, and other furnace accessories
- *Calciners*: Rotary kilns for calcining and production of high-purity alumina, calcining of chromic iron ores to produce ferrochrome, production of nickel and cobalt oxides and reclamation of spent nickel catalysts from petrochemical industries
- *Chemical/Petrochemical*: (a) Production of hydrogen via a new steam reformer technology, (b) production of phenol from benzene via a new and cheaper process, and (c) pig tails in refinery reformer
- *Automotive*: Catalytic support systems, glow plugs, exhaust gas flaps
- *Nuclear Industry*: Vitrification of nuclear waste
- *Metallurgy*: Direct reduction of iron ore technology to produce sponge iron
- Many others

A detailed description of some of these applications along with pictures of the various components fabricated from alloy 602CA are presented in the open literature [1, 13–15, 42, 43].

7.12 FABRICATION OF AQUEOUS CORROSION-RESISTANT ALLOYS

The nickel alloys for the modern chemical process and petrochemical industries not only have to resist uniform corrosion, localized corrosion, and stress corrosion cracking as characterized by their alloy content, but they must essentially maintain these same corrosion resistance characteristics during manufacture, fabrication, and assembly of various components used in these industries. The industry has experienced some costly mishaps and economic penalties due to premature failures of the component either during fabrication (cracking) or accelerated corrosion during service. This problem in many cases has been traced back to the lack of fundamental understanding of the physical and mechanical metallurgy of the different alloy systems and practice of technically wrong procedures during welding, hot- and cold-working operations, and stress relieving/solution annealing treatments.

The different nickel base alloys of the Ni–Mo and Ni–Cr–Mo families exhibit different physical metallurgical behavior as relates to various phase precipitations during thermomechanical processing and hence the need to follow somewhat different sets of guidelines and precautions in comparison to mild steel and in some cases even stainless steels during various fabrication procedures.

The do's and don'ts of various fabrication techniques such as hot forming vs. cold forming, the need to use proper weld consumables during welding, whether to anneal or not to anneal after welding, is stress relieve heat treatment applicable, the type and degree of cleanliness required for welding, other weld parameters, the heat treatment of formed components, and the heat treatment parameters, the question of iron contact and contamination during fabrication—these and many other such questions generally come up when working with the above alloy classes. The effect of the fabrication parameters on corrosion resistance and toughness is a frequently raised concern. In certain instances weld overlay is employed to combat corrosion, and this technology generates its own set of concerns and questions. Some of these questions are addressed below.

7.12.1 Welding

The welding of nickel base alloys requires some basic precautions regarding cleanliness of the weld zone, and one should take into account the more sluggish nature of welding products along with the low penetration in the base metal. Cleanliness of the weld joint is the single most important parameter for producing a sound weldment. Lack of thorough cleaning has accounted for a majority of the problems encountered in the industry, that is, cracking, porosity and accelerated corrosion. The contaminants to watch out for prior to welding are carbon, oxides, sulfur, lead, phosphorous, and other elements that form low-melting-point eutectics with nickel such as tin, zinc, bismuth, antimony, and arsenic. These could come from a variety of sources such as marking crayons, temperature-indicating sticks, machining oil, grease, oil mist from compressors, shop dirt, and other sources. The importance of thorough cleaning prior to welding cannot be overemphasized. Cleaning of the base metal in the weld area (both sides of the weld joint) should be carried with acetone or other suitable cleaners. Nowadays, due to environmental restrictions, the use of trichloroethylene (TRI), perchloroethylene (PER), and carbon tetrachloride (TETRA) is prohibited in many countries. Grinding with a clean alumina grinding wheel, at least 2 in. from either side of the weld joint has also been used for cleaning prior to welding. It is very important that the grinding wheels used are brand new and have not been previously used on either stainless steel or carbon steel.

Due to the viscous nature of molten nickel–alloy weld metal, increased joint angle and using relatively thinner land to compensate for low penetration are often necessary to produce good-quality welds. A misconception by welders unfamiliar with nickel alloys is to attempt to make the weld-metal flow by increasing the amperage (heat input) above the recommended range. Excessive amperage will not improve the welding characteristics or flowability of the weld metal but may cause mechanically unsound and metallurgically poor corrosion-resistant welds. Table 7.19 gives the basic guidelines of the weld parameters for various welding

TABLE 7.19 Welding Parameter Guidelines for Alloy 59 and Other Alloys

Sheet/ Plate Thickness (mm)	Welding Process	Filler Metal		Welding Parameters Root Pass		Intermediate and Final Passes		Welding Speed (cm/min)	Flux/ Shielding Gas Rate (L/min)	Plasma-Gas/ Rate (L/min)	Plasma/ Nozzle Diameter (mm)
		Diameter (mm)	Speed (m/min)	A	V	A	V				
3.0	Manual GTAW	2.0		90	10	110–120	11	10–15	Ar W3[a] 8–10		
6.0	Manual GTAW	2.0–2.4		100–110	10	120–130	12	10–15	Ar W3[a] 8–10		
8.0	Manual GTAW	2.4		110–120	11	130–140	12	10–15	Ar W3[a] 8–10		
10.0	Manual GTAW	2.4		110–120	11	130–140	12	10–15	Ar W3[a] 8–10		
3.0	Autom. GTAW	1.2	0.5	Manual		150	10	25	Ar W3[a] 15–20		
5.0	Autom. GTAW	1.2	0.5	Manual		150	10	25	Ar W3[a] 15–20		
2.0	Hot wire GTAW	1.0	0.3			180	10	80	Ar W3[a] 15–20		

10.0	Hot wire GTAW	1.2	0.45	Manual		250	12	40	Ar W3[a] 15–20		
4.0	Plasma arc	1.2	0.5	165	25			25	Ar W3[a] 30	Ar W3[a] 3.0	3.2
6.0	Plasma arc	1.2	0.5	190–200	25			25	Ar W3[a] 30	Ar W3[a] 3.5	3.2
8.0	MIG/ MAG[b]	1.0	Approx. 8	GTAW		130–140	23–27	24–30	Ar W3[a] 18–20		
10.0	MIG/ MAG[b]	1.2	Approx. 5	GTAW		130–150	23–27	20–26	Ar W3[a] 18–20		
6.0	SMAW	2.5		40–70	Approx. 21	40–70	Approx. 21				
8.0	SMAW	2.5–3.25		40–70	Approx. 21	70–100	Approx. 22				
16.0	SMAW	4.0				90–130	Approx. 22				

[a] Argon or argon + max. 3% hydrogen.
[b] For MAG Welding use of the shielding gas Cronigon He30S is recommended. (Cronigon He30S = argon + 30helium + 2hydrogen + 0.05 carbon dioxide). In all gas-shielded welding operations, ensure adequate back shielding. These figures are only a guide and are intended to facilitate setting of the welding machines and should not be taken as absolute.

TABLE 7.20 Guidelines for Energy Input per Unit Length During Welding

Welding Process	Energy Input[a]/Unit Length During Welding (kJ/cm)
GTAW (manual, automatic)	10 max
Hot wire GTAW	6 max
Plasma arc	10 max
GMAW (MIG/MAG)—manual, automatic	11 max
SMAW (manual metal arc)	7 max

[a]Heat input = $\frac{U \times I \times 60}{V \times 1000}$ kJ/cm, where U = arc voltage, volts; I = welding current, amps; V = welding speed cm/min.

processes, where as Table 7.20 give the basic guidelines for heat input during these processes. It must be noted that these are just basic guidelines and will have to be optimized for individual processes, positions, machines, and operators during any weld procedure qualifications.

Due to the need to manipulate the weld metal in the weld joint, nickel alloys require more openness (wider root gaps and larger included angles) to permit the use of a slight weaving technique, which should not exceed three times the diameter of the weld wire. These parameters are necessary in comparison to welding of mild steel because of lower thermal conductivity and higher thermal expansion characteristics of nickel alloys. Generally, welding can be done by all the conventional processes such as GTAW (TIG), GMAW (MIG), and SMAW (coated electrode). Submerged arc welding can be used, but, depending on the flux used, the danger of carbon and silicon pick up or chrome depletion in the weldment exists that could lower the corrosion resistance. Plasma arc welding has also been successfully used for welding nickel alloys. Another new technique in recent use is GTAW hot-wire welding, which uses a 2% hydrogen addition to the argon shielding gas. This has resulted in a significant increase of welding speed, thereby reducing the cost and increasing production efficiency. This process has been very successfully used in welding roll-clad steel with alloy 59 for FGD systems of large coal-fired power plants. Due to the extremely low iron content of alloy 59 filler metal, the iron dilution is kept at very low levels thus fully maintaining the corrosion resistance of the weld joint.

Another welding process, a variation of GMAW process known as the MAG process (metal active gas) is gaining popularity. In this the shielding gas contains an active gas component such as carbon dioxide (0.05%) in argon plus 30% helium and 2% hydrogen. This gas mixture is known as CronigonHe30S and is very popular in Germany.

Welding products suitable for welding Ni–Cr–Mo-type alloys are matching filler metal or overalloyed filler metal such as alloy 59, which is designated under AWS/ANSI A5.14 as ERNiCrMo-13 and ENiCrMo-13 in AWS A5.11. For welding the superaustenitic SS (6% Mo SS), and other low alloys such as alloy 825 or 904L, it is recommended that a minimum 9% Mo containing alloy such as alloy 625 (ERNiCrMo-3) be used because autogenous welding will significantly

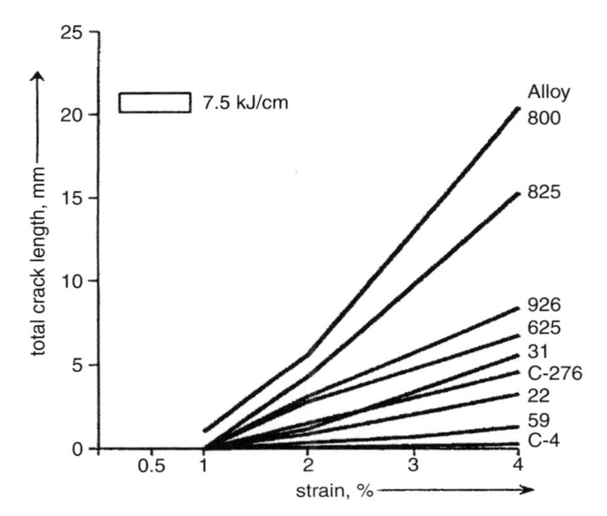

FIGURE 7.7 Hot-cracking sensitivity of various alloys as measured in a modified varestraint test.

lower the corrosion resistance through microsegregation of molybdenum during solidification. Since alloy 625 is a niobium-containing Ni–Cr–Mo alloy (9% Mo), its high hot cracking sensitivity has been known to cause some problems during welding (Fig. 7.7). Also, there has been an embrittling tendency when using this alloy as a filler metal for welding duplex and superduplex stainless steels, due to formation of niobium nitrides. Also, weldments of this alloy in the power industry and other high-temperature applications have been known to embrittle, due to formation of gamma double prime Ni_3Nb at temperatures above 1000 F. A new improved niobium-free version alloy of this Ni–Cr–Mo family, with even higher Mo content, has been developed by Krupp VDM to overcome the limitations of alloy 625 mentioned above. This alloy is known as alloy 50 with a UNS N06650. Its coverage in AWS and ASTM specifications is pending. Due to its higher Mo content with additions of some tungsten, alloy 50 (Ni bal, Mo 12%, W 2%, Fe 13.5%) would be an excellent filler metal, not only for welding 6% Mo alloys, duplex, and superduplex stainless steels, but also for low nickel alloys such as alloys 825, 904L, 20, and even stainless steels. Details on this alloy are published elsewhere [46, 47].

The alloy in the initial testing has shown promising results in weld overlay of superheater tubes in the power boilers of coal-fired power plants and refuse to energy power plants. Alloys C-276 and 59 have also been successfully used for welding these superaustenitic alloys. For welding of Ni–Mo alloys only the matching filler metals of the Ni–Mo family are recommended.

Neither preheating nor postweld heat treatment is required for these alloys, which are designed to be used in the as-welded condition due to their very low carbon contents. No major problems have been reported from the field if the welding was done properly. For dissimilar welding of these alloys to carbon steel, stainless steel, and other nickel alloys, the welding practices are not significantly different except for the selection of a proper filler metal, which is a very important parameter. Also, when joining these alloys to carbon or low-alloy steels, the arc may have a tendency to migrate on to the steel side of the weld joint. Hence, proper grounding procedures and techniques, short arc length, and torch/electrode manipulation capabilities are essential to counteract and compensate for this problem. When welding carbon steel/low-alloy steel to any stainless or superstainless steel, or Ni–Cr–Mo alloy, it is advisable to use either alloy C-276 or alloy 59, which fall in the category of fully austenitic overalloyed filler metals. Some fabrication shops have also used alloy 625 (ERNiCrMo-3) filler metal depending on the intended service. Under above conditions, when one of the base metals is of the Ni–Mo family, such as alloy B-2, then the recommended filler metals are alloys B-2 or B-4. As a general rule the choice of the filler metal will be greatly influenced by the intended corrosive service. It has been shown that coated electrode welding of Ni–Mo alloys is difficult and is not recommended for Ni–Mo alloys.

Figure 7.7 shows the sensitivity to hot cracking for various alloys as measured in a modified varestraint test. As is obvious, the sensitivity to hot cracking of tungsten and columbium free alloys, such as alloy 59 and C-4, is low compared to the tungsten and columbium containing alloys in the Ni–Cr–Mo family such as alloys C-276, 22, G-3, and 625. Another often-raised question with these alloys is whether to remove the heat tint, produced during welding. On the higher alloys as those discussed in the Ni–Cr–Mo and Ni–Mo family, it may not be necessary to remove the heat tint if proper welding procedures have been followed. Extensive field case histories in FGD systems of many coal-fired power plants in United States have shown that removal of heat tint (discoloration) is not necessary for the high alloys like C-276 but is necessary for lower alloys like 904L and stainless steels. Lab tests in simulated FGD environments have also confirmed this conclusion [48]. If excessive heat input or inadequate shielding has resulted in an inferior weld and excessive heat tint, then other remedial measures need to be taken, which will depend upon the specific condition in question.

Information on these remedial measures and answers to specific questions on welding are also available from the producers of the various Ni–Cr–Mo, Ni–Mo, and superaustenitic stainless alloys.

One question has been frequently raised on the manufacturing process of tubes and or pipes of Ni–Cr–Mo, Ni–Mo alloys, and lower grades of Ni–Cr–Mo alloys. The question relates to the need for a "full draw of the as-welded tube" prior to final anneal or the adequacy of the "welded or welded and bead-worked annealed tubes." Discussions with many people in this field and tests run have shown that although this requirement of welded and fully drawn and annealed tubes holds true for standard 18-8 variety stainless steel tubes for critical chemical services, this is not so for superaustenitics or high alloys of the Ni–Cr–Mo family. A very prominent and respected engineer, Dillon [49] states, "in the CPI the requirement to fully draw a longitudinally welded tube prior to annealing has been applied only to the standard 18-8 stainless steels. There have been no reports of selective weld corrosion in either the super-austenitic grades or high Ni–Cr–Mo alloys like C-276. Likewise, no cold-work is needed. In fact many of the field welds in this Ni–Cr–Mo alloy C-276 are left in the as-welded condition without the benefit of any solution annealing, and these too in the industry have performed well. I know of no instances in which welds in tubing, pipe or vessels have been selectively attacked, except in the case of defective welds or if the environment has been to severe for alloy C-276 base metal itself. Hence the bead worked and annealed tubes of these alloys, as opposed to fully drawn and annealed tubes, are totally adequate for the CPI corrosive environments." Not only are these tubes adequate for the service but they are significantly cheaper than the fully drawn and annealed tubes.

7.12.2 Cold Forming

Stainless steels and nickel base alloys are readily cold formed because they possess excellent ductility in the as-supplied condition (solution annealed). However, in comparison to mild steel, these alloys have somewhat higher strength and hence the need to have forming equipment with sufficient power commensurate with the mechanical characteristics of these alloys. Table 7.21 gives the mechanical properties of some of these alloys. Another point to consider is that nickel alloys have high strain-hardening coefficients and hence are subject to rapid strain-hardening, but because of their excellent ductility and uniform elongation, are capable of being reduced or drawn extensively without rupturing. However, due to the high strain-hardening the number of reductions between anneals is limited. Hence the nickel alloys require more frequent anneals in a progressive forming sequence in comparison to mild steel or standard stainless steels.

Since forming is very often accomplished with the same tools and equipment normally used for mild steels, it is very important and necessary to remove any

TABLE 7.21 Typical Room Temperature Mechanical Properties of Various Alloys

Alloy	UNS #	UTS (N/mm^2)	0.2% Yield (N/mm^2)	% El
316L	S31603	538	241	55
904L	N08904	586	283	50
926	N08926	655	310	50
31	N08031	717	352	50
G-3	N06985	676	310	50
625	N06625	862	448	45
C-276	N10276	759	365	60
22	N06022	731	359	60
59	N06059	772	379	60
B-2	N10665	847	426	61
B-4	N10629	860	425	66

embedded iron particles from the surface, if any, and the need for cleaning the formed part prior to any heat treatment or welding. Lubrication is generally not required for simple "U" bends as on a press brake. However, severe or progressive die forming operations require heavy-duty lubricants with good surface wetting characteristics and high film strength such as found in metallic soaps and chlorinated or sulfo-chlorinated oils. It is very important to remove all traces of these lubricants prior to any heat treatment or welding due to danger of the carbon pick-up and lowering of the corrosion resistance by forming complex carbides. Parts formed with zinc alloy dies should be flash pickled to prevent liquid metal embrittlement during heat treating.

A frequently raised concern is whether to anneal after forming prior to putting in service or prior to any welding. No specific guidelines can be given but from experience some general observations can be made:

- For most materials cold forming to a degree of 15% or less (maximum strain in the outer fiber) is permissible without subsequent heat treatment. However, in certain cases depending on the intended service, a full solution heat treatment may be necessary. Care should be exercised when annealing components with 5–10% reduction due to the danger of excessive irregular grain growth known as "orange peel effect." This may occur on materials that are severely bent and then annealed.
- In Ni–Mo alloys, if any welding is to be performed after forming, it is advisable to give a full solution anneal prior to welding. Prior to annealing of cold-formed heads, a shot peening operation with clean sand is necessary with alloy B-2. However, this shot peening may not be necessary with alloy B-4. However, in the author's opinion, until more data is generated, solution annealing prior to any welding of a cold-formed head or part is advisable for both alloys B-2 and B-4. The metallurgical reasons for this phenomenon is clearly explained in published studies [10, 11]. This phenomenon relates to the kinetics of formation of an embrittling body centered tetragonal β phase (Ni_4Mo). Higher iron content, as in controlled chemistry B-2 and B-4 significantly retards the formation of this detrimental phase, which appears to have been responsible for the cracking behavior

In Ni–Cr–Mo alloys such as alloy C-276 and 59, the corrosion resistance does not suffer as a result of cold work. This has been proven and well documented in successful use of alloy C-276 in deep sour gas wells in the cold reduced condition. However, if any welding is to be done on these Ni–Cr–Mo alloys, which have less than 15% cold work, it can be done without any heat treatment. The superaustenitic stainless steels can be treated similar as the Ni–Cr–Mo alloys regarding this aspect.

7.12.3 Hot Forming

Hot forming generally is carried out in a temperature range between the solidus temperature and start of recrystallization temperature. For most nickel base alloys,

TABLE 7.22 Hot Working and Solution Annealing Temperature Range for Various Alloys

Alloy	Hot Working Temperature Range (°C)	Solution Annealing Temperature Range (°C)
316L	1100–850	1080–1030
904L	1200–900	1150–1080
926	1200–900	1150–1100
31	1200–900	1150–1100
G-3	1170–950	1150–1100
625[a]	1180–1020	1150[a]–927[a]
C-276	1200–950	1140–1100
22	1200–950	1140–1100
59	1180–950	1140–1100
B-2	1200–980	1090–1050
B-4	1200–980	1090–1050

[a] Alloy 625 can be heat treated between these temperature ranges depending on specification needs. Generally the range is 950–1050°C.

these lie between 2250°F (1232°C) and 1600°F (871°C). Due to their alloying elements these alloys have higher high-temperature strength, lower thermal conductivity, higher strain-hardening coefficient, higher strain rate sensitivity, and undergo rapid increase in strength with falling temperatures in the hot-working temperature range. Due to these characteristics, the hot-forming temperature range tends to be narrow. The upper temperature range is used for either reducing the cross-section thickness or forming or a combination thereof, and lower temperature ranges are used to develop specific properties after solution annealing. Table 7.22 gives the hot-forming and solution-annealing temperatures for the various alloys. For low degrees of hot deformation (less than 20%), the forming temperatures should be as close as possible to the lower limit, and the finishing temperature should be sufficiently lower than the final solution heat treatment temperature so as to develop proper microstructure and mechanical properties after annealing. For greater degrees of deformation, higher temperatures are recommended with appropriate reheats. Care must be taken during the last reheat to ensure that there would be enough reduction after the last reheat and that the hot finish temperature would be sufficiently below the solution annealing temperature.

After all hot working, it is recommended to give a full solution anneal and water quench for the Ni–Mo, Ni–Cr–Mo, and superaustenitic stainless steels to provide optimum corrosion resistance. Figure 7.8 shows the time–temperature–sensitization diagram for most of the Ni–Cr–Mo alloys, as determined by ASTM Method G-28A. As is evident, alloy 59 is the most thermally stable alloy in comparison to other Ni–Cr–Mo alloys like C-276, C-4, and alloy 22.

For Ni–Mo alloys, Fig. 7.9 shows the effect of higher iron content on retarding the kinetics of formation of embrittling ß phase, as measured by isoimpact curves. The new alloy B-4 (UNS N10629) has solved both the cracking problem during

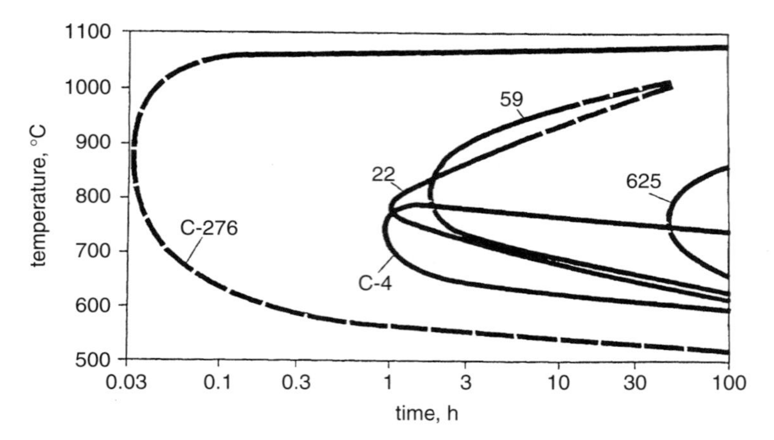

FIGURE 7.8 Time–temperature–sensitization diagram for Ni–Cr–Mo alloys as measured by ASTM G-28A test.

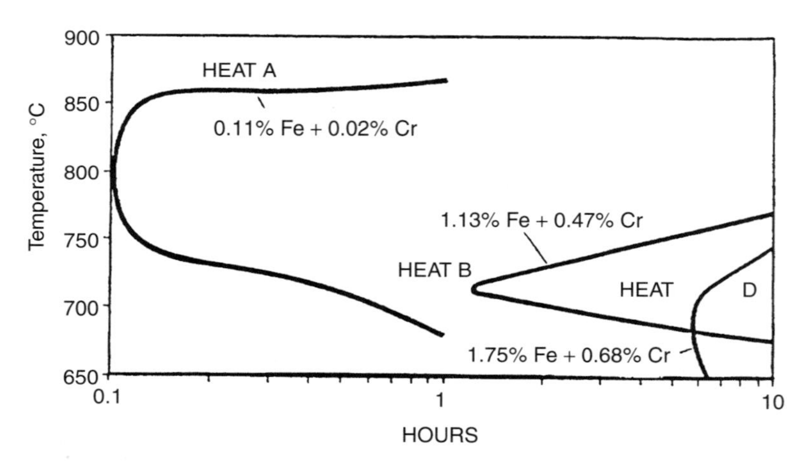

FIGURE 7.9 Kinetics of β phase (Ni_4Mo) formation in different iron-containing Ni–Mo alloys as indicated by ISO impact—100 J curves.

fabrication, as used to be the case with uncontrolled iron chemistry of alloy B-2, and stress corrosion cracking problems in specific environments. Again this SCC has been related to the formation of embrittling ß phase, which does not form in alloy B-4 during normal thermomechanical and welding processes.

7.12.4 Heat Treatment

Heat treatment of Ni–Mo, Ni–Cr–Mo, and superaustenitic stainless steels is relatively simple and straightforward but does require close attention to temperature, rate of heat up, time at temperature, cleanliness, and method of cooling.

The annealing is designed to optimize both mechanical and corrosion resistance characteristics. Following all hot-working/forming and most cold-forming operations, a full solution anneal is done. Stress relief at intermediate temperatures is generally not recommended due to danger of precipitation, which can be detrimental to mechanical behavior and/or corrosion resistance. Prior to annealing, cleanliness (i.e., removing grease, oil, and carboneous materials and foreign substances) is a must because of the possible diffusion of carbon, thus lowering the corrosion resistance. It is recommended to charge the alloy component (specially Ni–Mo alloys) in a hot furnace as opposed to charging cold and gradually raising the furnace temperature, unless there is a compelling reason to do so. Holding time of 5–10 min at temperature is sufficient, and this "counting of time at temperature" should only be started after the entire cross section of the part being heat treated is at the proper heat treatment temperature. Hence proper placement of thermocouple or thermocouples to ensure this is very critical. It is also important to avoid any stagnant condition during the heat treatment due to the possibility of accelerated scale formation and high-temperature attack especially in Ni–Mo alloys, which do not contain chromium. Rapid cooling is very important and essential to prevent any detrimental phase precipitation. Depending on section thickness, the component can be either rapid air-cooled or water quenched. Table 7.22 gives the solution annealing temperatures for various alloys.

7.12.5 Descaling and Pickling

After the heat treatment removal of oxide film, which is more tenacious and adherent than compared to stainless steel can be done chemically or mechanically. Chemical treatment may consist of molten caustic descaling salt treatment followed by nitric–hydrofluoric acid bath mixtures (20–25% nitric acid plus 3–5% hydrofluoric acid) at 120–160°F. For Ni–Mo alloys, since they do not contain any chromium, it is very important not to leave the material in the acid bath for more than 60 s followed by an immediate water rinse. This should be repeated as necessary. If sulfuric–hydrochloric acid baths are used, then this precaution is not that critical for Ni–Mo alloys.

Sand, shot, or vapor blasting can also be used for descaling with proper care.

7.12.6 Weld Overlay/Wallpapering

Overlay welding of carbon steel or low-alloy steel with the high-performance corrosion-resistant nickel alloys has seen increased activity in recent years such as in overlay welding of thick-walled tube sheets and flanges. Since the weld contains microsegregation of alloy constituents, its corrosion resistance may be slightly inferior to a fully wrought structure. This inferiority can be further compounded by base metal dilution or use of the improper flux. In recent years, electro slag overlay welding has been successfully employed providing excellent results. Other methods are submerged arc strip cladding and gas metal pulse arc overlay welding. Each method had its own advantages and disadvantages and is discussed in greater detail elsewhere [50]. Use of two- or three-layer deposits to achieve a desirable deposited chemistry is needed for optimum corrosion resistance and in most cases depends on the process and welding parameters used. However, with very low iron containing Ni–Cr–Mo alloys, the same desired chemistry and corrosion resistance can be obtained with one layer as is the case with Ni–Cr–Mo alloy 59, which has an iron content of typically less than 1%. However this will depend on the intended application, and it may be necessary to use two-layer deposit if the environment is very corrosive.

If large areas are to be protected, a technique known as wallpapering has been successfully used in the industry, especially in the power plant flue gas desulfurization systems. In these systems a 0.063-in.-thick sheet of corrosion-resistant alloy of the Ni–Cr–Mo family is applied over carbon steel structure by the "wallpapering" technique, which has proven to be very reliable and cost effective. Details of this method are described elsewhere [51].

7.12.7 Grinding and Machining

When very close tolerances are required, grinding is the preferred method. Grinding wheels from reputable manufacturers such as Norton, Carborundum, and others have been successfully used. Recommended wheels and coolants for these alloys are available from the manufacturers of these wheels. Nickel alloys are readily machinable by the conventional methods. Tungsten carbide and recently ceramic-tipped tools have been successfully used for machining these alloys. High-speed steel tools have also been used. During machining some of the high-nickel alloys work harden rapidly, generate high heat during cutting, may weld to the cutting tool surface, and offer high resistance to metals removal due to their high shear strength in comparison to stainless steels and/or carbon steels. The following points should be kept in mind when machining nickel alloys:

- Machine should have sufficient power and be rigid.
- Workpiece and cutting tool should be held rigid without any chatter. Vibration should not be present during machining operation.
- Tools should be sharp at all times and should be changed as the need arises. A 0.015-in. wear land is considered a "dull tool." Use positive rake angle tools for most machining operations. Negative rake angle tools can be used for intermittent cuts and heavy stock removal.
- Use heavy constant feed to maintain positive cutting action, that is, the tool should be constantly engaged with low cutting speed and higher feed rate. If feed slows and the tool dwells in the cut, work hardening will occur, tool life will lessen and close tolerances will be difficult to achieve.
- Lubrication is desirable. Water-based cutting oils are recommended to avoid overheating of the cutting tools. However, operations such as threading, tapping, or drilling require sulfonated or chlorinated lubricants. However, all traces of these fluids must be removed prior to any heat treatment or high-temperature service.

7.13 APPLICABLE SPECIFICATIONS

Just about all the alloys mentioned in this chapter are covered by the various national standards organizations such as ASTM, NACE, AWS, ASME, and international organizations in European countries, Japan, and others. The most prevalent standards covering these materials are of ASTM, AWS, NACE, and ASME.

7.14 SUMMARY

An attempt has been made to describe nickel and its alloys of the unary, binary, ternary, and quaternary systems for applications in both aqueous corrosion and high-temperature applications. Some corrosion data, mechanical data, and general information on fabrication has also been provided. Some of the newer alloys, including the 6% Mo alloys, developed in the last 50 years have been discussed, specially the chronology of the various alloys of the C family with both their advantages and limitations. These newer alloys such as alloy 59 have shown increased acceptance in the industry due to its superior properties over the current workhorse of the Ni–Cr–Mo alloy family, alloy C-276. Alloy C-22, an alloy introduced in the 1980s, has now been superseded by the newer alloys of the 1990s, that is, alloy 59, alloy 686, and alloy C-2000. Alloys 686 and C-2000 have yet to find major commercial applications. Alloy C-4 has found some applications, but mostly in European countries. Hence from the current portfolio of the C family alloys, it is obvious that alloy C-276 will continue to be the workhorse of this alloy family, followed by alloy 59, which has superior properties. Alloy 59 has and will continue to replace alloy C-276 in those extreme and severe corrosive media where alloy C-276 is either inadequate or marginal in nature or where the industry, due to "safety and reliability" considerations, seeks a better alloy than C-276. Alloy 59 will continue to fulfill these specific needs of the CPI and other industries.

BIBLIOGRAPHY

Krupp VDM, *High Temperature and Corrosion Resistant Alloys Product Literature*, Krupp VDM GmbH, P.O.Box 1820, Werdohl, Germany, 58778.

Haynes High Temperature and Corrosion Resistant Alloys Product Literature, Haynes International, Kokomo, IN.

Special Metals High Temperature and Corrosion Resistant Alloys Product Literature, Special Metals, Hungtington, WVA.

Publications of Nickel Development Institute on Nickel Alloys, Nickel Development Institute on Nickel, Toronto, Canada.

NACE International, Various Publications, NACE International, Houston, TX.

ASTM Book of Standards, Vol. 02.04, *Non-Ferrous Alloys*; Vol. 03.02, *Wear & Erosion*, ASTM, West Conshohocken, PA, 2003.

REFERENCES

1. U. Brill and M. Rockel, High-Temperature Alloys from Krupp VDM for Industrial Engineering, VDM Report # 25, Krupp VDM GmbH, P.O.Box 1820, Werdohl, Germany, 58778.
2. C. T. Sims et al. (Eds.), *Superalloys II, High-Temperature Materials for Aerospace and Industrial Power*, Wiley, New York, 1987.
3. G. Lai, High Temperature Corrosion of Engineering Alloys, ASM International, Metals Park, OH, 1990.
4. D. C. Agarwal and U. Brill, *Ind. Heating*, October, 55–60 (1994).
5. U. Brill, Krupp VDM High Temperature Alloys and Their Use in Furnace Construction, VDM Report #15, Krupp VDM GmbH, P.O.Box 1820, Werdohl, Germany, 58778.
6. W. Z. Friend, *Corrosion of Nickel and Nickel Base Alloys*, Wiley, New York, 1980, pp. 32–90, 292–367.
7. M. Fontana and N. Greene, *Corrosion Engineering*, McGraw-Hill, New York, 1967.
8. NACE Basic Corrosion Course, NACE International, Houston, TX, 1984.
9. Bulletin CEB-2, "Corrosion Resistance of Nickel and Nickel Containing Alloys in Caustic Soda and Other Alkalies," *Special Metals*, Hungtington, WV, 1973.
10. D. C. Agarwal et al., *Mat. Performance*, **33**(10), 64–68 (1994).
11. D. C. Agarwal, *Corrosion/2002*, Paper #119, NACE International, Houston, TX, 2002.
12. M. Köhler et al., *Corrosion/98*, Paper # 481, NACE International, Houston, TX, 1998.
13. D. C. Agarwal and U. Brill, *Corrosion/2000*, Paper # 521, NACE International, Houston, TX, 2000.
14. D. C. Agarwal and U. Brill, *Corrosion/2001*, Paper # 382, NACE International, Houston, TX, 2001.
15. D. C. Agarwal and U. Brill, *Corrosion/2002*, Paper # 372, NACE International, Houston, TX, 2002.
16. J. Klöwer et al., *Corrosion/97*, Paper # 139, NACE International, Houston, TX, 1997.
17. D. C. Agarwal and U. Brill, *Corrosion/1993*, Paper # 209, NACE International, Houston, TX, 1993.
18. D. C. Agarwal et al., "An Advanced High Chromium–High Molybdenum Super-Austenitic Stainless Steel Solves Corrosion Problems in Plants Producing Phosphoric Acid," *Corrosion: Its Mitigation & Preventive Maintenance*, Conference Proceedings, CORCON 2000, Mumbai, India, Nov, 2000.
19. D. C. Agarwal et al., *Corrosion/2000*, Paper # 574, NACE International, Houston, TX, 2000.
20. Data Sheet Alloy 31, Krupp VDM GmbH, P.O. Box 1820, Werdohl, Germany, 58778.
21. M. Koehler et al., KVDM Report # 24, *Nicrofer 3033–Alloy 33, A New Corrosion Resistant Austenitic Material for Many Applications*, Krupp VDM GmbH, P.O. Box 1820, Werdohl, Germany, 58778.
22. Data Sheet Alloy 33, Krupp VDM GmbH, P.O. Box 1820, Werdohl, Germany, 58778.
23. F. T. McCurdy, *Proc. Am. Soc. Testing Material*, **39**, 698 (1939).
24. Metals Handbook, 9th ed., Vol. 13, *Corrosion*, ASM International, Metals Park, OH, 1987, pp. 641–657.

25. I. Class, H. Gräfen, and E. Scheil, *Z. Metallkunde*, **53**, 283 (1962).
26. M. A. Streicher, *Corrosion*, **19**, 272 (1963); **32**, 79 (1976).
27. R. B. Leonard, *Corrosion*, **25**, 222 (1969).
28. R. W. Kirchner and F. G. Hodge, *Werkst. u. Korrosion*, **24**, 1042–1049 (1973).
29. F. G. Hodge and R. W. Kirchner, *Corrosion*, **32**, 332 (1976).
30. P. E. Manning et al., Paper no. 21, *Corrosion/83*, NACE Annual Meeting in Anaheim, California, April 1983.
31. P. E. Manning et al., presented in ACHEMA 85, International Meeting on Chemical Engineering, held in Frankfurt am Main, June 9–15, 1985.
32. N. Sridhar et al., "New Development in Corrosion Resistant Ni-Cr-Mo Alloys," *J. Metals* (1985).
33. D. C. Agarwal and W. R. Herda, "Alloying Effects and Innovations in Nickel Base Alloys for Combating Aqueous Corrosion," VDM Report # 23, 1995, Krupp VDM, P.O. Box 1820, Werdohl, Germany, D-58778.
34. V. Yanish, in *Proceedings of the Second International Conference on Heat Resistant Materials*, Gatlinburg, TN, 11–14 Sept., 1995, ASM, Metals Park, OH, 1985, pp. 655–656.
35. M. Morinaga, et al., "New Phacomp and Its Application to Alloy Design," in *Superalloys*, ASM, Metals Park, OH, 1984.
36. R. Kirchheiner, M. Köhler, and U. Heubner, *Corrosion/90*, Paper #90, NACE International, Houston, TX, 1990.
37. D. C. Agarwal et al., *Corrosion/91*, Paper #179, NACE International, Houston, TX, 1991.
38. U. Heubner, "Nickel Based Alloys," in R. W. Cahn, P. Haasen, and E. J. Kramer (Eds.), *Materials Science and Technology, A Comprehensive Treatment*, VCH, Germany.
39. D. C. Agarwal et al., *Corrosion/2000*, Paper #574, NACE International, Houston, TX, 2000.
40. D. C. Agarwal et al., *Corrosion/2001*, Paper #177, NACE International, Houston, TX, 2001.
41. D. C. Agarwal et al., *Corrosion/2000*, Paper #501, NACE International, Houston, TX, 2000.
42. Krupp VDM GmbH, *Material Data Sheet Nicrofer 6025H/6025HT—Alloy 602CA*, Material Data Sheet No. 4037, latest edition dated November 2001.
43. *Rolled Alloys Data Sheet on Alloy 602CA*, Form #1602, Feb. 2001, Rolled Alloys, 125 West Sterns Road, Temperance, MI 48182-9546.
44. U. Brill and D. C. Agarwal, *Corrosion/2002*, Paper # 373, NACE International, Houston, TX.
45. U. Brill, in K. Natesan and D. J. Tillac (Eds.), *Mathematical Description of Carburization Behavior of Various Commercial Heat-Resistant Alloys, Heat Resistant Materials, Conference Proceedings*, ASM International, Materials Park, OH, 1991, pp. 203–210.
46. U. Brill and J. Heinemann, *Corrosion/2001*, Paper #483, NACE International, Houston, TX, 2001.
47. U. Brill and G. K. Grossmann, *Corrosion/2001*, Paper #170, NACE International, Houston, TX, 2001.

48. W. L. Silence and L. H. Flasche, *Corrosion/1986*, Paper #358, NACE International, Houston, TX, 1986.
49. C. P. Dillon, private communication, August 2001.
50. U. Heubner et al., "Overlay Welding of Corrosion Resistant Nickel Superalloys," *Materials Weldability Symposium*, Materials Week 90, ASM International, Metals Park, OH, 1990.
51. VDM Report #17, June 1991, " 'Wallpaper' Installation Guidelines and Other Fabrication Procedures for FGD Maintenance, Repair and New Construction with VDM High Performance Nickel Alloys," Krupp VDM, Germany, 1991.

CHAPTER 8

Titanium Alloys

F. H. (SAM) FROES

Institute for Materials and Advanced Processes (IMAP), University of Idaho, Moscow, Idaho 83844

Handbook of Advanced Materials Edited by James K. Wessel
ISBN 0-471-45475-3

8.1 INTRODUCTION

Titanium has often been referred to as the "wonder metal" with excellent strength, ductility, and fracture resistant characteristics in combination with superior environmental resistance. However, significant difficulties in obtaining titanium from its ores (mainly rutile and ilmenite), combined with stringent processing requirements (both factors implying high cost), greatly slowed commercialization. However, today there is a vibrant titanium industry potential poised for breakthrough into the high-volume, cost-competitive automobile marketplace.

Titanium is a metal element of Group IVB of the periodic table, with a melting point of 1675°C, an atomic weight of 47.9, and a density of 4.5 g/cm^3. The element is the fourth most abundant structural metal in Earth's crust (behind Al, Fe, and Mg), occurring mainly as rutile (TiO_2) and ilmenite ($FeTiO_3$).

Metallic titanium use can be divided into two main categories: corrosion resistance (essentially titanium alloyed to a minor extent) and structural use (for which titanium is more highly alloyed to increase the strength level while maintaining usable levels of other mechanical properties such as ductility). While the market for metallic titanium is showing a generally upward trend, the major use of titanium, as TiO_2 a white compound with high refractive index, is as a pigment "whitener" in paints, paper, rubber, plastics, and the like at about 20$\times$ the use level of metallic titanium.

8.2 HISTORY

Titanium has been recognized as an element for more than 200 years since it was first identified in 1790 by a Cornish (UK) clergyman and named "titan" by a German chemist in 1795. Early reduction processes were both expensive and generally yielded a product of a purity level that was unsuitable for use of the metal itself, although the oxide was used from the early 1900s as a pigment. However, it is only in the last 50 years that metallic titanium has gained strategic importance since an economic extraction process was developed. In that time, commercial production of titanium and titanium alloys in the United States has increased from zero to a peak of more than 27 million kg/yr [1–9].

The catalyst for the remarkable growth of the metal was the development by Dr. Wilhelm J. Kroll of a relatively safe, economical method to produce titanium metal in the late 1930s. Kroll's process involved reduction of titanium tetrachloride ($TiCl_4$), first with sodium and calcium, and later with magnesium, under an inert gas atmosphere [1]. Research by Kroll and many others continued through World War II. By the late 1940s, the mechanical properties, physical properties, and alloying characteristics of titanium were defined as the commercial importance of the metal was apparent. The first titanium for actual flight was ordered from Remington Arms (later Rem-Cru, and still later, Crucible Steel) in the United States by Douglas Aircraft in 1949. Other early U.S. entrants to the titanium field included Mallory-Sharon (later IMI) and TMCA (later Timet). In the United Kingdom, ICI Metals (later IMI and recently Timet Europe) began sponge production in 1948, with other involvement from continental Europe a few years later. Recognizing the military potential of titanium, the Soviets began sponge production in 1954. In Japan, sponge production was initiated by Osaka Titanium in 1952, generally to supply other countries and internal corrosion-resistant applications.

The U.S. government invested large sums of money to develop the science and technology of titanium and its alloys. This included an investment of in excess of a quarter of a billion dollars for a sponge stockpile up to 1964 and establishment of a titanium laboratory at Battelle, Columbus, in 1955 [1].

The vast influx of money resulted in rapid development of a sound technology with very good scientific underpinning. The double consumable vacuum arc melting (VAR) technique was developed by Armor Research Foundation in 1953. Problems relating to hydrogen embrittlement and hot salt stress corrosion cracking were recognized and circumvented.

Alloy development progressed rapidly from about 1948 with people at Remington Arms recognizing the beneficial effects of aluminum additions. The "workhorse" Ti-6Al-4V* alloy was introduced in 1954, with the disputed patent assigned to Rem-Cru. This alloy soon became by far the most important titanium alloy because of its excellent combination of mechanical properties and "forgiving" processability. The first beta titanium alloy (Ti-13V-11Cr-3Al) was developed by Rem-Cru in the mid–late 1950s, with this high-strength heat-treatable alloy seeing extensive use on the high-speed surveillance aircraft the SR71. Alloy development in the United Kingdom, driven by Rolls-Royce, was concentrated more on elevated temperature alloys for use in engines.

Aircraft manufacturers have used a generally increasing amount of titanium in airframe applications for heavily stressed demanding components. Engine components using titanium began in the United States with the Pratt and Whitney J57 in 1954, including discs, blades, and spacers in the compressor section, and in the United Kingdom with the Rolls-Royce Avon engine in 1954.

Work in the 1950s also indicated the excellent corrosion resistance of titanium and its alloys, and early commercial applications included use in anodizing, wet chlorine, and nitric acid equipment.

*Throughout the text, all terminal alloys are given in wt %, intermetallics in at %.

During the late 1950s the shipment of titanium mill products in the United States dropped from a high of 4.5 million kg (9.9 million lb) in 1957 with a change in emphasis from aircraft to missiles and concerns over the cost of titanium.

Mill shipments in the United States tripled during the 1960s from 4.5 million kg (9.9 million lb) in 1960, with aerospace use accounting for 90% of the market in 1970. This increase resulted mainly from use in nonmilitary engines and the wide-body jets, the Boeing 747, the DC-10, and the L-1011. Advances also occurred in melting practices and expanded use in nonaerospace markets such as ships and heat exchanger tubing. The 1971 cancellation of the March 3 Supersonic Transport (SST), which was slated to use considerable amounts of titanium, was a blow to the titanium industry with mill products in the United States dropping from 12 million kg (26.5 million lb) in 1970 to 9 million kg (19.8 million lb) in 1971.

Mill product shipments increased in the 1970's in large part due to increased use in the large commercial transports and their high bypass engines, new military airframes with 20–35% of their structural weight produced from titanium products, and nonaerospace use, a result of the corrosion resistance of titanium. The U.S. industry set a new record in 1980–81 of about 23 million kg (51 million lb), but this figure then dropped because of hedge buying by the aerospace industry. This cyclic nature of the titanium industry will only smooth out if nonaerospace use increases. In Europe, and to a greater extent in Japan, industrial applications exceed 50% of the total use (see Section 8.11)

Titanium mill shipments in the United States increased steadily during the Reagan years, and the early years of the Bush administration with the buildup in military hardware, peaking in 1990 at a record 25 million kg (55 million lb) [4–6]. When "peace broke out," symbolized by the dismantling of the Berlin Wall, titanium shipments fell precipitously soon thereafter to about 16 million kg (35 million lb) per year, a level that was suggested at the time to possibly be the "norm"; however, this has proved not to be true, see below. This left a great overcapacity in the United States and painful "right-sizing" by the titanium industry occurred.

In the past decade, a steady growth of titanium shipments in the United States has occurred [3–7] fueled by increased commercial aerospace orders (the Boeing 777 has almost 10% in its airframe) and the amazing golf club phenomenon, the latter helped by the appearance of the extremely popular Tiger Woods on the golf scene [4–6].

Further expansion of the titanium market is now very critically dependent on reducing cost for a variety of applications [3–7]. Addressing this need, lower cost alloys are being introduced into the marketplace that utilize Al-Fe master alloys to reduce cost rather than the AI-V master alloy needed for alloys such as Ti-6AI-4V. These include the Ti-6AI-1. Fe-0. 1Si (Timetal 62S) and Ti-4. 5Fe-6.8Mo-1.5AI. [Timetal LCB (low-cost beta)] alloys (5.6). Attention is also being given to lower cost processes such as near-net-shape powder metallurgy (PM) and permanent mold casting approaches [4–7].

Recent realignments include the Allegheny-Technologies acquisition of Oremet Wah-Chang and that of IMI Titanium and Cezus by Timet. There are also rumblings of some "giants" getting into the titanium business, with some new approaches to reducing cost.

The effect of low-cost product from the former USSR, where the peak capacity is estimated to have been four times that of the United States [i.e., as much as 90 million kg (200 million lb) of mill products per year] has not yet caused any major problems with the U.S. production capacity, but this situation could change as VSMPO in Salda, Russia, strives to increase exports, particularly to the United States.

The current uses of titanium and its alloys are discussed in Section 8.11.

8.3 GENERAL CHARACTERISTICS

Titanium alloys may be divided into two major categories: corrosion-resistant and structural alloys [8, 9]. The corrosion-resistant alloys are generally based on the single-phase α with dilute additions of solid solution strengthening and α-stabilizing elements like oxygen, palladium, ruthenium, and aluminum. These alloys are used in the chemical, energy, paper, and food processing industries to produce highly corrosion-resistant tubings, heat exchangers, valve housings, and containers. The single-phase α alloys provide excellent corrosion resistance, good weldability, and easy processing and fabrication but at a relatively low strength.

The structural alloys can be divided into four categories: the near-α alloys, the $\alpha + \beta$ alloys, the β alloys, and the titanium aluminide intermetallics, which will be discussed later.

8.4 PHYSICAL PROPERTIES

Titanium in its natural form is a dark gray color; however, it is easily anodized to give a very attractive array of colors leading to use in jewelry and various other applications where appearance is important, including some buildings, the latter use particularly in Japan. The metal and its alloys are low density at approximately 60% of the density of steel. Titanium is nonmagnetic and has good heat transfer characteristics. Its coefficient of thermal expansion is somewhat lower than that of steel and less than half that of aluminum. The melting point of titanium and its alloys are higher than that of steel, but the maximum use temperature is much lower than would be anticipated based on this characteristic alone. A summary of the physical (and a few mechanical properties) of titanium are given in Table 8.1.

TABLE 8.1 Physical and Mechanical Properties of Elemental Titanium

Property	Value
Atomic number	22
Atomic weight	47.90
Atomic volume	10.6 weight/density
Covalent radius	0.132 nm
First ionization energy	661.5 MJ/kg · mol
Thermal neutron absorption cross section	560 fm^2/atom
Crystal structure	Alpha: close-packed hexagonal $\leq$ 1156 K Beta: body-centered cubic $\geq$ 1156 K
Color	Dark gray
Density	4510 kg/m^3
Melting point	1941 $\pm$ 285 K
Solidus/liquidus	1998 K
Boiling point	3533 K
Specific heat (at 298 K)	0.518 J/kg · K
Thermal conductivity	21 W/m · K
Heat of fusion	440 kJ/kg
Heat of vaporization	9.83 MJ/kg
Specific gravity	4.5
Hardness	HRB 70–74
Tensile strength	241 GPa
Modulus of elasticity	102.7 GPa
Young's modulus of elasticity	102.7 GPa
Poisson's ratio	0.41
Coefficient of friction	0.8 at 40 m/min 0.68 at 300 m/min
Specific resistance	0.554 $\mu\Omega \cdot m$
Coefficient of thermal expansion	8.64×10^{-6}/K
Electrical conductivity	3% IACS (copper 100%)
Electrical resistivity	0.478 $\mu\Omega \cdot m$
Electronegativity	1.5 Pauling's ratio
Temperature coefficient of electrical resistance	0.0026/K
Magnetic susceptibility	180×10^{-6}
Machinability rating	40 (equivalent to 3/4 hard stainless steel)

8.5 ALLOYING, ALLOYS, AND PHASE DIAGRAMS

8.5.1 Alloying Behavior

Titanium exists in two crystalline states: a low-temperature alpha (α) phase, which has a close-packed hexagonal crystal structure, and a high-temperature beta (β) phase, which has a body-centered cubic structure (Fig. 8.1) [8]. This allotropic transformation occurs at 880°C (1620°F) in nominally pure titanium. Titanium has certain features that make it very different from other light metals such as aluminum and magnesium [9]. The allotropic transformation allows the opportunity for formation of alloys composed of α, β, or α/β microstructures, in addition

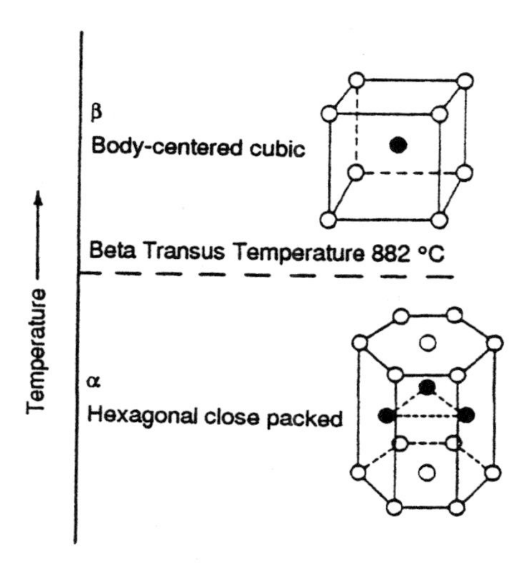

FIGURE 8.1 Two allotropic forms of titanium. Transition from low-temperature α-phase to the high-temperature β-phase occurs at 882°C (1620°F) [8].

to compound formation in certain alloys. Because of its electronic structure as a transition element, titanium can form solid solutions with most substitutional elements having a size factor within 20%, giving the opportunity for many alloying possibilities. Titanium also reacts strongly with interstitial elements such as nitrogen, oxygen, and hydrogen at temperatures below its melting point. When reacting with other elements. titanium may form solid solutions and compounds with metallic, covalent, or ionic bonding.

The choice of alloying elements is determined by the ability of the element to stabilize either the α or β phases (Fig. 8.2) [10]. Aluminum, oxygen, nitrogen, gallium, and carbon are the most common α-stabilizing elements. Zirconium, tin, and silicon are viewed as neutral in their ability to stabilize either phase. Elements that stabilize the β phase can either form binary systems of the β-isomorphous-type or the β-eutectoid type (see next section). Elements forming the isomorphous-type binary system include Mo, V, and Ta, while Cu, Mn, Cr, Fe, Ni, Co, and H are eutectoid formers in which compounds may form. The β-isomorphous alloying elements, which do not form intermetallic compounds, have traditionally been preferred to the eutectoid-type elements as additional to α-β or β alloys to improve hardenability and increase response to heat treatment.

8.5.2 Phase Diagrams

There have been a number of attempts to categorize titanium alloy phase diagrams [10, 11], all agreeing there are two major divisions: α-stabilized and β-stabilized systems. Of these probably the most convenient is that developed by Molchanova [10] (Fig. 8.2). Here the alpha stabilizers are divided into those having complete stability, in which the alpha phase can coexist with the liquid (e.g., Ti-O and

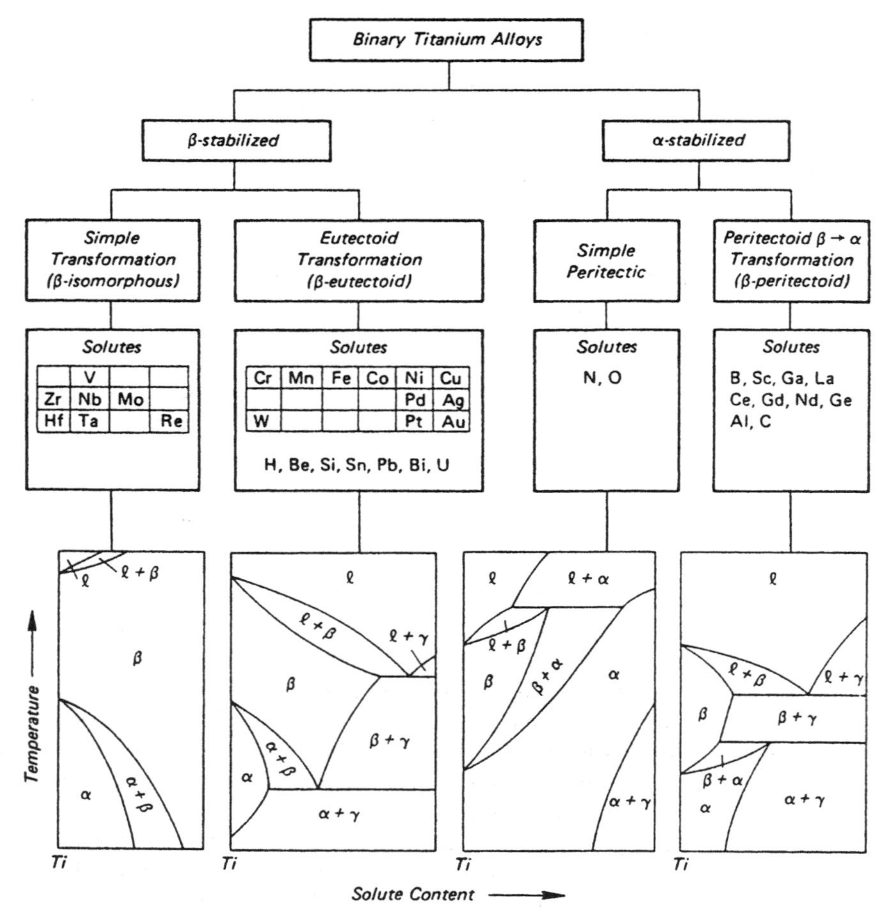

FIGURE 8.2 Classification scheme for binary titanium alloys [10].

Ti-N), and there is a simple peritectic reaction, and those that have limited alpha stability in which, with decreasing temperature, decomposition of the alpha takes place by a peritectoid reaction into beta plus a compound (beta peritectoid). Examples of the latter type of system are Ti-B, Ti-C, and Ti-AI. Molchanova [10] also divides the β stabilizers into two categories, β isomorphous and β eutectoid. In the former system an extensive β-solubility range exists with only a limited α-solubility range. Examples are Ti-Mo, Ti-Ta, Ti-V, with elements such as Zr and Hf occupying an intermediate position since they have complete mutual solubility in both the α and β phases. For the β-eutectoid systems the β phase has a restricted solubility range and decomposes into α and a compound (e.g., Ti-Cr and Ti-Cu). This class can also be further subdivided depending on whether the β transformation is rapid (the "active" eutectoid formers such as Ti-Si, Ti-Cu, and Ti-Ni) or slow (the "sluggish" eutectoid formers such as Ti-Cr and Ti-Fe).

8.5.3 Alloy Classes

Titanium alloys are categorized into one of four groups: alpha (α), alpha-beta (α-β), beta alloys (β), and the intermetallics (Ti_xAl, where $x = 1$ or 3). Titanium alloys for aerospace application contain α- and β-stabilizing elements to achieve required mechanical properties such as tensile strength, creep, fatigue, fatigue crack propagation resistance, fracture toughness, stress-corrosion cracking, and resistance to oxidation [12]. Once the chemistry is selected, optimization of mechanical properties is achieved by working (deformation) to control the size, shape, and dispersion of first the β phase and later the α phase.

Beta-isomorphous alloying elements (e.g., Mo, V, Nb), which do not form intermetallic compounds, have traditionally been preferred to "eutectoid-type" elements (e.g., Cr, Cu, Ni). However, some β-eutectoid-type compound formers are added to α-β or β alloys to improve hardenability and increase the response to heat treatment.

α Alloys The α alloys contain predominantly α phase at temperatures up to well above 540°C (1000°F). A major class of α alloys is the unalloyed titanium family of alloys that differ in the amount of oxygen and iron in each alloy. Alloys with higher interstitial content are higher in strength, hardness, and transformation temperature compared to high purity alloys. Approximately every 0.01 wt% oxygen gives a 10.5-MPa (1.5-ksi) increase in strength level [12]. Other α alloys contain additions such as Al and Sn (e.g., Ti-5Al-2.5Sn and Ti-6Al-2Sn-4Zr-2Mo).

Generally, α-rich alloys are more resistant to high-temperature creep than α-β or β alloys, and α alloys exhibit little strengthening by heat treatment. These alloys are usually annealed or recrystallized to remove stresses from cold working, and they have good weldability and generally inferior forgeability in comparison to α-β or β alloys.

α-β Alloys α-β Alloys contain one or more of the α and β stabilizers. These alloys retain more β after final heat treatment than the near α alloys and can be strengthened by solution treating and aging, although they are generally used in the annealed condition. Solution treatment is usually performed high in the α-β phase field followed by aging at lower temperature to precipitate α, giving a mixture of fine α in an α-β matrix. The solution treating and aging can increase the strength of these alloys by up to 80% [12]. Alloys with low amounts of β stabilizer (e.g., Ti-6Al-4V) have poor hardenability and must be rapidly quenched for subsequent strengthening. A water quench of Ti-6Al-4V will adequately harden sections only less than 25 mm (1 in.).

β Alloys β Alloys have more β-stabilizer content and less α stabilizer than α-β alloys. These alloys have high hardenability with the β phase retained completely during air cooling of thin sections, and water quenching of thick sections. β Alloys have good forgeability and good cold formability in the solution-treated condition. After solution treatment, aging is performed to transform some β phase to α. The strength level of these alloys is greater than α-β alloys, a result of the finely dispersed α particles in the β phase. These alloys have relatively higher densities and generally lower creep strengths than the α-β alloys. The fracture toughness of aged β alloys at a given strength level is generally higher than that of an aged α-β alloy, although crack growth rates can be faster [12].

Titanium Aluminides To increase the efficiency of gas turbine engines, higher operating temperatures are necessary, requiring alloys with enhanced mechanical properties at elevated temperatures. The family of titanium alloys showing potential for applications at temperatures as high as 900°C (1650°F) are the titanium aluminide intermetallic compounds $Ti_3Al(\alpha_2)$ and $TiAl(\gamma)$ [13–15]. The major disadvantage of this alloy group is low ambient temperature ductility. However, it has been found that niobium, or niobium with other β-stabilizing elements, in combination with microstructure control, can increase room temperature ductility in the Ti_3Al alloys up to as much as 26% elongation. Recently, by careful control of the microstructure the ambient temperature ductility of two-phase TiAl $(\gamma + \alpha_2)$ has been raised to levels as high as 5% elongation. The TiAl compositions (e.g., Ti-48Al-2Cr-2Nb) have now reached a stage of maturity where they are serious contenders for use in advanced gas turbine engines and automobiles [3–7].

8.5.4 Microstructural Development

In addition to chemistry, the mechanical properties of titanium alloys are strongly influenced by the microstructure [12]. In turn the microstructure is critically dependent on the processing, particularly whether this is carried out above or below the β-transus temperature, the temperature below which the α phase is stable. In general terms two microstructural features are important in commercially important terminal alloys: (a) the β-grain size and shape and (b) the morphology of the α phase within the β grains. Similar features strongly influence the properties of the intermetallics, but a discussion of these features is beyond the scope of this chapter, the interested reader is referred to [13–16].

β Grains Control of the β-grain size is dependent on two factors: recrystallization (when this occurs because of sufficient working) and subsequent grain growth [12]. A number of techniques have been developed for recrystallization of the β grains in α and α-β alloys by working followed by high β-field annealing.

The metastable β alloys require careful thermomechanical processing to achieve the required final microstructure. This controlled processing involves, first, the worked or recrystallized condition and, then, if recrystallized, the grain size. Under most melting conditions, the structure that occurs in an ingot ranges from small equiaxed β grains at the surface, to elongated columnar grains, to large equiaxed grains at the center of the ingot. Recently, it was shown that there is a supratransus "processing window" through which the alloy can be taken to result in a final fine equiaxed β-grain structure [12]. This processing window is relatively wide for the leaner (in β-stabilizer content) alloys and for heavy amounts of deformation.

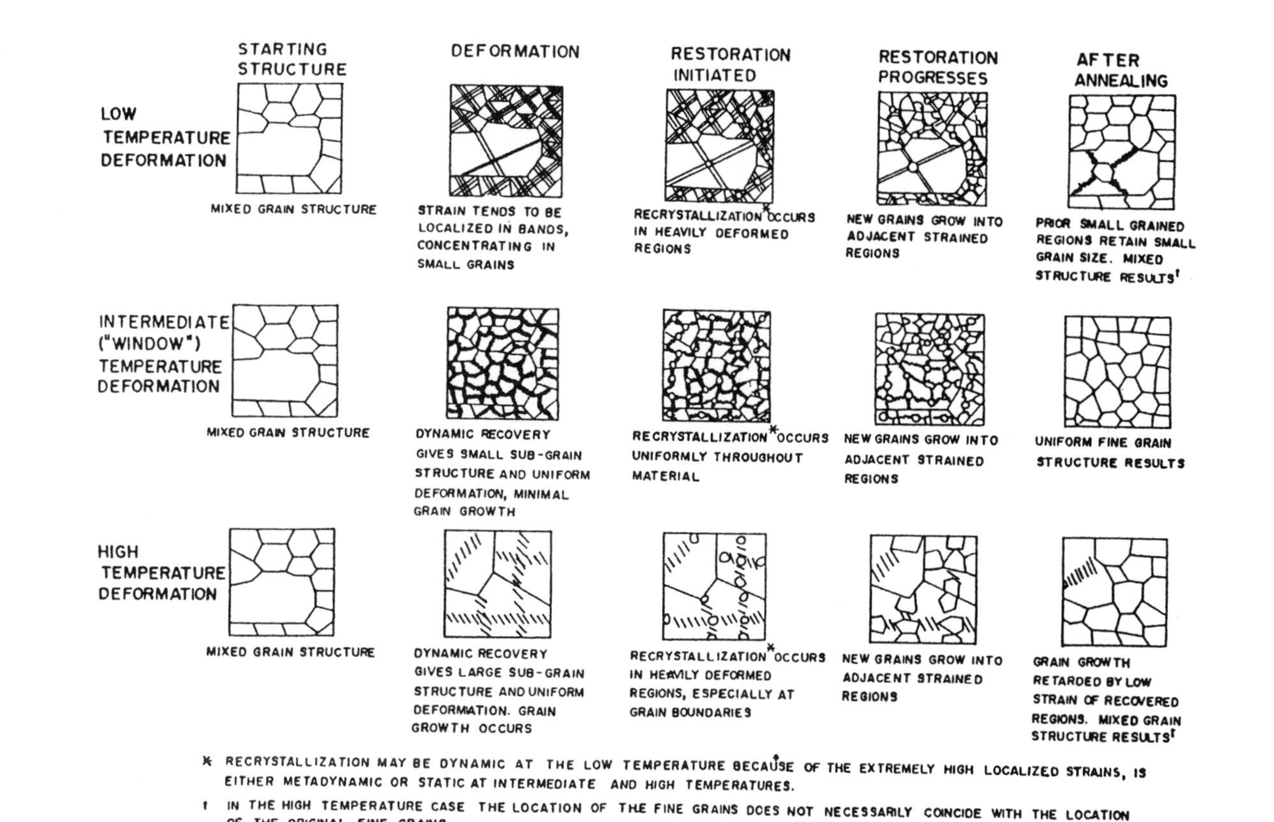

FIGURE 8.3 Restoration process by which strain in titanium grains is reduced during annealing treatment [12]:.

However, it is much more constricted for the rich β alloys and for lighter amounts of deformation, making control of the β grains much more difficult in these richer alloys. The mechanism by which the restoration to a low strain condition occurs is suggested schematically in Fig. 8.3 [12]. In general, a fine β-grain structure is promoted by working below the β-transus temperature and then heating through the transus.

Recrystallization follows the typical sigmoidal behavior, which is a function of temperature and prior deformation. The rate of grain boundary migration decreases inversely with annealing time, indicating a concurrent recovery process obeying second-order kinetics.

Grain growth follows the relationship

$$D^{1/n} - \mathrm{D}_0^{1/n} = At,$$

where D is the grain size after annealing at temperature for a time t, D_0 is the apparent initial grain size at $t = 0$, and n and A are constants. Generally, the kinetics of grain growth are not influenced by the prior grain size or amount of deformation, unless both recovered and recrystallized grains are measured, in which case a critical growth phenomenon occurs at low deformation levels (7–12%) [12].

8.5.5 α Morphology

The processing route determines the α morphology, which can vary quite considerably within the β matrix. This morphology in turn strongly influences the mechanical properties [12]. Two basic processing options are available: (1) β processing, carried out completely above the β transus or in which the processing is high enough that very little α phase is present, or (2) α-β processing, carried out below the β-transus temperature in the presence of the α phase. Subsequent annealing below the β-transus temperature within about 175°C (315°F) of the transus temperature results in a distribution of primary α, which is related to the processing sequence and annealing temperature. With β processes material, a lenticular α morphology occurs, while with α-β processing (and a sufficient amount of deformation) the primary α-β becomes globular during the subsequent heat treatment (Fig. 8.4) [8, 17].

The change in morphology of α from lenticular to globular is a direct result of the prior deformation of the α. Sufficient strain energy in the α causes it to recrystallize or relax to a lower-surface-energy globular configuration. The transformation of lenticular α to globular α is a function of annealing temperature and time and the amount of working the α has received; that is, lightly worked α will remain essentially lenticular while a heavily worked α will become globular.

Strength is virtually unaffected by the shape of the primary α but other properties such as fracture toughness and elevated-temperature flow characteristics (particularly creep, superplastic forming, and diffusion bonding) are strongly influenced. High fracture toughness is associated with α having a high aspect ratio (i.e., lenticular), while lower fracture toughness values at the same strength

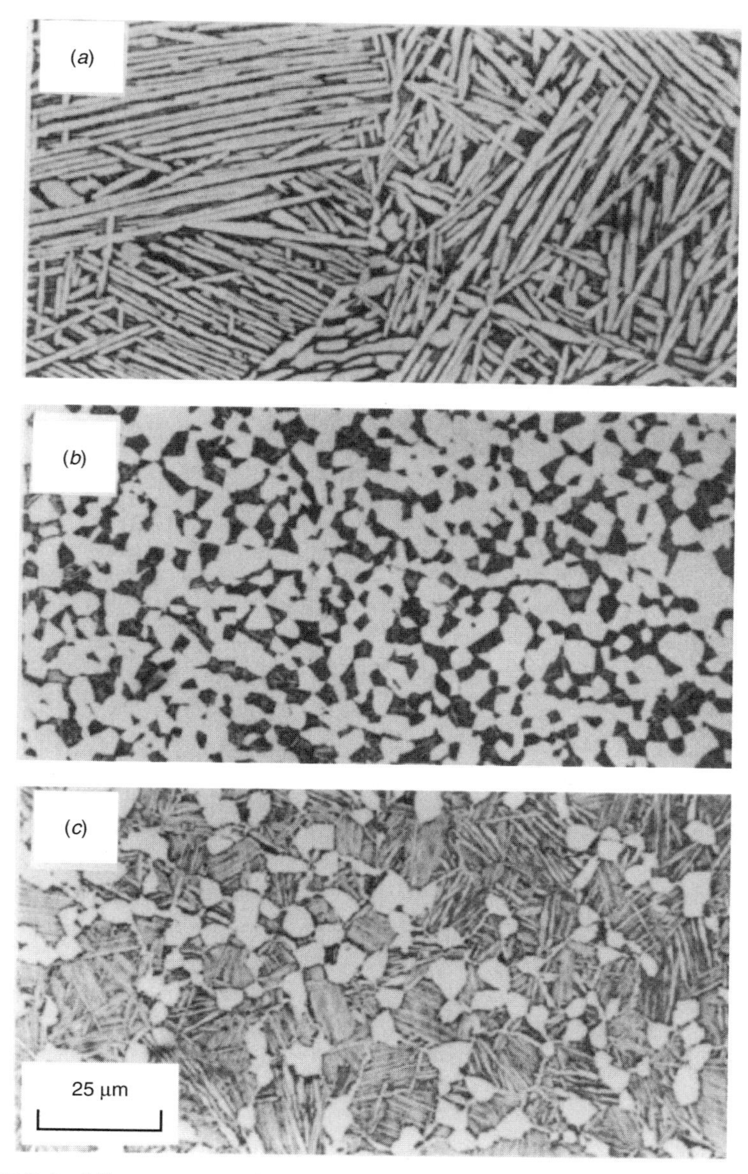

FIGURE 8.4 Microstructure of Ti-6Al-2Sn-4Zr-2Mo: (*a*) β worked followed by α-β anneal to produce lenticular α morphology, (*b*) α-β worked and α-β annealed to give predominantly an equiaxed α shape and (*c*) α-β worked followed by duplex anneal: just below the β-transus temperature [reduced volume fraction of equiaxed α compared to (*b*)], and significantly below the β-transus temperature (to form the lenticulary α between equiaxed regions) [8, 17].

level corresponded to α having a low aspect ratio (i.e., globular) [8, 12]. A similar trend occurs with fatigue crack growth rate. However, optimum superplastic forming and diffusion bonding is found in material with a globular microstructure; while creep performance is favored by lenticular α. Low-cycle fatigue behavior is optimized with a globular α morphology.

In β alloys, thermomechanical processing affects not only the microstructure but also the decomposition kinetics of the metastable β phase during aging. The increased dislocation density after working β alloys leads to extensive heterogeneous nucleation of the equilibrium α phase, which can suppress formation of the brittle ω phase [12].

8.6 PROCESSING/FABRICATION

8.6.1 Extraction

The commercial production of titanium metal involves the chlorination of rutile (TiO_2) in the presence of coke or other form of carbon. The most important chemical reaction involved is

$$TiO_2(s) + 2Cl_2(g) + 2C(s) \longrightarrow TiCl_4(g) + 2CO(g).$$

The resulting $TiCl_4$ ("tickle") is purified by distillation and chemical treatments and subsequently reduced to titanium sponge using either Mg (Kroll process) of Na (Hunter process). The basic reaction involved in the Kroll process is

$$2Mg(s) + TiCl_4(I) \longrightarrow Ti(s) + 2MgCl_2(I).$$

With either process the sponge produced is vacuum distilled, swept with an inert gas, or acid leached to reduce the remnant salt content. A number of alternate processes have been evaluated for sponge production, including electrolytic, molten salt, and plasma processes but none have reached commercial status [12].

8.6.2 Ingot

The starting stock for titanium ingot production may be either titanium sponge or reclaimed scrap or a combination of the two. In either case, stringent specifications must be met for control of ingot composition. Modern melting techniques remove volatile substances from sponge, so that ingot of high quality can be produced regardless of which method is used for the sponge production. However for critical aerospace use, especially in engines, melting must be carried out to virtually eliminate any type of defects. This has led to development of melting techniques in which the time–temperature for which the metal is molten is increased (e.g., electron beam and plasma cold hearth techniques) compared to conventional vacuum arc consumable electrode methods [5, 6, 12].

Recycling of titanium scrap (revert) is an important facet of cost-effective production of titanium product. The revert, which is recycled, includes cut sheet, reject castings, machine turning, and chips.

Ingots remain the major source of titanium mill products.

8.6.3 Castings

Castings are an attractive approach to the fabrication of titanium components since this technique allows production of relatively low-cost parts [5, 6, ,12]. Basically a near net shape is produced by allowing molten titanium to solidify in a graphite, ceramic, or metal mold. Use of a ceramic mold, generally produced by the "lost-wax" process, allows production of large, relatively high integrity, complex shapes. The metal mold process is capable of less complex and smaller parts, but cost can be only 50% of the ceramic mold process [5, 6]. Enhanced mechanical properties in combination with increased size and shape-making capabilities, have resulted in greatly increased use of titanium castings in both engine and airframe applications. The shipment of titanium castings has increased by a factor of 3 over the past 15 years to a level of about 400,000 kg/year (882,000 lb/year).

8.6.4 Powder Metallurgy

A number of powder metallurgy (PM) approaches have been evaluated for the titanium system including the blended elemental (BE), prealloyed (PA), rapid solidification (RS), mechanical alloying (MA), and vapor deposition (VD) techniques [12, 18].

Using a press-and-sinter technique, the BE approach allows fabrication of low-cost components from elemental and/or master alloy additions. However, because of the porosity resulting from this method, a result of the inherent salt from the Kroll or Hunter processes [1], generally initiation-related properties such as S–N fatigue are inferior to cast-and-wrought product.

The PA approach yields mechanical properties at least equivalent to those of ingot product. However, less than desirable cost advantages, in combination with a fear of the PM approach by design engines has resulted in few applications.

The powder metallurgy/rapid solidification (PM/RS) technique is a "far from equilibrium" approach that allows extension of alloying levels and much more refined microstructures than are possible using the ingot metallurgy (IM) technique. The greatly increased chemistry/microstructure "window" can lead to enhanced mechanical and physical properties in a variety of metallic systems.

Mechanical alloying (MA) with heavy working of powder particles results in intimate alloying by repeated welding and fracturing. This technique allows dispersoids to be produced, solubility extension, novel phase production, and microstructural refinement.

Production of alloys directly from the vapor allows even greater flexibility in microstructural development than RS or MA [20, 21]. A semicommercial scale electron beam vapor deposition process has been constructed to produce alloys that are not possible by ingot methods or even rapid solidification. One example is the production of low-density Ti-Mg alloys. Mg boils below the melting point of titanium, making production of a liquid alloy impossible by conventional methods.

None of the three processes discussed above has yet progressed from the laboratory.

8.6.5 Joining

Adhesive bonding, brazing, mechanical fastening, and diffusion bonding are all used routinely and successfully to join titanium and its alloys [12]. Welding of various types, including tungsten inert gas (TIG), electron beam, and plasma, is also used very successfully with titanium and its alloys. In all types of welds, contamination by interstitial impurities such as oxygen and nitrogen must be minimized to maintain useful ductility in the weldment. Thus, welding must be done under strict environmental controls to avoid pickup of interstitials that can embrittle the weld metal.

8.6.6 Wrought Product Processing

This section addresses the primary processing of wrought (ingot) product to mill products. The following section will address the fabrication of these mill products into final components. Mill products include billet, bar, plate, sheet, strip, foil, extrusions, tubing, and wire. Besides the reduction of section size, and shaping, the other objective of primary processing is control (generally refinement) of the microstructure to optimize final mechanical property combinations [8, 12]. In general terms as processing proceeds the temperature of the processing is decreased. The β-transus temperature, below which the α phase can be present, is the critical temperature for control of the microstructure. In many cases titanium is processed on the same equipment used for steel, with appropriate special auxiliary equipment [8].

Other concerns with the processing of titanium alloys include the high reactivity of titanium at elevated temperatures and the strain rate sensitivity; especially for the β alloys, strength decreases as the strain rate is reduced.

Billet product from an ingot starts above the β-transus temperature and proceeds at progressively decreasing temperatures. In some cases the β-grain size is reduced by a recrystallization treatment well above the β-transus temperature. However, minimization of grain boundary α, control of the morphology of the α phase, and refinement of the transgranular α can necessitate working below the β-transus temperature.

Bar, plate, sheet, and foil products are produced on a relatively routine basis. Generally, the processing is done hot, although the very high ductility of the metastable β alloys allows finishing of strip and foil by cold rolling.

Forging is a very common method for producing titanium alloy components. It allows both control of the part shape and manipulation of the microstructure and hence mechanical properties. Generally, titanium alloys are considerably more difficult to forge than aluminum alloys and alloy steels, particularly when processing at temperatures below the β-transus temperature is required.

Extrusions, tubing, and wire titanium products are also produced routinely with the same caveats regarding microstructural control as for the product forms discussed above.

8.6.7 Wrought Product Fabrication

Wrought products (mill products) are fabricated to desired configurations with the same concerns regarding microstructural control as discussed in the previous section. Examples of forming of wrought product include isothermal/hot forging, sheet metal forming, foil production, rod and wire, and superplastic forming/diffusion bonding.

Isothermal and hot forging are special forging operations in which the die temperatures are close to the metal temperature, that is, much higher than in conventional forging. This reduces chill effects and allows close to net-shape production. Strain rates are much lower than normal, contributing to the near-net-shape capability. The metastable β alloys, with a low β-transus temperature, are particularly amenable to the isothermal forging process.

Sheet metal forming is conducted either hot, which generally allows larger, more precise amounts of deformation, or cold, which is lower cost. Hot forming of titanium alloys is conducted in the range 595–815°C (1105–1500°F) with increased formability and reduced spring back. Formability increases with increasing temperature, but at the higher temperatures contamination can become a problem, sometimes necessitating an inert atmosphere or a coating. Beta alloys are easier to cold form than α and α-β alloys. The high degree of spring back exhibited by titanium alloys sometimes requires hot sizing after cold forming. This reduces internal stresses and restores a compressive yield strength.

Superplastic forming/diffusion bonding makes use of the fact that fine-grained material can deform extremely large amounts, especially at very low strain rates (0.0001–0.01 s^{-1}). Superplastic forming (SPF) is the propensity of sheet material to sustain very large amounts of deformation, without unstable deformation (tensile necking); for example-fine-grained (<10 μm) Ti-6Al-4V can be deformed >1000% in tension at 927°C (1700°F). Diffusion bonding (DB) is a solid-state bonding process in which a combination of pressure and temperature allow production of a metallurgically sound bond. Superplastic forming is now used routinely as a commercial sheet metal fabrication process for reduced cost and production of complex shapes, generally using gas pressure. The combined SPF/DB process has seen less commercial use than initially anticipated, predominantly because of problems in inspecting the integrity of the bond region.

8.6.8 Machining

Previous sections of this chapter have discussed a number of approaches to reduce the cost of titanium components, particularly near-net-shape methods. However, most titanium parts are still produced by conventional techniques involving a significant amount of machining [12]. As a result the machining of titanium and its alloys has been extensively evaluated, and well-defined procedures for various types of machining operations have been defined including turning, end milling, drilling, reaming, tapping, sawing, and grinding [12].

In many instances considerable amounts of machining are required for the production of complex components from mill products such as forgings, plate and bar, that is, a high buy-to-fly ratio (BFR). Titanium is chemically reactive leading to a tendency for welding to the tool, chipping, and premature failure. Other problems involve the low heat conductivity of titanium, which adversely affects tool life, and the ease of damaging the titanium surface. The latter effect is of particular concern because surface integrity strongly influences crack-initiation-related properties such as fatigue.

The machining of unalloyed titanium is similar to $\frac{1}{4}-\frac{1}{2}$ hard austenitic stainless steel. High-quality sharp tools, carbides for high productivity, and high-speed tool steels for more difficult operations are required for titanium. This, in combination with slow speeds, heavy feeds, and the correct cutting fluids, generally results in good machining behavior for titanium. The cutting fluids recommended are oil–water emulsions and water-soluble waxes at high cutting speeds, low viscosity sulfurized oils, and chlorinated oils at low speeds; in all cases the cutting fluids should be removed after machining, especially before heat treatment, to avoid potential stress-corrosion cracking problems.

8.6.9 Metal Matrix Composites

The success exhibited by organic matrix composites has led to parallel efforts to develop engineered metals including metal matrix composites (MMC's).

The CermeTi family of titanium alloy matrix composites, fabricated using the blended elemental approach, incorporates particulate ceramic (TiC or TiB_2) or intermetallic (TiAl) as a reinforcement [19]. These composites exhibit minimal particle–matrix interaction while maintaining the integrity of the essentially 100% dense homogeneously dispersed particles within the matrix.

An innovative method (XD) for the production of *in situ* discontinuous titanium-based composites has resulted in interesting property combinations in alloys such as the intermetallic γ [19], but to date there are no applications.

Reinforcement with continuous ceramic composites enhances the strength and modulus of terminal titanium alloys, such as Ti-6Al-4V [22, 23]. However, control of the reaction zone between the fiber and the matrix, inferior transverse properties, and very high cost remain major concerns. Innovative fabrication techniques such as plasma spray deposition and electron beam vapor deposition may help in controlling cost, while the design lessons learned with nonisotropic polymeric composites should be applicable in engineering metal composite structures.

The potential weight savings obtainable by replacing much heavier superalloys in both engine and airframes has resulted in considerable work being conducted on α_2 and γ MMCs [24]. Recently, increased attention has been given to the richer Nb varieties of Ti_3Al-Nb known as the "orthorhombic" alloys (22–27 at % Nb) as matrix materials [7, 14].

8.7 MECHANICAL PROPERTIES

8.7.1 Cast and Wrought Terminal Alloys

The mechanical properties of titanium alloys depend not only on the chemistry but are also strongly influenced by the microstructure as pointed out earlier, the latter in turn being dependent on the processing. The tensile properties of selected cast and wrought terminal titanium alloys are summarized in Table 8.2 [9].

TABLE 8.2 Compositions, Relative Densities, and Typical Room Temperature Tensile Properties of Selected Wrought Titanium Alloys[a]

Common Designations	Al	Sn	Zr	Mo	V	Si	Other	Relative Density	Condition	0.2% Proof Stress (MPα)	Tensile Strength (MPα)	Elongation (%)
α Alloys												
CP Ti 99.5% IMI 115, Ti-35A							0	4.51	Annealed 675°C	170	240	25
CP Ti 99.0% IMI 155, Ti-75A							0	4.51	Annealed 675°C	480	550	15
IMI 260							0.2 Pd	4.51	Annealed 675°C	315	425	25
IMI 317	5	2.5						4.46	Annealed 900°C	800	860	15
IMI 230							2.5 Cu	4.56	ST (α), duplex aged 400 and 475°C	630	790	24
Near-α alloys												
8-1-1	8			1	1			4.37	Annealed 780°C	980	1060	15
IMI 679	2.25	11	5	1		0.25		4.82	ST (α + β) aged 500°C	990	1100	15
IMI 685	6		5	0.5		0.25		4.49	ST (β) aged 550°C	900	1020	12
6-2-4-2S	6	2	4	2		0.2		4.54	ST (α + β) annealed 590°C	960	1030	15
Ti-11	6	2	1.5	1		0.1	0.35 Bi	4.45	ST (β) aged 700°C	850	940	15
IMI 829	5.5	3.5	3	0.3		0.3	1 Nb	4.61	ST (β) aged 625°C	860	960	15
α-β Alloys												
IMI 318, 6-4	6				4			4.46	Annealed 700°C	925	990	14
									ST (α + β) aged 500°C	1100	1170	10

TABLE 8.2 ***(Continued)***

Common Designations	Al	Sn	Zr	Mo	V	Si	Other	Relative Density	Condition	0.2% Proof Stress (MPα)	Tensile Strength (MPα)	Elongation (%)
IMI 550	4	2		4		0.5		4.60	ST (α + β) aged 500°C	1000	1100	14
IMI 680	2.25	11		4		0.2		4.86	ST (α + β) aged 500°C	1190	1310	15
6-6-2	6	2		6			0.7(Fe,Cu)	4.54	ST (α + β) aged 550°C	1170	1275	10
6-2-4-6	6	2	4	6				4.68	ST (α + β) annealed 590°C	1170	1270	10
IMI 551	4	4		4		0.5		4.62	ST (α + β) aged 500°C	1200	1310	13
Ti-8 Mn							8 Mn	4.72	Annealed 700°C	860	945	15
β-alloys 13-11-3	3				13		11 Cr	4.87	ST (β) aged 480°C	1200	1280	8
Beta III		4.5	6	11.5				5.07	ST (β) duplex aged 480 and 600°C	1315	1390	10
8-8-2-3	3			8	8		2 Fe	4.85}		1240	1310	8
Transage 129	2	2	11		11			4.81	ST (β) aged 580°C	1280	1400	6
Beta C	3		4	4	8		6 Cr	4.82	ST (β) aged 540°C	1130	1225	10
10-2-3	3				10		2 Fe	4.65	ST (β) aged 580°C	1250	1320	8

[a] ST (α), ST (α + β), and ST (β) correspond to solution treatment in the α, α + β, and β-phase fields, respectively. Annealing treatments normally involve shorter times than aging treatments

The influence of α morphology was discussed earlier, where it was pointed out that a lenticular shape favors high fracture toughness (K_{Ic}) while a globular morphology optimizes ductility. The effect of α morphology and section size on tensile properties and fracture toughness are demonstrated in Tables 8.3 and 8.4 [25] and illustrated in Fig. 8.5 [12]; as strength increases, fracture toughness decreases, and vice versa. Chemistry, particularly the interstitial content (e.g., O_2) (Figs. 8.6 and 8.7) [25], influences fracture toughness with high values of K_{Ic} associated with low O_2 values; and texture can also have an effect [25].

TABLE 8.3 Yield Strength and Plane-Strain Fracture Toughness of Various Titanium Alloys

Alloy	α Morphology or Processing Method	Yield Strength (MPa)	Plane-Strain Fracture Toughness (K_{IC}) (MPa$\sqrt{m}$)
Ti-6Al-4V	Equiaxed	910	44–66
	Transformed	875	88–110
	α-β Rolled + mill annealed[a]	1095	32
Ti-6Al-6V-2Sn	Equiaxed	1085	33–55
	Transformed	980	55–77
Ti-6Al-2Sn-4Zr-6Mo	Equiaxed	1155	22–23
	Transformed	1120	33–55
Ti-6Al-2Sn-4Zr-2Mo forging	α + β Forged, solution treated and aged	903	81
	β Forged, solution treated, and aged	895	84
Ti-17	α-β Processed	1035–1170	33–50
	β Processed	1035–1170	53–88

[a] Standard oxygen (<0.20 wt %).

TABLE 8.4 Relation of Tensile Strength of Solution-Treated and Aged Titanium Alloys to Size

Alloy	Tensile Strength of Square Bar in Section Size of:					
	13 mm (MPa)	25 mm (MPa)	50 mm (MPa)	75 mm (MPa)	100 mm (MPa)	150 mm (MPa)
Ti-6Al-4V	1105	1070	1000	930	—	—
Ti-6Al-6V-2Sn (Cu + Fe)	1205	1205	1070	1035	—	—
Ti-6Al-2Sn-4Zr-6Mo	1170	1170	1170	1140	1105	—
Ti-5Al-2Sn-2Zr-4Mo-4Cr (Ti-17)	1170	1170	1170	1105	1105	1105
Ti-10V-2Fe-3Al	1240	1240	1240	1240	1170	1170
Ti-13V-11Cr-3Al	1310	1310	1310	1310	1310	1310
Ti-11.5Mo-6Zr-4.5Sn (Beta III)	1310	1310	1310	1310	1310	—
Ti-3Al-8V-6Cr-4Zr-4Mo (Beta C)	1310	1310	1240	1240	1170	1170

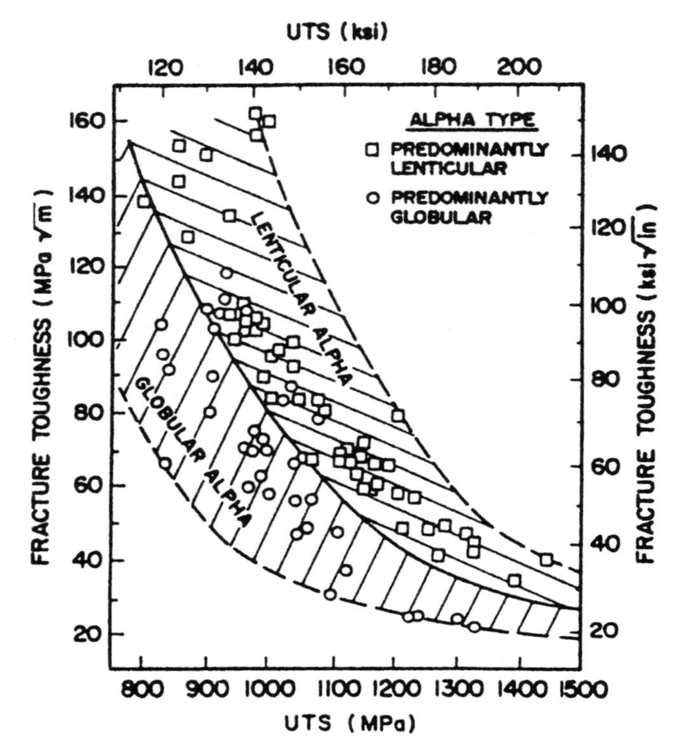

FIGURE 8.5 Variation of fracture toughness with strength level for CORONA 5 (Ti-4.5Al-5Mo-1.5Cr). At given strength level, lenticular α gives high fracture toughness than a globular morphology [12].

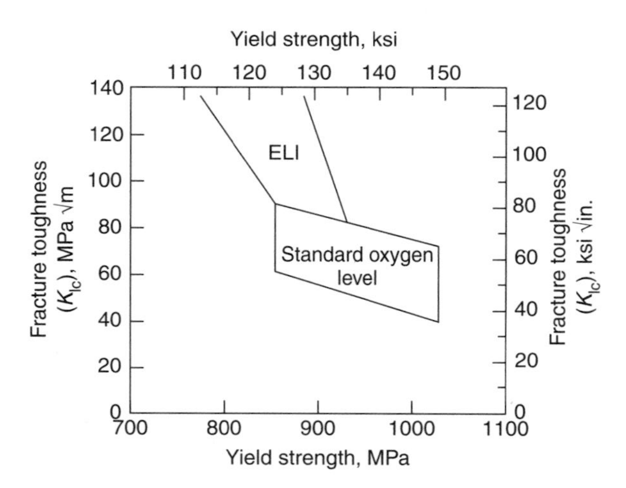

FIGURE 8.6 Range of yield strength and fracture toughness for Ti-6Al-4V alloy. ELI is extra low interstitial oxygen (<0.13 wt %), standard oxygen is <0.20 wt % [25].

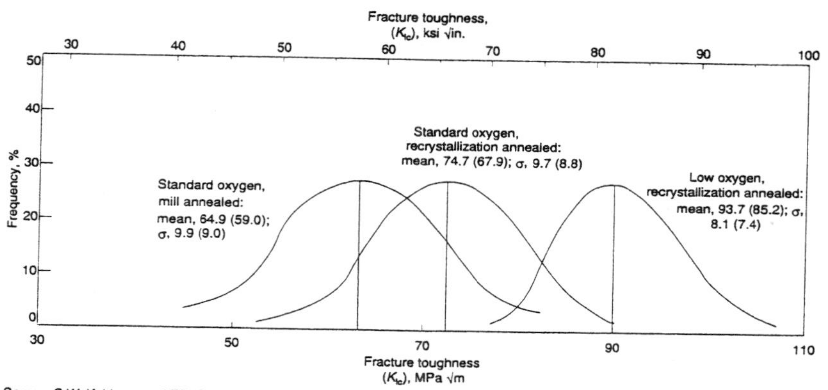

FIGURE 8.7 Effect of oxygen level and heat treatment on fracture toughness of Ti-6Al-4V [25].

TABLE 8.5 Strain Control Low-Cycle Fatigue Life of Ti-6242S at 480°C

		Number of Cycles to Failure	
Test Frequency (cycles/min)	Total Strain Range(%)	Acicular Structure	Equiaxed α Structure
0.4	1.2	1,196	10,500[a]
10	1.2	3,715	31,000[a]
0.4	2.5	273	722
10	2.5	353	1,166

[a] Run out.

The fatigue behavior of titanium alloys can be divided into S–N fatigue and fatigue crack growth rate (FCGR, or da/dn vs. δK). Within S–N fatigue a further subdivision can be made between low-cycle fatigue (LCF) and high-cycle fatigue (HCF). For LCF, failure occurs in 10^4 cycles or less, while HCF failure occurs at greater than 10^4 cycles.

Different uses favor different techniques for determining LCF, specifically strain-controlled and load-controlled tests, Table 8.5 and Fig. 8.8 [26]. Both notch concentration (K_t) and overall surface condition can strongly influence LCF. The beneficial effect of relatively gentle surface conditioning is shown in Fig. 8.9 [26]; more severe working of the surface can result in the formation of cracks and a degraded LCF behavior. The effect of K_t and crack propagation of LCF life on preloaded Ti-6Al-4V at 205°C is shown in Fig. 8.10 [26].

Surface condition can also strongly influence HCF (Fig. 8.11) [26]. The fatigue endurance limit is relatively flat to at least 315°C (600°F) (Fig. 8.12) [26]; with benefits apparent for titanium alloys over steels.

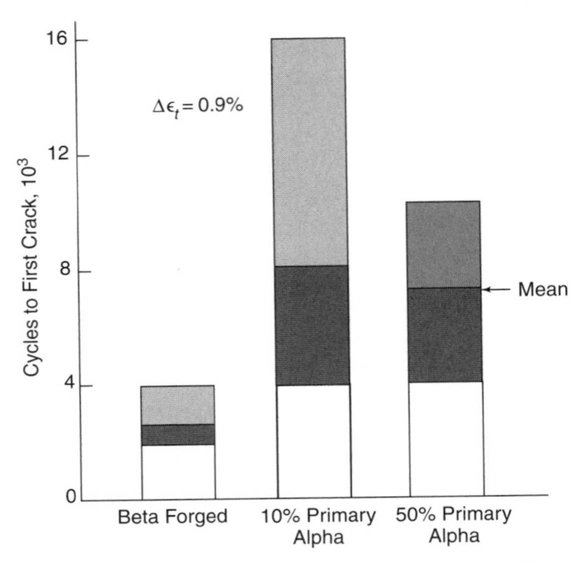

FIGURE 8.8 Low-cycle fatigue (LCF) life of Ti-6Al-4V with different structures [26].

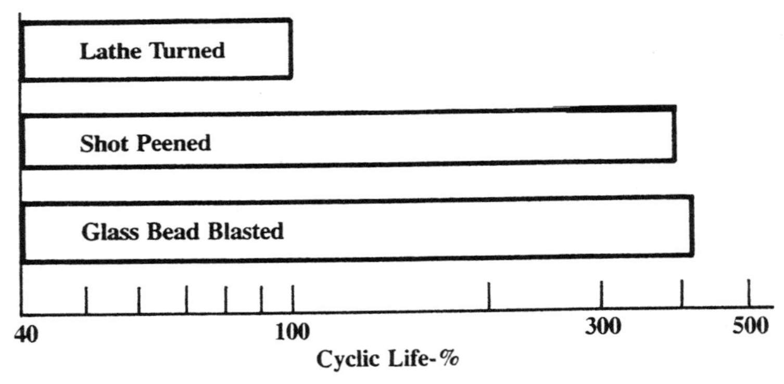

FIGURE 8.9 Effects of surface condition on LCF life of Ti-6Al-4V at 21°C (70°F) [26].

The FCGR performance generally parallels fracture toughness, with the caveat that severe corrosive environments (such as 3.5% NaCl solution) can adversely affect the FCGR by an order of magnitude. An example of the strong influence of microstructure on FCGR for the Ti-6Al-4V alloy is shown in Fig. 8.13 [26], with the β-2 annealed condition significantly better than the mill-annealed material.

The α and near-α alloys generally exhibited superior high-temperature behavior (Fig. 8.14) [25]. The reason why these alloys have replaced steels in advanced jet engines is clearly demonstrated in Fig. 8.15 [25], with the titanium alloys now used to about 600°C (1110°F). Generally these alloys contain Si for enhanced creep behavior (Fig. 8.16) [25].

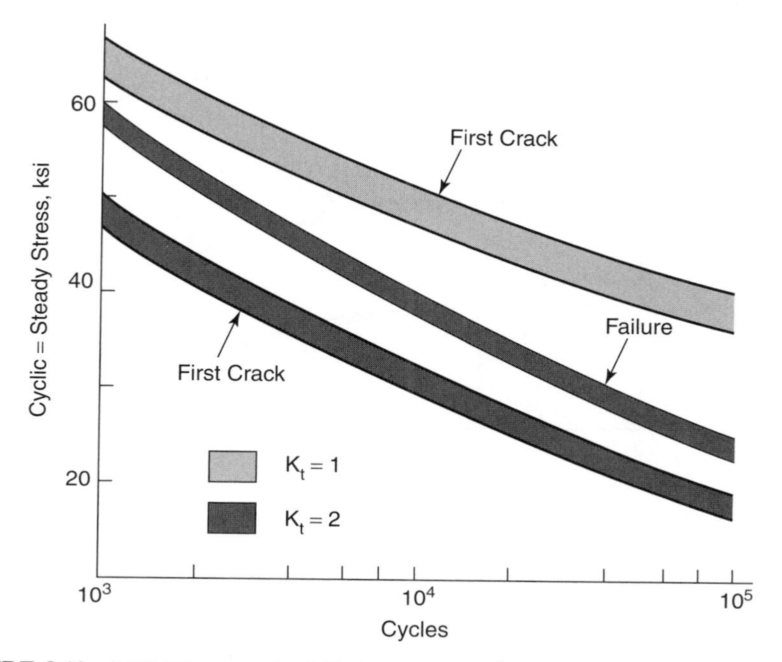

FIGURE 8.10 LCF life strength of Ti-6Al-4V at 205°C (400°F), showing effect of K_t (notch concentration) and crack propagation rate on life [26].

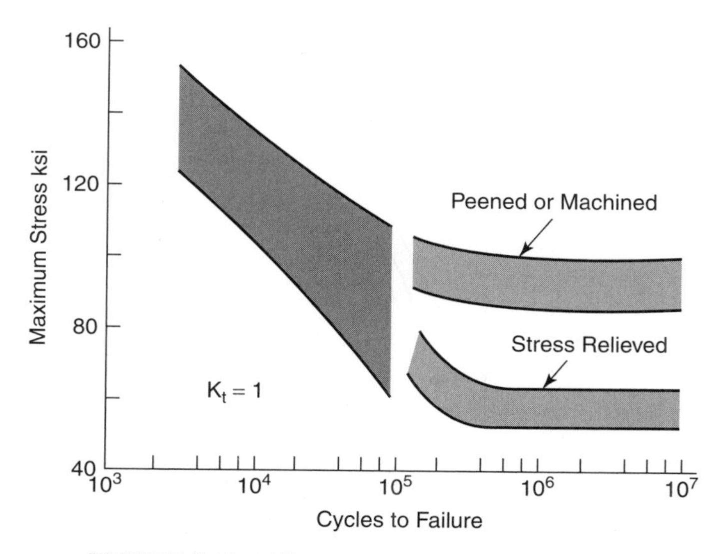

FIGURE 8.11 Effect of surface finish on LCF [26].

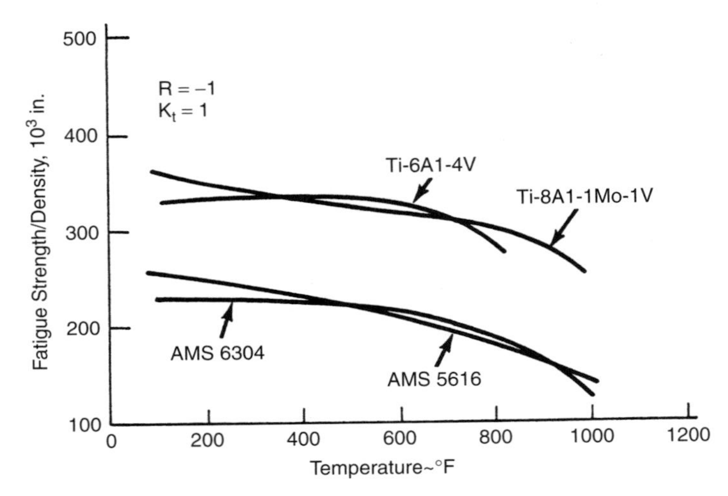

FIGURE 8.12 High-cycle fatigue (HCF) at 5×10^7 cycles, showing the benefits of titanium over steel [26].

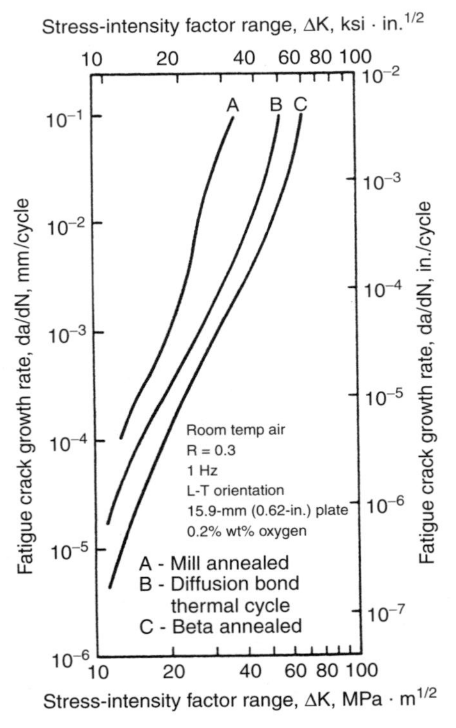

FIGURE 8.13 Effect of microstructure on the fatigue crack growth rate (FCGR) for the Ti-6Al-4V alloy [26]:.

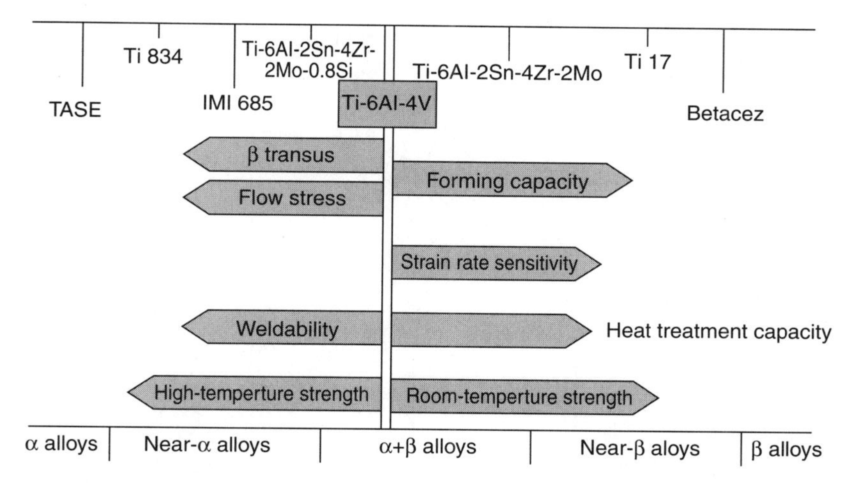

FIGURE 8.14 Major characteristics of three terminal classes of titanium alloys [25].

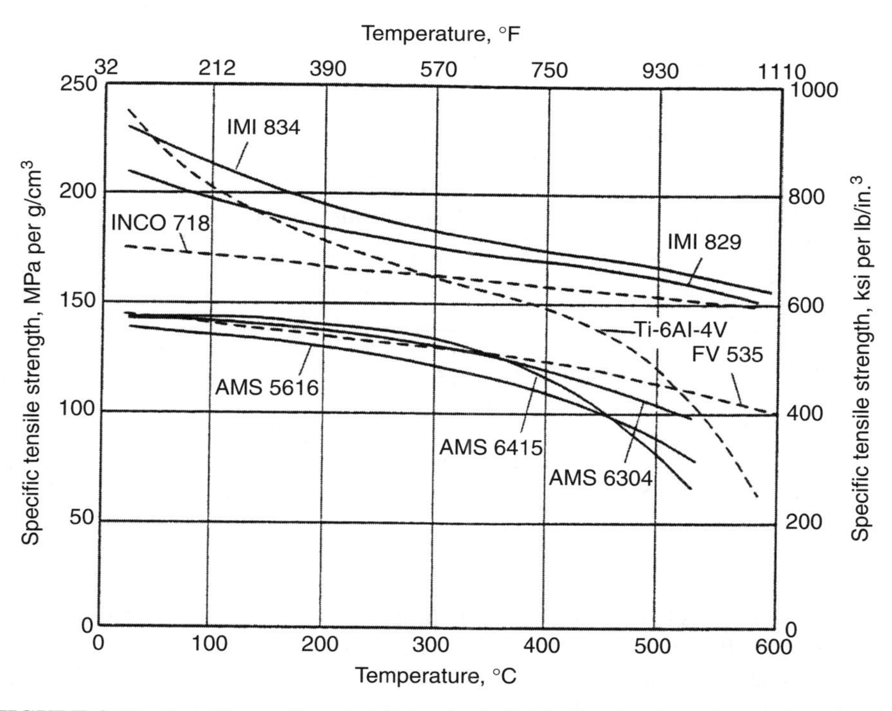

FIGURE 8.15 Specific tensile strength (strength divided by density) of various titanium alloys, compared with steels once used in gas turbine engines [25].

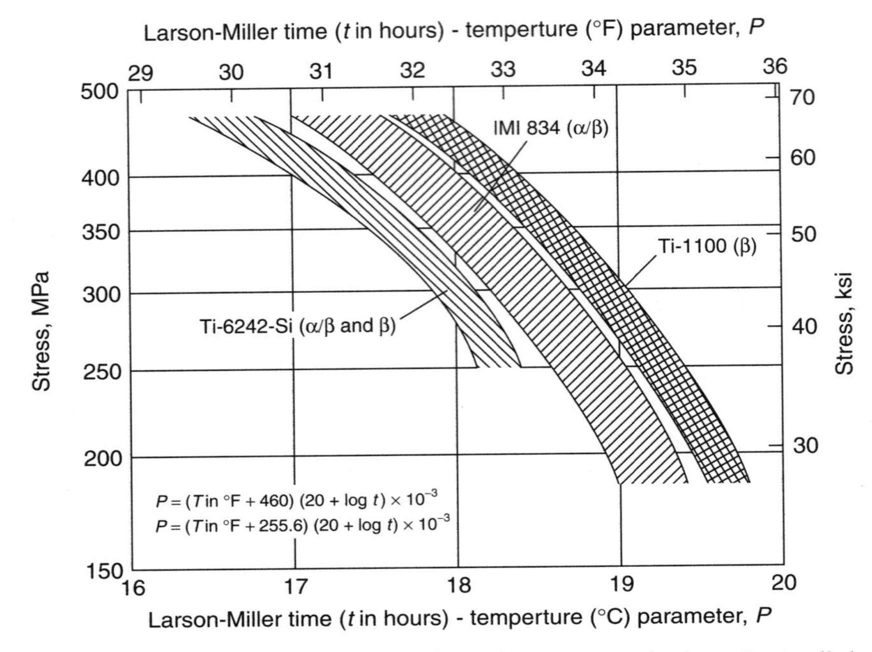

FIGURE 8.16 Creep behavior of three elevated temperature titanium alloys; all three alloys contain Si for enhanced performance. The alloys were either α-β or β processed [25].

TABLE 8.6 Typical β Titanium Alloy Compositions (wt %)

	COMPOSITION									
Alloy	Al	Sn	Zr	V	Fe	Mo	Cr	Nb	Si	Ti
10-2-3	2	—	—	10	3	—	—	—	—	Bal.
15-3	3	3	—	15	—	—	3	—	—	Bal.
Beta C	3	4	—	8	—	4	6	—	—	Bal.
Beta21S	3	—	—	—	—	15	—	2.6	0.2	Bal.
Beta III	—	4.5	6	—	—	11.5	—	—	—	Bal
Timetal LCB (low-cost beta)	1.5	—	—	—	4.5	6.8	—	—	—	Bal.

Titanium alloys have good cryogenic properties, with the α alloy Ti-5Al-2.5Sn and the α-β alloy Ti-6Al-4V seeing extensive use.

A number of β alloys have been developed over the years with current activity being concentrated on the alloys shown in Table 8.6. Generally these alloys exhibit higher strength–toughness combinations than the α-β alloys such as Ti-6Al-4V [25, 27].

The Timetal 62S (Ti-6Al-2Fe-0. 1Si) is a low-cost alloy (formulated with a lower cost Al-Fe master alloy than Ti-6Al-4V, which uses an Al-V master alloy) which has received consideration for armor applications.

8.7.2 Intermetallic Alloys

Intermetallics will only be discussed briefly here, with emphasis on indicating how both chemistry and processing/microstructure can be adjusted to control mechanical properties.

Typical properties of the $Ti_3Al(\alpha_2)$ type of titanium aluminide are shown in Table 8.7 [13–15]. A good example of how microstructural control can lead to tailoring of the mechanical properties is the very high ambient temperature ductility obtained by special processing to produce an optimum amount of equiaxed primary α_2.

Recently, control of the microstructure in TiAl(γ) alloys, especially those in the two phase $\gamma + \alpha_2$ phase field, has led to interesting mechanical property combinations, including an ambient temperature ductility of approximately 5%. A lenticular microstructure resulted in high toughness/low ductility, while the duplex microstructures give the opposite combination of properties (Table 8.8) [13, 14].

8.7.3 Cast Alloys

Because cast alloys are produced near-net shape directly from the molten state, they inherit a microstructure that cannot be modified by the thermomechanical processing used with cast and wrought (ingot) material [12]. Additionally, a number of defects can occur in castings, such as porosity, which can degrade mechanical properties.

TABLE 8.7 Typical Properties of Ti_3Al-Type Titanium Aluminides[a]

Alloy	UTS (MPa)	YS (MPa)	El. (%)	K_{Ic} (MPa$\sqrt{m}$)	Creep Rupture[b]
Ti-25Al	538	538	0.3	—	—
Ti-24Al-11Nb	824	787	0.7	—	44.7
Ti-25Al-10Nb-3V-1Mo	1042	825	2.2	13.5	360
Ti-24Al-14Nb-3V-0.5Mo	—	—	26.0[c]	—	—
Ti-24.5Al-17Nb	1010	952	5.8	28.3	62
	940	705	10.0	—	—
Ti-25Al-17Nb-1Mo	1133	989	3.4	20.9	476
Ti-15Al-22.5Nb	963	860	6.7	42.3	0.9

[a] Compositions in at %.
[b] Hours at 650°C/38 MPa
[c] Specially processed.

TABLE 8.8 Microstructure and Mechanical Properties in Ti-46.5Al-2.5V-1Cr TiAl-Type Titanium Aluminide[a]

	FL	NL	DM	NG
YS (MPa)	360	430	440–450	387
UTS (MPa)	400	480	505–538	468
El. (%)	0.5	2.3	3.3–4.8	1.7
K_{Ic} (MPa$\sqrt{m}$)	21	17	12	12

[a] Compositions in at %. FL, fully lamellar; NL, near lamellar; DM, duplex; NG, near gamma.

TABLE 8.9 Typical Room Temperature Tensile Properties of Several Cast Titanium Alloys

Alloy	Condition	Tensile Strength (MPa)	Yield Strength (MPa)	Elongation (%)	Reduction in Area (%)
Commercially pure titanium	As-cast or annealed	550	450	17	32
Ti-6Al-4V	As-cast or annealed	1035	890	10	19
Ti-6Al-2Sn-4Zr-2Mo	Duplex annealed	1035	895	8	16
Ti-5Al-2.5Sn-ELI	Annealed	805	745	11	—

The microstructure of cast products, for example, in the Ti-6Al-4V alloy, consists of large β grains, extensive grain boundary α, and elongated coarse intragranular α, which can occur in colonies (of similarly aligned plates) or in a Widmanstatten morphology. This leads to strength, fracture toughness, fatigue crack growth rate, and creep behavior, which are at a relatively high level (Table 8.9) [26]. However, ductility and S–N fatigue are lower than cast and wrought product (Fig. 8.17) [12]. Both ductility and S–N fatigue can be enhanced by use of either innovative heat treatments or the use of hydrogen as a temporary alloying element (thermohydrogen processing, THP) to refine the microstructure (Fig. 8.18) [28, 29]. The high-cycle fatigue of alloys such as Ti-6Al-4V can be enhanced by hot isostatic pressing (HIPing) [7, 30, 31].

Casting of titanium alloys other than the conventional Ti-6Al-4V alloy is also possible. An example is the Ti-3Al-8V-6Cr-4Zr-4Mo alloy (38-6-44 or beta C), which exhibits excellent tensile properties and impressive fatigue behavior, with an endurance limit 85% above the average value typical of the Ti-6Al-4V alloy [12]. Recently, cast γ alloys have been successfully produced and could see use in automobile and advanced gas turbine jet engines [3–17].

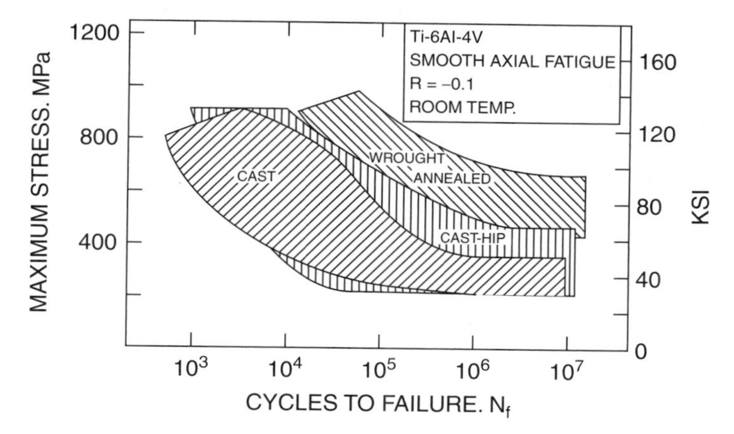

FIGURE 8.17 Room temperature smooth axial fatigue versus maximum cyclic stress for Ti-6Al-4V. Data scatterbands for cast, cast and hot isostatically pressed, and annealed wrought (ingot) material [12].

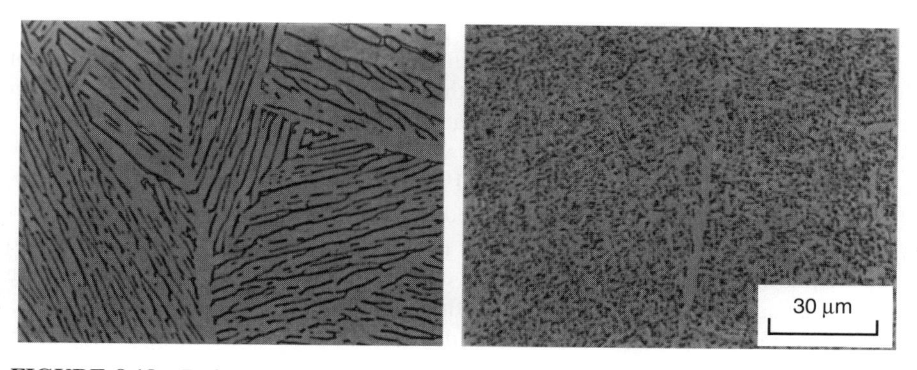

FIGURE 8.18 Refinement of microstructure of cast Ti-6Al-4V (left) by use of the thermohydrogen processing technique (right) [19, 28, 29]:.

8.7.4 Powder Metallurgy Alloys

The tensile properties of blended elemental (BE) product (elemental additions) meet typical minimum wrought specification properties (Table 8.10). However, because of the remnant salt (from the extraction process) and associated porosity, fatigue behavior is below wrought levels. However, this behavior may be enhanced in a similar fashion to cast product by use of innovative heat treatments or THP. At a cost penalty, properties may also be improved by using higher priced salt-free titanium sponge or a newly available hydride powder produced using a calcium process [5–7].

TABLE 8.10 Typical Tensile Properties of Blended Elemental Ti-6Al-4V Compacts Compared to Mill-Annealed Wrought Products

Material	0.2% YS (MPa)	UTS (MPa)	Elongation (%)	RA (%)
Cold isostatic press and HIP (CHIP)	827	917	13	26
Press and sinter (no HIP ing)	868	945	15	25
Wrought mill anneal	923	978	16	44
Typical minimum properties (MIL-T-9047)	827	896	10	25

TABLE 8.11 Properties of Ti-6Al-4V Prealloyed Powder Compacts

0.2%, YS (MPa)	UTS (MPa)	Elongation (%)	RA (%)	K_{Ic} (MPa$\sqrt{m}$)
930	992	15	33	77

The tensile and fracture toughness properties of prealloyed (PA) material are at levels at least equivalent to wrought product (Table 8.11) [18, 19], and with adequate precautions to avoid contamination of the powder S–N fatigue behavior is also at least at ingot levels. And as with BE product, S–N fatigue can be further improved by use of innovative heat treatments or THP.

Rapidly solidified (RS) titanium alloys containing rare-earth additions show some improvements in creep behavior [19] but have not yet seen commercial use. The RS approach can also be used to produce high-strength alloys such as the normally segregation prone Ti-1Al-8V-5Fe [19]. Little advantage has been achieved for the intermetallics using RS; with the caveat that the near-net-shape processing may offer an advantage for the very difficult to fabricate γ compositions.

Development of mechanically alloyed (MA) titanium alloys is at a very early stage with virtually no mechanical properties available [19]. Early indications, however, suggest that improved dispersions of second-phase particles and enhanced strength–ductility combinations may occur, the latter in very fine grained nanostructured material.

8.7.5 Welded Components

Welding generally increases strength and hardness and decreases ductility [12]. Welds in unalloyed titanium grades 1, 2, and 3 do not require postweld treatment unless the materials will be highly stressed in a strongly reducing atmosphere [26]. Welding of the α class of alloys and leaner α-β alloys such as Ti-6Al-4V can be accomplished with relative ease. Welds in more β-rich α-β alloys such as Ti-6Al-6V-2Sn have a high likelihood of fracturing with little or no plastic straining. Weld ductility can be improved by postweld heat treatment consisting of slow cooling from a high annealing temperature. And at the other end of the spectrum rich β-stabilized alloys can be welded, and such welds exhibit good

TABLE 8.12 Mechanical Properties of Titanium Metal Matrix Composites

System	UTS (MPa) L	UTS (MPa) T	E (GPa) L
Ti-6Al-4V	890	890	120
SCS-6(SiC)/Ti-6Al-4V	1455	340	240

ductility. However, here the aging kinetics of the weld metal may be substantially different from that of the parent metal.

The intermetallic Ti_3Al and TiAl are difficult to weld because of their low inherent ductility and the microstructures that develop in the weld region. Controlled energy inputs into the weld region using EB techniques allow manipulation of the heat input and elimination of solid-state cracking in the fusion zone [32].

8.7.6 Machined Components

The surface of titanium alloys may be damaged during machining and grinding; and this damage can lead to a degradation in fatigue strength and stress corrosion resistance [12]. Shallow compressive stresses can enhance fatigue behavior.

8.7.7 Metal Matrix Composites

The various grades of Cerme Ti offer higher elevated temperature strength, increased hardness, and improved modulus over the monolithic titanium alloy while maintaining both the fracture toughness and machinability of a metal (albeit more difficult), as opposed to those of a brittle ceramic [19].

The mechanical behavior of Ti-6Al-4V reinforced with continuous SiC fibers is shown in Table 8.12 [7, 33]. Greatly enhanced specific strength is obtained in α_2-type titanium aluminide/SiC composites compared to conventional superalloys, with dramatic weight savings of up to 75% by replacing a conventional disc-and-spacer assembly with a titanium aluminide reinforced ring configuration [34].

As a matrix, the richer orthorhombic α_2 alloys exhibit increased ambient temperature ductility and enhanced oxidation resistance. However, they are more costly and higher density [13, 14, 19].

8.8 CHEMICAL PROPERTIES/CORROSION BEHAVIOR

Titanium is used in aerospace and commercial applications because of its high strength-to-density ratio, good fracture characteristics, and generally outstanding corrosion resistance. Titanium's excellent resistance to most environments is the result of its stable, tightly adherent, protective surface film [12]. This film consists basically of TiO_2 at the metal–environment interface with underlying thin layers of Ti_2O_3 and TiO. This film forms naturally and is maintained when the metal and its alloys are exposed to moisture or air. In general, anhydrous conditions such as provided by chlorine or methanol as well as uninhibited

TABLE 8.13 Acid Concentration Limits for ASTM Grades 2,[a] 7,[a] and 12[a] Titanium in Pure Reducing Acids

Acid/Temperature	Acid-Concentration Limit (wt %)[b]		
	Grade 2	Grade 7	Grade 12
HCl			
24°C	6	25	9
Boiling	0.6	4.6	1.3
H_2SO_4			
24°C	5	48	10
Boiling	0.5	7	1.5
H_3PO_4			
24°C	30	80	40
Boiling	0.7	3.5	2

[a] Grade 2, Ti-50A; Grade 7, Ti-0.2Pd; Grade 12, Ti-0.3Mo-0.8Ni.
[b] For a corrosion rate of about 5 mil per year.

reducing conditions, should be avoided. The passive film formed in air may not be adequately stable and may not be regenerated if it is damaged during exposure to these environments.

General (uniform) corrosion rate information for titanium and many of its alloys exposed to a wide variety of environments is available [12]. Broadly, commercial purity titanium is resistant to natural environments, including sea, fresh, brackish, and mine waters; food products; crude oils; body fluids; and waste materials. The outstanding resistance of unalloyed titanium and the Ti-0.2 Pd alloy (ASTM Grades 2 and 7; see Table 8.13) in chloride-containing, aqueous environments is well established. With few exceptions, unalloyed titanium performs well when exposed to oxidizing inorganic acids (e.g., nitric and chromic acid), aqueous ammonia, anhydrous ammonia, molten sulfur, pure hydrocarbons, aqua regia, hydrogen sulfide, wet chlorine, most organic acids, dilute caustic solutions, and chlorine dioxide.

Titanium is not particularly resistant to pure reducing inorganic acids (i.e., those that generate hydrogen during the metal–acid reaction) such as sulfuric, hydrochloric, and phosphoric acids. The metal is dissolved rapidly by hydrofluoric acid. Other environments that should be avoided include fluoride-containing solutions (e.g., ammonium fluoride), hot concentrated caustics, certain organic acids (e.g., oxalic, concentrated citric and trichloroacetic, and nonaerated boiling formic), and powerful oxidizing agents [e.g., anhydrous liquid and gaseous chlorine, liquid and gaseous oxygen, anhydrous red fuming nitric acid (RFNA), anhydrous nitrogen tetroxide, and liquid bromine]. Powerful oxidizers are especially to be avoided because, under certain conditions such as impact, the reaction can be pyrophoric.

Figure 8.19 [12] shows that the use of titanium can be extended into the "reducing acid" region by alloying the metal with small amounts of a noble

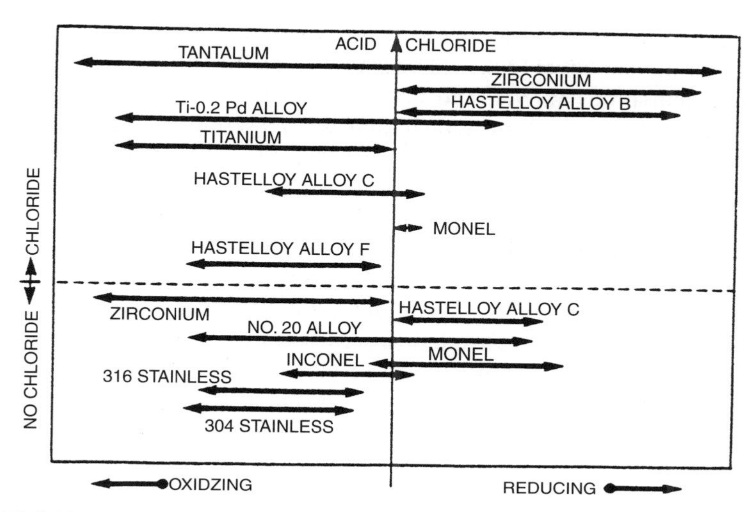

FIGURE 8.19 General corrosion behavior of commercially pure titanium and Ti-Pd alloys compared to other metals and alloys in oxidizing and reducing acids; with and without chloride ions. Each metal or alloy can generally be used for those environments below its respective solid lines [12].

metal such as 0.2 wt% palladium; Table 8.13 [12]. Similar but smaller benefit can be achieved by small alloying additions of nickel and molybdenum (e.g., 0.3 wt% Mo and 0.8 wt% Ni; ASTM Grade 12 titanium).

Unalloyed titanium is especially useful for applications where essentially no corrosion products can be tolerated in the process fluid. The metal is used extensively in the fabrication of food, drug, and dye processing equipment where even trace amounts of metal ion contamination could adversely affect the quality, color, and/or taste of the product produced.

Titanium, in most environments, is an effective cathode, thus coupling the metal to a less noble metal can result in a high galvanic corrosion current and rapid dissolution of the anodic material; and the titanium may absorb hydrogen.

Titanium and its alloys are susceptible to inhibitor-type concentration-cell corrosion when, for example, oxidizing heavy-metal ions are used to inhibit general corrosion and crevices exist. Concentration-cell corrosion of titanium can be mitigated in some cases by using either ASTM Grades 7 or 12 titanium (see Table 8.13) for fabricating entire components or just in local crevice zones. Palladium and nickel in these alloys, respectively, provide improved passivity (i.e., anodic protection) in the crevices.

Titanium's resistance to chloride-induced pitting attack is a primary reason for using this material (e.g., replacing type 316L stainless steel in petroleum refinery processes). However, under certain conditions, titanium is susceptible to pitting attack and has been reported to pit in the hot 130°C (270°F) brine solutions in salt evaporators. Pitting attack can also be mitigated by using ASTM Grades 7 and 12.

Titanium alloys are susceptible to stress-corrosion cracking (SCC) in a number of environments, including anhydrous methanol containing trace quantities of halides, anhydrous RFNA, and hot chloride-containing salts. Several of the alloys and unalloyed titanium (containing relatively high oxygen contents) are known to crack in ambient temperature seawater if the materials contain preexisting cracks.

The strong influence of microstructure on SCC has been demonstrated in the metastable β-titanium alloy Beta III (Ti-11.5Mo-6Zr-4.5Sn) [35]. This work demonstrated that the alloy was susceptible to SCC when equiaxed β grains and continuous grain boundary α were present, but a worked material in which no such microstructural features were observed was immune to SCC.

Although there have been no known service failures related to hot salt stress-corrosion cracking (HSSCC), HSSCC is a potential limitation to the long duration exposure of highly stressed titanium alloys at temperatures above about 220°C (430°F).

The near immunity of relatively high strength titanium alloys to corrosion fatigue in chloride-containing solutions allows these materials to be used in many hostile environments (e.g., body fluids) where other alloys have failed when subjected to cyclic stresses.

The tenacious passive film that forms naturally on titanium and its alloys provides excellent resistance to erosion corrosion. For turbine-blade applications where the components are impinged by high-velocity water droplets, unalloyed titanium has been shown to have superior resistance compared to conventional blade alloys (e.g., austenitic stainless steels and Monel).

It is known that the fatigue behavior of titanium and its alloys is surface condition sensitive; surface damage by fretting can adversely affect the ability of these materials to withstand cyclic stress. For example, fretting corrosion can reduce the fatigue strength of a titanium alloy, such as Ti-6Al-4V by more than 50%.

8.9 ALLOY SELECTION

As discussed in Section 8.5 there are four classes of titanium alloys: the near α alloys, the α-β alloys, the β alloys, and the intermetallic titanium aluminides. A separate categorization can be made into those alloys used predominantly for corrosion resistance and those used in load-bearing structural applications. A partial listing of the currently most significant titanium alloys is given in Table 8.14.

Within the corrosion-resistant alloy category are the commercially pure grades and those alloys containing specific additions to enhance corrosion behavior (e.g., the platinum group metals such as Pt, Pd, and Ru).

A broad separation of the structural alloys can be made between those which are used predominantly at ambient temperatures and those intended for elevated temperature use [to 600°C (1100°F) for the terminal alloys, to as high as 900°C (1650°F) for the intermetallics based on the equiatomic TiAl].

TABLE 8.14 Table of Most Significant Titanium Alloys[a]

Alloy Composition	UNS Number	ASTM Designation	Comments
Unalloyed titanium	R50250	Grade 1	Grades 2–4 have increased strength due to higher oxygen contents BT1-0 (purer) and BT1-0 Russian
Ti-0.2Pd	R52400 and R52250	Grade 7 and 11	Corrosion-resistant Ru substitution reduces cost
Ti-0.3Mo-0.8Ni	R53400	Grade 12	Corrosion-resistant
Ti-3Al-2.5V	R56320	Grade 9	Formable, tubing
Ti-5Al-2.5Sn	R54520		Weldable, cryogenic use Russian BT5-1
Ti-6Al-2Sn-4Zr-2Mo-0.1Si	R54620		Creep resistant
Ti-8Al-1Mo-1V	R54810		High modulus
Ti-6Al-2.7Sn-4Zr-0.4Mo-0.45Si			Timetal 1100 Use to 600°C (1100°F)
Ti-2.5Cu			IMI-230
Ti-5Al-3.5Sn-0.3Zr-1Nb-0.3Si			IMI-829
Ti-5.8Al-4Sn-3.5Zr-0.7Nb-0.5Mo-0.35Si			IMI-834
Ti-4.3Al-1.4Mn			Russian OT4 structural
Ti-6.7Al-3.3Mo-0.3Si			Russian BT8 high-temperature
Ti-6.4Al-3.3Mo-1.4Zr-0.28Si			Russian BT9 high-temperature
Ti-7.7Al-0.6Mo-11Zr-1.0Nb-0.12Si			Russian BT18 high-temperature
Ti-6Al-4V	R56400	Grade 5	Workhorse alloy Russian BT6
Ti-6Al-4VELI	R56401		Low interstitial, damage tolerant
Ti-6Al-6V-2Sn	R56620		Higher strength than Ti-6Al-4V
Ti-4Al-4Mo-4Sn-0.5Si			IMI 551
Ti-4.5Al-3V-2Mo-2Fe			Superplastic alloy SP-700
Ti-6Al-1.7Fe-0.1Si			Timetal 62S lowcost alloy
Ti-5Al-2Sn-2Zr-4Mo-4Cr	R58650		Ti-17, high strength, moderate temperature
Ti-6Al-2Sn-4Zr-6Mo	R56260		Moderate temperature, strength and long-term creep
Ti-3Al-8V-6Cr-4Mo-4Zr	R58640		Beta C (38-6-44)
Ti-10V-2Fe-3Al			Ti-10-2-3, high-strength forgings
Ti-15V-3Al-3Cr-3Sn			Ti-15-3, high strength and strip processible
Ti-3Al-7.4Mo-10.5Cr			Russian BT15 structural
Ti-1.5Al-5.5Fe-6.8Mo			Timetal LCB low-cost beta
Ti15-Mo-3Al-2.7Nb-0.25Si	R58210		Timetal 21S
Alpha-2 (Ti3Al) aluminide			Experimental intermetallics
Gamma (TiAl)			Two-phase alloys ($\alpha_2 + \gamma$) look best, semicommercial
Ti-Ni			Shape memory alloy
Cerme Ti			Particle reinforced, with TiC, TiB2, or TiAI

[a] IMI, Imperial Metals Industries (now part of Timet).

8.10 DESIGN ASPECTS

Designing with titanium and its alloys can conveniently be divided into the two areas discussed in Section 8.9: corrosion-resistant use and structural applications.

8.10.1 Corrosion-Resistant Design

As discussed in Section 8.8 the highly adherent oxide film that forms on the surface of titanium and its alloys offers exceptional resistance to a broad range of acids and alkalis, as well as natural salt and polluted waters. Titanium alloys are especially resistant to corrosion in oxidizing environments, and this behavior can be extended into the reducing regime with the addition of platinum group metals. A summary of corrosion environments where titanium's oxide film provides resistance are shown in Table 8.15.

8.10.2 Structural Design

With its high strength-to-density ratio, excellent fracture-related properties (fracture toughness, fatigue, and fatigue crack growth rate), and superior environmental resistance titanium is the material of choice for many aerospace and terrestrial structural (load-bearing) applications [36].

Selection of titanium for both airframes and engines is based upon its specific properties: weight reduction (due to the high strength-to-density ratio), coupled with exemplary reliability attributable to its outstanding corrosion resistance and general mechanical properties.

Highly efficient gas turbine engines are possible through the use of titanium alloy components such as fan blades, compressor blades, rotors, discs, hubs, and numerous nonrotor parts like inlet guide vanes. Titanium is the most common material for engine parts that operate up to 1100°F (593°C) because of its strength and ability to tolerate the moderate temperatures in the cooler parts of the engine. Other key advantages of titanium-based alloys include low density (which translates to fuel economy) and good resistance to creep and fatigue. The development of titanium aluminides should allow the use of titanium in even hotter sections of a new generation of engines.

Titanium alloys have replaced nickel and steel alloys in nacelles and landing gear components in the Boeing 777. This includes investment cast parts that allow complex shapes to be made at relatively low cost. For example, heat shields that protect wing components from engine exhaust are cast from titanium. Cold hearth melting leads to production of essentially clean metal for structural applications while controlling costs. Superplastic forming/diffusion bonding and powder metallurgy have helped to increase the use of titanium alloys in new airframe designs, by lowering the cost of machining and the amount of waste material produced (revert).

TABLE 8.15 Corrosive Environments Where Titanium Oxide Film Provides Resistance

Chlorine and other halides	*Inorganic salt solutions*
Fully resistant to moist chlorine and its compounds.	Highly resistant to chlorides of calcium, copper, iron, ammonia, manganese, and nickel.
Fully resistant to solutions of chlorites, hypochlorites, perchlorates, and chlorine dioxide.	Highly resistant to bromide salts.
Resistance to moist bromine gas, iodine, and their compounds is similar to chlorine resistance.	Highly resistant to sulfides, sulfates, carbonates, nitrates, chlorates, and hypochlorites.
Water	*Organic acids*
Immune to corrosion in all natural, sea, brackish, and polluted waters.	Generally very resistant to acetic, terephthalic, adipic, citric, formic, lactic, stearic, tartaric, and tannic acids.
Immune to microbiologically influenced corrosion (MIC).	
Oxidizing mineral acids	*Organic chemicals*
Highly resistant to nitric, chromic, perchloric, and hypochlorous (wet chlorine gas) acids.	Corrosion-resistant in organic process streams of alcohols, aldehydes, esters, ketones, and hydrocarbons, with air or moisture.
Gases	*Alkaline media*
Corrosion-resistant to sulfur dioxide, ammonium, carbon dioxide, carbon monoxide, hydrogen sulfide, and nitrogen.	Low corrosion rates in hydroxides of sodium, potassium, calcium, magnesium, and ammonia.

Information provided as an overview. Before specifying titanium in any aggressive environment, consult corrosion experts. Adapted from James S. Grauman and Brent Willey, "Shedding New Light on Titanium in CPI Construction," *Chemical Engineering*, August 1998.

8.11 APPLICATIONS

Titanium alloy markets, and product requirements can be described by three major market segments—jet engines, airframes, and industrial applications (see Table 8.16) [2] and Fig. 8.20 [6, 7] [2]. The first two of these segments are related to the broad aerospace market, which in the United States, dominates the use of titanium and consumes about equal amounts in engines and airframes. These two applications are based primarily on titanium's high specific strength (strength-to-density ratio). The third, and smallest, market segment in the United

TABLE 8.16 Titanium Alloys—Markets and Product Requirements

Market Segment	Market USA	Share Europe	Product Requirements
Jet engines	42%	37%	Elevated temp. tensile strength Creep strength Elevated temp. stability Fatigue strength Fracture toughness
Airframes	38%	33%	High tensile strength Fatigue strength Fracture toughness Fabricable
Industrial	20%	30%	Corrosion-resistant Adequate strength Fabricable Cost competitive
Total	100%	100%	
1990 Consumption, kg × 10^6	23.6	9.1	

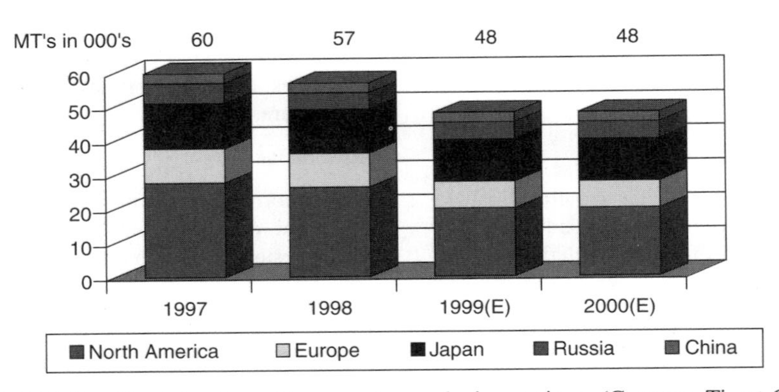

FIGURE 8.20 Total titanium shipments by producing regions. (Courtesy Timet Corp.)

States is industrial, which is based on titanium's excellent corrosion resistance in salt and other aggressive environments. As indicated in Table 8.16, these market segments have similar proportions in both the United States and Europe, although the total U.S. market is about 2.5 times that of Europe based on 1990 data [2]. In Japan, a majority of the titanium is for nonaerospace use. The titanium capacity of the former Soviet Union was estimated to be about 90 million kg (200 million lb) per year in 1990, a capacity that could totally change the western marketplace with low-cost products.

The product requirements for titanium alloys in each market segment are based on the specific needs for the particular application. For example, jet engine requirements are focused primarily on high-temperature tensile and creep strength and thermal stability at elevated temperatures. Second-tier property considerations are fatigue strength and fracture toughness. Airframe applications require high tensile strength combined with good fatigue strength and fracture toughness. Ease of fabricability of components is also an important consideration. Industrial applications emphasize good corrosion resistance in a variety of media as a primary consideration as well as adequate strength, fabricability, and competitive cost, relative to other types of corrosion-resistant alloys.

Jet engine applications include discs and fan blades (Figs. 8.21 and 8.22). Airframe components produced from titanium vary from small parts to large main landing gear support beams, the aft section of a fuselage, and truck beam forgings (Figs. 8.23–8.25) [1–7].

Traditional nonaerospace applications cover tubing in heat transfer equipment (Fig. 8.26) and watches (Fig. 8.27). They also include sporting goods (Fig. 8.28), corrosion prevention covers on seawater piers (Fig. 8.29), and roofs of buildings (Fig. 8.30) [1–7].

FIGURE 8.21 Ti-6Al-4V fan disc forgings for General Electric's CF6 series engine. Each forging is 90 cm (35 in.) in diameter and weighs 250 kg (550 lb). (Courtesy Wyman-Gordon Company.)

FIGURE 8.22 Titanium fan blades for jet engines. (Courtesy RMI Titanium Company.)

FIGURE 8.23 Ti-6Al-4V main landing gear support beam forging for Boeing 747. Each forging is 6.2 m long, 97 cm wide, 28 cm thick (20 ft × 38in. × 11in.) and weighs over 1600 kg (3525 lb). (Courtesy Wyman-Gordon Company.)

FIGURE 8.24 Aft Ti-6Al-4V/Ti-8Mn "boat-tail" section of fuselage of F-5. Section of plane experiences heating due to its proximity to engine. (Courtesy Northrop-Grumman Corporation, Aircraft Division.)

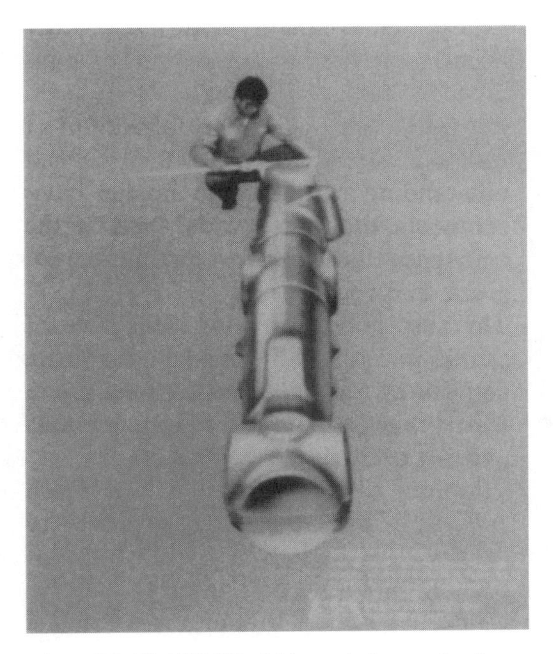

FIGURE 8.25 Boeing 777 Ti-10V-2Fe-3Al truck beam forging; a welded assembly about 10 m (33 ft) long. (Courtesy Boeing Commercial Airplane Company.)

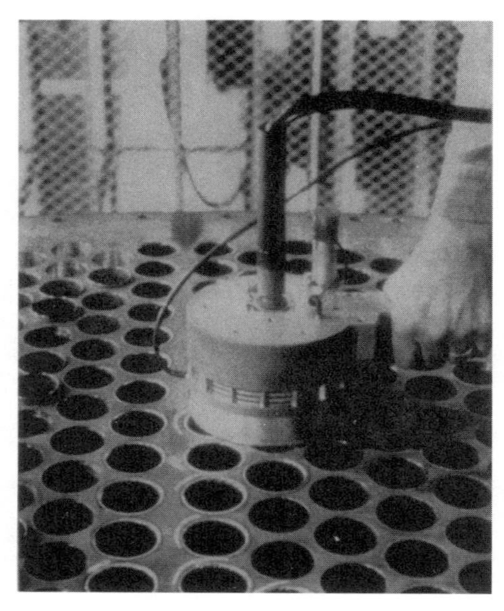

FIGURE 8.26 Titanium tubing in heat transfer equipment. (Courtesy RMI Titanium Company.)

FIGURE 8.27 Titanium alloy watch case produced using powder injection-molding (PIM) process. (Courtesy Hitachi Metals Precision/Casio Computer Co.)

FIGURE 8.28 Susan Abkowitz of Dynamet Technology holding titanium softball bat.

FIGURE 8.29 Titanium corrosion prevention cover on alluvial observation pier. (Courtesy Nippon Steel.)

FIGURE 8.30 Seam-welded titanium roof, Futtsu Technology Center, Japan. (Courtesy Nippon Steel).

TABLE 8.17 Airframe Weight Percentage of Titanium

System	Early Design	Final Concept
C5(Cargo)	24	3
B1(Bomber)	42	22
F15(Fighter)	50	34

The high cost of titanium alloy often limits use. For example, Table 8.17 compares the amount of titanium slated for use in three U.S. Air Force systems, expressed as airframe weight percentage, with early design figures shown for comparison [1]. Thus much work has concentrated on reducing component cost while maintaining acceptable mechanical property levels; approaches including near-net-shape techniques and lower cost alloy formulation.

An area for expansion for titanium is in automobiles with about 16 million cars and light trucks produced in the United States alone each year. Thus just 1.8 kg (4 lb) of titanium per vehicle could more than double titanium yearly consumption in the United States albeit with a dramatic effect on the titanium infrastructure [5, 6]. An "all-titanium" automobile was actually produced in the mid-1950s (Fig. 8.31). However, widespread use in large-volume production automobiles

FIGURE 8.31 All-titanium 1956 GM Titanium Firebird 2 automobile.

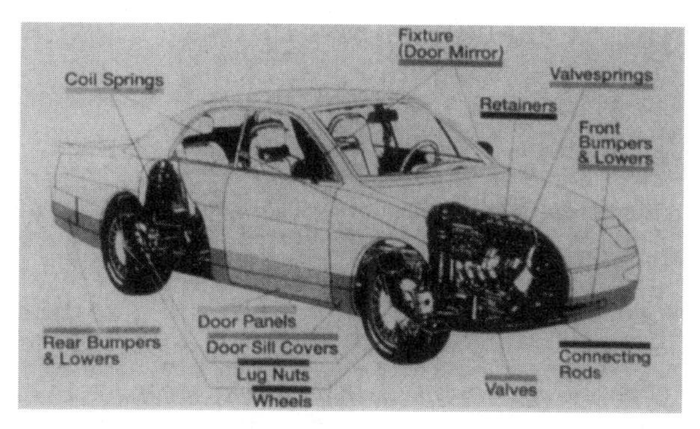

FIGURE 8.32 Prime components for titanium substitution in large production volume automobile. (Courtesy Japan Titanium Society).

FIGURE 8.33 Titanium metal "woods." (Courtesy TaylorMade Golf.)

(Fig. 8.32) will require a cost-effective product [1–7]. Recently, the attractive ballistic behavior of titanium has led to use in military armored vehicles [37, 38].

A new development has been the use of titanium in golf clubs, particularly metal "woods" (Fig. 8.33) [5, 6]. Also the requirement to reduce harmful defects [such as type I, hard, high interstitial (O_2 and N_2) defects], and the possibility of single melting billets closer to final configuration (hence reducing cost) has resulted in increased present and planned hearth melting facilities [5, 6].

8.12 CONCLUDING REMARKS AND FUTURE THOUGHTS

At the end of the 1990s is the titanium marketplace was in a minor downturn resulting from the decrease in commercial airplanes, especially by the Boeing Company. However, the marketplace strengthened in the late-1990s [39] and in combination with ever-increasing nonaerospace applications of titanium will lead to a continuation of the general growth trend shown in Fig. 8.20 [6, 7].

The major factor restricting much more widespread use of titanium and its alloys is cost. However, we have now seen the significant breakthrough of titanium use in the family automobile—the Toyota Altezza only on sale in Japan currently. If rumors of a lower cost extraction "reverse electrolytic" process [40] are true, then dramatically increased use of titanium and its alloys will occur.

REFERENCES

1. H. B. Bomberger, F. H. (Sam) Froes, and P. H. Morten, in F. H. (Sam) Froes, D. Eylon, and H. B. Bomberger (Eds.), *Titanium Technology: Present Status and Future Trends*, Titanium Development Association, Dayton, OH, 1985, pp. 3–18.
2. S. R. Seagle and J. R. Wood, in F. H. (Sam) Froes and T. Khan (Eds.), *Synthesis, Processing and Modelling of Advanced Materials*, Vol. 77–78, Trans. Tech. Publications, Brookville, VT, 1993, pp. 91–102.
3. F. H. (Sam) Froes, *Light Metal Age*, **53**(9 & 10), 6–9 (1995).
4. F. H. (Sam) Froes, *Third ASM Intl. Paris Conf. on Synthesis, Processing and Modeling of Advanced Materials*, ASM Intl., Materials Park, OH, 1997, pp. 3–38.
5. F. H. (Sam) Froes, P. G. Allen, and M. Niiomi, in F. H. (Sam) Froes, P. G. Allen, and M. Niiomi (Eds.), *Non-aerospace Applications of Titanium*, TMS, Warrendale, PA, 1998, pp. 3–20.
6. R. P. Pawlek and F. H. Froes, *Light Metal Age*, **57**(9 & 10), 55–61 (1999).
7. F. H. (Sam) Froes, *Light Metal Age*, **58**(3 & 4), 76–79 (2000).
8. S. S. Joseph and F. (Sam) H. Froes. *Light Metal Age*, **4–6**(11–12), 5–12 (1988).
9. I. J. Polmear, *Light Alloys, Metallurgy of the Light Metals*, 3rd Ed., Edward Arnold, London, 1996.
10. E. K. Molchanova, *Phase Diagrams of Titanium Alloys* [transl. of *Atlas Diagram Sostoyaniya Titanovyk Splavov*], Israel Program for Scientific Translations, Jerusalem, 1995.
11. H. Margolin and H. Neilson, "Titanium Metallurgy," in H. H. Hauser (Ed.), *Modern Materials, Advances in Development and Applications*, Vol. 2, Academic Press, Princeton, NJ, 1960, pp. 225–325.
12. F. H. (Sam) Froes, D. Eylon, and H. B. Bomberger (Eds.), *Titanium Technology: Present Status and Future Trends*, Titanium Development Association, Dayton, OH (now the International Titanium Association, Bolder, CO), 1985.
13. F. H. (Sam) Froes, C. Suryanarayana, and D. Eliezer, *ISIJ Intl.*, **31**, 1235–1248 (1991).
14. F. H. (Sam) Froes, C. Suryanarayana, and D. Eliezer, *J. Mater. Sci.*, **27**, 5113–5140 (1992).

15. F. H. (Sam) Froes, *J. Mater. Sci. Technol.*, **10**, 251–262 (1994).
16. C. M. Ward, *Int. Mats. Rev.*, **38**(2), 79–101 (1993).
17. B. H. Kear, *Scientific Am.*, **255**(4), 158–167 (1986).
18. F. H. (Sam) Froes and D. Eylon. *Int. Mats. Rev.*, **35**(3), 162–182 (1990).
19. F. H. (Sam) Froes and C. Suryanarayana, in A. Bose, R. M. German, and A. Lawley (Eds.), *Reviews in Particulate Materials*, Vol. 1, MPIF, Princeton, NJ, 1993, pp. 223–275.
20. F. H. (Sam) Froes, C. Suryanarayana, K. C. Russell, and C. M. Ward-Close, in J. Singh and S. M. Copely (Eds.), *Int. Conf. on Novel Techniques in Synthesis and Processing of Advanced Materials*, TMS, Warrendale, PA, 1995, pp. 1–22.
21. C. M. Ward-Close and F. H. (Sam) Froes, *JOM*, **46**(1), 28–31 (1994).
22. R. MacKay, P. Bindley, and F. H. (Sam) Froes, *JOM*, **43**(5), 23–29 (1992).
23. D. Upadhyaya, M. Wood, C. M. Ward-Close, P. Tsakiropoulos, and F. H. (Sam) Froes, *JOM*, Nov., 62–67 (1994).
24. F. H. (Sam) Froes, *Light Metal Age*, **49**(5/6), 6–11 (1991).
25. R. R. Boyer, G. Welsch, and E. W. Collings, *Materials Properties Handbook: Titanium Alloys*, ASM Int., Materials Park, OH, 1994.
26. M. J. Donachie, Jr. *Titanium, A Technical Guide*, ASM International, Materials Park, OH, 1988.
27. R. R. Boyer and J. A. Hall, in F. H. (Sam) Froes and I. L. Caplan (Eds.), *Titanium '92, Science and Technology*, Vol. 1, TMS, Warrendale, PA, 1993, pp. 77–88.
28. F. H. (Sam) Froes and D. Eylon, in N. R. Moody and A. W. Thompson (Eds.), *Hydrogen Effects on Material Behavior*, TMS, Warrendale, PA, 1990, pp. 261–284.
29. O. N. Senkov, J. J. Jonas, and F. H. (Sam) Froes, *JOM*, July, 42–47 (1996).
30. F. H. (Sam) Froes and J. Hebeisen, in L. Delaey and H. Tas (Eds.), *Hot Isostatic Pressing '93*, Elsevier, London, UK, 1994, pp. 71–90.
31. F. H. (Sam) Froes, R. Widmer, and J. Hebeisen, in F. H. (Sam) Froes, F. Widmer, and J. Hebeisen (Eds.), *Proc. Conf. Hot Isostatic Pressing '96*, ASM Intl., Materials Park, OH, 1996, pp. 3–18.
32. F. H. (Sam) Froes and W. A. Baeslack, in J. C. Danko and E. E. Noltin (Eds.), *High Energy Electron Beam Welding and Materials Processing*, AWS, Miami, FL, 1993, pp. 219–241.
33. P. R. Smith and F. H. (Sam) Froes, *JOM*, **36**(3), 19–26 (1984).
34. D. Driver, in E. Bachelet et al. (Eds.), *High Temperature Materials for Power Engineering*, Kluwer Academic, New York, NY, Part II, 1990, pp. 883–903.
35. J. B. Guernsey, V. C. Petersen, and F. H. (Sam) Froes, *Met. Trans.*, **3**, 339–341 (1972).
36. P. H. Morton (Ed.), *Designing with Titanium*, Inst. Metals, London, 1986.
37. J. C. Fanning, P. A. Blenkinsop, W. J. Evans and H. M. Flower (Eds.), *Titanium '95*, Vol. II, Inst. Materials, London, 1996, pp. 1688–1695.
38. F. H. (Sam) Froes, *Light Metal Age*, **55**(1 and 2), 66–73 (1997).
39. J. Ott and P. Proctor, *Aviation Week and Space Tech.*, **152**(2), 44 and 50 (2000).
40. C. M. Ward-Close, Defense Evaluation and Research Agency (DERA), Farnborough, UK, private communication, Feb. 2000.

CHAPTER 9

Aluminum and Aluminum Alloys

J. RANDOLPH KISSELL*

The TGB Partnership, 1325 Farmview Road, Hillsborough, North Carolina 27278

SYROS G. PANTELAKIS†

Department of Mechanical Engineering and Aeronautics, University of Patras, Patras, Greece

G. N. HAIDEMENOPOULOS†

Department of Mechanical and Industrial Engineering, University of Thessaly, Volos, Greece

*Sections 9.1– 9.5
†Section 9.6

Handbook of Advanced Materials Edited by James K. Wessel
ISBN 0-471-45475-3

9.1 DESCRIPTION

When a 6-lb aluminum cap was placed at the top of the Washington Monument upon its completion in 1884, aluminum was so rare it was considered a precious metal. In less than 100 years, however, aluminum became the most widely used metal after iron. While all aluminum alloys are recent discoveries compared to metals such as iron, copper, lead, and tin, the aluminum industry continues to develop new alloys and applications. Understanding of the new trends, though, is enhanced by an understanding of aluminum's history.

In nature, aluminum is found tightly combined with other elements, mainly oxygen and silicon, in red claylike bauxite deposits near the Earth's surface. Because it is so difficult to extract pure aluminum from its natural state, it wasn't until 1807 that it was identified by Sir Humphry Davy of England, who named it aluminum after alumine, the metal the Romans believed was present in clay. Davy had successfully produced small, relatively pure amounts of potassium but failed to isolate aluminum.

In 1825 Hans Oersted of Denmark finally produced a small lump of aluminum by heating potassium amalgam with aluminum chloride. Napoleon III of France, intrigued with possible military applications of the metal, promoted research leading to Sainte-Claire Deville's improved production method in 1854, which used less costly sodium in place of potassium. Deville named the aluminum-rich deposits near Les Baux in southern France "bauxite" and changed Davy's spelling to "aluminium." Probably because of the leading role played by France in the metal's early development, Deville's spelling was adopted around the world, including Davy's home country; only in the United States and Canada is the metal called "aluminum" today.

These chemical reaction recovery processes remained too expensive for widespread, practical application, however. In 1886, Charles Martin Hall of Oberlin, Ohio, and Paul L. T. Héroult in Paris, working independently, discovered virtually simultaneously the electrolytic process now used for the commercial production of aluminum. The Hall–Héroult process begins with aluminum oxide (Al_2O_3), a fine white material known as alumina and produced by chemically refining bauxite. Alumina is dissolved in a molten salt called cryolite in large, carbon-lined cells. A battery is set up by passing direct electrical current from the cell lining acting as the cathode and a carbon anode suspended at the center of the cell, separating the aluminum and oxygen. The molten aluminum produced is drawn off and cooled into a solid. Hall went on to patent this process and help found, in nearby Pittsburgh in 1888, what became the Aluminum Company of America, now called Alcoa. The success of this venture was aided by the discovery of Germany's Karl Joseph Bayer about this time of a practical process that bears his name for refining bauxite into alumina.

9.1.1 Attributes

Although aluminum is the most abundant metal in the Earth's crust, it costs more than some less plentiful metals because of the energy needed to extract the metal from ore. Its widespread use is due to the aluminum's qualities, which include:

High Strength-to-Weight Ratio Aluminum is the lightest metal other than magnesium, with a density about one-third that of steel. The strength of aluminum alloys, however, rivals that of mild carbon steel, and can approach 100 ksi (700 MPa). This combination of high strength and light weight makes aluminum especially well suited to transportation vehicles such as ships, rail cars, aircraft, rockets, trucks, and, increasingly, automobiles, as well as portable structures such as ladders, scaffolding, and gangways.

Ready Fabrication Aluminum is one of the easiest metals to form and fabricate, including operations such as extruding, bending, roll-forming, drawing, forging, casting, spinning, and machining [9]. In fact, all methods used to form other metals can be used to form aluminum. Aluminum is the metal most suited to extruding. This process (by which solid metal is pushed through an opening outlining the shape of the resulting part, like squeezing toothpaste from the tube) is especially useful since it can produce parts with complex cross sections in one operation. Examples include aluminum fenestration products such as window frames and door thresholds, and mullions and framing members used in curtainwalls, the outside envelope of many buildings.

Corrosion Resistance The aluminum cap placed at the top of the Washington Monument in 1884 is still there today. Aluminum reacts with oxygen very rapidly, but the formation of this tough oxide skin prevents further oxidation of the metal. This thin, hard, colorless oxide film tightly bonds to the aluminum surface and quickly reforms when damaged [24].

High Electrical Conductivity Aluminum conducts twice as much electricity as an equal weight of copper, making it ideal for use in electrical transmission cables.

High Thermal Conductivity Aluminum conducts heat three times as well as iron, benefiting both heating and cooling applications, including automobile radiators, refrigerator evaporator coils, heat exchangers, cooking utensils, and engine components.

High Toughness at Cryogenic Temperatures Aluminum is not prone to brittle fracture at low temperatures and has a higher strength and toughness at low temperatures, making it useful for cryogenic vessels.

Reflectivity Aluminum is an excellent reflector of radiant energy; hence its use for heat and lamp reflectors and in insulation.

Nontoxic Because aluminum is nontoxic, it is widely used in the packaging industry for food and beverages, as well as piping and vessels used in food processing and cooking utensils.

Recyclability Aluminum is readily recycled; about 30% of U.S. aluminum production is from recycled material. Aluminum made from recycled material requires only 5% of the energy needed to produce aluminum from bauxite.

Often a combination of the properties of aluminum plays a role in its selection for a given application. An example is gutters and other rain-carrying goods, made of aluminum because it can be easily roll-formed with portable equipment on site and it is so resistant to corrosion from exposure to the elements. Another is beverage cans, which benefit from aluminum's light weight for shipping purposes, and its recyclability.

9.1.2 The Aluminum Association

In the United States the Aluminum Association, founded in 1933, is composed of the primary American aluminum producers. The Aluminum Association is the main source of information, standards, and statistics concerning the U.S. aluminum industry. Contacts for the Aluminum Association are:

Mail: 900 19th Street, N.W., Suite 300, Washington, DC, 20006
Phone: 202-862-5100
Fax: 202-862-5164
Internet: `www.aluminum.org`

The Aluminum Association publishes standards on aluminum alloy and temper designations and tolerances for aluminum mill products. Publications offered by the association also provide information on applications of aluminum such as automotive body sheet and electrical conductors. Other parts of the world are served by similar organizations such as the European Aluminum Association in Brussels.

9.1.3 Alloy and Temper Designation System

Metals enjoy relatively little use in their pure state. The addition of one or more elements to a metal results in an alloy that often has significantly different properties than the unalloyed material. While the addition of alloying elements to aluminum sometimes degrades certain characteristics of the pure metal (such as corrosion resistance or electrical conductivity), this is acceptable for certain applications because other properties (such as strength) can be so markedly enhanced. About 15 alloying elements are used with aluminum, and, though they usually comprise less than 10% of the alloy by weight, they can dramatically affect material properties.

Aluminum alloys are divided into two categories: *wrought alloys*, those that are worked to shape, and *cast alloys*, those that are poured in a molten state into a mold that determines their shape. The Aluminum Association maintains a widely recognized designation system for each category, described in ANSI H35.1, *Alloy and Temper Designations for Aluminum* [15], discussed below.

9.1.3.1 Wrought Alloys

In the Aluminum Association's designation system for aluminum alloys, a four-digit number is assigned to each alloy registered with the association. The first number of the alloy designates the primary alloying element, which produces a group of alloys with similar properties. The last two digits are assigned sequentially by the association. The second digit denotes a modification of an alloy.

TABLE 9.1 Wrought Alloy Designation System and Characteristics

Series Number	Primary Alloying Element	Relative Corrosion Resistance	Relative Strength	Heat Treatment
1xxx	None	Excellent	Fair	Non-heat-treatable
2xxx	Copper	Fair	Excellent	Heat treatable
3xxx	Manganese	Good	Fair	Non-heat-treatable
4xxx	Silicon	—	—	Non-heat-treatable
5xxx	Magnesium	Good	Good	Non-heat-treatable
6xxx	Magnesium and silicon	Good	Good	Heat-treatable
7xxx	Zinc	Fair	Excellent	Heat-treatable

For example, 6463 is a modification of 6063 with slightly more restrictive limits on certain alloying elements such as iron, manganese, and chromium to obtain better finishing characteristics. The primary alloying elements and the properties of the resulting alloys are listed below and summarized in Table 9.1:

1xxx This series is for *commercially pure aluminum*, defined in the industry as being at least 99% aluminum. Alloy numbers are assigned within the 1xxx series for variations in purity and which elements comprise the impurities, the main ones being iron and silicon. The primary uses for alloys of this series are electrical conductors and chemical storage or processing because the best properties of the alloys of this series are electrical conductivity and corrosion resistance. The last two digits of the alloy number denote the two digits to the right of the decimal point of the percentage of the material that is aluminum. For example, 1060 denotes an alloy that is 99.60% aluminum. The strength of pure aluminum is relatively low.

2xxx The primary alloying element for this group is *copper*, which produces high strength but reduced corrosion resistance. These alloys were among the first aluminum alloys developed and were originally called duralumin. Alloy 2024 is perhaps the best known and most widely used alloy in aircraft. The original aluminum–copper alloys were not very weldable, but designers have overcome this obstacle in subsequently developed alloys of this series.

3xxx *Manganese* is the main alloying element for the 3xxx series, increasing the strength of unalloyed aluminum by about 20%. The corrosion resistance and workability of alloys in this group, which primarily consists of alloys 3003, 3004, and 3105, are good. The 3xxx series alloys are well suited to architectural products such as rain-carrying goods and roofing and siding.

4xxx *Silicon* is added to alloys of the 4xxx series to reduce the melting point for welding and brazing applications. Silicon also provides good flow characteristics, which in the case of forgings provide more complete filling of complex die shapes. Alloy 4043 is commonly used for weld filler wire.

5xxx The 5xxx series alloys contain *magnesium*, resulting in high strength and corrosion resistance. Alloys of this group are used in ship hulls and other marine applications, weld wire, and welded storage vessels. The strength of alloys in this series is directly proportional to the magnesium content, which ranges up to about 6%.

6xxx Alloys in this group contain *magnesium and silicon* in proportions that form magnesium silicide (Mg_2Si). These alloys have a good balance of corrosion resistance and strength. Alloy 6061 is one of the most popular of all aluminum alloys and has a yield strength comparable to mild carbon steel. The 6xxx series alloys are also very readily extruded, so they comprise the majority of extrusions produced and are used extensively in building, construction, and other structural applications.

7xxx The primary alloying element of this series is *zinc*. The 7xxx series includes two types of alloys—the aluminum–zinc–magnesium alloys (such as 7005) and the aluminum–zinc–magnesium–copper alloys (such as 7075 and 7178). The alloys of this series include some of the strongest aluminum alloys such as 7178, which has a minimum tensile ultimate strength of 84 ksi (580 MPa), and are used in aircraft frames and structural components. The corrosion resistance of those 7xxx series alloys alloyed with copper is less, however, than the 1xxx, 3xxx, 5xxx, or 6xxx series. The 7xxx alloys without copper are corrosion-resistant, and some (such as 7008 and 7072) are used as cladding to cathodically protect less corrosion-resistant aluminum alloys.

8xxx The 8xxx series is reserved for alloying elements other than those used for series 2xxx through 7xxx. Iron and nickel are used to increase strength without significant loss in electrical conductivity, such as in conductor alloys like 8017.

9xxx This series is not currently used.

Experimental alloys are designated in accordance with the above system, but with the prefix X until they are no longer experimental. Producers may also offer proprietary alloys to which they assign their own designation numbers or brand names.

The chemical composition limits in percent by weight for common wrought alloys are given in Table 9.2.

National variations of these alloys may be registered by other countries under this system. Such variations are assigned a capital letter following the numerical designation (e.g., 6005A, a variation on 6005 used in Europe). The chemical composition limits for national variations are similar to the Aluminum Association limits, but vary slightly. Some standards-writing organizations of other countries have their own designation systems that differ from the Aluminum Association system. A comparison of some alloy designations is given in Table 9.3.

The 2xxx and 7xxx series are sometimes referred to as aircraft alloys, but they are also used in other applications, including fasteners used in buildings. The 1xxx, 3xxx, and 6xxx series alloys are sometimes referred to as "soft," while the 2xxx, 5xxx, and 7xxx series alloys are called "hard." This description refers to the ease of extruding the alloys—hard alloys are more difficult to extrude, requiring higher capacity presses and are thus more expensive.

TABLE 9.2 Chemical Composition Limits of Wrought Aluminum Alloys① ②

AA Designation	Silicon	Iron	Copper	Manganese	Magnesium	Chromium	Nickel	Zinc	Titanium	Others㉒		Aluminum Min.④
										Each⑳	Total③	
1050	0.25	0.40	0.05	0.05	0.05	—	—	0.05	0.03	0.03⑨	—	99.50
1060	0.25	0.35	0.05	0.03	0.03	—	—	0.05	0.03	0.03⑨	—	99.60
1100	0.95 Si + Fe		0.05–0.20	0.05	—	—	—	0.10	—	0.05⑯	0.15	99.00
1145⑧	0.55 Si + Fe		0.05	0.05	0.05	—	—	0.05	0.03	0.03⑨	—	99.45
1175⑦	0.15 Si + Fe		0.10	0.02	0.02	—	—	0.04	0.02	0.02⑲	—	99.75
1200	1.00 Si + Fe		0.05	0.05	—	—	—	0.10	0.05	0.05	0.15	99.00
1230⑦	0.70 Si + Fe		0.10	0.05	0.05	—	—	0.10	0.03	0.03⑨	—	99.30
1235	0.65 Si + Fe		0.05	0.05	0.05	—	—	0.10	0.06	0.03⑨	—	99.35
1345	0.30	0.40	0.10	0.05	0.05	—	—	0.05	0.03	0.03⑨	—	99.45
1350⑥	0.10	0.40	0.05	0.01	—	0.01	—	0.05	—	0.03⑬	0.10	99.50
2011	0.40	0.7	5.0–6.0	—	—	—	—	0.30	—	0.05⑩	0.15	Remainder
2014	0.50–1.2	0.7	3.9–5.0	0.40–1.2	0.20–0.8	0.10	—	0.25	0.15	0.05	0.15	Remainder
2017	0.20–0.8	0.7	3.5–4.5	0.40–1.0	0.40–0.8	0.10	—	0.25	0.15	0.05	0.15	Remainder
2018	0.9	1.0	3.5–4.5	0.20	0.45–0.9	0.10	1.7–2.3	0.25	—	0.05	0.15	Remainder
2024	0.50	0.50	3.8–4.9	0.30–0.9	1.2–1.8	0.10	—	0.25	0.15	0.05	0.15	Remainder
2025	0.50–1.2	1.0	3.9–5.0	0.40–1.2	0.05	0.10	—	0.25	0.15	0.05	0.15	Remainder
2036	0.50	0.50	2.2–3.0	0.10–0.40	0.30–0.6	0.10	—	0.25	0.15	0.05	0.15	Remainder
2117	0.8	0.7	2.2–3.0	0.20	0.20–0.50	0.10	—	0.25	—	0.05	0.15	Remainder
2124	0.20	0.30	3.8–4.9	0.30–0.9	1.2–1.8	0.10	—	0.25	0.15	0.05	0.15	Remainder
2218	0.9	1.0	3.5–4.5	0.20	1.2–1.8	0.10	1.7–2.3	0.25	—	0.05	0.15	Remainder
2219	0.20	0.30	5.8–6.8	0.20–0.40	0.02	—	—	0.10	0.02–0.10	0.05⑱	0.15	Remainder
2319	0.20	0.30	5.8–6.8	0.20–0.40	0.02	—	—	0.10	0.10–0.20	0.05⑱	0.15	Remainder
2618	0.10–0.25	0.9–1.3	1.9–2.7	—	1.3–1	—	0.9–1.2	0.10	0.04–0.10	0.05	0.15	Remainder

3003	0.6	0.7	0.05–0.20	1.0–1.5	—	—	—	0.10	—	0.05	0.15	Remainder
3004	0.30	0.7	0.25	1.0–1.5	0.8–1.3	—	—	0.25	—	0.05	0.15	Remainder
3005	0.6	0.7	0.30	1.0–1.5	0.20–0.6	0.10	—	0.25	0.10	0.05	0.15	Remainder
3105	0.6	0.7	0.30	0.30–0.8	0.20–0.8	0.20	—	0.40	0.10	0.05	0.15	Remainder
4032	11.0–13.5	1.0	0.50–1.3	—	0.8–1.3	0.10	0.50–1.3	0.25	—	0.05	0.15	Remainder
4043	4.5–6.0	0.8	0.30	0.05	0.05	—	—	0.10	0.20	0.05⑯	0.15	Remainder
4045⑪	9.0–11.0	0.8	0.30	0.05	0.05	—	—	0.10	0.20	0.05	0.15	Remainder
4047⑪	11.0–13.0	0.8	0.30	0.15	0.10	—	—	0.20	—	0.05⑯	0.15	Remainder
4145⑪	9.3–10.7	0.8	3.3–4.7	0.15	0.15	0.15	—	0.20	—	0.05⑯	0.15	Remainder
4343⑪	6.8–8.2	0.8	0.25	0.10	—	—	—	0.20	—	0.05	0.15	Remainder
4643	3.6–4.6	0.8	0.10	0.05	0.10–0.30	—	—	0.10	0.15	0.05⑯	0.15	Remainder
5005	0.30	0.7	0.20	0.20	0.50–1.1	0.10	—	0.25	—	0.05	0.15	Remainder
5050	0.40	0.7	0.20	0.10	1.1–1.8	0.10	—	0.25	—	0.05	0.15	Remainder
5052	0.25	0.40	0.10	0.10	2.2–2.8	0.15–0.35	—	0.10	—	0.05	0.15	Remainder
5056	0.30	0.40	0.10	0.05–0.20	4.5–5.6	0.05–0.20	—	0.10	—	0.05	0.15	Remainder
5083	0.40	0.40	0.10	0.40–1.0	4.0–4.9	0.05–0.25	—	0.25	0.15	0.05	0.15	Remainder
5086	0.40	0.50	0.10	0.20–0.7	3.5–4.5	0.05–0.25	—	0.25	0.15	0.05	0.15	Remainder
5154	0.25	0.40	0.10	0.10	3.1–3.9	0.15–0.35	—	0.20	0.20	0.05	0.15	Remainder
5183	0.40	0.40	0.10	0.50–1.0	4.3–5.2	0.05–0.25	—	0.25	0.15	0.05⑯	0.15	Remainder
5252	0.08	0.10	0.10	0.10	2.2–2.8	—	—	0.05	—	0.03⑨	0.10	Remainder
5254	0.45Si + Fe		0.05	0.01	3.1–3.9	0.15–0.35	—	0.20	0.05	0.05	0.15	Remainder
5356	0.25	0.40	0.10	0.05–0.20	4.5–5.5	0.05–0.20	—	0.10	0.06–0.20	0.05⑯	0.15	Remainder
5454	0.25	0.40	0.10	0.50–1.0	2.4–3.0	0.05–0.20	—	0.25	0.20	0.05	0.15	Remainder
5456	0.25	0.40	0.10	0.50–1.0	4.7–5.5	0.05–0.20	—	0.25	0.20	0.05	0.15	Remainder
5457	0.08	0.10	0.20	0.15–0.45	0.8–1.2	—	—	0.05	—	0.03⑨	0.10	Remainder
5554	0.25	0.40	0.10	0.50–1.0	2.4–3.0	0.05–0.20	—	0.25	0.05–0.20	0.05⑯	0.15	Remainder
5556	0.25	0.40	0.10	0.50–1.0	4.7–5.5	0.05–0.20	—	0.25	0.05–0.20	0.05⑯	0.15	Remainder
5652	0.40Si + Fe		0.04	0.01	2.2–2.8	0.15–0.35	—	0.10	—	0.05	0.15	Remainder
5654	0.45Si + Fe		0.05	0.01	3.1–3.9	0.15–0.35	—	0.20	0.05–0.15	0.05⑯	0.15	Remainder
5657	0.08	0.10	0.10	0.03	0.6–1.0	—	—	0.05	—	0.02⑲	0.05	Remainder

TABLE 9.2 ***(Continued)***

AA Designation	Silicon	Iron	Copper	Manganese	Magnesium	Chromium	Nickel	Zinc	Titanium	Others(22)		Aluminum Min.(4)
										Each(20)	Total(3)	
6003(7)	0.35–1.0	0.6	0.10	0.8	0.8–1.5	0.35	—	0.20	0.10	0.05	0.15	Remainder
6005	0.6–0.9	0.35	0.10	0.10	0.40–0.6	0.10	—	0.10	0.10	0.05	0.15	Remainder
6053	(15)	0.35	0.10	—	1.1–1.4	0.15–0.35	—	0.10	—	0.05	0.15	Remainder
6061	0.40–0.8	0.7	0.15–0.40	0.15	0.8–1.2	0.04–0.35	—	0.25	0.15	0.05	0.15	Remainder
6063	0.20–0.6	0.35	0.10	0.10	0.45–0.9	0.10	—	0.10	0.10	0.05	0.15	Remainder
6066	0.9–1.8	0.50	0.7–1.2	0.6–1.1	0.8–1.4	0.40	—	0.25	0.20	0.05	0.15	Remainder
6070	1.0–1.7	0.50	0.15–0.40	0.40–1.0	0.50–1.2	0.10	—	0.25	0.15	0.05	0.15	Remainder
6101(12)	0.30–0.7	0.50	0.10	0.03	0.35–0.8	0.03	—	0.10	—	0.03(17)	0.10	Remainder
6105	0.6–1.0	0.35	0.10	0.15	0.45–0.8	0.10	—	0.10	0.10	0.05	0.15	Remainder
6151	0.6–1.2	1.0	0.35	0.20	0.45–0.8	0.15–0.35	—	0.25	0.15	0.05	0.15	Remainder
6162	0.40–0.8	0.50	0.20	0.10	0.7–1.1	0.10	—	0.25	0.10	0.05	0.15	Remainder
6201	0.50–0.9	0.50	0.10	0.03	0.6–0.9	0.03	—	0.10	—	0.03(17)	0.10	Remainder
6253(7)	(15)	0.50	0.10	—	1.0–1.5	0.04–0.35	—	1.6–2.4	—	0.05	0.15	Remainder
6262	0.40–0.8	0.7	0.15–0.40	0.15	0.8–1.2	0.04–0.14	—	0.15	0.15	0.05(5)	0.15	Remainder
6351	0.7–1.3	0.50	0.10	0.40–0.8	0.40–0.8	—	—	0.20	0.20	0.05	0.15	Remainder
6463	0.20–0.6	0.15	0.20	0.05	0.45–0.9	—	—	0.05	—	0.05	0.15	Remainder
6951	0.20–0.50	0.8	0.15–0.40	0.10	0.40–0.8	—	—	0.20	—	0.05	0.15	Remainder
7005	0.35	0.40	0.10	0.20–0.7	1.0–1.8	0.06–0.20	—	4.5–5.0	0.01–0.06	0.05(14)	0.15	Remainder
7008(7)	0.10	0.10	0.05	0.05	0.7–1.4	0.12–0.25	—	4.5–5.5	0.05	0.05	0.10	Remainder
7049	0.25	0.35	1.2–1.9	0.20	2.0–2.9	0.10–0.22	—	7.2–8.2	0.10	0.05	0.15	Remainder
7050	0.12	0.15	2.0–2.6	0.10	1.9–2.6	0.04	—	5.7–6.7	0.06	0.05(21)	0.15	Remainder
7072(7)	0.7 Si + Fe		0.10	0.10	0.10	—	—	0.8–1.3	—	0.05	0.15	Remainder
7075	0.40	0.50	1.2–2.0	0.30	2.1–2.9	0.18–0.28	—	5.1–6.1	0.20	0.05	0.15	Remainder
7175	0.15	0.20	1.2–2.0	0.10	2.1–2.9	0.18–0.28	—	5.1–6.1	0.10	0.05	0.15	Remainder
7178	0.40	0.50	1.6–2.4	0.30	2.4–3.1	0.18–0.28	—	6.3–7.3	0.20	0.05	0.15	Remainder
7475	0.10	0.12	1.2–1.9	0.06	1.9–2.6	0.18–0.25	—	5.2–6.2	0.06	0.05	0.15	Remainder

8017	0.10	0.55–0.8	0.10–0.20	—	0.01–0.05	—	—	0.05	—	0.03㉓	0.10	Remainder
8030	0.10	0.30–0.8	0.15–0.30	—	0.05	—	—	0.05	—	0.03㉔	0.10	Remainder
8176	0.03–.015	0.40–1.0	—	—	—	—	—	0.10	—	0.05㉕	0.15	Remainder
8177	0.10	0.25–0.45	0.04	—	0.04–0.12	—	—	0.05	—	0.03㉖	0.10	Remainder

Source: Aluminum Association [5, 6].

Note: Listed herein are designations and chemical composition limits for some wrought unalloyed aluminum and for wrought aluminum alloys registered with The Aluminum Association. This list does not include all alloys registered with The Aluminum Association. A complete list of registered designations is contained in the "Registration Record of International Alloy Designations and Chemical Composition Limits for Wrought Aluminum and Wrought Aluminum Alloys." These lists are maintained by the Technical Committee on Product Standards of the Aluminum Association.

①Composition in percent by weight maximum unless shown as a range or a minimum.

②Except for "Aluminum" and "Others," analysis normally is made for elements for which specific limits are shown. For purposes of determining conformance to these limits, an observed value or a calculated value obtained from analysis is rounded off to the nearest unit in the last right-hand place of figures used in expressing the specified limit, in accordance with ASTM Recommended Practice E 29.

③The sum of those "Other" metallic elements 0.010% or more each, expressed to the second decimal before determining the sum.

④The aluminum content for unalloyed aluminum not made by a refining process is the difference between 100.00% and the sum of all the other metallic elements present in amounts of 0.010% or more each, expressed to the second decimal before determining the sum.

⑤Also contains 0.40–0.7% each of lead and bismuth.

⑥Electric conductor. Formerly designated EC.

⑦Cladding alloy.

⑧Foil.

⑨Vanadium 0.05% maximum.

⑩Also contains 0.20–0.6% each of lead and bismuth.

⑪Brazing alloy.

⑫Bus conductor.

⑬Vanadium plus titanium 0.02% maximum; boron 0.05% maximum; gallium 0.03% maximum.

⑭Zirconium 0.08–0.20.

⑮Silicon 45–65% of actual magnesium content.

⑯Beryllium 0.0008 maximum for welding electrode and welding rod only.

⑰Boron 0.06% maximum.

⑱Vanadium 0.05–0.15; zirconium 0.10–0.25.

⑲Gallium 0.03% maximum; vanadium 0.05% maximum.

⑳In addition to those alloys referencing footnote⑯, a 0.0008 weight % maximum beryllium is applicable to any alloy to be used as welding electrode or welding rod.

㉑Zirconium 0.08–0.15.

㉒"Others" includes listed elements for which no specific limit is shown as well as unlisted metallic elements. The producer may analyze samples for trace elements not specified in the registration or specification. However, such analysis is not required and may not cover all metallic "Other" elements. Should any analysis by the producer or the purchaser establish that an "Others" elements exceeds the limit of "Total," the material shall be considered nonconforming.

㉓Boron 0.04% maximum; lithium 0.003% maximum.

㉔Boron 0.001–0.04.

㉕Gallium 0.03% maximum.

㉖Boron 0.04% maximum.

TABLE 9.3 Foreign Alloy Designations and Similar AA Alloys

Foreign Alloy Designation	Equivalent or Similar AA Alloy
Austria (Önorm)[1]	
Al99	1200
Al99,5	1050
E-Al	1350
AlCuMg1	2017
AlCuMg2	2024
AlCuMg0,5	2117
AlMg5	5056
AlMgSi0,5	6063
E-AlMgSi	6101
AlZnMgCu1,5	7075
Canada (CSA)[2]	
990C	1100
CB60	2011
CG30	2117
CG42	2024
CG42 Alclad	Alclad 2024
CM41	2017
CN42	2018
CS41N	2014
CS41N Alclad	Alclad 2014
CS41P	2025
GM31N	5454
GM41	5083
GM50P	5356
GM50R	5056
GR20	5052
GS10	6063
GS11N	6061
GS11P	6053
MC10	3003
S5	4043
SG11P	6151
SG121	4032
ZG62	7075
ZG62 Alclad	Alclad 7075
France (NF)[3]	
A5/L	1350
A45	1100
A-G1	5050
A-G0.6	5005
A-G4MC	5086
A-GS	6063
A-GS/L	6101
A-M1	3003
A-M1G	3004
A-U4G	2017
A-U2G	2117
A-U2GN	2618
A-U4G1	2024
A-U4N	2218
A-U4SG	2014
A-S12UN	4032
A-Z5GU	7075
Germany	
E-A1995[4] 3.0257[5]	1350
AlCuBiPb[4] 3.1655[5]	2011
AlCuMg0.5[4] 3.1305[5]	2117
AlCuMg1[4] 3.1325[5]	2017
AlCuMg2[4] 3.1355[5]	2024
AlCuSiMn[4] 3.1255[5]	2014
AlMg4.5Mn[4] 3.3547[5]	5083
AlMgSi0.5[4] 3.3206[5]	6063
AlSi5[4] 3.2245[5]	4043
E-AlMgSi0.5[4] 3.3207[5]	6101
AlZnMgCu1.5[4] 3.4365[5]	7075
Great Britian (BS)[6]	
1E	1350
91E	6101
H14	2017
H19	6063
H20	6061
L.80, L.81	5052
L.86	2117
L.87	2017
L.93, L.94	2014A
L.95, L.96	7075
L.97, L.98	2024
2L.55, 2L.56	5052
2L.58	5056
3L.44	5050
5L.37	2017
6L.25	2218
N8	5083
N21	4043
Great Britian (DTD)[7]	
150A	2017
324A	4032
372B	2017
717, 724, 731A 745, 5014, 5084	2618
5090	2024
5100	Alclad 2024

TABLE 9.3 (*Continued*)

Foreign Alloy Designation	Equivalent or Similar AA Alloy
Italy (UNI)⑧	
P-AlCu4MgMn	2017
P-AlCu4.5MgMn	2024
P-AlCu4.5MgMnplacc.	Alclad 2024
P-AlCu2.5MgSi	2117
P-AlCu4.4SiMnMg	2014
P-AlCu4.4SiMnMgplacc.	Alclad 2014
P-AlMg0.9	5657
P-AlMg1.5	5050
P-AlMg2.5	5052
P-AlSi0.4 Mg	6063
P-AlSi0.5 Mg	6101
Spain (UNE)⑨	
Al99.5E	1350
L-313	2014
L-314	2024
L-315	2218
L-371	7075
Switzerland (VSM)⑩	
Al-Mg-Si	6101
Al1.5 Mg	5050
Al-Cu-Ni	2218
Al3.5Cu0.5 Mg	2017
Al4Cu1.2 Mg	2027
Al-Zn-Mg-Cu	7075
Al-Zn-Mg-Cu-pl	Alclad 7075
ISO⑪	
Al99.0Cu	1100
AlCu2Mg	2117
AlCu4Mg1	2024
AlCu4SiMg	2014
AlCu4MgSi	2017
AlMg1	5005
AlMg1.5	5050
AlMg2.5	5052
AlMg3.5	5154
AlMg4	5086
AlMg5	5056
AlMn1Cu	3003
AlMg3Mn	5454
AlMg4.5Mn	5083
AlMgSi	6063
AlMg1SiCu	6061
AlZn6MgCu	7075

Source: Aluminum Association [5, 6].

①Austrian Standard M3430.
②Canadian Standards Association.
③Normes Françaises.
④Deutsche Industrie-Norm.
⑤Werkstoff -Nr.
⑥British Standard.
⑦Directorate of Technical Development.
⑧Unificazione Nazionale Italiana.
⑨Una Norma Espanol.
⑩Verein Schweizerischer Maschinenindustrieller.
⑪International Organization for Standardization.

Some alloys are provided with a thin coating of pure aluminum or corrosion-resistant aluminum alloy (such as 7072); the resulting product is called alclad. This cladding is metallurgically bonded to one or both sides of sheet, plate, 3003 tube, or 5056 wire, and may be 1.5–10% of the overall thickness. The cladding alloy is chosen so it is anodic to the core alloy and so protects it from corrosion. Any corrosion that occurs proceeds only to the cladding–core interface and then spreads laterally, making cladding very effective in protecting thin materials. Because the coating generally has a lower strength than the base metal, alclad alloys have slightly lower strengths than nonalclad alloys of the same thickness. Alloys used in clad products are given in Table 9.4.

9.1.3.2 Cast Alloys

Casting alloys contain larger proportions of alloying elements than wrought alloys. This results in a heterogeneous structure that is generally less ductile than the more homogeneous structure of the wrought alloys. Cast alloys also contain more silicon than wrought alloys to provide the fluidity necessary to make a casting.

TABLE 9.4 Components of Clad Products

Designation	Component Alloys①		Total Specified Thickness of Composite Product (in.)	Sides clad	Cladding Thickness per Side Percent of Composite Thickness		
						Average②	
	Core	Cladding			Nominal	min.	max.
Alclad 2014 Sheet and Plate	2014	6003	Up thru 0.024	Both	10	8	—
			0.025–0.039	Both	7.5	6	—
			0.040–0.099	Both	5	4	—
			0.100 and over	Both	2.5	2	3③
Alclad 2024 Sheet and Plate	2024	1230	Up thru 0.062	Both	5	4	—
			0.063 and over	Both	2.5	2	3③
$1\frac{1}{2}$% Alclad 2024 Sheet and Plate	2024	1230	0.188 and over	Both	1.5	1.2	3③
Alclad One Side 2024 Sheet and Plate	2024	1230	Up thru 0.062	One	5	4	—
			0.063 and over	One	2.5	2	3③
$1\frac{1}{2}$% Alclad One Side 2024 Sheet and Plate	2024	1230	0.188 and over	One	1.5	1.2	3③
Alclad 2219 Sheet and Plate	2219	7072	Up thru 0.039	Both	10	8	—
			0.040–0.099	Both	5	4	—
			0.100 and over	Both	2.5	2	3③
Alclad 3003 Sheet and Plate	3003	7072	All	Both	5	4	6③
Alclad 3003 Tube	3003	7072	All	Inside	10	—	—
			All	Outside	7	—	—
Alclad 3004 Sheet and Plate	3004	7072	All	Both	5	4	6③
Alclad 5056 Rod and Wire	5056	6253	Up thru 0.375	Outside	20	16 (of total cross-sectional area)	
Alclad 6061 Sheet and Plate	6061	7072	All	Both	5	4	6③

Alclad 7050 Sheet and Plate	7050	7072	Up thru 0.062	Both	4	3.2	—
			0.063 and over	Both	2.5	2	—
7108 Alclad 7050 Sheet and Plate	7050	7108	Up thru 0.062	Both	4	3.2	—
			0.063 and over	Both	2.5	2	—
Alclad 7075 Sheet and Plate	7075	7072	Up thru 0.062	Both	4	3.2	—
			0.063–0.187	Both	2.5	2	—
			0.188 and over	Both	1.5	1.2	3③
$2\frac{1}{2}$% Alclad 7075 Sheet and Plate	7075	7072	0.188 and over	Both	2.5	2	4③
Alclad One Side 7075 Sheet and Plate	7075	7072	Up thru 0.062	One	4	3.2	—
			0.063–0.187	One	2.5	2	—
			0.188 and over	One	1.5	1.2	3③
$2\frac{1}{2}$% Alclad One Side 7075 Sheet and Plate	7075	7072	0.188 and over	One	2.5	2	4③
7008 Alclad 7075 Sheet and Plate	7075	7008	Up thru 0.062	Both	4	3.2	—
			0.063–0.187	Both	2.5	2	—
			0.188 and over	Both	1.5	1.2	3③
7011 Alclad 7075 Sheet and Plate	7075	7011	Up thru 0.062	Both	4	3.2	—
			0.063–0.187	Both	2.5	2	—
			0.188 and over	Both	1.5	1.2	3③
Alclad 7178 Sheet and Plate	7178	7072	Up thru 0.062	Both	4	3.2	—
			0.063–0.187	Both	2.5	2	—
			0.188 and over	Both	1.5	1.2	3③
Alclad 7475 Sheet	7475	7072	Up thru 0.062	Both	4	3.2	—
			0.063–0.187	Both	2.5	2	—
			0.188–0.249	Both	1.5	1.2	—
No. 7 Brazing Sheet	3003	4004	Up thru 0.024	One	15	12	18
			0.025–0.062	One	10	8	12
			0.063 and over	One	7.5	6	9

TABLE 9.4 ***(Continued)***

Designation	Component Alloys①		Total Specified Thickness of Composite Product (in.)	Sides clad	Cladding Thickness per Side Percent of Composite Thickness		
						Average②	
	Core	Cladding			Nominal	min.	max.
No. 8 Brazing Sheet	3003	4004	Up thru 0.024	Both	15	12	18
			0.025–0.062	Both	10	8	12
			0.063 and over	Both	7.5	6	9
No. 11 Brazing Sheet	3003	4343④	Up thru 0.063	One	10	8	12
			0.064 and over	One	5	4	6
No. 12 Brazing Sheet	3003	4343④	Up thru 0.063	Both	10	8	12
			0.064 and over	Both	5	4	6
No. 23 Brazing Sheet	6951	4045	Up thru 0.090	One	10	8	12
			0.091 and over	One	5	4	6
No. 24 Brazing Sheet	6951	4045	Up thru 0.090	Both	10	8	12
			0.091 and over	Both	5	4	6
Clad 1100 Reflector Sheet	1100	1175	Up thru 0.064	Both	15	12	18
			0.065 and over	Both	7.5	6	9
Clad 3003 Reflector Sheet	3003	1175	Up thru 0.064	Both	15	12	18
			0.065 and over	Both	7.5	6	9

Source: Aluminum Association [5, 6].
Note: This table does not include all clad products registered with The Aluminum Association.
①Cladding composition is applicable only to the aluminum or aluminum alloy bonded to the alloy ingot or slab preparatory to processing to the specified composite product. The composition of the cladding may be subsequently altered by diffusion between the core and cladding due to thermal treatment.
②Average thickness per side as determined by averaging cladding thickness measurements taken at a magnification of 100 diameters on the cross-section of a transverse sample polished and etched for microscopic examination.
③Applicable for thickness of 0.500 in. and greater.
④The cladding component, in lieu of 4343 alloy, may be 5% 1xxx Clad4343.

While the Aluminum Association cast alloy designation system uses four digits like the wrought alloy system, most similarities end there. The cast alloy designation system has three digits, followed by a decimal point, followed by another digit. The first digit indicates the primary alloying element. The second two digits designate the alloy, or in the case of commercially pure casting alloys, the level of purity. The last digit indicates the product form—1 or 2 for ingot (depending on impurity levels) and 0 for castings. A modification of the original alloy is designated by a letter prefix (A, B, C, etc.) to the alloy number. The primary alloying elements are:

1xx.x These are the *commercially pure aluminum* cast alloys; an example of their use is cast motor rotors.

2xx.x The use of *copper* as the primary alloying element produces the strongest cast alloys. Alloys of this group are used for machine tools, aircraft, and engine parts. Alloy 203.0 has the highest strength at elevated temperatures and is suitable for service at 400°F (200°C).

3xx.x *Silicon*, with *copper* and/or *magnesium*, are used in this series. These alloys have excellent fluidity and strength and are the most widely used aluminum cast alloys. Alloy 356.0 and its modifications are very popular and used in many different applications. High silicon alloys have good wear resistance and are used for automotive engine blocks and pistons.

4xx.x The use of *silicon* in this series provides excellent fluidity in cast alloys as it does for wrought alloys, and so these are well suited to intricate castings such as typewriter frames and they have good general corrosion resistance. Alloy A444.0 has modest strength but good ductility.

5xx.x Cast alloys with *magnesium* have good corrosion resistance, especially in marine environments (e.g., 514.0), good machinability, and can be attractively finished. They are more difficult to cast than the 200, 300, and 400 series, however.

6xx.x This series is unused.

7xx.x Primarily alloyed with *zinc*, this series is difficult to cast and so is used where its finishing characteristics or machinability is important. These alloys have moderate or better strengths and good general corrosion resistance, but are not suitable for elevated temperatures.

8xx.x This series is alloyed with about 6% *tin* and primarily used for bearings, being superior to most other materials for this purpose. These alloys are used for large rolling mill bearings and connecting rods and crankcase bearings for diesel engines.

9xx.x This series is reserved for castings alloyed with elements other than those used in the other series.

The chemical composition limits for common cast alloys are given in Table 9.5.

9.1.3.3 Tempers

Aluminum alloys are tempered by *heat-treating* or *strain-hardening* to further increase strength beyond the strengthening effect of adding alloying elements.

TABLE 9.5 Chemical Composition Limits for Commonly Used Sand and Permanent Mold Casting Alloys[a][b]

Alloy	Product[c]	Silicon	Iron	Copper	Manganese	Magnesium	Chromium	Nickel	Zinc	Titanium	Others	
											Each	Total[k]
201.0	S	0.10	0.15	4.0–5.2	0.20–0.50	0.15–0.55	—	—	—	0.15–0.35	0.05[h]	0.10
204.0	S&P	0.20	0.35	4.2–5.0	0.10	0.15–0.35	—	0.05	0.10	0.15–0.30	0.05[i]	0.15
208.0	S&P	2.5–3.5	1.2	3.5–4.5	0.50	0.10	—	0.35	1.0	0.25	—	0.50
222.0	S&P	2.0	1.5	9.2–10.7	0.50	0.15–0.35	—	0.50	0.8	0.25	—	0.35
242.0	S&P	0.7	1.0	3.5–4.5	0.35	1.2–1.8	0.25	1.7–2.3	0.35	0.25	0.05	0.15
295.0	S	0.7–1.5	1.0	4.0–5.0	0.35	0.03	—	—	0.35	0.25	0.05	0.15
296.0	P	2.0–3.0	1.2	4.0–5.0	0.35	0.05	—	0.35	0.50	0.25	—	0.35
308.0	P	5.0–6.0	1.0	4.0–5.0	0.50	0.10	—	—	1.0	0.25	—	0.50
319.0	S&P	5.5–6.5	1.0	3.0–4.0	0.50	0.10	—	0.35	1.0	0.25	—	0.50
328.0	S	7.5–8.5	1.0	1.0–2.0	0.20–0.6	0.20–0.6	0.35	0.25	1.5	0.25	—	0.50
332.0	P	8.5–10.5	1.2	2.0–4.0	0.50	0.50–1.5	—	0.50	1.0	0.25	—	0.50
333.0	P	8.0–10.0	1.0	3.0–4.0	0.50	0.05–0.50	—	0.50	1.0	0.25	—	0.50
336.0	P	11.0–13.0	1.2	0.50–1.5	0.35	0.7–1.3	—	2.0–3.0	0.35	0.25	0.05	—
354.0	S&P	8.6–9.4	0.20	1.6–2.0	0.10	0.40–0.6	—	—	0.10	0.20	0.05	0.15
355.0	S&P	4.5–5.5	0.6[d]	1.0–1.5	0.50[d]	0.40–0.6	0.25	—	0.35	0.25	0.05	0.15
C355.0	S&P	4.5–5.5	0.20	1.0–1.5	0.10	0.40–0.6	—	—	0.10	0.20	0.05	0.15
356.0	S&P	6.5–7.5	0.6[d]	0.25	0.35[d]	0.20–0.45	—	—	0.35	0.25	0.05	0.15
A356.0	S&P	6.5–7.5	0.20	0.20	0.10	0.25–0.45	—	—	0.10	0.20	0.05	0.15
357.0	S&P	6.5–7.5	0.15	0.05	0.03	0.45–0.6	—	—	0.05	0.20	0.05	0.15
A357.0	S&P	6.5–7.5	0.20	0.20	0.10	0.40–0.7	—	—	0.10	0.04–0.20	0.05[e]	0.15
359.0	S&P	8.5–9.5	0.20	0.20	0.10	0.50–0.7	—	—	0.10	0.20	0.05	0.15
443.0	S&P	4.5–6.0	0.8	0.6	0.50	0.05	0.25	—	0.50	0.25	—	0.35
B443.0	S&P	4.5–6.0	0.8	0.15	0.35	0.05	—	—	0.35	0.25	0.05	0.15
A444.0	P	6.5–7.5	0.20	0.10	0.10	0.05	—	—	0.10	0.20	0.05	0.15
512.0	S	1.4–2.2	0.6	0.35	0.8	3.5–4.5	0.25	—	0.35	0.25	0.05	0.15
513.0	P	0.30	0.40	0.10	0.30	3.5–4.5	—	—	1.4–2.2	0.20	0.05	0.15
514.0	S	0.35	0.50	0.15	0.35	3.5–4.5	—	—	0.15	0.25	0.05	0.15
520.0	S	0.25	0.30	0.25	0.15	9.5–10.6	—	—	0.15	0.25	0.05	0.15

535.0	S&P	0.15	0.15	0.05	0.10–0.25	6.2–7.5	—	—	—	0.10–0.25	0.05 (f)	0.15
705.0	S&P	0.20	0.8	0.20	0.40–0.6	1.4–1.8	0.20–0.40	—	2.7–3.3	0.25	0.05	0.15
707.0	S&P	0.20	0.8	0.20	0.40–0.6	1.8–2.4	0.20–0.40	—	4.0–4.5	0.25	0.05	0.15
710.0	S	0.15	0.50	0.35–0.65	0.05	0.6–0.8	—	—	6.0–7.0	0.25	0.05	0.15
711.0	P	0.30	0.7–1.4	0.35–0.65	0.05	0.25–0.45	—	—	6.0–7.0	0.20	0.05	0.15
712.0	S	0.30	0.50	0.25	0.10	0.50–0.65	0.40–0.6	—	5.0–6.5	0.15–0.25	0.05	0.20
713.0	S&P	0.25	1.1	0.40–1.0	0.6	0.20–0.50	0.35	0.15	7.0–8.0	0.25	0.10	0.25
771.0	S	0.15	0.15	0.10	0.10	0.8–1.0	0.06–0.20	—	6.5–7.5	0.10–0.20	0.05	0.15
850.0	S&P	0.7	0.7	0.7–1.3	0.10	0.10	—	0.7–1.3	—	0.20	— (g)	0.30
851.0	S&P	2.0–3.0	0.7	0.7–1.3	0.10	0.10	—	0.30–0.7	—	0.20	— (g)	0.30
852.0	S&P	0.40	0.7	1.7–2.3	0.10	0.6–0.9	—	0.9–1.5	—	0.20	— (g)	0.30

Source: Aluminum Association [12].

(a) The alloys listed are those that have been included in Federal Specifications QQ-A-596d, ALUMINUM ALLOYS PERMANENT AND SEMIPERMANENT MOLD CASTINGS, QQ-A-601E, ALUMINUM ALLOY SAND CASTINGS, and Military Specification MIL-A-21180c, ALUMINUM ALLOY CASTINGS, HIGH STRENGTH. Other alloys are registered with The Aluminum Association and are available. Information on these should be requested from individual foundries or ingot suppliers.

(b) Except for "Aluminum" and "Others," analysis normally is made for elements for which specific limits are shown. For purposes of determining conformance to these limits, an observed value or calculated value obtained from analysis is rounded off to the nearest unit in the last right-hand place of figures used in expressing the specified limit, in accordance with the following:

When the figure next beyond the last figure or place to be retained is less than 5, the figure in the last place retained should be kept unchanged.
When the figure next beyond the last figure or place to be retained is greater than 5, the figure in the last place retained should be increased by 1.
When the figure next beyond the last figure or place to be retained is 5 and

(1) there are no figures or only zeros, beyond this 5, if the figure in the last place to be retained is odd, it should be increased by 1; If even, it should be kept unchanged;

(2) If the 5 next beyond the figure in the last place to be retained is followed by any figures other than zero, the figure in the last place retained should be increased by 1; whether odd or even.

(c) S = Sand Cast P = Permanent Mold Cast
(d) If iron exceeds 0.45%, manganese content shall not be less than one-half the iron content.
(e) Also contains 0.04–0.07% beryllium.
(f) Also contains 0.003–0.007% beryllium, boron 0.005% maximum.
(g) Also contains 5.5–7.0% tin.
(h) Also contains 0.40–1.0% silver.
(i) Also contains 0.05 max. % tin.
(k) The sum of those "Others" metallic elements 0.010% or more each, expressed to the second decimal before determining the sum.

Alloys are divided into two groups based on whether their strengths can be increased by heat-treating or not. Both *heat-treatable* and *non-heat-treatable* alloys can be strengthened by strain-hardening, also called cold working. The alloys that are not heat treatable may only be strengthened by cold working. Whether or not an alloy is heat treatable depends on its alloying elements. Alloys in which the amount of alloying element in solid solution in aluminum increases with temperature are heat treatable. The 1xxx, 3xxx, 4xxx, and 5xxx series wrought alloys are not heat treatable, while the 2xxx, 6xxx, and 7xxx wrought series are, with a few exceptions. Strengthening methods are summarized in Table 9.6.

Non-heat-treatable alloys may also be heat treated, but this treatment is only used to stabilize properties so that strengths don't decrease over time (behavior called *age softening*) and is only required for alloys with an appreciable amount of magnesium (the 5xxx series). Heating to 225–350°F (110–180°C) causes all the softening to occur at once and thus is used as the stabilization heat treatment.

Before tempering, alloys begin in the annealed condition, the weakest but most ductile condition. Tempering, while increasing the strength, decreases ductility and therefore decreases workability. To reduce material to the annealed condition, the typical annealing treatments given in Table 9.7 can be used.

Strain-hardening is achieved by mechanical deformation of the material at ambient temperature. In the case of sheet and plate, this is done by reducing its thickness by rolling. As the material is worked, it becomes resistant to further deformation and its strength increases. The effect of this work on the yield strength of some common non-heat-treatable alloys is shown in Fig. 9.1. Two heat treatments can be applied to annealed condition heat-treatable alloys. First, the material can be *solution heat treated*. This allows soluble alloying elements to enter into solid solution; they are retained in a supersaturated state upon *quenching*, a controlled rapid cooling usually performed using air or water. Next, the material may undergo a *precipitation heat treatment*, also called *artificial aging*, by which constituents are precipitated from solid solution to increase the strength. An example of this process is the production of 6061-T6 sheet. From its initial condition, 6061-O annealed material is heated to 990°F (530°C) as rapidly as possible (solution heat treated), then cooled as rapidly as possible (quenched), which renders the temper T4. Then the material is heated to 320°F (160°C) and

TABLE 9.6 Strengthening Methods

<table>
<tr><td>Pure Aluminum

1xxx</td><td>Alloying

2xxx–Cu
6xxx–Mg, Si
7xxx–Zn</td><td>Heat Treatment

Solution heat treatment;
natural aging or artificial aging</td><td>Strain Hardening
(Cold Working)</td><td>-T
tempers</td></tr>
<tr><td></td><td>Alloying

3xxx–Mn
5xxx–Mg</td><td colspan="2">Strain Hardening
(Cold Working)</td><td>-H
tempers</td></tr>
</table>

TABLE 9.7 Typical Annealing Treatments for Aluminum Alloys

Alloy	Metal Temperature (°F)	Approx. Time at Temperature (h)	Temper Designation
1060	650	①	O
1100	650	①	O
1145	650	①	O
1235	650	①	O
1345	650	①	O
1350	650	①	O
2014	775②	2–3	O
2017	775②	2–3	O
2024	775②	2–3	O
2117	775②	2–3	O
2219	775②	2–3	O
3003	775	①	O
3004	650	①	O
3005	775	①	O
3105	650	①	O
5005	650	①	O
5050	650	①	O
5052	650	①	O
5056	650	①	O
5083	650	①	O
5086	650	①	O
5154	650	①	O
5254	650	①	O
5454	650	①	O
5456	650	①	O
5457	650	①	O
5652	650	①	O
6005	775②	2–3	O
6053	775②	2–3	O
6061	775②	2–3	O
6063	775②	2–3	O
6066	775②	2–3	O
7072	650	①	O
7075	775③	2–3	O
7175	775③	2–3	O
7178	775③	2–3	O
7475	775③	2–3	O
Brazing sheet:			
Nos. 11 & 12	650	①	O
Nos. 23 & 24	650	①	O

①Time in the furnace need not be longer than necessary to bring all parts of load to annealing temperature. Rate of cooling is unimportant.

②These treatments are intended to remove effects of solution heat treatment and include cooling at a rate of about 50°F per hour from the annealing temperature to 500°F. The rate of subsequent cooling is unimportant. Treatment at 650°F, followed by uncontrolled cooling, may be used to remove the effects of cold work, or to partially remove the effects of heat treatment.

③This treatment is intended to remove the effects of solution heat treatment and includes cooling at an uncontrolled rate to 400°F or less, followed by reheating to 450°F for 4 h. Treatment at 650°F, followed by uncontrolled cooling, may be used to remove the effects of cold work or to partially remove the effects of heat treatment.

held for 18 h (precipitation heat treated); upon cooling to room temperature the temper is T6.

Solution heat-treated aluminum may also undergo *natural aging*. Natural aging, like artificial aging, is a precipitation of alloying elements from solid solution, but because it occurs at room temperature, it occurs much more slowly (over a period of days and months rather than hours) than artificial aging. Both aging processes produce an increase in strength and a corresponding decrease in ductility. Material that will be subjected to severe forming operations (like cold heading wire to make rivets or bolts) is often purchased in a T4 temper, formed, and then artificially aged or allowed to naturally age. Care must be taken to form the material before too long a period of time elapses, or natural aging of the material will cause it to harden and decrease its workability. Sometimes T4 material is refrigerated to prevent natural aging if cold forming required for fabrication into a product such as a fastener won't be performed soon after solution heat treatment.

The temper designation system is the same for both wrought and cast alloys, although cast alloys are only heat treated and not strain hardened, with the exception of some 85 $\times$ .0 casting alloys. The temper designation follows the alloy

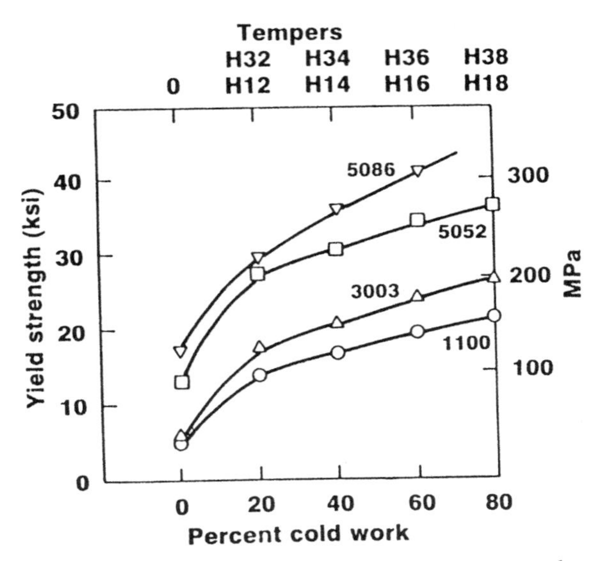

FIGURE 9.1 Effect of cold work on yield strength.

designation, the two being separated by a hyphen (e.g., 5052-H32). Basic temper designations are letters. Subdivisions of the basic tempers are given by one or more numbers following the letter.

The basic temper designations are:

F **As fabricated.** Applies to the products of shaping processes in which no special control over thermal conditions or strain-hardening is employed. For wrought products, there are no mechanical property limits.

O **Annealed.** Applies to wrought products that are annealed to obtain the lowest strength temper, and to cast products that are annealed to improve ductility and dimensional stability. The O may be followed by a number other than zero.

H **Strain hardened.** (wrought products only). Applies to products that have their strength increased by strain-hardening, with or without supplementary thermal treatments to produce some reduction in strength. The H is always followed by two or more numbers.

W **Solution heat treated.** An unstable temper applicable only to alloys that spontaneously age at room temperature after solution heat treatment. This designation is specific only when the period of natural aging is indicated; for example, W$\frac{1}{2}$ h.

T **Thermally treated to produce stable tempers other than F, O, or H.** Applies to products that are thermally treated, with or without supplementary strain-hardening, to produce stable tempers. The T is always followed by one or more numbers.

For strain-hardened tempers, the first digit of the number following the H denotes:

H1 **Strain hardened only.** Applies to products that are strain hardened to obtain the desired strength without supplementary thermal treatment. The number following this designation indicates the degree of strain-hardening. Example: 1100-H14.

H2 **Strain hardened and partially annealed.** Applies to products that are strain hardened more than the desired final amount and then reduced in strength to the desired level by partial annealing. For alloys that age soften at room temperature, the H2 tempers have the same minimum ultimate tensile strength as the corresponding H3 tempers. For other alloys, the H2 tempers have the same minimum ultimate tensile strength as the corresponding H1 tempers and slightly higher elongation. The number following this designation indicates the strain hardening remaining after the product has been partially annealed. Example: 3005-H25.

H3 **Strain hardened and stabilized.** Applies to products that are strain hardened and whose mechanical properties are stabilized either by a low-temperature thermal treatment or as a result of heat introduced during fabrication. Stabilization usually improves ductility. This designation is applicable only to those alloys that, unless stabilized, gradually age soften at room temperature. The number following this designation indicates the degree of strain-hardening remaining after the stabilization has occurred. Example: 5005-H34.

H4 **Strain hardened and lacquered or painted.** Applies to products that are strain hardened and subjected to some thermal operation during subsequent painting or lacquering. The number following this designation indicates the degree of strain-hardening remaining after the product has been thermally treated as part of the painting or lacquering curing. The corresponding H2X or H3X mechanical property limits apply.

The digit following the designation H1, H2, H3, or H4 indicates the degree of strain-hardening. Number 8 is for the tempers with the highest ultimate tensile strength normally produced. Number 4 is for tempers whose ultimate strength is approximately midway between that of the O temper and the HX8 temper. Number 2 is for tempers whose ultimate strength is approximately midway between that of the O temper and the HX4 temper. Number 6 is for tempers whose ultimate strength is approximately midway between that of the HX4 temper and the HX8 temper. Numbers 1, 3, 5, and 7 similarly designate intermediate tempers between those defined above. Number 9 designates tempers whose minimum ultimate tensile strength exceeds that of the HX8 tempers by 2 ksi (15 MPa) or more.

The third digit, when used, indicates a variation in the degree of temper or the mechanical properties of a two-digit temper. An example is pattern or embossed sheet made from the H12, H22, or H32 tempers; these are assigned

H124, H224, or H324 tempers, respectively, since the additional strain-hardening from embossing causes a slight change in the mechanical properties.

For heat-treated tempers, the number 1 through 10 following the T denotes:

T1 **Cooled from an elevated temperature shaping process and naturally aged to a substantially stable condition.** Applies to products that are not cold worked after cooling from an elevated temperature shaping process, or in which the effect of cold work in flattening or straightening may not be recognized in mechanical property limits. Example: 6005-T1 extrusions.

T2 **Cooled from an elevated temperature shaping process, cold worked, and naturally aged to a substantially stable condition.** Applies to products that are cold worked to improve strength after cooling from an elevated temperature shaping process or in which the effect of cold work in flattening or straightening is recognized in mechanical property limits.

T3 **Solution heat treated, cold worked, and naturally aged to a substantially stable condition.** Applies to products that are cold worked to improve strength after solution heat treatment or in which the effect of cold work in flattening or straightening is recognized in mechanical property limits. Example: 2024-T3 sheet.

T4 **Solution heat treated and naturally aged to a substantially stable condition.** Applies to products that are not cold worked after solution heat treatment or in which the effect of cold work in flattening or straightening may not be recognized in mechanical property limits. Example: 2014-T4 sheet.

T5 **Cooled from an elevated temperature shaping process and then artificially aged.** Applies to products that are not cold worked after cooling from an elevated temperature shaping process or in which the effect of cold work in flattening or straightening may not be recognized in mechanical property limits. Example: 6063-T5 extrusions.

T6 **Solution heat treated and then artificially aged.** Applies to products that are not cold worked after solution heat treatment or in which the effect of cold work in flattening or straightening may not be recognized in mechanical property limits. Example: 6063-T6 extrusions.

T7 **Solution heat treated and then overaged/stabilized.** Applies to wrought products that are artificially aged after solution heat treatment to carry them beyond a point of maximum strength to provide control of some significant characteristic. Applies to cast products that are artificially aged after solution heat treatment to provide dimensional and strength stability. Example: 7050-T7 rivet and cold heading wire and rod.

T8 **Solution heat treated, cold worked, and then artificially aged.** Applies to products that are cold worked to improve strength or in which the effect of cold work in flattening or straightening is recognized in mechanical property limits. Example: 2024-T81 sheet.

T9 **Solution heat treated, artificially aged, and then cold worked.** Applies to products that are cold worked to improve strength after artificial aging. Example: 6262-T9 nuts.

T10 **Cooled from an elevated temperature shaping process, cold worked, and then artificially aged.** Applies to products that are cold worked to improve strength or in which the effect of cold work in flattening or straightening is recognized in mechanical property limits.

Additional digits may be added to designations T1 through T10 for variations in treatment. Stress-relieved tempers follow the format T_5, which may be followed by additional numbers.

Typical heat treatments for wrought alloys are given in Table 9.8 Heat treatments for cast alloys are given in Table 9.9.

TABLE 9.8 Typical Heat Treatments for Aluminum Alloy Mill Products①

		Solution Heat Treatment②		Precipitation Heat Treatment		
Alloy	Product	Metal Temperature③ (°F)	Temper Designation	Metal Temperature③ (°F)	Approx. Time at Temperature④ (h)	Temper Designation
2011	Rolled or Cold Finished Wire, Rod & Bar	975	T3⑤	320	14	T8⑤
			T4	—	—	—
			T451⑥	—	—	—
	Drawn Tube	960	T3⑤	310	14	T8⑤
			T4511⑥	—	—	—
2014⑦	Flat Sheet	935	T3⑤	—	—	—
			T42	320	18	T62
	Coiled Sheet	935	T4	320	18	T6
			T42	320	18	T62
	Plate	935	T451⑥	320	18	T651⑥
			T42	320	18	T62
	Rolled or Cold Finished Wire, Rod & Bar	935	T4	320⑧	18	T6
			T451⑥	320⑧	18	T651⑥
			T42	320⑧	18	T62
	Extruded Wire, Rod, Bar, Profiles (Shapes) & Tube	935	T4	320⑧	18	T6
			T4510⑥	320⑧	18	T6510⑥
			T4511⑥	320⑧	18	T6511⑥
			T42	320⑧	18	T62
	Drawn Tube	935	T4	320⑧	18	T6
			T42	320⑧	18	T62
	Die Forgings	935⑨	T4	340	10	T6

TABLE 9.8 (*Continued*)

		Solution Heat Treatment②		Precipitation Heat Treatment		
Alloy	Product	Metal Temperature③ (°F)	Temper Designation	Metal Temperature③ (°F)	Approx. Time at Temperature④ (h)	Temper Designation
2014⑦	Hand Forgings and Rolled Rings	935⑨	T4㊴	340	10	T6
			T452⑩㊴	340	10	T652⑩
2017	Rolled or Cold Finished Wire, Rod & Bar	935	T4	—	—	—
			T451⑥	—	—	—
			T42	—	—	—
2018	Die Forgings	950⑪	T4㊴	340	10	T61
2024⑦	Flat Sheet	920	T3⑤	375	12	T81⑤
			T361⑤	375	8	T861⑤
			T42	375	9	T62
			T42	375	16	T72
	Coiled Sheet	920	T4	—	—	—
			T42	375	9	T62
			T42	375	16	T72
	Plate	920	T351⑥	375	12	T851⑥
			T361⑤	375	8	T861⑤
			T42	375	9	T62
	Rolled or Cold Finished Wire, Rod & Bar	920	T351⑥	375	12	T851⑥
			T36⑤	—	—	—
			T4	375	12	T6
			T42	375	16	T62
	Extruded Wire, Rod, Bar Profiles (Shapes) & Tube	920	T3⑤	375	12	T81⑤
			T3510⑥	375	12	T8510⑥
			T3511⑥	375	12	T8511⑥
			T42	—	—	—
	Drawn Tube	920	T3⑤	—	—	—
			T42	—	—	—
2025	Die Forgings	960	T4㊴	340	10	T6
2036	Sheet	930	T4	—	—	—
2117	Rolled or Cold Finished Wire and Rod	935	T4	—	—	—
			T42	—	—	—
2124	Plate	920	T351⑥	375	12	T851⑥
2218	Die Forgings	950⑪	T4㊴	340	10	T61
		950⑫	T41㊴	460	6	T72
2219⑦	Flat Sheet	995	T31⑤	350	18	T81⑤
			T37⑤	325	24	T87⑤
			T42㊴	375	36	T62

TABLE 9.8 (*Continued*)

Alloy	Product	Solution Heat Treatment②		Precipitation Heat Treatment		
		Metal Temperature③ (°F)	Temper Designation	Metal Temperature③ (°F)	Approx. Time at Temperature④ (h)	Temper Designation
2219⑦	Plate	995	T37⑤	350	18	T87⑤
			T351⑥	350	18	T851⑥
			T42㊴	375	36	T62
	Rolled or Cold Finished Wire, Road & Bar	995	T4㊴	375	36	T6
			T351㊴	375	18	T851⑥
	Extruded Rod, Bar, Profiles (Shapes) & Tube	995	T31⑤	375	18	T81⑤
			T3510⑥	375	18	T8510⑥
			T3511⑥	375	18	T8511⑥
			T42㊴	375	36	T62
	Die Forgings and Rolled Rings	995	T4㊴	375	26	T6
	Hand Forgings	995	T4㊴	375	26	T6
			T352⑩ ㊴	350	18	T852⑩
2618	Forgings and Rolled Rings	985⑪	T4㊴	390	20	T61
4032	Die Forgings	950⑨	T4㊴	340	10	T6
6005	Extruded Rod, Bar, Profiles (Shapes) & Tube	㉚	T1	350	8	T5
6053	Rolled or Cold Finished Wire and Rod	945	T4㊴	355	8	T61
	Die Forgings	970	T4㊴	340	10	T6
6061⑦	Sheet	990	T4	320	18	T6
			T42	320	18	T62
	Plate	990	T4㉑	320	18	T6㉑
			T451⑥	320	18	T651⑥
			T42	320	18	T62
	Rolled or Cold finished Wire, Rod & Bar	990	T4	320⑬	18	T6
			T3㊴	320⑬	18	T89⑤
			T4	320⑬	18	T913⑭
			T4	320⑬	18	T94⑭
			T451⑥	320⑬	18	T651⑥
			T42	320⑬	18	T62
	Extruded Rod, Bar, Profiles (Shapes) and Tube	㉚	T1	350	8	T51
		990⑮	T4	350	8	T6
			T4510⑥	350	8	T6510⑥
			T4511⑥	350	8	T6511⑥
		990	T42	350	8	T62

TABLE 9.8 (*Continued*)

Alloy	Product	Solution Heat Treatment[2]		Precipitation Heat Treatment		
		Metal Temperature[3] (°F)	Temper Designation	Metal Temperature[3] (°F)	Approx. Time at Temperature[4] (h)	Temper Designation
6061[7]	Structural Profiles (Shapes)	990[15]	T4[39]	350	8	T6
	Pipe	990[15]	T4[39]	350	8	T6
	Drawn Pipe	990	T4	320[13]	18	T6
			T42	320[13]	18	T62
	Die and Hand Forgings	990	T4[39]	350	8	T6
	Rolled Rings	990	T4[39] T452[10] [39]	350 350	8 8	T6 T652[10]
6063	Extruded Rod, Bar, Profiles (Shapes) & Tube	[30]	T1 T1	360[16] 360	3 3	T5 T52
		950[13]	T4	350[17]	8	T6
		970	T42	350[17]	8	T62
	Drawn Tube	970	T4 T3[5] [15] [39] T3[5] [15] [39] T3[5] [15] [39]	350 350 350 350	8 8 8 8	T6 T83[5] T831[5] T832[5]
			T42	350	8	T62
	Pipe	970[15]	T4[39]	350[17]	8	T6
6066	Extruded Rod, Bar, Profiles (Shapes) & Tube	990	T4 T4510[6] T4511[6]	350 350 350	8 8 8	T6 T6510[6] T6511[6]
			T42	350	8	T62
	Drawn Tube	990	T4	350	8	T6
			T42	350	8	T62
	Die Forgings	990	T4[39]	350	8	T6
6070	Extruded Rod, Bar Profiles (Shapes) & Tube	1015[15]	T4[39]	320	18	T6
		1015	T42[39]	320	18	T62
6101	Extruded Rod, Bar, Tube, Pipe and Structural Profiles (Shapes)	970[15]	T4[39] T4[39] T4[39] T4[39] T4[39]	390 440 410 535 430	10 5 9 7 3	T6 T61 T63 T64 T65
6105	Extruded Rod, Bar Profiles (Shapes) and Tube	[30]	T1	350	8	T5

TABLE 9.8 (*Continued*)

Alloy	Product	Solution Heat Treatment[2]		Precipitation Heat Treatment		
		Metal Temperature[3] (°F)	Temper Designation	Metal Temperature[3] (°F)	Approx. Time at Temperature[4] (h)	Temper Designation
6151	Die Forgings	960	T4[39]	340	10	T6
	Rolled Rings	960	T4[39]	340	10	T6
			T452[10] [39]	340	10	T652[10]
6162	Extruded Rod, Bar, Profiles (Shapes) & Tube	[30]	T1[39]	350	8	T5
			T1510[6] [39]	350	8	T5510[6]
			T1511[6] [39]	350	8	T5511[6]
		980[15]	T4[39]	350	8	T6
			T4510[6] [39]	350	8	T6510[6]
			T4511[6] [39]	350	8	T6511[6]
6201	Wire	950	T3[5] [39]	320	4	T81[5]
6262	Rolled or Cold Finished Wire, Rod and Bar	1000	T4[39]	340	8	T6
			T4[39]	340	12	T9[14]
			T451[6] [39]	340	8	T651[6]
			T42[39]	340	8	T62
	Extruded Rod, Bar, Profiles (Shapes) and Tube	1000[15]	T4[39]	350	12	T6
			T4510[6] [39]	350	12	T6510[6]
			T4511[6] [39]	350	12	T6511[6]
		1000	T42[39]	350	12	T62
	Drawn Tube	1000	T4[39]	340	8	T6
			T4[39]	340	8	T9[14]
			T42[39]	340	8	T62
6351	Extruded Rod, Bar and Profiles (Shapes)	[30]	T1	250	10	T54
		[30]	T1	350	8	T5
		985	T4	350	8	T6
6463	Extruded Rod, Bar and Profiles (Shapes)	[30]	T1	400	1	T5
		970[15]	T4[39]	350[17]	8	T6
		970	T42[39]	350[17]	8	T62
6951[29]	Sheet	985	T42	320	18	T62
7005	Extruded Rod, Bar and Profiles (Shapes)	[30]	T1[39]	[22]	[22]	T53
7049	Die Forgings	875[9]	W	[35]	[35]	T73
	Hand Forgings	875[9]	W	[35]	[35]	T73
			W52[10]	[35]	[35]	T7352[10]
7050	Plate	890	W51[6]	[31]	[31]	T7451[6]
			W51[6]	[32]	[32]	T7651[6]
	Rolled or Cold Finished Wire and Rod	890	W	[38]	[38]	T7

TABLE 9.8 (*Continued*)

Alloy	Product	Solution Heat Treatment② Metal Temperature③ (°F)	Solution Heat Treatment② Temper Designation	Precipitation Heat Treatment Metal Temperature③ (°F)	Precipitation Heat Treatment Approx. Time at Temperature④ (h)	Precipitation Heat Treatment Temper Designation
7050	Extruded Rod, Bar and Profiles (Shapes)	890	W510⑥	㊱	㊱	T73510⑥
			W510⑥	㊲	㊲	T74510⑥
			W510⑥	㉝	㉝	T76510⑥
			W511⑥	㊱	㊱	T73511⑥
			W511⑥	㊲	㊲	T74511⑥
			W511⑥	㉝	㉝	T76511⑥
	Die Forgings	890	W	㉞	㉞	T74
	Hand Forgings	890	W52㉞	㉞	㉞	T7452⑩
7075⑦	Sheet	900㉓	W	250⑱	24	T6
			W	⑳ ㉔	⑳ ㉔	T73㉗
			W	㉘	㉘	T76㉗
			W	250⑱	24	T62
	Plate	900㉓ ㊵	W51⑥	250⑱	24	T651⑥
			W51⑥	⑳ ㉔	⑳ ㉔	T7351⑥ ㉗
			W51⑥	㉘	㉘	T7651⑥ ㉗
			W	250⑱	24	T62
	Rolled or Cold finished Wire, Rod and Bar	915㉓ ㊵	W	250	24	T6
			W	⑳ ㉔	⑳ ㉔	T73㉗
			W	250	24	T62
			W51⑥	250	24	T651⑥
			W51⑥	⑳ ㉔	⑳ ㉔	T7351⑥ ㉗
	Extruded Rod, Bar and Profiles (Shapes)	870	W	250⑲	24	T6
			W	⑳ ㉔	⑳ ㉔	T73㉗
			W	㉘	㉘	T76㉗
			W	250⑲	24	T62
			W510⑥	250⑲	24	T6510⑥
			W510⑥	⑳ ㉔	⑳ ㉔	T73510⑥ ㉗
			W510⑥	㉘	㉘	T76510⑥ ㉗
			W511⑥	250⑲	24	T6511⑥
			W511⑥	⑳ ㉔	⑳ ㉔	T73511⑥ ㉗
			W511⑥	㉘	㉘	T76511⑥ ㉗
	Extruded Tube	870	W	250⑲	24	T6
			W	⑳ ㉔	⑳ ㉔	T73㉗
			W	250⑲	24	T62
			W510⑥	250⑲	24	T6510⑥
			W510⑥	⑳ ㉔	⑳ ㉔	T73510⑥ ㉗
			W511⑥	250⑲	24	T6511⑥
			W511⑥	⑳ ㉔	⑳ ㉔	T73511⑥ ㉗
	Drawn Tube	870	W	250	24	T6
			W	⑳ ㉔	⑳ ㉔	T73㉗
			W	250	24	T62

TABLE 9.8 (*Continued*)

Alloy	Product	Solution Heat Treatment②		Precipitation Heat Treatment		
		Metal Temperature③ (°F)	Temper Designation	• Metal Temperature③ (°F)	Approx. Time at Temperature④ (h)	Temper Designation
7075⑦	Die Forgings	880⑨	W	250	24	T6
			W	⑳	⑳	T73㉗
			W52⑩	⑳	⑳	T7352⑩ ㉗
	Hand Forgings	880⑨	W	250	24	T6
			W	⑳	⑳	T73㉗
			W52⑩	250	24	T652⑩
			W52⑩	⑳	⑳	T7352⑩ ㉗
	Rolled Rings	880	W	250	24	T6
7178⑦	Sheet	875	W	250	24	T6
			W	⑳	㉕	T76㉗
			W	250	24	T62
	Plate	875	W51⑥	250	24	T651⑥
			W51⑥	㉕	㉕	T7651⑥ ㉗
			W	250	24	T62
	Rolled or Cold Finished Wire and Rod	870	W	250	24	T6
	Extruded Rod, Bar and Profiles (Shapes)	870	W	250	24	T6
			W	㉖	㉖	T76㉗
			W	250	24	T62
			W510⑥	250	24	T6510⑥
			W510⑥	㉖	㉖	T76510⑥ ㉗
			W511⑥	250	24	T6511⑥
			W511⑥	㉖	㉖	T76511⑥ ㉗
7475	Sheet	900㊸	W	㊹	㊹	T61
			W	㊺	㊺	T761
	Plate	900㊸	W51⑥	240	24	T651
			W51⑥	㊶	㊶	T7351
			W51⑥	㊷	㊷	T7651
	Rod	900	W	㊻	㊻	T62

Source: Aluminum Association [5].

①The times and temperatures shown are typical for various forms, sizes and methods of manufacture and may not exactly describe the optimum treatment for a specific item.

②Material should be quenched from the solution heat-treating temperature as rapidly as possible and with minimum delay after removal from the furnace. Unless otherwise indicated, when material is quenched by total immersion in water, the water should be at room temperature and suitably cooled to remain below 100°F during the quenching cycle. The use of high-velocity high-volume jets of cold water is also effective for some materials. For additional details on aluminum alloy heat treatment and for recommendations on such specifics as furnace solution heat treat soak time see military Specification MIL-H-6088 or ASTM B597.

③The nominal metal temperatures should be attained as rapidly as possible and maintained ±10°F of nominal during the time at temperature.

④The time at temperature will depend on time required for load to reach temperature. The times shown are based on rapid heating, with soaking time measured from the time the load reached within 10°F of the applicable temperature.

⑤Cold work subsequent to solution heat treatment and, where applicable, prior to any precipitation heat treatment is required to attain the specified mechanical properties for these tempers.

TABLE 9.8 (*Continued*)

⑥Stress-relieved by stretching. Required to produce a specified amount of permanent set subsequent to solution heat treatment and, where applicable, prior to any precipitation heat treatment.
⑦These heat treatments also apply to alclad sheet and plate in these alloys.
⑧ An alternative treatment comprised of 8 hours at 350°F also may be used.
⑨ Quench after solution treatment in water at 140°F to 180°F.
⑩ Stress-relieved by 1–5 percent cold reduction subsequent to solution heat treatment and prior to precipitation heat treatment.
⑪ Quench after solution heat treatment in water at 212°F.
⑫ Quench after solution heat treatment in air blast at room temperature.
⑬ An alternative treatment comprised of 8 hours at 340°F also may be used.
⑭ Cold working subsequent to precipitation heat treatment is necessary to secure the specified properties for this temper.
⑮ By suitable control of extrusion temperature, product may be quenched directly from extrusion press to provide specified properties for this temper. Some products may be adequately quenched in air blast at room temperature.
⑯ An alternate treatment comprised of 1–2 hours at 400°F also may be used.
⑰An alternate treatment comprised of 6 hours at 360°F also may be used.
⑱ An alternate two-stage treatment comprised of 4 hours at 205°F followed by 8 hours at 315°F also may be used.
⑲ An alternate three-stage treatment comprised of 5 hours at 210°F followed by 4 hours at 250°F followed by 4 hours at 300°F also may be used.
⑳ Two -stage treatment comprised of 6 to 8 hours at 225°F followed by a second-stage of:

(a) 24–30 hours at 325°F for sheet and plate
(b) 8–10 hours at 350°F for rolled or cold-finished rod and bar.
(c) 6–8 hours at 350°F for extrusions and tube.
(d) 8–10 hours at 350°F for forgings in T73 temper and 6–8 hours at 350°F for forgings in T7352 temper.

㉑Applies to tread plate only.
㉒Held at room temperature for 72 hours followed by two stage precipitation heat-treatment of 8 hours at 225°F plus 16 hours at 300°F.
㉓With optimum ingot homogenization, heat-treating temperatures as high as 928°F are sometimes acceptable.
㉔An alternate two-stage treatment for sheet, plate, tube and extrusions comprised of 6 to 8 hours at 225°F followed by a second stage of 14–18 hours at 335°F may be used providing a heating-up rate of 25°F per hour is used. For rolled or cold-finished rod and bar the alternate treatment is 10 hours at 350°F.
㉕A two-stage treatment comprised of 3–5 hours at 250°F followed by 15–18 hours at 325°F.
㉖A two-stage treatment comprised of 3–5 hours at 250°F followed by 18–21 hours at 320°F.
㉗The aging of aluminum alloys 7075 and 7178 from any temper to the T73 (applicable to alloy 7075 only) or T76 temper series requires closer than normal controls on aging practice variables such as time, temperature, heating-up rates, etc., for any given item. In addition to the above, when reaging material in the T6 temper series to the T73 to T76 temper series, the specific condition of the T6 temper material (such as its property level and other effect of processing variables) is extremely important and will affect the capability of the re-aged material to conform to the requirements specified for the applicable T73 and T76 temper series.
㉘ The aging practice will vary with the product, size, nature of equipment, loading procedures and furnace control capabilities. The optimum practice for a specific item can be ascertained only by actual trial treatment of the item under specific conditions. Typical procedures involve a two-stage treatment comprised of 3–30 hours at 250°F followed by 15–18 hours at 325°F for extrusions. An alternate two-stage treatment of 8 hours at 210°F followed by 24–28 hours at 325°F may be used.
㉙ Core alloy in No. 21, 22, 23 and 24 brazing sheet.
㉚ Quenched directly from the extrusion press. Some extrusions may be adequately quenched using a room temperature air blast.
㉛ A two-stage treatment comprised of 3–6 hours at 250°F followed by 24–30 hours at 330°F.
㉜ A two-stage treatment comprised of 3–6 hours at 255°F followed by 12–15 hours at 330°F.
㉝ A two-stage treatment comprised of 4 hours at 250°F followed by 18–22 hours at 320°F.
㉞ A multi-stage treatment comprised of 8 hours at 225°F followed by 8 hours at 250°F followed by 4–10 hours at 350°F.
㉟ Held at room temperature for a minimum of 48 hours followed by a two-stage treatment comprised of 24 hours at 250°F followed by 10–16 hours at 330°F.
㊱ A two-stage treatment comprised of 24 hours at 250°F followed by 10–14 hours at 345°F.
㊲ A two-stage treatment comprised of 24 hours at 250°F followed by 8–10 hours at 345°F.
㊳ A two-stage treatment comprised of 4 hours at 250°F followed by 6–8 hours at 355°F.
㊴ By definition, this temper designation is that which would apply after natural aging even though mechanical properties for this alloy-temper product have not been registered.
㊵ For plate thickness over 4 inches and for rod diameters or bar thicknesses over four inches, a maximum temperature of 910°F is recommended to avoid eutectic melting.
㊶ A two-stage treatment comprised of 4–8 hours at 210°F followed by 24–30 hours at 320°F.
㊷ A two-stage treatment comprised of 4–8 hours at 250°F followed by 26–32 hours at 310°F.
㊸ Without adequate thermal pretreatment, melting may occur at this temperature.
㊹ A two-stage treatment comprised of 250°F for 3 hours plus 320°F for 3 hours.
㊺ A two-stage treatment comprised of 250°F for 3 hours plus 325°F for 10 hours.
㊻ A two-stage treatment comprised of 250°F for 3 hours plus 325°F for 3 hours.

TABLE 9.9 Recommended Times and Temperatures for Heat-Treating Commonly Used Aluminum Sand and Permanent Mold Castings

The heat treat times and temperatures given in this standard are those in general use in the industry. The times and temperatures shown for solution heat treatment are critical. Quenching must be accomplished by complete immersion of the castings with a minimum delay after the castings are removed from the furnace.

Under certain conditions complex castings which might crack or distort in the water quench can be oil or air blast quenched. When this is done the purchaser and the foundry must agree to the procedure and also agree on the level of mechanical properties which will be acceptable. Aging treatments can be varied slightly to attain the optimum treatment for a specific casting or to give agreed upon slightly different levels of mechanical properties.

Temper designations for castings are as follows:

F As cast—cooled naturally from the mold in room temperature air with no further heat treatment.

0 Annealed. Usually the weakest, softest, most ductile and most dimensionally stable condition.

T4 Solution heat treated and naturally aged to substantially stable condition. Mechanical properties and stability may change over a long period of time.

T5 Naturally cooled from the mold and then artificially aged to attain improved mechanical properties and dimensional stability.

T6 Solution heat treated and artificially aged to attain optimum mechanical properties and generally good dimensional stability.

T7 Solution heat treated and overaged for improved dimensional stability, but usually with some reduction from the optimum mechanical properties.

The T5, T6, and T7 designations are sometimes followed by one or more numbers which indicate changes from the originally developed treatment.

			Solution Heat Treatment(b)		Aging Treatment	
Alloy	Temper	Product(a)	Metal Temperatures ±10°F(c)	Time (h)	Metal Temperatures ±10°F(c)	Time (h)
201.0	T6	S	950–960 then 980–990	2 14–20	Room Temperature then 310	14–24 20
201.0	T7	S	950–960 then 980–990	2 14–20	Room Temperature then 370	12–14 5
204.0	T4	S or P	970	10	Room Temperature	5 days
208.0	T4	P	940	4–12	—	—
208.0	T6	P	940	4–12	310	2–5
208.0	T7	P	940	4–12	500	4–6
222.0	O(i)	S	—	—	600	3
222.0	T61	S	950	12	310	11
222.0	T551	P	—	—	340	16–22
222.0	T65	P	950	4–12	340	7–9
242.0	O(k)	S	—	—	650	3
242.0	T571	S	—	—	400	8
242.0	T77	S	960	5(d)	625–675	2 minimum
242.0	T571	S or P	—	—	400	7–9
242.0	T61	S or P	960	4–12(d)	400–450	3–5
295.0	T4	S	960	12	—	—
295.0	T6	S	960	12	310	3–6
295.0	T62	S	960	12	310	12–24
295.0	T7	S	960	12	500	4–6
296.0	T6	P	950	8	310	1–8
319.0	T5	S	—	—	400	8
319.0	T6	S	940	12	310	2–5

TABLE 9.9 ***(Continued)***

Alloy	Temper	Productⓐ	Solution Heat Treatmentⓑ		Aging Treatment	
			Metal Temperatures ±10°Fⓒ	Time (h)	Metal Temperatures ±10°Fⓒ	Time (h)
319.0	T6	P	940	4–12	310	2–5
328.0	T6	S	960	12	310	2–5
332.0	T5	P	—	—	400	7–9
333.0	T5	P	—	—	400	7–9
333.0	T6	P	940	6–12	310	2–5
333.0	T7	P	940	6–12	500	4–6
336.0	T551	P	—	—	400	7–9
336.0	T65	P	960	8	400	7–9
354.0	—	ⓔ	980–995	10–12	ⓕ	ⓕ
355.0	T51	S or P	—	—	440	7–9
355.0	T6	S	980	12	310	3–5
355.0	T7	S	980	12	440	3–5
355.0	T71	S	980	12	475	4–6
355.0	T6	P	980	4–12	310	2–5
355.0	T62	P	980	4–12	340	14–18
355.0	T7	P	980	4–12	440	3–9
355.0	T71	P	980	4–12	475	3–6
C355.0	T6	S	980	12	310	3–5
C355.0	T61	P	980	6–12	Room Temperature then 310	8 minimum 10–12
356.0	T51	S or P	—	—	440	7–9
356.0	T6	S	1000	12	310	3–5
356.0	T7	S	1000	12	400	3–5
356.0	T71	S	1000	10–12	475	3
356.0	T6	P	1000	4–12	310	2–5
356.0	T7	P	1000	4–12	440	7–9
356.0	T71	P	1000	4–12	475	3–6
A356.0	T6	S	1000	12	310	3–5
A356.0	T61	P	1000	6–12	Room Temperature then 310	8 minimum 6–12
357.0	T6	P	1000	8	330	6–12
A357.0	—	ⓔ	1000	8–12	ⓕ	ⓕ
359.0	—	ⓔ	1000	10–14	ⓕ	ⓕ
A444.0	T4	P	1000	8–12	—	—
520.0	T4	S	810	18ⓖ	—	—
535.0	T5ⓚ	S	—	—	750	5
705.0	T5	S	—	—	Room Temperature or 210	21 days 8
705.0	T5	P	—	—	Room Temperature or 210	21 days 10
707.0	T7	S	990	8–16	350	4–10
707.0	T7	P	990	4–8	350	4–10
710.0	T5	S	—	—	Room Temperature	21 days
711.0	T1	P	—	—	Room Temperature	21 days
712.0	T5	S	—	—	Room Temperature or 315	21 days 6–8
713.0	T5	S or P	—	—	Room Temperature or 250	21 days 16
771.0	T5	S	—	—	355	3–5ⓗ

TABLE 9.9 (*Continued*)

Alloy	Temper	Product(a)	Solution Heat Treatment(b) Metal Temperatures ±10°F(c)	Solution Heat Treatment(b) Time (h)	Aging Treatment Metal Temperatures ±10°F(c)	Aging Treatment Time (h)
771.0	T51	S	—	—	405	6
771.0	T52(i)	S	—	—	330	6–12(h)
771.0	T53(i)	S	—	—	360	4(h)
771.0	T6	S	1090	6(h)	265	3
771.0	T71	S	1090	6(i)	285	15
850.0	T5	S or P	—	—	430	7–9
851.0	T5	S or P	—	—	430	7–9
851.0	T6	P	900	6	430	4
852.0	T5	S or P	—	—	430	7–9

Source: Aluminum Association [12].
(a)S = Sand cast, P = Permanent mold cast.
(b)Unless otherwise noted, quench in water at 150–212F.
(c)Temperature range unless otherwise noted.
(d)Use air blast quench.
(e)Casting process varies, sand, permanent mold, or composite to obtain desired mechanical properties.
(f)Solution heat treat as indicated then artificially age by heating uniformly for the time and temperature necessary to obtain the desired mechanical properties.
(g)Quench in water at 150–212 for 10–20 seconds only.
(h)Cool in still air outside furnace to room temperature.
(i)Stress relieve for dimensional stability in following manner: (1) Hold at 775 ± 25°F for 5 hrs. Then (2) furnace cool to 650°F for 2 or more hrs. Then (3) furnace cool to 450°F for not more than 1/2 hr. Then (4) furnace cool to 250°F for approximately 2 hr. Then (5) cool to room temperature in still air outside the furnace.
(k)No quench required. Cool in still air outside furnace

9.2 PROPERTIES

9.2.1 Physical Properties

Physical properties include all properties other than mechanical properties. The physical properties of most interest to material designers include density, melting point, electrical conductivity, thermal conductivity, and coefficient of thermal expansion. While these properties vary among alloys and tempers, average values can be useful to the designer.

Density doesn't vary much by alloy (since alloying elements make up such a small portion of the composition) ranging from 0.092 to 0.103 lb/in.^{3} and averaging around 0.1 lb/in.^{3} (2700 kg/m^3). This compares to 0.065 for magnesium, 0.16 for titanium, and 0.283 lb/in.^{3} for steel. Density is calculated as the weighted average of the densities of the elements comprising the alloy; the 5xxx and 6xxx series alloys are the lightest of the common alloys since magnesium is the lightest of the main alloying elements. Densities for common wrought aluminum alloys are listed in Table 9.10. Densities for cast alloys are given in Table 9.11. The density of a casting is less than that of the cast alloy because some porosity cannot be avoided in producing castings. The density of castings is usually about 95–100% of the theoretical density of the cast alloy.

The *melting point* also varies by alloy. While pure aluminum melts at about 1220°F (660°C), the addition of alloying elements depresses the melting point to between about 900 and 1200°F (500 and 650°C) and produces a melting range since the different alloying elements melt at different temperatures. Most aluminum alloys' mechanical properties are significantly degraded well below their melting point. Few alloys are used above 400°F (200°C), although some, like 2219, have applications in engines up to about 600°F (300°C).

Thermal and electrical conductivity also vary widely by alloy. The purer grades of aluminum have the highest conductivities, up to a *thermal conductivity* of about

TABLE 9.10 Nominal Densities of Aluminum and Aluminum Alloys

Alloy	Density (lb/in.^{3})	Specific Gravity	Alloy	Density (lb/in.^{3})	Specific Gravity
1050	.0975	2.705	5252	.096	2.67
1060	.0975	2.705	5254	.096	2.66
1100	.098	2.71	5356	.096	2.64
1145	.0975	2.700	5454	.097	2.69
1175	.0975	2.700	5456	.096	2.66
1200	.098	2.70	5457	.097	2.69
1230	.098	2.70	5554	.097	2.69
1235	.0975	2.705	5556	.096	2.66
1345	.0975	2.705	5652	.097	2.67
1350	.0975	2.705	5654	.096	2.66
2011	.102	2.83	5657	.097	2.69
2014	.101	2.80	6003	.097	2.70
2017	.101	2.79	6005	.097	2.70
2018	.102	2.82	6053	.097	2.69
2024	.100	2.78	6061	.098	2.70
2025	.101	2.81	6063	.097	2.70
2036	.100	2.75	6066	.098	2.72
2117	.099	2.75	6070	.098	2.71
2124	.100	2.78	6101	.097	2.70
2218	.101	2.81	6105	.097	2.69
2219	.103	2.84	6151	.098	2.71
2618	.100	2.76	6162	.097	2.70
3003	.099	2.73	6201	.097	2.69
3004	.098	2.72	6262	.098	2.72
3005	.098	2.73	6351	.098	2.71
3105	.098	2.72	6463	.097	2.69
4032	.097	2.68	6951	.098	2.70
4043	.097	2.69	7005	.100	2.78
4045	.096	2.67	7008	.100	2.78
4047	.096	2.66	7049	.103	2.84
4145	.099	2.74	7050	.102	2.83
4343	.097	2.68	7072	.098	2.72
4643	.097	2.69	7075	.101	2.81
5005	.098	2.70	7175	.101	2.80
5050	.097	2.69	7178	.102	2.83

TABLE 9.10 (*Continued*)

Alloy	Density (lb/in.^{3})	Specific Gravity	Alloy	Density (lb/in.^{3})	Specific Gravity
5052	.097	2.68	7475	.101	2.81
5056	.095	2.64	8017	.098	2.71
5083	.096	2.66	8030	.098	2.71
5086	.096	2.66	8176	.098	2.71
5154	.096	2.66	8177	.098	2.70
5183	.096	2.66			

Source: Aluminum Association [5].
Note: Density and specific gravity are dependent upon composition, and variations are discernible from one cast to another for most alloys. The nominal values shown below should not be specified as engineering requirements but are used in calculating typical values for weight per unit length, weight per unit area, covering area, etc. The density values are derived from the metric and subsequently rounded. These values are not to be converted to the metric. X.XXX0 and X.XXX5 density values and X.XX0 and X.XX5 specific gravity values are limited to 99.35 percent or higher purity aluminum.

1625 Btu · in./ft^2 h°F(234 W/m · K) and an *electrical conductivity* of 62% of the International Annealed Copper Standard (IACS) at 68°F (20°C) for equal volume, or 204% of IACS for equal weight.

The *coefficient of thermal expansion*, the rate at which material expands as its temperature increases, is itself a function of temperature, being slightly higher at greater temperatures. Average values are used for a temperature range, usually from room temperature [68°F (20°C)] to water's boiling point [212°F (100°C)]. A commonly used number for this range is 13×10^{-6}/°F(23×10^{-6}/°C). This compares to 18 for copper, 15 for magnesium, 9.6 for stainless steel, and 6.5×10^{-6}/°F for carbon steel.

For wrought alloys, typical physical properties are given in Table 9.12. Typical physical properties of cast alloys are given in Table 9.11.

9.2.2 Mechanical Properties

Mechanical properties are properties related to the behavior of material when subjected to force. Most are measured according to standard test methods provided by the American Society for Testing and Materials (ASTM). The mechanical properties of interest for aluminum and ASTM test methods by which they are measured given in Table 9.13.

Mechanical properties are a function of the alloy and temper as well as, in some cases, product form. For example, 6061-T6 extrusions have a minimum tensile ultimate strength of 38 ksi (260 MPa), while 6061-T6 sheet and plate have a minimum tensile ultimate strength of 42 ksi (290 MPa).

TABLE 9.11 Typical Physical Properties of Commonly Used Sand and Permanent Mold Casting Alloys

Alloy	Temper	Specific Gravity(a)	Density(a) (lb/in.^{3})	Approximate Melting Range (°F)	Electrical Conductivity (% IACS)	Thermal Conductivity at 25°C, CGS(b)	Coeff. of Thermal Expansion, per °F × 10^{-6}	
							68–212 °F	68–572 °F
201.0	T6	2.80	0.101	1060–1200	27–32	0.29	19.3	24.7
	T7	2.80	0.101	1060–1200	32–34	0.29	—	—
204.0	T4	—	—	985–1200	—	—	—	—
208.0	F	2.79	0.101	970–1160	31	0.30	12.4	13.4
222.0	T61	2.95	0.107	965–1155	33	0.31	12.3	13.1
242.0	T571(c)	2.81	0.102	990–1175	34	0.32	12.6	13.6
	T77	2.81	0.102	990–1175	38	0.36	12.6	13.6
295.0	T6	2.81	0.102	970–1190	35	0.33	12.7	13.8
296.0	T6(c)	2.80	0.101	970–1170	33	0.31	12.2	13.3
308.0	F	2.79	0.101	970–1135	37	0.35	11.9	12.9
319.0	F	2.79	0.101	960–1120	27	0.26	11.9	12.7
328.0	F	2.70	0.098	1025–1105	30	0.29	11.9	12.9
332.0	T5(c)	2.76	0.100	970–1080	26	0.25	11.5	12.4
333.0	F(c)	2.77	0.100	960–1085	26	0.25	11.4	12.4
	T5(c)	2.77	0.100	960–1085	29	0.28	11.4	12.4
	T6(c)	2.77	0.100	960–1085	29	0.28	11.4	12.4
	T7(c)	2.77	0.100	960–1085	35	0.33	11.4	12.4
336.0	T551(c)	2.72	0.098	1000–1050	29	0.28	11.0	12.0
354.0	T61	2.71	0.098	1000–1105	32	0.30	11.6	12.7
355.0	T51	2.71	0.098	1015–1150	43	0.40	12.4	13.7
	T6	2.71	0.098	1015–1150	36	0.34	12.4	13.7
	T6(c)	2.71	0.098	1015–1150	39	0.36	12.4	13.7
	T61	2.71	0.098	1015–1150	37	0.35	12.4	13.7
	T62(c)	2.71	0.098	1015–1150	38	0.35	12.4	13.7
	T7	2.71	0.098	1015–1150	42	0.39	12.4	13.7
	T71	2.71	0.098	1015–1150	39	0.36	12.4	13.7
C355.0	T61	2.71	0.098	1015–1150	39	0.36	12.4	13.7
356.0	T51	2.68	0.097	1035–1135	43	0.40	11.9	12.9
	T6	2.68	0.097	1035–1135	39	0.36	11.9	12.9
	T6(c)	2.68	0.097	1035–1135	41	0.38	11.9	12.9
	T7	2.68	0.097	1035–1135	40	0.37	11.9	12.9
	T7(c)	2.68	0.097	1035–1135	43	0.40	11.9	12.9
A356.0	T61	2.67	0.097	1035–1135	39	0.36	11.9	12.9
357.0	F	2.67	0.097	1035–1135	39	0.36	11.9	12.9
A357.0	T61	2.67	0.097	1035–1135	39	0.36	11.9	12.9
359.0	T6	2.67	0.097	1045–1115	35	0.33	11.6	12.7
443.0	F	2.69	0.097	1065–1170	37	0.35	12.3	13.4
B443.0	F	2.69	0.097	1065–1170	37	0.35	12.3	13.4
A444.0	F	2.68	0.097	1070–1170	41	0.38	12.1	13.2
512.0	F	2.65	0.096	1090–1170	38	0.35	12.7	13.8
513.0	F(c)	2.68	0.097	1075–1180	34	0.32	13.4	14.5
514.0	F	2.65	0.096	1110–1185	35	0.33	13.4	14.5
520.0	T4	2.57	0.093	840–1120	21	0.21	13.7	14.8
535.0	F	2.62	0.095	1020–1165	23	0.23	13.1	14.8
705.0	F	2.76	0.100	1105–1180	25	0.25	13.1	14.3
707.0	F	2.77	0.100	1085–1165	25	0.25	13.2	14.4
710.0	F	2.81	0.102	1105–1195	35	0.33	13.4	14.6
711.0	F(c)	2.84	0.103	1120–1190	40	0.37	13.1	14.2
712.0	F	2.81	0.101	1135–1200	35	0.33	13.7	14.8(d)
713.0	F	2.81	0.100	1100–1180	30	0.29	13.4(d)	14.6(d)

TABLE 9.11 (*Continued*)

Alloy	Temper	Specific Gravity(a)	Density(a) (lb/in.^{3})	Approximate Melting Range (°F)	Electrical Conductivity (% IACS)	Thermal Conductivity at 25°C, CGS(b)	Coeff. of Thermal Expansion, per °F × 10^{-6} 68–212 °F	68–572 °F
771.0	F	2.81	0.102	1120–1190	37	0.33	13.7	14.8(d)
850.0	T5(c)	2.88	0.104	435–1200	47	0.43	13.0	(c)
851.0	T5(c)	2.83	0.103	440–1165	43	0.40	12.6	(c)
852.0	T5(c)	2.88	0.104	400–1175	45	0.41	12.9	(c)

Reference: *Aluminum, Volume I. Properties, Physical Metallurgy and Phase Diagrams*, American Society for Metals, Metals Park, Ohio (1967). Data for alloy 771.0 supplied by the U.S. Reduction Company, East Chicago, Indiana.
Note: These typical properties are not guaranteed, and should not be used for design purposes but only as a basis for general comparison of alloys and tempers with respect to any given characteristic.
(a) Assuming solid (void-free) metal. Since some porosity cannot be avoided in commercial castings, the actual values will be slightly less than those given.
(b) Cgs units equals calories per second per square centimeter per centimeter of thickness per degree centigrade.
(c) Chill cast samples; all other samples cast in green sand mold.
(d) Estimated value.
(e) Exceeds operating temperature.

9.2.2.1 Minimum and Typical Mechanical Properties

There are several bases for mechanical properties. A *typical property* is an average property. A *minimum property* is defined by the aluminum industry as the value that 99% of samples will equal or exceed with a probability of 95%. (The U.S. military calls such minimum values "A" values and also defines "B" values as those that 90% of samples will equal or exceed with a probability of 95%, a slightly less stringent criterion that yields higher values.) Typical mechanical properties are given in Table 9.14. Some minimum mechanical properties are given in ASTM and other specifications [17]; more are given in Table 9.15 for wrought alloys and Table 9.16 for cast alloys. Minimum mechanical properties are called "guaranteed" when product specifications require them to be met and are called "expected" when they are not required by-product specifications.

Structural design of aluminum components is usually based on minimum strengths. The rules for such design are given in the Aluminum Association's Specification for Aluminum Structures, part of the *Aluminum Design Manual* [2]. Safety factors given there, varying from 1.65 to 2.64 by type of structure, type of failure (yielding or fracture), and type of component (member or connection) are applied to the minimum strengths to determine the safe capacity of a component [25]. Typical strengths should be used to determine the capacity of fabrication equipment (e.g., the force required to shear a piece) or the strength of parts designed to fail at a given force to preclude failure of an entire structure. (Pressure-relieving panels are an example of this, called frangible design.) Maximum ultimate strengths are specified for some aluminum products (usually in softer tempers), but these materials are usually intended to be cold worked into final use products, changing their strength.

9.2.2.2 Strengths

While the stress–strain curve of aluminum is approximately linear in the elastic region, aluminum alloys do not exhibit a pronounced yield point like mild carbon

TABLE 9.12 Typical Physical Properties of Wrought Alloys

Alloy	Average① Coefficient of Thermal Expansion 68–212°F per°F	Melting Range②③ Approx. °F	Temper	Thermal Conductivity at 77°F English Units④	Electrical at 68°F Percent of International Annealed Copper Standard		Electrical Resistivity at 68°F Ohm—Cir. Mil/Foot
					Equal Volume	Equal Weight	
1060	13.1	1195–1215	O	1625	62	204	17
			H18	1600	61	201	17
1100	13.1	1190–1215	O	1540	59	194	18
			H18	1510	57	187	18
1350	13.2	1195–1215	All	1625	62	204	17
2011	12.7	1005–1190⑥	T3	1050	39	128	27
			T8	1190	45	142	23
2014	12.8	945–1180⑤	O	1340	50	159	21
			T4	930	34	108	31
			T6	1070	40	127	26
2017	13.1	955–1185⑤	O	1340	50	159	21
			T4	930	34	108	31
2018	12.4	945–1180⑥	T61	1070	40	127	26
2024	12.9	935–1180⑤	O	1340	50	160	21
			T3, T4, T361	840	30	96	35
			T6, T81, T861	1050	38	122	27
2025	12.6	970–1185⑤	T6	1070	40	128	26
2036	13.0	1030–1200⑥	T4	1100	41	135	25
2117	13.2	1030–1200⑥	T4	1070	40	130	26
2124	12.7	935–1180⑤	T851	1055	38	122	27
2218	12.4	940–1175⑤	T72	1070	40	126	26
2219	12.4	1010–1190⑤	O	1190	44	138	24
			T31, T37	780	28	88	37
			T6, T81, T87	840	30	94	35
2618	12.4	1020–1180	T6	1020	37	120	28
3003	12.9	1190–1210	O	1340	50	163	21
			H12	1130	42	137	25
			H14	1100	41	134	25
			H18	1070	40	130	26
3004	13.3	1165–1210	All	1130	42	137	25
3105	13.1	1175–1210	All	1190	45	148	23
4032	10.8	990–1060⑤	O	1070	40	132	26
			T6	960	35	116	30
4043	12.3	1065–1170	O	1130	42	140	25
4045	11.7	1065–1110	All	1190	45	151	23
4343	12.0	1070–1135	All	1250	47	158	25
5005	13.2	1170–1210	All	1390	52	172	20
5050	13.2	1155–1205	All	1340	50	165	21
5052	13.2	1125–1200	All	960	35	116	30
5056	13.4	1055–1180	O	810	29	98	36
			H38	750	27	91	38
5083	13.2	1095–1180	O	810	29	98	36
5086	13.2	1085–1185	All	870	31	104	33
5154	13.3	1100–1190	All	870	32	107	32
5252	13.2	1125–1200	All	960	35	116	30
5254	13.3	1100–1190	All	870	32	107	32

TABLE 9.12 (*Continued*)

Alloy	Average① Coefficient of Thermal Expansion 68–212°F per°F	Melting Range②③ Approx. °F	Temper	Thermal Conductivity at 77°F English Units④	Electrical at 68°F Percent of International Annealed Copper Standard		Electrical Resistivity at 68°F Ohm—Cir. Mil/Foot
					Equal Volume	Equal Weight	
5356	13.4	1060–1175	O	810	29	98	36
5454	13.1	1115–1195	O	930	34	113	31
			H38	930	34	113	31
5456	13.3	1055–1180	O	810	29	98	36
5457	13.2	1165–1210	All	1220	46	153	23
5652	13.2	1125–1200	All	960	35	116	30
5657	13.2	1180–1215	All	1420	54	180	19
6005	13.0	1125–1210⑥	T1	1250	47	155	22
			T5	1310	49	161	21
6053	12.8	1070–1205⑥	O	1190	45	148	23
			T4	1070	40	132	26
			T6	1130	42	139	25
6061	13.1	1080–1205⑥	O	1250	47	155	22
			T4	1070	40	132	26
			T6	1160	43	142	24
6063	13.0	1140–1210	O	1510	58	191	18
			T1	1340	50	165	21
			T5	1450	55	181	19
			T6, T83	1390	53	175	20
6066	12.9	1045–1195⑤	O	1070	40	132	26
			T6	1020	37	122	28
6070		1050–1200⑤	T6	1190	44	145	24
6101	13.0	1150–1210	T6	1510	57	188	18
			T61	1540	59	194	18
			T63	1510	58	191	18
			T64	1570	60	198	17
			T65	1510	58	191	18
6105	13.0	1110–1200⑥	T1	1220	46	151	23
			T5	1340	50	165	21
6151	12.9	1090–1200⑥	O	1420	54	178	19
			T4	1130	42	138	25
			T6	1190	45	148	23
6201	13.0	1125–1210⑥	T81	1420	54	180	19
6253		1100–1205	—	—	—	—	—
6262	13.0	1080–1205⑥	T9	1190	44	145	24
6351	13.0	1030–1200	T6	1220	46	151	23
6463	13.0	1140–1210	T1	1340	50	165	21
			T5	1450	55	181	19
			T6	1390	53	175	20
6951	13.0	1140–1210	O	1480	56	186	19
			T6	1370	52	172	20
7049	13.0	890–1175	T73	1070	40	132	26
7050	12.8	910–1165	T74⑧	1090	41	135	25
7072	13.1	1185–1215	O	1540	59	193	18
7075	13.1	890–1175⑦	T6	900	33	105	31

TABLE 9.12 (*Continued*)

Alloy	Average① Coefficient of Thermal Expansion 68–212°F per°F	Melting Range②③ Approx. °F	Temper	Thermal Conductivity at 77°F English Units④	Electrical at 68°F Percent of International Annealed Copper Standard Equal Volume	Equal Weight	Electrical Resistivity at 68°F Ohm—Cir. Mil/Foot
7175	13.0	890–1175⑦	T74	1080	39	124	26
7178	13.0	890–1165⑦	T6	870	31	98	33
7475	12.9	890–1175	T61, T651	960	35	116	30
			T76, T761	1020	40	132	26
			T7351	1130	42	139	25
8017	13.1	1190–1215	H12, H22	—	59	193	18
			H212	—	61	200	17
8030	13.1	1190–1215	H221	1600	61	201	17
8176	13.1	1190–1215	H24	1600	61	201	17

Source: Aluminum Association [5].
Note: The following typical properties are not guaranteed, since in most cases they are averages for various sizes, product forms and methods of manufacture and may not be exactly representative of any particular product or size. These data are intended only as a basis for comparing alloys and tempers and should not be specified as engineering requirements or used for design purposes.
① Coefficient to be multiplied by 10^{-6}. Example: $12.2 \times 10^{-6} = 0.0000122$.
② Melting ranges shown apply to wrought products of 1/4 inch thickness or greater.
③ Based on typical composition of the indicated alloys.
④ English units = btu-in./ft²hr°F.
⑤ Eutectic melting is not eliminated by homogenization.
⑥ Eutectic melting can be completely eliminated by homogenization.
⑦ Homogenization may raise eutectic melting temperature 20–40°F but usually does not eliminate eutectic melting.
⑧ Although not formerly registered, the literature and some specifications have used T736 as the designation for this temper.

TABLE 9.13 ASTM Test Methods for Mechanical Properties of Aluminum Alloys

Property	ASTM Test Method
Tensile strength	B557
Shear strength	B565
Plane strain fracture toughness	B645

steels. Therefore, an arbitrary definition for the *yield strength* has been adopted by the aluminum industry: a line parallel to a tangent to the stress–strain curve at its initial point is drawn, passing through the 0.2% strain intercept on the x (strain) axis. The stress where this line intersects the stress–strain curve is defined as the yield stress. The shape of the stress–strain curve for H, O, T1, T2, T3, and T4 tempers has a less pronounced knee at yield when compared to the shape of the curve for the T5, T6, T7, T8, and T9 tempers. (This causes the inelastic buckling strengths of these two groups of tempers to differ, since inelastic buckling strength is a function of the shape of the stress–strain curve after yield.)

Ultimate strength is the maximum stress the material can sustain. All stresses given in aluminum product specifications are engineering stresses; that is, they are calculated by dividing the force by the original cross-sectional area of

TABLE 9.14 Typical Mechanical Properties①②

	Tension				Hardness	Shear	Fatigue	Modulus
	Strength (ksi)		Elongation (percent in 2 in.)					
Alloy and Temper	Ultimate	Yield	1/16 in. Thick Specimen	1/2 in. Diameter Specimen	Brinell Number 500 kg load 10 mm ball	Ultimate Shearing Strength (ksi)	Endurance③ Limit (ksi)	Modulus④ of Elasticity (ksi × 10^3)
1060-O	10	4	43	—	19	7	3	10.0
1060-H12	12	11	16	—	23	8	4	10.0
1060-H14	14	13	12	—	26	9	5	10.0
1060-H16	16	15	8	—	30	10	6.5	10.0
1060-H18	19	18	6	—	35	11	6.5	10.0
1100-O	13	5	35	45	23	9	5	10.0
1100-H12	16	15	12	25	28	10	6	10.0
1100-H14	18	17	9	20	32	11	7	10.0
1100-H16	21	20	6	17	38	12	9	10.0
1100-H18	24	22	5	15	44	13	9	10.0
1350-O	12	4	—	—⑤	—	8	—	10.0
1350-H12	14	12	—	—	—	9	—	10.0
1350-H14	16	14	—	—	—	10	—	10.0
1350-H16	18	16	—	—	—	11	—	10.0
1350-H19	27	24	—	—⑥	—	15	7	10.0
2011-T3	55	43	—	15	95	32	18	10.2
2011-T8	59	45	—	12	100	35	18	10.2
2014-O	27	14	—	18	45	18	13	10.6
2014-T4, T451	62	42	—	20	105	38	20	10.6
2014-T6, T651	70	60	—	13	135	42	18	10.6

TABLE 9.14 (*Continued*)

	Tension				Hardness	Shear	Fatigue	Modulus
	Strength (ksi)		Elongation (percent in 2 in.)					
Alloy and Temper	Ultimate	Yield	1/16 in. Thick Specimen	1/2 in. Diameter Specimen	Brinell Number 500 kg load 10 mm ball	Ultimate Shearing Strength (ksi)	Endurance③ Limit (ksi)	Modulus④ of Elasticity (ksi × 10^3)
Alclad 2014-O	25	10	21	—	—	18	—	10.5
Alclad 2014-T3	63	40	20	—	—	37	—	10.5
Alclad 2014-T4, T451	61	37	22	—	—	37	—	10.5
Alclad 2014-T6, T651	68	60	10	—	—	41	—	10.5
2017-O	26	10	—	22	45	18	13	10.5
2017-T4, T451	62	40	—	22	105	38	18	10.5
2018-T61	61	46	—	12	120	39	17	10.8
2024-O	27	11	20	22	47	18	13	10.6
2024-T3	70	50	18	—	120	41	20	10.6
2024-T4, T351	68	47	20	19	120	41	20	10.6
2024-T361⑦	72	57	13	—	130	42	18	10.6
Alclad 2024-O	26	11	20	—	—	18	—	10.6
Alclad 2024-T3	65	45	18	—	—	40	—	10.6
Alclad 2024-T4, T351	64	42	19	—	—	40	—	10.6
Alclad 2024-T361⑦	67	63	11	—	—	41	—	10.6
Alclad 2024-T81, T851	65	60	6	—	—	40	—	10.6
Alclad 2024-T861⑦	70	66	6	—	—	42	—	10.6
2025-T6	58	37	—	19	110	35	18	10.4
2036-T4	49	28	24	—	—	—	18⑨	10.3

2117-T4	43	24	—	27	70	28	14	10.3
2124-T851	70	64	—	8	—	—	—	10.6
2218-T72	48	37	—	11	95	30	—	10.8
2219-O	25	11	18	—	—	—	—	10.6
2219-T42	52	27	20	—	—	—	—	10.6
2219-T31, T351	52	36	17	—	—	—	—	10.6
2219-T37	57	46	11	—	—	—	—	10.6
2219-T62	60	42	10	—	—	—	15	10.6
2219-T81, T851	66	51	10	—	—	—	15	10.6
2219-T87	69	57	10	—	—	—	15	10.6
2618-T61	64	54	—	10	115	38	18	10.8
3003-O	16	6	30	40	28	11	7	10.0
3003-H12	19	18	10	20	35	12	8	10.0
3003-H14	22	21	8	16	40	14	9	10.0
3003-H16	26	25	5	14	47	15	10	10.0
3003-H18	29	27	4	10	55	16	10	10.0
5254-O	35	17	27	—	58	22	17	10.2
5254-H32	39	30	15	—	67	22	18	10.2
5254-H34	42	33	13	—	73	24	19	10.2
5254-H36	45	36	12	—	78	26	20	10.2
5254-H38	48	39	10	—	80	28	21	10.2
5254-H112	35	17	25	—	63	—	17	10.2
5454-O	36	17	22	—	62	23	—	10.2
5454-H32	40	30	10	—	73	24	—	10.2
5454-H34	44	35	10	—	81	26	—	10.2
5454-H111	38	26	14	—	70	23	—	10.2
5454-H112	36	18	18	—	62	23	—	10.2
5456-O	45	23	—	24	—	—	—	10.3
5456-H25	45	24	—	22	—	—	—	10.3
5456-H321, H116	51	37	—	16	90	30	—	10.3

TABLE 9.14 (*Continued*)

Alloy and Temper	Tension				Hardness	Shear	Fatigue	Modulus
	Strength (ksi)		Elongation (percent in 2 in.)					
	Ultimate	Yield	1/16 in. Thick Specimen	1/2 in. Diameter Specimen	Brinell Number 500 kg load 10 mm ball	Ultimate Shearing Strength (ksi)	Endurance③ Limit (ksi)	Modulus④ of Elasticity (ksi × 10^3)
5457-O	19	7	22	—	32	12	—	10.0
5457-H25	26	23	12	—	48	16	—	10.0
5457-H38, H28	30	27	6	—	55	18	—	10.0
5652-O	28	13	25	30	47	18	16	10.2
5652-H32	33	28	12	18	80	20	17	10.2
5652-H34	38	31	10	14	68	21	18	10.2
5652-H36	40	35	8	10	73	23	19	10.2
5652-H38	42	37	7	8	77	24	20	10.2
5657-H25	23	20	12	—	40	12	—	10.0
5657-H38, H28	28	24	7	—	50	15	—	10.0
6061-O	18	8	25	30	30	12	9	10.0
6061-T4, T451	35	21	22	25	65	24	14	10.0
6061-T6, T651	45	40	12	17	95	30	14	10.0
Alclad 6061-O	17	7	25	—	—	11	—	10.0
Alclad 6061-T4, T451	33	19	22	—	—	22	—	10.0
Alclad 6061-T6, T651	42	37	12	—	—	27	—	10.0
6063-O	13	7	—	—	25	10	8	10.0
6063-T1	22	13	20	—	42	14	9	10.0
6063-T4	25	13	22	—	—	—	—	10.0

6063-T5	27	21	12	—	60	17	10	10.0
6063-T6	35	31	12	—	73	22	10	10.0
6063-T83	37	35	9	—	82	22	—	10.0
6063-T831	30	27	10	—	70	18	—	10.0
6063-T832	42	39	12	—	95	27	—	10.0
6066-O	22	12	—	18	43	14	—	10.0
6066-T4, T451	52	30	—	18	90	29	—	10.0
6066-T6. T651	57	52	—	12	120	34	16	10.0
6070-T6	55	51	10	—	—	34	14	10.0
6101-H111	14	11	—	—	—	—	—	10.0
6101-T6	32	28	15⑧	—	71	20	—	10.0
6262-T9	58	55	—	10	120	35	13	10.0
6351-T4	36	22	20	—	—	—	—	10.0
6351-T6	45	41	14	—	95	29	13	10.0
6463-T1	22	13	20	—	42	14	10	10.0
6463-T5	27	21	12	—	60	17	10	10.0
6463-T6	35	31	12	—	74	22	10	10.0
7049-T73	75	65	—	12	135	44	—	10.4
7049-T7352	75	63	—	11	135	43	—	10.4
7050-T73510, T73511	72	63	—	12	—	—	—	10.4
7050-T7451⑩	76	68	—	11	—	44	—	10.4
7050-T7651	80	71	—	11	—	47	—	10.4
7075-O	33	15	17	16	60	22	—	10.4
7075-T6, T651	83	73	11	11	150	48	23	10.4
Alclad 3003-O	16	6	30	40	—	11	—	10.0
Alclad 3003-H12	19	18	10	20	—	12	—	10.0
Alclad 3003-H14	22	21	8	16	—	14	—	10.0
Alclad 3003-H16	26	25	5	14	—	15	—	10.0
Alclad 3003-H18	29	27	4	10	—	16	—	10.0

TABLE 9.14 (*Continued*)

	Tension				Hardness	Shear	Fatigue	Modulus
	Strength (ksi)		Elongation (percent in 2 in.)					
Alloy and Temper	Ultimate	Yield	1/16 in. Thick Specimen	1/2 in. Diameter Specimen	Brinell Number 500 kg load 10 mm ball	Ultimate Shearing Strength (ksi)	Endurance(3) Limit (ksi)	Modulus(4) of Elasticity (ksi × 10^3)
3004-O	26	10	20	25	45	16	14	10.0
3004-H32	31	25	10	17	52	17	15	10.0
3004-H34	35	29	9	12	63	18	15	10.0
3004-H36	38	33	5	9	70	20	16	10.0
3004-H38	41	36	5	6	77	21	16	10.0
Alclad 3004-O	26	10	20	25	—	16	—	10.0
Alclad 3004-H32	31	25	10	17	—	17	—	10.0
Alclad 3004-H34	35	29	9	12	—	18	—	10.0
Alclad 3004-H36	38	33	5	9	—	20	—	10.0
Alclad 3004-H38	41	36	5	6	—	21	—	10.0
3105-O	17	8	24	—	—	12	—	10.0
3105-H12	22	19	7	—	—	14	—	10.0
3105-H14	25	22	5	—	—	15	—	10.0
3105-H16	28	25	4	—	—	16	—	10.0
3105-H18	31	28	3	—	—	17	—	10.0
3105-H25	26	23	8	—	—	15	—	10.0

4032-T6	55	46	—	9	120	38	16	11.4
5005-O	18	6	25	—	28	11	—	10.0
5005-H12	20	19	10	—	—	14	—	10.0
5005-H14	23	22	6	—	—	14	—	10.0
5005-H16	26	25	5	—	—	15	—	10.0
5005-H18	29	28	4	—	—	16	—	10.0
5005-H32	20	17	11	—	36	14	—	10.0
5005-H34	23	20	8	—	41	14	—	10.0
5005-H36	26	24	6	—	46	15	—	10.0
5005-H38	29	27	5	—	51	16	—	10.0
5050-O	21	8	24	—	36	15	12	10.0
5050-H32	25	21	9	—	46	17	13	10.0
5050-H34	28	24	8	—	53	18	13	10.0
5050-H36	30	26	7	—	58	19	14	10.0
5050-H38	32	29	6	—	63	20	14	10.0
5052-O	28	13	25	30	47	18	16	10.2
5052-H32	33	28	12	18	60	20	17	10.2
5052-H34	38	31	10	14	68	21	18	10.2
5052-H36	40	35	8	10	73	23	19	10.2
5052-H38	42	37	7	8	77	24	20	10.2
5056-O	42	22	—	35	65	26	20	10.3
5056-H18	63	59	—	10	105	34	22	10.3
5056-H38	60	50	—	15	100	32	22	10.3
5083-O	42	21	—	22	—	25	—	10.3
5083-H321, H116	46	33	—	16	—	—	23	10.3

TABLE 9.14 (*Continued*)

	Tension				Hardness	Shear	Fatigue	Modulus
	Strength (ksi)		Elongation (percent in 2 in.)					
Alloy and Temper	Ultimate	Yield	1/16 in. Thick Specimen	1/2 in. Diameter Specimen	Brinell Number 500 kg load 10 mm ball	Ultimate Shearing Strength (ksi)	Endurance③ Limit (ksi)	Modulus④ of Elasticity (ksi × 10^3)
5086-O	38	17	22	—	—	23	—	10.3
5086-H32, H116	42	30	12	—	—	—	—	10.3
5086-H34	47	37	10	—	—	27	—	10.3
5086-H112	39	19	14	—	—	—	—	10.3
5154-O	35	17	27	—	58	22	17	10.2
5154-H32	39	30	15	—	67	22	18	10.2
5154-H34	42	33	13	—	73	24	19	10.2
5154-H36	45	36	12	—	78	26	20	10.2
5154-H38	48	39	10	—	80	28	21	10.2
5154-H112	35	17	25	—	63	—	17	10.2
5252-H25	34	25	11	—	68	21	—	10.0
5252-H38, H28	41	35	5	—	75	23	—	10.0
Alclad 7075-O	32	14	17	—	—	22	—	10.4
Alclad 7075-T6, T651	76	67	11	—	—	46	—	10.4
7175-T74	76	66	—	11	135	42	23	10.4
7178-O	33	15	15	16	—	—	—	10.4
7178-T6, T651	88	78	10	11	—	—	—	10.4
7178-T76, T7651	83	73	—	11	—	—	—	10.3
Alclad 7178-O	32	14	16	—	—	—	—	10.4
Alclad 7178-T6, T651	81	71	10	—	—	—	—	10.4

7475-T61	82	71	11	—	—	—	—	10.2
7475-T651	85	74	—	13	—	—	—	10.4
7475-T7351	72	61	—	13	—	—	—	10.4
7475-T761	75	65	12	—	—	—	—	10.2
7475-T7651	77	67	—	12	—	—	—	10.4
Alclad 7475-T61	75	66	11	—	—	—	—	10.2
Alclad 7475-T761	71	61	12	—	—	—	—	10.2
8176-H24	17	14	15	—	—	10	—	10.0

Source: Aluminum Association [5].

Note: The following typical properties are not guaranteed, since in most cases they are averages for various sizes, product forms and methods of manufacture and may not be exactly representative of any particular product or size. These data are intended only as a basis for comparing alloys and tempers and should not be specified as engineering requirements or used for design purposes.

(1) The mechanical property limits are listed by major product in the "Standards Section" of this manual.

(2) The indicated typical mechanical properties for all except 0 temper material are higher than the specified minimum properties. For 0 temper products typical ultimate and yield values are slightly lower than specified (maximum) values.

(3) Based on 500,000,000 cycles of completely reversed stress using the R.R. Moore type of machine and specimen.

(4) Average of tension and compression moduli. Compression modulus is about 2% greater than tension modulus.

(5) 1350-O wire will have an elongation of approximately 23% in 10 inches.

(6) 1350-H19 wire will have an elongation of approximately $1\frac{1}{2}$% in 10 inches.

(7) Tempers T361 and T861 were formerly designated T36 and T86, respectively.

(8) Based on $\frac{1}{4}$ in. thick specimen.

(9) Based on 10^7 cycles using flexural type testing of sheet specimens.

(10) T7451, although not previously registered, has appeared in literature and in some specifications as T73651.

TABLE 9.15A Minimum Mechanical Properties for Aluminum Alloys

Alloy and Temper	Product	Thickness Range (in.)	Tension		Compression	Shear		Compressive Modulus of Elasticity‡ E (ksi)
			F_{ty}† (ksi)	F_{ty}† (ksi)	F_{cy} (ksi)	F_{su} (ksi)	F_{sy} (ksi)	
1100-H12	Sheet, Plate Rolled Rod & Bar	All	14	11	10	9	6.5	10,100
-H14		All	16	14	13	10	8	10,100
2014-T6	Sheet	0.040–0.249	66	58	59	40	33	10,900
-T651	Plate	0.250–2.000	67	59	58	40	34	10,900
-T6, T6510, T6511	Extrusions	All	60	53	52	35	31	10,900
-T6, T651	Cold Finished Rod & Bar, Drawn Tube	All	65	55	53	38	32	10,900
Alclad 2014-T6	Sheet	0.025–0.039	63	55	56	38	32	10,800
-T6	Sheet	0.040–0.249	64	57	58	39	33	10,800
-T651	Plate	0.250–0.499	64	57	56	39	33	10,800
3003-H12	Sheet & Plate	0.017–2.000	17	12	10	11	7	10,100
-H14	Sheet & Plate	0.009–1.000	20	17	14	12	10	10,100
-H16	Sheet	0.006–0.162	24	21	18	14	12	10,100
-H18	Sheet	0.006–0.128	27	24	20	15	14	10,100
-H12	Drawn Tube	All	17	12	11	11	7	10,100
-H14	Drawn Tube	All	20	17	16	12	10	10,100
-H16	Drawn Tube	All	24	21	19	14	12	10,100
-H18	Drawn Tube	All	27	24	21	15	14	10,100
Alclad 3003-H12	Sheet & Plate	0.017–2.000	16	11	9	10	6.5	10,100
-H14	Sheet & Plate	0.009–1.000	19	16	13	12	9	10,100
-H16	Sheet	0.006–0.162	23	20	17	14	12	10,100
-H18	Sheet	0.006–0.128	26	23	19	15	13	10,100
Alclad 3003-H14	Drawn Tube	0.025–0.259	19	16	15	12	9	10,100
-H18	Drawn Tube	0.010–0.500	26	23	20	15	13	10,100
3004-H32	Sheet & Plate	0.017–2.000	28	21	18	17	12	10,100
-H34	Sheet & Plate	0.009–1.000	32	25	22	19	14	10,100
-H36	Sheet	0.006–0.162	35	28	25	20	16	10,100
-H38	Sheet	0.006–0.128	38	31	29	21	18	10,100
3004-H34	Drawn Tube	0.018–0.450	32	25	24	19	14	10,100
-H36	Drawn Tube	0.018–0.450	35	28	27	20	16	10,100
Alclad 3004-H32	Sheet	0.017–0.249	27	20	17	16	12	10,100
-H34	Sheet	0.009–0.249	31	24	21	18	14	10,100
-H36	Sheet	0.006–0.162	34	27	24	19	16	10,100
-H38	Sheet	0.006–0.128	37	30	28	21	17	10,100
-H131, H241, H341	Sheet	0.024–0.050	31	26	22	18	15	10,100
-H151, H261, H361	Sheet	0.024–0.050	34	30	28	19	17	10,100
3005-H25	Sheet	0.013–0.050	26	22	20	15	13	10,100
-H28	Sheet	0.006–0.080	31	27	25	17	16	10,100
3105-H25	Sheet	0.013–0.080	23	19	17	14	11	10,100
5005-H12	Sheet & Plate	0.017–2.000	18	14	13	11	8	10,100
-H14	Sheet & Plate	0.009–1.000	21	17	15	12	10	10,100

TABLE 9.15A (*Continued*)

Alloy and Temper	Product	Thickness Range (in.)	Tension F_{ty}† (ksi)	Tension F_{ty}† (ksi)	Compression F_{cy} (ksi)	Shear F_{su} (ksi)	Shear F_{sy} (ksi)	Compressive Modulus of Elasticity‡ E (ksi)
5005-H16	Sheet	0.006–0.162	24	20	18	14	12	10,100
-H32	Sheet & Plate	0.017–2.000	17	12	11	11	7	10,100
-H34	Sheet & Plate	0.009–1.000	20	15	14	12	8.5	10,100
-H36	Sheet	0.006–0.162	23	18	16	13	11	10,100
5050-H32	Sheet	0.017–0.249	22	16	14	14	9	10,100
-H34	Sheet	0.009–0.249	25	20	18	15	12	10,100
-H32	Cold Fin. Rod & Bar† Drawn Tube	All	22	16	15	13	9	10,100
-H34	Cold Fin. Rod & Bar† Drawn Tube	All	25	20	19	15	12	10,100
5052-O	Sheet & Plate	0.006–3.000	25	9.5	9.5	16	5.5	10,200
-H32	(Sheet & Plate, Cold Fin. Rod & Bar, Drawn Tube)	All	31	23	21	19	13	10,200
-H34	(Sheet & Plate, Cold Fin. Rod & Bar, Drawn Tube)	All	34	27	24	20	15	10,200
-H36	Sheet	0.006–0.162	37	29	26	22	17	10,200
5083-O	Extrusions	up thru 5.000	39	16	16	24	9	10,400
-H111	Extrusions	up thru 0.500	40	24	21	24	14	10,400
-H111	Extrusions	0.501–5.000	40	24	21	23	14	10,400
-O	Sheet & Plate	0.051–1.500	40	18	18	25	10	10,400
-H116	Sheet & Plate	0.188–1.500	44	31	26	26	18	10,400
-H321	Sheet & Plate	0.188–1.500	44	31	26	26	18	10,400
-H116	Plate	1.501–3.000	41	29	24	24	17	10,400
-H321	Plate	1.501–3.000	41	29	24	24	17	10,400
5086-O	Extrusions	up thru 5.000	35	14	14	21	8	10,400
-H111	Extrusions	up thru 0.500	36	21	18	21	12	10,400
-H111	Extrusions	0.501–5.000	36	21	18	21	12	10,400
-O	Sheet & Plate	0.020–2.000	35	14	14	21	8	10,400
-H112	Plate	0.250–0.499	36	18	17	22	10	10,400
-H112	Plate	0.500–1.000	35	16	16	21	9	10,400
-H112	Plate	1.001–2.000	35	14	15	21	8	10,400
-H112	Plate	2.001–3.000	34	14	15	21	8	10,400
-H116	Sheet & Plate	All	40	28	26	24	16	10,400
-H32	Sheet & Plate Drawn Tube	All	40	28	26	24	16	10,400
-H34	Sheet & Plate Drawn Tube	All	44	34	32	26	20	10,400
5154-H38	Sheet	0.006–0.128	45	35	33	24	20	10,300
5454-O	Extrusions	up thru 5.000	31	12	12	19	7	10,400
-H111	Extrusions	up thru 0.500	33	19	16	20	11	10,400
-H111	Extrusions	0.501–5.000	33	19	16	19	11	10,400
-H112	Extrusions	up thru 5.000	31	12	13	19	7	10,400
-O	Sheet & Plate	0.020–3.000	31	12	12	19	7	10,400
-H32	Sheet & Plate	0.020–2.000	36	26	24	21	15	10,400
-H34	Sheet & Plate	0.020–1.000	39	29	27	23	17	10,400
5456-O	Sheet & Plate	0.051–1.500	42	19	19	26	11	10,400
-H116	Sheet & Plate	0.188–1.250	46	33	27	27	19	10,400

TABLE 9.15A (*Continued*)

Alloy and Temper	Product	Thickness Range (in.)	Tension		Compression	Shear		Compressive Modulus of Elasticity‡ E (ksi)
			F_{ty}† (ksi)	F_{ty}† (ksi)	F_{cy} (ksi)	F_{su} (ksi)	F_{sy} (ksi)	
5056-H321	Sheet & Plate	0.188–1.250	46	33	27	27	19	10,400
-H116	Plate	1.251–1.500	44	31	25	25	18	10,400
-H321	Plate	1.251–1.500	44	31	25	25	18	10,400
-H116	Plate	1.501–3.000	41	29	25	25	17	10,400
-H321	Plate	1.501–3.000	41	29	25	25	17	10,400
6005-T5	Extrusions	up thru 1.000	38	35	35	24	20	10,100
6061-T6, T651	Sheet & Plate	0.010–4.000	42	35	35	27	20	10,100
-T6, T6510, T6511	Extrusions	All	38	35	35	24	20	10,100
-T6, T651	Cold Fin. Rod & Bar	up thru 8.000	42	35	35	25	20	10,100
-T6	Drawn Tube	0.025–0.500	42	35	35	27	20	10,100
-T6	Pipe	All	38	35	35	24	20	10,100
6063-T5	Extrusions	up thru 0.500	22	16	16	13	9	10,100
-T5	Extrusions	0.500–1.000	21	15	15	12	8.5	10,100
-T6	Extrusions & Pipe	All	30	25	25	19	14	10,100
6066-T6, T6510, T6511	Extrusions	All	50	45	45	27	26	10,100
6070-T6, T62	Extrusions	up thru 2.999	48	45	45	29	26	10,100
6105-T5	Extrusions	up thru 0.500	38	35	35	24	20	10,100
6351-T5	Extrusions	up thru 1.000	38	35	35	24	20	10,100
6463-T6	Extrusions	up thru 0.500	30	25	25	19	14	10,100

Source: Aluminum Association [2].

† F_{tu} and F_{ty} are minimum specified values (except F_{ty} for 1100-H12, -H14 Cold Finished Rod and Bar and Drawn Tube, Alclad 3003-H18 Sheet and 5050-H32, -H34 Cold Finished Rod and Bar which are minimum expected values); other strength properties are corresponding minimum expected values.

‡ Typical values. For deflection calculations an average modulus of elasticity is used; this is 100 ksi lower than values in this column.

TABLE 9.15B Minimum Mechanical Properties for Aluminum Alloys (SI)

Alloy and Temper	Product	Thickness Range (mm)	Tension		Compression	Shear		Compressive Modulus of Elasticity‡ E (MPa)
			F_{tu}† (MPa)	F_{ty}† (MPa)	F_{cy} (MPa)	F_{su} (MPa)	F_{sy} (MPa)	
1100-H12	Sheet, Plate Rolled Rod & Bar	All	95	75	70	62	45	69,600
-H14		All	110	95	90	70	55	69,600
2014-T6	Sheet	1.00–6.30	455	400	405	275	230	75,200
-T651	Plate	6.30–50.00	460	405	400	275	235	75,200
-T6, T6510, T6511	Extrusions	All	415	365	360	240	215	75,200
-T6, T651	Cold Finished Rod & Bar, Drawn Tube	All	450	380	365	260	220	75,200
Alclad								
2014-T6	Sheet	0.63–1.00	435	380	385	260	220	74,500
-T6	Sheet	1.00–6.30	440	395	400	270	230	74,500
-T651	Plate	6.30–12.50	440	395	385	270	230	74,500

TABLE 9.15B ***(Continued)***

Alloy and Temper	Product	Thickness Range (mm)	Tension F_{tu}† (MPa)	Tension F_{ty}† (MPa)	Compression F_{cy} (MPa)	Shear F_{su} (MPa)	Shear F_{sy} (MPa)	Compressive Modulus of Elasticity‡ E (MPa)
3003-H12	Sheet & Plate	0.40–50.00	120	85	70	75	48	69,600
-H14	Sheet & Plate	0.20–25.00	140	115	95	85	70	69,600
-H16	Sheet	0.15–4.00	165	145	125	95	85	69,600
-H18	Sheet	0.15–3.20	185	165	140	105	95	69,600
-H12	Drawn Tube	All	120	85	75	75	48	69,600
-H14	Drawn Tube	All	140	115	110	85	70	69,600
-H16	Drawn Tube	All	165	145	130	95	85	69,600
-H18	Drawn Tube	All	185	165	145	105	95	69,600
Alclad								
3003-H12	Sheet & Plate	0.40–50.00	115	80	62	70	45	69,600
-H14	Sheet & Plate	0.20–25.00	135	110	90	85	62	69,600
-H16	Sheet	0.15–4.00	160	140	115	95	85	69,600
-H18	Sheet	0.15–3.20	180	160	130	105	90	69,600
Alclad								
3003-H14	Drawn Tube	0.63–6.30	135	110	105	85	62	69,600
-H18	Drawn Tube	0.25–12.50	180	160	140	105	90	69,600
3004-H32	Sheet & Plate	0.40–50.00	190	145	125	115	85	69,600
-H34	Sheet & Plate	0.20–25.00	220	170	150	130	95	69,600
-H36	Sheet	0.15–4.00	240	190	170	140	110	69,600
-H38	Sheet	0.15–3.20	260	215	200	145	125	69,600
3004-H34	Drawn Tube	0.45–11.50	220	170	165	130	95	69,600
-H36	Drawn Tube	0.45–11.50	240	190	185	140	110	69,600
Alclad								
3004-H32	Sheet	0.40–6.30	185	140	115	110	85	69,600
-H34	Sheet	0.20–6.30	215	165	145	125	95	69,600
-H36	Sheet	0.15–4.00	235	185	165	130	110	69,600
-H38	Sheet	0.15–3.20	255	205	195	145	115	69,600
-H131, H241, -H341	Sheet	0.60–1.20	215	180	150	125	105	69,600
-H151, H261, -H361	Sheet	0.60–1.20	235	205	195	130	115	69,600
3005-H25	Sheet	0.32–1.20	180	150	140	105	90	69,600
-H28	Sheet	0.15–2.00	215	185	170	115	110	69,600
3105-H25	Sheet	0.32–2.00	160	130	115	95	75	69,600
5005-H12	Sheet & Plate	0.40–50.00	125	95	90	75	55	69,600
-H14	Sheet & Plate	0.20–25.00	145	115	105	85	70	69,600
-H16	Sheet	0.15–4.00	165	135	125	95	85	69,600
-H32	Sheet & Plate	0.40–50.00	120	85	75	75	48	69,600
-H34	Sheet & Plate	0.20–25.00	140	105	95	85	59	69,600
-H36	Sheet	0.15–4.00	160	125	110	90	75	69,600
5050-H32	Sheet	0.40–6.30	150	110	95	95	62	69,600
-H34	Sheet	0.20–6.30	170	140	125	105	85	69,600
-H32	Cold Fin. Rod & Bar† Drawn Tube	All	150	110	105	90	62	69,600
-H34	Cold Fin. Rod & Bar† Drawn Tube	All	170	140	130	105	85	69,600

Source: Aluminum Association [2].

† F_{tu} and F_{ty} are minimum specified values (except F_{ty} for 1100-H12, -H14 Cold Finished Rod and Bar and Drawn Tube, Alclad 3003-H18 Sheet and 5050-H32, -H34 Cold Finished Rod and Bar which are minimum expected values); other strength properties are corresponding minimum expected values.

‡Typical values. For deflection calculations an average modulus of elasticity is used; this is 700 MPa lower than values in this column.

TABLE 9.16A Mechanical Property Limits for Commonly Used Aluminum Sand Casting Alloys[a]

Alloy	Temper[b]	Minimum Properties					Typical Brinell Hardness[e]
		Tensile Strength				% Elongation	
		Ultimate		Yield (0.2% Offset)		2 in. or 4 times diameter	500—kgf load 10—mm ball
		ksi	(MPa)	ksi	(MPa)		
201.0	T7	60.0	(414)	50.0	(345)	3.0	110–140
204.0	T4	45.0	(310)	28.0	(193)	6.0	—
208.0	F	19.0	(131)	12.0	(83)	1.5	40–70
222.0	0	23.0	(159)	—	—	—	65–95
222.0	T61	30.0	(207)	—	—	—	100–130
242.0	0	23.0	(159)	—	—	—	55–85
242.0	T571	29.0	(200)	—	—	—	70–100
242.0	T61	32.0	(221)	20.0	(138)	—	90–120
242.0	T77	24.0	(165)	13.0	(90)	1.0	60–90
295.0	T4	29.0	(200)	13.0	(90)	6.0	45–75
295.0	T6	32.0	(221)	20.0	(138)	3.0	60–90
295.0	T62	36.0	(248)	28.0	(193)	—	80–110
295.0	T7	29.0	(200)	16.0	(110)	3.0	55–85
319.0	F	23.0	(159)	13.0	(90)	1.5	55–85
319.0	T5	25.0	(172)	—	—	—	65–95
319.0	T6	31.0	(214)	20.0	(138)	1.5	65–95
328.0	F	25.0	(172)	14.0	(97)	1.0	45–75
328.0	T6	34.0	(234)	21.0	(145)	1.0	65–95
354.0	[c]	—	—	—	—	—	—
355.0	T51	25.0	(172)	18.0	(124)	—	50–80
355.0	T6	32.0	(221)	20.0	(138)	2.0	70–105
355.0	T7	35.0	(241)	—	—	—	70–100
355.0	T71	30.0	(207)	22.0	(152)	—	60–95
C355.0	T6	36.0	(248)	25.0	(172)	2.5	75–105
356.0	F	19.0	(131)	—	—	2.0	40–70
356.0	T51	23.0	(159)	16.0	(110)	—	45–75
356.0	T6	30.0	(207)	20.0	(138)	3.0	55–90
356.0	T7	31.0	(214)	29.0	(200)	—	60–90
356.0	T71	25.0	(172)	18.0	(124)	3.0	45–75
A356.0	T6	34.0	(234)	24.0	(165)	3.5	70–105
357.0	[c]	—	—	—	—	—	—
A357.0	[c]	—	—	—	—	—	—
359.0	[c]	—	—	—	—	—	—
443.0	F	17.0	(117)	7.0	(49)	3.0	25–55
B433.0	F	17.0	(117)	6.0	(41)	3.0	25–55
512.0	F	17.0	(117)	10.0	(69)	—	35–65
514.0	F	22.0	(152)	9.0	(62)	6.0	35–65
520.0	T4[f]	42.0	(290)	22.0	(152)	12.0	60–90
535.0	F or T5	35.0	(241)	18.0	(124)	9.0	60–90
705.0	F or T5	30.0	(207)	17.0	(117)	5.0	50–80
707.0	T5	33.0	(228)	22.0	(152)	2.0	70–100
707.0	T7	37.0	(255)	30.0	(207)	1.0	65–95
710.0	F or T5	32.0	(221)	20.0	(138)	2.0	60–90
712.0	F or T5	34.0	(234)	25.0	(172)	4.0	60–90
713.0	F or T5	32.0	(221)	22.0	(152)	3.0	60–90
771.0	T5	42.0	(290)	38.0	(262)	1.5	85–115

TABLE 9.16A (*Continued*)

Alloy	Temper(b)	Minimum Properties					Typical Brinell Hardness(e)
		Tensile Strength				% Elongation	
		Ultimate		Yield (0.2% Offset)		2 in. or 4 times diameter	500—kgf load 10—mm ball
		ksi	(MPa)	ksi	(MPa)		
771.0	T51	32.0	(221)	27.0	(186)	3.0	70–100
771.0	T52	36.0	(248)	30.0	(207)	1.5	70–100
771.0	T53	36.0	(248)	27.0	(186)	1.5	—
771.0	T6	42.0	(290)	35.0	(241)	5.0	75–105
771.0	T71	48.0	(331)	45.0	(310)	2.0	105–135
850.0	T5	16.0	(110)	—	—	5.0	30–60
851.0	T5	17.0	(117)	—	—	3.0	30–60
852.0	T5	24.0	(165)	18.0	(124)	—	45–75

Source: Aluminum Association [12].

(a)Values represent properties obtained from separately cast test bars and are derived from ASTM B-26, Standard Specification for Aluminum-Alloy Sand Castings; Federal Specification QQ-A-601e, Aluminum Alloy Sand Castings; and Military Specification MIL-A-21180c, Aluminum Alloy Castings, High Strength. Unless otherwise specified, the average tensile strength, average yield strength and average elongation values of specimens cut from castings shall be not less than 75 percent of the tensile and yield strength values and not less than 25 percent of the elongation values given above. The customer should keep in mind that (1) some foundries may offer additional tempers for the above alloys, and (2) foundries are constantly improving casting techniques and, as a result, some may offer minimum properties in excess of the above.

(b)F indicates "as cast" condition; refer to AA-CS-M11 for recommended times and temperatures of heat treatment for other tempers to achieve properties specified.

(c)Mechanical properties for these alloys depend on the casting process. For further information consult the individual foundries.

(e)Hardness values are given for information only: not required for acceptance.

(f)The T4 temper of Alloy 520.0 is unstable; significant room temperature aging occurs within life expectancy of most castings. Elongation may decrease by as much as 80 percent.

TABLE 9.16B Mechanical Property Limits for Commonly Used Aluminum Permanent Mold Casting Alloys(a)

Alloy	Temper(b)	Minimum Properties					Typical Brinell Hardness(c)
		Tensile Strength				% Elongation	
		Ultimate		Yield (0.2% Offset)		2 in. or 4 times diameter	500—kgf load 10—mm ball
		ksi	(MPa)	ksi	(MPa)		
204.0	T4	48.0	(331)	29.0	(200)	8.0	—
208.0	T4	33.0	(228)	15.0	(103)	4.5	60–90
208.0	T6	35.0	(241)	22.0	(152)	2.0	75–105
208.0	T7	33.0	(228)	16.0	(110)	3.0	65–95
222.0	T551	30.0	(207)	—	—	—	100–130
222.0	T65	40.0	(276)	—	—	—	125–155
242.0	T571	34.0	(234)	—	—	—	90–120
242.0	T61	40.0	(276)	—	—	—	95–125
296.0	T6	35.0	(241)	—	—	2.0	75–105
308.0	F	24.0	(165)	—	—	—	55–85
319.0	F	28.0	(193)	14.0	(97)	1.5	70–100
319.0	T6	34.0	(234)	—	—	2.0	75–105
332.0	T5	31.0	(214)	—	—	—	90–120
333.0	F	28.0	(193)	—	—	—	65–100
333.0	T5	30.0	(207)	—	—	—	70–105
333.0	T6	35.0	(241)	—	—	—	85–115

TABLE 9.16B ***(Continued)***

Alloy	Temper[b]	Minimum Properties					Typical Brinell Hardness[c]
		Tensile Strength				% Elongation	
		Ultimate		Yield (0.2% Offset)		2 in. or 4 times diameter	500—kgf load 10—mm ball
		ksi	(MPa)	ksi	(MPa)		
333.0	T7	31.0	(214)	—	—	—	75–105
336.0	T551	31.0	(214)	—	—	—	80–120
336.0	T65	40.0	(276)	—	—	—	110–140
354.0	T61	48.0	(331)	37.0	(255)	3.0	—
354.0	T62	52.0	(359)	42.0	(290)	2.0	—
355.0	T51	27.0	(186)	—	—	—	60–90
355.0	T6	37.0	(255)	—	—	1.5	75–105
355.0	T62	42.0	(290)	—	—	—	90–120
355.0	T7	36.0	(248)	—	—	—	70–100
355.0	T71	34.0	(234)	27.0	(186)	—	65–95
C355.0	T61	40.0	(276)	30.0	(207)	3.0	75–105
356.0	F	21.0	(145)	—	—	3.0	40–70
356.0	T51	25.0	(172)	—	—	—	55–85
356.0	T6	33.0	(228)	22.0	(152)	3.0	65–95
356.0	T7	25.0	(172)	—	—	3.0	60–90
356.0	T71	25.0	(172)	—	—	3.0	60–90
A356.0	T61	37.0	(255)	26.0	(179)	5.0	70–100
357.0	T6	45.0	(310)	—	—	3.0	75–105
A357.0	T61	45.0	(310)	36.0	(248)	3.0	85–115
359.0	T61	45.0	(310)	34.0	(234)	4.0	75–105
359.0	T62	47.0	(324)	38.0	(262)	3.0	85–115
443.0	F	21.0	(145)	7.0	(49)	2.0	30–60
B443.0	F	21.0	(145)	6.0	(41)	2.5	30–60
A444.0	T4	20.0	(138)	—	—	20.0	—
513.0	F	22.0	(152)	12.0	(83)	2.5	45–75
535.0	F	35.0	(241)	18.0	(124)	8.0	60–90
705.0	T5	37.0	(255)	17.0	(117)	10.0	55–85
707.0	T7	45.0	(310)	35.0	(241)	3.0	80–110
711.0	T1	28.0	(193)	18.0	(124)	7.0	55–85
713.0	T5	32.0	(221)	22.0	(152)	4.0	60–90
850.0	T5	18.0	(124)	—	—	8.0	30–60
851.0	T5	17.0	(117)	—	—	3.0	30–60
851.0	T6	18.0	(124)	—	—	8.0	—
852.0	T5	27.0	(186)	—	—	3.0	55–85

Source: Aluminum Association [12].

[a] Values represent properties obtained from separately cast test bars and are derived from ASTM B-108, Standard Specification for Aluminum-Alloy Permanent Mold Castings; Federal Specification QQ-A-596d, Aluminum Alloy Permanent and Semi-Permanent Mold Castings; and Military Specification MIL-A-21180c, Aluminum Alloy Castings, High Strength. Unless otherwise specified, the average tensile strength, average yield strength and average elongation values of specimens cut from castings shall be not less than 75 percent of the tensile and yield strength values and not less than 25 percent of the elongation values given above. The customer should keep in mind that (1) some foundries may offer additional tempers for the above alloys, and (2) foundries are constantly improving casting techniques and, as a result, some may offer minimum properties in excess of the above.

[b] F indicates "as cast" condition; refer to AA-CS-M11 for recommended times and temperatures of heat treatment for other tempers to achieve properties specified.

[c] Hardness values are given for information only; not required for acceptance.

the specimen, rather than the actual cross-sectional area under stress. The actual area is less than the original area since necking occurs after yielding; thus the engineering stress is slightly less than the actual stress.

When strengths are not available, relationships between the unknown strength and known properties may be used. The tensile ultimate strength (F_{tu}) is almost always known and the tensile yield strength (F_{ty}) is usually known, so other properties are related to these:

$$F_{cy} = 0.9F_{ty} \quad \text{(for cold-worked tempers)}$$

$$F_{cy} = F_{ty} \quad \text{(for heat-treatable alloys and annealed tempers)}$$

$$F_{sy} = 0.6F_{ty}$$

$$F_{su} = 0.6F_{tu}$$

These relationships are approximate but usually accurate enough for design purposes.

Tensile ultimate strengths vary widely among common alloys and tempers, from a minimum of 8 ksi (55 MPa) for 1060-O and 1350-O to a maximum of 84 ksi (580 MPa) for 7178-T62. For some tempers (usually the annealed temper) of certain alloys, strengths are also limited to a maximum value to ensure workability without cracking.

The strength of aluminum alloys is a function of temperature. Most alloys have a plateau of strength between roughly −150°F(−100°C) and 200°F (100°C), with higher strengths below this range and lower strengths above it. Ultimate strength increases 30–50% below this range, while the yield strength increase at low temperatures is not so dramatic, being on the order of 10%. Both ultimate and yield strengths drop rapidly above 200°F, dropping to nearly zero at 750°F (400°C). Some alloys (such as 2219) retain useful (albeit lower) strengths as high as 600°F (300°C). Figure 9.2 shows the effect of temperature on strength for various alloys.

Heating tempered alloys also has an effect on strength. Heating for a long enough period of time reduces the condition of the material to the annealed state, which is the weakest temper for the material. The higher the temperature, the briefer the period of time required to produce annealing. The length of time of high-temperature exposure causing no more than a 5% reduction in strength is given in Table 9.17 for 6061-T6. Since welding introduces heat to the parts being welded, welding reduces their strength. This effect is discussed in Section 9.4.1, and minimum reduced strengths for various alloys are given there.

Under a constant stress, the deformation of an aluminum part may increase over time, a behavior known as creep. Creep effects increase as the temperature increases. At room temperature, very little creep occurs unless stresses are near the tensile strength. Creep is usually not significant unless stresses are sustained at temperatures over about 200°F (95°C).

9.2.2.3 Modulus of Elasticity, Modulus of Rigidity, and Poisson's Ratio

The *modulus of elasticity (E)* (also called Young's modulus) is the slope of the stress–strain curve in its initial, elastic region before yielding occurs. The modulus is a measure of a material's stiffness (or resistance to elastic deformation) and its buckling strength, and varies slightly by alloy since it is a function of the alloying elements. It can be estimated by averaging the moduli of the alloying

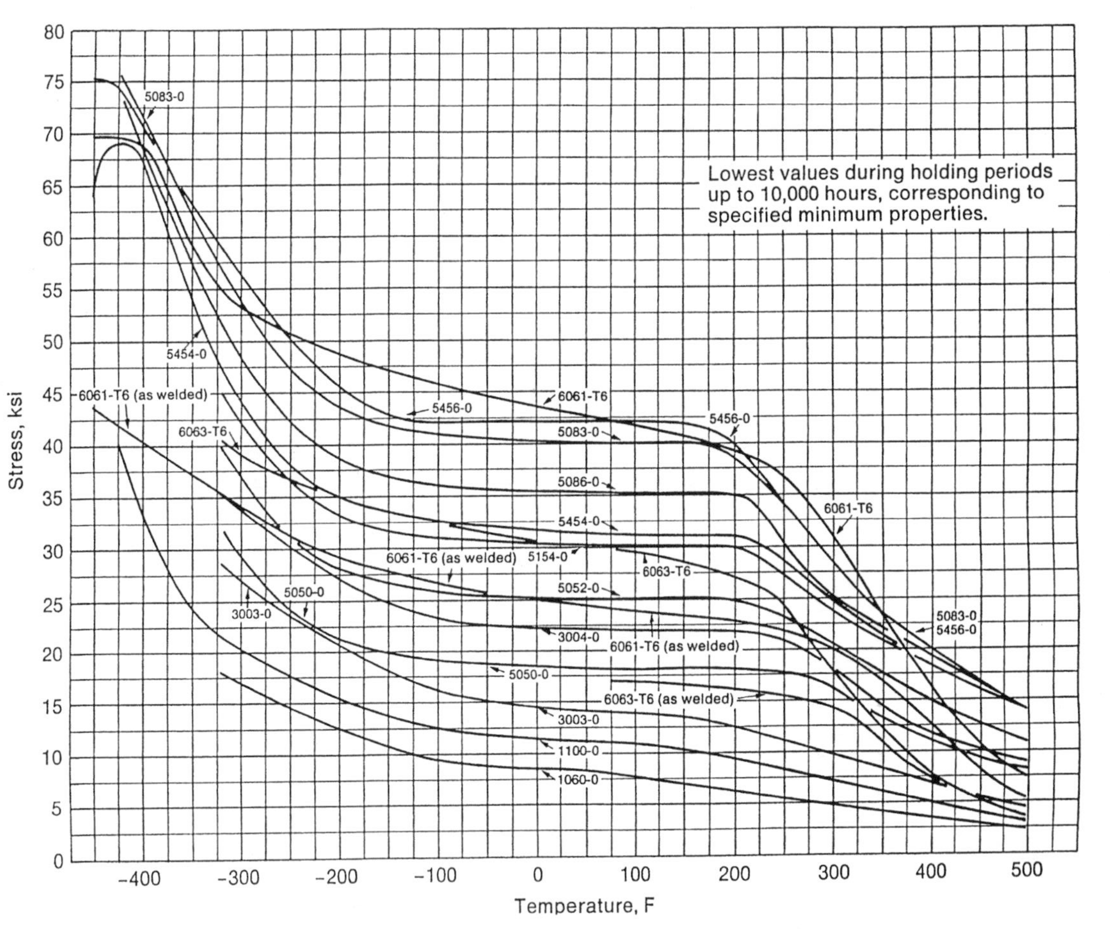

FIGURE 9.2 Typical tensile strengths of some aluminum alloys at various temperatures.

TABLE 9.17 Maximum Time at Elevated Temperatures—6061-T6[a]

Elevated Temperature		
°F	°C	Maximum Time
450	230	5 min
425	220	15 min
400	205	30 min
375	190	2 h
350	175	10 h
325	165	100 h

[a]Loss of strength will not exceed 5% at these times.

elements according to their proportion in the alloy, although magnesium and lithium tend to have a disproportionate effect. An approximate value of 10,000 ksi (69,000 MPa) is sometimes used, but moduli range from 10,000 ksi for pure aluminum (1xxx series), manganese (3xxx series), and magnesium–silicon alloys (6xxx series) to 10,800 ksi (75,000 MPa) for the aluminum–copper alloys and 11,200 ksi (77,200 MPa) for 8090, an aluminum–lithium alloy. Moduli of elasticity for various alloys are given in Table 9.14. This compares to 29,000 ksi (200,000 MPa) for steel alloys (about three times that of aluminum) and to 6500 ksi (45,000 MPa) for magnesium.

For aluminum, the tensile modulus is about 2% less than the compressive modulus. An average of tensile and compressive moduli is used to calculate bending deflections; the compressive modulus is used to calculate buckling strength.

Aluminum's modulus of elasticity is a function of temperature, increasing about 10% around −300°F(−200°C) and decreasing about 30% at 600°F (300°C).

At strains beyond yield, the slope of the stress–strain curve is called the tangent modulus and is a function of stress, decreasing as the stress increases. Values for the tangent modulus or the Ramberg–Osgood parameter n define the shape of the stress–strain curve in this inelastic region and are given in the *U.S. Military Handbook on Metallic Materials and Elements for Aerospace Structures* (MIL HDBK 5) [23] for many aluminum alloys. The Ramberg–Osgood equation is

$$\varepsilon = \frac{\sigma}{E} + 0.002\left(\frac{\sigma}{F_y}\right)^n,$$

where ε = strain
σ = stress
F_y = yield strength

The *modulus of rigidity (G)* is the ratio of shear stress in a torsion test to shear strain in the elastic range. The modulus of rigidity is also called the shear modulus. An average value for aluminum alloys is 3800 ksi (26,000 MPa).

Poisson's ratio (ν) is the negative of the ratio of transverse strain that accompanies longitudinal strain caused by axial load in the elastic range. Poisson's ratio is approximately 0.33 for aluminum alloys, similar to the ratio for steel. While the ratio varies slightly by alloy and decreases slightly as temperature decreases, such variations are insignificant for most applications. Poisson's ratio can be used to relate the modulus of rigidity (G) and the modulus of elasticity (E) through the formula

$$G = \frac{E}{2(1 + \nu)}.$$

9.2.2.4 Fracture Toughness and Elongation

Fracture toughness is a measure of a material's resistance to the extension of a crack. Aluminum has a face-centered cubic crystal structure and so does not exhibit a transition temperature (like steel) below which the material suffers a significant loss in fracture toughness. Furthermore, alloys of the 1xxx, 3xxx, 4xxx, 5xxx, and 6xxx series are so tough that their fracture toughness cannot be readily measured by the methods commonly used for less tough materials and is rarely of concern. Alloys of the 2xxx and 7xxx series are less tough and when they are used in fracture critical applications such as aircraft, their fracture toughness is of interest to the designer.

The plane strain fracture toughness (K_{Ic}) for some products of the 2xxx and 7xxx alloys can be measured by ASTM B645. For those products whose fracture toughness cannot be measured by this method (such as sheet, which is too thin for applying B645), nonplane strain fracture toughness (K_c) may be measured by ASTM B646. Fracture toughness limits established by the Aluminum Association are given in Table 9.18. Fracture toughness is a function of the orientation of the specimen and the notch relative to the part, and so toughness is identified by two letters: L for the length direction, T for the width (long transverse) direction, and S for the thickness (short transverse) direction. The first letter denotes the specimen direction perpendicular to the crack, and the second letter the direction of the notch.

Ductility, the ability of a material to absorb plastic strain before fracture, is related to elongation. *Elongation* is the percentage increase in the distance between two gage marks of a specimen tensile tested to fracture. All other things being equal, the greater the elongation, the greater the ductility. The elongation of aluminum alloys tends to be less than mild carbon steels; while A36 steel has a minimum elongation of 20%, the comparable aluminum alloy, 6061-T6, has a minimum elongation requirement of 8 or 10%, depending on the product form. An alloy that is not ductile may fracture at a lower tensile stress than its minimum ultimate tensile stress because it is unable to deform plastically at local stress concentrations. Instead, brittle fracture occurs at a stress raiser, leading to premature failure of the part.

The elongation of annealed tempers is greater than that of strain-hardened or heat-treated tempers, while the strength of annealed tempers is less. Therefore, annealed material is more workable and able to undergo more severe forming operations without cracking.

TABLE 9.18 Fracture Toughness Limits

Alloy and Temper	Thickness (in.)	K_{Ic}, ksi$\sqrt{\text{in.}}$ min		
		L-T	T-L	S-L
*Fracture Toughness Limits for Plate*①				
2124-T851	1.500–6.000	24	20	18
7050-T7451②③	1.000–2.000	29	25	—
	2.001–3.000	27	24	21
	3.001–4.000	26	23	21
	4.001–5.000	25	22	21
	5.001–6.000	24	22	21
7050-T7651②	1.000–2.000	26	24	—
	2.001–3.000	24	23	20
7475-T651	1.250–1.500	30	28	—
7475-T7351	1.250–2.499	40	33	—
	2.500–4.000	40	33	25
7475-T7651	1.250–1.500	33	30	—
*Fracture Toughness Limits for Sheet*④				
7475-T61	0.040–0.125	—	75	—
	0.126–0.249	60	60	—
7475-T761	0.040–0.125	—	87	—
	0.126–0.249	—	80	—

Source: Aluminum Association [5].

①When tested per ASTM Test Method E399 and ASTM Practice B645.

②Thickness for K_{Ic} specimens in the T-L and L-T test orientations: Up thru 2 in. (ordered, nominal thickness) use full thickness; over 2 thru 4 in. use 2-in. specimen thickness, centered at T/2; over 4 in. use 2-in specimen thickness centered at T/4. Test location for K_{Ic} specimens in the S-L test orientation: locate crack at T/2.

③T74 type tempers, although not previously registered, have appeared in the literature and in some specifications as T736 type tempers.

④When tested per ASTM Practice B646 and ASTM Practice E561.

Elongation values are affected by the thickness of the specimen, being higher for thicker specimens. For example, typical elongation values for 1100-O material are 35% for a $\frac{1}{16}$-in.-thick specimen, and 45% for a $\frac{1}{2}$-in.-diameter specimen. Elongation is also very much a function of temperature, being lowest at room temperature and increasing at both lower and higher temperatures.

9.2.2.5 Hardness

The hardness of aluminum alloys can be measured by several methods, including Webster hardness (ASTM B647), Barcol hardness (ASTM B648), Newage hardness (ASTM B724), and Rockwell hardness (ASTM E18). The Brinell hardness (ASTM E10) for a 500-kg load on a 10-mm ball is used most often and is given in Tables 9.14 and 9.16. Hardness measurements are sometimes used for quality assurance purposes on temper. The Brinell hardness number (BHN) is approximately related to minimum ultimate tensile strength: BHN $= 0.556F_{tu}$; this relationship can be useful to help identify material or estimate its strength based on a simple hardness test. The relationship between hardness and strength is not as dependable for aluminum as for steel, however.

9.2.2.6 Fatigue Strength

Tensile strengths established for metals are based on a single application of load at a rate slow enough to be considered static. The repeated application of loads causing tensile stress may result in fracture at a stress less than the static tensile strength. This behavior is called fatigue. The fatigue strength of aluminum alloys varies by alloy and temper, but this variation is more marked when the number of load cycles is small, which corresponds with high stress ranges (Fig. 9.3). When the number of load cycles is high, designers often consider fatigue strength to be independent of alloy and temper [28].

The fatigue strengths of the various aluminum alloys can be compared based on the endurance limits given in Table 9.14. These endurance limits are the stress range required to fail an R. R. Moore specimen in 500 million cycles of completely reversed stress. Endurance limits are not useful for designing components, however, because the conditions of the test by which endurance limits are established are rarely duplicated in actual applications. Also, endurance limit test specimens are small compared to actual components, and fatigue strength is a function of size, being lower for larger components. This is because fatigue failure initiates at local discontinuities such as scratches or weld inclusions and the probability that a discontinuity will be present is greater the larger the part.

Fatigue strength is strongly influenced by the number of cycles of load and the geometry of the part. Geometries such as connections that result in stress concentrations due to abrupt transitions such as sharp corners or holes have lower fatigue strengths than plain metal without such details. Therefore, for design purposes, applications are categorized by the severity of the detail, from A (being least severe, such as base metal in plain components) to F (being most severe, such as fillet weld metal). Design strengths in fatigue can be found in Table 9.19 by substituting parameters given there into the equation

$$S_{rd} = \frac{C_f}{N^{1/m}}$$

where S_{rd} = allowable stress range, which is the algebraic difference between the minimum and maximum stress (tension is positive, compression is negative)

C_f = constant from Table 9.19

N = number of cycles of load
m = constant from Table 9.19

This equation is set so that there is a 95% probability that 97.7% of components subjected to fatigue will be strong enough to withstand the stress range given by the equation [27].

This equation shows that fatigue strength decreases rapidly as the number of load cycles increases. For loads of constant amplitude, however, it is believed that the fatigue strength of aluminum alloys does not decrease once the number of cycles reaches approximately 5 million. The fatigue strength predicted by the above equation for $N = 5$ million is called the constant amplitude fatigue limit (CAFL, or simply fatigue limit), and is given in Table 9.19. Loads may also

TABLE 9.19 Fatigue Strengths of Aluminum Alloys

Category	C_f (ksi)	m	Fatigue Limit (ksi)	Category Examples
A	96.5	6.85	10.2	Plain metal
B	130	4.84	5.4	Members with groove welds parallel to the direction of stress
C	278	3.64	4.0	Groove welded transverse attachments with transition radius 24 in. $< R \leq 6$ in. (610 mm $< R \leq 150$ mm)
D	157	3.73	2.5	Groove welded transverse attachments with transition radius 6 in. $< R \leq 2$ in. (150 mm $< R \leq 50$ mm)
E	160	3.45	1.8	Base metal at fillet welds
F	174	3.42	1.9	Fillet weld metal

TABLE 9.20 Relative Hot Extrudability

Alloy	Extrudability (% of 6063)	Alloy	Extrudability (% of 6063)
1060	135	5456	20
1100	135	6061	60
1350	160	6063	100
2011	35	6066	40
2014	20	6101	100
2024	15	6151	70
3003	120	6253	80
5052	80	6351	60
5083	20	6463	100
5086	25	7001	7
5154	50	7075	10
5254	50	7079	10
5454	50	7178	7

TABLE 9.21 Cross-Sectional Dimension Tolerances—Profiles (Shapes)①

EXCEPT FOR T3510, T4510, T6510, T73510, T76510 AND T8510 TEMPERS⑦

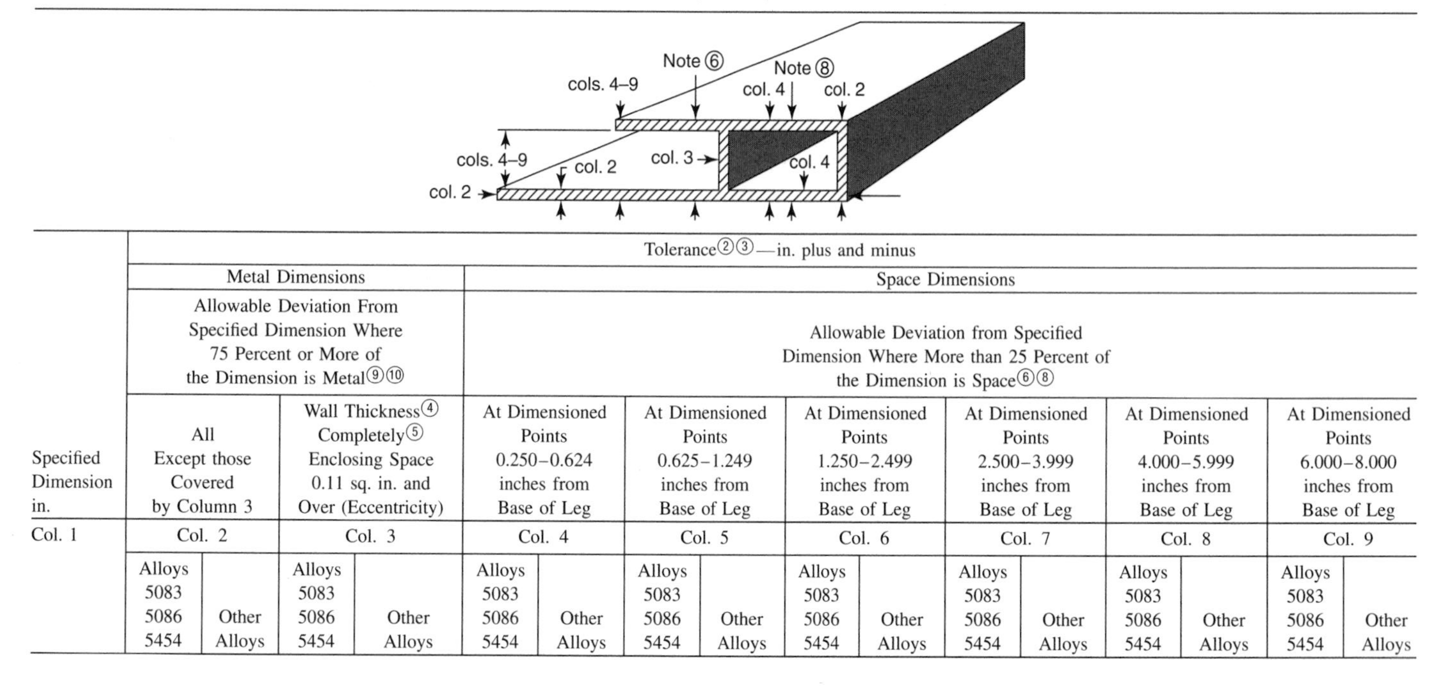

Specified Dimension in.	Tolerance②③—in. plus and minus															
	Metal Dimensions				Space Dimensions											
	Allowable Deviation From Specified Dimension Where 75 Percent or More of the Dimension is Metal⑨⑩				Allowable Deviation from Specified Dimension Where More than 25 Percent of the Dimension is Space⑥⑧											
	All Except those Covered by Column 3		Wall Thickness④ Completely⑤ Enclosing Space 0.11 sq. in. and Over (Eccentricity)		At Dimensioned Points 0.250–0.624 inches from Base of Leg		At Dimensioned Points 0.625–1.249 inches from Base of Leg		At Dimensioned Points 1.250–2.499 inches from Base of Leg		At Dimensioned Points 2.500–3.999 inches from Base of Leg		At Dimensioned Points 4.000–5.999 inches from Base of Leg		At Dimensioned Points 6.000–8.000 inches from Base of Leg	
Col. 1	Col. 2		Col. 3		Col. 4		Col. 5		Col. 6		Col. 7		Col. 8		Col. 9	
	Alloys 5083 5086 5454	Other Alloys	Alloys 5083 5086 5454	Other Alloys	Alloys 5083 5086 5454	Other Alloys	Alloys 5083 5086 5454	Other Alloys	Alloys 5083 5086 5454	Other Alloys	Alloys 5083 5086 5454	Other Alloys	Alloys 5083 5086 5454	Other Alloys	Alloys 5083 5086 5454	Other Alloys

Circumscribing Circle Sizes Less Than 10 in. in Diameter																
Up thru 0.124	0.009	0.006	±15% of specified dimension; ±0.090 max. ±0.015 min.	±10% of specified dimension; ±0.060 max. ±0.010 min.	0.013	0.010	0.015	0.012	—	—	—	—	—	—	—	—
0.125–0.249	0.011	0.007			0.016	0.012	0.018	0.014	0.020	0.016	—	—	—	—	—	—
0.250–0.499	0.012	0.008			0.018	0.014	0.020	0.016	0.022	0.018	0.024	0.020	—	—	—	—
0.500–0.749	0.014	0.009			0.021	0.016	0.023	0.018	0.025	0.020	0.027	0.022	—	—	—	—
0.750–0.999	0.015	0.010			0.023	0.018	0.025	0.020	0.027	0.022	0.030	0.025	0.035	0.030	—	—
1.000–1.499	0.018	0.012			0.027	0.021	0.029	0.023	0.032	0.026	0.036	0.030	0.041	0.035	—	—
1.500–1.999	0.021	0.014			0.031	0.024	0.033	0.026	0.038	0.031	0.043	0.036	0.049	0.042	0.057	0.050
2.000–3.999	0.036	0.024			0.046	0.034	0.050	0.038	0.060	0.048	0.069	0.057	0.080	0.068	0.092	0.080
4.000–5.999	0.051	0.034			0.061	0.044	0.067	0.050	0.081	0.064	0.095	0.078	0.111	0.094	0.127	0.110
6.000–7.999	0.066	0.044			0.076	0.054	0.084	0.062	0.104	0.082	0.121	0.099	0.142	0.120	0.162	0.140
8.000–9.999	0.081	0.054			0.091	0.064	0.101	0.074	0.127	0.100	0.147	0.120	0.182	0.145	0.197	0.170
Circumscribing Circle Sizes 10 in. in Diameter and Over																
Up thru 0.124	0.021	0.014	±15% of specified dimension; ±0.090 max. ±0.025 min.	±15% of specified dimension; ±0.090 max. ±0.015 min.	0.025	0.018	0.027	0.020	—	—	—	—	—	—	—	—
0.125–0.249	0.022	0.015			0.026	0.019	0.029	0.022	0.035	0.028	—	—	—	—	—	—
0.250–0.499	0.024	0.016			0.028	0.020	0.032	0.024	0.038	0.030	0.058	0.050	—	—	—	—
0.500–0.749	0.025	0.017			0.030	0.022	0.035	0.027	0.049	0.040	0.068	0.060	—	—	—	—
0.750–0.999	0.027	0.018			0.031	0.023	0.039	0.030	0.057	0.050	0.079	0.070	0.099	0.090	—	—
1.000–1.499	0.028	0.019			0.033	0.024	0.043	0.034	0.069	0.060	0.089	0.080	0.109	0.100	—	—
1.500–1.999	0.036	0.024			0.046	0.034	0.056	0.044	0.082	0.070	0.102	0.090	0.122	0.110	0.182	0.170
2.000–3.999	0.051	0.034			0.061	0.044	0.071	0.054	0.097	0.080	0.117	0.100	0.137	0.120	0.197	0.180
4.000–5.999	0.066	0.044			0.076	0.054	0.086	0.064	0.112	0.090	0.132	0.110	0.152	0.130	0.212	0.190
6.000–7.999	0.081	0.054			0.091	0.064	0.101	0.074	0.127	0.100	0.147	0.120	0.167	0.140	0.227	0.200
8.000–9.999	0.096	0.064			0.106	0.074	0.116	0.084	0.142	0.110	0.162	0.130	0.182	0.150	0.242	0.210
10.000–11.999	0.111	0.074			0.121	0.084	0.131	0.094	0.157	0.120	0.177	0.140	0.197	0.160	0.257	0.220
12.000–13.999	0.126	0.084			0.136	0.094	0.146	0.104	0.172	0.130	0.192	0.150	0.212	0.170	0.272	0.230
14.000–15.999	0.141	0.094			0.151	0.104	0.161	0.114	0.187	0.140	0.207	0.160	0.227	0.180	0.287	0.240
16.000–17.999	0.156	0.104			0.166	0.114	0.176	0.124	0.202	0.150	0.222	0.170	0.242	0.190	0.302	0.250
18.000–19.999	0.171	0.114			0.181	0.124	0.191	0.134	0.217	0.160	0.237	0.180	0.257	0.200	0.317	0.260
20.000–21.999	0.186	0.124			0.196	0.134	0.206	0.144	0.232	0.170	0.252	0.190	0.272	0.210	0.332	0.270
22.000–24.000	0.201	0.134			0.211	0.144	0.221	0.154	0.247	0.180	0.267	0.200	0.287	0.220	0.347	0.280

TABLE 9.21 (*Continued*)

①These Standard Tolerances are applicable to the average profile (shape); wider tolerances may be required for some profiles (shapes) and closer tolerances may be possible for others.
②The tolerance applicable to a dimension composed of two or more component dimensions is the sum of the tolerances of the component dimensions if all of the component dimensions are indicated.
③When a dimension tolerance is specified other than as an equal bilateral tolerance, the value of the standard tolerance is that which applies to the mean of the maximum and minimum dimensions permissible under the tolerance for the dimension under consideration.
④Where dimensions specified are outside and inside, rather than wall thickness itself, the allowable deviation (eccentricity) given in Column 3 applies to mean wall thickness. (Mean wall thickness is the average of two wall thickness measurements taken at opposite sides of the void.)
⑤In the case of Class 1 Hollow Profiles (Shapes) the standard wall thickness tolerance for extruded round tube is applicable. (A Class 1 Hollow Profile (Shape) is one whose void is round and one inch or more in diameter and whose weight is equally distributed on opposite sides of two or more equally spaced axes.)
○6At points less than 0.250 inch from base of leg the tolerances in Col. 2 are applicable.
○7Tolerances for extruded profiles (shapes) in T3510, T4510, T6510, T73510, T76510 and T8510 tempers shall be as agreed upon between purchaser and vendor at the time the contract or order is entered.
○8The following tolerances apply where the space is completely enclosed (hollow profiles (shapes)); For the width (*A*), the balance is the value shown in Col. 4 for the depth dimension (*D*). For the depth (*D*), the tolerance is the value shown in Col. 4 for the width dimension (*A*). In no case is the tolerance for either width or depth less than the metal dimensions (Col. 2) at the corners.

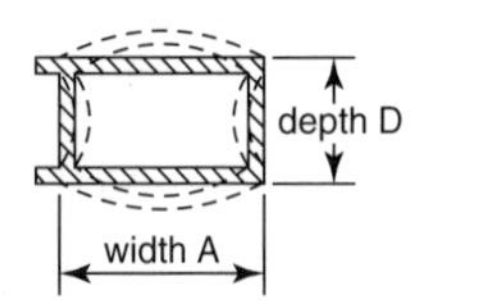

Example–Alloy 6061 hollow profile (shape) having 1×3 rectangular outside dimensions; width tolerance is ± 0.021 inch and depth tolerance ± 0.034 inch. (Tolerances at corners, Col. 2, metal dimensions, are ± 0.024 inch for the width and ± 0.012 inch for the depth.) Note that the Col. 4 tolerance of 0.021 inch must be adjusted to 0.024 inch so that it is not less than the Col. 2 tolerance.
○9These tolerances do not apply to space dimensions such as dimensions *X* and *Z* of the example (right), even when *Y* is 75 percent or more of *X*. For the tolerance applicable to dimensions *X* and *Z*, use Col. 4, 5, 6, 7, 8 or 9, dependent on distance *A*.

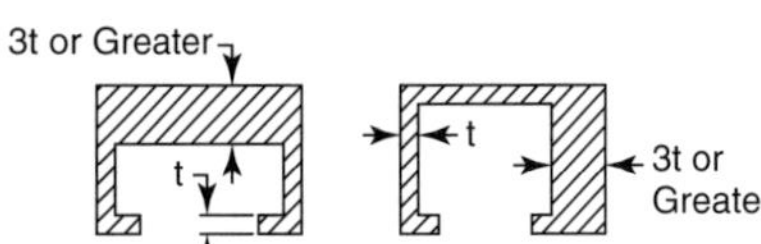

○10The wall thickness tolerance for hollow or semihollow profiles (shapes) shall be as agreed upon between purchaser and vendor at the time the contract or order is entered when the nominal thickness of one wall is three times or greater than that of the opposite wall.

have variable amplitudes, such as the loads on a beam in a bridge carrying traffic composed of cars and trucks of various weights. For variable amplitude loads, no lower bound on the fatigue strength is believed to exist, but some design codes use one half of the constant amplitude fatigue limit as the fatigue limit for variable amplitude loading.

Fatigue strengths of aluminum alloys are 30–40% those of steel under similar circumstances of loading and severity of the detail.

Fatigue is also affected by environmental conditions. The fatigue strength of aluminum in corrosive environments such as salt spray can be considerably less than the fatigue strength in laboratory air. This may be because corrosion sites such as pits act as points of initiation for cracks, much like flaws such as dents or scratches. The more corrosion-resistant alloys of the 5xxx and 6xxx series suffer less reduction in fatigue strength in corrosive environments than the less corrosion-resistant alloys such as those of the 2xxx and 7xxx series. On the other hand, fatigue strengths are higher at cryogenic temperature than at room temperature. There isn't enough data on these effects to establish design rules, so designers must test specific applications to determine the magnitude of environmental factors on fatigue strength.

The fatigue strength of castings is less than that of wrought products, and no fatigue design strengths are available for castings.

9.2.3 Property Ratings

Ratings for properties such as corrosion resistance, weldability, and machinability are given in tables for wrought (Table 9.20) and cast alloys (Table 9.21). See Chapter 12 for more information about corrosion.

9.3 PRODUCTS

Ingot (large, unfinished bars of aluminum) is the result of primary aluminum production. This material is not in a useful form for purchasers, so it is wrought into semifabricated mill products that include flat rolled products (foil, sheet, or plate, depending on thickness), rolled elongated products (wire, rod, or bar, depending on dimensions), drawn tube, extrusions, and forgings. Ingot is also furnished to foundries that remelt it to produce castings.

9.3.1 Wrought Products

Aluminum mills produce wrought products. ASTM specifications provide minimum requirements for wrought aluminum alloys by product (e.g., sheet and plate) rather than by alloy. Each ASTM specification addresses all the alloys that may be used to make that product. For example, ASTM B209 *Sheet and Plate* includes 3003, 6061, and all other alloys used to make sheet.

Tolerances on the dimensional, mechanical, and other properties of wrought alloys are given in the Aluminum Association's *Aluminum Standards and Data* [5] and ANSI H35.2 *Dimensional Tolerances for Aluminum Mill Products* [16]. The two are the same, although they are revised at different dates and so may not match exactly until the next revision. The tolerances are standard tolerances (also called commercial or published tolerances). Special (either more strict or less strict than standard) tolerances may be met by agreement between the purchaser and the supplier. Tolerances approximately one-half of standard tolerances can usually be met if the purchaser so specifies.

9.3.1.1 Flat Rolled Products (Foil, Sheet, and Plate)

Flat rolled products are produced in rolling mills, where cylindrical rolls reduce the thickness and increase the length of ingot. The process begins with huge heated ingots 6 ft wide, 20 ft long, and more than 2 ft thick and weighing over 20 tons that are rolled back and forth in a breakdown mill to reduce the thickness to a few inches. Plate of this thickness can be heated, quenched, stretched (to straighten and relieve residual stresses), aged at temperature, and shipped, or may be coiled and sent on to a continuous mill to further reduce its thickness. Before further rolling at the continuous mill, the material is heated to soften it for cold rolling. Heat treatments and stretching are also applied there after rolling to the desired thickness. Alternatively, sheet can be produced directly from molten metal rather than ingot in the continuous casting process, by which molten aluminum passes through water-cooled casting rollers and is solidified. The thickness is then further reduced by cold rolling.

The resulting flat rolled products are rectangular in cross section and are called foil, sheet, or plate, depending on their thickness (Table 9.22). *Foil* has a thickness less than 0.006 in. (up through 0.15 mm). *Sheet* has a thickness less than 0.25 in., but not less than 0.006 in. (over 0.15–6.3 mm). *Plate* is 0.25 in. (over 6.3 mm) thick or more. The use of the term "flat" in flat rolled products is to distinguish these from rolled wire, rod, or bar, which are discussed in Section 9.3.1.3.

Foil is produced as thin as 0.00017 in. thick, in alloys 1100, 1145, 1235, 2024, 3003, 5052, 5056, 6061, 8079, and 8111, in both rolls and sheets, and in the annealed and H19 tempers. The H19 variety is called hard foil because it is fully strain hardened. Uses for foil include the cores of aluminum honeycomb panels used in aircraft, capacitors (ASTM B373 *Foil for Capacitors*), and for packaging

TABLE 9.22 Relative Cold Extrudability of Annealed Alloys

Alloy	Relative Cold Extrusion Pressure
1100	1.0
3003	1.2
6061	1.6
2014	1.8
7075	2.3

(ASTM B479 *Foil for Flexible Barrier Applications*). Standard household foil is approximately 0.0006 in. thick, or about $\frac{1}{4}$ the thickness of the paper used in this book, while capacitor foil is four times thinner. In many packaging applications, foil is laminated to paper or plastic films for strength.

Fin stock is coiled sheet or foil in specific alloys (1100, 1145, 3003, and 7072), tempers, and thicknesses (0.004–0.030 in.) used for the manufacture of fins for heat exchangers.

Sheet is one of the most widely used aluminum products and is produced in more alloys than any other aluminum product. Sheet is available rolled into coils with slit edges or in flat sheets with sheared, slit, or sawed edges. Circular blanks from coil or flat sheet are available. Panel flat sheet is flat sheet with a tighter tolerance on flatness than flat sheet. The maximum weight of a coil is about 9900 lb (4500 kg). Coils are available in standard widths such as 24, 30, 36, 48, 60, and 72 in., and up to about 108 in. (2740 mm). Commonly available widths for flat sheet are 96, 120, and 144 in. Aluminum sheet gages are different from steel sheet gages, so it's better to identify aluminum sheet by thickness rather than gage number to avoid confusion.

Plate is available in thicknesses up to about 8 in. (200 mm) in certain alloys; single plates may weigh up to 7900 lb (3600 kg). Circular plate blanks are available. Common widths and lengths for plate are similar to those for sheet.

Sheet and plate tolerances have been established for width, length, deviation of edges from straight, squareness, and flatness. Thickness tolerances are different for alloys specified for aerospace applications (including 2014, 2024, 2124, 2219, 2324, 2419, 7050, 7150, 7178, and 7475) than for other alloys. The ASTM specification for sheet and plate is B209, *Sheet and Plate*.

A number of sheet and plate products are made for specific applications by the primary producers. These include tread plate (ASTM B632 *Rolled Tread Plate*) made with a raised diamond pattern on one side to provide improve traction, duct sheet, and roofing and siding, available in corrugated and other shapes.

9.3.1.2 Extrusions

Extrusions are products formed by pushing heated metal, in a log-shaped form called a billet, through an opening called a die, the outline of which defines the cross-sectional shape of the extrusion. Some examples are shown in Fig. 9.3. Thousands of pounds of pressure are exerted by the extrusion press as it forces the aluminum through the die and onto a runoff table, where the extrusion is straightened by stretching and cut to length. Artificial aging heat treatment may then be applied to extrusions made of heat-treatable alloys.

Extruding was developed in the 1920s and replaced rolling around 1950 for producing standard shapes such as I beams and angles. Today it is used for virtually all aluminum shapes. Shapes, called profiles in the aluminum industry, are all products that are long relative to their cross-sectional dimensions and are not sheet, plate, tube, wire, rod, or bar. Commonly available shapes include I beams, tees, zees, channels, and angles; a great variety of custom shapes can be produced with modest die costs and short lead times. In the late 1960s, the

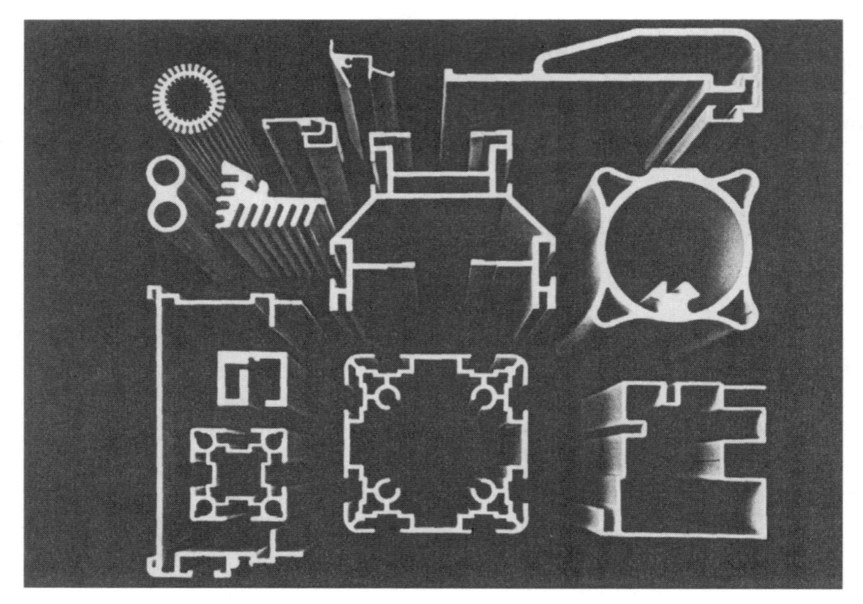

FIGURE 9.3 Extruded shapes.

Aluminum Association developed standard I beam and channel sizes that are structurally efficient and have flat (as opposed to tapered) flanges for convenient connections (ASTM B308 *6061-T6 Standard Structural Shapes*).

Profiles are sized by the smallest diameter circle that encloses their cross section, called the circumscribing circle size. The size of an extruded shape is limited by the capacity and size of the extrusion press used to produce the shape. Presses that are used for common applications are usually limited to those profiles that fit within about a 15-in. (375 mm) diameter circle; presses for military and aerospace applications can handle shapes as large as 31 in. (790 mm) in circle size. Standard I beams range up to 12 in. (305 mm) in depth because the entire shape can fit in a 15-in.-diameter circle. Standard channels are as large as 15 in. deep.

Some alloys are more difficult to extrude than others; generally, the stronger the alloy, the more difficult it is to extrude. The relative extrudability of some alloys is given in Table 9.20. Alloy 6063 is the benchmark for extrudability because it's easy to extrude and widely used; about 75% of all extrusions are 6xxx series alloys. The 1xxx, 3xxx, and 6xxx series alloys are called soft alloys, while the 2xxx, 5xxx, and 7xxx are called hard alloys, based on the relative difficulty of extruding.

Hollow shapes are extruded with hollow billets or with solid billets through porthole or bridge dies. The billet metal must divide and flow around the support for the die outlining the inside surface of the hollow extrusion and then weld itself back together as it exits the die. Such dies are more expensive than those for solid shapes. Hollow shapes produced with porthole or bridge dies are not considered seamless and are not used for parts designed to hold internal pressure. (See Section 9.3.1.4 for information on tubes used to contain internal pressure.)

A profile's shape factor (the ratio of the perimeter of the profile to its area) is an approximate indication of its extrudability. The higher the ratio, the more difficult the profile is to extrude. In spite of this, profiles that are thin and wide are extrudable, but the trade-off between the higher cost of extruding and the additional cost of using a thicker section should be considered.

Sometimes a thicker part is less expensive in spite of its greater weight because it is more readily extruded. Parts as thin as 0.04 in. (1 mm) can be extruded, but as the circle size increases, the minimum extrudable thickness also increases.

Other factors also affect cost. Unsymmetric shapes are more difficult to extrude, as are shapes with sharp corners. Providing a generous fillet or rounded corner decreases cost. Profiles with large differences in thickness across the cross section or with fine details are also more difficult to extrude. Consulting extruders when designing a shape can help reduce costs [14].

Metal flows fastest at the center of the die, so as shapes become larger, it becomes more difficult to design and construct dies that keep the metal flowing at a uniform rate across the cross section. For this reason, larger shapes have larger dimensional tolerances. Tolerances on cross-section dimensions, length, straightness, twist, flatness, and other parameters are given in *Aluminum Standards and Data* [5]. Tolerances on cross-sectional dimensions are also given here in Table 9.21.

Cold extrusions (also called impact extrusions or impacts) are also produced, an effective method for making tubes or cuplike pieces that are hollow with one end partially or totally closed [22]. The metal is formed at room temperature, and any heating of the metal is a consequence of the conversion of deformation energy to heat. The slug or preform is struck by a punch and deformed into the shape of a die, resulting in a wrought product with tight tolerances and zero draft and no parting lines. The five most commonly cold-extruded alloys and their relative cold extrudability are shown in Table 9.22. An example of a cold extrusion is irrigation tubing, which can be produced up to 6 in. in diameter, 0.058-in. wall thickness, and 40 ft in length.

9.3.1.3 Wire, Rod, and Bar

Wire, rod, and bar are defined as products that have much greater lengths than cross-section dimensions (ASTM B211 *Bar, Rod, and Wire*). *Wire* is rectangular (with or without rounded corners), round, or a regular hexagon or octagon in cross section and with one perpendicular distance between parallel faces less than 0.375 in. (10 mm or less). Material that has such a dimension of 0.375 in. or greater (greater than 10 mm) is *bar*, if the section is rectangular or a regular hexagon or octagon, and *rod*, if the section is round (see Table 9.23).

Rod and bar can be produced by hot rolling long, square ingot or hot extruding; the product may also be subsequently cold finished by drawing through a die. Wire can be hot extruded or drawn (pulled through a die or series of dies that

TABLE 9.23 Wire, Rod, and Bar

Width or Diameter	Square, Rectangular, Hexagon, or Octagon	Circular
<0.375 in. (≤10 mm)	Wire	Wire
≥0.375 in. (>10 mm)	Bar	Rod

TABLE 9.24 Minimum Strengths of Rivet and Cold Heading Wire and Rod

		Tensile Strength		
Alloy–Temper	Diameter (in.)	Ultimate (ksi)	Yield (ksi)	Shear Ultimate Strength (ksi)
2017-T4	0.063 to 1.000	55	32	33
2024-T42	0.063 to 0.124	62	—	37
2024-T42	0.125 to 1.000	62	40	37
2117-T4	0.063 to 1.000	38	18	26
2219-T6	0.063 to 1.000	55	35	30
6053-T61	0.063 to 1.000	30	20	20
6061-T6	0.063 to 1.000	42	35	25
7050-T7	0.063 to 1.000	70	58	39
7075-T6	0.063 to 1.000	77	66	42
7075-T73	0.063 to 1.000	68	56	41
7178-T6	0.063 to 1.000	84	73	46
7277-T62	0.500 to 1.250	60	—	35

define its cross-sectional shape) or flattened by roll-flattening round wire into a rectangular shape with rounded corners. Drawing and cold finishing result in much tighter tolerances on thickness than rolling, which in turn produces more precise dimensions than extruding (ASTM B221, *Extruded Bars, Rods, Wire, Shapes, and Tubes*).

The round products (wire and rod), when produced for subsequent forming into fasteners such as rivets or bolts, are called *rivet and cold-heading wire and rod* (ASTM B316). Rivet and cold-heading wire and rod is produced in the alloys and with the strengths shown in Table 9.24; these alloys have good machinability for threading. Rivet and cold heading wire and rod, and the fasteners produced from it, shall upon proper heat treatment be capable of developing the properties presented in Table 9.24.

9.3.1.4 Tubes

Tube is a product that is hollow and long in relation to its cross section, which may be round, a regular hexagon, a regular octagon, an ellipse, or a rectangle, and has uniform wall thickness. Seamless tubes, common in pressure applications, are made without a metallurgical weld resulting from the method of manufacture; such tubes can be produced from a hollow ingot or by piercing a solid ingot. Pipe is tube that is made in standardized diameter and wall thickness combinations. Tube is produced by several different methods.

Drawn tube is made by pulling material through a die (ASTM B210, *Drawn Seamless Tubes* and B483, *Drawn Tubes for General Purpose Applications*). Drawn tube is available in straight lengths or coils, but coils are generally available only as round tubes with a wall thickness of 0.083 in. (2 mm) or less and only in non-heat-treatable alloys. Drawn seamless tubes are used in surface condensers, evaporators, and heat exchangers, in wall thicknesses up to 0.200 in. and diameters up to 2.00 in., and in alloys 1060, 3003, 5052, 5454, and 6061 (ASTM B234, *Drawn Seamless Tubes for Condensers and Heat Exchangers* and B404, *Seamless Condenser and Heat-Exchanger Tubes with Integral Fins*). Heat exchanger tube is very workable and is tested for leak tightness and marked "HE."

Welded tube is produced from sheet or plate that is rolled into a circular shape and then longitudinally welded by gas tungsten or gas metal arc welding. Tube from ASTM B547, *Formed and Arc-Welded Round Tube*, is available in diameters from 9 to 60 in. (230 to 1520 mm) in wall thicknesses from 0.125 to 0.500 in. (3.15 to 12.5 mm). Tube from ASTM B313, *Round Welded Tubes*, are made in wall thicknesses from 0.032 in. (0.80 mm) to 0.125 in. (3.20 mm).

Extruded tube (ASTM B241, *Seamless Pipe and Seamless Extruded Tube*, B345, *Seamless Pipe and Seamless Extruded Tube for Gas and Oil Transmission and Distribution Piping Systems*, B429, *Extruded Structural Pipe and Tube*, and B491, *Extruded Round Tubes for General Purpose Applications*) is made by the extrusion process, discussed above. Tube may be extruded and then drawn to minimize ovality, a process sometimes called sizing.

9.3.1.5 Forgings

Forgings are one of the oldest wrought products, since they can be produced by simply hammering a hot lump of metal into the desired shape. A hammer, hydraulic press, mechanical press, upsetter, or ring roller is used to form the metal. Both castings and forgings can be used to produce parts with complex shapes; forgings are more expensive than castings but have more uniform properties and better ductility [3].

There are two types of forgings: open die forgings and closed die forgings (Table 9.25). *Open die forgings* (also called hand forgings) are produced without lateral confinement of the material during the forging operation. Minimum mechanical properties are not guaranteed for open die forgings unless specified by the customer, so they don't tend to be used for applications where structural integrity is critical.

Closed die forgings (also called die forgings) are more common and are produced by pressing the forging stock (made of ingot, plate, or extrusion) between a counterpart set of dies. Popular uses of closed die forgings are automotive and aerospace applications; they have been made up to 23 ft (7 m) long and 3100 lb (1400 kg) in weight. Die forgings are divided into four categories described below, from the least intricate, lowest quantity forgings, and lowest cost to the

TABLE 9.25 Mechanical Property Limits—Die Forgings⑥

Alloy and Temper	Specified Thickness	Specimen Axis Parallel to Direction of Grain Flow				Specimen Axis Parallel Not to Direction of Grain Flow			Brinell Hardness⑥
		Tensile Strength (ksi min.)		Elongation (percent min. in 2 in. or 4D)③		Tensile Strength (ksi min.)		Elongation (percent min. in 2 in. or 4D)③	500-kg Load—10 mm ball min.
		Ultimate	Yield	Coupon	Forging	Ultimate	Yield	Forging	
1100-H112④	Up thru 4.000	11.0	4.0	25	18	—	—	—	20
2014-T4	Up thru 4.000	55.0	30.0	16	11	—	—	—	100
2014-T6	Up thru 1.000	65.0	56.0	8	6	64.0	55.0	3	125
	1.001–2.000	65.0	56.0	①	6	64.0	55.0	2	125
	2.001–3.000	65.0	55.0	①	6	63.0	54.0	2	125
	3.001–4.000	63.0	55.0	①	6	63.0	54.0	2	125
2018-T61	Up thru 4.000	55.0	40.0	10	7	—	—	—	100
2025-T6	Up thru 4.000	52.0	33.0	16	11	—	—	—	100
2218-T61	Up thru 4.000	55.0	40.0	10	7	—	—	—	100
2218-T72	Up thru 4.000	38.0	29.0	8	5	—	—	—	85
2219-T6	Up thru 4.000	58.0	38.0	10	8	56.0	36.0	4	100
2618-T61	Up thru 4.000	58.0	45.0	6	4	55.0	42.0	4	115
3003-H112④	Up thru 4.000	14.0	5.0	25	18	—	—	—	25
4032-T6	Up thru 4.000	52.0	42.0	5	3	—	—	—	115
5083-H111④	Up thru 4.000	42.0	22.0	—	14	39.0	20.0	12	—
5083-H112④	Up thru 4.000	40.0	18.0	—	16	39.0	16.0	14	—
5456-H112	Up thru 4.000	44.0	20.0	—	16	—	—	—	—

6053-T6	Up thru 4.000	36.0	30.0	16	11	—	—	—	75
6061-T6	Up thru 4.000	38.0	35.0	10	7	38.0	35.0	5	80
6066-T6	Up thru 4.000	50.0	45.0	12	8	—	—	—	100
6151-T6	Up thru 4.000	44.0	37.0	14	10	44.0	37.0	6	90
7049-T73⑦	Up thru 1.000	72.0	62.0	10	7	71.0	61.0	3	135
	1.001–2.000	72.0	62.0	10	7	70.0	60.0	3	135
	2.001–3.000	71.0	61.0	10	7	70.0	60.0	3	135
	3.001–4.000	71.0	61.0	10	7	70.0	60.0	2	135
	4.001–5.000	70.0	60.0	10	7	68.0	58.0	2	135
7050-T74⑧⑨	Up thru 2.000	72.0	62.0	—	7	68.0	56.0	5	—
	2.001–4.000	71.0	61.0	—	7	67.0	55.0	4	—
	4.001–5.000	70.0	60.0	—	7	66.0	54.0	3	—
	5.001–6.000	71.0	59.0	—	7	66.0	54.0	3	—
7075-T6	Up thru 1.000	75.0	64.0	10	7	71.0	61.0	3	135
	1.001–2.000	74.0	63.0	①	7	71.0	61.0	3	135
	2.001–3.000	74.0	63.0	①	7	70.0	60.0	3	135
	3.001–4.000	73.0	62.0	①	7	70.0	60.0	2	135
7075-T73⑦	Up thru 3.000	66.0	56.0	—	7	62.0	53.0	3	125
	3.001–4.000	64.0	55.0	—	7	61.0	52.0	2	125
7075-T7352⑦	Up thru 3.000	66.0	56.0	—	7	62.0	51.0	3	125
	3.001–4.000	64.0	53.0	—	7	61.0	49.0	2	125
7175-T74⑧⑩	Up thru 3.000	76.0	66.0	—	7	71.0	62.0	4	—
7175-T7452⑧⑩	Up thru 3.000	73.0	63.0	—	7	68.0	55.0	4	—
7175-T7454⑧⑩	Up thru 3.000	75.0	65.0	—	7	70.0	61.0	4	—

TABLE 9.25 (*Continued*)

Source: Aluminum Association [5].

① When separately forged coupons are used to verify acceptability of forgings in the indicated thicknesses, the properties shown for thicknesses "Up thru 1 inch," including the test coupon elongation, apply.

② As-forged thickness. When forgings are machined prior to heat treatment, the properties will also apply to the machined heat treat thickness, provided the machined thickness is not less than one-half the original (as-forged thickness).

③ D equals specimen diameter.

④ Properties of H111 and H112 temper forgings are dependent on the equivalent cold work in the forgings. The properties listed should be attainable in any forging within the prescribed thickness range and may be considerably exceeded in some cases.

⑤ For information only: The Brinell Hardness is usually measured on the surface of a heat-treated forging using a 500 kg load and a 10-mm penetrator ball.

⑥ The database and criteria upon which these mechanical property limits are established are outlined elsewhere.

⑦ Material in this temper, 0.750 in. and thicker, when tested in accordance with ASTM G47 in the short transverse direction at a stress level of 75 percent of the specified minimum yield strength, will exhibit no evidence of stress corrosion cracking. Capability of individual lots to resist stress corrosion is determined by testing the previously selected tensile test sample in accordance with the applicable lot acceptance criteria outlined elsewhere.

⑧ T74 type tempers, although not previously registered, have appeared in the literature and in some specifications as T736 type tempers.

⑨ Material in this temper when tested at any plane in accordance with ASTM G34-72 will exhibit exfoliation less than that shown in Category B, Figure 2 of ASTM G34-72. Also, material, 0.750 in. and thicker, when tested in accordance with ASTM G47 in the short transverse direction at a stress level of 35 ksi, will exhibit no evidence of stress corrosion cracking. Capability of individual lots to resist exfoliation corrosion and stress corrosion cracking is determined by testing the previously selected tensile test sample in accordance with the applicable lot acceptance criteria outlined elsewhere.

⑩ Material in this temper, 0.750 in. and thicker, when tested in accordance with ASTM G47 in the short transverse direction at a stress level of 35 Ksi, will exhibit no evidence of stress corrosion cracking. Capability of individual lots to resist stress corrosion is determined by testing the previously selected tensile test sample in accordance with the applicable lot acceptance criteria outlined elsewhere.

most sharply detailed, highest quantity type, and highest cost. Less intricate forgings are used when quantities are small because it is more economical to incur machining costs on each of a few pieces than to incur higher one-time die costs. The most economical forging for a particular application depends on the dimensional tolerances and quantities required.

Blocker-type forgings have large fillet and corner radii and thick webs and ribs so that only one set of dies is needed; generally two squeezes of the dies are applied to the stock. Fillets are about two times the radius of conventional forgings, and corner radii about 1.5 times that of conventional forgings. Usually, all surfaces must be machined after forging. Blocker-type forgings may be selected if tolerances are so tight that machining would be required in any event, or if the quantity to be produced is small (typically up to 200 units). Blocker-type forgings can range in size from small to very large.

Finish only forgings also use only one set of dies, like blocker-type forgings, but typically one more squeeze than blocker-type forgings is applied to the part. Because of the additional squeezes, the die experiences more wear than for other forging types, but the part can be forged with tighter tolerances and reduced fillet and corner radii and web thickness. Fillets are about 1.5 times the radius of conventional forgings, and corner radii about the same as that of conventional forgings. The average production quantity for finish only forgings is 500 units.

Conventional forgings are the most common of all die forging types. Conventional forgings require two to four sets of dies; the first set produces a blocker forging that is subsequently forged in finishing dies. Fillet and corner radii and web and rib thicknesses are smaller than for blocker-type or finish-only forgings. Average production quantities are 500 or more.

Precision forgings, as the name implies, are made to closer than standard tolerances and include forgings with smaller fillet and corner radii and thinner webs and ribs.

There are other ways to categorize forgings. *Can* and *tube forgings* are cylindrical shapes that are open at one or both ends; these are also called extruded forgings. The walls may have longitudinal ribs or be flanged at one open end. *No-draft forgings* require no slope on vertical walls and are the most difficult to make. *Rolled ring forgings* are short cylinders circumferentially rolled from a hollow section.

Die forging alloys and their mechanical properties are listed in Table 9.25. Alloys 2014, 2219, 2618, 5083, 6061, 7050, 7075, and 7178 are the most commonly used. The ASTM specification for forgings is B247 *Die Forgings, Hand Forgings, and Rolled Ring Forgings*.

9.3.1.6 Electrical Conductors

Aluminum is used as a conductor because of its excellent electrical conductivity. Alloys 1350, [formerly known as EC (electrical conductor) grade] 5005, 6201, 8017, 8030, 8176, and 8177 are used in the form of wire, and alloys 1350 and 6101 are produced as bus bar (ASTM B317 *Extruded Bar, Rod, Pipe, and Structural Shapes for Electrical Purposes (Bus Conductors)* and ASTM B236 *Bars for Electrical Purposes (Bus Bars)*), made by extruding, rolling, or sawing from plate or sheet. The minimum conductivity of aluminum conductors is about 60% of the International Annealed Copper Standard (IACS).

In power transmission lines, the necessary strength for long spans is obtained by stranding aluminum wire around a high-strength galvanized or aluminized steel core. This product is called aluminum conductor, steel reinforced (ACSR). The resulting strength-to-weight ratio is about twice that of copper of equal conductivity.

9.3.2 Cast Products

The first aluminum products, including the first commercial application (a tea kettle), were castings, made by pouring molten aluminum into a mold. They are useful for making complex shapes and are produced by a number of methods [26]. Common methods and their ASTM specifications are:

B26 *Sand Castings*
B85 *Die Castings*
B108 *Permanent Mold Castings*
B618 *Investment Castings*

Castings are made in foundries. Usually the aluminum to be cast is received in ingot form, but foundries located next to a smelter may receive molten aluminum directly from the reduction plant, and some foundries use recycled material. Castings make up about one half the aluminum used in automotive applications.

The minimum mechanical properties of separately cast test bars of cast alloys are given in Table 9.16. The average tensile ultimate strength and tensile yield strength of specimens cut from castings need only be 75% of the minimum strengths given in Table 9.16, and 25% of the minimum elongation values given in Table 9.16. The values for specimens cut from castings should be used in design because they are more representative of the actual strength of the casting.

9.3.2.1 Casting Types

Sand castings are made with a sand mold that is used only once. This method is used for larger castings without intricate details and that are produced in small quantities. The mold material is sometimes referred to as green sand or dry sand. Aluminum sand castings as large as 7000 lb (3000 kg) have been produced.

Permanent mold castings are made in reusable molds; sometimes the flow is assisted by a small vacuum but otherwise is gravity induced. Permanent mold castings are more expensive than sand castings but can be held to tighter tolerances and finer details, including wall thicknesses as small as 0.09 in. (2 mm). Semipermanent molds made of sand or other material are used when the geometry of the casting makes it impossible to remove the mold in one piece from the solidified part.

Die castings are made by injecting the molten metal under pressure into a reusable steel die at high velocity. Solidification is rapid, so high production rates are possible. Die castings are usually smaller and may have thinner wall thicknesses and tighter tolerances than either sand or permanent mold castings.

Investment castings are made by surrounding (investing) an expendable pattern (usually wax or plastic) with a refractory slurry that sets at room temperature. The pattern is then removed by heating and the resulting cavity is filled with molten metal.

Not all casting alloys are appropriate for all production methods, but some may be produced by multiple methods. Fewer alloys are suitable for die casting than the other methods.

New methods, such as squeeze casting and thixocasting, are showing promise in producing high-strength, ductile castings but have not been proven in aluminum yet. Thixocasting has more recently been called semisolid forming and may be thought of as a cross between casting and forging. Semisolid forming stock has a special globular crystal structure that behaves as a solid until sufficient shearing forces are applied during forming, upon which the material flows like a viscous liquid.

9.3.2.2 Casting Quality

Foundries only hold those tolerances that are specified by the purchaser. This is unlike the case for wrought products, for which mills will meet standard mill tolerances as a minimum. The dimensions of castings can be difficult to control because it is sometimes difficult to predict the shrinkage during solidification and the warping that may be produced by nonuniform cooling. The quality of cast material may also vary widely, and any inspection methods must be specified by the purchaser. The most commonly used inspection techniques are radiography and penetrant methods. Radiography is performed by X-raying the part to show discontinuities such as gas holes, shrinkage, and foreign material. These discontinuities are then rated by comparing them to reference radiographs shown in ASTM E155. The ratings are then compared to inspection criteria agreed to beforehand by the customer and the foundry. The inspection criteria for quality and frequency of inspection can be selected and then specified from the Aluminum Association's casting quality standard AA-CS-M5-85, which provides seven quality levels and four frequency levels from which to choose. The penetrant inspection method is only useful for detecting surface defects. Two techniques are available. The fluorescent penetrant procedure is to apply penetrating oil to the part, remove the oil, apply developer to absorbed oil bleeding out of surface discontinuities, and then inspect the casting under ultraviolet light. The dye penetrant method uses a color penetrant, enabling inspection in normal light. Frequency levels are given in AA-CS-M5-85 for penetrant testing also.

A test bar cast with each heat is also useful. It can be tested and the results compared directly to minimum mechanical properties listed for the alloy in Table 9.16.

9.3.3 Aluminum Powder

There are many uses for aluminum powder particles, which can be as small as a few microns thick. Larger particles are used in the chemical and metal production industries; one of the first uses of aluminum was as particles to remove oxygen from molten steel during its production. Finer particles are used as an explosive in fireworks and flares and as a solid fuel for rockets. Each launch of the space shuttle uses 350,000 16 (160,000 kg) of aluminum powder. Powder is also flattened into flakes in a rotating mill and used as a constituent for paints to provide a metallic finish. Finally, aluminum powder may be pressed into parts, referred to as powder metallurgy, competing with conventionally cast aluminum parts.

9.4 WELDING, BRAZING, AND SOLDERING

9.4.1 Welding

Welding is the process of uniting parts by either heating, applying pressure, or both. When heat is used to weld aluminum (as is usually the case), it reduces the strength of all tempers other than annealed material, and this must be taken into account where strength is a consideration [13].

Aluminum's affinity for oxygen, which quickly forms a thin, hard oxide surface film, has much to do with the welding process. This oxide is nearly as hard as diamonds, attested to by the fact that aluminum oxide grit is often used for grinding. It has a much higher melting point than aluminum itself [3725°F (2050°C), versus 1220°F (660°C)], so trying to weld aluminum without first removing the oxide melts the base metal long before the oxide. The oxide is also chemically stable; fluxes to remove it require corrosive substances that can damage the base metal unless they are fully removed after welding. Finally, the oxide is an electrical insulator and porous enough to retain moisture. For all these reasons, the base metal must be carefully cleaned and wire brushed immediately before welding, and the welding process must remove and prevent reformation of the oxide film during welding.

The metal in the vicinity of a weld can be considered as two zones: the weld bead itself, a casting composed of a mixture of the filler and the base metal, and the heat affected zone (HAZ) in the base metal outside the weld bead. The extent of the HAZ is a function of the thickness and geometry of the joint, the welding process, the welding procedure, and preheat and interpass temperatures, but rarely exceeds 1 in. (25 mm) from the centerline of the weld. The strength of the metal near a weld is graphed in Fig. 9.4. Smaller welds and higher welding speeds tend to have a smaller HAZ. As the base metal and filler metal cool after freezing, if the joint is restrained from contracting and its strength at the elevated temperature is insufficient, hot cracking may occur.

The magnitude of the strength reduction from welding varies: for non-heat-treatable alloys, welding reduces the strength to that of the annealed (O) temper of the alloy; for heat-treatable alloys, the reduced strength is slightly greater than that of the solution heat treated but not artificially aged temper (T4) of the alloy. Minimum strengths across groove-welded aluminum alloys are given in Table 9.26. These strengths are the same as those required to qualify a welder or weld procedure in accordance with the American Welding Society (AWS)

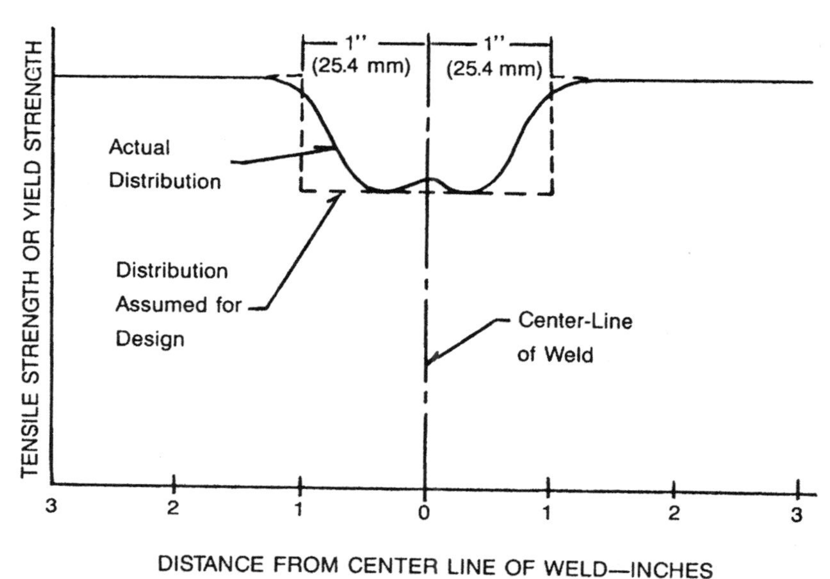

FIGURE 9.4 Strength near a weld.

D1.2 *Structural Welding Code—Aluminum* [21] and the American Society of Mechanical Engineers (ASME) *Boiler and Pressure Vessel Code, Section IX* [7]. They are based on the most common type of welding (gas-shielded arc, discussed next), and as long as a recommended filler alloy is used, they are independent of filler. Yield strengths for welded material are also given in the Aluminum Association's *Aluminum Design Manual* [2], but they must be multiplied by 0.75 to obtain the yield strength of the weld-affected metal because the association's yield strengths are based on a 10-in. (250-mm) long gage length, and only about 2 in. (50 mm) of that length is heat-affected metal.

Fillet weld shear strengths are a function of the filler used; minimum shear strengths for the popular filler alloys are given in Table 9.27. Fillet welds transverse (perpendicular) to the direction of force are generally stronger than fillet welds longitudinal (parallel) to the direction of force. This is because transverse welds are in a state of combined shear and tension and longitudinal welds are in shear, and tension strength is greater than shear strength.

Heat-treatable base metal alloys welded with heat-treatable fillers can be heat treated after welding to recover strength lost by heat of welding. This postweld heat treatment can be a solution heat treatment and aging or just aging (see 9.1.3.3). While solution heat-treating and aging will recover more strength than aging alone, the rapid quenching required in solution heat-treating can cause distortion of the weldment because of the residual stresses that are introduced. Natural aging will also recover some of the strength; the period of time required is a function of the alloy. The fillet weld strengths for 4043 and 4643 in Table 9.27 are based on 2–3 months of natural aging.

TABLE 9.26 Minimum Strengths of Welded Aluminum Alloys

Alloy	Product	Thickness (in.)	Tensile Ultimate Strength (ksi)	Tensile Yield Strength (ksi)[a]
1060	Sheet and plate	Up thru 3.000	8	2.5
1060	Extrusion	All	8.5	2.5
1100	All	Up thru 3.000	11	3.5
2219	All	All	35	—
3003	All	Up thru 3.000	14	5
Alclad 3003	Tube	All	13	4.5
Alclad 3003	Sheet and plate	Up to 0.500	13	4.5
Alclad 3003	Plate	0.500 to 3.000	14	5
3004	All	Up thru 3.000	22	8.5
Alclad 3004	Sheet and plate	Up to 0.500	21	8
Alclad 3004	Plate	0.500 to 3.000	22	8.5
5005	All	Up thru 3.000	15	5
5050	All	Up thru 3.000	18	6
5052	All	Up thru 3.000	25	9.5
5083	Forging	All	39	16
5083	Extrusion	All	39	16
5083	Sheet and plate	Up thru 1.500	40	18
5083	Plate	>1.500, thru 3.000	39	17
5083	Plate	>3.000, thru 5.000	38	16
5083	Plate	>5.000, thru 7.000	37	15
5083	Plate	>7.000, thru 8.000	36	14
5086	All	Up thru 2.000	35	14
5086	Extrusion	>2.000, thru 5.000	35	14
5086	Plate	>2.000, thru 3.000	34	14
5154	All	Up thru 3.000	30	11
5254	All	Up thru 3.000	30	11
5454	All	Up thru 3.000	31	12
5456	Extrusion	Up thru 5.000	41	19
5456	Sheet and plate	Up thru 1.500	42	19
5456	Plate	>1.500, thru 3.000	41	18
5456	Plate	>3.000, thru 5.000	40	17
5456	Plate	>5.000, thru 7.000	39	16
5456	Plate	>7.000, thru 8.000	38	15
5652	All	Up thru 3.000	25	9.5
6005	Extrusion	Up thru 1.000	24	—
6061	All	All	24	—
Alclad 6061	All	All	24	—
6063	Extrusion	Up thru 1.000	17	—
6351	Extrusion	Up thru 1.000	24	—
7005	Extrusion	Up thru 1.000	40	—
356.0	Casting	All	23	—
443.0	Casting	All	17	7
A444.0	Casting	All	17	—
514.0	Casting	All	22	9
535.0	Casting	All	35	18

[a] Yield strengths are for 2 in. gage length.

TABLE 9.27 Minimum Shear Strengths of Filler Alloys

Filler Alloy	Longitudinal Shear Strength (ksi)	Transverse Shear Strength (ksi)
1100	7.5	7.5
2319	16	16
4043	11.5	15
4643	13.5	20
5183	18.5	—
5356	17	26
5554	17	23
5556	20	30
5654	12	—

Prior to 1983, the ASME *Boiler and Pressure Vessel Code, Section IX, Welding and Brazing Qualifications* [7] was the only widely available standard for aluminum welding. Many aluminum structures other than pressure vessels were welded in accordance with the provisions of the *Boiler and Pressure Vessel Code*, therefore, due to the lack of an alternative standard. In 1983, the American Welding Society's (AWS) D1.2 *Structural Welding Code—Aluminum* [21] was introduced as a general standard for welding any type of aluminum structure (e.g., light poles, space frames, etc.). In addition to rules for qualifying aluminum welders and weld procedures, D1.2 includes design, fabrication, and inspection requirements. There are other standards that address specific types of welded aluminum structures, such as ASME B96.1 *Welded Aluminum-Alloy Storage Tanks*, AWS D15.1 *Railroad Welding Specification—Cars and Locomotives*, and AWS D3.7 *Guide for Aluminum Hull Welding*.

9.4.1.1 Gas-Shielded Arc Welding

Before World War II, shielded metal arc welding (SMAW) using a flux-coated electrode was one of the few ways aluminum could be welded. This process, however, was inefficient and often produced poor welds. In the 1940s, inert gas-shielded arc welding processes were developed that used argon and helium instead of flux to remove the oxide and quickly became more popular. Other methods of welding aluminum are used (and will be discussed below), but today most aluminum welding is by the gas-shielded arc processes.

There are two gas-shielded arc methods: gas metal arc welding (GMAW), also called metal inert gas welding, or MIG, and gas tungsten metal arc welding (GTAW), also called tungsten inert gas welding, or TIG. MIG welding uses an electric arc between the base metal being welded and an electrode filler wire. The electrode wire is pulled from a spool by a wire-feed mechanism and delivered to the arc through a gun. In TIG welding, the base metal and, if used, the filler metal are melted by an arc between the base metal and a nonconsumable tungsten electrode in a holder.

Tungsten is used because it has the highest melting point of any metal [6170°F (3410°C)] and reasonably good conductivity, about one-third that of copper. In each case, the inert gas removes the oxide from the aluminum surface and protects the molten metal from oxidation, allowing coalescence of the base and filler metals.

Tungsten inert gas welding was developed before MIG welding and was originally used for all metal thicknesses. Today, however, TIG is usually limited to material $\frac{1}{4}$ in. (6 mm) thick or less. TIG welding is slower and does not penetrate as well as MIG welding. In MIG welding, the electrode wire speed is controlled by the welding machine and once adjusted to a particular welding procedure does not require readjustment, so even manual MIG welding is considered to be semiautomatic. MIG welding is suitable for all aluminum material thicknesses.

The weldability of wrought alloys depends primarily on the alloying elements, discussed below for the various alloy series:

1xxx Pure aluminum has a narrower melting range than alloyed aluminum. This can cause a lack of fusion when welding, but generally the 1xxx alloys are very weldable. The strength of pure aluminum is low, and welding decreases the strength effect of any strain hardening, so welded applications of the 1xxx series are used mostly for their corrosion resistance.

2xxx The 2xxx alloys are usually considered poor for arc welding, being sensitive to hot cracking, and their use in the aircraft typically has not required welding. However, alloy 2219 is readily weldable, and 2014 is welded in certain applications.

3xxx The 3xxx alloys are readily weldable, but have low strength and so are not used in structural applications unless their corrosion resistance is needed.

5xxx The 5xxx alloys retain high strengths even when welded and are free from hot cracking and are very popular in welded plate structures such as ship hulls and storage vessels.

6xxx The 6xxx alloys can be prone to hot cracking if improperly designed and lose a significant amount of strength due to the heat of welding, but are successfully welded in many applications. Postweld heat treatments can be applied to increase the strength of 6xxx weldments. The 6xxx series alloys (like 6061 and 6063) are often extruded and combined with the sheet and plate products of the 5xxx series in weldments.

7xxx The low copper content alloys (such as 7004, 7005, and 7039) of this series are weldable; the others are not, losing considerable strength and suffering hot cracking when welded.

Some cast alloys are readily welded and some are postweld heat-treated because they are usually small enough to be easily placed in a furnace. The condition of the cast surface is key to the weldability of castings; grinding and machining are often needed to remove contaminants prior to welding. The weldability of the 355.0, 356.0, 357.0, 443.0, and A444.0 alloys is considered excellent.

Filler alloys can be selected based on different criteria, including: resistance to hot cracking, strength, ductility, corrosion resistance, elevated temperature performance, MIG electrode wire feedability, and color match for anodizing. Recommended selections are given in Table 9.28, and a discussion of some fillers is given below. Material specifications for these fillers are given in AWS A5.10, *Specification for Bare Aluminum and Aluminum Alloy Welding Electrodes and Rods* [19]. There is no ASTM specification for aluminum weld filler.

Filler alloys 5356, 5183, and 5556 were developed to weld the 5xxx series alloys, but they have also become useful for welding 6xxx and 7xxx alloys. Alloy 5356 is the most commonly used filler due to its good strength, compatibility with many base metals, and good MIG electrode wire feedability. Alloy 5356 also is used to weld 6xxx series alloys because it provides a better color match with the base metal than 4043 when anodized. Alloy 5183 has slightly higher strength than 5356, and 5556 higher still. Because these alloys contain more than 3% magnesium and are not heat treatable, however, they are not suitable for elevated temperature service or postweld heat-treating. Alloy 5554 was developed to weld alloy 5454, which contains less than 3% magnesium so as to be suitable for service over 150°F(66°C).

Alloy 5654 was developed as a high-purity, corrosion-resistant alloy for welding 5652, 5154, and 5254 components used for hydrogen peroxide service. Its magnesium content exceeds 3% so it is not used at elevated temperatures.

Alloy 4043 was developed for welding the heat-treatable alloys, especially those of the 6xxx series. Its has a lower melting point than the 5xxx fillers and so flows better and is less sensitive to cracking. Alloy 4643 is for welding 6xxx base metal parts over 0.375 in. (10 mm) to 0.5 in. (13 mm) thick that will be heat treated after welding. Alloys 4047 and 4145 have low melting points and were developed for brazing but are also used for some welds; 4145 is used for welding 2xxx alloys and 4047 is used instead of 4043 in some instances to minimize hot cracking and increase fillet weld strengths.

Alloy 2319 is used for welding 2219; it's heat treatable and has higher strength and ductility than 4043 when used to weld 2xxx alloys that are postweld heat treated.

Pure aluminum alloy fillers are often needed in electrical or chemical industry applications for conductivity or corrosion resistance. Alloy 1100 is usually satisfactory, but for even better corrosion resistance (due to its lower copper level), 1188 may be used. These alloys are soft and sometimes have difficulty when fed through MIG conduit.

The filler alloys used to weld castings are castings themselves (C355.0, A356.0, and A357.0), usually a $\frac{1}{4}$-in. (6-mm) rod used for TIG welding. They are mainly used to repair casting defects. More recently, wrought versions of C355.0 (4009), A356.0 (4010), and A357.0 (4011) have been produced so that they can be produced as MIG electrode wire. (Alloy 4011 is only available as rod for GTAW, however, since its beryllium content produces fumes too dangerous for MIG welding.) Like 4643, 4010 can be used for postweld heat-treated 6xxx weldments.

TABLE 9.28 Guide to Choice of Filler Metal for General-Purpose Welding

Base Metal	319.0, 333.0, 354.0, 355.0, C355.0, 380.0	356.0, A356.0, A357.0, 359.0, 413.0, A444.0, 443.0	511.0, 512.0, 513.0, 514.0	7005[k], 7039, 710.0, 711.0, 712.0	6070	6061, 6063, 6101, 6201, 6151, 6351, 6951	5456	5454	5154, 5254[a]	5086	5083	5052, 5652[a]	5005, 5050	3004, Alc.3004	2219, 2519	2014, 2036	1100, 3003, Alc.3003	1060, 1070, 1080, 1350
1060, 1070, 1080, 1350	4145[c,i]	4043[i,f]	4043[e,i]	4043[i]	4043[i]	4043[i]	5356[c]	4043[i]	4043[e,i]	5356[c]	5356[c]	4043[i]	1100[c]	4043	4145	4145	1100[c]	1188[j]
1100, 3003, Alclad 3003	4145[c,i]	4043[i,f]	4043[e,i]	4043[i]	4043[i]	4043[i]	5356[c]	4043[e,i]	4043[e,i]	5356[c]	5356[c]	4043[e,i]	4043[e]	4043[e]	4145	4145	1100[c]	
2014, 2036	4145[g]	4145			4145	4145									4145[g]	4145[g]		
2219, 2519	4145[g,c,i]	4145[c,i]	4043[i]	4043[i]	4043[f,i]	4043[f,i]	4043	4043[i]	4043[i]	4043	4043	4043[i]	4043	4043	2319[c,f,i]			
3004, Alclad 3004	4043[i]	4043[i]	5654[b]	5356[e]	4043[e]	4043[b]	5356[e]	5654[b]	5654[b]	5356[e]	5356[e]	4043[e,i]	4043[e]	4043[e]				
5005, 5050	4043[i]	4043[i]	5654[b]	5356[e]	4043[e]	4043[b]	5356[e]	5654[b]	5654[b]	5356[e]	5356[e]	4043[e,i]	4043[d,e]					
5052, 5652[a]	4043[i]	4043[b,i]	5654[b]	5356[e]	5356[b,c]	5356[b,c]	5356[b]	5654[b]	5654[b]	5356[e]	5356[e]	5654[a,b,c]						
5083		5356[c,e,i]	5356[e]	5183[e]	5356[e]	5356[e]	5183[e]	5356[e]	5356[e]	5356[e]	5183[e]							
5086		5356[c,e,i]	5356[e]	5356[e]	5356[e]	5356[e]	5356[e]	5356[b]	5356[b]	5356[e]								
5154, 5254[a]		4043[b,i]	5654[b]	5356[b]	5356[b,c]	5356[b,c]	5356[b]	5654[b]	5654[a,b]									
5454	4043[i]	4043[b,i]	5654[b]	5356[b]	5356[b,c]	5356[b,c]	5356[b]	5554[c,e]										
5456		5356[c,e,i]	5356[e]	5556[e]	5356[e]	5356[e]	5556[e]											
6061, 6063, 6351, 6101, 6201, 6151, 6951	4145[c,i]	4043[f,i]	5356[b,c]	5356[b,c,i]	4043[b,i]	4043[b,i]												
6070	4145[c,i]	4043[f,i]	5356[c,e]	5356[c,e,i]	4043[e,i]													

7005[k], 7039, 710.0, 711.0, 712.0	4043[i]	4043[b,i]	5356[b]	5356[e]
511.0, 512.0, 513.0, 514.0		4043[b,i]	5654[b,d]	
356.0, A356.0, A357.0, 359.0, 413.0, A444.0, 443.0	4145[c,i]	4043[d,i]		
319.0, 333.0, 354.0, 355.0, C355.0, 380.0	4145[d,c,i]			

1. Service conditions such as immersion in fresh or saltwater, exposure to specific chemicals or a sustained high-temperature (over 150°F) may limit the choice of filler metals. Filler alloys 5356, 5183, 5556, and 5654 are not recommended for sustained elevated temperature service.
2. Recommendations in this table apply to gas shielded are welding processes. For gas welding, only 1100, 1188, and 4043 filler metals are ordinarily used.
3. All filler metals are listed in AWS specification A5.10.

[a]Base metal alloys 5652 and 5254 are used for hydrogen peroxide service. 5654 filler metal is used for welding both alloys for low temperature service (150°F and below)
[b]5183, 5356, 5554, 5556, and 5654 may be used. In some cases they provide: (1) improved color match after anodizing treatment; (2) highest weld ductility, and (3) higher weld strength. 5554 is suitable for elevated temperature service.
[c]4043 may be used for some applications.
[d]Filler metal with the same analysis as the base metal is sometimes used.
[e]5183, 5356, or 5556 may be used.
[f]4145 may be used for some applications.
[g]2319 may be used for some applications.
[i]4047 may be used for some applications.
[j]1100 may be used for some applications.
[k]This refers to 7005 extrusions only (X-prefix still applies to sheet and plate).

4. Where no filler metal is listed, the base metal combination is not recommended for welding.

Weld quality may be determined by several methods. *Visual inspection* detects incorrect weld sizes and shapes (such as excessive concavity of fillet welds), inadequate penetration on butt welds made from one side, undercutting, overlapping, and surface cracks in the weld or base metal. *Dye penetrant inspection* uses a penetrating dye and a color developer and is useful in detecting defects with access to the surface. *Radiography* (making X-ray pictures of the weld) can detect defects as small as 2% of the thickness of the weldment, including porosity, internal cracks, lack of fusion, inadequate penetration, and inclusions. *Ultrasonic inspection* uses high-frequency sound waves to detect similar flaws, but is expensive and requires trained personnel to interpret the results. Its advantage over radiography is that it is better suited to detecting thin planar defects parallel to the X-ray beam. Destructive tests, such as bend tests, fracture (or nick break) tests, and tensile tests are usually reserved for qualifying a welder or a weld procedure. Acceptance criteria for the various methods of inspection and tests are given in AWS D1.2 [21] and other standards for specific welded aluminum components or structures.

9.4.1.2 Other Arc Welding Processes

Stud welding (SW) is a process used to attach studs to a part. Two methods are used for aluminum: arc stud welding, which uses a conventional welding arc over a timed interval, and capacitor discharge stud welding, which uses an energy discharge from a capacitor. Arc stud welding is used to attach studs ranging from $\frac{1}{4}$ in. (6 mm) to $\frac{1}{2}$ in. (13 mm) in diameter, while capacitor discharge stud welding uses studs $\frac{1}{16}$ in. (1.6 mm) to $\frac{1}{4}$ in. (6 mm) in diameter. Capacitor discharge stud welding is very effective for thin sheet [as thin as 0.040 in. (1.0 mm)], because it uses much less heat than arc stud welding and does not mar the appearance of the sheet on the opposite side from the stud. Studs are inspected using bend, torque, or tension tests. Stud alloys are the common filler alloys. Stud welding requirements are included in AWS D1.2 [21].

Plasma arc welding with variable polarity (PAW-VP) [also called variable polarity plasma arc (VPPA) welding] is an outgrowth of TIG welding and uses a direct current between a tungsten electrode and either the workpiece or the gas nozzle. Polarity is constantly switched from welding to oxide cleaning modes at intervals tailored to the joint being welded. Two gases, a plasma gas and a shielding gas, are provided to the arc. Welding speed is slower than MIG welding, but often fewer passes are needed, single pass welds in metal up to $\frac{5}{8}$ in. (16 mm) thick having been made. The main disadvantage is the cost of the required equipment.

Plasma–MIG welding is a combination of plasma arc and MIG welding, by which the MIG electrode is fed through the plasma coaxially, superimposing the arcs of each process. Higher deposition rates are possible, but equipment costs are also higher than for conventional MIG welding.

Arc spot welding uses a stationary MIG arc on a thin sheet held against a part below, fusing the sheet to the part. The advantage over resistance welding (discussed below) is that access to both sides of the work is unnecessary.

Problems with gaps between the parts, overpenetration, annular cracking, and distortion have limited the application of this method. It has been used to fuse aluminum to other metals such as copper, aluminized steel, and titanium for electrical connections.

Shielded metal arc welding (SMAW) is an outdated, manual process that uses a flux-coated filler rod, the flux taking the place of the shielding gas in removing oxide. Its only advantage is that it can be performed with commonly used shielded metal arc steel welding equipment. Shielded metal arc welding is slow, prone to porosity [especially in metal less than $\frac{3}{8}$ in. (10 mm) thick], susceptible to corrosion if the slightest flux residue is not removed, and produces spatter (especially if rods are exposed to moisture) and requires preheating for metal 0.10 in. thick and thicker. Only 1100, 3003, and 4043 filler alloys are available for this process; see AWS A5.3, *Specification for Aluminum and Aluminum Alloy Electrodes for Shielded Metal Arc Welding* [18] for more information. For these reasons, gas-shielded arc welding is preferred.

9.4.1.3 Other Fusion Welding Processes

Fusion welding is any welding method that is performed by melting of the base metal or base and filler metal. It includes the arc welding processes mentioned above, and several others discussed below as they apply to aluminum.

Oxyfuel gas welding (OFW), or oxygas welding, was used to weld aluminum prior to development of gas-shielded arc welding. The fuel gas, which provides the heat to achieve coalescence, can be acetylene or hydrogen, but hydrogen gives better results for aluminum. The flux can be mixed and applied to the work prior to welding, or flux-coated rods used for shielded metal arc welding can be used to remove the oxide. Oxyfuel gas welding is usually confined to sheet metal of the 1xxx and 3xxx alloys. Preheating is needed for parts over $\frac{3}{16}$ in. (5 mm) thick. Problems include large heat-affected zones, distortion, flux residue removal labor and corrosion, and the high degree of skill required. The only advantage is the low cost of equipment; so oxyfuel gas welding of aluminum is generally limited to less developed countries where labor is inexpensive and capital is lacking.

Electrogas welding (EGW) is a variation on automatic MIG welding for single pass, vertical square butt joints such as in ship hulls and storage vessels. It has not been widely applied for aluminum because the sliding shoes needed to contain the weld pool at the root and face of the joint have tended to fuse to the molten aluminum and tear the weld bead.

Electroslag welding uses electric current through a flux without a shielding gas; the flux removes the oxide and provides the welding heat. This method has only been experimentally applied to aluminum for vertical welds in plate.

Electron beam welding (EBW) uses the heat from a narrow beam of high-velocity electrons to fuse plate. The result is a very narrow heat-affected zone and suitability for welding closely fitted, thick parts [even 6 in. (150 mm) thick] in one pass. A vacuum is needed or the electron beam is diffused; also, workers must be protected from X-rays resulting from the electrons colliding with the work. Thus electron beam welding must be done in a vacuum chamber or with a sliding seal vacuum and a lead-lined enclosure.

Laser beam welding (LBW) is an automatic welding process that uses a light beam for heat; for aluminum, a shielding gas is used also. Equipment is costly.

Thermit welding uses an exothermic chemical reaction to heat the metal and provide the filler; the process is contained in a graphite mold. Its application to aluminum is for splicing high-voltage aluminum conductors. These conductors must be kept dry because the copper and tin used in the filler have poor corrosion resistance when exposed to moisture.

9.4.1.4 Arc Cutting

Arc cutting is not a joining process, but rather a cutting process, but is included in this section on joining because it is similar to welding in that an arc from an electrode is used. Plasma arc cutting is the most common arc cutting process used for aluminum. It takes the place of flame cutting (such as oxyfuel gas cutting) used for steel, a method unsuited to aluminum because aluminum's oxide has such a high melting point relative to the base metal that flame cutting produces a very rough severing.

In plasma arc cutting, an arc is drawn from a tungsten electrode and ionized gas is forced through a small orifice at high velocity and temperature, melting the metal and expelling it and in so doing cutting through the metal. To cut thin material, a single gas (air, nitrogen, or argon) may act as both the cutting plasma and to shield the arc, but to cut thick material, two separate gas flows (nitrogen, argon, or, for the thickest cuts, an argon–hydrogen mix) are used. Cutting can be done manually, usually on thicknesses from 0.040 to 2 in. (1 to 50 mm) or by machine, more appropriate for material $\frac{1}{4}$ to 5 in. (6 to 125 mm) thick.

Arc cutting leaves a heat-affected zone and microcracks along the edge of the cut. Thicker material is more prone to cracking, since thick metal provides more restraint during cooling. The cut may also have some roughness and may not be perfectly square in the through thickness direction. The specification for aluminum structures requires therefore that plasma-cut edges be machined to a depth of $\frac{1}{8}$ in. (3 mm). The quality of the cut is a function of alloy (6xxx series alloys cut better than 5xxx), cutting speed, arc voltage, and gas flow rates.

9.4.1.5 Resistance Welding

Resistance welding is a group of processes that use the electrical resistance of an assembly of parts for the heat required to weld them together. Resistance welding includes both fusion and solid-state welding, but it's useful to consider the resistance welding methods as their own group. Because aluminum's electrical conductivity is higher than steel's, it takes more current to produce enough heat to fuse aluminum by resistance welding than for steel.

Resistance spot welding (RSW) produces a spot weld between two or more parts that are held tightly together by briefly passing a current between them. It is useful for joining aluminum sheet and can be used on almost every aluminum alloy, although annealed tempers may suffer from excessive indentation due to their softness. Its advantages are that it is fast, automatic, uniform in appearance, not dependent on operator skill, strong, and minimizes distortion of the parts. Its disadvantages are that it only applies to lap joints, is limited to parts no thicker than $\frac{1}{8}$ in. (3 mm), requires access to both sides of the work, and the equipment is costly and not readily portable. Tables are available that provide the minimum weld diameter, minimum spacing, minimum edge distance, minimum overlap, and shear strengths as a function of the thickness of the parts joined. Proper cleaning of the surface by etching or degreasing and mechanical cleaning is needed for uniform quality.

Weld bonding is a variation on resistance spot welding in which adhesive is added at the weld to increase the bond strength.

Resistance roll spot welding is similar to resistance spot welding except that the electrodes are replaced by rotating wheel electrodes. Intermittent seam welding has spaced welds; seam welding has overlapped welds and is used to make liquid or vapor-tight joints.

Flash welding (FW) is a two-step process: Heat is generated by arcing between two parts and then the parts are abruptly forced together. The process is automatically performed in special-purpose machines, producing very narrow welds. It has been used to make miter and butt joints in extrusions used for architectural applications and to join aluminum to copper in electrical components.

High-frequency resistance welding uses high-frequency welding current to concentrate welding heat at the desired location and for aluminum is used for longitudinal butt joints in tubular products. The current is supplied by induction for small-diameter aluminum tubing and through contacts for larger tubes.

9.4.1.6 Solid-State Welding

Solid-state welding is a group of welding processes that produce bonding by the application of pressure at a temperature below the melting temperatures of the base metal and filler.

Explosion welding (EXW) uses a controlled detonation to force parts together at such high pressure that they coalesce. Explosion welding has two applications for aluminum: It has been used to splice natural-gas distribution piping in rural areas where welding equipment and skilled labor is hard to come by and to bond aluminum to other metals like copper, steel, and stainless steel to make bimetallic plates.

Ultrasonic welding (USW) produces coalescence by pressing overlapping parts together and applying high-frequency vibrations that disperse the oxide films at the interface. Ultrasonic welding is very well suited to aluminum: Spot welds join aluminum wires to themselves or to terminals, ring welds are used to seal containers, line and area welds are used to attach mesh, and seam welds are used to join coils for the manufacture of aluminum foil. Welds between aluminum and copper are readily made for solid-state ignition systems, automotive starters, and small electric motors. The advantages of the process are that it requires less surface preparation than other methods, is automatic, fast (usually requiring less than a second), and joint strengths approach that of parent material. Joint designs are similar to resistance spot welds, but edge distance and spot spacing requirements are much less restrictive.

Diffusion welding uses pressure, heat, and time to cause atomic diffusion across the joint and produce bonding, usually in a vacuum or inert gas environment. Pressures can reach the yield strength of the alloys and times may be in the range of a minute. Sometimes a diffusion aid such as aluminum foil is inserted in the joint. Diffusion welding has been useful to join aluminum to other metals or to join dissimilar aluminum alloys. Welds are of high quality and leak tightness.

Pressure welding uses pressure to cause localized plastic flow that disperses the oxide films at the interface and causes coalescence. When performed at room temperature, it is called cold welding (CW); when at elevated temperature, it is termed hot pressure welding (HPW). Cold welding is used for lap or butt joints. Butt welds are made in wire from 0.015 in. (0.4 mm) to $\frac{3}{8}$ in. (10 mm) in diameter, rod, tubing, and simple extruded shapes. Lap welds can be made in thicknesses from foil to $\frac{1}{4}$ in. (6 mm). The 5xxx alloys with more than 3% magnesium, 2xxx and 7xxx alloys, and castings fracture before a pressure weld can be made and so are not suitable for this process. Hot pressure welding is used to make alclad sheet.

9.4.2 Brazing

Brazing is the process of joining metals by fusion using filler metals with a melting point above 840°F (450°C) but lower than the melting point of the base metals being joined [1]. *Soldering* also joins metals by fusion, but filler metals for soldering have a melting point below 840°F (450°C). Brazing and soldering differ from welding in that no significant amount of base metal is melted during the fusion process. Ranking the temperature of the process and the strength and the corrosion resistance of the assembly, from highest to lowest, are welding, brazing, and then soldering.

Brazing's advantage is that it is very useful for making complex and smoothly blended joints, using capillary action to draw the filler into the joint. A disadvantage is that it requires that the base metal be heated to a temperature near the melting point; since yield strength decreases drastically at such temperatures, parts must often be supported to prevent sagging under their own weight. Another disadvantage is the corrosive effect of flux residues, which can be overcome by using vacuum brazing or chloride-free fluxes.

Brazing can be used on lap, flange, lock-seam, and tee joints to form smooth fillets on both sides of the joint. Joint clearances are small, ranging from 0.003 in. (0.08 mm) to 0.025 in. (0.6 mm) and depend on the type of joint and the brazing process.

Non-heat-treatable alloys 1100, 3003, 3004, and 5005, heat-treatable alloys 6061, 6063, and 6951, and casting alloys 356.0, A356.0, 357.0, 359.0, 443.0, 710.0, 711.0, and 712.0 are the most commonly brazed of their respective categories. The melting points of 2011, 2014, 2017, 2024, and 7075 alloys are too low to be brazed, and 5xxx alloys with more than 2% magnesium are not very practically brazed because fluxes are ineffective in removing their tightly adhering oxides. Brazing alloys are shown in Table 9.29, and brazing sheet (cladding on sheet) parameters are given in Table 9.30.

TABLE 9.29 Common Brazing Filler Alloys and Forms

Brazing Alloy Designation	AWS Classification Number	Nominal Composition (%)			Melting Range °F (°C)	Normal Brazing °F (°C)	Available As:				Brazing Process			Remarks
		Si	Cu	Mg			Rod	Sheet	Clad[a]	Powder	Torch	Furnace	Dip	
4343	BA1Si-2	7.5	—	—	1070–1135 (517–613)	1110–1150 (599–621)		X	X			X	X	
4145	BA1Si-3	10	4	—	970–1085 (521–585)	1060–1120 (571–604)	X	X			X	X	X	Desirable where control of fluidity is necessary
4047	BA1Si-4	12	—	—	1070–1080 (577–582)	1080–1120 (582–604)	X	X		X	X	X	X	Fluid in entire brazing range
4045	BA1Si-5	10	—	—	1070–1095 (577–591)	1090–1120 (588–604)		X	X			X	X	
4004	BA1Si-7	10	—	1.5	1030–1105[b] (554–596)[b]	1090–1120 (588–604)			X			X		Vacuum furnace brazing
4147	BA1Si-9	12	—	2.5	1044–1080[b] (562–582)[b]	1080–1120 (582–604)			X			X		Vacuum furnace brazing
4104[c]	BA1Si-11	10	—	1.5	1030–1105[b] (554–596)[b]	1090–1120 (588–604)			X			X		Vacuum furnace brazing
4044	—	8.5	—	—	1070–1115[b] (577–602)[b]	1100–1135 (593–613)			X			X	X	

[a] As a cladding on aluminum brazing sheet (Table 7.30)
[b] The melting range temperatures shown for this filler were obtained in air. These temperatures are different in vacuum.
[c] Also contains 0.10 Bi.

TABLE 9.30 Some Standard Brazing Sheet Products

Commercial Brazing Sheet Designation	Number of Sides Cladding	Core Alloy	Cladding Composition	Thickness			Brazing Range °F(°C)
				Sheet		% Cladding on Each Side	
				in.	mm		
No. 7	1	3003	4004	0.024 and less	0.61 and less	15	1090-1120
No. 8	2			0.025 to 0.062	0.62 to 1.59	10	(588–604)
				0.063 and over	1.60 and over	7.5	
No. 11	1	3003	4343	0.063 and less	1.60 and less	10	1100-1150
No. 12	2			0.064 and over	1.62 and over	5	(593–621)
No. 13	1	6951	4004	0.024 and less	0.61 and less	15	1090-1120
No. 14	2			0.025 to 0.062	0.62 to 1.59	10	(588–604)
				0.063 and over	1.60 and over	7.5	
No. 21	1	6951	4343	0.090 and less	2.29 and less	10	1100-1150
No. 22	2			0.091 and over	2.3 and over	5	(593–621)
No. 23	1	6951	4045	0.090 and less	2.29 and less	10	1090-1120
No. 24	2			0.091 and over	2.3 and over	5	(588–604)
No. 33	1	6951	4044	All	All	10	1100-1135
No. 34	2						(593–613)
No. 44	[a]	6951	4044/7072	All	All	15/5	1100-1135 (593–613)

[a] This product is Clad with 4044 on one side and 7072 on the other side for resistance to corrosion.

Brazing fluxes are powders that are mixed with water or alcohol to make a paste that removes the oxide film from the base metal upon heating. Chloride fluxes have traditionally been used, but their residue is corrosive to aluminum. More recently, fluoride fluxes, which are not corrosive and thus do not require removal, have come into use. They are useful where flux removal is difficult, such as in automobile radiators.

Brazing can be done by several processes. Torch brazing uses heat from an oxyfuel flame and can be manual or automatic. Furnace brazing is most common, and is used for complex parts like heat exchangers where torch access is difficult. Assemblies are cleaned, fluxed, and sent through a furnace on a conveyor. Dip brazing is used for complicated assemblies with internal joints. The assemblies are immersed in molten chloride flux; the coating on brazing sheet or preplaced brazing wire, shims, or powder supply the filler. Vacuum brazing does not require fluxes and is done in a furnace; it's especially useful for small matrix heat exchangers, which are difficult to clean after fluxing.

Upon completion of brazing, the assembly is usually water quenched to provide the equivalent of solution heat treatment and to assist in flux removal and subsequently may be naturally or artificially aged to gain strength.

Minimum requirements for fabrication, equipment, material, procedure, and quality for brazing aluminum are given in the American Welding Society's publication C3.7 *Specification for Aluminum Brazing* [20].

9.4.3 Soldering

Soldering is the process of joining metals by fusion, but filler metals for soldering have a melting point below 840°F (450°C) [4]. (*Brazing*, described in Section 9.4.2, uses filler metals with a melting point above 840°F (450°C) but lower than the melting point of the base metals being joined.)

Soldering is much like brazing but conducted at lower temperatures. Soldering is limited to aluminum alloys with no more than 1% magnesium or 4% silicon, because higher levels produce alloys that have poor flux wetting characteristics. Alloys 1100 and 3003 are suitable for soldering, as are clad alloys of the 2xxx and 7xxx series. Alloys of zinc, tin, cadmium, and lead are used to solder aluminum; they are classified by melting temperature and described in Table 9.31.

Soldering fluxes are classified as organic and inorganic. Organic fluxes are used for low temperature [300–500°F (150–260°C)] soldering and usually need not be removed, being only mildly corrosive. Inorganic fluxes are used for intermediate [500–700°F (260–370°C)] and high-temperature [700–840°F (370–450°C)] soldering. Inorganic flux must be removed since it is very corrosive to aluminum. Both fluxes produce obnoxious fumes that must be properly ventilated.

Like brazing, soldering can be performed by several processes. Soldering with a hot iron can be done on small wires and sheet less than $\frac{1}{16}$ in. (1.6 mm) thick. Torch soldering can be performed in a much wider variety of cases, including automatic processes used to make automobile air-conditioning condensers. Torch soldering can also be done without flux by removing the aluminum oxide from the work by rubbing with the solder rod, called abrasion soldering. Abrasion soldering can also be performed with ultrasonic means. Furnace and dip soldering are much like their brazing counterparts. Resistance soldering is well suited to spot or tack soldering; flux is painted on the base metal, the solder is placed, and current is passed through the joint to melt the solder.

Soldered joint shear strengths vary from 6 to 40 ksi (40 to 280 MPa) depending on the solder used. Corrosion resistance is poor if chloride containing flux residue remains and the joint is exposed to moisture. Zinc solders have demonstrated good corrosion resistance, even for outdoor exposure.

TABLE 9.31 Classification of Aluminum Solders

Type	Melting Range °F (°C)	Common Constituents	Ease of Application	Wetting of Aluminum	Relative Strength	Relative Corrosion Resistance
Low Temp.	300–500 (149–260)	Tin or lead plus zinc and/or cadmium	Best	Poor to fair	Low	Low
Intermediate Temp.	500–700 (260–371)	Zinc base plus cadmium or zinc–tin	Moderate	Good to excellent	Moderate	Moderate
High Temp.	700–840 (371–449)	Zinc base plus aluminum, copper, etc.	Most difficult	Good to excellent	High	Good

9.5 RECENT DEVELOPMENTS

In the United States, about 21 billion pounds of aluminum worth $30 billion was produced in 1995, about 23% of the world's production. (To put this in perspective, about $62 billion of steel is shipped each year). Of this, about 25% is consumed in transportation applications, 25% in packaging, 15% in the building and construction market, and 13% in electrical products. Other markets include consumer durables such as appliances and furniture, machinery and equipment for use in petrochemical, textile, mining, and tool industries, reflectors, and powders and pastes used for paint, explosives, and other products.

The current markets for aluminum have developed over the relatively brief history of industrial production of the metal. Commercial production became practical with the invention of the Hall–Héroult process in 1886 and the birth of the electric power industry, a requisite because of the energy required by this smelting process. The first uses of aluminum were for cooking utensils in the 1890s, followed by electrical cable shortly thereafter. Shortly after 1900, methods to make aluminum stronger by alloying it with other elements (such as copper) and by heat treatment were discovered, opening new possibilities. Although the Wright brothers used aluminum in their airplane engines, it wasn't until World War II that dramatic growth in aluminum use occurred, driven largely by the use of aluminum in aircraft. Following the war, building and construction applications of aluminum boomed due to growth in demand and the commercial advent of the extrusion process, an extremely versatile way to produce prismatic members. The next big market for aluminum was packaging. Between the late 1960s and the 1980s, the aluminum share of the U.S. beverage can market went from zero to nearly 100%. The most recent growth market for aluminum has been in automobiles and light trucks; over 220 lb of aluminum were used, on average, in each car produced in North America in 1996. In the 1990s, aluminum use grew at a mean rate of about 3% annually in the United States.

After the emergence of the initial aluminum alloys in the first half of the twentieth century, the development of aluminum alloys became more narrowly focused on specific applications. This has reduced the likelihood of alloy crossover from one market to another but not eliminated it. Also, new alloys are being developed both for mature markets such as aircraft and developing markets such as automobiles. These circumstances combine to offer opportunities for designers to employ aluminum in new ways.

9.5.1 Aluminum–Lithium Alloys

Lithium is the lightest metallic element, and since the density of an alloy is the weighted average of the density of its constituents, lithium is attractive as an alloying element. But lithium has additional benefits—in addition to a 3% decrease in density for every 1% of lithium added (up to the solubility limit of 4.2%), the elastic modulus increases by 5–6%. Aluminum–lithium alloys are also heat treatable. These advantages are offset by the reactivity of lithium, which necessitates the use of an inert gas atmosphere when adding the liquid metal to the alloy. Al–Li alloys are often alloyed with copper, magnesium, zirconium, or other elements to improve properties. Since there is no aluminum–lithium alloy series, when lithium is the greatest alloying element, the designation number is 8xxx (see Section 9.1.3.1). When other alloying elements are in greater proportion than lithium, the designation number is based on the element in greatest proportion (such as 2195, which contains 4% copper and 1% lithium).

The Germans developed the first aluminum–lithium alloy in the 1920s, but the first Al–Li alloys to win commercial application were those developed for aircraft between the 1950s and 1970s. Alloy 2020 was used for compression wing skins of the RA5C Vigilante, but its registration was discontinued in 1974. Applications were hampered by low ductility and fracture toughness.

The second phase of Al–Li alloy development, which occurred in the 1980s, used relatively high levels of lithium (over 2%) in order to maximize property improvements. Alloys 2090 and 8090, typical of this phase, had some success but were limited by anisotropic behavior and relatively low corrosion resistance. Finally, in the late 1980s and 1990s, work done at Martin Marietta yielded the Weldalite Al–Li alloys, which appear destined to achieve significant success in aerospace and aircraft applications. These alloys are weldable, as the name implies, and use copper as the primary alloy, with modest amounts of lithium (slightly over 1%), and about 0.4% magnesium and 0.4% silver.

The most promising application for experimental, extremely light, and strong materials is space launch vehicles, where the cost of attaining low Earth orbit is about $8000/ kg and the number of reuses is limited. The U.S. space shuttle external fuel tank is a good example. The first application of the Weldalite-type alloys was the use of 2195 to replace 2219, a weldable aluminum–copper alloy, for the shuttle's liquid hydrogen and liquid oxygen tanks, producing a weight savings of 3500 kg. Alloy 2197 is now being used to refurbish F-16 fighter jet bulkheads, improving the range and performance of the aircraft. Commercial aircraft applications are anticipated next. As consumption has increased, Al–Li alloy material costs have fallen from a premium of 20 times that of common alloys to less than 4 times.

9.5.2 New Aluminum Automotive Alloys

The need to reduce emissions while enhancing performance and adding features has driven manufacturers to use more aluminum in automobiles and light trucks. This effort has been accompanied by the development of new aluminum alloys specifically tailored for these applications. These alloys are too new to be listed in ASTM specifications or *Aluminum Standards and Data*, so detailed information is given here.

Since automobiles and light trucks undergo a paint bake cycle at temperatures high enough to affect the temper of both heat-treatable and non-heat-treatable aluminum alloys, the automotive alloys are provided in the -T4 (solution heat treated) and -O (annealed) tempers, respectively. Both have the best formability in these tempers for the cold working they undergo in the process of being formed into body panels. The forming operation increases strengths through cold working. The subsequent paint bake artificially ages the heat-treatable alloys, which can additionally increase their strength, but reanneals the non-heat-treatable alloys, erasing any strength increase due to cold work. High strength is not necessarily important in this application, however.

The automotive alloys fall in three groups:

- 2xxx series (aluminum–copper alloys), including 2008, 2010, and 2036. Alloys 2008 and 2010 were developed to provide improved formability over 2036. Alloy 2036 has more copper than 2008 and 2010, giving it about 40% higher strength but less corrosion resistance. These alloys are heat treatable.
- 5xxx series (aluminum–magnesium alloys), including 5182 and 5754. Alloy 5182 was developed for the ends of beverage cans. It has high magnesium content, providing high strength but also sensitivity to corrosion if exposed to temperatures above 150°F for extended periods. Alloy 5754 is a variant on 5454, with slightly more magnesium (3.1 vs. 2.7%), lower strength, but better formability.
- 6xxx series (aluminum–magnesium–silicon alloys), including 6009, 6111, and 6022. These alloys are heat treatable and can attain fairly high strengths during the paint bake cycle. The newest of these alloys, 6022, is used in the Plymouth Prowler body panels.

Extrusions have not seen significant automotive use, but some alloys such as 7029 have been used in bumpers for some time.

9.5.3 Aluminum Foam

Closed-cell aluminum foam is made by bubbling gas or air through aluminum alloys or aluminum metal matrix composites (see Section 9.5.4) to create a strong but lightweight product. The foam's density is 2–20% that of solid aluminum. Foamed aluminum's advantages include fire-retardant properties, a high strength-to-weight ratio, rigidity, and energy absorbency. Current applications include sound insulation panels. Standard size blocks as well as parts with complex shapes can be cast.

9.5.4 Aluminum Metal Matrix Composites

A relatively new product, aluminum metal matrix composites (MMCs) consist of an aluminum alloy matrix with carbon, metallic, or, most commonly, ceramic reinforcement. Of all metals, aluminum is the most commonly used matrix material in MMCs. MMCs combine the low density of aluminum with the benefits of ceramics such as strength, stiffness (by increasing the modulus of elasticity), wear resistance, and high-temperature properties. They can be formed from both solid and molten states into forgings, extrusions, sheet and plate, and castings. Disadvantages include decreased ductility and higher cost; MMCs cost about three times more than conventional aluminum alloys. Yet even though they're still being developed, MMCs have been applied in automotive parts such as diesel engine pistons, cylinder liners, drive shafts, and brake components such as rotors.

Reinforcements are characterized as continuous or discontinuous depending on their shape and make up 10–70% of the composite by volume. Continuous fiber or filament reinforcements (designated f) include graphite, silicon carbide (SiC), boron, and aluminum oxide (Al_2O_3). Discontinuous reinforcements include SiC whiskers (designated w), SiC or Al_2O_3 particles (designated p), or short or chopped (designated c) Al_2O_3 or graphite fibers. The Aluminum Association standard designation system for aluminum MMCs identifies each as:

matrix material / reinforcement material / reinforcement volume %, form

For example, 2124/SiC/25 w is aluminum alloy 2124 reinforced with 25% by volume of silicon carbide whiskers; 6061/Al_2O_3/10p is aluminum alloy 6061 reinforced with 10% by volume of aluminum oxide particles. Chapter 5 has additional information on metal matrix composites.

9.5.5 Friction Stir Welding

Friction stir welding (FSW) is a new technique by which a nonconsumable tool is rotated and plunged into the joint made by abutting parts. The tool then moves along the joint, plasticizing the material to join it. No filler or shielding gas is needed, nor is there any need for current or voltage controls, and no welding fumes are produced. It has been applied to 2xxx, 5xxx, 6xxx and 7xxx alloys, in thicknesses up to 1 in. (25 mm). Friction stir welding produces uniform welds with little heat input and attendant distortion and loss of strength. The disadvantage is that high pressures must be brought to bear on the work, so the work must be properly supported and parts designed with this in mind. Commercial applications include rocket fuel tanks and ship decks. FSW is especially suited to making butt welds in long joints that are mass produced.

9.5.6 Hydrotalcite Coatings

Although aluminum is often used without coatings, coatings are sometimes necessary for appearance or corrosion protection. Many coatings do not adhere well to aluminum, however, without a surface pretreatment. The most effective pretreatment for many years has been a chromate coating, but the oxidizing solutions used to make chromate coatings and the coatings themselves contain hexavalent chromium (Cr^{6+}), a carcinogen. Cyanide and other toxic substances are also commonly used in chromate coating operations. Anodizing is an alternate chromate coatings, but it is more expensive.

Environmental and worker safety issues have led to the search for alternate coating pretreatment methods. Recently, a hydrotalcite coating has been developed at the University of Virginia to take the place of chromate coatings. The hydrotalcite coating costs less than chromate, is nontoxic, and is effective with the low-copper (3xxx, 5xxx, and 6xxx series) alloys.

9.6 CORROSION AND HYDROGEN EMBRITTLEMENT

9.6.1 Introduction

Corrosion presents a major threat to the structural integrity of aging aircraft structures. As the time of an aircraft structure in service increases, there is a growing probability that corrosion will interact with other forms of damage, such as single fatigue cracks or multiple-site damage in the form of widespread cracking at regions of high-stress gradients; it can result in loss of structural integrity and may lead to fatal consequences. Thus, the effect of corrosion on the damage tolerance ability of advanced aluminum alloys calls for a very diligent consideration of the problems associated with the combined effect of corrosion and embrittling mechanisms. There has been, recently, an increasing attention of basic research and development concerning structural integrity taking into account the related corrosion aspects[29–31, 36]. It has been realized that the establishment of damage functions for quantifying the simultaneous accumulation of corrosion and fatigue-induced damage is very complex and difficult. Therefore, despite the advancements in modeling fatigue crack growth [32–35] and multiple-site damage phenomena [29–31], the assessment of structural degradation in aging aircraft is still relying heavily on test data. To face the corrosion-induced structural degradation issue, available data usually refer to accelerated laboratory corrosion tests and, more rarely, to in-nature atmospheric or marine exposure corrosion tests.

With the exception of the atmospheric corrosion test where, according to the relevant specification the tensile properties of corroded specimens are measured as well, these tests are used for evaluating the corrosion susceptibility of the materials by measuring weight loss and characterizing depth and type of corrosion attack. The above methodology toward understanding corrosion susceptibility of a material does not relate corrosion to their effect on the materials mechanical behavior and residual properties. Yet, it is exactly these missing data that are needed to face structural integrity problems of corroded aircraft components. Corrosion-induced mechanical degradation studies have been based mainly on the results of stress corrosion cracking tests [37, 38] or, more rarely on the results of fatigue tests performed in the presence of a corrosive environment [29, 30]. Both types of tests provide useful results; they refer, however, to the case where a material is loaded in a corrosive environment but not to situations where a corroded material is subjected to mechanical loads. Present-day considerations of the corrosion-induced structural degradation relate the presence of corrosion with a decrease of the load-bearing capacity of the corroded structural member [36, 39].

This decrease is associated with the presence of corrosion notches that lead to local increase of stress promoting fatigue crack initiation as well; in addition, corrosion-induced reduction of the members' load-bearing thickness which, in the case of the thin alloy skin sheets, may be essential, can lead to appreciable increase of stress gradients [39]. Corrosion-induced material embrittlement is not accounted for. The above consideration of the corrosion-induced structural integrity issue is consistent with the classical understanding of the corrosion attack of aluminum alloys as the result of complex oxidation processes at the materials surface [40].

Regarding corrosion-induced material embrittlement, Pantelakis et al. [41, 42] claimed that hydrogen embrittlement could be responsible for the dramatic degradation of toughness and ductility of 2091 and 8090 Al–Li alloys as well as conventional 2024-T3 alloy in several types of accelerated corrosion tests. In other alloy systems there is mounting evidence connecting embrittlement and stress corrosion cracking to hydrogen penetration. Speidel [37] reviews recent results, mainly for Al–Mg–Zn alloys. Studies by Scamans et al. [43] of Al embrittlement in humid air, point to the major role of hydrogen. In particular, the intergranular crack path and the reversibility of the phenomenon (recovery of ductility after degassing) support a hydrogen, rather than an anodic dissolution, mechanism. Also, Scamans and Tuck [44] measured H_2 permeability and stress corrosion resistance of the Al–Mg–Zn alloy, as functions of quench rate and aging treatment, and found similar trends. However, the stress-corrosion-resistant Al–Mg–Si alloy does not allow hydrogen permeation through its matrix, though the volume of hydrogen produced by surface reaction with the water in humid air is even higher than that of the Al–Mg–Zn alloy [44]. It has been suggested [37] that hydrogen plays a major role in stress corrosion cracking of aluminum alloys exposed to aqueous solutions as well. An indication in favor of this argument is provided by measurement, in Al–Mg–Zn alloys, of hydrogen permeation [45] and stress corrosion crack growth rates [46]. These parameters are found to vary similarly as functions of the electrode potential. Despite the lack of a universally accepted hydrogen embrittlement mechanism, a generally recognized common feature is that some critical concentration of hydrogen must buildup at potential crack sites, for failure to initiate. Thus, the distribution of hydrogen inside the metal and its pattern of migration are of paramount importance in understanding the phenomena and designing alloys with improved behavior.

It has been shown [47, 48] that lattice defects (vacancies, dislocations, grain boundaries) and precipitates provide a variety of trapping sites for diffusing hydrogen. Hydrogen traps have mechanistically been classified by Pressouyre [49] as reversible and irreversible, depending on the steepness of the energy barrier needed to be overcome by hydrogen to escape from the trap. For example, during a degassing experiment reversible traps will release hydrogen continuously, while irreversible ones will do so only after a critical temperature has been reached. This is the temperature at which the probability of a single jump out of the steep trap becomes nonnegligible. Reversible and irreversible traps may play different roles during an actual experiment [50]. In particular, irreversible traps will always act as sinks for hydrogen, whereas reversible traps may act as sinks or sources depending on initial hydrogen charging of the lattice. A uniform distribution of irreversible traps is believed to provide a beneficial effect in alloy behavior under embrittling conditions, by arresting diffusing hydrogen and thus delaying its buildup at the crack sites [51]. When crack nucleation and growth is along the grain boundaries, boundary chemistry may be playing an important role. Various studies on Al–Mg–Zn alloys [52–54] have indicated that alloying elements (and in particular Mg) are segregated on the grain boundary. Tuck [55] proposed that Mg hydride forms at grain boundaries and is responsible for material embrittlement. In an effort to explain the connection between Mg–H interaction and material embrittlement, Song et al. [56] recently showed that stress corrosion and fatigue crack growth rates increase with the concentration of solid solution Mg on grain boundaries. The same authors theoretically calculated a decrease in the intergranular fracture work with both Mg and H segregation.

Useful insight in the nature and intensity of hydrogen traps can be offered by studying the temperature needed to break these bonds. Thus, thermal analysis techniques have been used for a variety of alloys [55, 57]. In particular, thermal desorption has been successfully used to study hydrogen partitioning in pure cast aluminum [58] and in Al–Cu and Al–Mg_2Si alloys [47] and hydrogen diffusion in Al–Li alloys [59]. Among other findings, these studies show that, for aluminum alloys, the energy of chemisorption is lower than the energy for lattice diffusion. Thus, the layer of passive oxide—formed on the surface of aluminum alloys—does not mask the bulk trapping states, and the results of thermal analysis are meaningful. Accelerated corrosion tests were recently used by Haidemenopoulos et al. [60] to characterize corrosion and hydrogen absorption in the less studied but widely used Al–Cu alloy 2024. In [61] hydrogen evolution from the corroded specimen of Al alloy 2024 was systematically measured as a function of temperature. The exfoliation test [62] was used as an accelerated corrosion method, and different exposure times were tested. The existence of multiple trapping states was verified and the quantity and evolution pattern of hydrogen is discussed.

The investigations summarized above indicate that characterization of corrosion susceptibility should involve information on the residual mechanical properties of a structural material following exposure to corrosive environment. In the following the effect of corrosion and hydrogen embrittlement on the mechanical behavior of aluminum aircraft alloys is discussed. The work is based on extended, new experimental data of the aircraft aluminum alloys 2024, 6013, 2091, and 8090. A short overview of the currently used corrosion resistance characterization procedures is first made along with a critical appraisal of their suitability for characterization of aluminum alloy corrosion susceptibility. Evaluation of corrosion resistance is performed based on the tensile behavior following several types of corrosion tests; to the evaluation conventional metallographic and stereoscopic characterization of corroded specimens is also employed. The results are referred to current considerations of the effect of corrosion on structural integrity analysis of aged aircrafts. Results on the fatigue behavior of corroded 2024 alloy specimens are presented as well. The results include stress-life (S–N) curves, as well as, fatigue crack growth tests for several R ratios. Finally, the obtained results are discussed under the viewpoint of hydrogen embrittlement. For investigating the possible links between material embrittlement and corrosion-induced hydrogen evolution, experimental determination of hydrogen uptake has been performed followed by controlled heating experiments in order to determine the hydrogen trapping states in the material.

9.6.2 Experimental Procedures

9.6.2.1 Materials and Specimens

In the present investigation the aluminum alloys 2024, 6013, 8090, and 2091 were used. Alloy 2024 is the most widely used aircraft structure aluminum alloy; alloy 6013 is weldable; the investigated aluminum–lithium alloys, in addition to their desirable high values of specific strength and specific modulus of elasticity, provide good creep resistance and are considered as candidate materials for the European new-generation civil supersonic aircraft. Chemical compositions of the alloys are given in Table 9.32; the selected materials refer to the aluminum alloy systems Al–Cu, Al–Si–Mg–Cu, and Al–Li. All alloys were received in sheet form of 1.6 mm nominal thickness in the following temper conditions: T3 for alloy 2024, T6 for alloy 6013, T81 for alloy 8090, and T3 for alloy 2091. Tensile specimens were machined according to the specification ASTM E8m-94a [63]; specimens were cut in both longitudinal (L) and long transverse (LT) direction. Prior to tensile testing the specimens were precorroded as described below. Corroded tensile specimens have been also used for the metallographic and stereoscopic corrosion characterization. A portion of the 2024-T351 alloy tensile specimens were first subjected to hard anodization coating and sealing according to MIL-A-8625E specification [64] and then corroded. The thickness of the anodization layer was 50 μm. Sealing was performed by immersing the anodized specimens in a hot aqueous 5% sodium dichromate solution. Fatigue specimens to derive S–N curves were machined according to the specification ASTM E466-82 [65]; and specimens to measure fatigue crack growth were machined according to ASTM E647-93 [66].

TABLE 9.32 Chemical Composition (in wt %) of Aluminum Alloys 2024, 6013, 8090, and 2091

Aluminum Alloy	Chemical Composition (in wt %)										
	Si	Fe	Cu	Mn	Mg	Cr	Zn	Ti	Ni	Zr	Li
2024	0.10	0.18	4.35	0.67	1.36	0.02	0.07	0.03	—	0.01	—
6013	0.25	—	0.90	0.35	0.95	—	—	—	—	—	—
8090	0.02	0.05	1.26	0.04	0.83	0.003	0.02	0.024	0.004	0.06	2.34
2091	0.044	0.034	2.02	—	1.25	—	—	0.025	—	0.085	1.97

9.6.2.2 Corrosion Characterization Procedures

Presently, in order to characterize corrosion susceptibility of aircraft aluminum alloys, accelerated laboratory corrosion tests as well as natural corrosion tests are used. The tests include exposure of the specimen into a certain corrosive environment as defined in the respective specification and then determination of weight loss as well as type and depth of corrosion attack. Some companies involve in their internal corrosion test evaluation procedures measurement of pitting density as well. The specifications for atmospheric corrosion tests include the determination of tensile properties after exposure of the specimens to atmospheric corrosion for certain time intervals. The most widely used laboratory corrosion tests will be shortly reviewed in this chapter. Yet, with the exception of the atmospheric corrosion test, any correlation of the above tests to the real service environment of aircraft components is relatively arbitrary. Most characteristic is the case of the widely used exfoliation corrosion test. It has been originally specified to evaluate the quality of a certain heat treatment procedure and not to simulate in the laboratory any operational corrosive environment. Even the salt spray test, which obviously relies on exposure close to the sea, should be interpreted very carefully. Presently, no models exist to correlate the accelerated corrosion attack of a specimen at laboratory conditions to the long-term gradually accumulated corrosion in the operating conditions of a structure. In addition, the mentioned corrosion susceptibility evaluation does not provide any information concerning the possible influence of progressing corrosion attack on the material's mechanical properties.

One should consider that current design specifications of aircraft components do not account for corrosion-induced structural degradation in service. Even in calculations of the residual strength of aged components, the material properties employed refer to the virgin material. In case studies discussed recently in the open literature [39], the presence of corrosion in an aged aluminum aircraft member is correlated to an increase of the stress gradients applied on the member; it is assumed that the corrosion-attacked surface layer of the material is not capable of carrying a load. Thus, the thickness of the material may be reduced essentially and the stress gradients increase, respectively. This approach accounts for the corrosion-induced decrease of yield and ultimate tensile stress. Yet, in several recent publications corrosion has been related to a dramatic material embrittlement as well [41, 42, 60, 61]; the phenomenon has been attributed to corrosion-induced hydrogen evolution [60], which is trapped at different states in the material [61]. The assumption that the corroded material surface layer is not capable of carrying a load does not account for corrosion-induced hydrogen embrittlement.

The remarks made above indicate that characterization of corrosion susceptibility should involve information on the residual mechanical properties of a structural material following exposure to a corrosive environment. In addition, there is a need to advance models capable of correlating the in-service expected long-term corrosion induced material property decrease to the decrease determined at accelerated corrosion attack conditions in the laboratory. This will allow as well to specify accelerated laboratory corrosion tests which simulate the in-service expected corrosion attack. The corrosion processes used to precorrode the specimens are described below; they refer to the most commonly used corrosion tests in today praxis.

Exfoliation Corrosion (EXCO) Test The EXCO test was conducted according to ASTM specification G34-90 [62]. Following the G34-90 specification the recommended duration of specimen exposure in the corrosive solution is 48 and 96 h, respectively. In order to determine with even greater precision the behavior of the 2024-T3 alloy for the entire test duration of 96 h and to locate the exact time for the initiation of pitting and exfoliation corrosion, additional specimens were used as well; they correspond to intermediate durations of exposure. For the corrosive solution, the following chemicals were diluted in 1 L distilled water to the indicated concentrations: NaCl(4.0 *M*), KNO_3 (0.5 *M*), and HNO_3 (0.1 *M*). The initial pH was 0.3. The required amount of the solution was 10 mL/cm^2 according to the specification G34-90. The specimens were exposed individually. The solution temperature was maintained constant at $25 \pm 3°C$ during the entire experiment by conducting the experiment in a controlled temperature chamber. During the specimen exposure, regular pH measurements of the solution were taken.

To determine weight loss, specimens were weighed before and after exposure. Specimen cleaning after removal from the corrosive solution was conducted by rinsing in distilled water, soaking in concentrated nitric acid, rinsing in distilled water, and thoroughly drying in a hot-air stream. Particular attention was payed to the removal of hydroxide deposits (white film) from the surface of the exposed coupons.

Alternate Immersion Test The alternate immersion (AI) test was conducted according to ASTM specification G44-94 [67]. As for the EXCO test, additional periods of exposure were selected in order to record the development of corrosion and to determine the exact appearance time of pitting. A 3.5-wt % NaCl solution in distilled water was used. The amount of solution according to ASTM G44-94 was 32 mL/cm^2. All specimens were exposed simultaneously to the solution. The total amount of solution was 29 L. The solution temperature was maintained constant at $25 \pm 3°C$ during the entire experiment. The solution pH was in the 6.4–7.2 range.

The alternate immersion apparatus consisted of two Plexiglas tanks and two transfer pumps suitable for corrosive solutions. The operation of the apparatus was controlled automatically. The specimens, suspended by polyvinyl chloride (PVC) beams, were submerged in the first tank. After 10 min exposure, the solution in the first tank was pumped into the second tank, where it remained for 50 min, before it was pumped back into the first tank. The time cycle (1 h) was constant for all experiments. As an example, the 30 day exposure corresponded to a total of 720 cycles. In order to determine and record weight loss, the specimens were weighed before and after the experiment. Specimen cleaning was conducted according to ASTM specification G1-90 [68].

Intergranular Corrosion The intergranular corrosion test was conducted according to ASTM G110-92 specification [69]. Prior to immersion in the test solution, each specimen was immersed for 1 min in an etching cleaner at 93°C. The etching cleaner was prepared by adding 50 mL of nitric acid, HNO_3, (70%) and 5 mL of hydrofluoric acid, HF (48%), to 945 mL of distilled water. The specimens were first rinsed in the reagent water and immersed in concentrated nitric acid (70%) for 1 min. The specimens were rinsed again in distilled water and dried into moving air. Then they were immersed into the test solution. The test solution consisted of 57 g of sodium chloride (NaCl) and 10 mL of hydrogen peroxide (H_2O_2), prepared just diluted to 1 L with reagent water. The test solution volume per exposure area was 8 mL/cm^2 of specimen surface area. The solution temperature was maintained at 30 ± 3°C. The exposure time was 6 h. After exposure, each specimen was rinsed with reagent water and allowed to dry. Examination of specimens was made following the specification ASTM G110-92.

Salt Spray Test The salt spray test was conducted according to ASTM B117-94 specification [70]. The salt solution was prepared by dissolving 5 parts by mass of sodium chloride in 95 parts of distilled water. The pH of the salt solution was such that when atomized at 35°C the collected solution was in the pH range from 6.5 to 7.2. The pH measurement was made at 25°C. The temperature at the exposure zone of the salt spray chamber was maintained at $35 + 1.1 - 1.7$°C. The test duration was 30 days. After the exposure the specimens were washed in clean running water to remove salt deposits from their surface and then immediately dried.

Cyclic Acidified Salt Fog Test The cyclic acidified salt fog test was performed according to ASTM G85-94 specification, Annex A2 [71]. The salt solution was prepared by dissolving 5 parts by weight of sodium chloride, NaCl, in 95 parts of distilled water. The pH value of the solution was adjusted to range between 2.8 and 3.0 by the addition of the acetic acid. The temperature in the saturation tower was 57 ± 1°C. The temperature in the exposure zone of the salt spray chamber was maintained at $49 + 1.2 - 1.7$°C. The specimens were subjected to 6 h repetitive corrosion cycles; they included: 45 min spray, 120 min dry-air purge, and 195 min soak at high relative humidity. The test duration was 30 days.

Atmospheric Corrosion Test The tests were made in accordance to ASTM G50-76 specification [72]. Exposure racks and framers were prepared according to the requirements of the above specification. A specimen exposure angle of 30° from the horizontal, facing south, was selected. Before exposure, corrosion specimens were weighed to the nearest 1×10^{-4} g. Prior to the exposure all data recommended in ASTM G33-88 specification [73] were recorded. Atmospheric factors were recorded continuously. Specimens were removed every 3 months starting from the third month of exposure. In the present work, the results cover an exposure period of 21 months; tests of longer exposure duration are ongoing. After the exposure, weight loss, depth of pits at skyward, and groundward surface were measured; pitting density was also determined using an image analysis facility.

9.6.2.3 Mechanical Testing

Following exposure in the corrosive environments described above, the corroded tensile specimens were subjected to tensile testing. All tensile tests carried out are summarized in Tables 9.33– 9.36. The test series performed included: (i) tensile tests on uncorroded specimens to derive the reference tensile behavior of the material, (ii) tensile tests on specimens subjected to accelerated corrosion tests, (iii) tensile tests on specimens exposed to the corrosive environment for different exposure times to determine the gradual tensile property degradation during corrosion exposure in accelerated laboratory tests as well as in atmospheric

TABLE 9.33 Tensile Tests Performed on Alloy 2024-T351

Test Series	Test Series Description	Corrosion Exposure Prior to Tensile Test	Specimen Direction	Number of Tests Performed
1	Tensile tests on reference materials	None	L/LT	3/3
2	Tensile tests after accelerated corrosion testing according to ASTM specifications	Exfoliation corrosion 48 h	L/LT	3/6
		Exfoliation corrosion 96 h	L/LT	3/2
		Intergranular corrosion 6 h	L/LT	3/3
		Alternate immersion 30 days	L/LT	3/3
		Salt spray 30 days	L/LT	3/2
		Cyclic acidified salt fog (mastmaasis) 30 days	L/LT	3/3
		Atmospheric corrosion 12 months	L/LT	4/4
3	Tensile tests after exfoliation or atmospheric corrosion	Exfoliation corrosion (exposure times: 0.3, 2.0, 24, 48, 72, 96 h)	L/LT	15/15
		Atmospheric corrosion (exposure times: 3, 6, 9, 12, 15, 18, 21 months)	L/LT	27/27
4	Tensile tests on reference specimens subjected to anodizing and sealing	None	L	3
5	Tensile tests after accelerated corrosion testing according to ASTM specifications	Exfoliation corrosion 24 h	L	2
		Exfoliation corrosion 48 h	L	3
		Alternate immersion 30 days	L	3
6	Tensile tests after exfoliation corrosion and mechanical removal of the corroded material surface layer	Exfoliation corrosion (exposure times: 48 and 72 h)	L/LT	4/4

TABLE 9.34 Tensile Tests Performed on Alloy 6013-T6

Test Series	Test Series Description	Corrosion Exposure Prior to Tensile Test	Specimen Direction	Number of Tests Performed
1	Tensile tests on reference materials	None	L/LT	4/4
2	Tensile tests after accelerated corrosion testing according to ASTM specifications	Exfoliation corrosion 48 h	L/LT	4/4
		Exfoliation corrosion 96 h	L/LT	4/4
		Alternate immersion 30 days	L/LT	3/3
3	Tensile tests after exfoliation corrosion for different exposure times	Exfoliation corrosion (exposure times: 24, 48, 72, 96 h)	L/LT	14/14
4	Tensile tests after exfoliation corrosion and mechanical removal of the corroded material surface layer	Exfoliation corrosion 48 h	L/LT	2/2
		Exfoliation corrosion 72 h	L/LT	2/2

(natural) environment, and (iv) tensile tests on specimens subjected to corrosion exposure and then to removal of the corrosion-affected surface layer by machining in order to investigate whether the corrosion-induced tensile property degradation is volumetric. For alloy 2024-T3 some specimens were subjected to hard anodizing and sealing prior to corrosion exposure. They were used to derive (i) the reference tensile behavior of uncorroded material subjected to anodizing and sealing and (ii) to evaluate this protective role of the coating when the material is exposed in corrosive environments. Indicated in Tables 9.33–9.36 are also the corrosion tests applied to precorrode the tensile specimens, the specimen direction, as well as the number of tests performed. The tensile tests were performed according to ASTM E8m-94a [63] specification. For the tests a 200 KN Zwick universal testing machine and a servohydraulic MTS 250 KN machine were used. The deformation rate was 10 mm/min.

The fatigue tests performed to obtain S–N curves on as received, on protected, as well as on corroded 2024 alloy are summarized in Table 9.37. The same material fatigue crack growth tests were performed for the R values 0.01, 0.1, 0.5, and 0.7. For each R value two tests were performed on the as-received specimens and two tests on specimens subjected to 36 h exfoliation corrosion exposure prior to fatigue crack growth test. The frequency for all fatigue crack growth tests was constant at 20 Hz. Fatigue crack growth measurements were

TABLE 9.35 Tensile Tests Performed on Alloy 2091-T84

Test Series	Test Series Description	Corrosion Exposure Prior to Tensile Test	Specimen Direction	Number of Tests Performed
1	Tensile tests on reference materials	None	L/LT	4/4
2	Tensile tests after accelerated corrosion testing according to ASTM specifications	Exfoliation corrosion 48 h	L/LT	4/4
		Exfoliation corrosion 96 h	L/LT	4/4
		Intergranular corrosion 6 h	L/LT	5/5
		Alternate immersion 30 days	L/LT	3/3
		Salt spray 30 days	L/LT	3/3
		Cyclic acidified salt fog (mastmaasis) 30 days	L/LT	3/3
		Atmospheric corrosion 12 months	L/LT	3/3
3	Tensile tests after exfoliation corrosion for different exposure times	Exfoliation corrosion (exposure times: 24, 48, 72, 96 h)	L/LT	15/15
4	Tensile tests after exfoliation corrosion and mechanical removal of the corroded material surface layer	Exfoliation corrosion 48 h	L/LT	2/2
		Exfoliation corrosion 72 h	L/LT	2/2

made using the potential drop method. All fatigue tests were performed on a servohydraulic MTS 250 KN machine.

9.6.2.4 Material Corrosion Characterization Procedure

The characterization of corroded specimens was conducted by stereoscopic and metallographic analysis. Stereoscopic analysis was conducted with a stereomicroscope at 60×. The analysis included observation of the corroded surfaces and determination of pit density (pits/m^2).

The preparation of specimens for metallographic analysis consisted of cutting, mounting, grinding, and polishing. Mounting was performed in a vacuum impregnation device in order to preserve the shape of pits and keep the corroded material in place during subsequent preparation. The specimens were chemically etched using Keller's reagent. Measurements of the maximum depth of attack were conducted as well.

9.6.2.5 Determination of Trapped Hydrogen

For the hydrogen measurements, strips 2.5 mm wide and 40 mm long were cut at right angles to the rolling direction and were then exposed to exfoliation

TABLE 9.36 Tensile Tests Performed on Alloy 8090-T81

Test Series	Test Series Description	Corrosion Exposure Prior to Tensile Test	Specimen Direction	Number of Tests Performed
1	Tensile tests on reference materials	None	L/LT	4/4
2	Tensile tests after accelerated corrosion testing according to ASTM specifications	Exfoliation corrosion 48 h	L/LT	4/4
		Exfoliation corrosion 96 h	L/LT	4/4
		Intergranular corrosion 6 h	L/LT	3/3
		Alternate immersion 30 days	L/LT	3/3
		Salt spray 30 days	L/LT	3/3
		Cyclic acidified salt fog (mastmaasis) 30 days	L/LT	3/3
		Atmospheric corrosion 12 months	L/LT	3/3
3	Tensile tests after exfoliation corrosion for different exposure times	Exfoliation corrosion (exposure times: 8, 16, 24, 48, 72, 96 h)	L/LT	20/20
4	Tensile tests after exfoliation corrosion and mechanical removal of the corroded material surface layer	Exfoliation corrosion 48 h	L/LT	2/2
		Exfoliation corrosion 72 h	L/LT	2/2

TABLE 9.37 Fatigue Tests Performed on 2024 Alloy to Derive S–N Curves; $f = 25$ Hz

Specimen Preparation Prior to Fatigue	K_t	$R = \sigma_{\min}/\sigma_{\max}$	Number of Tests to Derive the S–N Curve
As received	1	0,1	17
Anodization coating and sealing	1	0,1	11
Anodization coating and sealing	2,5	0,1	11
Anodization coating, sealing, and then exposure to exfoliation corrosion	1	0,1	11
Anodization coating, sealing, and then exposure to exfoliation corrosion	2,5	0,1	12

corrosion. The large surface-to-volume ratio of the specimen was chosen with a view to decreasing the hydrogen evolution time and increasing the sensitivity of the measurements.

Hydrogen evolved from the corroded specimen with controlled heating in an inert atmosphere and was measured by a gas chromatograph. The specimen was placed in a 10-mm-diameter quartz tube and was held in place by an inert porous bottom (quartz wool). The tube was inserted in a temperature-controlled (±1°C) vertical furnace and was heated at a controlled rate. A continuous, high-purity nitrogen flow was maintained through the tube at a rate of 20 mL/min and was then driven to a gas chromatograph equipped with a TCD detector. Calibration runs were performed using standard H_2-N_2 samples. Blind experiments were conducted with an empty tube heated to 600°C and no hydrogen was detected.

The rate of heating is an important parameter in the experiments. A very slow rate (typically 0.5°C/min), followed by extended constant-temperature ramps, was used initially to accurately detect the temperatures at which hydrogen first appears. This procedure was adopted to avoid overlap of desorption fields from different trapping states. However, quantitative estimate of total amount evolved at each stage is awkward with this setup; small quantities of hydrogen evolve for extended periods of time, and the integration of the concentration vs. time curve is not reliable. Thus, after detection of the temperatures at which hydrogen first appears, the bulk of the experiments was conducted at an optimum heating rate (5–6°C/min), which permitted separation of the peaks and reliable integration.

9.6.3 Results and Discussion

9.6.3.1 Microstructural Characterization of Corrosion

Performed metallographic and stereoscopic corrosion analysis has shown that in all accelerated laboratory corrosion tests, corrosion developed gradually from pitting into intergranular attack. In contrast, outdoor exposure for durations up to 24 months did not lead to appreciable corrosion damage. Protection by anodizing and sealing decreased significantly the rate of corrosion attack. Characteristic results are shown for the EXCO test as well as the alternate immersion test for alloy 2024-T3. Pitting density (in pits/m^2) and maximum depth of attack for the EXCO test are given in Fig. 9.5. Between 5 and 24 h, the pitting density increased slowly. Between 24 and 36 h, pitting density increased even more, and after 48 h exfoliation corrosion commenced. The depth of attack increased with exposure time (Fig 9.5) and reached a value of 0.35 mm after 96 h of exposure. Characteristic micrographs of sectioned specimens are shown in Fig. 9.6 *a*, and *b*. Intergranular attack (cracks) (Fig. 9.6*a*) developed in the area adjacent to the corrosion pits. For long exposure times (>72 h), the cracks run parallel to the specimen surface (parallel to rolling direction) resulting in macroscopic exfoliation of the specimen (Fig 9.6*b*). The derived maximum depth of attack of all alloys investigated in this study to exfoliation corrosion solution are summarized in Table 9.38. These values were used as reference to machine the corrosion-attacked material surface for performing tensile tests to investigate whether the corrosion-induced tensile property degradation is volumetric. For specimens exposed at the alternate immersion test pitting density (in pits/m^2) the maximum depth of attack is given in Fig. 9.7. The rate of increase in pitting density was lower than in the EXCO test, but it was roughly constant

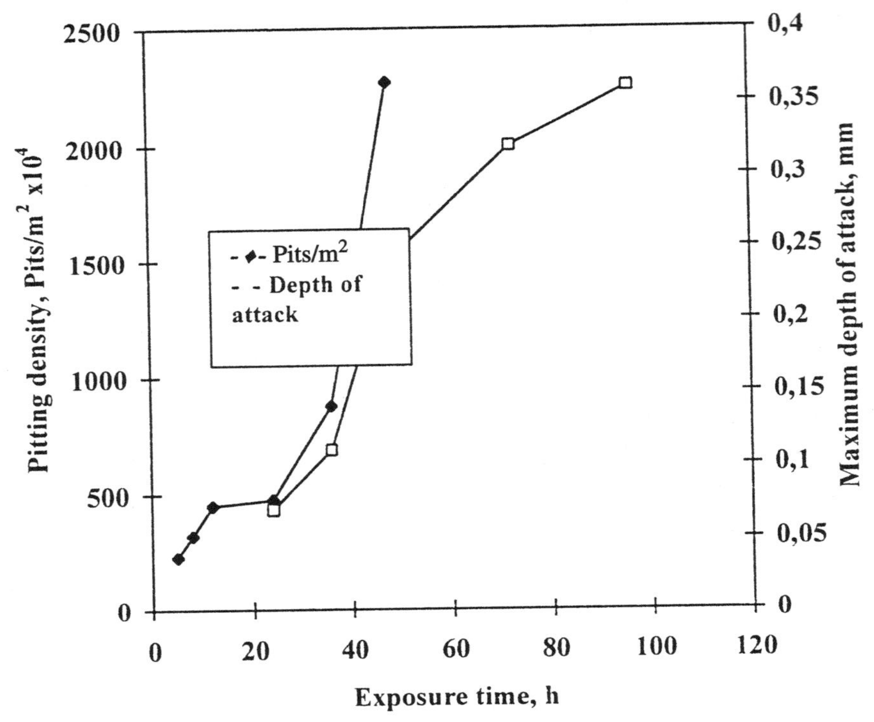

FIGURE 9.5 Variation of pitting density and maximum depth of attack with exposure time during EXCO test.

throughout the entire exposure period of 90 days. It should be noticed that the alternate immersion test resulted in a much higher final pitting density than the EXCO test. However, the pits were at the same time finer. The depth of attack showed a similar behavior to the exfoliation test in that it increased slowly in the beginning of exposure and proceeded faster with longer exposure. Alternate immersion resulted in a higher depth of attack than the EXCO test. Figure 9.8 shows a characteristic micrograph of a sectioned specimen. As in the EXCO test, a network of intergranular attack developed in areas adjacent to the corrosion pits. However, intergranular attack parallel to the rolling direction was not observed.

9.6.3.2 Tensile Behavior of Corroded Specimens

Reference Materials The tensile properties of all reference materials are summarized in Table 9.39. The drop of elongation to failure and energy density obtained on alloy 2024-T3 following anodization coating and sealing is significant; coated 2024 material is referred to as "ref. 2." Investigations on this latter material were limited to the L direction.

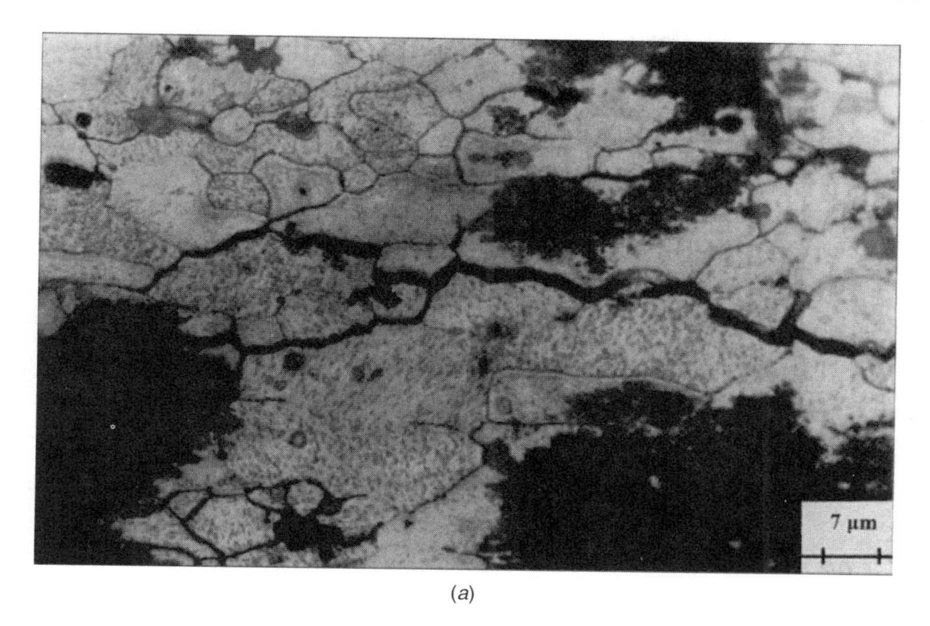

(a)

(b)

FIGURE 9.6 Micrographs of sectioned specimens exposed to the EXCO test for 96 h: (a) network intergranular cracks and (b) exfoliation crack.

TABLE 9.38 Depth of Attack after 48-h Exposure to Exfoliation Corrosion Solution

Alloy	Depth of Attack
2024	0.33 mm
6013	0.49 mm
2091	0.35 mm
8090	0.38 mm

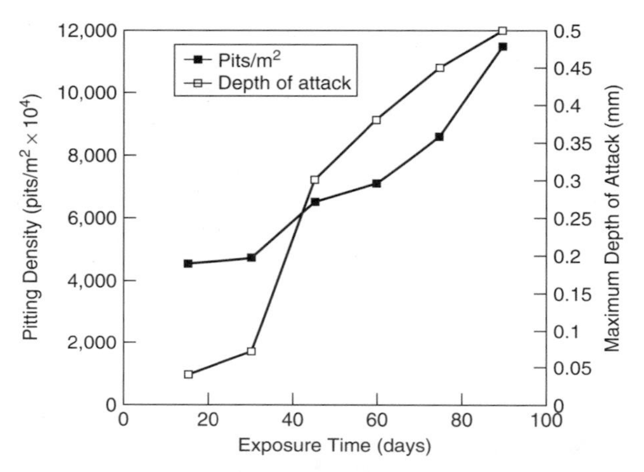

FIGURE 9.7 Variation of pitting density and maximum depth of attack with exposure time during alternate immersion test.

FIGURE 9.8 Micrograph of sectioned specimen exposed to alternate immersion test for 60 days.

TABLE 9.39 Tensile Properties of Reference Materials

Material	Yield Stress S_y (MPa)		Ultimate Tensile Stress R_m (MPa)		Elongation to Failure $A_{50}(\%)$		Energy Density $W(\mathrm{MJ/m^3})$	
	L	LT	L	LT	L	LT	L	LT
2024-T351	396	339	520	488	18	18	86.7	81.7
2024-T351 "ref. 2"	387		458		12.72		56.64	
6013-T6	371	345	407	401	11.25	11.62	44.7	45.14
2091-T84	355	359	439	470	9.5	11.8	40.2	53.2
8090-T81	347	302	437	423	8.9	12.5	36.9	50.6

Tensile Behavior after Exposure to Various Corrosion Tests Figures 9.9–9.13 show the variation of the tensile properties degradation for the investigated accelerated corrosion tests as compared to the tensile properties of the reference material; Figs. 9.9– 9.13 stand for the alloys 2024, 2024 "ref. 2," 6013, 8090, and 2091, respectively. In all figures the values of the residual tensile properties are given in percent of the respective properties of the reference material. Referring to the data in Figs. 9.5 and 9.7, it is evident that the corrosion attack causes a decrease of ultimate tensile stress and yield stress. The effect is small for most corrosion processes considered and becomes appreciable with increasing aggressiveness of the corrosive solutions. This observation does not apply for material 6013; the loss of yield and ultimate tensile stress for this material following exfoliation corrosion was remarkable. On the other hand a dramatic drop has always been determined on elongation to fracture and energy density. Tensile ductility reduction was found to be appreciable even for the 2024-T3 "ref. 2" specimens, which, prior to corrosion tests, were subjected to anodizing and sealing. Particularly strong was the influence of exfoliation corrosion; it leads to extremely low values of remaining tensile ductility. The metallographic characterization of the materials has shown serious corrosion attack in this aggressive environment, and the measured depth of attack has reached values up to 0.49 mm for the worst case of the 6013 alloy. For the 2024 alloy the measured depth of attack was under 0.33 mm. Mechanical degradation has been associated with the initiation of corrosion defects that grow and lead to early failure of the material [29, 30]. The results of the performed metallographic evaluation may explain the determined degradation of yield and ultimate tensile stress but not the dramatic embrittlement of all materials investigated. Even more difficult to explain is the measured appreciable tensile ductility drop after a short outdoor exposure time. For the same outdoor exposure conditions, the yield and ultimate tensile stress, practically, do not decrease. Notice that stereoscopic analysis of the respective tensile specimens (performed after the outdoor exposure and classical metallographic characterization following the tensile tests) could not prove the occurrence of significant corrosion. Yet, the exposure of the materials during the corrosion processes in hydrogen-rich environments

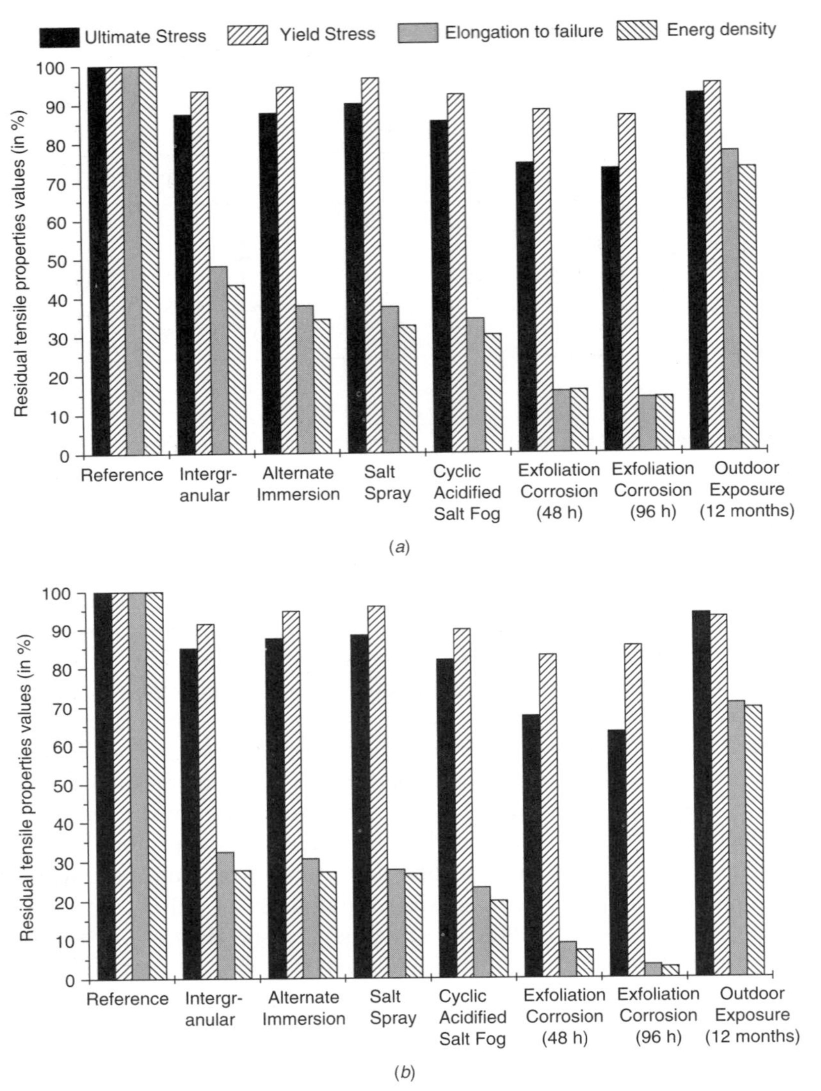

FIGURE 9.9 Tensile behavior of alloy 2024 following corrosion exposure according to relevant specifications for accelerated corrosion tests: (*a*) L direction and (*b*) LT direction.

is not considered when evaluating corrosion susceptibility; obviously, conventional metallographic corrosion characterization cannot detect the possible impact of hydrogen on the material microstructure. Although, hydrogen embrittlement may be the cause for the determined dramatic decrease of the tensile ductility properties.

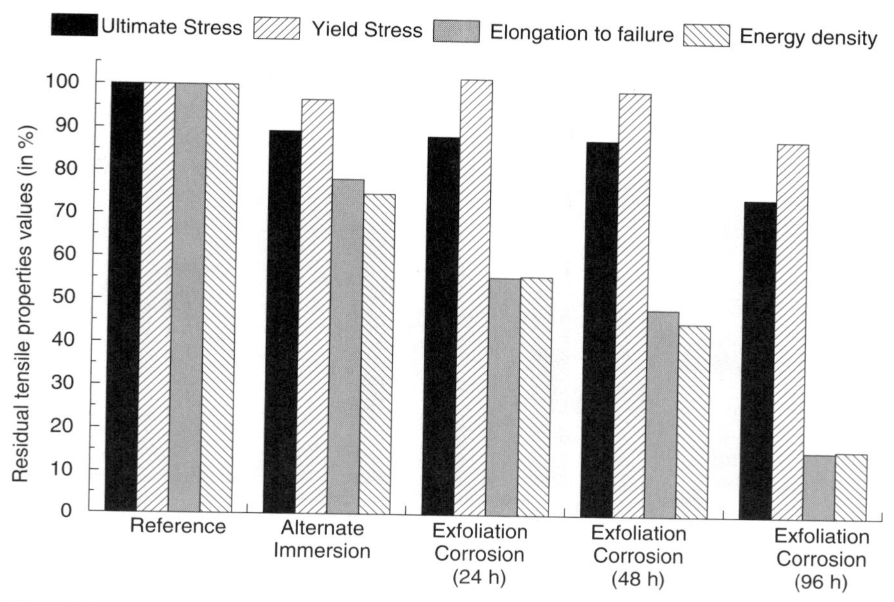

FIGURE 9.10 Tensile behavior of alloy 2024 "ref. 2," for L direction, following corrosion exposure according to relevant specifications for accelerated corrosion tests.

Influence of Exposure Time on Tensile Behavior The determined tensile property degradation occurs gradually. It is shown in Figs. 9.14– 9.17; they display the trends of property decrease versus the exposure time into exfoliation corrosion solution for the alloys 2024, 6013, 8090, and 2091, respectively. For the alloy 2024 the gradual property degradation with exposure time was determined for atmospheric corrosion tests as well Fig. 9.18. As seen in these figures all properties are decreasing nonlinearly with the exposure time. By defining with $\overline{P}(t) = [\Delta P(t)/P_{\text{initial}} \times 10^2]$ the loss in percent of the respective property during the exposure, the results of Figs. 9.14– 9.17 were used to formulate expressions of the gradual decrease of the material properties. Applied functions as well as the values of the respective fitting parameters used for the case of exposure in exfoliation corrosion solution are summarized in Table 9.40. For atmospheric corrosion, exposure duration is still too short to formulate reliable functions. The yield and ultimate stress drop during exfoliation corrosion exposure are best approximated by applying power time functions. The power exponents were found for all alloys <1, that is, the obtained property loss occurs with a decreasing rate. The reported behavior is consistent with the experimental observation of the progressive growth of a protective oxide film on the materials surface [74]. The growth of this film has been found to follow power time functions; the rate of attack during exposure decreases at very low rates or it ceases. The drop of elongation to failure and

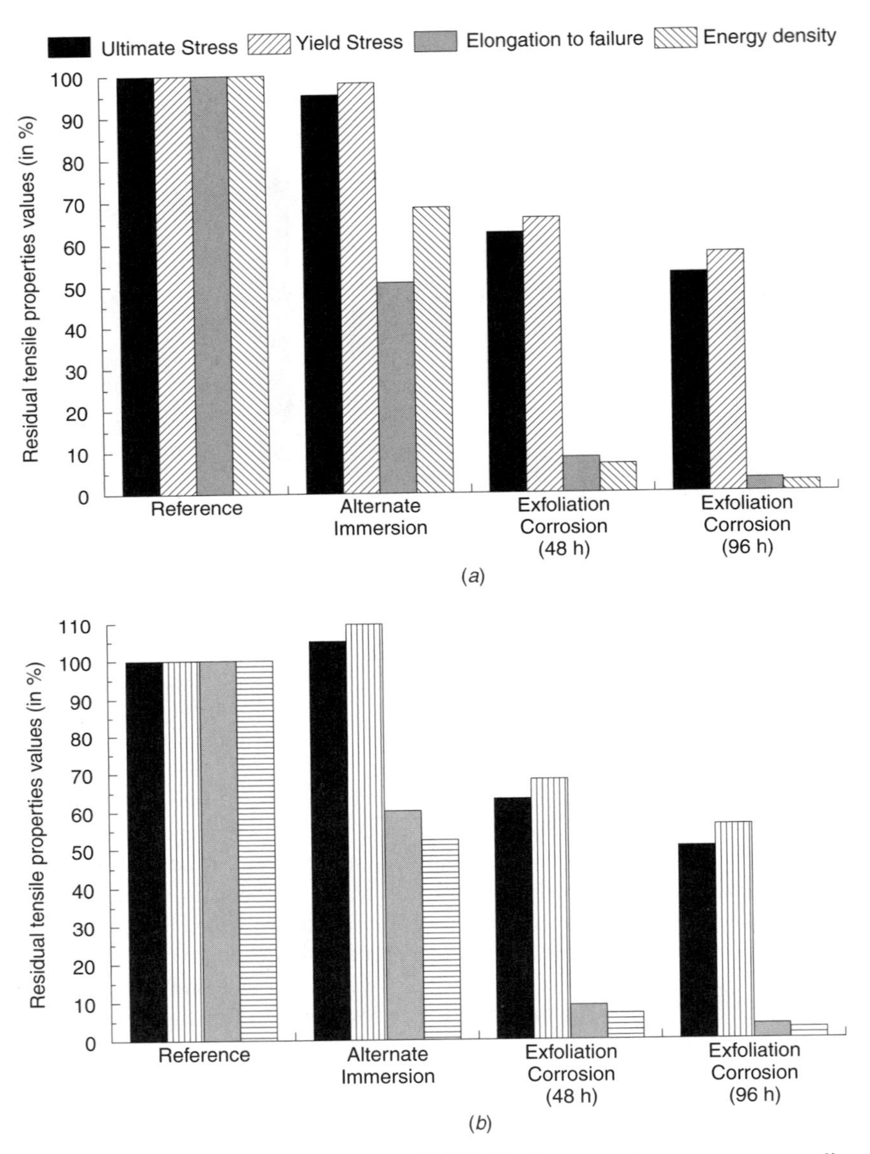

FIGURE 9.11 Tensile behavior of alloy 6013 following corrosion exposure according to relevant specifications for accelerated corrosion tests: (*a*) L direction and (*b*) LT direction.

energy density were fitted well by exponential time functions. In the equations for tensile ductility degradation, $\overline{A}_{50(f)}$ and $\overline{W}_f$ stand for the final value of the respective property loss. Hence, they may be interpreted to reflect the susceptibility of the property to the corrosive environment. The parameters β and k indicate the rate of property decrease in the corrosive solution. The exponential form of these

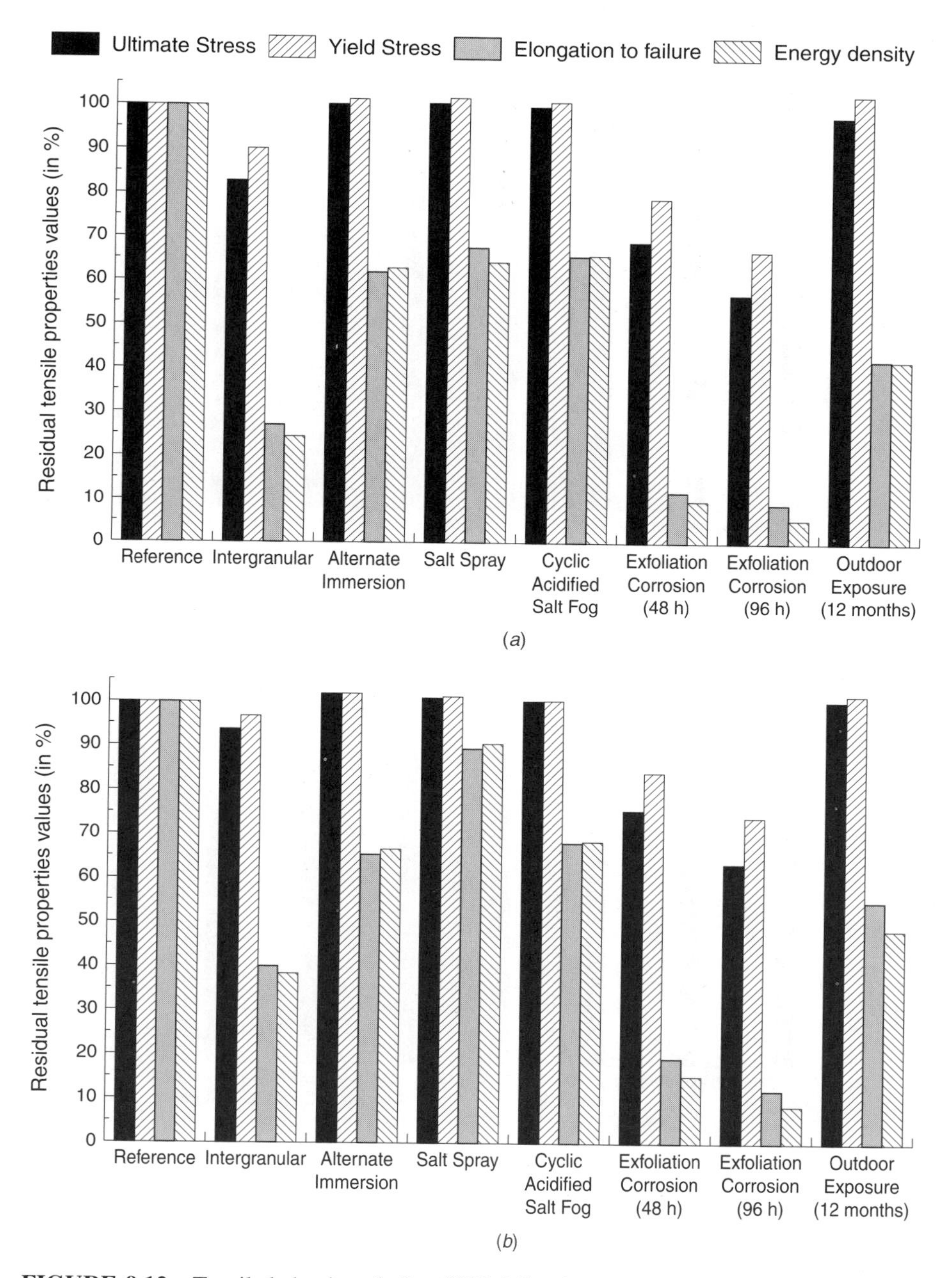

FIGURE 9.12 Tensile behavior of alloy 8090 following corrosion exposure according to relevant specifications for accelerated corrosion tests: (*a*) L direction and (*b*) LT direction.

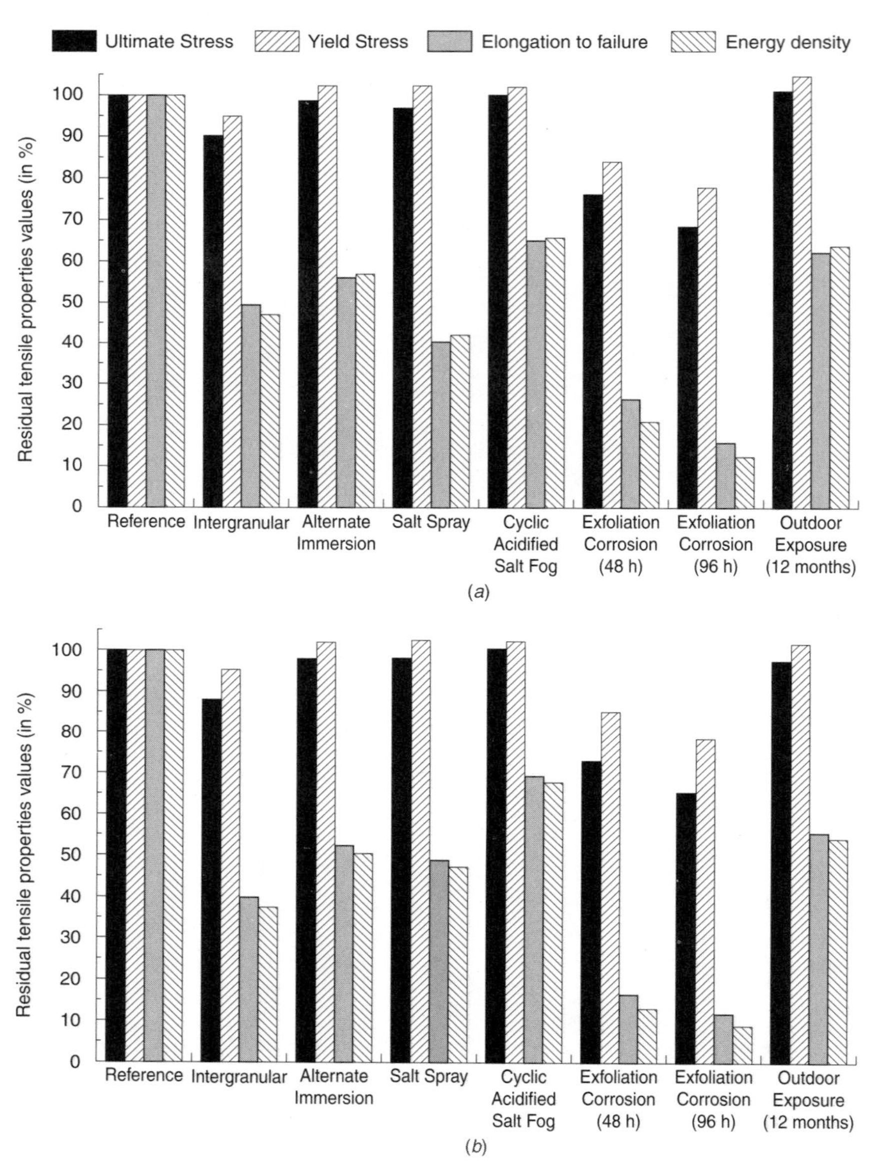

FIGURE 9.13 Tensile behavior of alloy 2091 following corrosion exposure according to relevant specifications for accelerated corrosion tests: (*a*) L direction and (*b*) LT direction.

equations leads to the association of diffusion-controlled processes. In addition, the occurrence of saturation values for the tensile ductility degradation is supporting the viewpoint of a volumetric, diffusion-controlled phenomenon. In [60] evaluation of exfoliation corrosion tests performed on 2024-T351 specimens has shown a good correlation between hydrogen evolution and the available surface

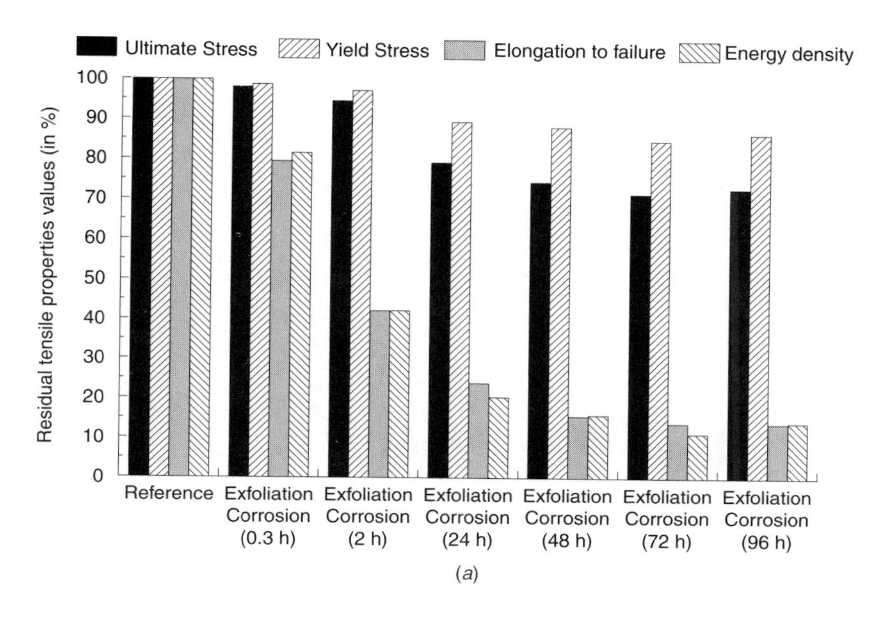

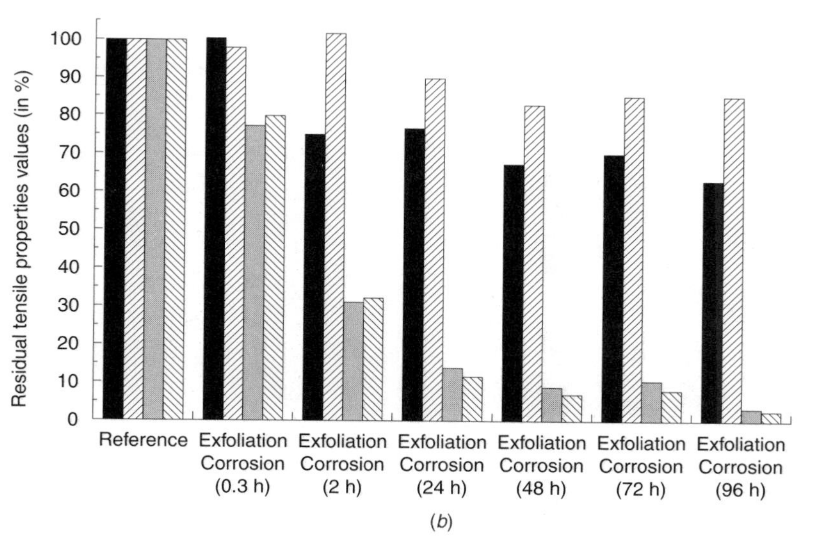

FIGURE 9.14 Gradual tensile property degradation for alloy 2024 during exposure in exfoliation corrosion solution: (*a*) L direction and (*b*) LT direction.

area for penetration. Indirect evidence was reported that hydrogen was initially absorbed chemically. Recall also present results of corrosion characterization; increasing pitting density and occurrence of intergranular attack when exposure time increases were reported. Referred results provide substantial explanations in favor of hydrogen absorption and bulk penetration.

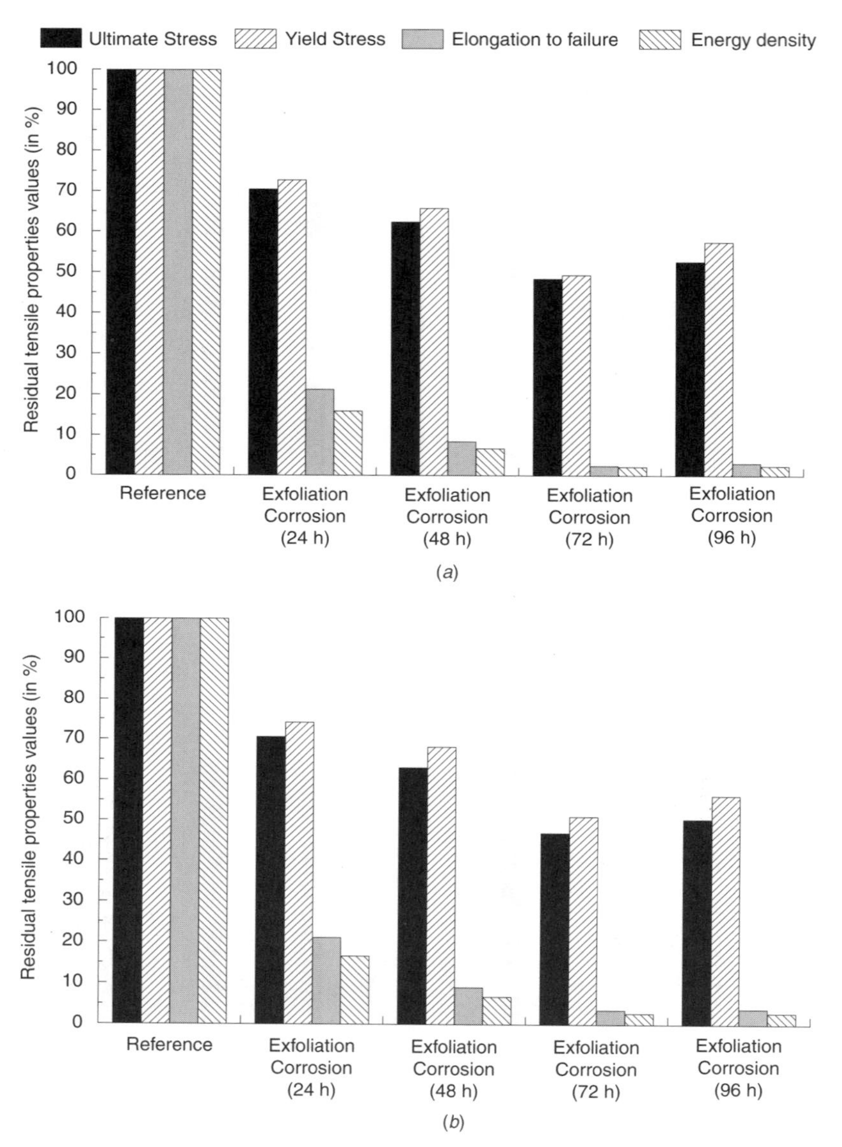

FIGURE 9.15 Gradual tensile property degradation for alloy 6013 during exposure in exfoliation corrosion solution: (*a*) L direction and (*b*) LT direction.

Tensile Behavior Following Exfoliation Corrosion and Mechanical Removal of Corrosion-Induced Surface Damage The results reported above are strengthening the hypothesis for bulk hydrogen embrittlement of the investigated aluminum alloy sheets. To investigate farther on this hypothesis, tensile specimens were subjected to mechanical removal of the corrosion-attacked surface layer

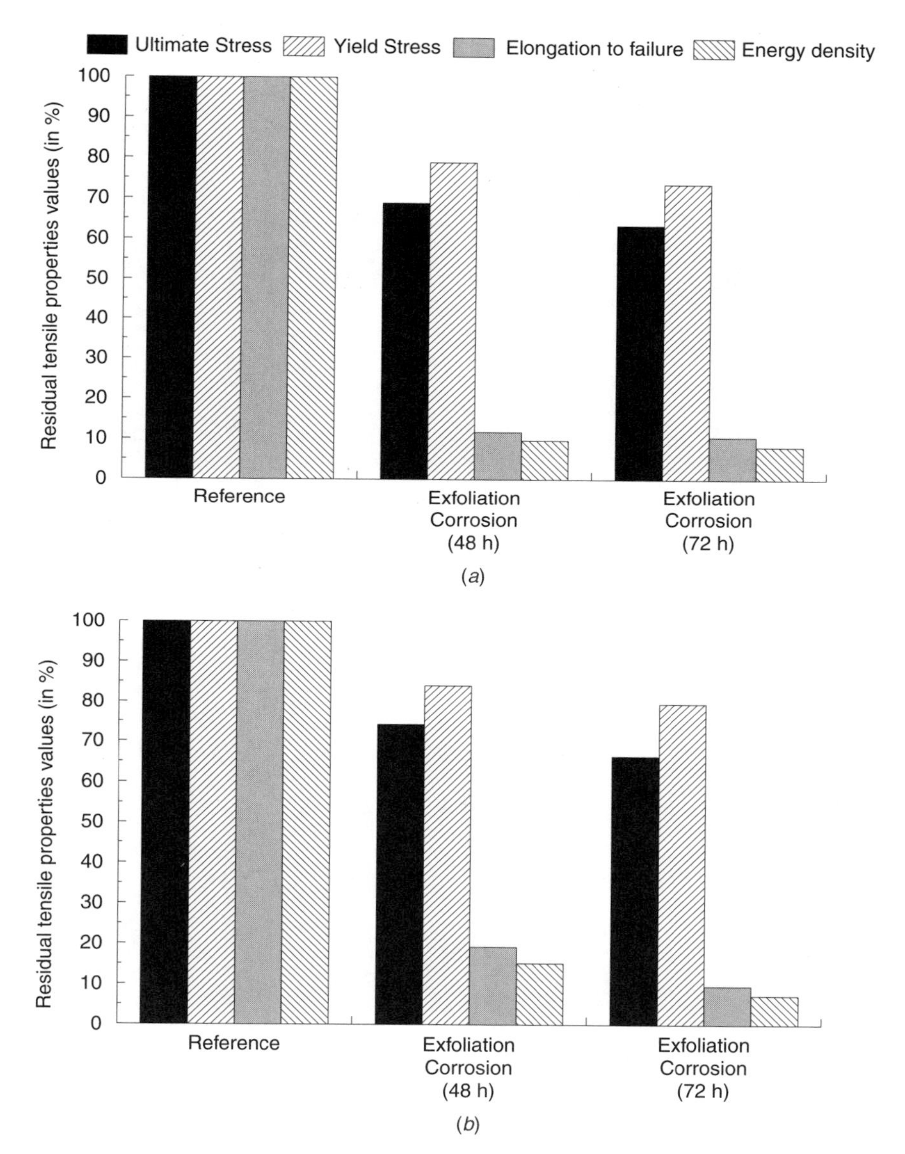

FIGURE 9.16 Gradual tensile property degradation for alloy 8090 during exposure in exfoliation corrosion solution: (*a*) L direction and (*b*) LT direction.

following their exposure in exfoliation corrosion. The depth of the removed layers was taken from Table 9.38 to reach the "uncorroded" material core. The obtained results are given in Figs. 9.19– 9.22 for the four alloys investigated. Machining of the corroded surface layer leads to an appreciable rewinning of yield and ultimate tensile stress but not into substantial recovery of the tensile ductility.

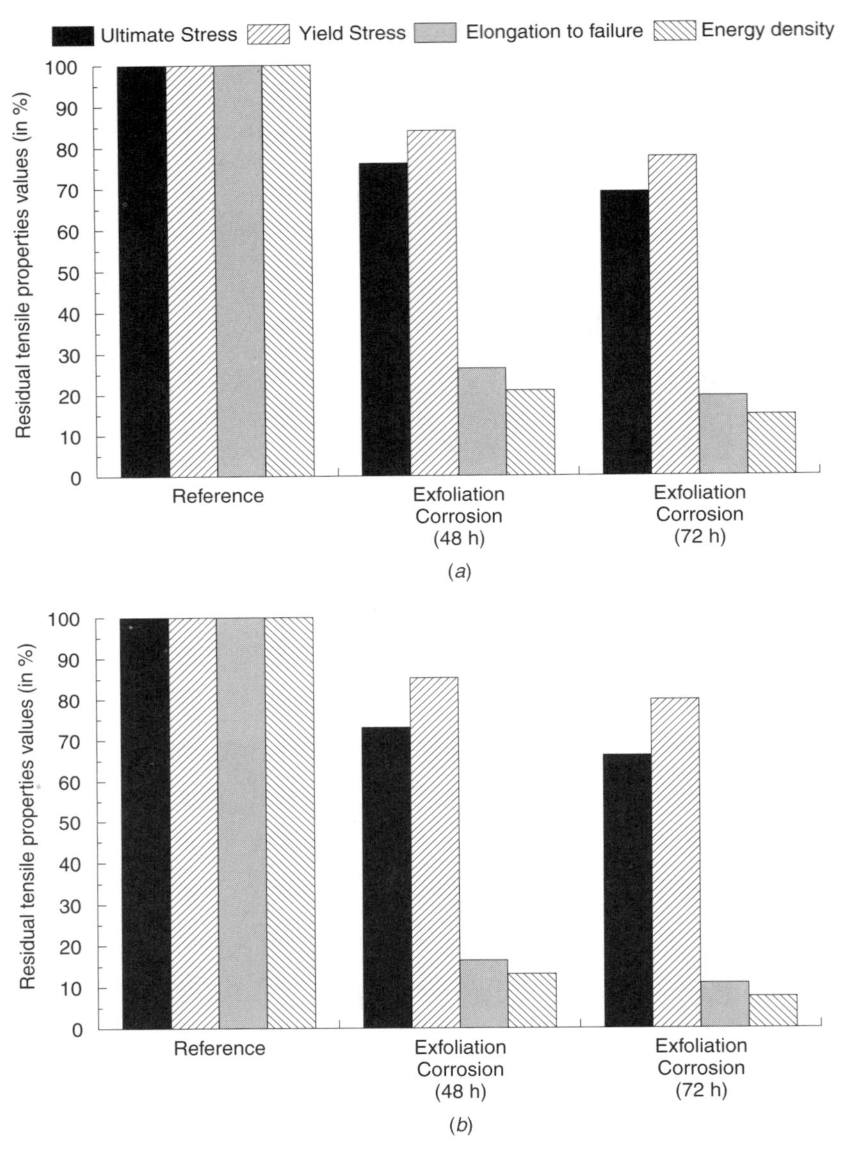

FIGURE 9.17 Gradual tensile property degradation for alloy 2091 during exposure in exfoliation corrosion solution: (*a*) L direction and (*b*) LT direction.

One should notice that the corrosion-damaged layer could be removed only across the width of the specimens. Yet, it is known that the severity of corrosion attack increases at free edges of the material. Hence, remaining edge corrosion notches at the sides of the specimen along the specimen length may explain the discrepancies between yield and ultimate tensile stress obtained following

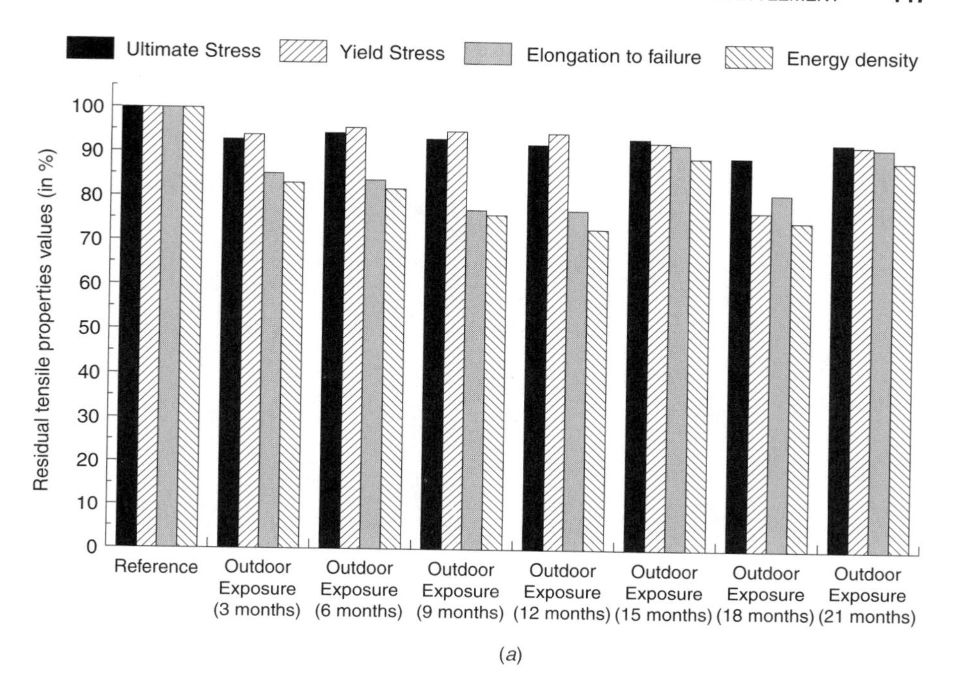

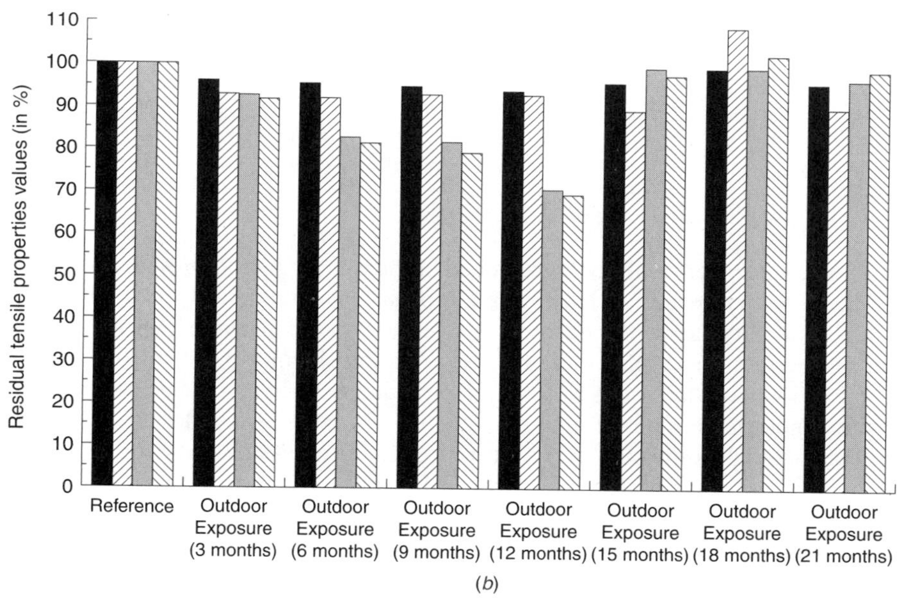

FIGURE 9.18 Gradual tensile property degradation for alloy 2024 during exposure in atmospheric corrosion: (*a*) L direction and (*b*) LT direction.

TABLE 9.40 Time Dependency of Tensile Property Degradation During Exfoliation Corrosion

			L		LT	
Property	Function	Material	$P(h^{-m})$	m	$P(h^{-m})$	m
Yield stress S_y	$\overline{S_y}(t) = P \cdot t^m$	2024	5.4	0.2	4.6	0.3
		6013	9.08	0.37	7.87	0.42
		2091	2.24	0.52	2.73	0.45
		8090	1.0	0.78	0.56	0.86
			$K(h^{-n})$	n	$K(h^{-n})$	n
Ultimate tensile stress R_m	$\overline{R_m}(t) = K \cdot t^n$	2024	9.41	0.25	9.6	0.3
		6013	8.65	0.37	6.55	0.43
		2091	7.01	0.33	8.19	0.32
		8090	3.90	0.53	2.29	0.62
			A_{50f}	$\beta(h^{-1})$	A_{50f}	$\beta(h^{-1})$
Elongation to failure A_{50}	$\overline{A_{50}}(t) = A_{50f}(1 - e^{-\beta t})$	2024	86.0	0.08	93	0.10
		6013	97.01	0.07	95.98	0.07
		2091	85.2	0.08	88.1	0.08
		8090	90.0	0.13	88.0	0.08
			W_f	$k(h^{-1})$	W_f	$k(h^{-1})$
Energy density W	$\overline{W}(t) = W_f(1 - e^{-kt})$	2024	88	0.08	95	0.10
		6013	97.8	0.08	96.72	0.08
		2091	87.0	0.08	91.5	0.08
		8090	92.0	0.15	91.5	0.08

the mechanical removal of the corroded surface layer and the respective values of the reference material. However, the persistent, dramatically low values for tensile ductility prove the occurrence of a volumetric phenomenon that cannot be related to surface oxidation processes. The results of present investigation in conjunction with the results in [60, 75], provide evidence that the determined dramatic embrittlement of the investigated alloys is caused by hydrogen absorption in the material during the corrosion processes. Present-day tests for evaluating corrosion resistance of aircraft structure aluminum alloys do not account for this important aspect.

9.6.3.3 *Fatigue Behavior of Corroded Specimens*

S–N Curves Following Exfoliation Corrosion Displayed in Fig. 9.23 are S–N curves derived for the material 2024 alloy (anodized and sealed) following exposure to exfoliation corrosion solution for 36 h for specimens without a hole ($K_t = 1$) or including a hole ($K_t = 2.5$). For comparison the S–N curves of the uncorroded alloys with and without anodizing and sealing are included as well.

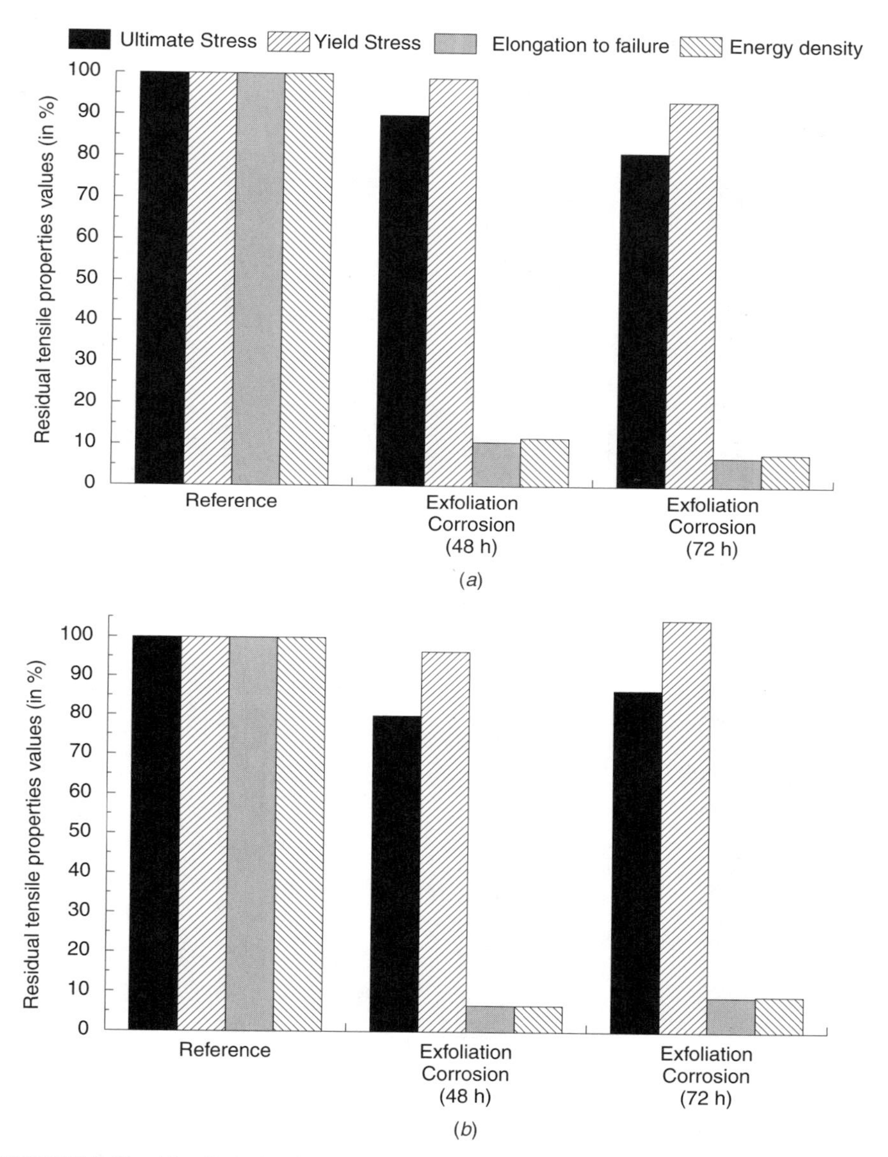

FIGURE 9.19 Tensile behavior following exfoliation corrosion and mechanical removal of corrosion-induced surface damage for alloy 2024: (*a*) L direction and (*b*) LT direction.

The results show that anodizing and sealing reduces fatigue resistance appreciably. Specimens including a hole ($K_t = 2.5$) show, as expected, reduced fatigue life as compared to specimens without a hole loaded at the same stress amplitude. The exponents of the Weibull distribution, describing the plots in Fig. 9.23, are summarized in Table 9.41; the respective curves are plotted in Fig. 9.23 as well.

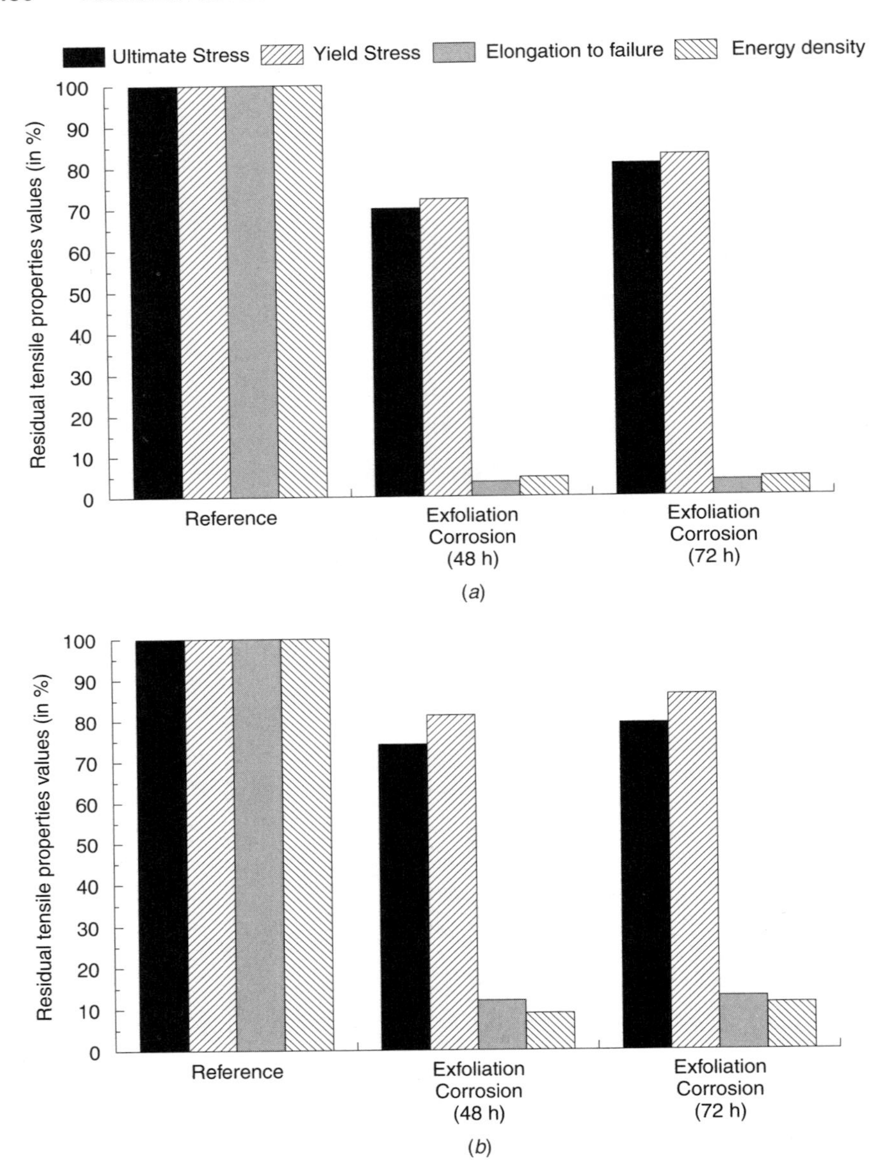

FIGURE 9.20 Tensile behavior following exfoliation corrosion and mechanical removal of corrosion-induced surface damage for alloy 6013: (*a*) L direction and (*b*) LT direction.

Fatigue Crack Growth Following Exfoliation Corrosion The above discussed corrosion-induced fatigue degradation was not reflected in the fatigue crack propagation tests performed for some material on both uncorroded and corroded anodized and sealed alloy specimens for four different $R = \sigma_{min}/\sigma_{max}$ values. The results are displayed in Fig. 9.24 where the influence of the R value on

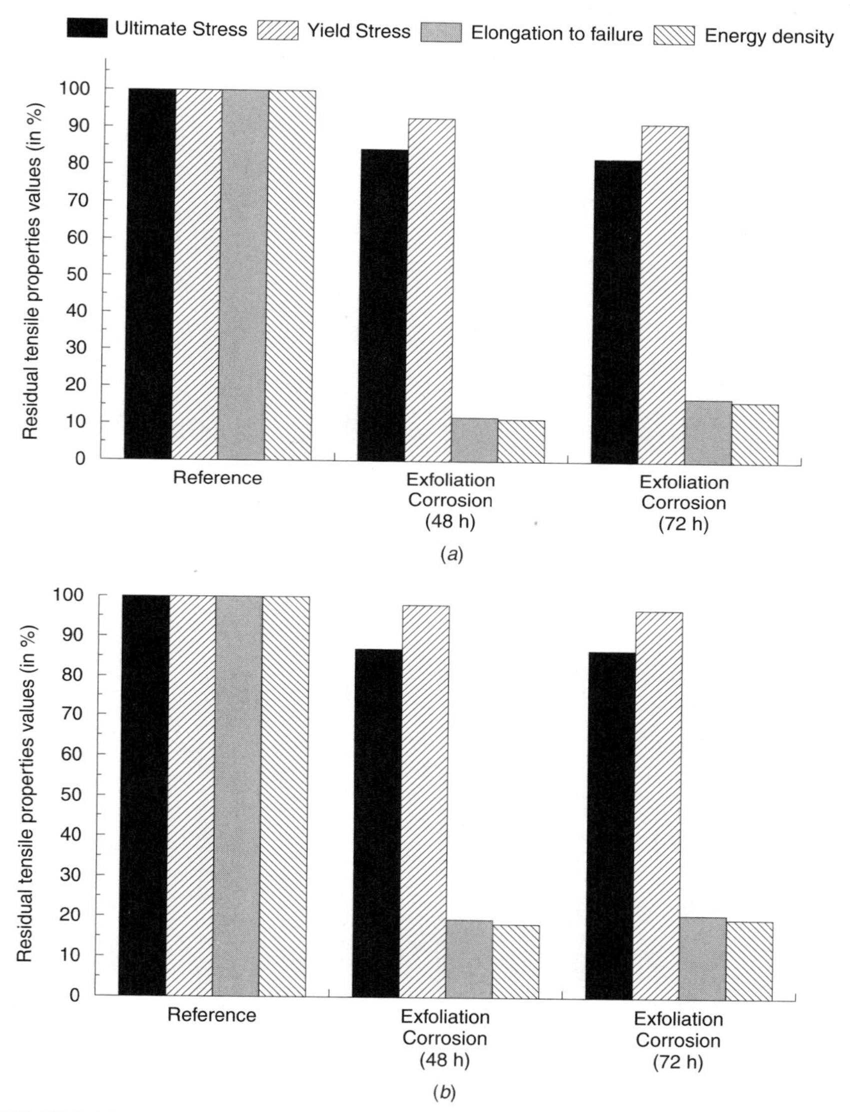

FIGURE 9.21 Tensile behavior following exfoliation corrosion and mechanical removal of corrosion-induced surface damage for alloy 8090: (*a*) L direction and (*b*) LT direction.

fatigue crack propagation rate is depicted. Corrosion caused by the exposure in exfoliation corrosion solution for 36 h has practically no influence on fatigue crack growth behavior. Summarized in Table 9.42 are the determined constants C and n of the Paris equation:

$$\frac{da}{dN} = C(\Delta K)^n$$

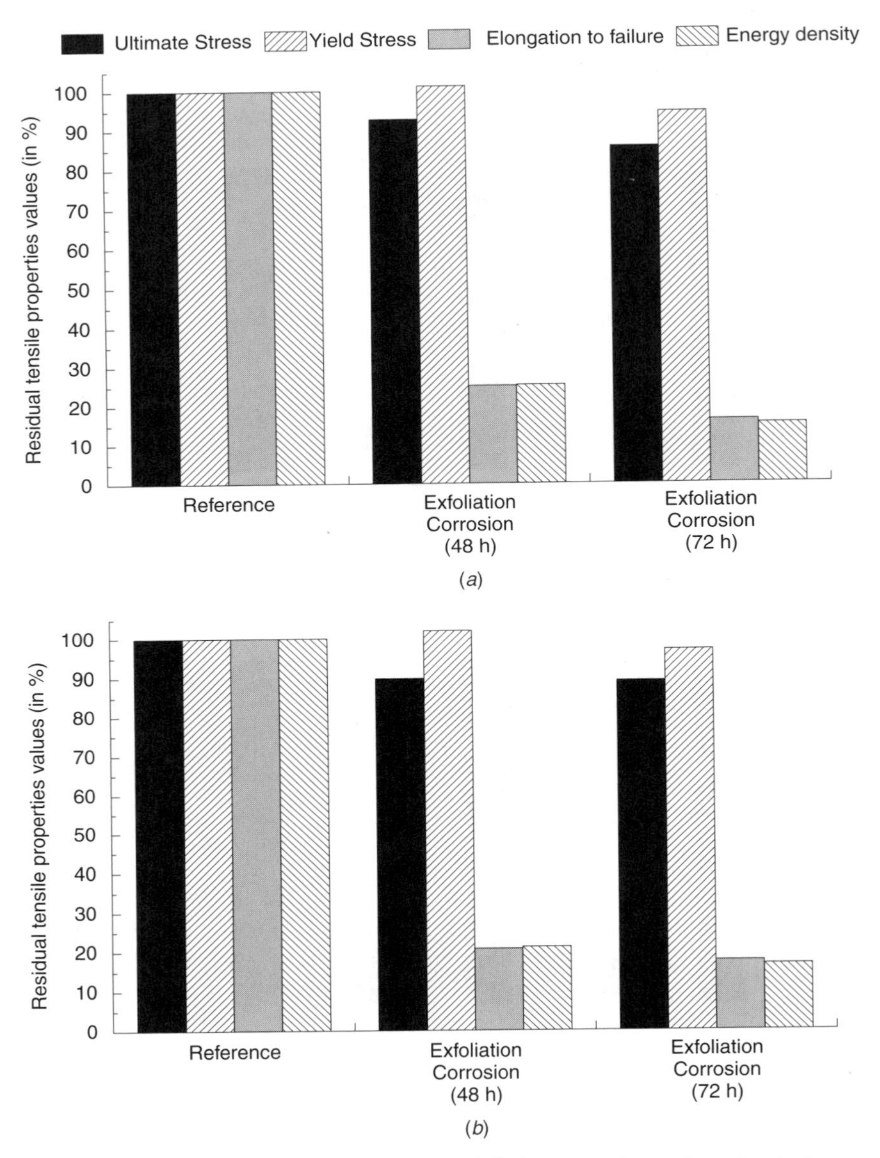

FIGURE 9.22 Tensile behavior following exfoliation corrosion and mechanical removal of corrosion-induced surface damage for alloy 2091: (*a*) L direction and (*b*) LT direction.

where ΔK stands for the stress intensity factor range. Derived results are consistent with experimental results from [76]. The fact that fatigue crack propagation is not affected by existing corrosion is a result that is not intuitively understandable. This result should not be misinterpreted as fatigue behavior not being related to corrosion. Recall that as discussed above, the fatigue life following corrosion

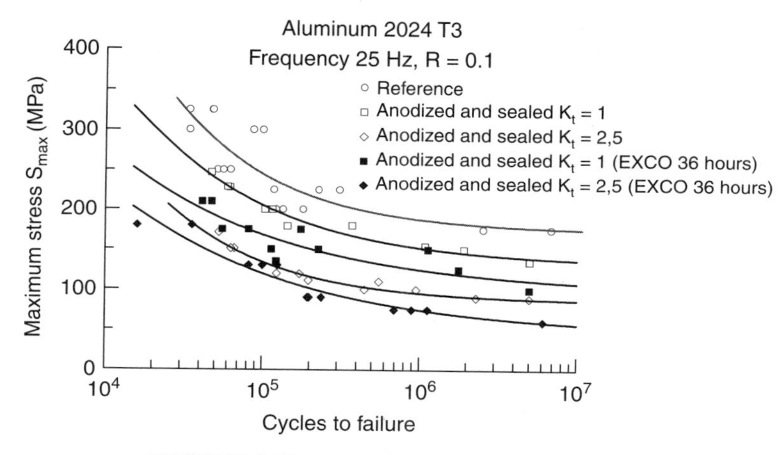

FIGURE 9.23 S–N curves for alloy 2024-T3.

TABLE 9.41 Derived Exponents of the Weibull Distribution for Alloy 2024

Specimen Preparation Prior to Fatigue	C_1	C_2	C_3	C_4
As received, $K_t = 1$	499	171.3	4.3	−8.8
Anodization coating and sealing, $K_t = 1$	450	129	4.16	−7.09
Anodization coating and sealing, $K_t = 2.5$	350	83.8	4.15	−8.41
Anodization coating and sealing and then exposure to exfoliation corrosion, $K_t = 1$	450	93.03	3.77	−5.11
Anodization coating and sealing and then exposure to exfoliation corrosion, $K_t = 2.5$	350	40.97	3.94	−5.14

decreases significantly. In addition, the discussed material embrittlement is diffusion controlled. Thus, large components with small amounts of free edges are expected to suffer less from obtained material embrittlement. Yet, locally (e.g., in the area of neighboring rivets where multiple-site damage (MSD) and widespread fatigue damage (WFD) may be developed as well) corrosion-induced embrittlement may become significant. Further investigation is needed to quantify whether the observed fatigue resistance drop due to corrosion may be explained solely by the presence of corrosion notches, which are reducing the fatigue crack initiation phase significantly, or may be related to determined ductility drop as well.

9.6.3.4 Hydrogen Characterization

The pattern of hydrogen evolution, with heating of the corroded specimen, has been shown to depend strongly on temperature. A plot of hydrogen concentration in the purge stream versus temperature is shown for some representative cases in Fig. 9.25. Different curves correspond to specimens with varying exposure time to the exfoliation solution. Multiple peaks are observed and are attributed to different trapping states.

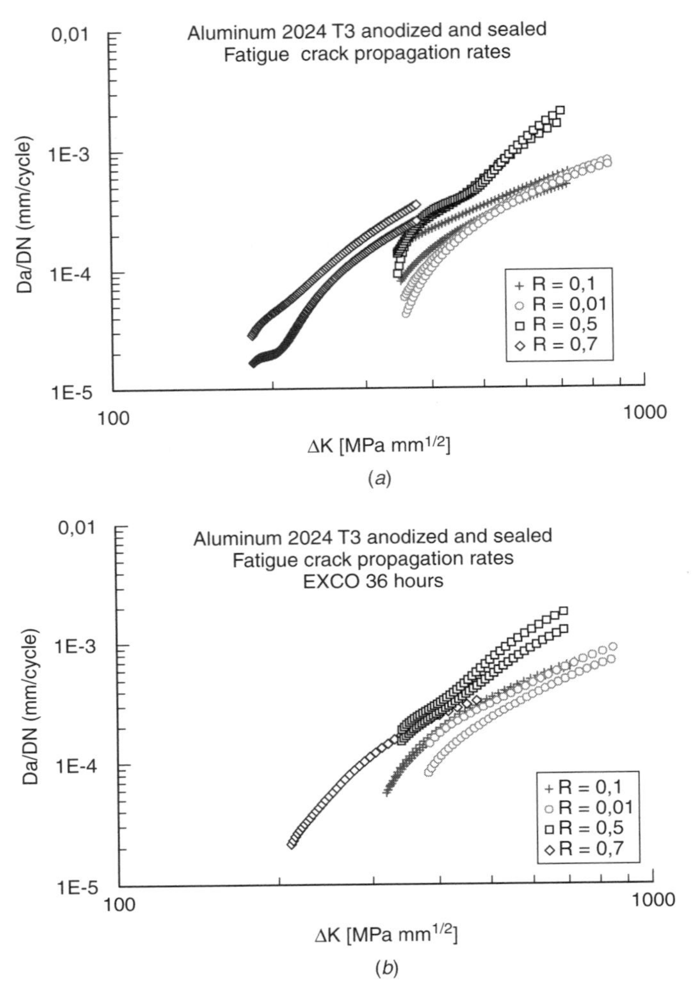

FIGURE 9.24 Fatigue crack growth rates for (*a*) anodized and sealed specimens and (*b*) anodized, sealed and then corroded 2024-T3 alloy specimens.

The onset of the peaks labeled as T2, T3, and T4 occur at 200, 410, and 500°C, respectively. These represent critical temperatures below which no hydrogen evolution from the respective states is observed, even if the specimen is exposed to a constant, lower temperature for an extensive time period. According to Pressouyre [49], the existence of a critical temperature classifies states T2, T3, and T4 as irreversible. Trapping state T1 is found to release hydrogen continuously at lower temperatures. Thus, T1 is considered a reversible trap. This trap corresponds to the low binding-energy state reported by Haidemenopoulos et al. [60], who observed hydrogen evolution with mild heating of corroded Al 2024-T3.

TABLE 9.42 Effect of R Values and Existing Corrosion on Paris Equation Constants for Alloy 2024

	"Ref 2" 2024		"Ref 2" 2024 Corroded	
Stress Ratio R	C	n	C	n
$R = 0.01$	2.87×10^{-10} 9.31×10^{-10}	2.20 2.01	2×10^{-10} 3.29×10^{-10}	2.24 2.20
$R = 0.1$	2.8747×10^{-9} 3×10^{-9}	1.876 1.87	8.73×10^{-11} 6.32×10^{-10}	2.42 2.12
$R = 0.5$	2.66×10^{-13} 4.899×10^{-13}	3.45 3.36	3.13×10^{-12} 5.90×10^{-13}	3.04 3.356
$R = 0.7$	4.46×10^{-13} 2.71×10^{-12}	3.47 3.108	6.03×10^{-11}	2.53

Total quantity of hydrogen in each bonding state is estimated by integrating the original concentration versus time data to calculate the area under each peak. Results for the four peaks are shown in Fig. 9.26 (*a, b, c,* and *d*), where the amount of hydrogen (expressed in ppmw relative to the specimen weight) is plotted as a function of exposure time in the exfoliation solution. The three strong traps T2, T3, and T4 share common features. Linear increase of the amount of hydrogen with exposure time is initially observed, followed by asymptotic approach to a constant value. This behavior is reminiscent of a saturation process by depletion of available active sites.

State T4 reaches a plateau concentration of 1200 ppmw after ~35 h exposure in the exfoliation solution, while states T2 and T3 saturate at concentrations 40 and 300 ppmw, respectively, after the elapse of ~60 h. The fact that hydrogen desorbs from state T4 at the highest temperature of all identified trapping states indicates that T4 is energetically favored. The fact that this is the first state to become saturated further supports the above result.

The amount of hydrogen in the T1 state (Fig. 9.26*a*) increases linearly with exposure time, and no saturation is evidenced up to an exposure of 120 h in the exfoliation solution. Since it is the lowest energy state observed, T1 could in principle be associated with adsorbed hydrogen. However, in aluminum alloys the energy of chemisorption is lower than the migration energy and no peak should appear.

The physical origin of the above trapping states is difficult to identify solely from thermal analysis. However, some speculations can be made based on information in the literature. Given the fact that T1 is a low-energy state and—when compared to the other states—relatively unsaturable, it would appear that this trapping state is related to hydrogen at interstitial sites. The continuous increase in the amount of hydrogen with exposure time to the exfoliation solution is attributed to the creation, by the corrosion process, of new penetration paths for hydrogen (intergranular cracks and surfaces), as reported by Haidemenopoulos et al. [60].

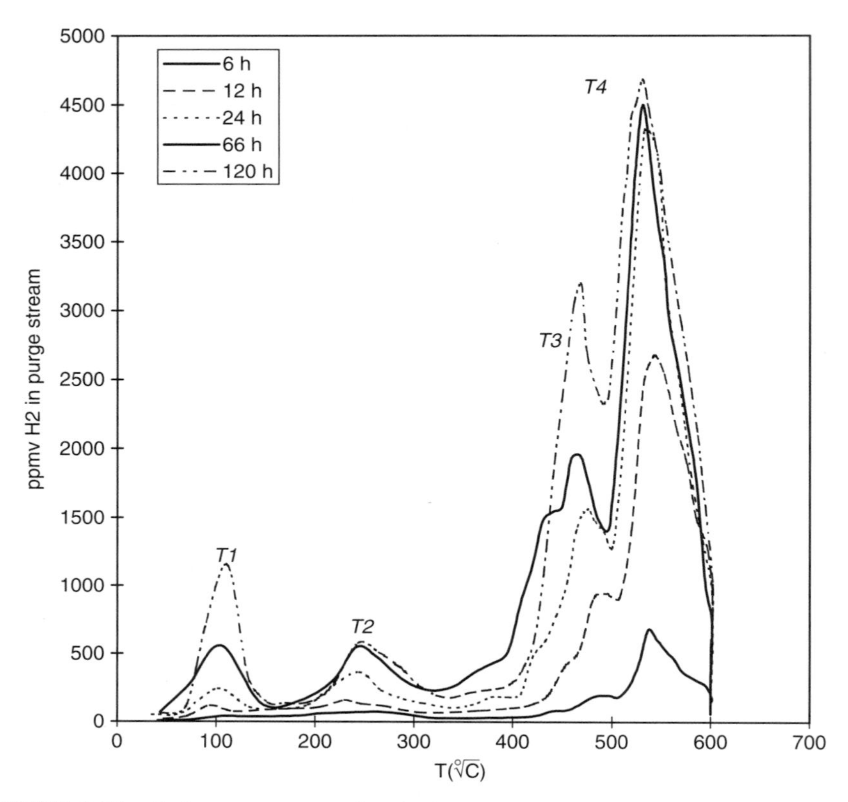

FIGURE 9.25 Hydrogen concentration in furnace purge steam as function of furnace temperature, during rapid heating. Curves correspond to different exposure times to exfoliation solution.

Trapping state T2 is an intermediate energy state that saturates with relatively little hydrogen. A possible physical origin of this trap is the interface between the Mg_2Si precipitate and the matrix lattice. The Mg_2Si precipitate is incoherent with the matrix and the interfacial dislocations that exist around it can trap hydrogen. This has been shown by Saitoh et al. [47], using tritium autoradiography.

The critical temperature of 410°C, below which no hydrogen evolves from state T3, compares favorably with the thermal decomposition temperature of MgH_2 as reported by Tuck [55] (450°C with a heating rate of 50°C/min, which should bias the peak to higher temperature). Thus, trapping state T3 could tentatively be associated with Mg hydride. It has been noted by Saitoh et al. [47] that Mg is bonded to Si by a strong ionic bond that precludes formation of MgH_2. However, Mg content in the 2024 alloy presently tested is in roughly 40% excess over the stoichiometric analogy with Si and—if saturated—results in a 500-ppmw concentration of hydrogen. This estimate compares favorably with the plateau of 300 ppmw shown in Fig. 9.26*c*.

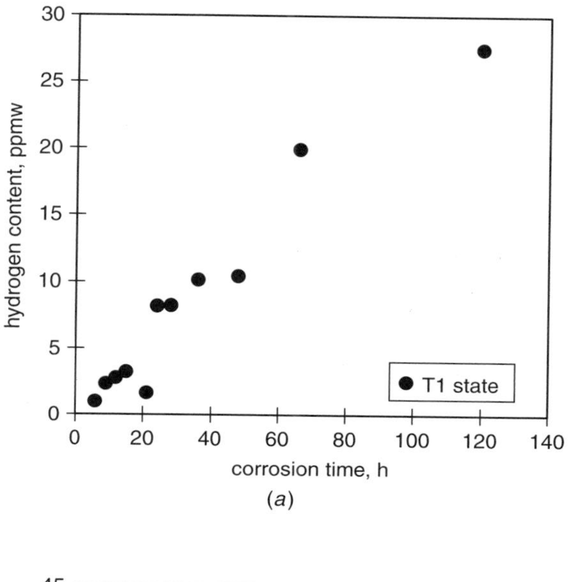

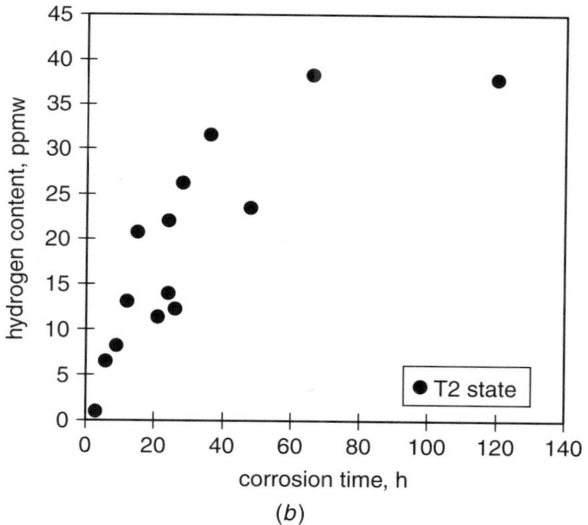

FIGURE 9.26 Total hydrogen content of corroded specimen as function of exposure time to exfoliation solution (a–d) corresponds to trapping states T1–T4, respectively.

The critical temperature of 500°C, which marks hydrogen evolution from state T4, coincides with the dissolution temperature of the Al_2Cu precipitate, as calculated by the computational alloy thermodynamics software ThermoCalc [77]. Thus, trapping state T4 should be associated with this phase. Indeed, it has been demonstrated by tritium autoradiography [47] that the bulk of the θ precipitate can serve as a hydrogen trap.

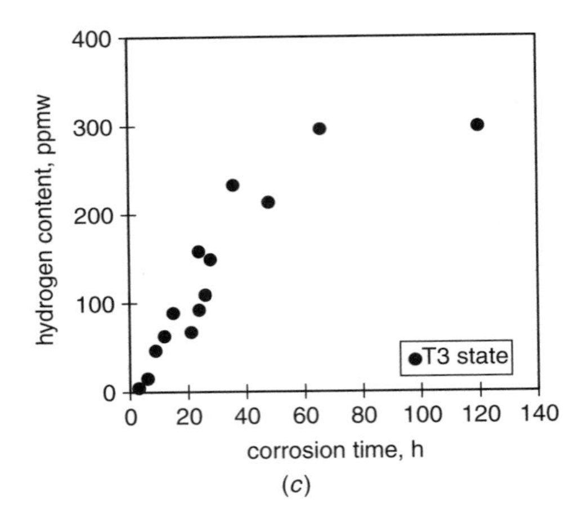

(*c*)

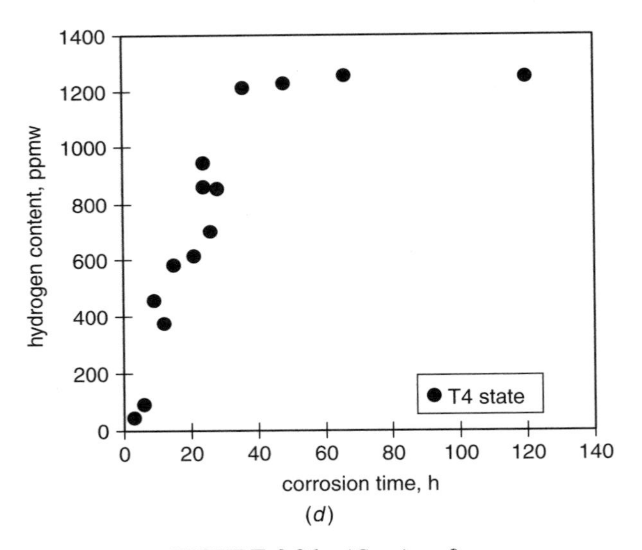

(*d*)

FIGURE 9.26 (*Continued*).

9.6.4 Concluding Remarks

The results presented show clearly the following:

1. There is a significant degradation of tensile ductility of Al alloys with exposure time in the corrosive environment. Controlled experiments (by removing the corroded surface layers) have shown that the observed embrittlement is not caused by a surface damage mechanism. It is rather related to a bulk embrittlement effect.

2. The fatigue life of corroded specimens decreases significantly as well. Yet, the fatigue crack growth behavior of the materials is not affected by existing corrosion.
3. There is a considerable buildup of hydrogen in the material with exposure time in the corrosive environment. Hydrogen is trapped at different states that are related to microstructural traps. Focusing our discussion on the exfoliation corrosion test, where the hydrogen measurements were performed, one can observe that both tensile ductility and hydrogen uptake follow similar time dependence. Figure 9.27 is an effort to link the degradation of tensile ductility to hydrogen. The energy density and hydrogen content for state T4 are plotted against exposure time in the exfoliation solution. State T4 was selected as the state associated with the highest amounts of hydrogen. The figure shows that the rapid degradation of energy density in the first 30–40 h of exposure is associated with a respective rapid buildup of the hydrogen state T4 in the material. Both energy density and hydrogen saturate to their final values at about 40–50 h of exposure. This similarity of behavior is a strong indication that indeed the observed tensile ductility degradation is caused by bulk hydrogen embrittlement mechanisms. Hydrogen states T2 and T3 also show similar behavior (rapid increase followed by saturation) to the T2 state. Controlled experiments are necessary to separate the different hydrogen states and to identify the state being responsible for the observed embrittlement.

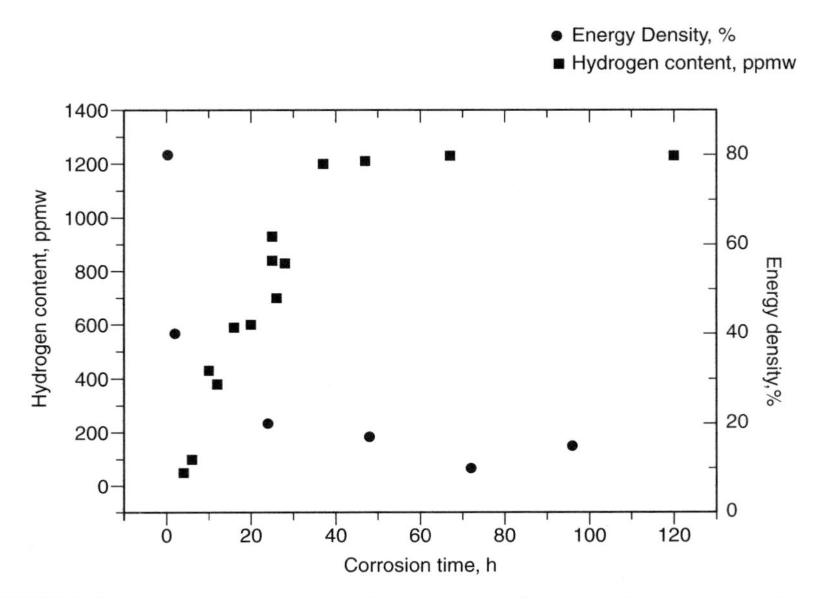

FIGURE 9.27 Energy density and hydrogen content for state T4 versus corrosion time in exfoliation solution for alloy 2024.

REFERENCES

1. Aluminum Association, *Aluminum Brazing Handbook*, Aluminum Association, Washington, DC, 1990.
2. Aluminum Association, *Aluminum Design Manual*, Aluminum Association, Washington, DC, 2000.
3. Aluminum Association, *Aluminum Forging Design Manual*, Aluminum Association, Washington, DC, 1995.
4. Aluminum Association, *Aluminum Soldering Handbook*, Aluminum Association, Washington, DC, 1996.
5. Aluminum Association, *Aluminum Standards and Data 2000*, Aluminum Association, Washington, DC, 2000.
6. Aluminum Association, *Aluminum Standards and Data 1998 Metric SI*, Aluminum Association, Washington, DC, 1998.
7. American Society of Mechanical Engineers, *Boiler and Pressure Vessel Code*, ASME, New York, 2001.
8. Aluminum Association, *Designation System for Aluminum Finishes*, Aluminum Association, Washington, DC, 1997.
9. Aluminum Association, *Forming and Machining Aluminum*, Aluminum Association, Washington, DC, 1988.
10. Aluminum Association, *Guidelines for Minimizing Water Staining of Aluminum*, Aluminum Association, Washington, DC, 1990.
11. Aluminum Association, *Guidelines for the Use of Aluminum With Food and Chemicals*, Aluminum Association, Washington, DC, 1994.
12. Aluminum Association, *Standards for Aluminum Sand and Permanent Mold Castings*, Aluminum Association, Washington, DC, 1992.
13. Aluminum Association, *Welding Aluminum: Theory and Practice*, Aluminum Association, Washington, DC, 1997.
14. Aluminum Association and Aluminum Extrusion Council, *The Aluminum Extrusion Manual*, Aluminum Association and Aluminum Extrusion Council, Washington, DC, 1998.
15. Aluminum Association, Secretariat, *ANSI H35.1 Alloy and Temper Designation Systems for Aluminum*, Aluminum Association, Washington, DC, 1997.
16. Aluminum Association, Secretariat, *ANSI H35.2 Dimensional Tolerances for Aluminum Mill Products*, Aluminum Association, Washington, DC, 1997.
17. American Society for Testing and Materials, Vol. 02.02 *Aluminum and Magnesium Alloys*, ASTM, Conshohoken, PA, 1999.
18. American Welding Society, *A5.3 Specification for Aluminum and Aluminum Alloy Electrodes for Shielded Metal Arc Welding*, AWS, Miami, FL, 1999.
19. American Welding Society, *A5.10 Specification for Bare Aluminum and Aluminum Alloy Welding Electrodes and Rods*, AWS, Miami, FL, 1992.
20. American Welding Society, *C3.7 Specification for Aluminum Brazing*, AWS, Miami, FL, 1993.
21. American Welding Society, *D1.2 Structural Welding Code—Aluminum*, AWS, Miami, FL, 1997.
22. ASM International, *Aluminum and Aluminum Alloys*, ASM, Materials Park, OH, 1993.

23. Department of Defense, MIL-HDBK-5 *Military Handbook Metallic Materials and Elements for Aerospace Vehicle Structures*, DOE, 1994.
24. H. P. Godard, et al., *The Corrosion of Light Metals*, Wiley, New York, 1967.
25. J. R. Kissell and R. L. Ferry, *Aluminum Structures*, 2nd Ed., Wiley, New York, 2002.
26. Non-Ferrous Founders Society, *The NFFS Guide to Aluminum Casting Design: Sand and Permanent Mold*, University of Northern Iowa, Des Plaines, IL, 1994.
27. M. L. Sharp, *Behavior and Design of Aluminum Structures*, McGraw-Hill, New York, 1993.
28. M. L. Sharp, G. E. Nordmark, and C. C. Menzemer, *Fatigue Design of Aluminum Components and Structures*, McGraw-Hill, New York, 1996.
29. *FAA–NASA Symposium on the Continued Airworthiness of Aircraft Structures*, FAA Center of Excellence in Computational Modeling of Aircraft Structures, Atlanta, Georgia, August 28–30, 1996, National Technical Information Service, Springfield, VA, 1997.
30. AGARD Workshop, Fatigue in the Presence of Corrosion, Oct. 5–9, 1998, Corfu, Greece, RTO Meeting, Proceedings 18.
31. Structural Integrity for the Next Millennium, ICAF '99 Conference, 12–16 July 1999, Bellevue, WA.
32. J. B. Chang, M. Szamossi, and K. W. Lin, in *Methods and Models for Predicting Fatigue Crack Growth under Random Loading*, ASTM STP 748, ASTM, Philadelphia, 1981, p. 115.
33. X. Zhang, A. S. L. Chan, and G. A. O. Davis, *Eng. Mech.*, **42**, 305 (1992).
34. G. C. Sih and D. Y. Jeong, *J. Theor. Appl. Fract. Mech.*, **14**, 1 (1990).
35. Sp. G. Pantelakis, Al. Th. Kermanidis, and P. G. Daglaras, *Theor. Appl. Fract. Mech.*, **28**, 1 (1997).
36. BRITE/EURAM No. 1053, Structural Maintenance of Ageing Aircraft (SMAAC), Final Report, CEC Brussels, 1998.
37. M. O. Speidel, in R. Gibala and R. F. Heheman (Eds.), *Hydrogen Embrittlement and Stress Corrosion Cracking*, ASM, Materials Park, OH, 1992, p. 271.
38. H. F. de Jong, *Aluminium*, **58**, 526 (1982).
39. M. E. Inman, R. G. Kelly, S. A. Willard, and R. S. Piascik, *Proc. of the FAA–NASA Symposium on the Continued Airworthiness of Aircraft Structures*, Atlanta, Georgia, August 28–30, 1996, National Technical Information Service, Springfield, VA, 1997, p. 129.
40. W. Wallace, D. W. Hoeppher, and P. V. Kandachar, *Aircraft Corrosion: Causes and Case Histories, AGARD Corrosion Handbook*, Vol. 1, AGARD AG 278, 1985.
41. Sp. G. Pantelakis, N. I. Vassilas, and P. G. Daglaras, *METALL*, **47**, 135 (1983).
42. Sp. G. Pantelakis, Th. B. Kermanidis, P. G. Daglaras, and Ch. Alk. Apostolopoulos, "The Effect of Existing Corrosion on the Structural Integrity of Aging Aircraft," in the Proc. of the Fatigue in the Presence of Corrosion, AGARD Workshop, Oct. 5–9, 1998, Corfu, Greece.
43. M. Scamans, R. Alani, and P. R. Swann, *Corrosion Sci.*, **16**, 443 (1976).
44. G. M. Scamans and C. D. S. Tuck, in A. Foroulis (Ed.), *Environment-Sensitive Fracture of Engineering Materials* Conf. Proceedings, the Metallurgical Soc. of AIME, Warrendale, PA, 1979, p. 464.

45. R. J. Gest and A. R. Troiano, *L' Hydrogene dans les metaux*, Science et Industrie, Paris, 1972, p. 427.
46. M. O. Speidel, *Hydrogen in Metals*, American Society for Metals, Materials Park, OH, 1974, p. 249.
47. H. Saitoh, Y. Iijima, and K. Hirano, *J. Mat. Sci.*, **29**, 5739 (1994).
48. G. Itoh, K. Koyama, and M. Kanno, *Scripta Mat.*, **35**, 695 (1996).
49. G. M. Pressouyre, *Met. Trans.*, **10A**, 1571 (1979).
50. G. M. Pressouyre, *Acta Met.*, **28**, 895 (1980).
51. G. A. Young and J. R. Scully, *Scripta Mat.*, **36**, 713 (1997).
52. A. Joshi, C. R. Shastry, and M. Levy, *Metall. Trans.*, **12A**, 1081 (1981).
53. G. M. Scamans, N. J. H. Holroyd, and C. D. S. Tuck, *Corrosion Sci*, **27**, 329 (1987).
54. Y. Iijima, P. Yoshida, H. Saitoh, H. Tanaka, and K. Hirano, *J. Mat. Sci.*, **27**, 5735 (1992).
55. C. D. S. Tuck, Proc. 3rd Int. Conf. Effects of Hydrogen on the Behaviour of Materials, Jackson, Wyoming, 1980, p. 503.
56. R. G. Song, M. K. Tseng, B. J. Zhang, J. Liu, Z. H. Jin, and K. S. Shin, *Acta Mater.*, **44**, 3241 (1996).
57. S. -M. Lee and J. -Y. Lee, *Metall. Trans.*, **17A**, 181 (1986).
58. R. A. Outlaw, D. T. Peterson, and F. A. Schmidt, *Metall. Trans.*, **12A**, 1809 (1981).
59. P. N. Anyalebechi, *Metall. Trans.*, **21B**, 649 (1990).
60. G. N. Haidemenopoulos, N. Hassiotis, G. Papapolymerou, and V. Bontozoglou, *Corrosion*, **54**, 73 (1997).
61. EPET 30, Damage Tolerance Behaviour of Corroded Aluminium Structures, Final Report, General Secretariat for Research and Technology, Athens, Greece, 1990.
62. G 34–90, Standard Test Method for Exfoliation Corrosion Susceptibility in 2xxx and 7xxx Series Aluminium Alloys (EXCO Test), in *Annual Book of A.S.T.M Standards*, Section 3, *Metals Test and Analytical Procedures*, Vol. 03.02, *Wear and Erosion*; Metal Corrosion, ASTM, West Conshohocken, PA, 1994, p. 129.
63. ASTM E8M-94a, *Annual Book of ASTM Standards*, Section 3, *Metal Test Methods and Analytical Procedures*, ASTM, West Conshohocken, PA, 1994, p. 81.
64. MIL-A-8625E, Type III-CLASS 1, *Anodic Coatings for Aluminum and Aluminum Alloys*, Dept. of Defense Systems Eng. Dept., Lakehurst, NJ, April 25 1988, p. 1.
65. ASTM E 466-82, Standard Practice for Conducting Constant Amplitude Axial Fatigue Tests of Metallic Materials, in *Annual Book of A.S.T.M. Standards*, Section 3, *Metals Test and Analytical Procedures*, Vol. 03.01, *Metals, Mechanical Testing; Elevated and Low-Temperature Tests; Metallography*, ASTM, West Conshohocken, PA, 1994, p. 465.
66. ASTM E647-93, Standard Test Method for Measurement of Fatigue Crack Growth Rates, in *Annual Book of A.S.T.M. Standards*, Section 3, *Metals Test and Analytical Procedures*, Vol. 03.01, *Metals, Mechanical Testing; Elevated and Low-Temperature Tests; Metallography*, ASTM, West Conshohocken, PA, 1994, p. 569.
67. ASTM G 44–94, Standard Practice for Evaluating Stress Corrosion Cracking Resistance of Metals and Alloy by Alternate Immersion in 3.5% Sodium Chloride Solution,

in *Annual Book of A.S.T.M Standards*, Section 3, *Metals Test and Analytical Procedures*, Vol 03.02, *Wear and Erosion; Metal Corrosion*, ASTM, West Conshohocken, PA, 1995, p. 172.

68. ASTM G1-90, Standard Practice for Preparing, Cleaning, and Evaluation Corrosion Test Specimens, in *Annual Book of A.S.T.M Standards*, Section 3, *Metals Test and Analytical Procedures*, Vol 03.02, *Wear and Erosion; Metal Corrosion, G1-90. Wear and Erosion; Metal Corrosion*, ASTM, West Conshohocken, PA, 1995, p. 25.
69. ASTM G110-92, Standard Practice for Evaluating Intergranular Corrosion Resistance of Heat Treatable Aluminium Alloys by Immersion in *Sodium Cloride + Hydrogen Peroxide Solution*, Section 3, *Metals Test and Analytical Procedures*, Vol 03.02, *Wear and Erosion; Metal Corrosion*, ASTM, West Conshohocken, PA, 1995, p. 457.
70. ASTM GB117-93, Standard Guide for Calculating and Reporting Measures of Prediction Using Data from Interlaboratory Wear or Erosion Tests, in *Annual Book of ASTM Standards*, Section 3, *Metal Test Methods and Analytical Procedures*, ASTM, West Conshohocken, PA, 1995, p. 1.
71. ASTM G85-94, Standard Practice for Modified Salt Spray (Fog) Testing, in *Annual Book of ASTM Standards*, Section 3, *Metal Test Methods and Analytical Procedures*, ASTM, West Conshohocken, PA, 1995, p. 351.
72. ASTM G50-76, Standard Practice for Conducting Atmospheric Corrosion Tests on Metals, in *Annual Book of ASTM Standards*, Section 3, *Metal Test Methods and Analytical Procedures*, ASTM, West Conshohocken, PA, 1995, p. 185.
73. ASTM G33-88, Standard Practice for Recording Data from Atmospheric Corrosion Tests of Metallic-Coated Steel Specimens, in *Annual Book of ASTM Standards*, Section 3, *Metal Test Methods and Analytical Procedures*, ASTM, West Conshohocken, PA, 1995, p. 111.
74. ASM Specialty Handbook, "Aluminium and Aluminum Alloys", J. R. Davis (ed.), ASM International, Materials Park, OH, 1993.
75. L. Kompotiatis, Investigation on the Behavior of Aluminium Alloys in Corrosive Environments, Doctor Engineer Thesis, Chemical Engineering Department, University of Patras, 1995.
76. J. P. Chubb, T. A. Morad, B. S. Hockenhull, and J. W. Bristow, "The Effect of Exfoliation Corrosion on the Fatigue Behaviour of Structural Aluminium Alloys," in S. N. Atluri, S. G. Sampath, and P. Tong (Eds.), *Structural Integrity of Aging Airplanes*, Springer Series in Computational Mechanics, Springer, Berlin, 1991, p. 87.
77. B. Sundman, B. Jansson, and J. O. Anderson, *Calphad*, **9**, 153 (1985).

CHAPTER 10

Functionally Graded Materials

IVAR E. REIMANIS

Metallurgical and Materials Engineering Department, Colorado School of Mines, Golden, Colorado 80401

Handbook of Advanced Materials Edited by James K. Wessel
ISBN 0-471-45475-3

10.1 DESCRIPTION

10.1.1 General

Functionally graded materials (FGMs) are materials that comprise a spatial gradation in structure and/or composition, tailored for a specific performance or function. FGMs are not technically a separate class of materials but rather represent an engineering approach to modify the structural and/or chemical arrangement of materials or elements. This approach is most beneficial when a component has diverse and seemingly contradictory property requirements, such as the necessity for high hardness *and* high toughness in wear-resistant coatings. Generally, it is very difficult to provide broad design guidelines for utilizing FGMs since the structures are complex and diverse. The purpose of this chapter is to give the reader an understanding of how specific gradations in structure and/or composition will affect specific material behavior. Because of the complexity of FGM systems, most of the information is qualitative and is meant to provide broad guidelines. Some of the descriptions are generic and apply to a large class of material and structural systems while others are quite specific, and only "work" for a small set of materials.

While the term *functionally graded materials* has only existed since the mid-1980s, the concept has been utilized in engineering for a relatively long time. For example, the concept behind surface hardening of steel by carburization has been understood for some 60 years and has been used for many hundreds of years. For a second example, as early as 1912 metal/glass sealing technologists developed graded structures for minimizing thermal residual stresses due to thermal expansion coefficients mismatch [1, 2]. A third example exists in graded band gap semiconductors for use in heterojunction bipolar transistors, introduced in 1957 [3, 4]. Finally, graded structures were introduced in structural composites in the 1970s [5]. Despite well-established guidelines for the latter applications, until recently no unified view of graded structures existed. By focusing research and development efforts on generic aspects of FGMs, further advancements may be made to understand which structures are desirable for specific applications. This chapter provides a current description and provides a framework for utilizing the FGM concept in engineering applications.

10.1.2 Classification

Perhaps FGMs are best classified according to processing, as illustrated in Fig. 10.1, which separates FGMs into those produced by constructive processes and those produced by transport-based processes [6]. In short, constructive processes rely on the placement of phases within a structure by the processing engineer. Transport-based processes rely on well-timed and designed *in situ* reactions or processes during material fabrication. Many protective coatings (e.g., thermal barrier coatings) fall into the former category. The carburization of steel

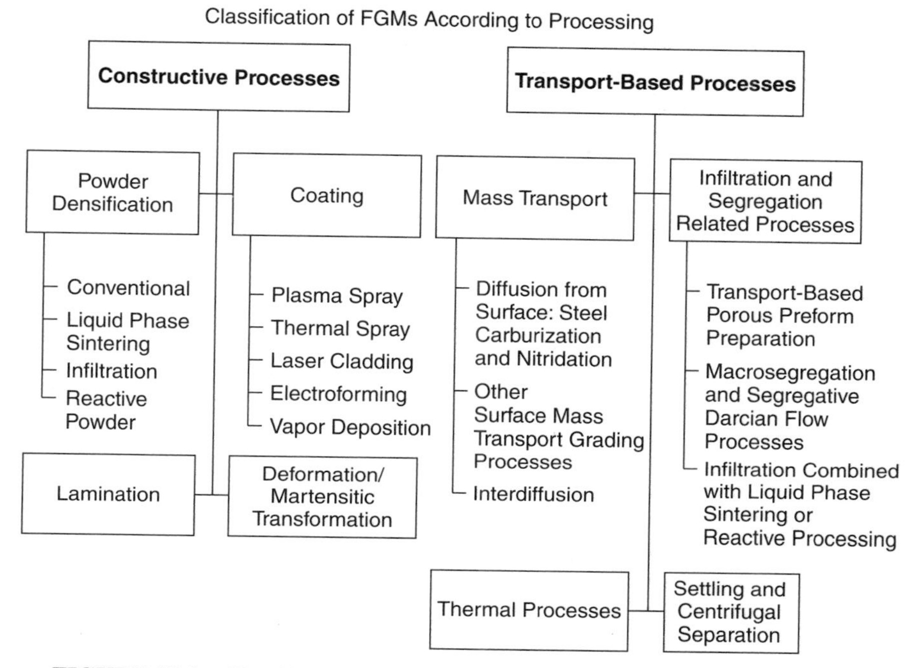

FIGURE 10.1 Classification of functionally graded materials according to [5].

falls into the latter category. Naturally, a design approach in which a gradation is formed *in situ*, by a transport-based process, would be simpler and generally more desirable than a constructive approach.

10.2 CHARACTERISTICS, BEHAVIOR, AND PERFORMANCE

It is inappropriate to assign a unique property to a functionally graded material since the local properties vary throughout the material. Instead, the approach required is to super impose the various material properties within the FGM to predict a specific type of behavior. Thus, utilization of FGMs relies on controlling the spatial variation in material properties of a component so that the desired spatial variance of performance may be achieved. The challenge is to devise models that predict the characteristics, behavior, and performance of graded materials as a function of the constituent properties and the graded architecture. At the current time, only a few such specific models exist. The approach taken here is to provide a more qualitative framework to build models that predict material response. Various characteristics of graded materials are discussed next. The emphasis is on mechanical behavior.

10.2.1 Thermal Residual Stress Modification

Historically, one of the first purposes of constructively processed FGMs was to reduce thermal residual stresses in joints between dissimilar materials. The thermal residual stress results from cooling the joint from elevated temperature: Thermal expansion mismatch results in differential thermal strains that give rise to potentially deleterious stresses. While for bimaterial joints the thermal stresses are relatively simple to predict, for layered and graded systems, the situation is more complex. The schematic in Fig. 10.2 indicates an interlayer region between two materials that have different thermal expansion coefficients. Ideally, this interlayer region exhibits a spatially varying thermal expansion coefficient (TEC) whose value at any point is intermediate between the two materials. Specific optimum design of this interlayer region (e.g., whether its gradation is linear, parabolic, or stepwise) depends on many details, including the overall geometry (in the simplest case, whether the joint is a plate, a cylinder, or a coating—see Fig. 10.3), the specific material constituent properties, whether or not plasticity plays a role in relaxing stresses, and the particular design-related constraints. Some generic models have been developed to examine trends in stress distribution in FGMs, and these models provide the design engineer with guidance [6]. Figure 10.4 illustrates the kind of benefit attained by incorporating an FGM interlayer. The dashed curve indicates the (in-plane) stress distribution for a sharp Ni/Al_2O_3 interface, while the solid line indicates the stress distribution for the same interface that contains an FGM interlayer that exhibits linearly varying elastic modulus and thermal expansion coefficient. Note the reduction in maximum tensile stress from approximately 125 MPa in the sharp joint to about 25 MPa in the graded joint. These results indicate the kind of benefit FGMs can provide to dissimilar material joints. Even though a detailed understanding of stress distributions is complex and depends on many parameters, some qualitative generalizations may be made. The stress distributions depend on the component geometry (Fig. 10.3) and the graded architecture (Fig. 10.5). The differences are summarized qualitatively for a few general cases in Tables 10.1 and 10.2. More detailed, quantitative information may be found elsewhere [7–9]. It is highly

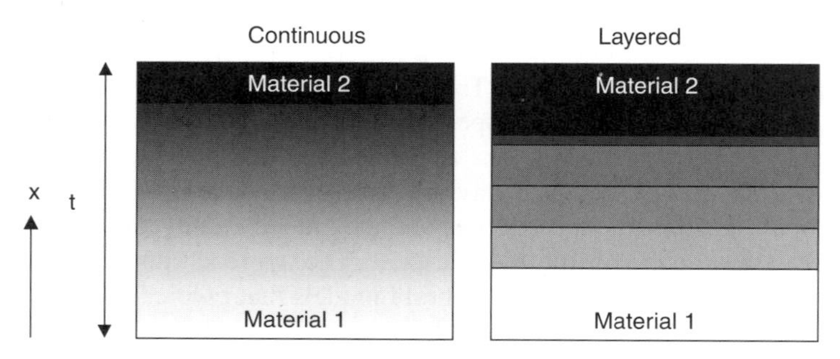

FIGURE 10.2 Schematic showing two graded joint geometries with material 1 (white) and material 2 (black). Architecture on left is layered or discrete while one on right is continuous.

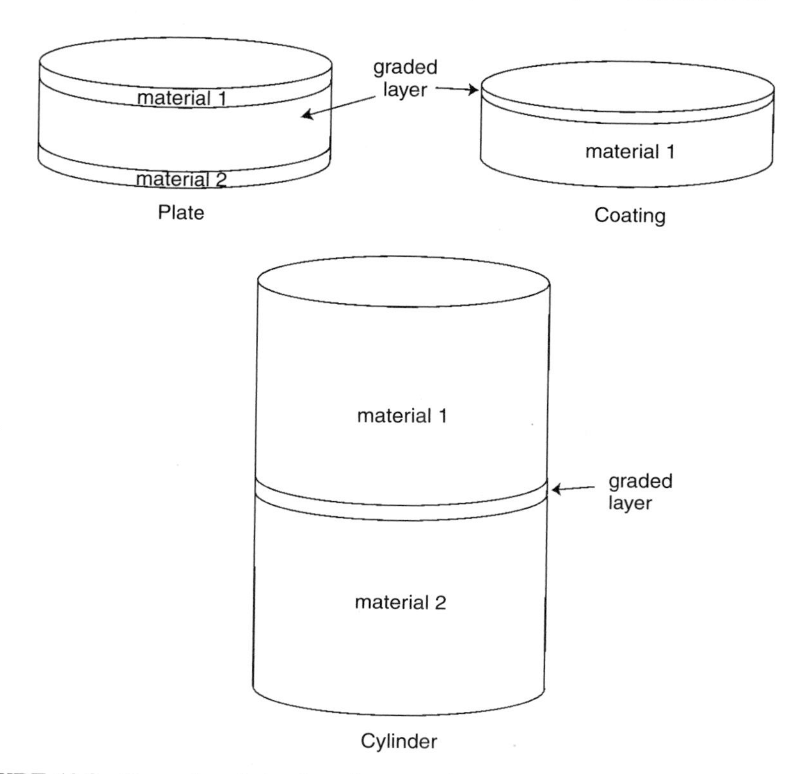

FIGURE 10.3 Examples of simple, axisymmetric geometries for which thermal residual stress distributions are very different. Table 10.1 gives qualitative information on effect of geometries in reference to stresses typically important in joints for structural applications (axial and shear stresses at edge and in-plane stresses in interior).

recommended that the engineer constructs a finite-element model for detailed stress distribution prediction.

Typically, the edges of a joint experience higher shear and normal thermal residual stresses than those predicted in the center of the joint. Because component edge regions frequently contain flaws from surface machining, these regions are particularly likely to experience high stress singularities that can result in premature crack propagation. Thus, special attention should be given to the development of stresses in edges and any singular regions. Work on multilayered brittle symmetric composites in which a material with lower TEC is placed between two materials with higher TCE has shown that upon cooling, the inner layers, which normally experience compressive axial stresses in the bulk, may experience tensile axial stresses at the edge [10]. The magnitude of tensile stresses depends on the layer thickness but has been observed to drive channeling cracks at the edge in ceramic/ceramic systems [10]. Generally, gradation in material properties is an effective method to reduce edge stresses in nonsymmetric

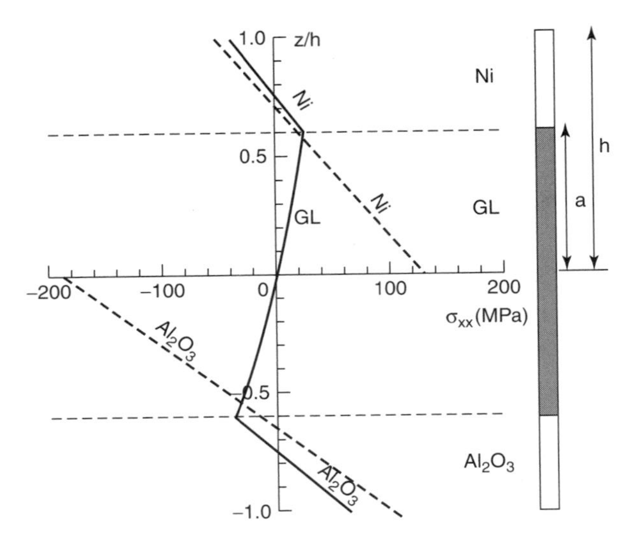

FIGURE 10.4 Stress distribution in nickel–alumina joint with (solid line) and without (dashed line) graded interlayer. Joint has experienced temperature drop of 100°C from stress-free temperature. (Figure reprinted from Suresh and Mortensen [6].)

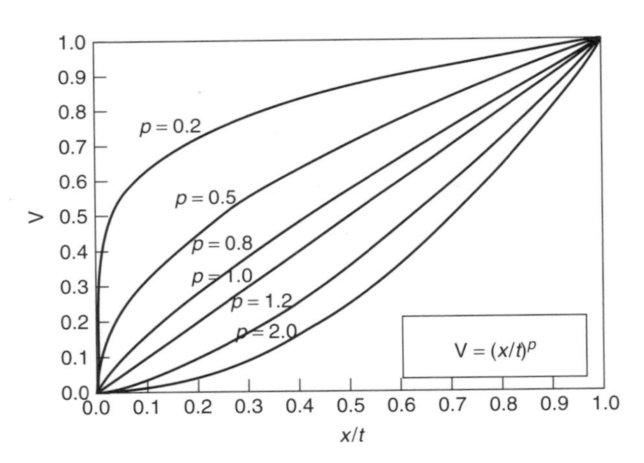

FIGURE 10.5 Possible graded architectures for joint between two materials. Volume fraction of material 2, f_2, as function of distance across graded joint normalized to graded region thickness, x/t. Linear gradient is defined as $p = 1$, parabolic as $p > 1$, and square root as $p < 1$. See Fig. 10.2 for definition of x and t.

joints [6, 11, 12]. However, stress reduction in one region may enhance stresses in another; furthermore, stress relaxation through plastic strain may not be desired for some applications [11]. Whether or not the presence of a graded interlayer is desired to reduce effects of thermal residual stress depends on the combination of component geometry and size and material properties.

TABLE 10.1 The Type of Geometry Alters the Stress Distribution[a,b]

Type of Geometry (see Figure 10.3)	Relative Level of Residual Stress in Material with Lowest Thermal Coefficient of Expansion (TCE)[c]
Plate	High
Cylinder	Low
Coating	Medium

[a]Three generic geometrical configurations are listed: 1. a plate; 2. a cylinder; and 3. a coating.
[b]Quantitative information concerning locations and magnitudes of stresses may be found in references [7, 8]
[c]In most metal/ceramic joints, the material with the lowest TCE is the most brittle.

TABLE 10.2 The Type of Gradation Alters the Stress Distribution[a,b]

Type of Gradation (see Figure 10.2 and Figure 10.5)	Relative Level of Residual Stress in Material with Lowest Thermal Coefficient of Expansion (TCE)[c]
Linear ($V = bx$)	Medium
Parabolic ($V = bx^2$)	High
Square-root ($V = bx^{1/2}$)	Low

[a]Three general types of gradients are listed: 1. the volume fraction of material 1, V, varies linearly with distance, x; 2. V varies parabolically with x; and 3. V exhibits a square root dependence on x.
[b]Quantitative information locations and magnitudes of stresses may be found in references [7–9]
[c]In most metal/ceramic joints, the material with the lowest TCE is the most brittle.

10.2.2 Mechanical Behavior

10.2.2.1 Elastic

If the constituent properties of materials comprising an FGM are known, then the overall elastic response of the FGM component can be predicted, either by analytic means in the simplest cases or by numerical methods. When the gradation is accomplished by discrete layering, then a prediction of the elastic response is usually a relatively simple mechanics problem. If each layer is considered to be an isotropic composite, then the elastic modulii of that layer should not depend on the fact that the layer is surrounded by other layers. The few reported cases where the modulii of a layer within an FGM are different than those of the same composition composite processed separately are due to slight processing variations that probably result in microstructural differences. For discrete layers where the microstructural features within the layer are much smaller than the layer thickness, classical plate and beam theories of continuum mechanics apply [6]. For example, to predict the bending of an inhomogeneous beam, a homogeneous beam of equivalent cross section should be constructed, and standard mechanics methods should be applied [13]. FGMs frequently experience thermal loading, and, in this case, a system of linear equations that contain the thermal strain and plate curvature should be developed and solved to predict elastic response [6]. The solutions to the thermal residual stress distribution in Fig. 10.4 was solved using these types of classical theories. Detailed application of these techniques to FGMs may be found in [6]. When gradation is continuous, elastic response prediction may be more challenging, and numerical procedures such as the finite-element method are likely the most efficient approaches.

10.2.2.2 Plastic Deformation and Fracture

Fracture mechanics concepts can be applied to FGMs in a way that is similar to monocomposition materials, though there may be some deviations. The linear elastic fracture mechanics approach of stress intensity factors can be applied to FGMs if the elastic constants are continuous and differentiable [14]. However, for layered materials in which a crack propagates between the layers, interface fracture mechanics may have to be invoked [15]. The two main differences between a crack that propagates in a graded material perpendicular to the direction of the gradient and one that propagates in a homogeneous material are the amount of shear loading on the crack tip and the precise level of the stress intensity factor [16]. For a given geometry and load application, a crack in a graded material will experience more shear stress and have a lower driving force for propagation than the same crack in a homogeneous material or at a bimaterial interface. The end result is that an FGM is slightly more resistant to crack propagation than a homogeneous material, all else except elastic properties being equal. The magnitudes of the differences in crack propagation driving force depend on the specific geometries, the material properties, and type of gradient but, in general, are in fact quite small. Little is currently known about crack propagation perpendicular to the direction of the gradient when plasticity is relatively large, though it appears that the effect of elastic mismatch is much stronger than plastic mismatch [17].

Numerical simulation studies have shown that as a crack propagates in a linear elastic graded material parallel to the direction of the gradient from a more compliant to a less compliant material, the driving force for its continued propagation (i.e., the mode I stress intensity factor) typically increases [18]. In other words, if the fracture toughness were the same everywhere, the FGM would "appear" less resistant to fracture than a homogeneous material with the same fracture toughness. When plasticity is considered, it has been shown that when a crack grows toward a material with a higher yield stress, the crack's driving force decreases [6]. Since the driving force must be compared with a fracture criterion to predict crack propagation, it is necessary to superimpose the elastic and plastic effects and to compare the resultant driving forces with values of the fracture resistance. More recent mechanics studies on ceramic/metal graded materials in which crack bridging occurs have shown that the fracture resistance increases as the crack propagates toward the metal [19]. In this case, the fracture resistance increases because the crack bridging forces reduce the driving force of the crack tip. Thus, the elastic, plastic, and fracture properties together dictate when the crack driving force will exceed the fracture resistance of the graded material. This complex interplay of loading, component geometry, and distribution of material properties makes it difficult to arrive at a set of general guidelines for design purposes.

10.2.2.3 Wear

The wear of material is a complex process that may occur by a variety of mechanisms that depend on several properties such as the fracture toughness and the hardness, as well as the environmental conditions. Many applications, for example, cutting tools, gears, and prosthetic implants, utilize concepts of wear to engineer a component that exhibits a hard surface and a tough interior. Since for most materials there is an inverse correlation between hardness and toughness, a gradient frequently produces the optimum structure. A compositionally graded approach has been applied to at least three different materials systems: Co–WC, TiC–NiMo, and diamond–SiC. These have been shown to exhibit superior wear resistance compared to monocomposition tools [3]. In the case of TiC–NiMo, optimum graded composites were fabricated in which the compositions ranged from 95 wt % TiC at the surface to 86 wt % at the transition site to plane steel [20]. In the case of diamond–SiC, a graded layer between the diamond chip and the SiC shank is formed through a powder metallurgy–reaction sintering approach.

Elastic modulus gradients are believed to alter wear resistance of materials [6, 21]. While it is difficult to separate the specific effects of elastic modulus gradient on wear, model studies conducted on alumina and glass suggest that wear resistance is improved when the surface is more compliant than the interior [22].

10.2.3 Thermomechanical Behavior

Thermomechanical behavior in graded materials refers to nonuniform deformation during heating or cooling induced by differential thermal expansion coefficients. Many of the thermomechanical descriptions for graded materials are only applicable for small deformations in the context of beam and plate theories of classical continuum mechanics [6]. Fortunately, these descriptions are not only simpler than large deformation models but are also applicable to a wider range of applications, since in many applications, large deformations are not acceptable. On the other hand, they do not include dynamic effects, thermal gradients, stress relaxation due to plasticity, and edge and singular effects. User-friendly computer programs have been developed for predicting this kind of deformation [23]. After a component has been exposed to a particular temperature excursion, the resulting deformation and residual stress (e.g., Fig. 10.4) are uniquely determined by the component geometry and material constitutive properties. Specific situations have been analyzed by several authors [24, 25].

The presence of plastic deformation alters the stress distribution from the elastic case above. The von Mises stress criterion is most commonly used in describing plasticity in graded materials. One of the challenges in obtaining accurate predictions is to specify this criterion as a function of position within the graded structure. Because of microstructural changes along the compositional gradation, the effective plastic flow stress may change as well (e.g., the degree of constraint around a plastic phase is a function of composition). The current models do not consider these microstructural variations.

10.2.4 Electrical and Optical Characteristics

In describing the electrical conduction of FGMs, it is necessary to consider whether the potential is applied parallel or perpendicular to the gradient. When considering a potential applied perpendicular to the gradient, the electrical conductivity through a particular compositional layer depends on the relative contiguity of the two or more phases within that layer. Applying percolation theory, the electrical conductivity of the two-phase composite Λ is proportional to the volume fraction of phase 1, V_1, and the percolation threshold V_c, which is the spatial location where phase 1 becomes continuous [26, 27]:

$$\Lambda \propto (V_1 - V_c)^n, \tag{10.1}$$

where n is an exponent with a value between 1.6 and 2.0 [26, 27]. The percolation threshold V_c depends on the relative particle sizes of the two phases and is approximately 0.16 when the particles are spheres of identical radius. The percolation threshold for the second phase may be different from V_c. In general, the spatial location of the percolation threshold for the two phases are not equivalent, implying that there exists a band of a three-dimensional interpenetrating composite between the two percolation thresholds. Thus, the electrical characteristics of FGMs may be dramatically altered by changes in the microstructure.

In the case where the electrical potential E is applied parallel to the gradient, the current through a graded structure with thickness t is [3]

$$I = \frac{SE}{\int_0^t [dx/\sigma(x)]},$$

where S is the cross-sectional area and $\sigma(x)$ is the conductivity as a function of distance parallel to the gradient x. In a similar manner, the capacitance across a graded material is given as [3]

$$C = \frac{S}{\int_0^t [dx/\varepsilon(x)]},$$

where $\varepsilon(x)$ is the dielectric permittivity as a function of distance x. If a compositional gradation is designed to spatially alter the Curie point of the capacitor, a broader temperature range of operation may be achieved. For a nonmagnetic transparent material, a gradation in the dielectric permittivity implies a gradation in the index of refraction. Designers of optical fibers have utilized the concept of gradation for a long time [28].

10.3 APPLICATIONS

10.3.1 Environmental Protection Systems

A large number of applications require component protection from elevated temperature or from a corrosive environment. These include aerospace vehicles, both for skin protection for reentry and for engine components such as turbine blades, stirrers, and nozzles for the molten glass and metal processing industry, nuclear reactor components (diverter and first-wall components), and for any cutting tools that experience harsh environments.

Thermal barrier coatings are commonly applied to turbine engine blades to protect against high-temperature and corrosive gases. A large variety of coating compositions, microstructures, and morphologies are possible, though any change to an existing turbine blade design is not trivial from the point of view of engine designers. The coating performance requirements are inherently multifunctional through the coating thickness, and thus it is logical to attempt to employ functionally graded materials.

Silicon carbide–carbon (SiC–C) and carbon–carbon (C–C) composites are widely used as protective shields on the outside of space reentry vehicles and also in the combustion chamber components. It has been shown that applying a graded SiC–C interlayer between a C–C component and a SiC coating improves the life time of a space vehicle nose cone exposed to 1900°C in an oxygen atmosphere at Mach 3. $MoSi_2$-based compositionally graded coatings have been used for protecting components against corrosion in molten glass environments [29], and improvements based on a functionally graded design have been suggested.

10.3.2 Wear-Resistant Coatings

Cutting tools typically require the exterior to be hard and the interior to be tough and strong. Because WC–Co cutting tools have been used extensively in the past, they have been used in the design of graded cutting tools, and it has been shown that resistance to wear may be achieved by appropriately grading the composition of the tool. The compositional gradient should be designed with two objectives in mind. First, the material with highest hardness should be at the surface to maximize hardness there. Second, if the tool is processed at an elevated temperature, the material with the lowest thermal expansion should be at the surface so that upon cooling, compressive stresses develop, thereby increasing the effective hardness. Fortunately, harder materials (e.g., ceramics) typically have lower thermal expansion coefficients than softer ones. Graded Ti–TiN coatings on titanium alloys are common for wear resistance in biological environments [3]. In this case, magnetron sputtering may be used to deposit Ti-rich material near the titanium alloy substrate and TiN-rich material at the coating surface. Chromium nitride films (of varying stoichiometry) produced by cathodic arc deposition are being evaluated for coatings on metal working dies [30]. An optimum combination of adhesion, wear resistance, and intrinsic residual stress has been achieved using a trilayer deposition of 50 nm Cr, 0.4 μm CrN-rich layer, and 3.6 μm Cr_2N-rich layer onto a tool steel substrate [30].

10.3.3 Joining

Joining of dissimilar materials is challenging due to differences in thermal expansion coefficients between different materials. Thermal residual stresses that develop due to this difference may result in premature joint failure. One of the most common solutions is to employ a braze, which is typically designed to promote good wetting and relieve stresses through plastic deformation of the interlayer. Another approach is to employ a graded interlayer so that the effect of the material mismatch is diffused over a greater distance. In addition, a graded layer may provide additional plastic deformation when metals are bonded to brittle materials. The technique of tailoring the glass thermal expansion coefficient in glass/metal seals has been utilized since 1912 [1, 2]. In this technique, the glass composition is varied so that the thermal expansion coefficient increases nearer the metal. Additionally, the thermal expansion coefficient of the metal may be chosen accordingly as well. General guidelines state that the thermal expansion of the glass should be no less than 10% smaller than that of the metal for glass/metal seals [31]. By forming a graded interlayer, the range of possible glasses increases. However, it is not possible to give more specific guidelines since the success of a glass/metal seal depends on many factors, including geometry, glass mechanical properties, and processing conditions among others. The most efficient approach is to develop a numerical model (e.g., utilizing the finite-element method) for the specific application. While the potential exists for utilizing graded materials to form joints for high-temperature structural materials, techniques have not been sufficiently developed. However, at room temperature, one application is in biological prosthesis [3]. The rationale behind using graded materials is that (1) material mismatch is minimized and (2) when graded porosity is created, the bone can grow into it and become an integral part of the prosthesis. In addition, it may be desirable to grade the aspect of the joint that accounts for biocompatibility. Dental implants have the additional requirement that the external surfaces must be hard and tough [3].

10.3.4 Energy Conversion

10.3.4.1 Thermoelectric and Thermionic Conversion

The effectiveness in the conversion of thermal energy to electricity for a thermoelectric device is typically expressed as a figure of merit, Z, where

$$Z = \frac{\alpha^2}{\rho\kappa},$$

where α is the thermoelectric power, ρ is the electrical resistivity, and κ is the thermal conductivity. The figure of merit Z is ultimately determined by the band gap E_g and the carrier concentration n and is also a function of temperature, exhibiting a maximum, Z_{max}, at a unique temperature. Thus, by grading E_g and n along the device, it is possible to create a gradation in Z_{max}. Since the device experiences a temperature gradient, one can tailor it such that Z_{max} is achieved for each location in the device. Thus, the tailoring corresponds to choosing materials with the best performance at a particular temperature, and then stacking those materials from one side of the device to the other. The ultimate efficiency of power conversion through thermoelectric techniques depends not only on the difference in temperature from one side of the device to the other but also on the figure of merit achieved at each location. Figure 10.6 shows the concept for a three-layer device in which Bi_2Te_3, PbTe, and SiGe are used. It has been

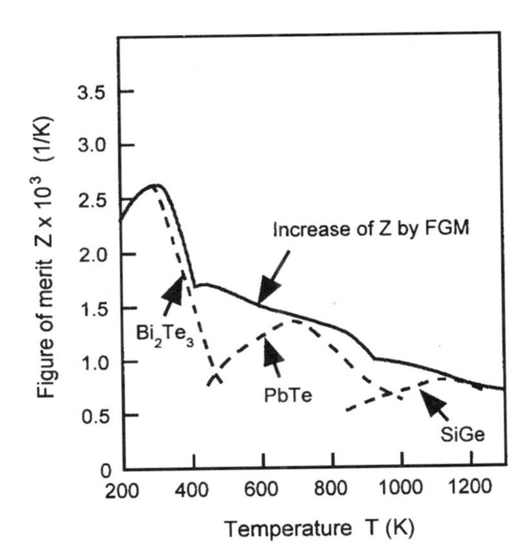

FIGURE 10.6 Figure of merit estimation for graded thermoelectric device (solid line) as function of temperature. Estimate is based on figure of merit for three compounds, also shown (dashed line). (Taken from [3].)

suggested that the efficiency of a graded thermoelectric device could be twice that of a homogeneous device. Thermionic conversion refers to the conversion of electrons emitted from a heated material into electrical current. Thermionic devices typically contain several dissimilar material joints, and thus graded materials are used to relieve thermal residual stresses.

10.3.4.2 Fuel Cells

Cathodic materials for solid oxide fuel cells have diverse property requirements: high electrical and ionic conductivity, high catalytic activity for oxygen reduction, chemical compatibility with the electrolyte and interconnects, compatibility of the thermal expansion coefficient with the other components in the fuel cell, stability in air at high temperatures, and the ability to be processed into thin films. By modifying the electrode main structure so that graded multilayer configuration is used, problems of poor adherence related to thermal expansion mismatch have been minimized, and the electrochemically active portion of the interface may be widened, increasing the efficiency of the cell.

10.3.5 Optical Fibers

Graded optical fibers have been used successfully for more than a decade to optimize the multiple transmission of light signals of different wavelength [32]. Figure 10.7 illustrates the benefits of graded index fibers. The multimode graded index design can transmit the widest bandwidth of any of the designs. Common glasses are based on silica, borosilicate, or soda-lime, but multicomponent glasses are also used. Methods to manufacture glass-graded optical fibers include chemical vapor deposition and the double-crucible method. Methods to manufacture polymer-graded optical fibers are based on either vapor-phase diffusion processes or monomer reaction processes [3]. The former involves diffusing a second monomer with a lower index of refraction from the outside to the inside of the polymer fiber, followed by copolymerization. The latter involves a copolymerization reaction where the two different monomers used have different reaction rates. By mixing the monomers in a glass tube, and polymerizing by application of ultraviolet (UV) radiation, the rates of polymerization are different in the outside of the tube from the inside, and thus the outside of the tube ends up with a different (lower) index of refraction. Details may be found elsewhere [3].

10.3.6 Electrical and Magnetic Behavior

The concept of using graded band-gap semiconductors has existed for more than 40 years. Nongraded heterojunctions exhibit sharp energy spikes corresponding to sharp interfaces; these sharp spikes may act as charge carrier barriers. The employment of compositional gradations in heterojunctions results in smooth band-gap energy transitions, avoiding spikes. Applications have included bipolar transistors, solar cell structures, and separate confinement heterostructures

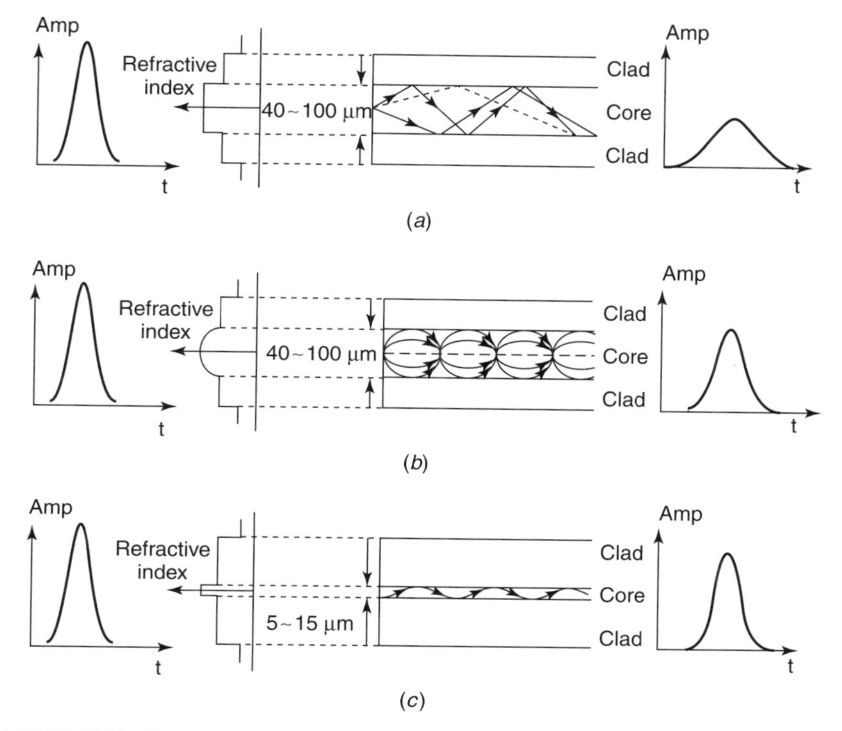

FIGURE 10.7 Schematic illustrating differences between (*a*) discrete architecture (multimode step index), (*b*) continuous architecture (multimode graded index), and (*c*) non-graded index (single-mode step index). (Taken from [3].)

in semiconductor lasers. Further details of these applications may be found elsewhere [3].

Materials in which magnetic properties are graded may be desired in applications where a magnetic sensor may be used to measure position [33]. A unique approach to processing graded magnetic steels is by rolling deformation of wedge-shaped samples [34]. In this technique, different sections of the wedge experience different strains (as measured by the rolling ratio) and thus exhibit different degrees of transformation from austenite (low saturation magnetization) to martensite (high saturation magnetization).

10.4 FABRICATION

10.4.1 General

It is useful to view fabrication methods as falling into one of the two categories shown in Fig. 10.1: constructive processes and transport-based processes. Constructive processes are those in which material is placed in the appropriate locations by some technique such as vapor deposition or powder metallurgy. Transport-based processes utilize the presence of a steep gradient to promote mass transport (e.g., atomic diffusion) from one location to another. A functionally graded material is in some sense inherently unstable since at a high enough temperature and long enough time, atomic or molecular diffusion occurs until there is no gradient. This natural tendency for the material to homogenize limits its use conditions. As with many material fabrication processes, the temperature of fabrication of FGM components should be higher than the service temperature or else structural and compositional evolution of the FGM may occur during service. However, the greater challenge with respect to the compositional or microstructural instability may be in processing the material because the processing conditions are typically more severe than the service conditions, as is true with many engineering materials. Furthermore, methods for producing many conventional materials seek to produce uniform composition and microstructure and therefore may not apply to graded material fabrication. Some methods not discussed here but used to produce FGMs include electrophoresis, electrodeposition, reaction synthesis (or self-propagating high-temperature synthesis), cladding, and sedimentation.

10.4.2 Thermal Spraying

Because of the ability to control composition with relative ease and the potential for batch manufacturing, thermal spraying is one of the easiest methods to fabricated graded coatings. In addition, it may be used to form bulk materials [35]. Generally, feedstock consists of powders, though rods or wire may also be used. The feedstock is mixed in the appropriate composition while it is fed into an intense heat source, such as a combustion plasma, an arc, or a laser beam. A torch is frequently used to generate the combustion plasma, and this is usually referred to as plasma spraying. The particles of material melt while they feed through the heat source, and they impinge on the substrate, typically undergoing rapid solidification. For more details on thermal spraying, the reader is referred to references [35–38]. One of the reasons composition may be easily controlled is that plasma spraying may involve multiple feeders that either feed into a single torch or multiple torches, thereby resulting in blended powders. Additionally, very dissimilar materials, such as refractories and metals may be simultaneously melted in the desired composition. One of the disadvantages is that the microstructure, including the texture, porosity, and presence of nonequilibrium phases is somewhat difficult to control. Some of these latter aspects may be affected by adjusting the cooling rate during deposition. Certain applications in which porosity gradients are desired (e.g., where increased compliance toward the coating surface is required) may utilize the ability to control porosity through the thickness of the coating by adjusting parameters such as the particle size, the plasma gas pressure, and the torch-to-substrate distance.

10.4.3 Vapor Deposition

10.4.3.1 Physical Vapor Deposition

Physical vapor deposition (PVD) provides a great degree of compositional control because multiple sources and targets may be employed. The most common method for producing graded coatings, electron beam physical vapor deposition (EB-PVD) utilizes an electron beam to heat the target material. Evaporation from the target to the substrate then occurs, the rate depending on a number of factors including temperature, constituent vapor pressures, and the geometrical conditions. The most successful method to produce graded compositions is to employ multiple targets using multiple electron guns. Single-gun EB-PVD units may be programmed so that the gun jumps back and forth between targets, at frequencies as high as kilohertz [3], varying whatever parameters (e.g., energy) are necessary to achieve the desired evaporation rates. Sputtering may also be employed to form films and coatings, though the rates are generally lower than EB-PVD. In sputtering, an inert gas is ionized using a high voltage. The resulting high-energy ions accelerate toward the cathodic target, and this results in the release of target material atoms that then deposit on the substrate. Sputtering is typically used to form wear-resistant coatings, such as in compositionally graded TiN-coated titanium alloys for surgical implants [3]. In this latter case, pure titanium is sputtered in an $Ar{-}N_2$ atmosphere, forming an amount of TiN that depends on the relative $Ar{-}N_2$ concentration. The $Ar{-}N_2$ concentration is varied during the deposition to produce a graded composition.

10.4.3.2 Chemical Vapor Deposition

Chemical vapor deposition (CVD) utilizes a source gas fed into a reaction chamber that excites the gas, typically by using heat, light, or a plasma. The gas reacts to form products that subsequently deposit onto a substrate. Control over composition of a CVD coating is achieved by modifying the gas mixture, the gas flow rate, the gas pressure, or the deposition temperature. Some common graded coatings fabricated by CVD include C–SiC, ZrC–C, and diamond–metal.

10.4.4 Powder Metallurgy

The techniques used to form graded materials by the sintering of powders are similar to those used for ceramics and metals, and the reader is encouraged to refer to other chapter of this handbook. A layered geometry is made by stacking powders of varying composition prior to consolidation. It may be difficult to achieve layer thickness uniformity when stacking powders of different compositions, particularly when the layer of each composition is thin (e.g., less than about 1 mm in the final dense part). Automated systems have been designed to maximize uniformity of layer thickness, allowing very large diameter (up to 300 mm) parts to be fabricated [39].

One of the most significant issues in processing graded materials by powder metallurgy concerns the residual stresses that may develop from differential shrinkage characteristics. This residual stress is distinctly different from that developed due to differential thermal expansion coefficients. Because different materials exhibit different (1) initial packing densities, (2) sintering start temperatures, and (3) sintering rates, mismatch strains develop during densification. Two methods used to minimize this mismatch include [40]: (1) adding different amounts of an organic binder phase to different regions of the FGM so that the initial packing densities are closely matched, and (2) altering the constituent particle sizes in different regions to produce different sintering start temperatures, sintering rates, and potentially initial packing densities. An example formulation is given for a Ni–Al_2O_3 system in Fig. 10.8 [40]. The same principle could be applied to other materials. The procedure is to conduct a set of experiments to determine the differential sintering characteristics (sintering rate and sintering start temperature) and the initial packing densities for each compositional layer. Once these are determined, matching the amount of shrinkage in each layer may be achieved by adding a binder phase that burns out and altering particle sizes, provided the latter does not change the resulting microstructure in an undesirable way. The same procedure may be used to intentionally produce gradients in porosity in a material, for example, for fabricating a graded porosity preform that is subsequently infiltrated; an example is described in [41]. Other techniques for producing compositional gradations during powder preparation include powder

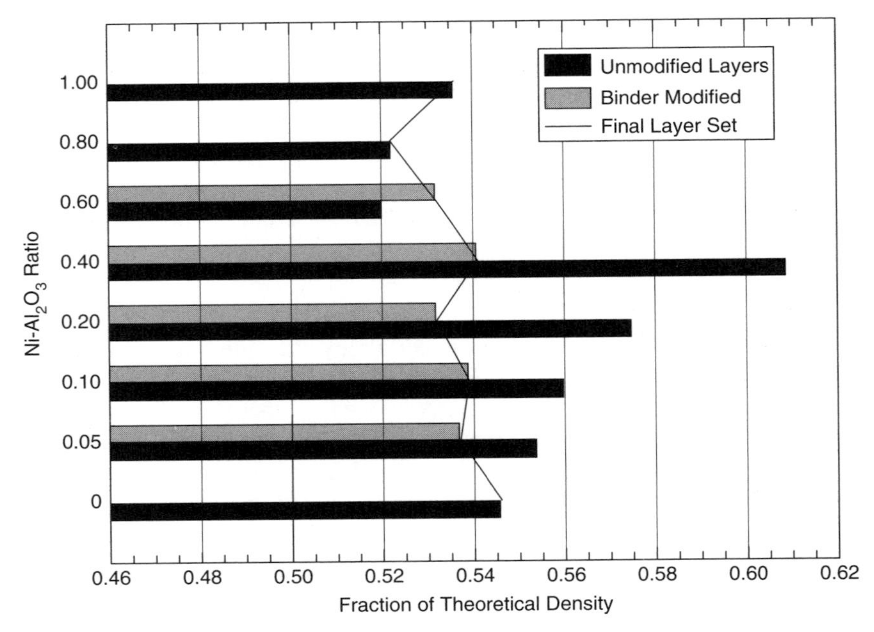

FIGURE 10.8 Example formulation for powder metallurgy processing of graded Ni–Al_2O_3 composites utilizing varying binder contents. (Reprinted from Winter [41].)

stacking under centrifugal forces [42, 43] and slip casting under pressure-induced flow [44–46].

Consolidation and sintering are also achieved in a manner similar to methods used for ceramics and metals. These include pressureless sintering, hot pressing, hot isostatic pressing, and microwave sintering. Gradients in temperature during densification may be used as a method to modify sintering rates. Spark plasma sintering (SPS) (also called pulse electric current sintering or plasma-activated sintering [47]) has been relatively recently shown to be an effective method for manufacturing graded materials, particularly when dissimilar materials must be sintered simultaneously [48]. In SPS an electric current is applied to the compact while it is pressed at elevated temperature. The electric current results in local "Joule" heating as well as heating from the creation of localized plasma between nearby powder particles, as indicated schematically in Fig. 10.9. Since the amount of Joule heating and plasma heating are highest in less dense areas of the compact, regions that would not sinter well under only radiative heat and pressure experience enhanced sintering rates.

10.4.5 Preform Methods

A second phase may be intruded by solid-, liquid-, or vapor-phase diffusion. Solid-state diffusion has been used for a long time to produce desired compositional and/or microstructural gradients, for example, in steel processing. Other examples of solid-state diffusion are discussed elsewhere [3]. In liquid- or

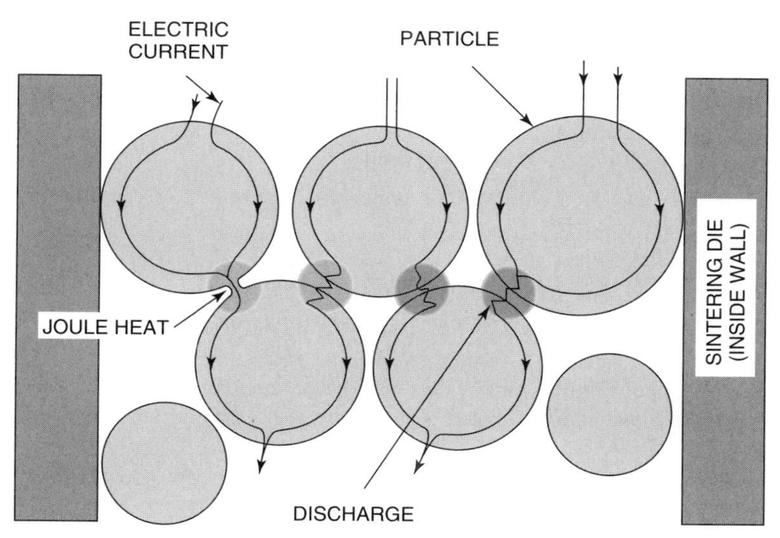

FIGURE 10.9 Schematic showing ways in which heat is generated in spark plasma sintering (SPS). (Courtesy of M. Tokita, Sumitomo Coal Mining Group at Izumi Technology Company.)

vapor-phase diffusion, a porous preform is usually required. Thus, the formation of a compositionally graded material is greatly facilitated if the preform porosity is graded. The techniques of infiltration are similar to those used in the fabrication of composites by preform/infiltration methods, though new techniques have been developed for FGMs [49].

REFERENCES

1. Keyes and Kraus, U. S. Patent 1,014,757 (1912).
2. L. J. Buttolph, *J. Optic. Soc. Am.*, **11**, 549–557 (1925).
3. Y. Miyamoto, W. A. Kaysser, B. H. Rabin, A. Kawasaki, R. G. Ford, (Eds.), *Functionally Graded Materials: Design, Processing and Applications*, Kluwer Academic, Boston, MA, 1999.
4. H. Kroemer, *RCA Rev.*, **18**(3), 332–342 (1957).
5. M. B. Bever and P. E. Duwez, *Mater. Sci. Eng.* **10**, 1–8 (1972).
6. S. Suresh and A. Mortensen, Fundamentals of Functionally Graded Materials, IOM Communications, London, 1998.
7. R. L. Williamson, B. H. Rabin, and J. T. Drake, *J. Appl. Phys.*, **74**(2), 1310–1320 (1993).
8. J. T. Drake, R. L. Williamson, and B. H. Rabin, *J. Appl. Phys.*, **74**(2), 1321–1326 (1993).
9. R. D. Torres, G. G. W. Mustoe, I. E. Reimanis, and J. J. Moore, "Evaluation of Residual Stresses Developed in a Functionally Graded Material Using the Finite Element Technique," in T. S. Srivatsan and J. J. Moore (Eds.), *Processing and Fabrication of Advanced Materials*, Vol. IV, TMS, Warrendale, PA, 1996.
10. S. Ho, C. Hillman, F. F. Lange, and Z. Suo, *J. Am. Ceram. Soc.*, **78**(9), 2353–2359 (1995).
11. J. Chapa and I. Reimanis, *J. Nucl. Mater.* **303**, 131–136 (2002).
12. F. Erdogan, *Comp. Eng.*, **5**, 753–770 (1995).
13. B. B. Muvdi and J. W. McNabb, *Engineering Mechanics of Materials*, 3rd ed., Springer, New York, 1991.
14. Z. -H. Jin and N. Noda, *ASME J. Appl. Mech.*, **62**, 738–740 (1994).
15. J. W. Hutchinson and Z. Suo, *Adv. Appl. Mech.*, **29**, 63–191 (1992).
16. P. Gu and R. J. Asaro, *Int. J. Solids Structures*, **34**(1), 1–17 (1997).
17. I. Reimanis, J. Chapa, A. N. Winter, W. Windes, and E. Steffler, "Fracture and Deformation in Ductile/Brittle Joints with Graded Structures," in K. Ichikawa (Ed.), *Proceedings of Functionally Graded Materials in the 21st Century*, Tsukuba, Japan, March 26–28, 2000.
18. F. Erdogan, A. C. Kaya, and P. F. Joseph, *J. Appl. Mech.*, **58**, 410–418 (1991).
19. Z.-H. Jin and R. C. Batra, *J. Mech. Phys. Solids*, **44**(8), 1221–1235 (1996).
20. C. F. Cline, "Preparation and Properties of Gradient TiC Cermet Cutting Tools," in B. Ilschner and N. Cherradi (Eds.), *Proceedings of the 3rd International Symposium on Structural and Functional Gradient Materials*, Presses Polytechniques et Universitaires Romandes, Lausanne, Switzerland, 1995, p. 595.

21. J. Jitcharoen, N. Padture, A. E. Giannakopoulos, and S. Suresh, *J. Am. Ceram. Soc.* **81**(9), 2301–308 (1998).
22. S. Suresh, M. Olsson, A. E. Giannakopoulos, N. P. Padture, and J. Jitcharoen, *Acta Mater.*, **47**(14), 3915–3926 (1999).
23. M. Finot and S. Suresh, MultiTherm, Version 1.4, a personal computer software for the thermomechanical analysis of layered and graded materials, MIT, 1994.
24. S. Suresh, A. E. Giannakopoulos, and M. Olsson, *J. Mech. Phys. Solids*, **42**, 979–1018 (1994).
25. L. B. Freund, *J. Mech. Phys. Solids*, **44**, 723–736 (1996).
26. C. -W. Nan, R. -Z. Yuan, and L. -M. Zhang. "The Physics of Metal/Ceramic Functionally Gradient Materials," in *Functionally Gradient Materials, Ceramic Transactions*, American Ceramic Society, Westerville, OH, 1993, pp. 75–82.
27. D. Stauffer and A. Aharony, *Introduction to Percolation Theory*, rev. 2nd ed., Taylor and Francis, New York, 1994.
28. J. F. Stroman, "Optical Fibers," in *Ceramics and Glasses*, Vol. 4, Engineered Materials Handbook, ASM International, Materials Park, OH, 1991, pp. 409–417.
29. J. J. Petrovic, Los Alamos National Laboratory, private communication (2000).
30. M. Peters, Development of a Model Graded Architecture for Chromium Nitride Coatings Deposited by Cathodic Arc Evaporation for Wear Resistant and Forming Applications, Ph.D. Thesis, Colorado School of Mines, 1999.
31. A. P. Tomsia, J. A. Pask, and R. E. Leohman, "Glass/Metal and Glass-Ceramic/Metal Seals," in *Engineered Materials Handbook, Ceramics and Glasses*, Vol. 4, ASM International, Materials Park, OH, 1991.
32. S. R. Nagel, "Telecommunications and Related Uses," in *Engineered Materials Handbook, Ceramics and Glasses*, Vol. 4, ASM International, Materials Park, OH, 1991.
33. Y. Watanabe, Y. Nakamura, Y. Fukui, and K. Nakanishi, *J. Mater. Sci. Lett.*, **12**, 326–328 (1993).
34. Y. Watanabe, S. H. Kang, J. W. Chan, and J. W. Morris, Jr., *Mater. Trans., JIM*, **40**(9), 961–966 (1999).
35. S. Sampath, H. Herman, N. Shimoda, and J. Saito, *MRS Bull.*, January, 27 (1995).
36. S. Kuroda, "Properties and Characterization of Thermal Sprayed Coatings—A Review of Recent Research Progress," in Proc. of the 15th International Thermal Spray Conference 25–29 May 1998, Nice, France.
37. M. L. Thorpe, *Adv. Mater. Proc. ASM International*, May, 68–89, 1993.
38. L. Pawlowski, *The Science and Engineering of Thermal Spray Coatings*, Wiley, New York, 1995.
39. Y. Nakayama, "Development of Automatic FGM Manufacturing System," in Proceedings of the NEDO International Symposium on Functionally Graded Materials, Mielparque, Tokyo, Japan, October 21–22, 1999, pp. 35–40.
40. A. N. Winter, B. A. Corff, I. E. Reimanis, and B. H. Rabin, *J. Amer. Ceram. Soc.* **83**(9), 2147–2154 (2000).
41. A. N. Winter, Ph.D. Thesis, Colorado School of Mines (1999).
42. R. K. Roeder, K. P. Trumble, and K. J. Bowman, *J. Am. Ceram. Soc.*, **80** 27–36 (1997).

43. D. P. Miller, J. J. Lannutti, and R. D. Nöbe, *J. Mater. Res.*, **8**(8), 2004–2013 (1993).
44. B. R. Marple and S. Tuffe, "Graded Casting for Producing Smoothly Varying Gradients in Materials," in I. Shiota and Y. Miyamoto (Eds.), *Proceedings of the 4th International Symposium on FGMs*, FGM96 Elsevier Science, Amsterdam, 1997, pp. 167–172.
45. S. Tuffe and B. R. Marple, *J. Am. Ceram. Soc.*, **78**(12) 3297–3303 (1995).
46. J. Chu et al., *J. Ceram. Soc. Jpn*, **101**(7), 818–820 (1993).
47. Y. Zhou, K. Hirao, M. Toriyama, and H. Tanaka, *J. Am. Ceram. Soc.*, **83**(3), 654–656 (2000).
48. M. Tokita, "Trends in Advanced SPS Systems and FGM Technology," in Proceedings of the NEDO International Symposium on Functionally Graded Materials, Mielparque, Tokyo, Japan, October 21–22, 1999, pp. 23–33.
49. K. J. McClellan, J. J. Petrovic, and I. E. Reimanis, U.S. Patent 5,928,583, filed May 29, 1997, received July 27, 1999.

CHAPTER 11

Corrosion of Engineering Materials

ROBERT AKID

Materials Research Institute, Sheffield Hallam University, Howard Street, Sheffield, United Kingdom S1 1WB

Handbook of Advanced Materials Edited by James K. Wessel
ISBN 0-471-45475-3

11.1 INTRODUCTION

Corrosion, *environmental degradation*, and *rusting* are terms commonly used to describe the processes that bring about a loss of performance of engineering materials. *Corrosion* is the term generally used to describe the chemical "wasting," which occurs when a metal or alloy reacts with the environment within which it is in contact. More specifically, *rusting* is a term describing the dissolution of iron-based alloys, notably steels. A study of corrosion requires a multidisciplinary approach, involving the disciplines of chemistry, physics, metallurgy, and mechanical engineering. This interdisciplinarity may well be the reason why, despite our 200+ years of awareness of the problems imposed by aggressive environments, corrosion incidents continue to occur on a daily basis.

This chapter will attempt to provide an "engineering" approach to the subject of corrosion, offering general guidance on the basic principles, forms of corrosion, methods of evaluating corrosion resistance, and design aspects. From this it is hoped that mechanical and design engineers and material selectors will be better positioned to improve the overall performance and, hence, lifetime of engineering components and structures.

11.2 ECONOMIC ASPECTS OF CORROSION

11.2.1 Cost of Corrosion

Records of corrosion failures date back well over 200 years, although at the time no assessments were made of the costs of such failures. This is of no surprise as much of the cost of corrosion relates to *indirect or hidden* costs, for example, general maintenance, loss of containment of product, pollution remediation, loss of production, and endangerment of life. Surveys have been conducted [1, 2] that identify the cost of corrosion for an industrialized nation to be close to 4% of gross national product (GNP). It is interesting to note that in the report by Hoar only 8% of the total corrosion costs was attributed to the oil and chemical industry, in comparison to 42% attributed to the power industry. To put some meaning on this figure, approximately 1 penny out of every pound (2 cents out of every dollar) is used to replace corroded metals. In terms of the quantity of iron produced worldwide, 30% is used simply to replace corroded iron [3].

More recent surveys are now awaiting publication. However, it would appear that new estimates of the cost of corrosion remain around 4% GNP. This is a sobering thought given that the fundamentals of corrosion and its prevention are now well understood. Such facts may well reflect society's view of corrosion, being that of an "unavoidable feature of life." However, a more realistic appraisal of society's failure to reduce the cost of corrosion lies in the subject's "interdisciplinary" nature, and corrosion is often therefore not treated as a mainstream subject.

11.2.2 Learning from the Past

Our prediction of the future performance of materials under specific operating conditions is often based upon previous past experience. However, what is often not appreciated is that a specified material can exhibit an entirely different performance as a result of its' manufacturing history and the nature of the design (see Section 11.9). Examples of galvanic (or bimetallic) corrosion are littered throughout the marine industry, ranging from loss of hull sheeting on frigates, steel bolt failures on copper end plates, to magnesium alloy wheel failures on jump jet harriers [4]. Such failures were predictable, and the risk of these failures could have been minimized by the application of existing knowledge. Similar examples can be found in the automotive, power, construction, and chemical industries [5–7].

11.3 CORROSION PROCESSES

11.3.1 Basic Concepts

Numerous definitions of the term *corrosion* have been cited in the literature, each having its own respective merits. Here corrosion is defined as the *degradation* of a metal or alloy arising from the interaction of the material with an *aggressive* environment. Degradation may be in the form of loss of aesthetic appearance or loss of structural integrity, either of which may result in the component or structure being deemed "unacceptable." Such a definition provides suitable discrimination between the *deterioration* and *transformation* definitions of corrosion [8]. The latter relates to the transformation of a metal to an oxide to form an adherent protective oxide film, for example, $Ti \rightarrow TiO_2$, $Al \rightarrow Al_2O_3$, etc. Corrosion will, unless otherwise stated, be considered as an electrochemical process with the specific requirements of an anode, a cathode, a conducting electrolyte, and a metallic path connecting the anode and cathode sites.

11.3.1.1 Definitions [9]

Anode The electrode of an electrolytic cell at which oxidation is the principal reaction leading to metal ions entering solution, that is, the electrode where corrosion occurs.

Cathode The electrode of an electrolytic cell at which reduction is the principle reaction.

Electrolyte An aqueous ionic conducting substance in which the anode and cathode are immersed.

Metallic Path The external circuit that connects the anode to the cathode, permitting the flow of electrons from the anode to the cathode.

A corrosion cell therefore consists of two electronically connected sites (an anodic and cathodic site) in contact with a conducting electrolyte. Figure 11.1 shows that corrosion cells can result from the connection of two metals or from the separation of anode and cathode sites on an individual metal or alloy.

11.3.1.2 Formation of Corrosion Cells

In aqueous corrosion reduction at the cathode (C) and oxidation at the anode (A) occurs involving the transfer of electrons (e); see Fig. 11.1. This process requires

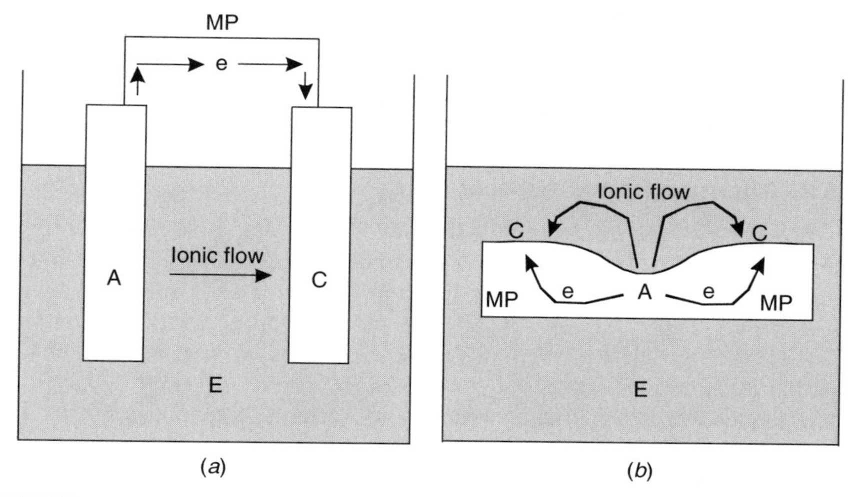

FIGURE 11.1 Examples of corrosion cells (*a*) schematic (bimetallic system), A = anode (e.g., zinc), C = cathode (e.g., copper), E = electrolyte (e.g., seawater), MP = metallic path (e.g., wire); (*b*) corrosion cell derived from single metal (e.g., pitting in stainless steel) A = anode, C = cathode, MP = metallic path, E = electrolyte.

electrical contact between the anode and cathode via a metallic path (MP) and the electronic circuit is completed via ionic flow (charged ions) through the presence of a conducting electrolyte (E). Classically corrosion cells are formed when two dissimilar metals are connected together within an electrolyte, (Fig. 11.1a). However, a number of other conditions may arise that lead to the formation of a corrosion cell upon an isolated metal or alloy (Fig. 11.1*b*). These include:

1. *Differences in Material Microstructure*
   a. Matrix versus grain boundary
   b. Difference in grain orientation between adjacent grains
   c. Second-phase particles within a solid solution matrix
2. *Chemical Heterogeneities within a Matrix* Nonmetallic inclusions, for example, oxides, sulfides, etc., which provide a cathodic/anodic site with respect to the matrix.
3. *Differential Aeration* Where metal surfaces experience differences in oxygen concentration; the site of lower oxygen concentration becoming the anode.
4. *Differential Concentration* In a similar manner to condition 3, differences in the concentration of a species, that is, metal ions, leads to the formation and separation of anodic and cathodic sites.
5. *Heat Treatment Effects* Areas that exhibit different microstructures due to heat treatments, for example, quenching, weld heat-affected zones, and so forth, lead to conditions where anode and cathodic sites may be established.

6. *Mechanical Working* Adjacent areas on a metal surface that have received different degress of strain (deformation) may lead to the formation of a corrosion cell, where the strained areas become anodic with respect to unstrained cathodic areas.

11.3.2 Thermodynamic Aspects

Very few metals exist in nature in their native form. That most metals are found either in the form of oxides, sulfates, carbonates, or other complex (mixed) compounds eludes to the fact that in their elemental state metals are inherently reactive. This tendency for the formation of natural metallic compounds is governed by the laws of thermodynamics, laws that govern whether or not a particular reaction is possible.

The tendency for a particular reaction to occur is determined by the free-energy change that takes place when reactants come into physical contact with each other, for example, a water droplet on a steel structure. Given that corrosion is an electrochemical process, consideration of the electrode potentials of the half-cell reactions (see below), which occur at the anodic and cathodic sites, provides a more direct approach to assessing the tendency of a reaction. A corrosion cell consists of two half-cell reactions, which may or may not occur on the same electrode (see Fig. 11.1). In the case of Fig. 11.1*a* the two half-cells A and C are connected to complete the corrosion cell via a metallic path (i.e., a wire). Each half-cell can be designated as E_a or E_c, relating to whether or not an oxidation (anodic) or reduction (cathodic) reaction occurs. The total cell potential when two half-cells are coupled together is given by

$$E_{cell} = E_c - E_a. \tag{11.1}$$

For a reaction to occur E_{cell} must be positive, that is, $E_{cell} = E_c - E_a > 0$ or $E_c > E_a$.

Faraday's law relates the free-energy change of a corrosion process with that of the cell potential and is given by

$$\Delta G = -nFE, \tag{11.2}$$

where ΔG is the free-energy change (J/mol), E is the cell potential (V), F is Faradays constant (96,494 C/mol), and n is the number of electrons (negative charge) transferred during the corrosion reaction.

This leads to the well-known Nernst equation:

$$E = E^0 - \frac{RT}{nF} \ln \frac{[\text{products}]}{[\text{reactants}]}. \tag{11.3}$$

Here E is the potential generated by the reaction, [ ] represents the concentration of either reactants or products, E^0 is designated the standard electrode potential, R is the gas constant (8.314 J/mol/K), T is temperature in Kelvin (298 K).

A simple example of corrosion consisting of two half-cell reactions is that of iron in an acid solution. Here the cathodic reaction is

$$2H^{+}{}_{(aq)} + 2e^{-} = H_2(g) \tag{11.4}$$

and the anodic reaction is

$$Fe_{(s)} = Fe^{2+}{}_{(aq)} + 2e^{-} \tag{11.5}$$

The application of the Nernst equation to corrosion studies was embedded in the work of Pourbaix [10] who constructed E versus pH diagrams that provided "maps" showing regimes of corrosion, immunity, or passivity for metals immersed in water.

Figure 11.2 presents a typical E versus pH (Pourbaix) diagram for the iron/water system. From this diagram three principle regions can be found. Region 1 (shaded) represents active "corrosion" where the concentration of metal ions in solution is $\geq 10^{-6}$ M. In the case where the ion concentration is less than 10^{-6} M the metal is considered to be in a condition of "immunity" (noncorroding). This represents the second region. Finally, a third region exists being that of "passivity". In this state the metal surface is covered by a film (usually an oxide), which prevents the metal from corroding.

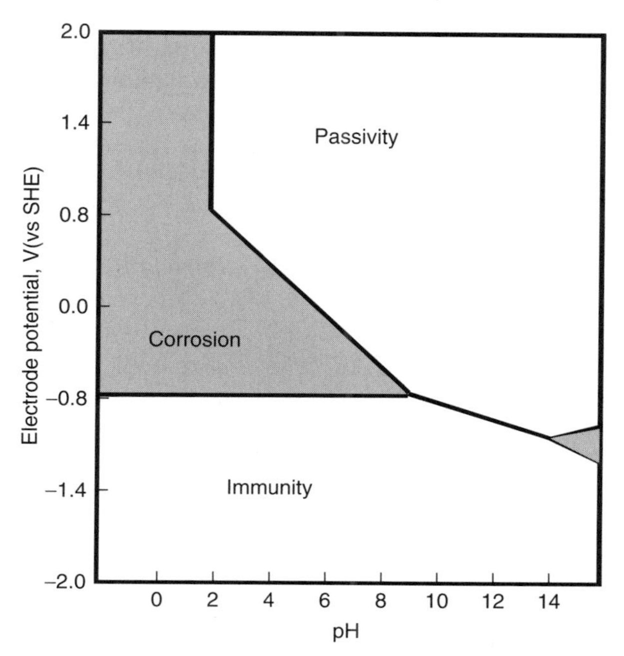

FIGURE 11.2 Pourbaix diagram for the iron–water system at 25°C (after [10]).

In Fig. 11.2 the potential scale (y axis) is given as a measure (in volts) versus the standard hydrogen electrode (SHE). It should be noted that all electrode potentials are quoted against a given reference electrode. The potential established when a metal is immersed in a solution arises from the release of positively charged cations together with the creation of a negatively charged metal. The potential difference created cannot be measured in absolute terms, but what may be measured is the potential difference between it and another electrode, the latter being a reference electrode (of designated electrode potential). The SHE is designated as having an electrode potential of 0.0 V. Other reference electrodes such as the silver/silver chloride and saturated calomel reference electrode have assigned potential values measured with respect to the SHE, in this case being 0.248 and 0.242 V at room temperature, respectively.

11.3.3 Kinetics (Rates) of Corrosion

The construction of thermodynamic corrosion (Pourbaix) diagrams provides valuable information on whether or not a particular reaction is feasible. However, these diagrams do not provide any information on the rate at which a reaction may occur. For example, the reaction of gaseous hydrogen and gaseous oxygen to form water is accompanied by a large decrease in free energy, some 200 + kJ at 25°C. However, the rate of this reaction is so slow that it might be regarded as not occurring at all. Furthermore, these diagrams, given normally at 25°C, do not provide information on a time basis. In this respect it is therefore necessary to consider the kinetic aspects of corrosion.

As previously discussed, a corrosion cell consists of two half-cell reactions that, in the case of steel immersed in an acidic solution, occur on the same electrode. The two half-cell reactions are

$$Fe_{(s)} \rightarrow Fe^{2+}{}_{(aq)} + 2e^{-}\,(\text{anodic}) \tag{11.6}$$

and

$$2H^{+}{}_{(aq)} + 2e^{-} \rightarrow H_{2(g)}\,(\text{cathodic}) \tag{11.7}$$

Overall

$$Fe_{(s)} + 2H^{+}{}_{(aq)} \rightarrow Fe^{2-}{}_{(aq)} + H_{2(g)} \tag{11.8}$$

The standard half-cell potential for the oxidation of iron, $E_{(a)}$, is 0.44 V while that for the cathodic reaction, $E_{(c)}$ is 0.0 V [11]. The total cell potential $E_{(\text{cell})}$ is given by $E_{(\text{cell})} = E_{(c)} - E_{(a)} = 0.0\text{ V} - 0.44\text{ V} = -0.44\text{ V}$ Based on Faraday's law, Eq. 11.2, a large negative free-energy value is obtained and the reaction proceeds spontaneously.

11.3.3.1 Mixed-Potential Theory and Evans Diagrams

The mixed-potential theory was developed to address the problem of several electrochemical reactions proceeding simultaneously at the same metal–solution interface. The rest potential (corrosion potential) of, for example, iron immersed in an acidic solution, is a mixed potential and lies between the equilibrium potentials of the two participating reactions (i.e., anodic dissolution and hydrogen evolution).

The Evans diagram is a kinetic diagram representing electrode potential in volts versus corrosion current in amperes per unit area. Figure 11.3 presents an Evans diagram for iron immersed in an acidic solution. Theoretically, four reactions are possible for this system, that is, iron dissolution to from ferrous ions (Fe^{2+}), the reverse of this process, that is, ferrous ions attaining electrons to form Fe, hydrogen ions in solution forming hydrogen gas or the reverse of this process. Hence four E vs. i lines are presented in Fig. 11.3. For clarity the two thermodynamically feasible reactions are given as solid lines. An extrapolation of the portions of the two solid lines gives rise to an intersection at the corrosion potential (E_{corr}) and corrosion current density (i_{corr}) for the given system. Evans diagrams provide kinetic information, that is, the rate of a corrosion reaction, as a function of the applied potential.

The corrosion potential E_{corr} is in effect the line (on the Pourbaix diagram) that differentiates the corrosion and immunity regions. That is, if the potential is held below E_{corr} in Fig. 11.3, the rate of iron dissolution decreases and the hydrogen generation reaction dominates. On the other hand if the potential is held above E_{corr}, the metal dissolution reaction dominates and the rate of corrosion increases. Maintaining the potential below that of E_{corr} is adopted in practice and is known as cathodic protection (see Section 11.7.3)

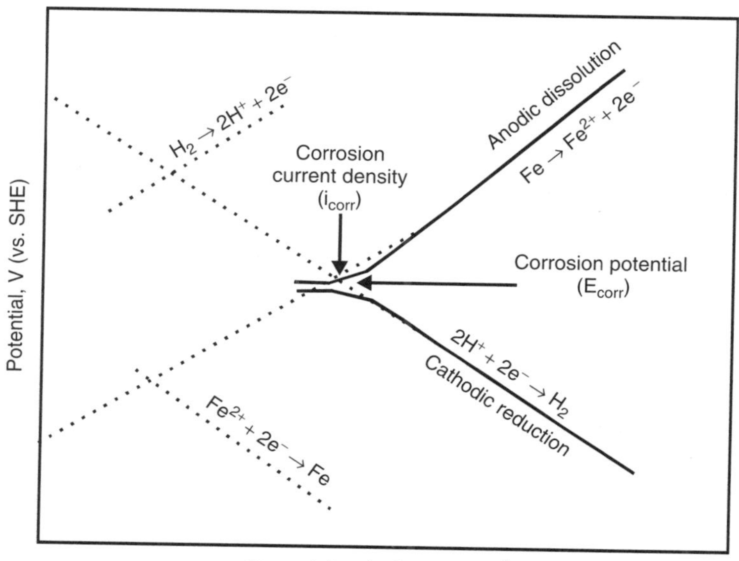

FIGURE 11.3 Schematic of the Evans diagram for iron in acid solution.

11.3.3.2 Determination of Corrosion Rate

It can be seen that by plotting the corrosion current as a function of potential (Figs. 11.3 and 11.4), four important parameters can be determined, that is, corrosion potential E_{corr}, corrosion current I_{corr}, and the anodic (β_a) and cathodic (β_c) Tafel constants. Figure 11.3 also shows that by moving the potential away from E_{corr} by a set value $+\Delta E (= E - E_{corr})$ a straight line is obtained:

$$\eta_a = \beta_a \log\left(\frac{i_{app}}{i_{corr}}\right), \tag{11.9}$$

where $\eta_a = E - E_{corr}$.

Similarly moving the potential by a value $-\Delta E$, a corresponding equation is obtained:

$$\eta_c = \beta_c \log\left(\frac{i_{app}}{i_{corr}}\right),$$

where a and b represent anodic and cathodic terms and η is designated as the overpotential or polarization. Hence the kinetics of the anodic or cathodic reaction may be obtained by plotting η versus log (i_{app}) for either reaction. It should be recognized that the above description of the Stern and Geary electrode kinetic expression is somewhat simplified and a full treatise can be found in reference [12].

This method of *polarizing* the sample and measuring the corresponding change in corrosion current (or current density for a known area) is commonly used to determine the corrosion rates of metals in given electrolytes. The plots obtained by this method are known as *polarization* curves.

Although polarization curves are commonly used to provide kinetic data for corrosion reactions, several complications arise due to both solution resistance and concentration polarization effects. The first of these is associated with the ohmic drop between the sample and reference electrode. This arises primarily due to (a) the resistivity of the solution. (b) the magnitude of the applied current, and (c) the location of the reference electrode. The result of ohmic errors is that the measured potential requires correction [13].

Concentration polarization occurs when the rate of removal of reacting species from the electrode surface, or rate of transfer of species to the electrode, becomes diffusion limited. such effects occur at large over potentials [14] and cause the anodic and cathodic Tafel lines to deviate from linearity. Figure 11.4 shows a typical polarization curve for the aluminum alloy 2024/NaCl system.

Figure 11.4 shows that at low values of overpotential, η the E versus log i curve deviates from a straight line. If, however, E versus I values are plotted on linear axes the following relationship holds [15]:

$$i_{corr} = \left(\frac{\Delta i}{2.3\Delta E}\right)\left(\frac{\beta_a\beta_c}{\beta_a + \beta_c}\right). \tag{11.10}$$

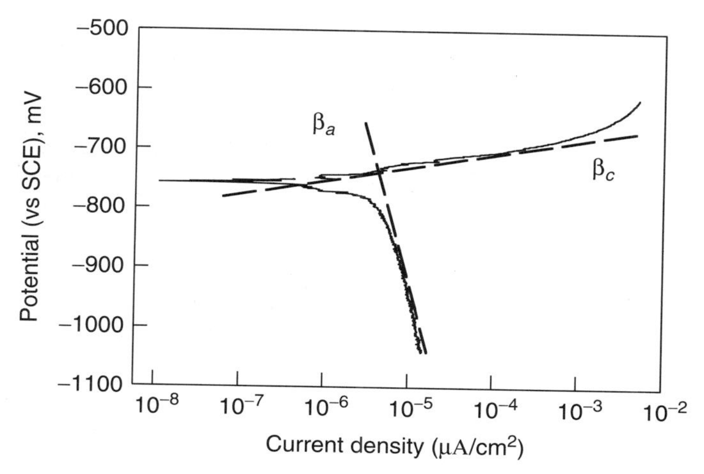

FIGURE 11.4 Typical anodic and cathodic polarization curves for aluminum alloy 2024 in 3.5% NaCl solution at 25°C.

Figure 11.5 shows a schematic of the typical E versus i relationship at low overpotentials for a corroding system. By determining the slope of the curve at $\Delta E = 0$, that is, E_{corr}, the polarization term R_p may be found and equation (10) can be rearranged to give

$$i_{\text{corr}} = \frac{1}{2.3R_p}\left(\frac{\beta_a\beta_c}{\beta_a + \beta_c}\right), \tag{11.11}$$

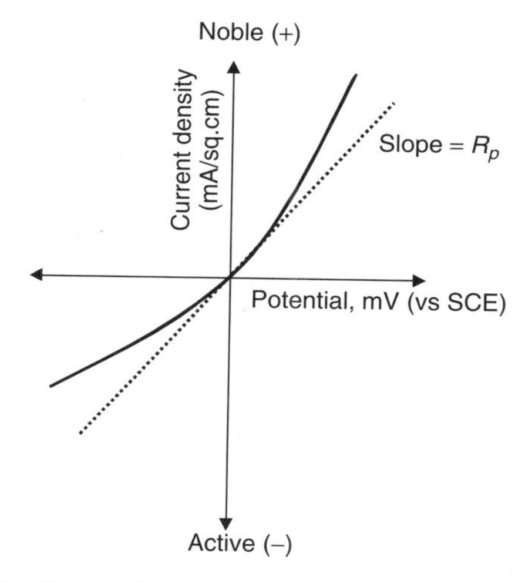

FIGURE 11.5 Schematic showing typical polarization resistance plot.

where $R_p = \Delta E/\Delta i$ and R_p is inversely proportional to the corrosion rate. The standard method for determining R_p and hence the corrosion rate is given in ASTM G59 Practice for Conducting Potentiodynamic Polarisation Resistance Measurements [16].

The above procedure is commonly used in industry although care must be taken that large overpotentials are not applied that invalidate the linearity of Eq. 11.10. In addition complications can arise when the open-circuit (free) corrosion potential changes over the period of the measurement. Finally the above procedure assumes that the anodic and cathodic corrosion reactions are not influenced by limitations in diffusion of species [17] and that the solution resistance of the electrolyte is low. Additional consideration of the above can be found in reference [18].

The practical considerations of evaluating corrosion rate are further detailed in Section 11.6.

11.4 FACTORS AFFECTING THE RATE OF CORROSION

As discussed previously, the corrosion of a metal proceeds via anodic and cathodic reactions. Hence any process that limits or interferes with one or both of these reactions will influence the overall rate of corrosion. Specific cases of corrosion control will be detailed in Section 11.7. However, some general comments will be given here to overview the factors affecting corrosion rate.

11.4.1 Environment

The rate of a corrosion reaction will be determined by the nature and concentration of reacting species of the environment (and the nature and form of the corrosion products). For atmospheric corrosion these include water content (humidity), the presence of chloride (Cl^-), or oxides of sulfur (SO_x) and nitrogen (NO_x), where x can vary from 1 to 3 depending upon the source reactions. Additionally carbon in the form of CO_2 influences the rate of corrosion by causing a drop in the pH of aqueous phases or in solid form by depositing soot particles on the surfaces of metals, which act as cathodic sites.

When the environment is predominantly aqueous in nature, the corrosion rate is influenced by solution conductivity, acidity (pH), dissolved gases and solids, and temperature. Natural waters vary considerably in their composition, being dependent upon the nature of the rock/soil composition through which they permeate following rainfall or being high in chloride concentration as in the case of seawater.

Typically natural waters vary from being very soft (low $CaCO_3$ content) to very hard. Table 11.1 provides a typical water analysis (in milligrams/liter or parts per million) for the above cases.

Dissolved gases are present in all waters, oxygen being the most important, as it often takes part in the cathodic reaction (oxygen reduction) for neutral

TABLE 11.1 Typical Composition/Properties of Soft and Hard Waters

	pH	Alkalinity[a]	$CaCO_3$[b]	Sulfate	Chloride	DS[c]
Very soft	6.1–6.4	<5	5–15	<10	4–8	<50
Very hard	6.9–7.2	450–500	>500	400–450	130–150	>1500

[a] $CaCO_3$ as determined against methyl orange indicator.
[b] $CaCO_3$ given as total hardness.
[c] Dissolved solids.

and alkaline solutions. The concentration of oxygen varies with temperature, being around 15 mg/L at 0°C and 8 mg/L at 25°C. Carbon dioxide can lead to a decrease in the pH of water, but its effect is dependent upon the bicarbonate (HCO_3^-) content of the water. Depending upon this relationship, temperature and other constituents within the water, precipitation of carbonate on the metal surface may occur leading to some degree of protection. A fuller explanation of this is given in reference [19].

Seawater environments are naturally more aggressive than natural waters due to the presence of a high concentration of salts, most notably chloride. Seawater is a complex electrolyte, the composition of which varies around the world. Table 11.2 presents a typical composition for seawater.

Seawater in equilibrium with atmospheric CO_2 is slightly alkaline with a pH of 8.1–8.3 and contains a number of dissolved gases, the most prevalent being O_2. Oxygen saturation in seawater is approximately 7–8 mg/L (ppm) although this value varies with depth being a minimum (<4 mg/L) at depths of 400–800 m.

The corrosion rates of metals in seawater are higher than those experienced in fresh (natural) water, typically being on the order of 0.65 mm/yr and 0.25 mm/yr, respectively.

11.4.2 Concentration, Temperature, and Solution Velocity

In general, increasing concentration of reactive species leads to an increase in the corrosion rate. For example, increasing the hydrogen ion concentration of a solution, that is, decreasing the pH, causes an increase in corrosion rate. However, where solution chemistry can lead to the formation of a passive film, that is, oxidizing acids such as nitric acid, a decrease in corrosion rate can occur.

TABLE 11.2 Typical Composition of Seawater

Element	Ion	Concentration (g/kg)
Sodium	Na^+	10.68
Potassium	K^+	0.40
Magnesium	Mg^{2++}	1.28
Chlorine	Cl^-	18.98
Sulfur	SO_4^{2-}	2.70

Increases in temperature also tend to increase the corrosion rate. For example, an increase in temperature of 10°C will tend to double the corrosion rate. Such effects are, however, not universal as increasing the temperature of a solution affects the concentration of gases dissolved in solution, notably oxygen. Increased corrosion rates are the result of increased mass transport, which for oxygen reduction controlled reactions, occurs at about 80°C [20].

Solution velocity effects the corrosion rate of metals by affecting the rate of mass transport of species to and from the reaction sites and also by affecting the stability of passive films. In addition, increased flow rates affect whether or not corrosion products remain at the anode/cathode sites, thereby affecting the ability of subsequent electron transfer processes. Oldfield and Todd [21] proposed the following equation to predict the corrosion rate of mild steel in flowing seawater:

$$\text{Corrosion rate (mm/y)} = 1.17 \times 10^{-2}\ \mathrm{CO_2}\ \mathrm{U}^{0.9}\ \frac{D^{0.75}}{\nu}, \tag{11.12}$$

where CO_2 = concentration of oxygen, ppb

U = flow rate, cm/s

D = diffusion coefficient, cm^2/s

ν = kinematic viscosity, cm^2/s

11.4.3 Metallurgical Effects

As previously discussed corrosion processes take place at the metal–electrolyte interface with specific reactions occurring at the metal surface. It seems quite reasonable, therefore, that the nature of the metal, its composition, and metallurgical structure will determine its corrosion resistance. These are in turn influenced by the manufacturing history and any thermal treatments, that is, heat treatment or welding.

Whether or not a metal will corrode uniformly or at specific sites (locally) will be dependent upon the relative stability of the components of the alloy, the presence of metalloids such as carbides, and local compositions within a single phase. Nonmetallic inclusions such as oxides and sulfides, for example, play an important role in the development of pitting. Corrosion may occur because of the difference in electrochemical potential between the matrix (grain) and grain boundary. Such "intergranular" corrosion is favored by alloys that exhibit an active–passive behavior, for example, stainless steels. A further description of intergranular corrosion is given in Section 11.5.

11.5 FORMS OF CORROSION

Corrosion may result from many different sets of circumstances. As previously mentioned, aqueous corrosion is electrochemical in nature and therefore requires

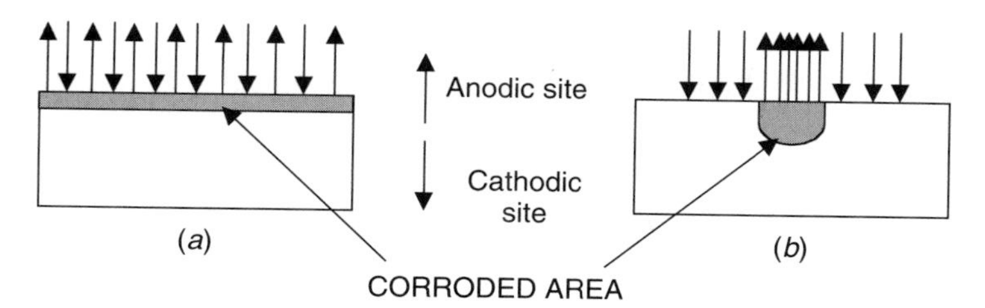

FIGURE 11.6 Schematic representations of (*a*) uniform and (*b*) localized corrosion.

electron transfer between anodic and cathodic sites. In the case where the anodic and cathodic processes are distributed homogeneously over the metal surface (see Fig. 11.6*a*), uniform corrosion is said to occur. Where oxidation and reduction reactions occur at fixed, spatially separated sites, localized corrosion results (Fig. 11.6*b*).

11.5.1 Uniform Corrosion

This type of corrosion is the most common form of attack for nonpassive alloys. In order that *uniform corrosion* occurs, the material should exhibit no major differences in microstructure. In addition to this, the environment should be of uniform composition at the metal–liquid interface, and conditions of solution temperature and flow should remain constant. A further requirement is that concerning the state of stress of the material, which should be uniform throughout the structure and the design of the structure/component should be such that no large stress concentrations exist. As all areas of the metal corrode at a similar rate, it is possible to apply the results obtained from polarization studies to predict the rate of metal loss over a given period of time. This allows the design engineer the opportunity to make use of a *corrosion allowance* for design purposes.

11.5.2 Localized Corrosion

Localized corrosion results when the anodic and cathodic reactions are fixed at sites spatially separated from each other. Localized corrosion may occur on both the microscopic and macroscopic scale and is the most destructive type of corrosion. The rate of corrosion can depend upon minute changes in local conditions, for example, MnS inclusions exposed at the steel surface and in contact with an aqueous environment can, via hydrolysis reactions, give rise to highly localized acidic conditions. In the case of localized corrosion, the rate of metal loss is often very unpredictable. This applies particularly to pitting in which the location, distribution, and size of pits depends upon the precise microstructure and environmental conditions prevailing.

A number of factors control the rate of localized corrosion and include:

1. *Anode/Cathode Area Ratio* The rate of reaction is governed by the relative surface areas of the anode and cathode and will be limited by the smallest

surface area. Local attack (dissolution) will therefore be more pronounced when the cathodic area is larger than the anodic area. This occurs because the anodic process is confined to a relatively small area, and therefore a large anodic current can be sustained through the availability of a large cathode area. Impurities in the form of segregation, for example, Cu and Fe in Zn give rise to local cathodes with low hydrogen overpotentials that subsequently affects the rate of the hydrogen reduction reaction. This causes an increase in corrosion rate of the metal in low pH solutions but has little effect in aerated neutral solutions where the cathodic reaction is that of oxygen reduction.

2. *Differential Aeration Cells* Differential aeration cells can arise when a metal is in contact with a solution in which the concentration of oxygen within the solution differs from one site to another. This may be the result of limited transport of oxygen due to poor solution agitation. The concentration of oxygen determines the corrosion rate and the site at which the cathodic reaction takes place, Note; corrosion rate is proportional to [pO (anode)/pO (cathode)], where pO is the partial pressure of oxygen. As oxygen is easily replaced where the electrolyte is exposed to the atmosphere, this area is favored for the cathodic reduction of O_2 to OH^-. The site of lower oxygen concentration, that is, below the water line, favors the formation of anodic sites.
3. *Changes in Solution pH* The pH value of a solution is dependent upon the concentration of H^+ (in the form of H_3O^+) or correspondingly of OH^- ions. Where dissolved oxygen or H^+ are involved in the corrosion reaction, the rate of the anodic and cathodic reactions will therefore depend upon the pH of the solution. For near neutral solutions the cathodic reaction (reduction of oxygen/water to hydroxide) involves a decrease in acidity, that is, an increase in pH, while the anodic reaction, via hydrolysis, leads to a decrease in pH, that is, an increase in H^+ concentration.
4. *Corrosion Products and Deposits* Corrosion products such as surface films, for example, oxide films/layers, often have the effect of reducing the overall rate of corrosion, for example, stainless steels depend for their corrosion resistance on a thin chromium oxide film. However, if these surface films are disrupted, for example, by cracking under stress, then localized corrosion may result. Where deposits lie on the surface of a metal, cathodic sites are often formed at the outer edge of the deposit. As the degree of aeration decreases away from this location, that is, toward the center of the deposit, corrosion activity (dissolution) is favored beneath the deposit.
5. *Active–Passive Cells* Passivity is a property exhibited by a material when an insoluble film forms on the corroding surface preventing metal–electrolyte contact, hence greatly reducing the corrosion rate. Where a surface has lost its passive film, that is, is depassivated, the local corrosion rate increases markedly. The electrode potential of the passive and active surfaces differ and an electrochemical cell is formed. The magnitude of

corrosion enhancement depends upon the ratio of active and passive surface areas and the nature of the electrolyte. Normally the passive (cathodic) surface is far larger than the active (anodic) surface and rapid localized corrosion results.

11.5.3 Galvanic (Bimetallic) Corrosion

This type of corrosion occurs when dissimilar metals are in electrical contact with an aqueous aggressive electrolyte (see Fig. 11.7). A guide to the tendency (driving force) for *galvanic corrosion* can be assessed from the galvanic series (see Table 11.3). The magnitude of galvanic corrosion is affected by:

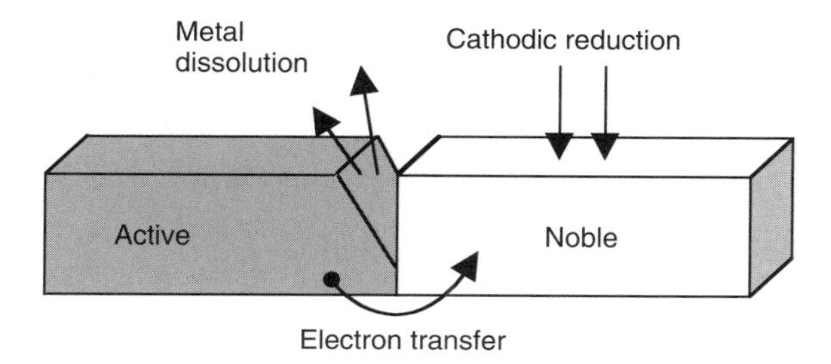

FIGURE 11.7 Schematic showing a galvanic corrosion cell.

TABLE 11.3 Galvanic Series for Metals/Alloys in Aerated Flowing Seawater (25°C, 0.25 m/s)

Metal/Alloy System	Electrode Potential in Volts (vs (SCE)
Pt	0.2
Ti	0.05 to −0.05
316/317 Stainless steel[a]	0.0 to −0.1 (*−0.4*)
304 Stainless steel[a]	−0.05 to −0.10 (*−0.5*)
Cu/Ni (70/30)	−0.15 to −0.25
430 Stainless steel[a]	−0.20 to −0.25 (*−0.55*)
Tin bronze	−0.25 to −0.30
Admiralty brass	−0.28 to −0.35
Cu	−0.3 to −0.35
Sn	−0.32 to −0.35
Brass (70/30)	−0.30 to −0.40
Low-alloy steel	−0.57 to −0.63
Mild steel	−0.60 to −0.72
Al alloys	−0.75 to −1.0
Zn	−0.95 to −1.05
Mg	−1.60 to −1.65

[a]Note values in (*italics*) represent deaerated seawater conditions.

1. The difference between the corrosion potentials of the uncoupled metals
2. The conductivity of the solution
3. Anode/cathode area ratio
4. Distance between the couple.

In general, a large difference in corrosion potentials, a large cathode/anode ratio, and small distance between the metals will give rise to the greatest corrosion rate.

This assessment of susceptibility assumes that the corresponding cathodic reactions are not influenced by polarization (overpotential) effects. For example, if we consider the galvanic couples of Mg–Pt and Mg–Hg, the rate of corrosion of Mg should be similar for each couple because E for Mg $= -1.6$ V (vs. SHE), and both Pt and Hg ≈ 0.2 V. However, the rate of corrosion for the Mg–Hg couple is extremely low because the hydrogen evolution overpotential on the Hg surface is 8 orders of magnitude lower than that at the Pt surface.

11.5.4 Flow-Induced Erosion–Corrosion

The flow of a gas or electrolyte across the surface of a metal generally enhances the rate of corrosion due to an increase in the rate of mass transport of gaseous and ionic species to the surface and solid products away from the surface. The accelerated corrosion will be uniform if the flow does not produce an uneven diffusion layer at the metal–environment interface. However, localized corrosion may take place because of turbulent flow, that is, *flow-induced erosion–corrosion* may result (Fig. 11.8). Erosion–corrosion is characterized by the appearance of grooves, gullies, waves, and rounded holes often having a distinct directional pattern. Cavitation corrosion falls within this group and proceeds through the collapse of a vapor bubble. Material loss may arise through mechanical and/or electrochemical mechanisms.

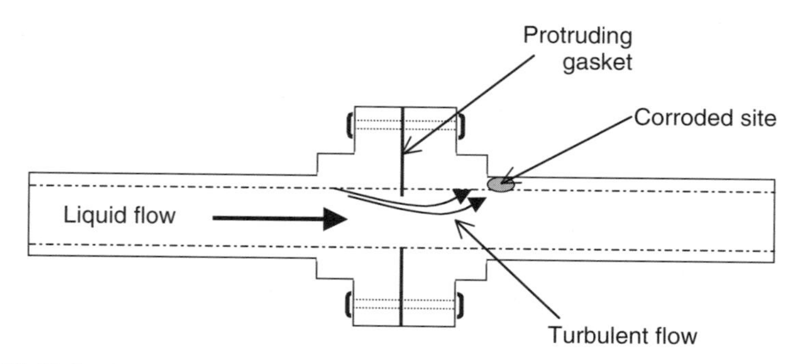

FIGURE 11.8 Schematic showing flow-induced corrosion in pipe wall due to turbulent flow.

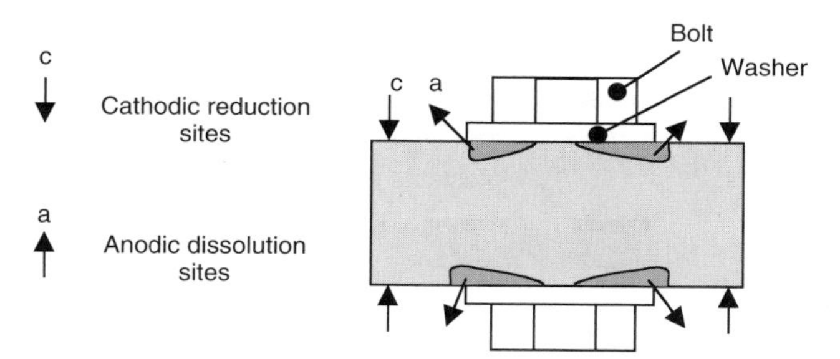

FIGURE 11.9 Crevice corrosion under washer on bolted component.

11.5.5 Crevice Corrosion

Crevice corrosion is a type of intense localized corrosion, which occurs within crevices and other shielded areas on the metal surface. This form of corrosion is often associated with the presence of Cl^- ions and small volumes of stagnant solution caused by holes, lap joints, and crevices under bolt and rivet heads (Fig. 11.9). The cathodic reaction (often oxygen reduction) occurs on the metal surface outside the crevice. Inside the crevice metal dissolution and acidification reactions occur that result in further oxygen depletion creating conditions that allow corrosion to continue. The rate of crevice corrosion is dependent upon the crevice geometry, which in turn influences the local solution chemistry, that is, pH, and O_2 concentration. Oldfield and Sutton [22] proposed the following mechanism for crevice corrosion: (i) deoxygenation of the solution in the crevice, (ii) accumulation of positive metal ions in the crevice, (iii) diffusion of Cl^- into the crevice, and (iv) increase in aggressivity of the crevice solution.

11.5.6 Pitting Corrosion

Pitting corrosion is often considered to be identical to that of crevice corrosion because the nature of the corrosion processes that occur are almost identical. The difference is, however, the nature by which each type of corrosion initiates. In the case of pitting local surface film damage or material heterogeneities lead to the formation of a pit, as opposed to differences in oxygen concentration, as is the case for crevice corrosion. Localized pitting may result from the preferential dissolution at the site of a nonmetallic inclusion or where the surface oxide film is broken, for example, by scratching the surface (see Fig. 11.10) [23].

Pitting is often associated with materials that show a high tendency to passivate, for example, stainless steels. Localized metal loss via pitting corrosion is governed by the following:

1. Spatial separation of anodic and cathodic sites
2. Presence of an oxidant, for example, dissolved oxygen and negative ions (e.g., SO_4^{2-}, Cl^-, but not OH^-)
3. Concentration of salts within the pit, that is, solubility of the corrosion product

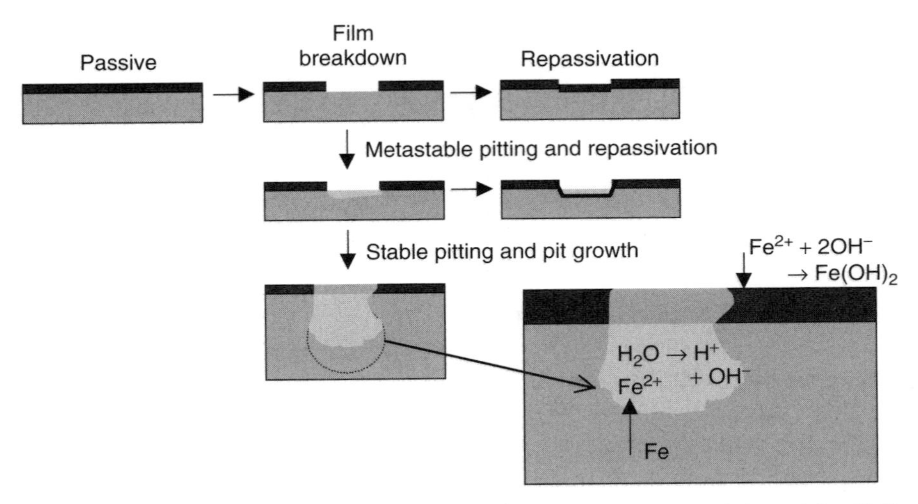

FIGURE 11.10 Pit development and growth following surface film breakdown (after [23]).

Stable pit propagation is the result of the inability of the pit surface to repassivate. This is due to the aggressivity of the pit solution. As can be seen from Fig. 11.10, metal dissolution and cation hydrolysis lead to the formation of a local low pH environment; see Eqs. 11.13 and 11.14:

$$Fe \rightarrow Fe^{2+} + 2e^- \tag{11.13}$$

$$Fe^{2+} + H_2O \rightarrow Fe\,(OH)^+ + H^+ \tag{11.14}$$

The aggressiveness of the environment will be determined by the pit depth and local metal dissolution current [24]. Oxygen depletion occurs inside the pit further reducing the lack of repassivation. The pitting mechanism is said to be *autocatalytic* with anodic dissolution, hydrolysis of metal ions, outward transport of metal ions, and inward transport of anions such as chloride. The processes by which this occurs include diffusion, convection, and ionic migration due to the potential gradient between pit and bulk solution. Where sulfur is present, as, for example, in the form of sulfide and oxysulfate oxysulfide inclusions, the pitting resistance of the metal is reduced [25], hence the steel-making trend toward that of low sulfur steels. In addition it has been reported that treatments that result in a redistribution of the sulfur at the surface of the metal, lead to an improvement in pitting resistance [26].

11.5.7 Selective Corrosion

Corrosion, which occurs at preferred sites on a metal surface, may be described as selective corrosion. As metals are rarely uniform in composition or structure and contain grain boundaries as a result of solute segregation during solidification, spatial separation of anodic and cathodic sites may readily occur. Typical examples of selective corrosion include (i) intergranular corrosion and (ii) selective leaching or demetallification.

Intergranular corrosion is caused primarily by a difference in the chemical activity between the atoms within the grain boundary region and the alloy–metal matrix. The concentration of impurities at the grain boundaries and enrichment/depletion of alloying elements in the grain boundary area therefore influence intergranular corrosion. Heat treatment processes that cause preferential migration of elements within the material can result in sensitization of the microstructure, for example, 304 stainless steel with carbon contents above 0.05%, heated in the temperature range 500–800°C suffer chromium carbide precipitation at the grain boundaries. Sensitization may be prevented by quenching quickly through the critical temperature range, using very low carbon content steels ($<0.05\%$) or adding carbide stabilizers such as Ti or Nb ($<1\%$).

Exfoliation of Al alloys is another example of selective grain boundary corrosion being especially dramatic in alloys where there is a strong rolled texture, that is, in wrought alloys, which do not recrystallize during heat treatment. Elongated pancake-shaped grains are susceptible as corrosion attack occurs along paths parallel to the working direction. The corrosion products formed have a volume greater than the original material, causing swelling and eventual mechanical separation of individual layers. Alloys that are susceptible to exfoliation, include, Al–Cu, Al–Zn–Mg, and Al–Mg. However, exfoliation is not encountered in alloys having an equiaxed grain structure.

Dezincification of brasses is a well-known example of selective corrosion. The area around a dezincified zone is mechanically weak and porous and therefore lowers the mechanical properties of the alloy. Susceptibility of the 70Cu/30Zn alloy may be decreased by the addition of 0.05% As (arsenic). *Graphitization* of cast-iron is another example of selective corrosion, gray cast-iron being most susceptible. In this case graphite flakes are more noble than the other phases in the alloy and act as cathode sites with the matrix acting as the anode. The addition of Ni (13–36%) gives rise to an austenitic matrix and improved resistance to corrosion. White cast-iron contains no graphite and therefore is not susceptible. In the case of spheroidal cast-iron the graphite particles are found as discrete spheroids and not interlinked as in the case graphitic cast-iron; therefore this type of cast-iron does not suffer from graphitization.

11.6 ENVIRONMENT-ASSISTED FAILURE

In many instances corrosion problems are exacerbated by synergistic effects caused by the joint action of mechanical stress and chemical reactions. There are three major processes by which the integrity of a component may be affected: (i) stress corrosion cracking (SCC), (ii) corrosion fatigue (CF), and (iii) hydrogen embrittlement (HE).

TABLE 11.4 Metal/Environment Systems Susceptible to SCC

Metal/Alloy	Environment
Al alloy	NaCl and H_2O solutions, H_2O vapor
Cu alloys	Ammonia solutions and vapors, water, amines
Mg alloys	NaCl, $NaCl/K_2Cr_2O_7$ solutions
Steels	NaOH, NaCl, and NO_3 solutions; mixed acids; H_2S gas/solutions
Stainless steels	Acidic NaCl solutions at elevated temperatures, $MgCl_2$, H_2S

11.6.1 Stress Corrosion Cracking

Stress corrosion cracking results in the brittle failure of normally ductile material. SCC is defined as cracking under the combined action of corrosion and a tensile stress where failure would not occur if either stress or environment were applied in isolation (see Fig. 11.11). There are a number of sources of tensile stress including external applied stresses (structural loads) and internal residual stress (from heat treatment or manufacturing processes such as welding or rolling etc.). The phenomenon of SCC was originally termed *season cracking*. The term came about as a result of the failure of brass ammunition cartridges used in hot, humid (ammoniacal) environments during the late 1800s. Since that time a whole range of metal/environment systems that suffer from SCC have been recognized, a few of which are given in Table 11.4.

Although SCC tends to result from the combination of tensile stress and a specific environment, the susceptibility of a material is very unpredictable because small changes in metal composition, heat treatment, environmental composition, or temperature can drastically alter the SCC behavior of the material. In addition the rate of application of stress/strain also affects the SCC susceptibility of a material.

11.6.1.1 Prevention of SCC

A simple approach to avoiding SCC can be deduced from Fig. 11.11, where SCC results from an appropriate combination of material type, environment, and stress. SCC can therefore be reduced or eliminated by: (i) removing or reducing the working or residual stresses within the component or inducing compressive residual stresses, (ii) controlling the environment, that is, composition, temperature,

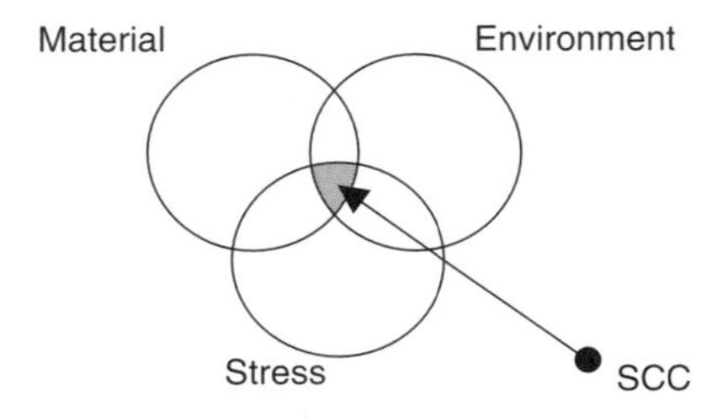

FIGURE 11.11 Components contributing to SCC.

pH, and potential, applying anodic or cathodic protection, and (iii) changing alloy composition or structure.

A reduction in the stress can have a significant effect on SCC if the cracking is caused by residual stresses, for example, a simple annealing treatment can be carried out. Furthermore prevention of mechanical damage, that is, hammering/machining, can elevate susceptibility. Environmental control is not always possible particularly when side reactions occur as a result of slight changes in process variables. However, if SCC occurs outside a set of well-known electrochemical parameters, that is, pH, temperature, concentration, and metal potential, careful monitoring and control of the environment can avoid possible cracking.

Material selection is perhaps the easiest of all the options available for an engineer and although corrosion-resistant materials are more expensive, it is often more economical, based on full life-cycle costs, to use such materials.

11.6.1.2 Mechanism of SCC Growth

SCC is currently thought to occur by one of three possible mechanisms.

Preexisting Active Path Mechanisms This mechanism presupposes the existence of regions within the metallurgical structure that are chemically reactive and susceptible to corrosion, for example, grain boundaries and elemental depleted zones in heat-treated materials. This can result in anodic dissolution at such active sites. The importance of stress in such instances is twofold: first, that it ensures the crack remains open and, second, it can result in the fracture of an existing protective surface film.

Film-Induced Cleavage This mechanism is based on the formation of a brittle film, which results from corrosion reactions, and its subsequent fracture. Once the film is broken, localized corrosion, that is, dissolution may occur allowing the crack to propagate for a short distance before arresting due to repassivation. This mechanism may repeat itself, allowing continued crack growth.

Adsorption-Related Mechanisms These mechanisms rely heavily on the stress component and associate crack propagation with a decrease in the integrity, in the region of the crack tip, due to the adsorption of chemical species. The most common failure type being that of hydrogen embrittlement, this process is discussed in more detail later.

11.6.1.3 Origins of SCC

Favorable development sites for SCC may arise either as a result of manufacturing or service installation practices or may form as a result of poor design. Table 11.5 illustrates a brief classification of some of these.

11.6.1.4 Assessment of SCC Susceptibility

A number of test methods have been devised to provide information regarding a material's susceptibility to SCC. These methods range from simple observation

TABLE 11.5 Initiation Sites for SCC

Source	Manufacture/Installation	SCC Origin in Service
Metallurgical	Nonmetallic inclusions	Pitting
	Grain boundary segregation	Preferential dissolution
	Heat treatment sensitization	
Mechanical	Residual stresses	Pitting
	Rolling	Crevice corrosion
	Machining	Preferential dissolution
	Welding	
	Hammer/stamp marks	
Design	Laps, seams, holes, notches, defective coatings	Crevice corrosion
		Pitting
	Galvanic coupling	Preferential dissolution

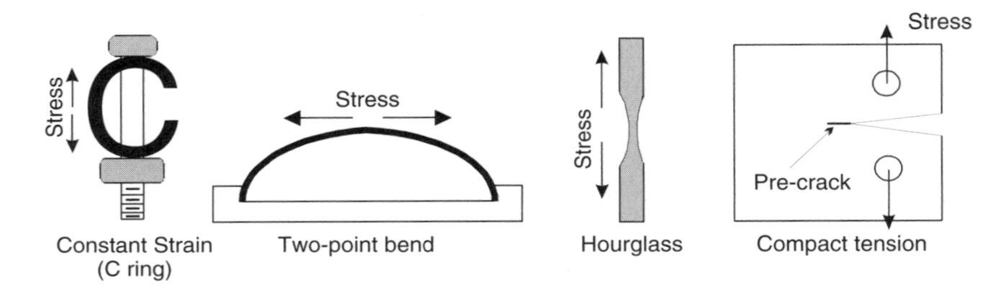

FIGURE 11.12 Specimen configurations used to evaluate SCC.

of static loaded specimens within an environment with and without the presence of a defect (crack) to slow strain rate tests on smooth/notched specimens and electrochemical polarization techniques. Examples of the types of test specimen used for SCC assessment are given in Fig. 11.12.

11.6.1.5 SCC Growth Rates

SCC growth rates are determined by adopting fracture mechanics test methods [27]. In this case compact tension specimens containing a preexisting defect (crack) are used (see Fig. 11.12). Crack extension is monitored as a function of time, and a plot of crack growth rate (da/dt) vs. stress intensity factor (K) is derived. This latter parameter is based upon Eq. 11.15:

$$K = \sigma_{app} Y \sqrt{(\pi a)}, \tag{11.15}$$

where σ_{app} is the applied tensile stress, Y is a factor dependent upon specimen geometry, and a is the crack length.

Figure 11.13 presents a typical SCC growth curve. The important features of this plot are (i) a threshold stress intensity K_{ISCC} value below which crack growth does not occur, (ii) a limited crack growth plateau where the crack velocity is

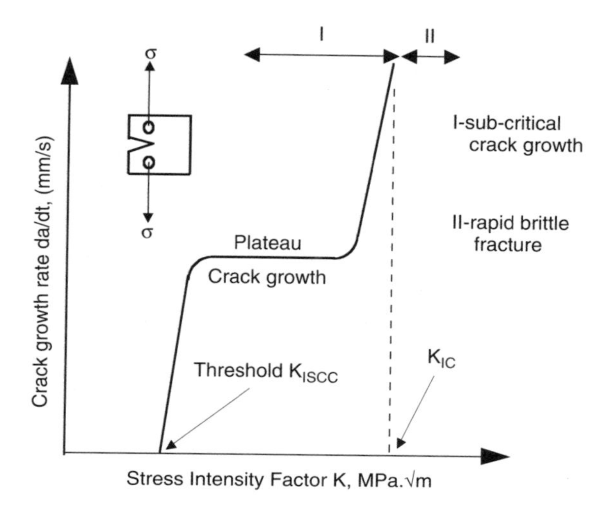

FIGURE 11.13 SCC growth rate curve.

controlled by the rate of chemical reactions at the crack tip, and (iii) rapid crack growth to fracture as stress intensity approaches K_{IC}, the critical stress intensity for rapid brittle fracture in air.

Growth rates vary depending upon the cracking mechanisms involved. For dissolution-type processes, moderate crack growth rates (ca. 10^{-3} mm/s) are limited by the rate at which metal can enter solution and be transported away from the crack tip. For hydrogen embrittlement, crack growth rates are much higher as hydrogen often needs only to diffuse a short distance ahead of the crack tip, and consequently rates of 1 mm/s may be encountered.

It should be noted that evaluation of SCC cracking behavior based on precracked specimens provides no information on the early development of SCC cracks, often known as the *initiation stage*. Caution should therefore be taken when extrapolating SCC linear elastic fracture mechanics (LEFM) data to small defects on smooth engineered surfaces as it is recognized that cracks can develop at stress intensities below the K_{ISCC} threshold level.

11.6.2 Corrosion Fatigue (CF)

Corrosion fatigue differs from SCC in that a fluctuating cyclic load exists while the component is in contact with an aggressive environment. A more alarming difference between SCC and CF is that the combination of cyclic stress and environment shows no specificity and CF failure of a material may occur over a wide range of stress levels and in a variety of different environments. Other features associated with CF include the transgranular nature of fracture, improved resistance to CF with increase in corrosion resistance of the material, and the lack of relationship between corrosion fatigue strength and tensile strength.

11.6.2.1 Prevention of CF

A number of methods are available for reducing a material's susceptibility to CF. However, the methods tend to reduce crack growth rates, that is, increase fatigue lifetimes, rather than eliminate CF entirely. These methods include:

- Increasing the corrosion resistance of the material, for example, by addition of Cr to steels
- Reducing the aggressivity of the environment (dilution) or addition of inhibitors
- Deaerating solutions (which avoid crevice-type conditions for passive alloys) and using lower solution temperatures
- Lowering the applied stress or induce surface compressive residual stresses (shot peening)
- Prevent metal/environment contact by applying coatings, for example, plating or painting
- Apply cathodic protection by reducing the potential or attach sacrificial anodes, for example, Zn or Al alloys

11.6.2.2 Mechanisms of CF Crack Growth

Although a number of mechanisms have been forwarded, it is still uncertain as to the exact mechanism by which corrosion fatigue cracks propagate. However, two main categories exist: anodic dissolution and absorption-related mechanism.

Anodic Dissolution This mechanism presupposes that crack advance occurs as a result of dissolution at the crack tip. As with SCC, local cells are set up either as a result of galvanic action, that is, cathodic matrix and anodic grain boundaries, or due to the formation of bare (crack tip) metal surfaces adjacent to protective films on the walls of the crack. Corrosion rates measured using static techniques suggest that crack advance cannot be wholly accounted for by anodic dissolution. However, the fact that environments within crack enclaves differ from bulk solution composition, along with changes in crack tip potential, and that transient increases in corrosion currents occur when bare metal surfaces are formed, suggests that a calculation of the environmental contribution via the Faradaic equation is possible [28] and,

$$\text{(Crack rate propagation)} \quad V = \frac{M i_a}{z F \rho} \tag{11.16}$$

where M is the molecular weight, i_a is the anodic corrosion current, z is the number of electrons involved in the reaction, F is Faraday's constant, and ρ is the density of the metal–alloy.

Adsorption-Related Mechanism This mechanism suggests that adsorption of hydrogen directly at the crack tip lowers the surface energy of the material and, subsequently, lowers the fracture stress of the material:

$$\text{(Fracture stress) } \sigma_{FR} = \frac{(E\gamma)^{1/2}}{b}, \tag{11.17}$$

where γ is the surface energy of material, b is the interatomic spacing, and E is Young's modulus.

Alternatively, hydrogen may diffuse ahead of the crack tip to locations of maximum stress triaxiality. An increase in crack growth can be assumed to be proportional to the amount of hydrogen generated (C_H) per load cycle:

$$\frac{da}{dN_{Cf}} = f(C_H)\, g(\Delta K, R) \tag{11.18}$$

In this case, $f(C_H)$ includes the influence of environmental composition, temperature, potential, and frequency, while $g(\Delta K, R)$ relates to the mechanical driving force, where R is the stress ratio ($\sigma_{min}/\sigma_{max}$).

11.6.2.3 Assessment of CF Susceptibility

Like the evaluation of SCC, both smooth and precracked specimens may be used to assess the CF resistance of a material. When using smooth specimens fatigue lifetime, that is, cycles to failure, (N_f) is plotted as a function of the applied stress range ($\Delta\sigma$). The resulting S–N_f curve obtained is typical of that shown in Fig. 11.14 [29].

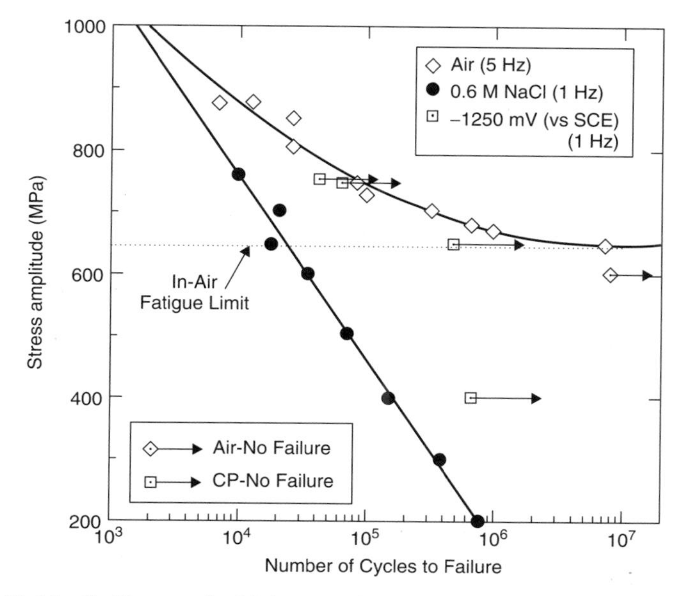

FIGURE 11.14 S–N curve for high-strength steel in 3.5% NaCl solution (after [29]).

One of the most notable features when comparing the in-air and corrosion fatigue resistance of a material is the loss of the *fatigue or endurance limit* for tests conducted in an aggressive environment. Figure 11.14 shows that fatigue failure continues to occur despite a reduction in the applied stress range to 70% of the in-air fatigue limit stress range. In addition, it can be seen that some restoration of fatigue strength occurs on applying a cathodic potential (CP), thereby eliminating the damaging anodic dissolution reaction that leads to pitting and early crack initiation.

11.6.2.4 CF Crack Growth (CFCG)

The vast majority of CF studies have involved the use of precracked specimens with corresponding results plotted in the form of crack growth rate (da/dN in, e.g., millimeters/cycle), versus stress intensity factor range, ΔK (MPa $\sqrt{m}$). This type of plot is also known as the *Paris curve* [30] (see Fig. 11.15).

Figure 11.15 shows that an enhancement in crack growth occurs for fatigue cycling within the environment, particularly in the mid-ΔK range. At high ΔK values, CFCG rates approach the air fatigue crack growth rate as mechanical effects dominate. At low ΔK values, the threshold value for crack growth in an environment may be lower or higher than that for air depending upon effects such as closure (i.e., corrosion product blocking the crack enclave). Alternatively anodic dissolution at the crack tip may create a notch-like feature, thereby increasing the stress concentration and making slip easier or conversely blunting

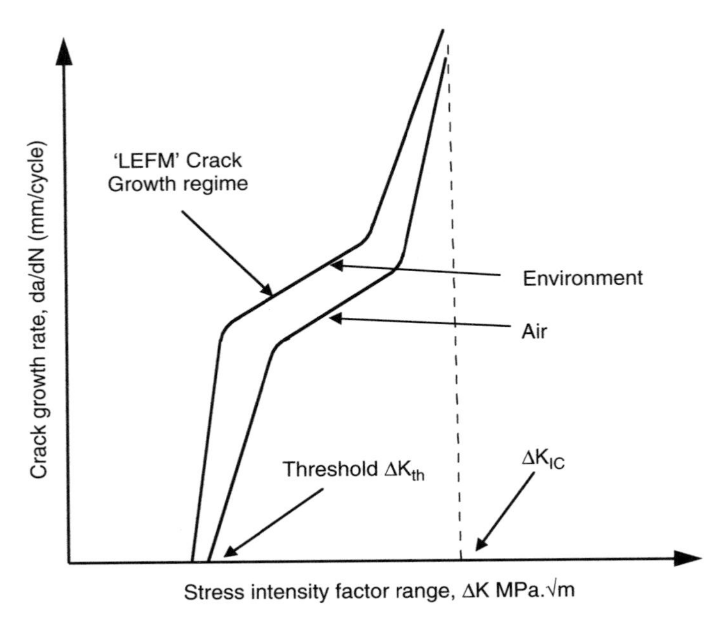

FIGURE 11.15 Schematic showing typical fatigue and corrosion fatigue crack growth curves.

of the crack tip may occur decreasing the effective crack tip stress intensity and reducing CFCG rate.

Effect of Variables on CFCG Rates of CFCG are affected by changes in both mechanical and electrochemical variables, for example.

FREQUENCY It is clear that if the CFCG rate is dependent upon both stress and chemical reactions occurring inside the crack, then crack growth rates will be influenced by the time allowed for chemical reactions to occur. Changes in frequency have two effects. First, there will be a change in crack tip strain rate with changes in frequency, and, second, the time interval during which the crack is fully open will also change. In general, a decrease in frequency causes an increase in crack growth rate.

ENVIRONMENTAL COMPOSITION In general, the corrosion fatigue crack growth rate increases as the aggressive nature of the environment increases. However, a general prediction of the effects of environment is difficult and trends may be restricted to specific environmental changes, for example, pH, oxygen concentration, relative humidity, and the like.

TEMPERATURE As corrosion rates increase with increasing temperature, that is, an approximate doubling of rate for every 10°C rise in temperature, it is not unexpected that crack growth rates will also increase as the environment temperature rises, particularly where mass transport/diffusion processes control crack growth rate. However, it should also be recognized that as temperature increases the solubility of oxygen decreases and where the cathodic reaction is that of oxygen reduction, a decrease in crack growth rate may occur.

LOADING WAVEFORM The type of loading experienced by the component can be one of a number of forms, that is, sinusoidal, square wave, triangular, and positive or negative sawtooth form. Where initial rise times are sufficiently slow, as in sinusoidal, triangular, and positive sawtooth waveforms, reactions at the crack tip can take place during the loading half-cycle leading to an effect on crack growth rate [31]. Conversely, negative sawtooth and square waveforms, which have fast rise times, result in a negligible effect on crack growth rate.

SCC CONTRIBUTIONS During corrosion fatigue cycling, it is possible that other static modes of fracture, that is, SCC, contribute to crack growth. For this to happen it has been argued that the material must exhibit an SCC tendency, and any perturbations to normal corrosion fatigue crack growth will only occur when the stress intensity at the crack tip, during the cycle, rises above the threshold stress intensity for the onset of SCC, that is, K_{ISCC}.

As a result of these factors, it is possible that the Paris curve may take on one of three forms [32]: (i) true corrosion fatigue (TCF), (ii) stress corrosion fatigue (SCF). In this case, perturbations occur when the ΔK value of the cycle is equal to or greater than K_{ISCC}, and (iii) true corrosion fatigue on stress corrosion fatigue. This type of behavior will occur when the material exhibits both CF and SCC.

11.6.3 Hydrogen Embrittlement

Hydrogen embrittlement may be categorized as one form of stress corrosion cracking because of the mechanism by which cracking occurs. However, a brief separate description is given in order to highlight some of its features. HE is widely associated with premature failure of high yield strength steels in "sour" oil wells and acid brine environments. Spring steels also suffer from failure by HE, particularly when the manufacturing process route includes pickling or electroplating treatments.

11.6.3.1 Mechanisms of HE

There are three postulated mechanisms responsible for hydrogen embrittlement:

1. *Internal pressure*. Here molecular hydrogen is said to accumulate at subcritical crack sites leading to a pressure enhancement of existing residual or applied tensile stresses. The subsequent effect is that the fracture stress is exceeded, that is, $\sigma + p \geq \sigma_0$, where σ is the applied/residual stress, p is the internal pressure, and σ_0 is unembrittled fracture stress [33]. This mechanism is generally invoked to explain the phenomenon of delayed fracture [34].
2. *Surface adsorption*. In this case adsorbed H atoms lower the surface energy of the cracks [35].
3. *Decohesion*. Absorbed hydrogen atoms at interstitial sites weaken the lattice cohesion causing a lowering of the cleavage fracture stress [36].

The basic process resulting in hydrogen adsorption involves hydrogen entering the metal interstitially via natural corrosion reactions or through cathodic polarization. Two possible pathways are available for the atomic H formed (Fig. 11.16): (i) interstitial diffusion into the metal lattice and (ii) the combination of adsorbed species to form molecular H, which leaves the metal surface in gaseous form.

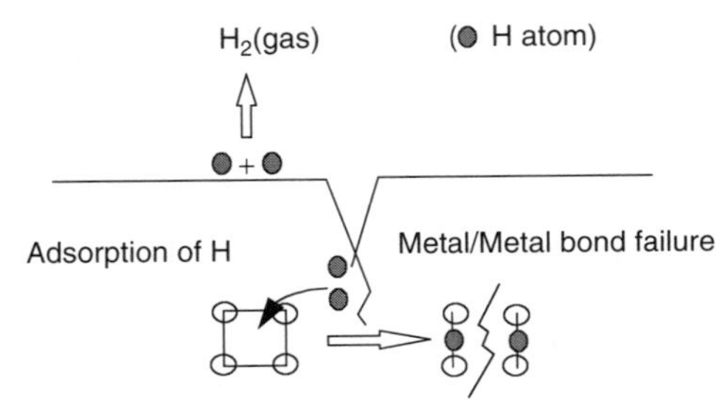

FIGURE 11.16 Possible pathways available for hydrogen atom recombination or interstitial adsorption.

Which of these reactions will dominate and the rate at which they proceed will depend upon factors such as alloy composition, temperature, metal potential, catalytic activity of the metal surface, and solution chemistry.

11.6.3.2 Prevention of HE

Measures used to avoid HE involve either removal of any diffused hydrogen or prevention of hydrogen adsorption on the metal surface. Metallic, inorganic, and organic coatings may be used to prevent hydrogen adsorption on the metal surface. Inhibitors and additives that reduce the corrosion rate or increase the rate of hydrogen gas evolution can be helpful. Furthermore as hydrogen diffusion increases with temperature, low-temperature heat treatments may be used to remove any diffused hydrogen, Modification of material chemistry is also possible, for example, in steels small amounts of Ti and V may be added that act as H traps reducing the diffusion rate of H in the steel.

11.7 CORROSION TESTING

11.7.1 Introduction

The following section will deal briefly with some of the common forms of corrosion testing carried out both in the research and development laboratory and in plant/field studies. Comments on assessing the susceptibility/performance of materials subject to the conjoint action of stress and corrosion have been dealt with under environment-assisted failure (Section 11.5.7).

11.7.2 Standard Laboratory Procedures

11.7.2.1 Pretreatment of Test Samples

Corrosion processes occur at the environment–metal interface and therefore quite naturally are dependent upon the state of the surface. Any sample preparation or pretreatment should provide a surface that is representative of that to be encountered in service. Often sample preparation involves wet polishing with emery papers followed by fine polishing to a micron surface finish using diamond paste media. The effect of these and other treatments, that is, cutting, grinding and so on, is one of including residual stresses into the sample surface. Care should therefore be taken during sample preparation not to change the natural characteristics of the surface.

11.7.3 Electrochemical Corrosion Tests

Electrochemical techniques are used for a variety of purposes, not least to provide the engineer with information of the corrosion rate of a metal/alloy in a given environment. Only a brief summary is possible here, and the reader is advised to refer to references [18 and 37], which provide further descriptions of the different types of corrosion tests currently available.

Assessment of corrosion rate may be conducted using standard polarization methods as described in Section 11.3.4. Alternatively, a technique based upon applying a small sinusoidal alternating current or potential perturbation to the corroding system may be employed. This is commonly known as electrochemical impedance spectroscopy (EIS) and has been applied to systems where a metal is coated or where the metal is passive, for example, stainless steel or aluminum [38, 39]. Impedance measurements have gained in popularity and are used to determine the polarization resistance value, hence corrosion rate, particularly in poorly conducting media and on coated samples, such as paints and lacquers. As with the polarization resistance methods alternating current (ac) impedance is used to evaluate the "general" uniform corrosion rate and provides no information on localized corrosion rates.

Galvanic corrosion may also be determined by superimposing individual anodic and cathodic polarization curves for the respective anodic and cathodic corrosion reactions that occur on each of the two metals making up the galvanic couple [40, 41]. An example of this is shown in Fig. 11.17.

Figure 11.18 presents an alternative experimental set up for determining the magnitude of galvanic corrosion (current) using zero resistance ammetry.

In addition polarization curves may also be used to determine the pitting resistance and potential regimes within which SCC may occur. Figure 11.19 presents the polarization curves for three different stainless steels: AISI 304L, 316L, and 2025 duplex stainless steels, in an artificial seawater environment [42]. Highlighted in Fig. 11.19 are the pitting potentials (→) and regions where SCC may be encountered (}). As can be seen from this figure the pitting resistance is in the order duplex > 316L > 304L.

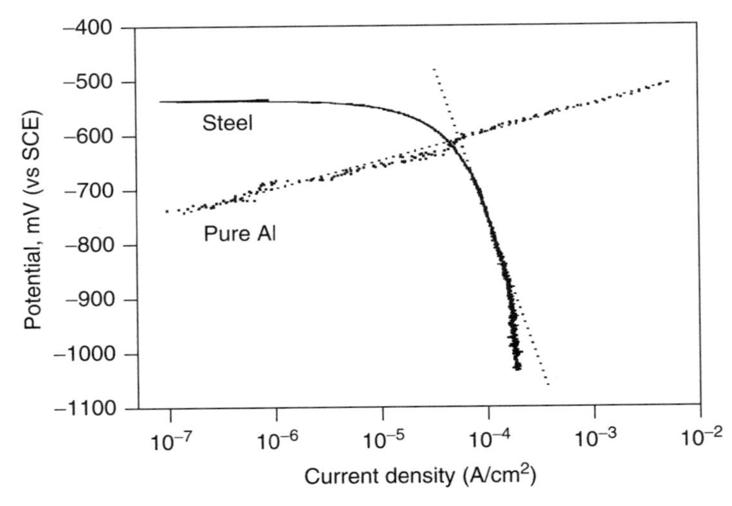

FIGURE 11.17 Composite polarization plot: pure aluminum (anodic branch) and steel (cathodic branch) (after [41]).

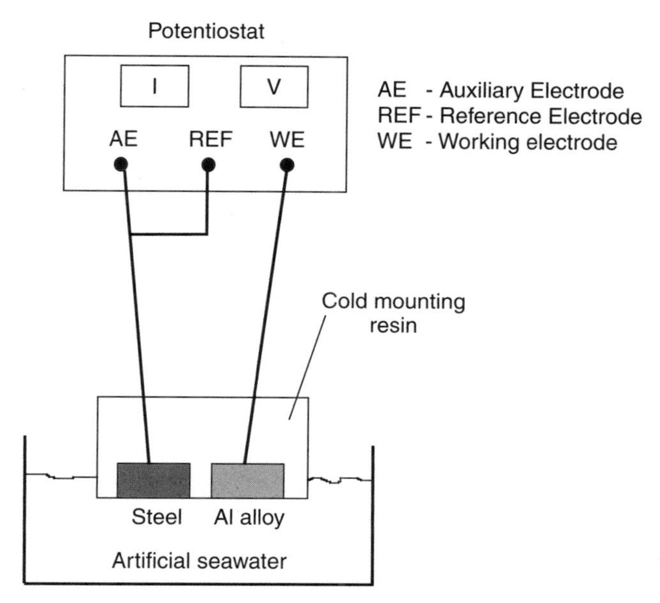

FIGURE 11.18 Application of potentiostat as a zero resistance ammeter for measurement of galvanic current (corrosion rate) from two different metals.

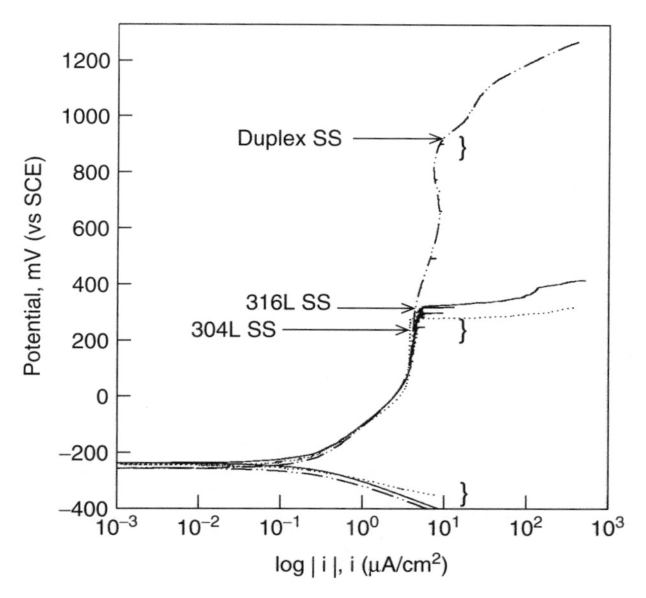

FIGURE 11.19 Polarization curves highlighting different pitting potentials for duplex, 316L, and 304L stainless steel grades.

As previously pointed out, the above techniques are primarily used to assess the general corrosion behavior of a material. Recent developments in scanning methods and rapid data acquisition have given rise to *electrochemical scanning probe* techniques. An example is the scanning reference electrode technique (SRET) [43–45]. These techniques are based upon measurement of corrosion activity via a twin platinum wire probe, which moves in discrete steps, from 0.5 μm, across the surface of a corroding sample. The result is a two-dimensional map of the localized corrosion activity on the surface. When scans are repeated over selected intervals in time, the result is a time-lapsed progression of corrosion activity. Figure 11.20*a* shows a typical SRET map showing galvanic corrosion. In this case the galvanic couple is that of an explosively bonded (kelocouple) joint containing steel/pure A1 and A1 alloy. It can be seen from Figure 11.20b that by changing the conductivity of the solution, corrosion activity (dark areas) moves from the A1 alloy to the pure A1 [46]. Such spatial information concerning corrosion activity cannot be derived from those conventional techniques previously described.

11.7.4 Service and Field Practice

Like the above laboratory-based tests, in-service and field tests are designed to evaluate a range of material degradation processes including erosion–corrosion, stress corrosion, and conventional corrosion processes. In addition information is required concerning material performance within a wide range of operating environments, including atmospheric, sea and fresh waters, high-temperature, in vivo, soils and concrete. In this respect only a brief survey of the relevant test methods can be made here.

By far the most extensive type of test, arising from its simplicity is that of outdoor atmospheric coupon testing. These tests consist of exposing appropriate sized coupons (bare metal or coated) within the desired atmospheric conditions. Normally these are graded as urban, rural, and industrial; other classifications of

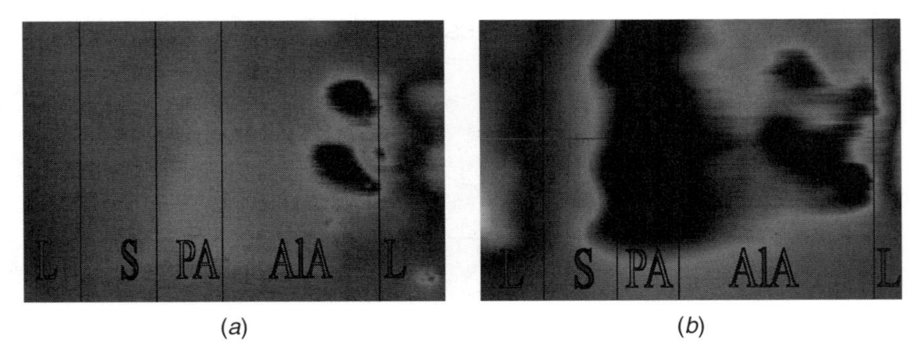

FIGURE 11.20 SRET maps illustrating shift in localized corrosion activity on a kelocouple joint due to changes in solution concentration: (*a*) = 4% solution and (*b*) = 0.4% solution. Note L = lacquer, S = steel, PA = pure aluminum, and A1A = aluminum alloy.

atmospheric conditions have also been derived [47]. Coupons are then examined periodically during which visual inspection is carried out and weight changes are monitored. This latter measurement provides an average corrosion rate for the selected metal/alloy in the given environment. Depending upon alloy type and the nature of the environment, corrosion rates can range from 0.5 μm/year to 300 + μm/year for highly concentrated chloride environments. Corrosion weight loss per unit area is based upon Faraday's law:

$$m = \frac{iMTS}{nF}, \tag{11.19}$$

where m is metal loss (g), i is corrosion current (A), M is molar mass of the metal (g/mol), T is time (s), S is area of metal involved (cm^2), n is number of electrons released in the dissolution reaction, and F is Faraday's constant, 96487 C (As). It should be noted that this equation only applies to uniformly corroding surfaces. Corrosion coupons can also be used for assessing corrosion resistance in fresh and seawater environments.

Online monitoring techniques are becoming increasingly popular, particularly where chemical treatments such as inhibitor and biocide dosing is used to minimize corrosion activity. Potential monitoring is a common online technique whereby measurements of free corrosion potential, as a function of time, are made. Typical examples include measurements of potential on cathodically protected structures such as offshore oil platforms and reinforced steel within concrete structures.

Electrical resistance probes are used in a similar manner to coupons but without the need to remove the sample from the environment. This has two advantages: (i) all corrosion processes remain continuous and (ii) the method can be automated to provide information on "instantaneous" corrosion rate. However, information relating to whether corrosion is uniform or localized is not provided using this method. Similarly electrochemical probes are used to provide online information of corrosion rate, normally via measurements of polarization resistance. The use of probes, like that of laboratory tests, relies on the aqueous media being relatively conductive, otherwise errors are incurred due to an internal resistance (IR) drop across the electrodes. Electrochemical noise (EN) in the form of potential and current measurements is finding increasing application, although its extensive use is restricted due to the complex statistical analysis required to identify the nature of the corrosion. EN is used primarily for evaluating localized corrosion (pitting) and SCC [48, 49].

11.8 METHODS OF CORROSION PREVENTION

It was shown earlier that a corrosion cell consists of four components: an anode, a cathode, a conducting electrolyte, and a metallic conducting circuit. This knowledge provides us with an understanding of how to minimize or prevent corrosion. Corrosion prevention strategies are therefore based on one of the following approaches:

1. Reduce the aggressiveness of the environment/adding inhibitors
2. Apply cathodic or anodic protection
3. Apply a protective coating to the substrate
4. Appropriate material selection
5. Application of good design principles

Approaches 4 and 5 will be dealt with in Sections 11.9 and 11.10.

11.8.1 Reducing the Aggressiveness of the Environment

Changes to the environment may be brought about to alter the factors that affect corrosion rate, namely temperature, pH, oxygen concentration, humidity, concentration of active species, presence of dissolved solids, and pollutants, for the case where the environment is predominantly gaseous, that is atmospheric corrosion. The aggressiveness of the environment may be reduced by lowering the temperature, lowering the relative humidity, and eliminating aggressive species such as chloride, sulfur, and nitrogen oxides and carbon dioxide. Also conditions should not be such that temperature differentials exist between surfaces such that dew point or condensation corrosion can be initiated.

With aqueous environments corrosion rates may be reduced by lowering the temperature, conductivity, and oxygen concentration. In addition, but dependent upon the nature of the metal/alloy, a change in the pH of the solution can reduce corrosion rate through the formation of passive films.

11.8.2 Inhibitors

Environmental change is often brought about by the use of inhibitors, which may, depending upon their chemical composition, influence either or both the rate of the anodic and cathodic reactions. Alternatively, an inhibitor may form a relatively thick film on the surface and thereby increase the electrolyte resistance of the circuit. Figure 11.21 presents schematic polarization curves illustrating how anodic, cathodic, and mixed inhibitors influence the rate of corrosion [50].

Classification of inhibitor systems is often based upon the following [50, 51]:

- Anodic or cathodic inhibitors—depending upon which electrode reaction is affected.
- Oxidizing or nonoxidizing inhibitors—depending upon the ability to passivate the metal. In the case of the latter, dissolved oxygen in the aqueous phase is required to form the passive film.
- Organic or inorganic inhibitors—depending upon the chemical nature of the inhibitor.

Examples of the application of different types of inhibitor are given in Table 11.6.

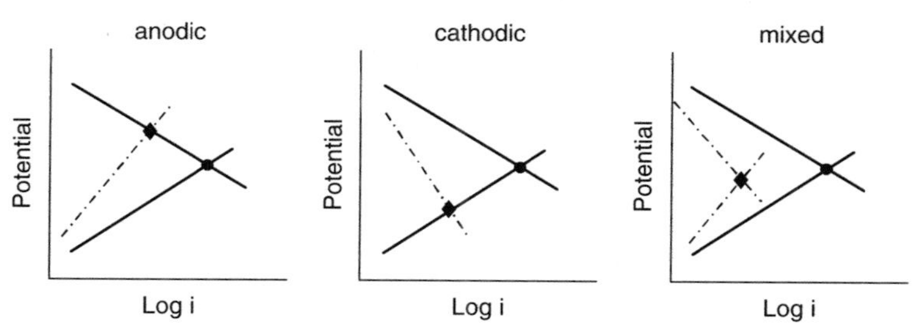

FIGURE 11.21 Influence of inhibitor type on corrosion rate.

TABLE 11.6 Examples of Inhibitors and Their Effectiveness for Use in Different Water Systems

Application	Type	Example	Metals	Concentration	Effectiveness[a]
Potable waters	Anodic	Silicate	Steel	4–10 ppm	R
			Zinc		R
			Cast-iron		R
			Copper alloys		R
			Aluminium alloys		R
Cooling systems (once through)	Cathodic	Polyphosphate	Steel	2–10 ppm	E
			Cast Fe		E
			Copper alloys		E
			Al alloys		V
Recirculation	Anodic	Chromate	Steel	300–500 ppm	E
			Cast Fe		E
			Copper alloys		E
			Al alloys		E
	Anodic	Nitrate	Steel	500 ppm	E
			Cast Fe		PE
			Cu alloys		PE
			Al alloys		PE
			Zn alloys		IE
Central heating system	Anodic	Nitrite/ benzoate	Steel	Nitrate/benzoate	E
			Cast Fe	1:7 ratio	E
			Lead joints		E
			Cu alloys		E

[a]R = reasonably effective, E = effective, V = variable effectiveness, PE = partially effective.

The concentration of the inhibitor will be dependent upon the metal being protected and the nature of the environment. A certain minimum concentration is required for an inhibitor to be fully effective. In fact inhibitors may be classed as *safe* or *dangerous*, for example, an insufficient concentration of a safe inhibitor can lead, at worst, to a rate equal to that of an uninhibited system. Conversely an insufficient concentration of a dangerous inhibitor can lead to accelerated localized attack that is, pitting. Anodic inhibitors are often classed as dangerous inhibitors because they rely on complete passivation of the metal surface for effective corrosion control.

Inhibitors may also be used in the absence of a water phase. These types of inhibitor are known as vapor-phase inhibitors (VPI) and are used mainly in closed systems and packaging. As their name suggests, they rely upon volatilization of a compound into the atmosphere around the product to be protected, the inhibitor then being absorbed onto the surface of the metal. Typical examples include cyclohexylamine nitrate, sodium benzoate, and dicyclohexylamine carbonates.

11.8.3 Anodic and Cathodic Protection

Examination of the Pourbaix diagram given in Fig. 11.2 provides information on the corrosion *state* for a given pH, as a function of the electrode potential. For example, Table 11.7 shows the various regions that exist at pH 7: This information provides the engineer with two options in order to reduce the rate of corrosion, namely (i) decrease the potential until the metal becomes "immune"—cathodic protection—and (ii) increase the potential until the metal becomes passive—anodic protection.

11.8.3.1 Cathodic Protection

As previously discussed corrosion degradation is the result of the oxidation of metal atoms to produce metal ions that enter the solution (Eq. 11.5). CP is therefore an attempt to convert the whole metal surface (structure to be protected) into a cathodic area. In this case the anodic reaction is moved to a more durable material, notably an inert anode (see Fig. 11.22). CP therefore involves a shift in the structure's potential from the active to the immune region of the Pourbaix

TABLE 11.7 Various Regions That Exist at pH 7 for Iron/Water System

Region	Electrode Potential (vs SHE)	Corrosion State
a	≤ -0.65	Immune
b	-0.65 to -0.1	Active corrosion
c	≥ -0.1	Passive[a]

[a] Above a given potential the passive film breaks down and pitting develops. The value of this potential is material and environment dependent.

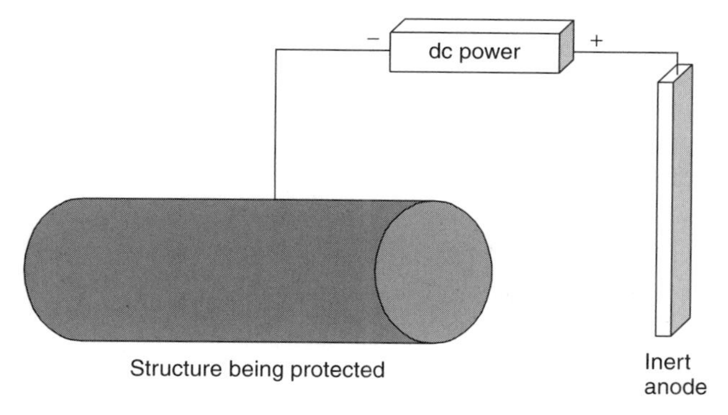

FIGURE 11.22 Schematic arrangement illustrating cathodic protection system.

diagram where unreacted metal is the most stable product. Overall the following processes take place:

- The potential of the metal is polarized below the free corrosion potential.
- Aggressive ions such as chloride (Cl^-) are repelled from the negatively charged metal surface (like charges repel).
- Oxygen is consumed at the metal surface.
- Hydroxyl ions are formed at the metal surface promoting the formation of a passive film (note for Al and Zn, alkali corrosion may occur at high pH values).

An alternative to the use of a direct current (dc) voltage to lower the potential of the metal to be protected is that of the application of a sacrificial anode. This method effectively creates a galvanic couple with the metal to be protected acting as the noble metal (cathode) while the sacrificial anode dissolves preferentially. Table 11.8 identifies various impressed current and sacrificial anodes used in CP systems. A full description of the principles and practice of cathodic protection can be found in references [52–54].

11.8.3.2 Anodic Protection

Anodic protection differs from that of CP in that its purpose is to raise the potential of a metal into the passivation domain of the Pourbaix corrosion diagram. In this case, the metal to be protected becomes the *anode* of the system, which is coupled to an inert counter electrode (cathode), for example, platinum. Anodic protection systems require the formation of a stable passive film and are therefore associated with the protection of metals that readily passivate, for example, stainless steels and titanium and its alloys. The majority of applications of anodic protection involve the manufacture, storage, and transport of sulfuric acid. Other applications include the protection of mild steel in paper mill acids, ammonia, ammonium nitrate, and other fertilizers. Typically for steel in contact with 93%

TABLE 11.8 Impressed Current and Sacrificial Anode Materials

Consumable Types		
(a) Cast-iron	Consumption	$\leq$10 kg/Ay
(b) Si cast-iron	Consumption	$\leq$1 kg/Ay (buried)
Nonconsumable		
(a) Lead alloys	Consumption	$\leq$0.1 kg/Ay
(b) Pt activated	Consumption	$\leq 1 \times 10^{-5}$ kg/Ay
Sacrificial Anodes		
(a) Zn alloy	Capacity	780 $Ahkg^{-1}$
(b) Al alloy	Capacity	2640 $Ahkg^{-1}$
(c) Mg alloy	Capacity	1230 $Ahkg^{-1}$

sulfuric acid, the corrosion rate of a protected system is around four times less than that of an unprotected system. Further information on the subject of anodic protection may be found in reference [55].

11.8.4 Surface Coatings

Surface coatings are an established anticorrosion strategy providing the opportunity to use "poor" corrosion resistance metals/alloys and offering the opportunity to aesthetically enhance a component or structure. Coating systems are extremely diverse, being applied to the substrate via a variety of techniques. In general, coatings may be classified as (i) metallic coatings, (ii) conversion coatings, and (iii) nonmetallic coatings.

The choice and application of a particular coating system will be dictated by the following considerations:

- Economics (cost per unit area or number of articles treated)
- Applicability to a given substrate
- Applicability of a given coating material
- Coating vs. durability consideration

11.8.4.1 Protection Mechanisms

Coating systems may be classified according to the process by which they afford protection, in this respect three broad categories may be applied; see Fig. 11.23.

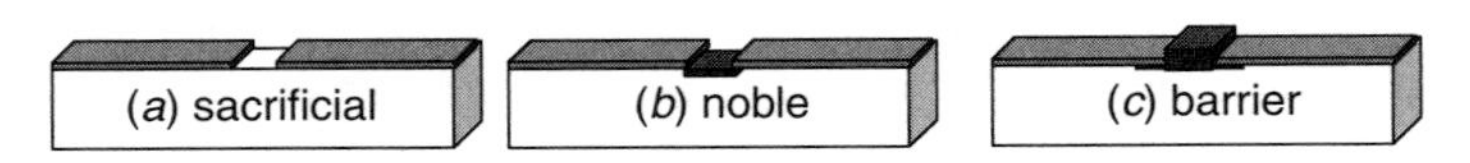

FIGURE 11.23 Categories of coating systems.

Sacrificial Coatings These coatings, as their name suggests are "sacrificed" to afford protection to the substrate, that is, the coating corrodes preferentially to the metal substrate (e.g., Zn, Cd).

Noble Coatings These coatings are more corrosion resistant than the substrate being protected. In this respect they are required to be free from cracks and pores, otherwise preferential corrosion of the substrate will occur at defects in the coatings (e.g., Ni, Cr).

Barrier Coatings These coatings are principally nonmetallic and nonconducting. Their primary mechanism of protection is one of separating the substrate from the corrosive environment (e.g., alkyd, epoxy).

The success of a coating system is largely dependent upon the surface preparation received by the substrate. Typically the following stages can be identified for a paint coating:

1. Chemical/physical cleaning with acids/alkalis, wire brushing, or blast cleaning
2. Chemical or physical pretreatment using chemical etchants, conversion coatings, or particle abrasion
3. Primary coat to form an adherent substrate/coating interface
4. Additional coats to build up the bulk properties of the coating
5. Top coat to provide additional resistance and aesthetic qualities

Metallic coatings applied via spraying or hot dipping may be restricted to stages 1 and 2. A brief description of some of the more common types of anticorrosion coatings will be given, although more extensive information may be found in references [53, 56, 57].

11.8.5 Metallic Coatings

Metallic coatings are commonly adopted being based upon the following application techniques:

- Metal spraying, for example, zinc and aluminum and their alloys. These metals are frequently applied to steel structures.
- Cladding, for example, aluminum and stainless steel applied to aluminum alloys and steels, respectively. This method produces a "sandwich"-type structure with the less corrosion-resistant substrate forming the inner layer of the sandwich.
- Hot dipping, for example, tin, zinc, and aluminum and their alloys. This method produces a coating with very good adhesion properties that can be made available in a variety of coating thickness.
- Electroplating, for example, copper, nickel, chromium, zinc, gold, and so forth. This method allows complex-shaped components to be coated with

a relatively thin layer of the metal of interest. This coating method lends itself to batch production processing.

- Vacuum/vapor deposition, for example, titanium and complex titanium alloys (TiAlN). This method is relatively expensive and requires high capital expenditure equipment; however, extremely hard coatings can be applied to complex objects.

11.8.6 Conversion Coatings

These coatings operate by chemical conversion of the metal surface. There are three main types:

1. Phosphate coatings—applied primarily to steel surfaces as a base for paint topcoats.
2. Chromate coatings—applied to aluminum, magnesium, and zinc-based alloys, providing an excellent base for subsequent paint overcoats.
3. Anodizing, this is, an electrolytic process whereby the oxide layer of a metal/alloy, notably aluminum, is increased in thickness providing some additional degree of protection but also providing an excellent base for paint overcoats.

11.8.7 Nonmetallic Barrier Coatings

Barrier coatings act, as their name suggests, by providing a barrier between the metal substrate and the environment. Barrier coatings may be found in the form of organic paints, enamels, varnishes and lacquers. Paints are by far the most common barrier coating in existence today, consisting essentially of three components: a solvent (organic or water based) providing fluidity, a pigment suspended in the solvent, and additives that promote drying and can provide additional properties. It has been estimated that over 50% of all metal surfaces for which an impervious, pore-free, adherent and attractive property is required are treated with paints of one sort or another [56].

11.8.7.1 Properties and Types of Paint Systems

The properties of a paint coating will primarily be dependent upon the bulk properties of the system, and in this respect paint coatings generally consist of a primary coat(s), undercoat(s), and a top coat. From a corrosion protection viewpoint, these systems have two essential roles: (a) to provide an adherent bond between primer and substrate, and (b) to restrict the access of air, moisture, and aggressive ions to the substrate. It should be recognized at this stage that all paint coatings are permeable to a greater or lesser extent. It is the pigment that limits the rate of diffusion through the paint film. The performance of a paint is, however, often limited because of poor substrate surface preparation and it is crucial that the structure being painted is free from dust, grease, or corrosion products. This can be achieved chemically or mechanically as discussed

TABLE 11.9 Typical Paint Types and Applications

Type of Coating	Uses and Comments
Alkyd	General metal finishing. Mild corrosive atmospheres. Not recommended for immersion conditions. Fast drying, economic. Poor resistance to alkalis.
Phenolic	Can and tank linings. Immersed structural steels or high-humidity atmospheres. Good chemical resistance.
Acrylic	Automotive top coats, coil coats. Protection of steel in mild corrosive environments.
Epoxy resins	Chemical processing. Good chemical resistance especially to alkalis. Good adherence properties. Surface deterioration (chalking) occurs in sunlight.
Urethane	Aircraft finishes. Low-temperature applications. Excellent abrasion and impact resistance. Variable corrosion resistance depending upon formulation. Good gloss retention.
Epoxy coatings	Air conditioners, tanks, heat exchangers. 'White goods.' Excellent corrosion resistance. Versatile, flexible, and good hardness.
Vinyl coatings	Used widely in the chemical industry for severe corrosive environments. Not recommended above 50–60°C. Can be used as thin or high-build film thickness deposits. Poor resistance to solvents.
Metal-rich primers (Zn)	Act sacrificially to protect the underlying steel substrate. Good abrasion and temperature resistance. Used in conjunction with other top-coat systems.

previously. Paints should thereby be highly impermeably and resistant to abrasion, be suitable for the desired temperature range, and be flexible, thereby resisting cracking on movement of the substrate.

Selection of the most suitable coating system is not trivial and, as previously noted, depends upon a number of considerations. One of the more important steps in coating selection will be a thorough evaluation of the environmental conditions under which the coating will operate. Table 11.9 provides a brief summary of some of the more common types of paint and their typical applications.

11.9 MATERIAL SELECTION

11.9.1 Introduction

Selection of the optimum material for construction is critical in terms of safety, performance, and economic considerations. The basic information used for determining the most appropriate material is derived from laboratory or pilot-plant tests, previous published corrosion data, and in many cases from past experience. It should be recognized that material selection is a compromise and rarely

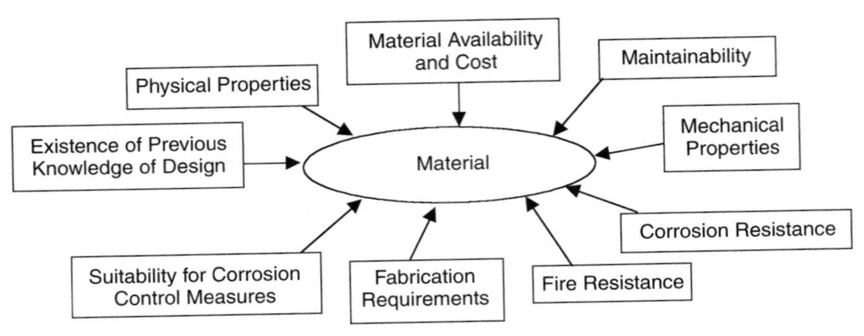

FIGURE 11.24 Factors affecting material selection.

is the highest corrosion-resistant material selected. Figure 11.24 highlights the type of information required to formulate decisions on material selection.

Material selection should be seen as an integral part of the design exercise and used in conjunction with a corrosion management strategy. In too many cases corrosion is considered only when damage has occurred; by this time remedial measures may be several times the cost of the original materials. In addition where poor design has led to a corrosion failure, alternative design and not alternative material selection may be the only effective solution.

The choice of materials available to the material selector or design engineer is extensive. In the limits of this chapter only a brief mention will be made of the "popular" engineering materials used extensively for construction. The reader is advised to consult the literature and seek advice from manufacturers and suppliers prior to the application of a given material.

11.9.2 Ferrous Materials (Steels)

Typically these materials may be classed as (i) carbon steels (mild steel), (ii) low-alloy steels (1–2% alloy additions), and, (iii) high-alloy steels (stainless steels).

Carbon Steels Carbon steel is by far the most widely used engineering alloy by virtue of its low cost, variability in mechanical properties, formability and ease of fabrication, and suitability for the application of protective coatings and cathodic protection methods.

The corrosion rate of carbon steel, like all materials, depends upon the chemical composition and previous history, that is, mechanical working, heat treatment, welding, surface condition and so forth. In general corrosion takes place when the relative humidity is greater than 60% where upon a liquid-phase may form on the metal surface. Once formed the rate of corrosion will depend upon the nature of the environment as discussed previously (Section 11.4). Table 11.10 presents a list of typical corrosion rates for carbon steels within different types of environments.

TABLE 11.10 Typical Corrosion Rates for Mild Steel in Various Environments

Environment	Corrosion Rate (mm/yr)
Atmospheric	
Rural	0.003–0.005
Suburban	0.050–0.08
Marine (onshore)	0.02–0.50
Marine (offshore)	0.05–0.15
Industrial	0.1–0.5
Aqueous	
Freshwater	0.01–0.05
Seawater	0.1–0.2
Soil	
Clay	0.02–0.4[a]
Marl	0.01–0.015
Alluvium	0.03–0.04

[a] Depending upon depth

Low-Alloy Steels Low-alloy steels with 1–2% alloying elements, for example, Cr, Ni, Cu, and Mn, offer improved mechanical strength with limited improvement in corrosion resistance. Increases in the Cr content have some marginal effects on the general corrosion resistance. Improvements in pitting resistance have been observed when Cu additions (0.2–0.3%) are made. However, where steels are used in indoor environments no significant effect of Cr and Cu additions have been noted when compared to the corrosion rates of a nonalloyed steel [58].

Stainless Steels Stainless steels vary widely in composition and are classified according to the metallurgical phase produced on solidification of the metal. Grades are designated as austenitic, ferritic, martensitic, and duplex (austenite and ferrite). In addition a further type of stainless steel is that known as precipitation hardening. The high corrosion resistance of these steels is derived from the ability of the metal to form a protective, self-repairing oxide film. This "self-repairing" ability is, however, subject to the availability of oxygen in the environments surrounding the steel; hence stainless steels are susceptible to corrosion under deaerated conditions, where upon localized corrosion (pitting and crevice corrosion) can occur. The corrosion resistance of stainless steel is the result of the addition of Cr, 11% (minimum). Chromium contents below this level provide limited improvement in corrosion resistance over that of the low-alloy counterparts. In addition, steels with the minimum Cr content and too high a carbon content ($\geq$0.03%) are susceptible to *sensitization* when the metal is welded or heat treated in the temperature range 500–800°C. In this case precipitation of chromium carbide ($Cr_{23}C_6$) takes place close to the austenite grain boundaries, resulting in a zone highly susceptible to corrosion. This type of corrosion has also been named *weld corrosion*: see Section 11.5.7. Resistance to weld corrosion is improved by lowering the carbon content, rapid cooling through the 500–800°C range, or adding carbide stabilizing elements such as Ti and Nb.

Other elements that have a major effect on the mechanical and corrosion properties of stainless steel include Ni, Mo, and N. Nickel is added in varying amounts and leads to the stabilization of the austenite phase. Molybdenum and nitrogen are important alloying additions as they lead to improved strength and resistance to localized corrosion, for example, stress corrosion cracking, pitting, and crevice corrosion. The pitting resistance of a stainless steel may be "indexed" based on the popular *pitting resistance equivalent number* (PREN) and is obtained from the following formula:

$$\text{PREN} = \%\text{Cr} + 3.3\%\text{Mo} + 16\%\text{N}.$$

From this formula it can quickly be seen that increasing the Cr, Mo, and N content of a stainless steel will lead to a higher, more pitting-resistant, alloy.

11.9.2.1 Grades of Stainless Steel

Austenitic Stainless Steels Austenitic grades are highly ductile and have good toughness properties. The alloys are nonmagnetic and have excellent weldability. Table 11.11 provides a summary of the important austenitic grades and their uses.

Ferritic Stainless Steels Ferritic grades, although having a lower corrosion resistance than the austenitic grades, have excellent structural strength at high temperatures. In addition, if suitable precautions are taken to avoid sensitization, these alloys have good resistance to SCC. However, the ductile/brittle transition temperature of the ferritic grades is above room temperature, and this limits the section size of these grades. Table 11.12 summarizes the types and uses of the ferritic grades.

TABLE 11.11 Examples of Austenitic Stainless Steel Grades and Their Uses

AISI Series	Comment and Uses
200 Series	Basic 18Cr/8Ni[a] type used in the food industry where minimal corrosion resistance required
300 Series	304 and 304L—food industry, transportation, chemical industry, heat exchangers, piping and tubing. L grade used where welding is required. Typical corrosion rates;[b] (mm/y) 0.025 (general), 0.25 (crevice).
Superaustenitic	254 SMO,[c] 6% Mo for improved pitting and crevice corrosion resistance. Marine and offshore applications

[a]Compositions given in weight percent, e.g., 18Cr/8Ni represents 18%Cr and 8%Ni.
[b]Unless stated corrosion rates are based upon contact with seawater.
[c]Manufactured by AVESTA.

TABLE 11.12 Examples of Ferritic Stainless Steel Grades and Their Uses

AISI Type	Comments and Uses
400 Series	430. Basic ferritic grade (17Cr)—Furnace parts below 850°C, heat exchangers, tubing and piping. Typical corrosion rates; (mm/y), 0.025 (general), 5.0 (crevice).
Superferritic	29Cr/4Mo. Severe corrosive environments, e.g., 10% boiling sulfuric acid, petrochemical industries.

Martensitic Stainless Steel Martensitic grades, unlike the austenitic and ferritic types, are amenable to hardening by heat treatment, like that of carbon steels. The corrosion resistance of these grades is generally lower than the austenitic and ferritic stainless steels. These grades are used where corrosion resistance is not of paramount importance, but other properties such as hardness or wear resistance are required. Table 11.13 summarizes the types and uses of these grades.

Duplex Stainless Steels Duplex grades have microstructures composed of both austenite and ferrite, although austenite/martensite or ferrite/martensite mixed microstructures, in principle, are also classed as duplex stainless steel. The duplex grades have an excellent combination of toughness, strength, weldability, and corrosion resistance: see Fig. 11.19. Typically the grades contain between 18 and 25%Cr, 4 and 8%Ni, and 2 and 4%Mo. Table 11.14 summarizes the types and uses of theses grades.

Precipitation–Hardening (PH) Stainless Steels PH stainless steels provide a range of grades with high strength and good hardness. Unfortunately, the PH grades have limited corrosion resistance and at best are parallel with that of the 304 grade. They are used in the aircraft industry where high strength is required.

TABLE 11.13 Examples of Martensitic Stainless Steel Grades and Their Uses

AISI Type	Comments and Uses
400 Series	410. Basic type (12Cr)—steam and water valves, pump and steam turbine parts.
	420. Higher carbon content for higher hardenability—cutlery and surgical instruments.
	431. Higher Cr and Ni content—food industry, valves and separators.

TABLE 11.14 Examples of Duplex Stainless Steel Grades and Their Uses

Class and Steel	Comments and Uses
Duplex	22Cr/5Ni/3Mo. Used widely in the offshore industry.
Super duplex	25Cr/7Ni/3Mo. Used where H_2S environments are encountered in 'sour' oil wells.

11.9.3 Nonferrous Materials

11.9.3.1 Aluminum and Its Alloys

The corrosion resistance of Al is reduced on alloying, and therefore pure Al has a higher corrosion resistance than its alloy counterparts. This decrease in corrosion resistance is offset by an increase in the mechanical properties. Al and its alloys gain their corrosion resistance from the natural formation of a protective oxide film (Al_2O_3). Corrosion resistance of the alloys is related to the thickness of this film, and recognition of this has led to the process of anodizing, during which the surface film is thickened to produce a hard, compact, and tightly adherent layer. Al and Al alloys are amphoteric in nature and are therefore susceptible to corrosion attack in both strong acids and strong alkalis. An exception to this is where the acids are strongly oxidizing, that is, nitric acid above 82%.

Typically Al alloys are divided into two groups, namely cast and wrought alloys, the latter group having two categories: those that can be hardened by heat treatment and nonhardenable types, which may only be strengthened by cold working. A compromise in mechanical properties and corrosion resistance can be gained by cladding a high-strength Al alloy with pure Al, this is known as Alclad. Table 11.15 provides a summary of the types and uses of common Al alloys.

11.9.3.2 Copper and Its Alloys

Like that of Al the corrosion resistance of copper decreases on increasing the alloy content. However, in its pure form copper has limited uses due to its softness and lack of strength. Copper alloys have excellent heat transfer properties and therefore find extensive use in heat exchanger applications. The alloys have excellent resistance to seawater environments and are used in marine applications

TABLE 11.15 Examples of Aluminum and Aluminum Alloy Grades and Their Uses

Grade/Type	Comment and Uses
Wrought Commercially pure	99Al. Acetic acid tanks pumps and handling equipment. Typical corrosion rates; (mm/y) 0.04 (general), <0.025 (crevice).
3003 (USA) N3 (UK)	1.2Mn. Stronger than commercially pure grade—used in tanks and heat exchangers.
5052 (USA) N4 (UK)	Higher strength usage—i.e., chemical tanks and pressure vessels.
2024	4.5Cu/1.5 Mg. Commonly used engineering alloy used in aerospace applications.
7075	Al–Li alloys, higher strength to weight ratio than 2024 alloys—use in aircraft fuselages. These alloys are weldable but can suffer from SCC.
Cast 356	7Si. Heat treatable to higher strength—general marine usage.

TABLE 11.16 Examples of Copper and Copper Alloy Grades and Their Uses

Grade/Type	Comment and Uses
Wrought Al Bronze	8 Cu/5-10Al/2.5Fe. Marine, heat exchanger and superheated valve and pump components. Typical corrosion rates; (mm/y) seawater 0.05 (general), <0.05 (crevice).
Brasses	Cu/10-30Zn. Used widely within mildly corrosive environments and atmospheres. Brasses are susceptible to dezincification, which is reduced on the addition of arsenic (0.04%), corrosion rates; (mm/y) 0.05 (general), <0.05 nickels (crevice).
Cupro-nickels	Cu/10-30Ni/0.5-1.0Fe. Used extensively within marine environments, having excellent corrosion resistance in aerated, nonpolluted seawater. Cupro nickels have excellent resistance to marine biofouling. Typical corrosion rates; (mm/y) 0.025 (general), <0.025 (crevice). Polluted seawater and stagnant conditions can lead to accelerated corrosion via the action of microbial sulfate reducing bacteria (SRBs) >5 mm/yr.

for pumps, piping, and sheathing. Copper alloys are susceptible to SCC in the presence of ammonia or amines. In addition selective corrosion (dezincification) of the brasses (preferential corrosion of the Zn phase) can occur in fresh and seawater environments. Like Al, copper alloys are also found in both the wrought and cast forms. Table 11.16 presents a summary of the main Cu alloy types and their typical uses.

11.9.3.3 Titanium and Its Alloys

Titanium and its alloys combine good mechanical properties of strength, toughness, and fatigue with high corrosion resistance, particularly in seawater environments. Titanium is, however, an expensive material. Ti and its alloys are finding increasing applications in deep seas, heat exchangers, drilling risers, pumping, and piping of fresh and seawater. A distinct advantage of Ti and its alloys is its high resistance to microbial-induced corrosion (MIC). Care, however, should be taken in high-carbonate-containing waters where carbonate deposits can build up and accelerated underdeposit corrosion can occur. The Ti/6%Al/4%V grades have good resistance to SCC in brackish or seawaters. Of added value, particularly for thin-wall applications, for example, heat exchangers, is the very high resistance to pitting of Ti and its alloys. Caution should be taken when coupling Ti and its alloys to other materials as in most cases Ti will act as the cathode and can be harmful where conditions can lead to the uptake of hydrogen. In the case of Ti used in conjunction with cathodic protection, potentials less than −0.9V (vs. SCE) should be used. Similarly, the application of Ti and its alloy should be restricted to pH ranges between 3 and 12 to avoid possible hydrogen damage. Crevice corrosion of Ti and Ti alloys will not occur in environments of any pH below 70°C [59].

11.10 DESIGN ASPECTS

Protection against corrosion should begin at the design stage, that is, adopting a proactive approach rather than taking reactive measures after corrosion has occurred. Clear consideration to limiting metal/electrolyte interfaces and operating protective systems under optimum conditions rather than selecting maximum protection can lead to a reduction of up to 20% in the costs arising from corrosion. During a project there are various stages at which corrosion control considerations should be adopted, as shown in Table 11.17.

11.10.1 Contributions to Safe Plant Construction and Operation

There is a common misconception that presupposes that the responsibility for optimum component or plant lifetime rests with the user. To maximize cost effectiveness in terms of capital outlay and minimize running costs, including maintenance costs, requires the cooperation of the designer, manufacturer, and customer (end user); see Fig. 11.25. The individual parties concerned should take a responsible role in ensuring an optimum corrosion control strategy is implemented. In this respect the following considerations should be made:

Designer:

- Prevent the use of geometrical shapes that can cause corrosion problems.
- Select the correct choice of materials.
- Based on the above take account of operational environment.
- Consider the application of protective coatings or the use of electrical protection methods.

Manufacturer:

- Ensure accurate reproduction of design.
- Restrict the use of materials to those specified at the design stage.
- Use correct heat treatments where appropriate.

TABLE 11.17 Identification of Corrosion Control Options Required at Different Stages of Project

Project Stage	Input
• Conceptual design	Corrosion control options considered
• Preliminary design	Select specific control options
• Appraisal of Cost	Estimate cost of options chosen
• Final design	Corrosion control specifications agreed
• Engineering design	Material selection/compatibility evaluated
	CP design confirmed
	Coating type, method, and application agreed
• Construction and commissioning	Checks on materials/CP systems/coatings, etc.
• Operation	Monitor corrosion control systems

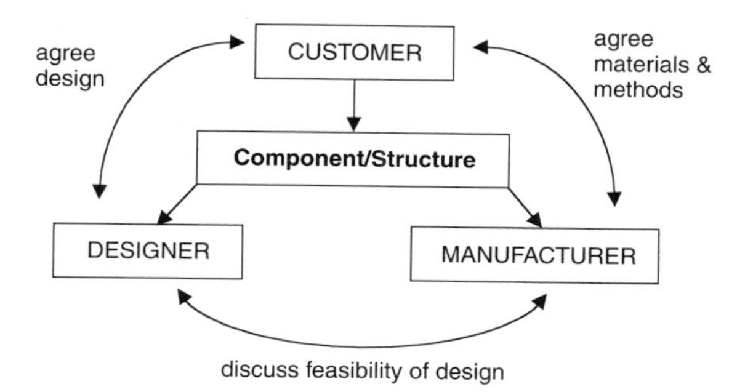

FIGURE 11.25 Interaction between designer/manufacturer and customer for maximum cost effectiveness and minimum corrosion risk.

- Ensure correct fabrication techniques are applied.
- Correctly apply any protective coating systems.

Customer:

- Ensure correct maintenance procedures are applied.
- Ensure the correct replacement of materials.
- Progressively monitor environmental conditions.
- Maintain protective coatings.
- Monitor the progress of any electrical protection systems.

11.10.2 Design Considerations

Successful plant operation requires an understanding of the factors that contribute to a decrease in the inherent corrosion resistance of a material. Summarized below

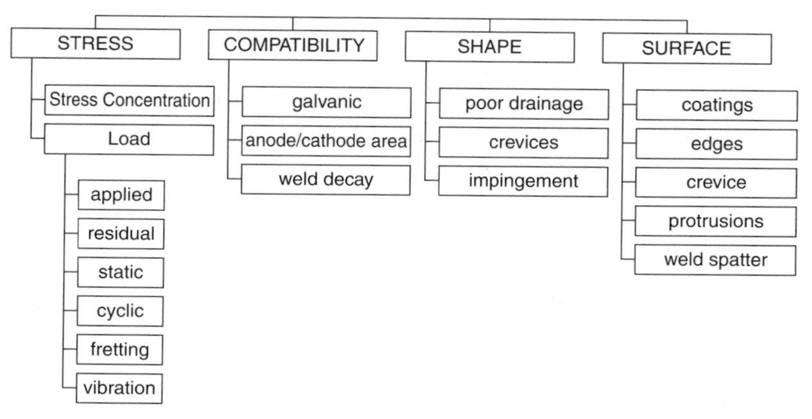

FIGURE 11.26 Factors to be considered during design, fabrication, and commissioning stages, which affect overall corrosion risk of component or structure.

are some of the factors that need to be considered from the initial design stage, through manufacture to the final end product (Fig 11.26).

11.10.3 Examples of Poor Design

Figure 11.26 identified a number of factors that can give rise to increased risk of corrosion. Figures 11.27–11.30 provide examples that highlight typical designs leading to this increased risk. Examples of simple modifications are also given, which lead to reduced risk. Examples are given under the following headings:

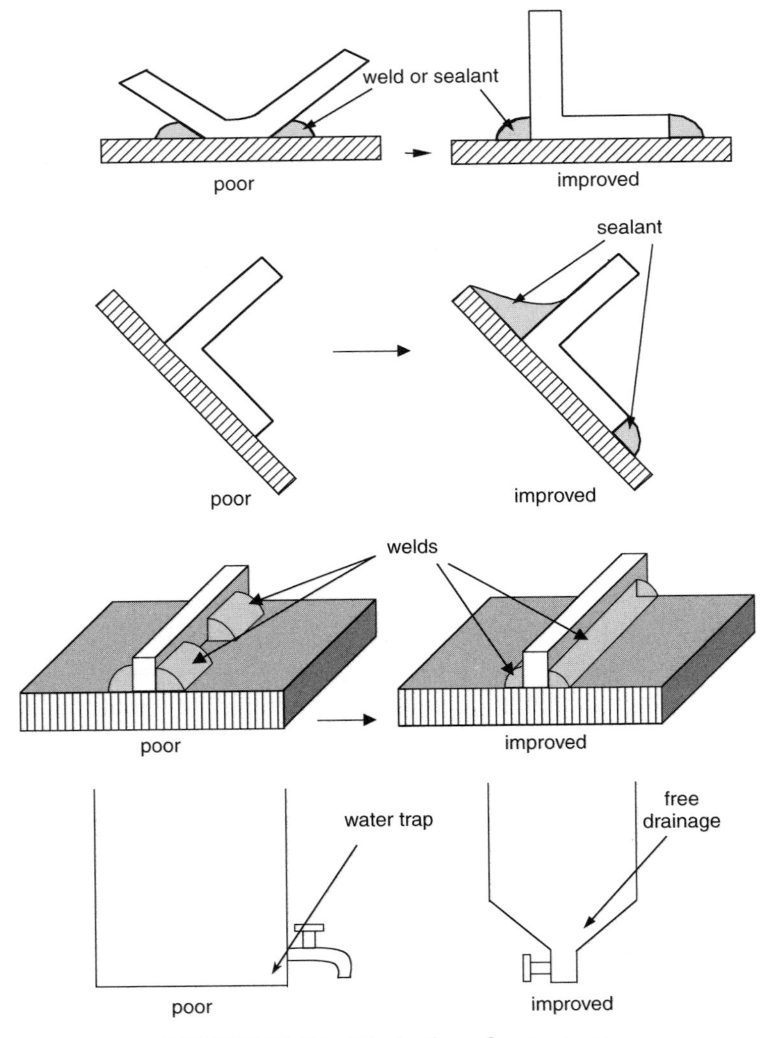

FIGURE 11.27 Elimination of water traps.

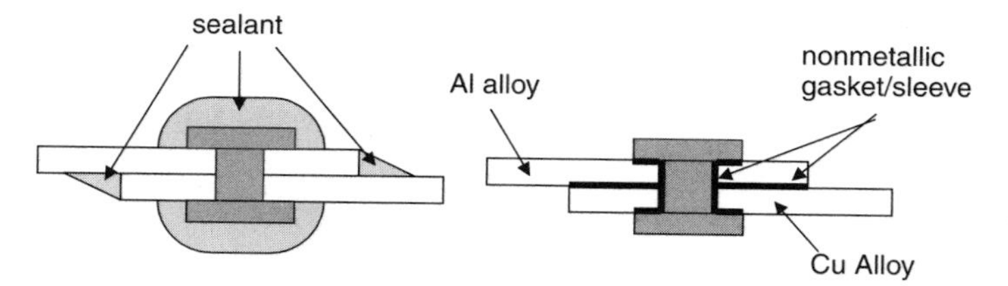

FIGURE 11.28 Elimination of galvanic coupling.

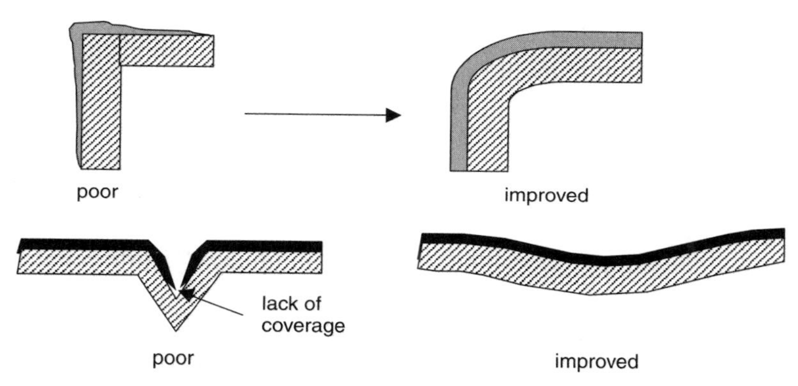

FIGURE 11.29 Elimination of poor profile (prior to coating).

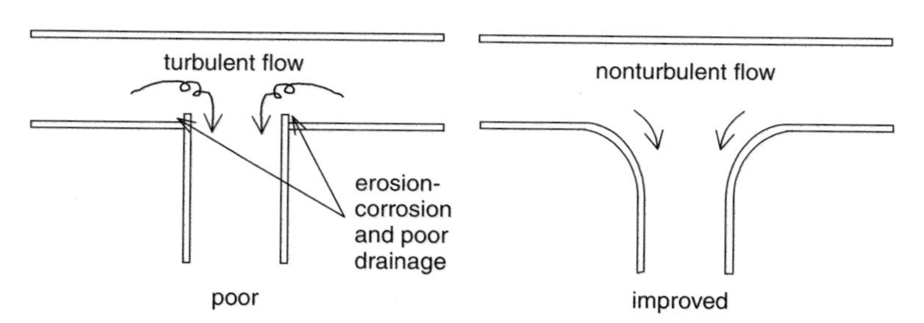

FIGURE 11.30 Elimination of erosion–corrosion.

1. Water traps (Fig. 11.27)
2. Galvanic (bimetallic) coupling (Fig. 11.28)
3. Profile effects affecting coating coverage/adhesion (Fig. 11.29)
4. Erosion (flow-induced) corrosion (Fig. 11.30)

11.10.4 Design Aspects — Summary

Avoiding or more realistically minimizing corrosion risk stems from a simple application of existing knowledge and a good deal of common sense. Following the comments made previously and adopting elementary design considerations, material lifetimes may be extended.

In summary some of the following design aspects should be considered:

- Prevent stress concentrations, changes in section and sites for solution accumulation.
- Avoid changes in pipe bore.
- Avoid galvanic coupling where less anodic material has unfavorable area ratio.
- Apply correct material selection.
- Prevent excessive operating conditions.
- CP or AP may be more economical than design changes but avoid excessive current densities.
- Avoid mechanical damage to coating systems.

REFERENCES

1. H. H. Uhlig, United Nations Scientific Conference on Conversation of Resources, *Corrosion*, **6** 29 (1950).
2. T. P. Hoar, *Report on the Committee on Corrosion and Protection*, HMSO, London, 1971.
3. Topic Study 3—Corrosion, The Open University, 1985.
4. K. R. Trethewey and J. Chamberlain, *Corrosion for Science and Engineering*, 2nd. ed., Longman Group, Essex, England, 1995.
5. H. McArthur *Corrosion Prevention and Control*, **28**(3), 5–10 (1981).
6. H. H. Krause, "Historical Perspective of Fireside Corrosion Problems in Refuse-Fired Boilers," in R. D. Gundry (Ed.), *Corrosion 93*, NACE International, Houston, 1993.
7. R. N. Parkins, *Engineering Solutions for Corrosion in Oil and Gas Applications*, NACE International, Houston, 1991.
8. H. H. Uhlig (Ed.), *The Corrosion Handbook*, Wiley, New York, and Chapman and Hall, London, 1948.
9. ASTM G15, Standard Terminology Relating to Corrosion and Corrosion Testing, *Annual Book of Standards*, ASTM, Philadelphia, 1993.
10. M. Pourbaix, *Lectures on Electrochemical Corrosion*, Plenum Press, New York, 1973.
11. D. L. Piron, *The Electrochemistry of Corrosion*, NACE, Houston, 1991, p. 42.
12. J. O. Bockris and A. K. N. Reddy, *Modern Electrochemistry*, Plenum/Rosetta, New York, 1970.
13. E. Gileadi, *Electrode Kinetics for Chemists, Chemical Engineers and Material Scientists*, VCH, New York, 1993.
14. J. O' M. Bockris and D. Drazic, *Electro-Chemical Science*, Taylor and Francis, London, 1972.
15. M. Stern and A. L. Geary, *J. Electrochem. Soc.* **104**, 56 (1957).
16. ASTM G59, Standard Practice for Conducting Potentiodynamic Polarisation Resistance Measurements, *Annual Book of Standards*, ASTM, Philadelphia, 1990.

17. I. Epelboin, C. Gabrielli, M. Keddam, and H. Takenouti, in F. Mansfeld and V. Bertocci (Eds.), *Electrochemical Corrosion Testing*, ASTM 727, ASTM, Philadelphia, 1981, p. 150.
18. R. Baboian, *Corrosion Tests and Standards, Application and Interpretation*, ASTM, Philadelphia, 1995.
19. L. L. Shreir, R. A. Jarman, and G. T. Burstein (Eds.), *Corrosion, Metal/Environment Reactions*, Vol. 1, Butterworth-Heinemann, Ltd, Oxford, UK, 1995, pp. 2.43–2.59.
20. F. Speller, *Corrosion*, McGraw-Hill, London, 1951.
21. J. W. Oldfield and P. Todd, *Desalination*, **31**, 365 (1979).
22. J. W. Oldfield and W. H. Sutton, *Br. Corr. J.*, **13**, 13–22 (1978).
23. G. T. Burstein, P. C. Pistorius, and S. P. Martin, *Corr. Sci.*, **35**, 57–62 (1993).
24. U. Steinsmo and H. S. Isaacs, *Corr. Sci.*, **35**, 83–88 (1993).
25. P. Marcus, A. Teisser, and J. Oudar, *Corr. Sci.*, **24**, 259–268 (1984).
26. J. Stewart and D. E. Williams, *Corr. Sci.*, **33**, 457–474 (1992).
27. British Standard BS6890 Pt 5, British Standards Institution, London (1990).
28. A. R. Despic, R. G. Raicheff, and J. O' M. Bockris, *J. Chem. Phys.* **49**, 926 (1968).
29. R. Akid, Y. Z. Wang, and U. Fernando, in T. Magnin, and J. M. Gras (Eds.), *Corrosion-Deformation Interactions*, Physique Les Ulis, Les Ulis Cedex A, France, France (1993), pp. 659–670.
30. D. Broek, *Elementary Fracture Mechanics*, 4th ed., Martinus Nijhoff, Dordrecht, The Netherlands, 1986.
31. J. M. Barsom, *Proc. Int. Conference on Corrosion Fatigue*, NACE, Houston, 1972 p. 424.
32. I. M. Austen and E. F. Walker, *The Influence of Environment on Fatigue*, Institution of Mechanical Engineers, London, 1977, pp. 1–10.
33. J. M. West, *Metal Sci.*, **11**, 534–540 (1980).
34. B. A. Bilby and J. Hewitt, *Acta Metall.* **10**, 587 (1962).
35. N. J. Petch and R. Stables, *Nature*, **169**, 842 (1952).
36. R. A. Oriano, Proc. Symp. on Fundamentals of SCC, National Association of Corrosion Engineers, NACE, Houston, 1969. p. 32.
37. E. Heitz, R. Henkhaus, and A. Rhamel (Eds.), *Corrosion Science—An Experimental Approach*, Ellis-Horwood, Chichester, England, 1992.
38. F. Mansfeld, *Corrosion*, **37**, 301–307 (1981).
39. K. H. Ladky, L. M. Callow, J. L. Dawson, *Br. Corr. J.* **15**, 20 (1980).
40. R. Baboian, *Electrochemical Techniques for Corrosion Engineering*, NACE, Houston, 1986, pp. 253–259.
41. R. Akid and D. Mills, *Corr. Sci.*, **43**, 1203–1216, (2001).
42. J. Gonzalez, Ph.D. Thesis, University of Sheffield, UK. (1999).
43. H. S. Isaacs and Y. Ishikawa, *Electrochemical Techniques for Corrosion Engineering*, NACE, Houston, 1986, pp. 17–23.
44. K. R. Trethewey, D. A. Sargeant, D. J. Marsh, and S. Haines, in K. R. Trethewey and P. R. Roberge (Eds.), *Modelling Aqueous Corrosion*, Kluwer Academic, Dordrecht, The Netherlands, (1994), pp. 417–442.
45. R. Akid, *Materials World*, **11**, 522–525 (1995).

46. R. Akid and D. Mills, *Proceedings EuroCorr 97*, September 1997, Trondheim, Norway, Vol. 1, 1997, pp. 757–763.
47. S. W. Dean, *Corrosion Testing and Evaluation: Silver Anniversary Volume ASTM STP 1000*, R. Baboian and S. W. Dean (Eds.), ASTM, Philadelphia, 1990, pp. 163–176.
48. K. H. Ladky and J. L. Dawson, *Corr. Sci.*, **22**, 231–237 (1982).
49. P. C. Searson and J. L. Dawson *J. Electrochemical. Soc.*, **135**, 1908–1915 (1988).
50. A. D. Mercer, in L. L. Shreir, R. A. Jarman, and G. T. Burstein (Eds.), *Corrosion*, Butterworth-Heinemann Ltd, Oxford, UK, 1995, p. 17:10.
51. G. Wranglen, *An Introduction to Corrosion and Protection of Metals*, Chapman and Hall, London 1985.
52. V. Ashworth and C. J. L. Booker (Eds.), *Cathodic Protection: Theory and Practice*. Ellis–Horwood, Chichester, UK, 1986.
53. K. A. Chandler and D. A. Bayliss (Eds.), *Corrosion Protection of Steel Structures*, Elsevier Applied Science, London, 1985.
54. V. Chaker (Ed.), *Corrosion Forms and Control for Infrastructure*, ASTM STP 1137, ASTM, Philadelphia (1992).
55. O. L. Riggs and C. E. Locke (Eds.), *Anodic Protection*, Plenum, New York, 1981.
56. D. R. Gabe, *Principles of Metal Surface Treatments & Protection*, Pergamon Press, 1978.
57. C. G. Munger, *Corrosion Prevention by Protective Coatings*, NACE, Houston, 1984.
58. J. F. Stanners, *J. Appl. Chem*, **10**, 461 (1960).
59. D. K. Peacock, Int. Conf. on Titanium in Practical Applications, Norwegian Assoc. of Corrosion Engineers, Trondheim, 1990.

CHAPTER 12

Standards and Codes for Advanced Materials

MICHAEL G. JENKINS

Department of Mechanical Engineering, University of Detroit Mercy, 4001 W. McNichols Rd., Detroit, Michigan 48219

Handbook of Advanced Materials Edited by James K. Wessel
ISBN 0-471-45475-3

12.1 INTRODUCTION

Thermomechanical behavior as well as physical attributes (and their subsequent characterization) of advanced materials [e.g., superalloys, intermetallics, functionally graded materials, advanced composites (polymer, metal, or ceramic matrix)] are currently the subject of extensive investigation worldwide. The determination of the properties and performance (mechanical, thermal, thermomechanical, physical, environmental, etc.) of these advanced materials is required for a number of reasons: (1) to provide basic characterization for purposes of materials development, quality control, and comparative studies, (2) to provide a research tool for revealing the underlying mechanisms of mechanical performance, and (3) to provide engineering performance prediction data for engineering applications and components design [1]. As prototype and trial products comprised for advanced materials reach the marketplace, the paucity of standards (i.e., test methods, classification systems, unified terminology, and reference materials) for these materials and the lack of design codes and their related databases specific to these materials are limiting factors in commercial diffusion and industrial acceptance [2] of advanced materials.

12.1.1 Standards

The term *standards* has many implications. To the researcher and the technical community, it may be fundamental test methodologies and units of measure. To the manufacturer or end-product user it may be materials specifications and tests to meet requirements. Commercial standards equate to the rules and terms of information transfer among designers, manufacturers, and product users [2]. There are even fundamental differences between levels of standards: company (internal use with only internal consensus); industry (trade/project use with limited organizational consensus); government (wide usage and varying levels of consensus); fullconsensus (broadest usage and greatest consensus).

At present, there are few—national or international—full-consensus standards for testing advanced materials such as advanced ceramics. This limited ability to test on a common-denominator basis hampers further material development [2]. For example, specific areas where standardization (or consensus) are required include terminology/nomenclature, test fixtures, test specimen geometries, specimen preparation, machining procedures and allowable tolerances, test specimen alignment, optimal straining/stressing rates, metrology (temperature and strain), testing environment, and identification of fracture and failure modes. These needs are particularly acute at elevated temperatures or in aggressive environments where test equipment and measurement techniques are often being developed simultaneous with the test material. Although considerable development may be required for standards for many advanced materials, rather than adopting entirely new or unconventional methods and techniques, test methods developed originally for the room temperature characterization of conventional materials are a good starting point to develop test methodologies for advanced materials in particular.

12.1.2 Design Codes and Databases

The meaning of the term *design code* is not generally well understood. As used in the following discussion *design code* is not a design manual (i.e., a "cookbook" design procedure resulting in a desired component or system. [3]. Instead, design codes are widely accepted but general rules for the construction of components or systems where safety is important. A primary objective is the reasonably certain protection of life and property for a reasonably long safe life of the design. Although needs of the users, manufacturers, and inspectors are recognized, the safety of the design can never be compromised.

By not imposing specific rules for design, the code allows flexibility for introducing new designs as required for performance, efficiency, usability, or manufacturability while still providing constraints for safety. The code must be wide ranging, incorporating figurative links between materials, general design (formulas, loads, allowable stress, permitted details), fabrication techniques, inspection, testing, certification by stamping and data reports, and finally quality control to ensure that the code has been followed. Thus, implicit in the design codes may be many of the standards previously discussed for materials testing, characterization, and quality control. In addition, unlike standards that provide no rules for compliance or accountability, codes require compliance through documentation, and certification through inspection and quality control.

A logical outcome of design codes is the incorporation of databases of material properties and performance "qualified" for inclusion in the code. These data are "qualified" because they have been attained through testing per the statistical requirement of the code as well as per the standards indicated in the code. Qualified databases often require a minimum numbers of test for (1) a particular batch of material and (2) multiple batches of material. In addition, databases may include primary summary data (e.g., mean, standard deviation, and numbers of tests) along with secondary data from the individual tests. Some databases may contain only numerical information while others may include graphical information (e.g., stress–strain curves, temperature profiles, or test specimen geometry). Databases are increasingly in electronic form to speed data retrieval and many are even web-based to provide instant access and frequent updatability.

Design codes and their databases may even be backed as legal requirements for implementing an engineering design [e.g., certification and compliance with the American Society of Mechanical Engineers (ASME), *Boiler and Pressure Vessel Code* [4] is a legal requirement in 48 of the 50 United States]. At present there are no national or international design codes allowing continuous fiber-reinforced ceramic composites (CFCCs) in any type of design. This situation may be hampering material utilization since designers cannot use a material directly in new designs but instead must (1) show evidence that the material meets the requirements of the code and (2) obtain special permission to used the material in the code design. In addition material development is impaired since without a demand for a new material, there is no incentive for further refinement.

This chapter concentrates on standards and codes for advanced materials. First, national and international standards bodies are briefly reviewed keying on important mechanical, thermal, and physical aspects that require standardization. Next, a similar brief review of national and international design codes and evolving databases for advanced materials is presented. Finally, the summary and conclusion section contains successes, lessons, and future directions for standards and codes for advanced materials.

12.2 STANDARDS

12.2.1 National

Most of the industrialized nations in the world have at least one national standards institute. Within the International Organization for Standardization (ISO), there are currently over 130 such institutes represented. The following sections provide a brief overview of some of these national standards writing bodies.

12.2.2 United States

The United States is the only industrialized nation in which the national standards institute is not part of or supported by the national government [5]. The National Institute for Standards and Technology (formerly National Bureau of Standards) (NIST), which is an entity within the U.S. Department of Commerce, conducts research on advanced materials and processing as well an initiating and promoting development of voluntary standards. NIST also develops standard reference materials (SRMs), which are used for "calibration" and verification of test instruments and procedures. However, the American National Standards Institute (ANSI) is the coordinating organization for the voluntary standards systems for the United States and is comprised of individuals and standards writing bodies. It certifies the standards-making processes of other organizations, initiates new standards-making projects, represents the United States within ISO, and examines the standards prepared by other organizations to determine whether they meet the requirements for consensus so as to be included as an ANSI standard [5]. Of the over 175 organizations accredited by ANSI to produce standards, the American Society for Testing and Materials (ASTM) has produced over half of the existing ANSI standards. The following discussion briefly discusses ASTM and its role in standards for advanced materials.

12.2.3 ASTM

ASTM has implemented committees (e.g., ASTM Committee C28, Advanced Ceramics) that are in the process of identifying and addressing the more pressing standardization needs for advanced materials, surveying which needs are satisfied by existing ASTM (or other) standards and establishing liaison with other organizations to enhance the standards-setting process through collaborative arrangements. One requirement for the success of these standardization efforts is that the resulting test methods are shaped with industrial users in mind. Often research-oriented test methods are not necessarily acceptable as commercial standards. To ensure commercial usefulness, industry must participate in the standards-setting process. The ideal standardized test method should be simple to conduct, should use small, easily fabricated and prepared specimens, and should be capable of measuring simultaneously elastic properties and performance (e.g., first matrix microcracking stress, ultimate strength as well as corresponding strains).

It is useful to look at the organization of ASTM to understand the standards development process. As shown in Fig. 12.1, the oversight of the ASTM standardization process is the Committee on Standards. Below this executive committee are the technical committees, for example, Committee C28. Committee C28 is further subdivided into subcommittees, for example Subcommittee C28.07 Ceramic Matrix Composites. Finally, the technical work of the subcommittee is accomplished on the task group level, of which there are currently eight in the example of Subcommittee C28.07. It is important to note that task group leaders are free (and encouraged) to recruit any appropriate experts (members and nonmembers of ASTM) who they deem necessary to complete the work of the task group. Note that in Fig. 12.1 that the flow of standards development is from the task groups to the executive committee. Thus, the technical aspects of the standard are the driving force, with the executive committee ensuring procedural, format, and layout conformance with the norms of the society, and not necessarily technical correctness.

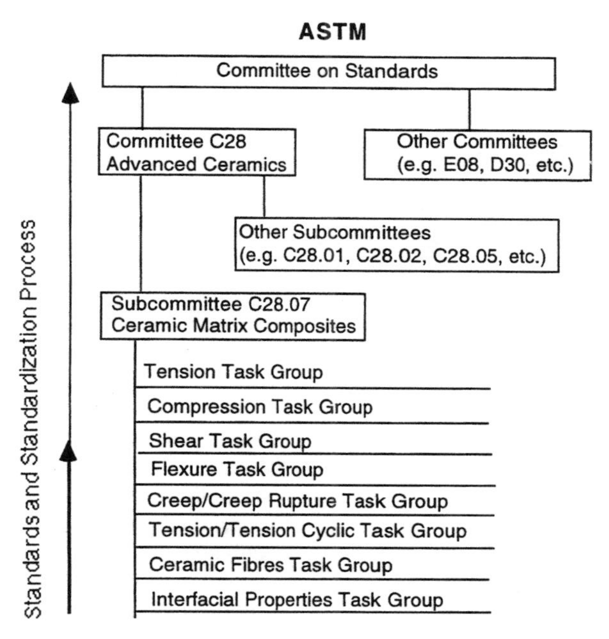

FIGURE 12.1 Standardization organization of ASTM.

Once a task group has developed a draft standard, it is submitted to all members of the subcommittee for ballot. Members vote and comment on the technical rigor as well as the usability (format, language, etc.) of the draft standard. Any negative ballots will stop the approval process until either the negative is dealt with (withdrawn or found nonpersuasive) or the item withdrawn from ballot. Once the draft standard has been approved on the subcommittee level, it is submitted to all members of the main committee for ballot. As in the subcommittee ballot, members vote and comment on the draft standard and any resulting negatives stop the approval process until they are resolved. Finally, after being approved at the sub and main committee levels, the draft standard is submitted for a "society review," which does not require each member of the society to vote, but instead only holds up a draft standard if a member votes negative after inquiring about the draft. Once the draft standard has passed these three levels of approval the Committee on Standards must approve that due process was followed. The ASTM standards produced by this process are described as "technically rigorous and of high quality."

12.2.3.1 ASTM Standards for Advanced Materials

The work of the task groups of the subcommittees may be either ad hoc or may follow a program of work. Because the standardization process of ASTM is voluntary, the work is accomplished only if willing volunteers [i.e., someone with an interest (technical, industrial, or profit)] are able to "champion" a standard through the full-consensus approval process. ASTM has six different types of standards that require this level of approval: test method (procedures for determining material properties or product performance), specification (concise statement of the requirements that need to be satisfied), terminology (definitions, nomenclature, symbols, initialisms, and acronyms), guide [options and instructions (e.g., other ASTM test methods) but does not require or recommend a specific course of action], and classifications (systematic arrangement or division of materials or products into groups). Table 12.1 provides a partial listing of those ASTM

TABLE 12.1 ASTM Committees on Advanced Materials

Committee	Materials
B-10 Reactive and Refractory Metals and Alloys	Titanium
C-5 Manufactured Carbon and Graphite Products	Carbon, graphite, carbon–carbon composites
C-21 Ceramic Whitewares and Related Products	Porcelains, aluminum oxide products
C-28 Advanced Ceramics	Monolithic and composite technical ceramics
D-30 Composite Materials	Polymer and metal matrix composites
F-1 Electronics	Electronic substrates, circuit boards, etc.
F-4 Medical and Surgical Materials and Devices	Biomaterials

committees (out of over 150 total committees) working on advanced materials as an obvious part of the committee charter. Other committees may also develop standards for advanced materials indirectly through other related work.

Not all advanced materials are represented by the committees of ASTM shown in Table 12.1. This does not mean standards for those missing advanced materials are not or will not be developed in the United States. It simply means that the "critical mass" of people and interest has not yet developed for a committee to be formed within ASTM on that topic. ASTM standards can be easily assessed via the *Annual Book of ASTM Standards* (either hard-copy form in printed books or in electronic form on compact disk) [6] or through the World Wide Web through ASTM's web page [7].

An example of problems addressed for elevated temperature of a particular advanced material (ceramic matrix composite) is shown in Fig. 12.2 where the temperature distributions in a test specimen being held in hot, warm, and cold grips are illustrated. As used here, grips can be located outside the furnace and either water-cooled (cold grips) or noncooled [or heated] (warm grips). Grips can also be located inside the furnace and exposed to the test temperature (hot grips). Thermal analyses such as those shown in Fig. 12.2 have indicated substantial temperature gradients when using cold grips ($\Delta T \approx 1200°C \text{in} \Delta L \approx 50$ mm). Such steep temperature gradients may introduce thermal stresses and thereby promote nongage section failures. Progressively less steep temperature gradients exist for warm and hot grips.

A failure outside the gage section, which would be considered unacceptable according to the criteria in current ASTM standards (e.g., C1275-95 [8] on tensile tests of CFCCs at room temperature), could be common for a material that

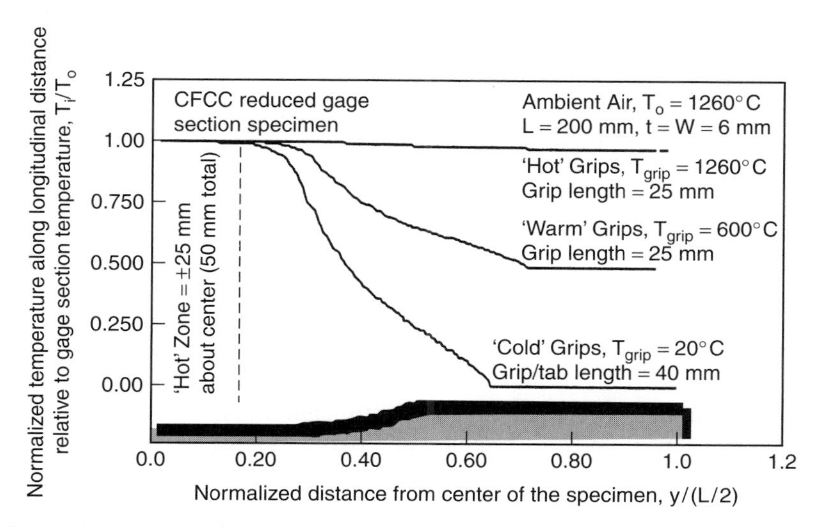

FIGURE 12.2 Temperature distributions in a ceramic matrix composite test specimen for various types of grips.

exhibits increasing strength with increasing temperature (e.g., SiC fiber-reinforced SiC matrix composites). Therefore, it is often necessary to define a gage section where the material is not only exposed uniformly to the test temperature but where it is also subjected to the maximum stress. This is accomplished by reducing the cross section of the specimen, thereby "forcing" the failures to occur in the uniformly heated and stressed region.

In addition, no comprehensive study addressing the reliability of temperature measurements at high temperatures for CFCCs has been reported to date. In addition, an in-depth study is required to compare optical and other contact-type temperature measuring devices and to determine their reliability especially as a function of time. It will be necessary to monitor the temperature at the extensometer contact points and along the gage section, including comparison of temperature measurements at the surface and at the interior of the specimen. Temperature control and its relation to temperature distribution in the furnace and in the specimen should be addressed, and, ideally, maximum allowed temperature differences along the gage length should be required for consistent tests. Possible recommendations of temperature uniformity are $\pm 5^\circ C$ at $<500^\circ C$ and $\pm 1\%$ of test temperature at $>500^\circ C$. Finally, temperature measurement must have an accuracy of $\pm 5^\circ C$.

12.2.4 Europe

Within the European continent each industrialized nation has a national standards institute. Some well-known examples of these are the United Kingdom's British Standards Institute (BSI), Germany's Deutsches Institut für Normung (DIN), and France's Association Française de Normalisation (AFNOR). Some lesser known examples are Hellenic Organization for Standardization (ELOT) in Greece and Instituto Portugues da Qualidade (IPQ) in Portugal.

The evolving aspects of the European Union (EU) have created the need for a unified vision of standards within the rapidly shrinking European continent. The Comité Européen de Normalisation (CEN) has evolved as a regional standardization forum for EU nations. In the CEN arena, once a technical area has been proposed for standardization within CEN, all standardization work in that area within CEN member nations must halt. The cessation of effort is intended to avoid duplication of efforts within Europe. Given this aspect of CEN, further discussion of European standardization efforts is deferred until a following section on regional efforts (CEN).

12.2.5 Asia

In Asia, as in Europe, each industrialized nation has a national standards institute. Some well-known examples of these are Japan's Japanese Industrial Standards Committee (JISC), People's Republic of China's Chinese Association for Standards (CAS), and Australia's Standards Australia (SA). Some lesser known examples are Bapan Standardardisasi Nasional (BSN) in Indonesia and Korea Standards Institute (KSI) in Korea.

The dominance of certain industrial nations such as Japan and the United States as well as the need to promote international trade and industry have provided a driving force for many Asian countries to adopt ASTM, JISC, CEN, or ISO standards as international standards. Nonetheless, a JISC is an example of strong Asian standards organization, which is discussed in the following section.

12.2.5.1 JISC

In Japan, industrial standardization is promoted at the national, industry association, and company level. JISC develops Japanese industrial standards (JIS) as voluntary national standards. JISC has implemented divisions (e.g., JISC committee R, Fine Ceramics) that develop processes and techniques for, among other things, providing a JIS mark that assures a certain level of quality. One requirement for the success of the JISC standardization efforts is that the resulting standards should be shaped with industrial users in mind. Seldom if ever are research-oriented standards acceptable as commercial standards. To ensure commercial usefulness, industry must participate in the standards-setting process. Various JISC divisions and the type of JIS standards that result are shown in Fig. 12.3. Note that the 19 divisions of JISC are far fewer in number than the over 150 committees of ASTM. Moreover, divisions within JISC specifically exclude medicines, agricultural chemicals, chemical fertilizers, silk yarn, and foodstuffs/agricultural/forest products. However, within each division, a myriad of activities and subactivities are present, often guided by national programs under the guidance of the Ministry on International Trade and Industry (MITI).

It is useful to look at the organization of JISC to understand the standards development process. As shown in Fig. 12.4, the oversight of the JISC standardization process is through its Divisional Committee and its Technical Division Council. Draft or requests for JIS standards can come from many directions: any interested party (e.g., industry) or a relevant governmental minister or governmental program. The technical scrutiny and progressive ballot/revision process such as that of ASTM is not obvious in the JISC system. Once a draft standard is submitted to JISC by a competent minister, one of the divisional councils is asked to deliberate it. If necessary, further deliberation may take place within of the technical committees. When the draft is considered appropriate and rationale, JISC reports this to the competent minister and the standard is announced and published as a JIS standard through official channels. No provisions for revision or reapproval are obvious.

12.2.5.2 JISC Standards for Advanced Materials

The work of the divisions within technical areas may be ad hoc or more often it is part of a mandated program of work. Because the standardization process of JISC is often mandated, the work is often accomplished through government-funded projects. JISC has three major domains of standards (see Fig. 12.3): substance or products (shape, dimensions, appearances, etc.), actions or methods (operations, procedures, or method), and basic (units, terminology, conditioning, classification, etc.). Figure 12.3 shows a very broad overview of divisions, some of which

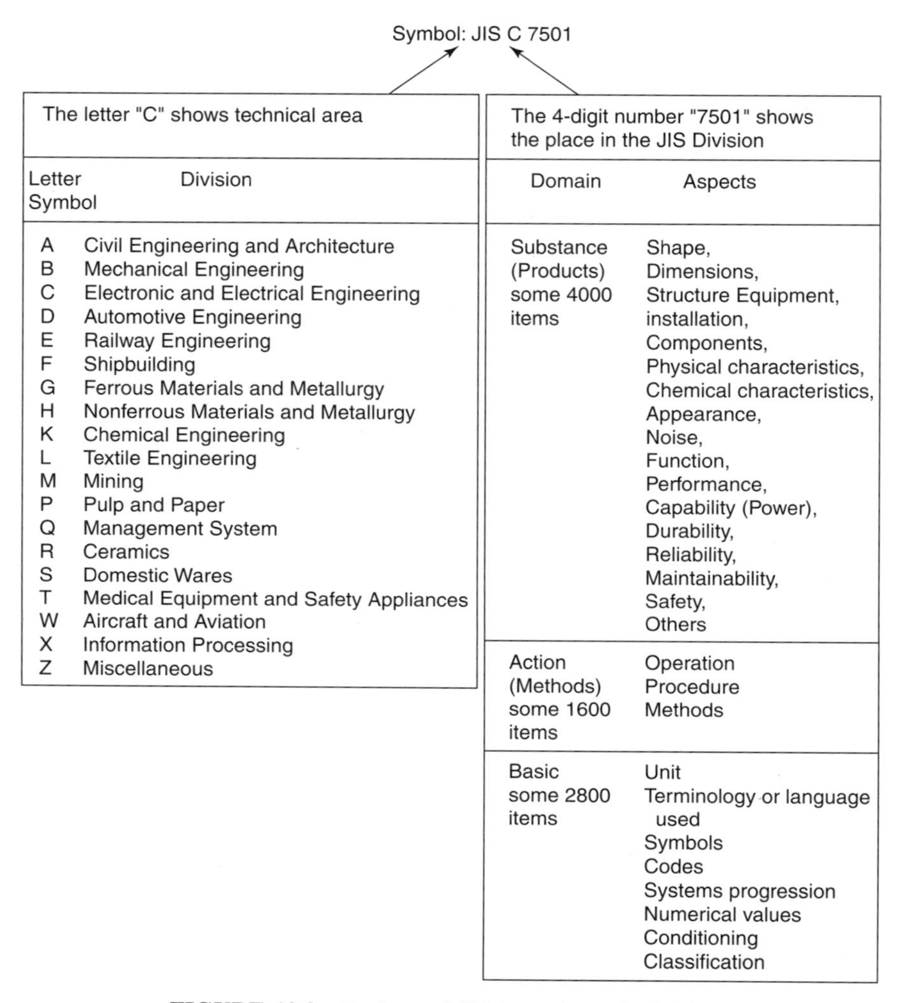

FIGURE 12.3 Designated JIS items in each division.

may deal directly with advanced materials. For example, Divisions H and R deal with nonferrous materials (e.g., titanium, etc.) and fine ceramic, respectively. However, Divisions A, B, D, and W may all deal with various aspects of polymeric composites. JISC standards can be accessed via the standards (either hard-copy form in printed books or in electronic form on compact disk) [10] or through the World Wide Web through JISC's web page [9].

12.2.6 International Standards

There is much debate as of late about what constitutes an "international" standard [11]. From a legal and trade standpoint, an obvious international

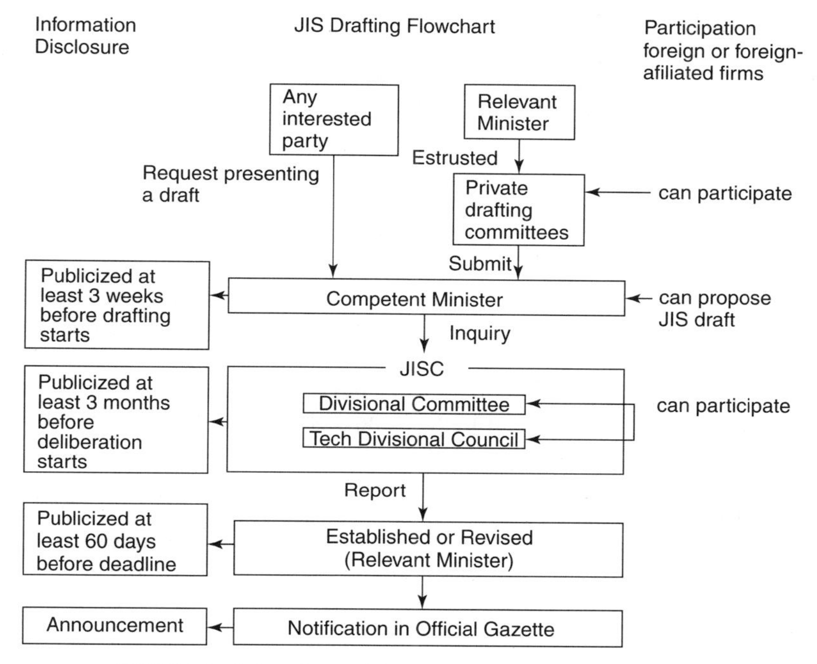

FIGURE 12.4 Flow of the JISC standardization process.

standard is one that has the word *international* in its name (e.g., International Organization for Standardization, i.e., ISO). However, from a practical and profit-motive standpoint, any standard that is accepted internationally in the marketplace, regardless of country of origin or country of use is an international standard [11].

International standards are needed because the existence of multiple nonharmonized standards for similar technologies in different countries or regions can contribute to so-called technical barriers to trade. Export-minded industries have long sensed the need to agree on world standards to help rationalized the international trading process. In the following discussion, *international* is used to refer to those standards institutions that are composed of entities or efforts from multiple nations.

12.2.7 Regional

Although various parts of the world are developing various regional efforts for defense, trade, and economic reasons [e.g. North Atlantic Treaty Organization (NATO), EC, World Trade Organization (WTO), Group of 7/8, North American Free Trade Agreement (NAFTA), etc.], there are few obvious and successful examples of regional efforts in regard to standards. The most evident recent example of a regional standards effort is CEN as discussed in the following discussion.

12.2.7.1 CEN

European Committee for Standardization (CEN), European Committee for Electrotechnical Standardization (CENELEC), and European Telecommunications Institute are the three European standardization bodies recognized competent in the area of voluntary technical standardization as listed in European Union Directive. Together they prepare European Standards (a.k.a. Norms) (EN) in specific sectors of activity. As in ASTM, most standards are prepared at the request of industry. However, as in JISC, a government entity, in this case the European Commission (EC), can also request that the standards bodies prepare standards in order to implement European legislation. Such standardization is "mandated" by the EC and becomes part of the program of work of each standardization body (Table 12.2).

The CEN consists of a "system" to carry out formal processes shared between national members, associates, affiliates, and correspondents. For the purposes of approving standards, only the national members can vote, with each nation having only one vote. It is useful to look at the organization of CEN to understand the standards development process. As shown in Fig. 12.5, the oversight of the CEN standardization process is controlled by the Technical Board of CEN. Once agreed to by the CEN national members in the Technical Board, the development of European standards with precise scopes, titles, and target dates for completion is conducted through one of three main routes:

1. *International Standardization (ISO)* This procedure (formalized as the Vienna Agreement) allows CEN to decide case by case and according to precise conditions to transfer the execution of European standards to ISO (and in a few cases, vice versa). The work is done according to specific ISO rules and CEN/ISO parallel procedures for public enquiry and formal vote. Under this procedure ISO may nominate a representative to sit in a CEN committee and vice versa.

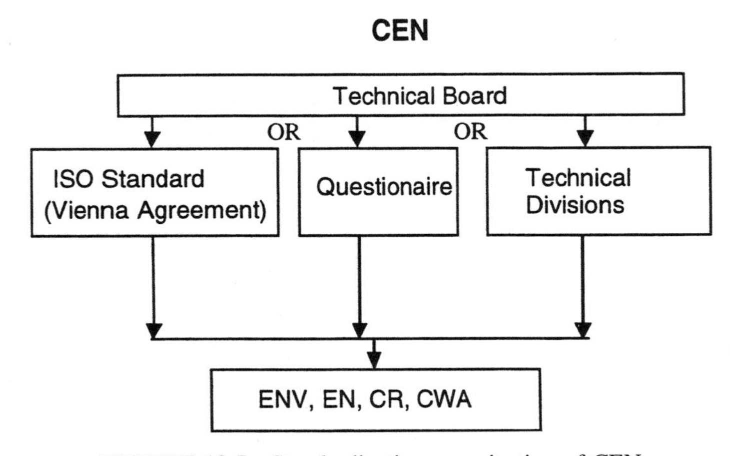

FIGURE 12.5 Standardization organization of CEN.

2. *Questionnaire Procedure (PQ)* This route is used when an appropriate "reference" document exists (often but not only an ISO item). This procedure does not require the formation a new technical committee.
3. *Technical Committee (TC)* When the first two cases are not possible, the CEN technical committees (TCs) gather the national delegations of experts convened by national members which must ensure that such delegations convey a national point of view that accounts for all interests affected by the work. Participation as observers of recognized European/international interests are authorized. TCs must take into account any relevant work (in ISO, for example) falling within its scope, as well as any data that may be supplied by national members and by other relevant European/international organizations. The results of this work can then be offered to ISO. Note: the third path (TC) can still use and modify ISO work. The first (ISO) and second (PQ) paths are used when the texts remain strictly the same.

Technical committees can be further subdivided into subcommittees if the scope of the work needs to be narrowed before the work on standards is actually implemented. Principal TCs are grouped as: biotechnology; heating, cooling, and ventilation; building and civil engineering; household goods, sports, and leisure; chemistry; information society standardization system; environment; materials; food; mechanical engineering; gas appliances; quality, measurement, and value analysis; health and safety at the workplace; services; health care; and transport and packaging.

The outcomes of this standardization process can take several forms as approved by formal votes of national members:

1. European standards (ENs) are usually the general rule because it is important that the national standards of members are identical wherever possible. Once an EN is implemented, members adopt them as the national standard.
2. European prestandards (ENVs) are established as prospective standards for provisional application in technical fields where the innovation rate is high or where there is an urgent need for guidance and primarily where the safety of persons or goods is not involved. ENVs do not have to be adopted by the members (but they must be announced and made available).
3. CEN reports (CRs) provide information and are adopted by the Technical Board.
4. CEN workshop agreements (CWAs) are consensus-based specifications, drawn up in an open workshop environment. Although CWAs are not formal standards, they can be produced on a rapid basis to meet market needs.

12.2.7.2 CEN Standards for Advanced Materials

As shown in Table 12.3 there is a limited number of technical committees specifically dealing with advanced materials, although work on advanced materials may take place within committees dealing with broader topics than just materials. CEN standards can be accessed via the national standards institute of each

TABLE 12.2 Type and Actual Members of CEN

Type of Member	Member [Nation (Standards Body)]
National member (from EU, EFTA, and the Czech Republic) and the representative expertise	Austria (ON), Belgium (IBN/BIN), Czech Republic (CSNI), Denmark (DS), Finland (SFS), France (AFNOR), Germany (DIN), Greece (ELOT), Iceland (STRÍ), Ireland (NSAI), Italy (UNI), Luxembourg (SEE), Netherlands (NNI), Norway (NSF), Portugal (IPQ), Spain (AENOR), Sweden (SIS), Switzerland (SNV), United Kingdom (BSI)
Associate	ANEC, European Association for the cooperation of consumer representation in standardization; CEFIC, European Chemical Industry Council; EUCOMED, European Confederation of Medical Devices Associations; FIEC, European Construction Industry Federation; NORMAPME, European Office of Crafts, Trades and Small and Medium-sized Enterprises for standardization; TUTB, European Trade Union Technical Bureau for Health and Safety
Affiliates (nations of Central and Eastern Europe expected to join CEN as full members in the coming years)	Albania (DPS), Bulgaria (CSM), Croatia (DZNM), Cyprus (CYS), Estonia (EVS), Hungary (MSZH), Latvia (LVS), Lithuania (LST), Malta (Malta Standardization Authority, Poland (PKN), Romania (IRS), Slovakia (UNMS), Slovenia (SMIS), Turkey (TSE)
Corresponding organizations	Egypt (EOS), South Africa (SABS), Ukraine (DSTU), Yugoslavia, (SZS)

national member or the *Catalogue of European Standards* (either hard-copy form in printed books or in electronic form on compact disk) [12] or through the World Wide Web through CEN's web page [13].

12.2.8 International Cooperation

International cooperation in standards often results only when it becomes obvious that differences in national and regional standards present serious technical barriers to engineering, economic, or trade efforts. The resulting cooperative efforts are aimed more at "harmonizing" (i.e., rationalizing and resolving similarities and differences) existing national/regional standards to create a mutually agreeable international standard rather than in creating new international standards without regard to existing standards. ISO is the most apparent example of international standardization efforts.

12.2.8.1 ISO

International Organization for Standardization is a worldwide federation of national standards bodies from some 130 nations. ISO is a nongovernmental organization with the mission of promoting the development of standardization and related activities in the world in order to facilitate the international exchange of goods and services. Just as in CEN, ISO exists because the nonharmonized standards can contribute to "technical barriers to trades."

The ISO is made up of three categories of members: national member bodies (e.g., ANSI in the United States), correspondent member and subscriber member. The first member category contains members who actively participate in the development of ISO standards. The second and third categories are for information purposes.

The technical work of ISO is highly decentralized, carried out in a hierarchy of some 2850 technical committees, subcommittees, and working groups. In these committees, qualified representatives of industry, research institutes, government authorities, consumer bodies, and international organizations from all over the world come together as equal partners in the resolution of global standardization problems. Some 30,000 experts participate in meetings each year. The major responsibility for administrating a standards committee is accepted by one of the national standards bodies that make up the ISO membership. The member body holding the secretariat of a standards committee normally appoints one or two persons to do the technical and administrative work. A committee chair assists committee members in reaching consensus. Generally, a consensus will mean that a particular solution to the problem at hand is the best possible one for international application at that time.

The Central Secretariat in Geneva acts to ensure the flow of documentation in all directions, to clarify technical points with secretariats and chairperson, and to ensure that the agreements approved by the technical committees are edited, printed, submitted as draft international standards to ISO member bodies for voting, and published. Meetings of technical committees and subcommittees are convened by the Central Secretariat, which coordinates all such meetings with the committee secretariats before setting the date and place. Figure 12.6 shows the organizational structure of ISO.

An international standard (IS) is the result of an agreement between the member bodies of ISO. It may be used as such, or it may be implemented through incorporation in national standards of different countries. ISs are developed by ISO technical committees (TC) and subcommittees (SC) by a six-step process:

Stage 1: Proposal stage
Stage 2: Preparatory stage [working draft (WD)]
Stage 3: Committee stage [committee draft (CD)]
Stage 4: Enquiry stage [draft international standard (DIS)]
Stage 5: Approval stage [final draft international standard (FDIS)]
Stage 6: Publication stage [international standard (IS)]

If a document with a certain maturity is available at the start of a standardization project, for example, a standard developed by another organization, it is possible to omit certain stages. In the so-called fast-track procedure, a document is

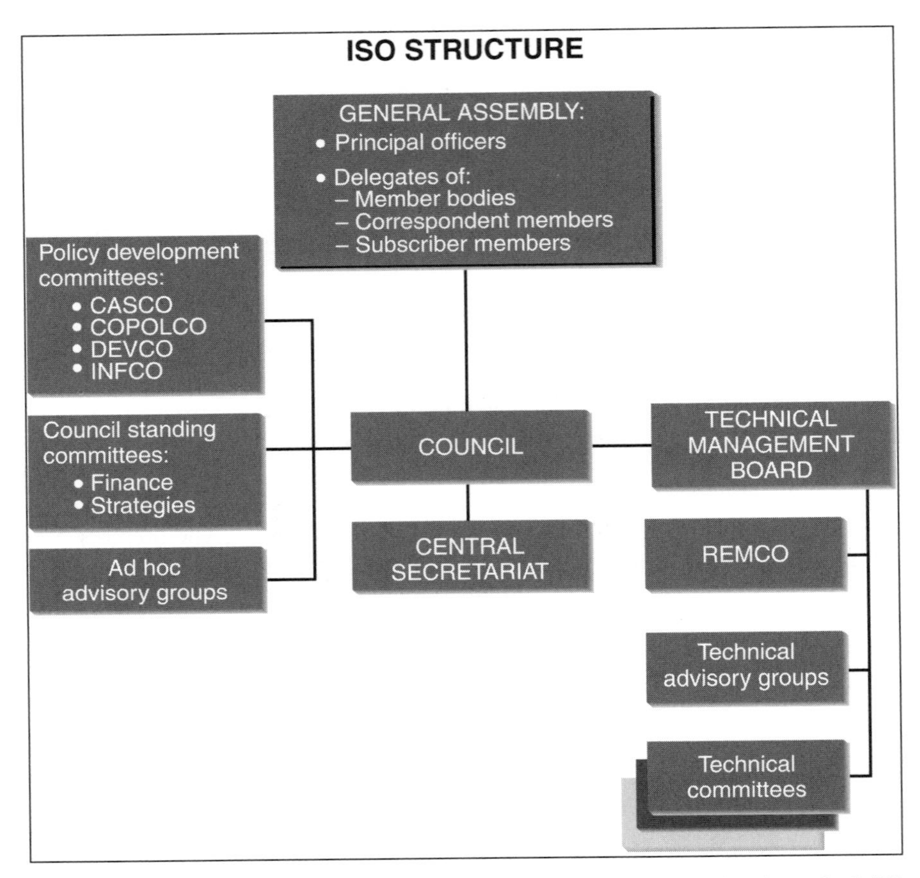

FIGURE 12.6 ISO administrative and technical structure, an international standard (IS).

submitted directly for approval as a DIS to the ISO member bodies (stage 4) or, if the document has been developed by an international standardizing body recognized by the ISO Council, as an FDIS (stage 5), without passing through the previous stages. ISs are reviewed at maximum intervals of 5 years by the relevant technical committee to determine whether they should be confirmed, revised, or withdrawn.

12.2.8.2 ISO Standards for Advanced Materials

As shown in Table 12.4 there are a limited number of technical committees specifically dealing with advanced materials, although work on advanced materials may take place within committees dealing with broader topics than just materials. ISO standards can be accessed via the national standards institute of each national member or the ISO *Catalogue* (either hard-copy form in printed books or in electronic form on compact disk) [14] or through the World Wide Web through ISO's web page [15].

TABLE 12.3 CEN Committees on Advanced Materials

Technical Committee (TC)	TC Category	Materials
TC 262 Protection of metallic materials against corrosion	Materials	Various
TC 249 Plastics	Materials	Polymers
TC 240 Thermal spraying and thermally sprayed coating related products	Materials	Various
TC 189 Geotextiles and geotextile-related products	Materials	Polymer, glass, etc.
TC 187 Refractory products and materials	Materials	Ceramics and high-temperature metals
TC 184 Advanced technical ceramics	Materials	Monolithic and composite advanced ceramics
TC 233 Biotechnology	Biotechnology	Biomaterials

12.3 DESIGN CODES AND DATABASES

As discussed in the introduction, in this discussion design codes are widely accepted but general rules for the construction of components or systems where safety is important. Often design codes, like standards, are developed as voluntary, consensus documents. Eventually widespread acceptability and utility of design codes may cause them to become legally binding in some circumstances.

Databases are often linked to design codes because engineers using the design codes need ready access to engineering materials that have been "qualified" for inclusion in the design code. This connection of databases and design codes does not always exist, and with the increasing popularity of relational, electronic database formats, stand-alone electronic databases for advanced materials are becoming more common.

In the following sections, national and international efforts at design codes and databases for advanced materials will be briefly discussed. Where appropriate illustrative examples will be included.

12.3.1 Codes

In the United States, numerous organizations have developed and maintain design codes. Some more common examples of these and their applications are contained in Table 12.5. For advanced materials, two of the more pertinent examples of design codes are the *ASME Boiler and Pressure Vessel Code* [4] (various types

TABLE 12.4 ISO Committees on Advanced Materials

Technical Committee (TC)	Materials
TC 77 Products in fiber-reinforced cement	Fiber-reinforced cement
TC 106 Dentistry	Biomaterials
TC 119 Powder metallurgy	Powder metals
TC 155 Nickel and nickel alloys	Nickel alloys
TC 166 Ceramic ware, glassware and glass ceramic ware in contact with food	Ceramics and glasses
TC 150 Implants for surgery	Biomaterials
TC 206 Fine ceramics	Monolithic and composite advanced ceramics

TABLE 12.5 Examples of Design Codes and Application in United States

Design Code	Application
Uniform Building Code (UBC)	Building, dwelling, etc.
Boiler and Pressure Vessel Code (ASME)	Pressure vessels and pressure equipment (fired and unfired)
Structural Welding Code (AWS)	Welding and weldments in metals
Uniform Fire Code (UFC)	Fire protection systems, insulation, etc.
Uniform Mechanical Code	HVAC, piping, etc.

of composites, ceramics, and high-temperature metals) and *Military Handbook 17 on Composite Materials* [16]. The following sections give brief overviews of these two efforts.

ASME Boiler and Pressure Vessel Code The *ASME Boiler and Pressure Vessel Code* is a good example of a design code where the material specifics of CFCCs must be incorporated before one of the potential applications (power generation) of CFCCs can be realized. As shown in Fig. 12.7, the code is divided into 11 sections, the 2 most important for CFCCs being Section I Power Boilers and Section II Materials, each detailed in the following discussion.

Section I dates to the adoption of the code in 1914 and is divided into eight subgroups each dealing with specific aspects of the care, piping, design, fabrication, and examination of power boilers. In addition, subgroups also deal with particular types of boilers, including firetube and electric. Generally, Section I applies to boilers with >103 kPa (15 psi) pressures external to the boiler itself

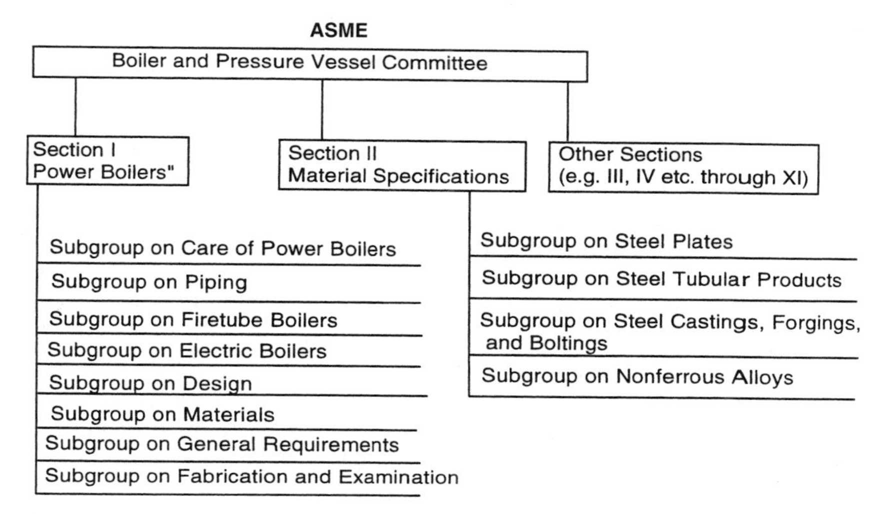

FIGURE 12.7 Organization of ASME Boiler and Pressure and Pressure Vessel Code.

and >1103 kPa (160 psi) pressures and/or >121°C(250°F) temperatures for high-temperature water boilers. The scope of jurisdiction of Section I is the boiler proper and the boiler external piping.

Of particular interest for advanced materials producers is the subgroup on materials in Section I, which limits materials to (1) those listed in Section II and (2) those listed in certain tables in Section I. In addition, certain materials are restricted in regard to usage. For example, an austenitic stainless steel might not be permitted for parts in water-wetted service but may be permitted for steam-touched service. Also included in Section I, are provisions for approval of new materials for code construction:

1. New materials shall have been previously adopted by ASTM (adoption of a new materials for the ASME code means adoption of the ASTM specification for the material).
2. Items to be furnished for evaluation by the code committee are:
   a. Such mechanical properties as ultimate tensile strength, yield strength, creep and rupture strength, heat treatment, toughness, etc.
   b. Stress–strain data for vessels designed for external pressure
   c. Weldability, including data for establishing the requirements of Section IX Welding
   d. Physical changes and resistance to effects of both elevated temperature and cryogenic temperature where applicable
   e. Availability of the material regarding patents and licensing

For design purposes, two approaches are prescribed in the code: design by rule and design by analysis. Section I relies primarily on design by rule, which is basically an empirical approach (what has proven to work successfully in the past) setting limits on: factors of safety (typically 4–5); design pressures and temperatures, minimum thicknesses [6.35 mm (0.25 in.)]; maximum pressures [no greater than 1.06 times the maximum allowable working pressure (use of safety valves required)]; loadings due to internal pressures (these set the minimum required thicknesses unless other loadings exceed 10% of the allowable working stress); thickness of cylindrical components under internal pressure as determined from formulas; openings and reinforcements; fatigue, fast-fracture, creep, and other failure mechanisms; and hydrostatic proof tests (1.5 times the maximum allowable working pressure). Note, that Section III Nuclear Power Plant Components uses the more sophisticated, but less historically based, design by analysis. The use of design by analysis in Section III is a precedent important for acceptance of CFCCs in the code since these materials may require recently developed reliability techniques employing computer algorithms for implementation in advanced designs.

Finally, Section I requires affixing a special code stamp and a proper nameplate to the components and resulting system, respectively, after complying with all the code requirements of design and construction. To document compliance, seven types of Manufacturers Data Report Forms must be completed.

Section II also has its roots in the original 1914 edition of the code. The original materials specifications were developed in a joint effort of ASME and ASTM for ferrous and nonferrous materials. In addition, joint specifications with the American Welding Society (AWS) have since been developed for welding rods, electrodes, and filler metals. Only those ASTM or AWS specifications required by ASME are addressed in Section II. The documentation of Section II is a four-part compendium of materials data: Part A—Ferrous Materials Specifications, Part B—Nonferrous Material Specifications, Part C—Specification for Welding Rods, Electrodes, and Filler Metals, and Part D—Properties. Part D lists material properties for all materials accepted by Sections I, III, and VIII. Not only are mechanical properties but also physical properties are contained in Part D. However, a major portion of the data contained in Part D are tables of stresses as functions of temperature.

Currently, certain advanced materials such as monolithic and composite advanced composites are not currently allowed materials in the ASME Boiler and Pressure Vessel Code. For a manufacturer wishing to implement CFCC materials in a code-certified power boiler design, several steps must be taken. First, the CFCC material of interest must receive a specification from ASTM, including materials analysis, physical properties, and mechanical properties. Second, the material must be accepted by the code as an allowable material. Third, the design-by-analysis approach used by the manufacturer must be shown to meet the minimum requirements of Section I. Fourth, the manufacturing process that includes the CFCC material must be certified as being acceptable under the code. Fifth, special nondestructive characterization procedures now being developed for CFCCs must be certified as under Section V Nondestructive Examination. Sixth, the system must be constructed using the approved methods and material, ultimately passing a pressurized proof test as per the code requirements before receiving official certification.

Military Handbook 17 *Mil-Hdbk-17* is the outgrowth of a collaborative effort on the part of industry (i.e., defense contractors) and the U.S. Department of Defense to clarify and codify issues involving the use of polymer matrix composites (PMCs) in advanced designs. *Mil-Hdbk-17* has been in existence in various forms since 1959 and has been relatively successful in creating common language, design philosophies, fabrication methods, maintenance approaches, and certification of advanced PMCs. Recent directives from the U.S. Department of Defense have established the format for developing offshoots of *Mil-Hdbk-17* for other advanced composites, namely, metal matrix composites (MMCs) in 1993 and ceramic matrix composites (CMCs) in 1996. The organization of the *Mil-Hdbk-17* effort for CMCs is shown in Fig. 12.8.

The vision of the *Mil-Hdbk-17* effort for CMCs is as follows: *Mil-Hdbk-17* is the primary and authoritative source for characterization, statistically based property, and performance data of current and emerging ceramic matrix composites. It reflects the best available data and methodologies for characterization, testing, analysis and design and usage guidelines in support of design methodologies for composites.

The objectives are:

- Development of a framework for the future, successful use of CMCs.
- Provide guidance to industry for the collection of statistically meaningful critical data that designers need to utilize CMCs.

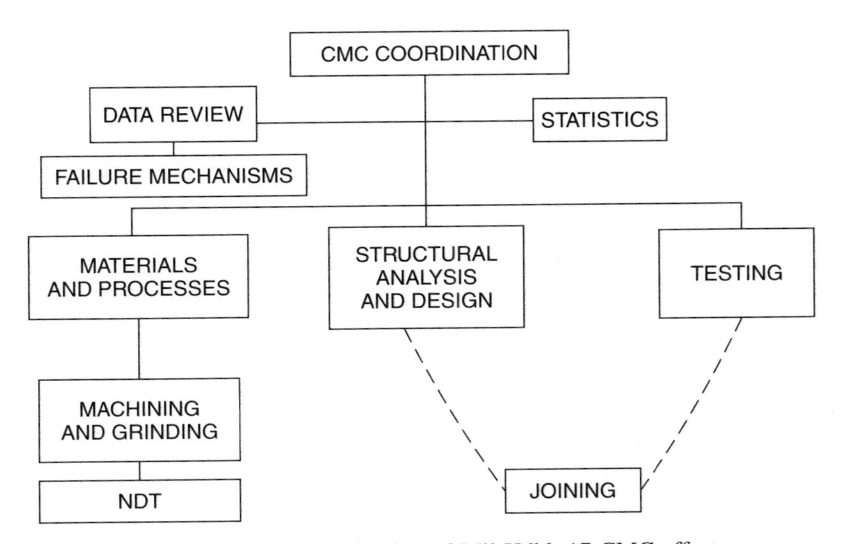

FIGURE 12.8 Organization of Mil-Hdbk-17 CMC effort.

- Based on the requirements from the design community. Identify appropriate properties and broadly accepted testing procedures—including consideration of the designation of the precision level and prioritization of properties required.
- Provide guidelines and recommendations for the characterization, testing, design, and utilization of CMC materials and structures.
- Provide the primary and authoritative sources for characterization, property, and performance data of current and emerging CMC systems.
- Provide recommendations for the statistical analysis of materials data and structures relativity.

Data Review Any data generated for inclusion in *Mil-Hdbk-17* must satisfy requirements for confidence bounds and statistical sample size within a batch and batch to batch. Because of the need to establish the CMC portion of the *Mil-Hdbk-17* as quickly as possible, database generation has been simultaneous with establishing design rules and guidelines. Note also in Fig. 12.8 that while the testing activity is separate, it can actually be thought of as performing a service role to the other activities. Moreover, the stated mission of the testing activity is not to develop standards (this is best left to the standards-writing bodies already discussed), but instead to identify and recommend those existing standards that are appropriate. Where appropriate standards do not exist, the testing activity will assist standards-writing bodies in the development of the standards.

12.3.2 Databases

Data bases can either stand separately or can be integrated into design codes. Separate databases for materials have existed for some time in book form and are often compilations of test results reported in the literature (e.g., qualified databases must be incorporated into design codes if materials are included in the requirements of the code (e.g. *ASME Boiler and Pressure Vessel Code* and *Mil-Hdbk-17*).

Conventional databases are often organized according to material system (e.g., ceramics, composites, nickel-based superalloy, etc.). The data is then often organized according to type of property such as thermal, mechanical, or physical. Finally, information is often summarized as a mean or range of values. Less frequently the standard deviation and number of tests is provided. Even less frequently the individual data points are provided. Usually a reference is provided for the original source of the information. Unless the information is being used to "qualify" a material, the information is seldom screened to determine if the tests were conducted in accordance with accepted test methods and practices.

Three predominant formats for databases have emerged in recent years: hard copy (usually book form); electronic–local access (compact disk, floppy disk, or magnetic tape) and electronic–remote access (Internet-based). Each of these formats has its advantages and disadvantages as discussed in the following sections.

An example of hard-copy database is *The Materials Selector*, which covers many advanced materials and is available in a multivolume set [17]. Hard-copy databases provide a permanent record of data, especially if it is qualified (e.g., recognized test methods were used to generate statistically significant data). Unfortunately, the time required to collect, organize, and publish the information makes the some of the data obsolete even before the information is published. In addition hard-copy databases are difficult to search, and any relational aspects must be manually implemented.

There are several examples of electronic–local access databases [18–20], which are increasingly available on stand-alone, searchable compact or floppy disk formats [19, 20]. Some electronic databases [18] can be accessed by engineering analysis software such as finite-element analysis (FEA) or computer-aided design (CAD) software packages to add even more usability to the database. Flexibility, searchability, and relational aspects are all advantages of the electronic–local access database. Unfortunately, unless some provision is added to provide periodic updates to the information, these databases suffer the same obsolescence drawback as the hard-copy database.

Several examples of electronic–remote access data bases [21, 22] reveal an exciting tread in information management. Web-based electronic databases can be accessed and searched using a variety of criteria. With the proper interface, even engineering analysis software such as FEA or CAD software packages can access these online databases. Flexibility, searchability, and relational aspects are all advantages of the electronic–remote access database. The final advantage is that the information contained in online databases can be constantly update as required. A possible drawback is that Internet access is required, but this requirement is usually easily met.

Several key aspects must be present in a successful database. Among these are completeness (recently over 130 separate items were identified for producing a qualified database for a ceramic composite in the Mil-Hdbk-17 effort), usability (the interface must be user-friendly and "intuitive" including the relational search capabilities), and up-to-date (data entry must be simple and self-guided). It is interesting to note that ASTM considers electronic databases so critical that a committee has been created (Committee E49 Computerized Systems and Chemical and Material Information) to aid not only in establishing standards for electronic databases, but also to assist other ASTM committees in reporting results in ways amenable to (and consistent with) the efforts of Committee E49. An example of a complete data set for an electronic database is shown in Fig. 12.9 where both quantified and graphical information are contained in the data sets. An example of an electronic–local access database interface is shown in Fig. 12.10.

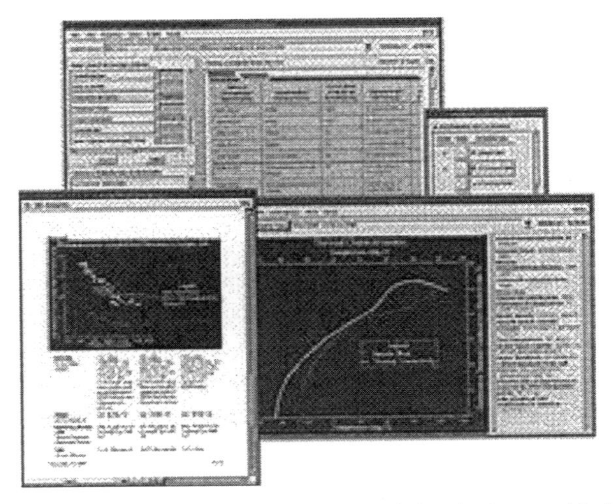

FIGURE 12.9 Complete electronic database containing both quantified and graphical information [19] (www.macsch.com/products/mvision).

Material Data

Fiber Matrix Material Adhesive Neat Resin Core

Type:

Subtype:

Id	Product Name	Type	Fiber	Matrix	Core	Layer
[illegible]	Glass/Vinyl ester (ASDV)	Laminate	WR 24	510A		Laminat
212	Glass/Vinyl ester (ASDV)	Laminate	WR 24	8084		Laminat
213	Glass/Epoxy (ASDV)	Laminate	WR 24	Tactix 1		Laminat
214	Glass/Polyester (ASDV)	Laminate	WR 24	8472		Laminat
215	Glass/Vinyl ester (ASDV)	Laminate	WR 24	8520		Laminat
217	Glass/Vinyl ester (ASDV)	Laminate	WR 24	8510		Laminat
218	Glass/Vinyl ester (ASDV)	Laminate	10oz Twill	8084		Laminat
219	Glass/Vinyl ester (ASDV)	Laminate	24oz Twill	8084		Laminat
220	Glass/Vinyl ester (ASDV)	Laminate	Chomarat	8084		Laminat
221	Glass/Vinyl ester (ASDV)	Laminate	DF1400	8084		Laminat
222	Glass/Vinyl ester (ASDV)	Laminate	7781	8084		Laminat
223	Glass/Vinyl ester (ASDV)	Laminate	Biaxial	8084		Laminat
224	Carbon/Epoxy (ASDV)	Laminate	1059(AS4	9405		Laminat

Record: 1 of 62

Approve Record

Refresh

Property Data

View Record

Edit Record

Delete Record

Print Record

Export Record

FIGURE 12.10 User interface for material database system (DBS) [20] www.trl.com/mds.

12.4 SUMMARY AND CONCLUSIONS

In this chapter, a broad overview of standards, design codes, and databases for advanced materials was provided. There are many standards bodies—nationally, regionally, and internationally—all producing standards (i.e., test methods, rules, and terms of information transfer) for advanced materials depending on industry demands or ad hoc programmes. Building on the work of the standards bodies are the design code writers who must incorporate standards and databases into the "general rules for design" while still fulfilling the requirements. Where infrastructures for creating standards and design codes for various materials (including advanced materials) are well established, no such infrastructure exists for databases. Databases are still being developed so as to be usable and consistent with modern engineering analysis tools. Development of useful databases requires effort both up and down stream of data collection. However, the current electronic–remote access databases show promise in being complete, usable, and updatable, not only for advanced materials but all engineering materials.

REFERENCES

1. D. C. Phillips and R. W Davidge, *Br. Ceram. Trans. J.*, **85**, 123–130 (1986).
2. S. J. Schneider and D. R. Bradley, *Adv. Ceram Mat.*, **3** (5) 442–449 (1988).
3. J. R. Farr, "The ASME Boiler and Pressure Vessel Code: Overview," in *Pressure Vessel Codes and Standards*, Elsevier Applied Science, New York, 1983, pp. 1–34.
4. American Society of Mechanical Engineers, ASME Boiler and Pressure Vessel Code, ASME, New York, 1983.
5. G. E. Dieter, *Engineering Design: A Materials and Processing Approach*, McGraw-Hill, New York, 1991.
6. American Society for Testing and Materials, Annual Book of ASTM Standards, ASTM, West Conshohoken, PA, 1999.
7. American Society for Testing and Materials www.astm.org, ASTM, West Conshohoken, PA, 1999.
8. ASTM, C1275, "Standard Test Method for Monotonic Tensile Strength Testing of Continuous Fibre-Reinforced Advanced Ceramics with Solid Rectangular Cross-Section Specimens at Ambient Temperatures," in *Annual Book of ASTM Standards*, V 15.01, ASTM, W. Conshohocken, PA, 1999.
9. Japanese Industrial Standards Committee, www.jisc.org, JISC, Tokyo, Japan, 1999.
10. Japanese Industrial Standards Committee, *Book of Standards*, JISC, Tokyo, Japan, 1999.
11. ASME Position Paper, ASHE, New York, 1998.
12. Austrian Standards Institute, "Catalogue of European Standards," Austrian Standards Institute (ON), Vienna, Austria, 1999.
13. Committee for European Standardization (CEN), www.cenorm.be, CEN, Brussels, Belgium, 1999.
14. International Organization for Standardization (ISO), *ISO Catalogue*, International Organization for Standardization (ISO), Geneva, Switzerland, 1999.
15. International Organization for Standardization (ISO), www.iso.ch, ISO, Geneva, Switzerland, 1999.
16. U.S. Department of Defense, Ceramic Matrix Composites Coordination Group for *Mil-Hdbk-17 Composite Materials Handbook*, U.S. Army Research Laboratory, Aberdeen Proving Ground, MD, 1999.

17. M. Waterman, *The Materials Selector*, CRC Press, Boca Raton, FL, 1996.
18. MacNeal-Schwendler Corp., MSC/Mvision, MacNeal-Schwendler Corp., Los Angeles, California, 1999.
19. Granta Design, Cambridge Materials Selection, Granta Design, Cambridge, UK, 1999.
20. Touchstone Research Laboratory, The Material Data Base System (DBS), Touchstone Research Laboratory, Triadelphia, WV, 1999.
21. MatWeb, www.matls.com, MatWeb, Christiansburg, VA, 1999.
22. MEMS Clearinghouse, http://mems.isi.edu/mems/materials/index.html, University of Southern California, Information Sciences Institute, Los Angeles, 1999.

CHAPTER 13

Nondestructive Evaluation of Structural Ceramics

WILLIAM A. ELLINGSON and CHRIS DEEMER

Energy Technology Division, Argonne National Laboratory, 9700 South Cass Avenue, Argonne, Illinois 60439

Handbook of Advanced Materials Edited by James K. Wessel
ISBN 0-471-45475-3

13.1 INTRODUCTION

Structural ceramics may be classified in many ways, but for this chapter structural ceramics will be limited to (1) monolithic materials developed for application in power generation systems, for example, gas turbines and diesels engines [1–3], and certain biomedical applications, for example, balls for artificial hips [4–7], and (2) ceramic matrix composites (CMC) being developed for power generation systems, for example, gas turbines and for industrial applications such as hot-gas filters in advanced coal-fired plants or circulating fans in very aggressive environmental conditions [8]. In most engineering applications where warranty issues play an important role, nondestructive evaluation (NDE) methods need to be in place to assess the condition of components at scheduled or unscheduled opportunities. For ceramics to be economically competitive, processing costs must be kept low while at the same time assure the user of a reliable product. One way to reduce costs is to improve yield. Yield can be improved through rejection of parts in the early processing steps prior to sintering, machining, and proof testing. There are several steps in the processing of most ceramic materials. Table 13.1 lists the steps often used to produce monolithic structural ceramics. These several processing steps can introduce "flaws" in the materials, not all failure causing. The critical flaw size is dependent upon the fracture toughness. Table 13.2 shows the wide range of material fracture toughness values of various ceramic materials, both monolithics and composites. Table 13.2 also notes the critical flaw sizes for many of these materials. It is the relatively small flaw sizes, 10–100 μm, in monolithic materials that have provided challenges to the NDE community. Ceramic composites with higher fracture toughness have critical flaw sizes that are much larger and the failure modes are completely different than for monolithics. Another important aspect about ceramic materials, from an NDE standpoint, is that the design methodology is based on probability statistics [9, 10]. This means that there is going to be variability in the base material itself.

13.2 BRIEF HISTORICAL PERSPECTIVE

Over the past 30 years there have been great strides made in the development of NDE technologies that can be used to characterize ceramic materials. These

TABLE 13.1 Common Processing Operations for Advanced Ceramics

Operation	Method	Examples
Powder preparation	Synthesis	SiC
	Sizing	Si_3N_4
	Granulating	ZrO_2
	Blending	
	Solution chemistry	Glasses
Forming	Slip casting	Combustors, stators
	Dry pressing	Cutting tools
	Extrusion	Tubing, honeycomb
	Injection-molding	Turbocharger rotors
	Tape casting	Capacitors
	Melting/casting	Glass ceramics
Densification	Sintering	Al_2O_3
	Reaction bonding	Si_3N_4
	Hot pressing	Si_3N_4, SiC, BN
	Hot isostatic pressing	Si_3N_4, SiC
Finishing	Mechanical	Diamond grinding
	Chemical	Etching
	Radiation	Laser, electron beam
	Electric	Electric discharge

Source: Office of Technology Assessment.

developments have been the focus of recent topical conferences [11, 12], and the reader is referred to these excellent sources. It will not be the intent of this chapter to thoroughly review the historical developments of the several NDE technologies, but it is instructive to look at some of the primary developments and closely related technologies, such as high-speed, large-capacity desk-top computers. Figure 13.1 shows schematically one historical perspective on the recent development of ceramic materials together with one perspective on parallel developments in NDE technology. To be noted is the fact that composite ceramic materials really were not available until very recently—after 1990. It is essential to highlight the personal impact computers have had on the development of NDE because this has allowed developments related to sensors, digital signal processing, and image processing.

Prior to the early 1970s, there was little worldwide development activity in the area of structural ceramics and certainly very little research directed toward NDE for ceramics. A great deal of the very earliest efforts directed toward NDE developments for ceramic materials for gas turbines occurred in the late 1970s and the early 1980s in Germany [13]. Efforts began in the late 1970s in the United States [14], and subsequently in the early 1980s efforts began in Japan [15]. It is interesting to note that in the summary of their 1983 report, Goebbels and Reiter [13] noted the following, which was directed toward monolithic materials:

TABLE 13.2 Fracture Toughness and Critical Flaw Sizes of Monolithic and Composite Ceramic Materials Compared with Metals[a]

Material	Fracture Toughness ($MPa \cdot m^{1/2}$)	Critical Flaw Size (μm)
Conventional microstructure		
Al_2O_3	3.5–4.0	25–33
Sintered SiC	3.0–3.5	18–25
Fibrous or interlocked microstructure		
Hot pressed Si_3N_4	4.0–6.0	33–74
Sintered Si_3N_4	4.0–6.0	33–74
SiAlON	4.0–6.0	33–74
Particulate dispersions		
Al_2O_5-TiC	4.2–4.5	36–41
SiC-TiB_2		
Si_3N_4-TiC	4.5	41
Transformation toughening		
ZrO_2-MgO	9–12	165–294
ZrO_2-Y_2O_3	6–9	74–165
Al_2O_3-ZrO_2	6.5–15	86–459
Whisker dispersions		
Al_2O_3-SiC	8–10	131–204
Fiber reinforcement[b]		
SiC in borosilicate glass	15–25	
SiC in LAS	15–25	
SiC in CVD SiC	8–15	
Aluminum[c]	33–44	
Steel[c]	44–66	

[a] Assumes a stress of 700 MPa (~100,000 psi). Al_2O_3, alumina; LAS, lithium aluminosilicate; CVD, chemical vapor deposition.
[b] The strength of these composites is independent of preexisting flaw size.
[c] The toughness of some alloys can be much higher.

"Further development is needed in high-resolution surface inspection (dye penetrant, photo-acoustic microscopy, surface acoustic microscopy) together with further developments in filmless radiography, resonant frequency methods, modal analysis of bulk waves and acoustic emission." These projections were correct and have been addressed over these past 15 plus years. Surface crack detection for monolithic materials has been largely accomplished through use of high-magnification microscopes with fluorescent penetrants, while internal features have been largely addressed through very significant advances in filmless high-resolution X-ray computed tomography [16–18]. These advances in NDE technology together with advances in higher fracture toughness ceramics have allowed structural monolithic ceramics to begin to be utilized in large quantities.

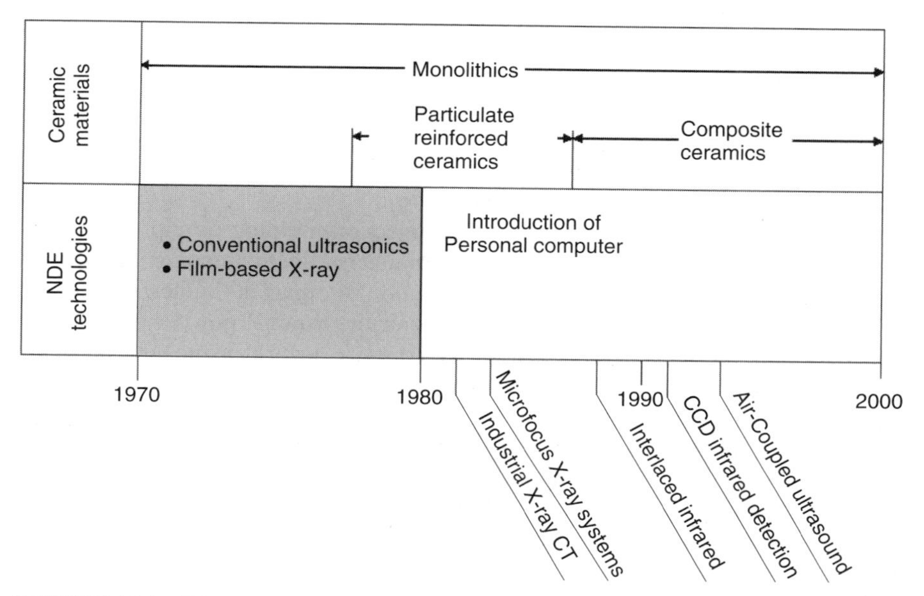

FIGURE 13.1 Diagram showing historical perspective on NDE technology development for structural ceramics.

13.3 CHARACTERISTICS OF INTEREST TO MONOLITHIC CERAMICS

There are several characteristics of monolithic ceramic materials that influence the properties and failure probability of as-produced components. These include uniformity of density, presence of internal and surface flaws, and regions of porosity. Common processing steps were listed in Table 13.1. Each step in the process has the potential to contribute to the introduction of a flaw. NDE technologies that can be applied in the different steps are discussed in the following sections. References are provided to allow the reader more complete information.

13.3.1 Green-State Information

13.3.1.1 Powders

The presence of impurities such as Fe and WC in the starting powders have been shown to be detrimental to ceramics and have been traced to ball milling and powder transfer operations [19, 20]. If present, these high-density particles can be carried on throughout the processing stages and become failure-causing flaws. Various NDE methods can be used to characterize these powders [20], but use of microfocus X-ray imaging [21, 22] has been shown to be able to quickly detect high-density impurities. It is important for radiographic imaging of these powders that small focal spot, microfocus, X-ray sources are used. These microfocus X-ray sources are important because they allow large magnification to be employed. Coupling a microfocus X-ray source to near real-time detectors allows this method to be implemented without high cost.

13.3.1.2 Injection-Molding/Slip Casting

High-volume production methods such as injection-molding or slip casting [23] are well established, and several casting parameters effect component uniformity. In the injection-molding process, the distribution of organics (binder, plasticizers, and mold-release agents) used as the carrier for the ceramic powder is important. The distribution effects homogeneity, green density, local densification rates, and mechanical properties. Yeh et al. [24] have listed typical defects and causes in injection-molded parts; see Table 13.3.

Two NDE methods have shown promise in determining the distribution of organics. These are X-ray computed tomographic imaging [16–18] and nuclear magnetic resonance (NMR) imaging [25, 26]. X-ray computed tomographic imaging with its inherent high sensitivity to small variations in density has been used to determine variations in organics. In a set of experiments, multicomponent polymer binder content was varied by using inserts in a cylinder. By examining the gray-scale values of the X-ray images and relating this to the organic content, the relation shown in Fig. 13.2 was established. Note that the linear relation holds regardless of whether the organic content was changed for either cold-pressed or injection-molding composition. Imaging solids by NMR presents problems. The main issue is the NMR signal to be detected decays rapidly (i.e., short T2) and requires special imaging methods. Special NMR instrumentation has been developed that combines high gradient field strengths (ca. 10 G/cm gradients) and short gradient switching times. Further details of this instrumentation are given by Ellingson et al. [25] and Carduner et al. [26].

TABLE 13.3 Injection-Molding Defects and Causes

Type of Defect	Cause
Incomplete part	Improper feed material
	Poor tool design
	Improper material and/or tool temperature
Large pores	Entrapped air
	Improper material flow and consolidation during injection
	Agglomerates
	Large pockets of organic binder-plasticizer due to incomplete mixing
Knit lines	Improper tool design or feed material
	Incorrect temperatures
Cracks	Stocking during removal from tool
	Improper tool design
	Improper extraction of binder/plasticizer

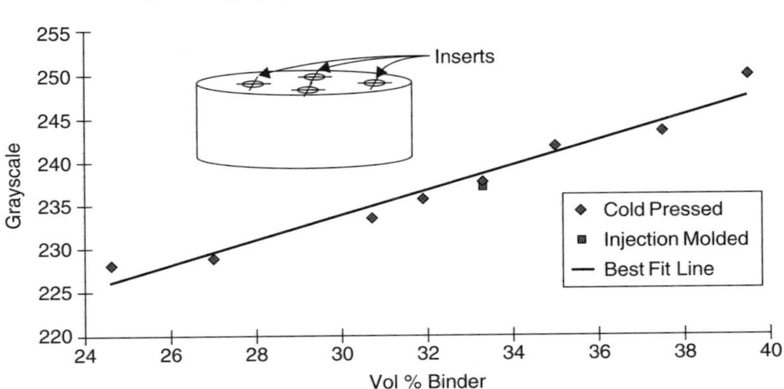

FIGURE 13.2 Determination of organic binder content in green-state ceramics by X-ray computed topographic imaging.

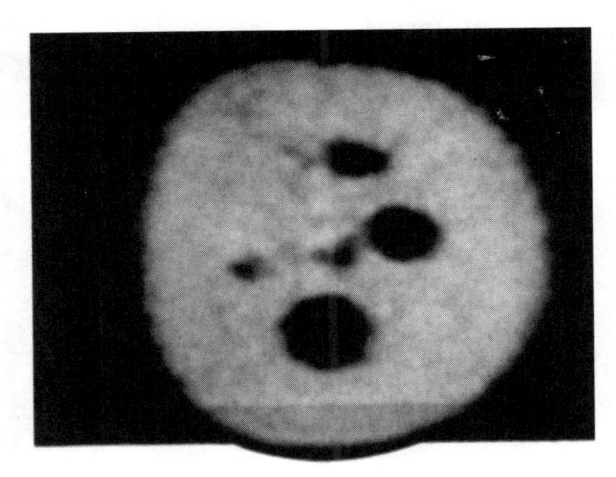

FIGURE 13.3 NMR image of cross section of 25-mm-diameter green-state (15 wt% binder) cold-pressed Si_3N_4 sample with intentional holes. Two holes at 1.1 mm, one at 2.2 mm, one at 3.2 mm, and one at 4.8 mm. Data acquisition time was 29 min.

Figure 13.3 shows a transaxial NMR image of a green, Si_3N_4 compact disk with 15 wt % organic material. In this sample, several holes were drilled, with diameters from 1.1 to 4.8 mm. Spatial resolution in the image is approximately 640 μm in either direction and the time to acquire the imaging data was 29.3 min.

It is possible to achieve images with higher spatial resolution and to observe lower concentrations of NMR-sensitive nuclei but with a significant increase in imaging time.

In the slip-casting process, the location of the cast surface/slip interface and the efficacy of molds depend on the amount of solids that fill the open pores as well as other parameters [26]. These features can be measured using magnetic resonance imaging (MRI) NDE methods because most slips are water based (hence an abundance of ^1H protons).

Hayashi and Kawashima [27] examined an Al_2O_3 slip in a plaster of paris cylindrical mold [20 mm outer diameter (OD) and 10 mm inner diameter (ID)]. The water content of the slip was 30 wt % and the experiment consisted of observing the cross section of the mold at different time intervals.

Figure 13.4 is a sequence of cross-sectional ^1H NMR images obtained as a function of time after the slip was poured into the mold. In each image, three

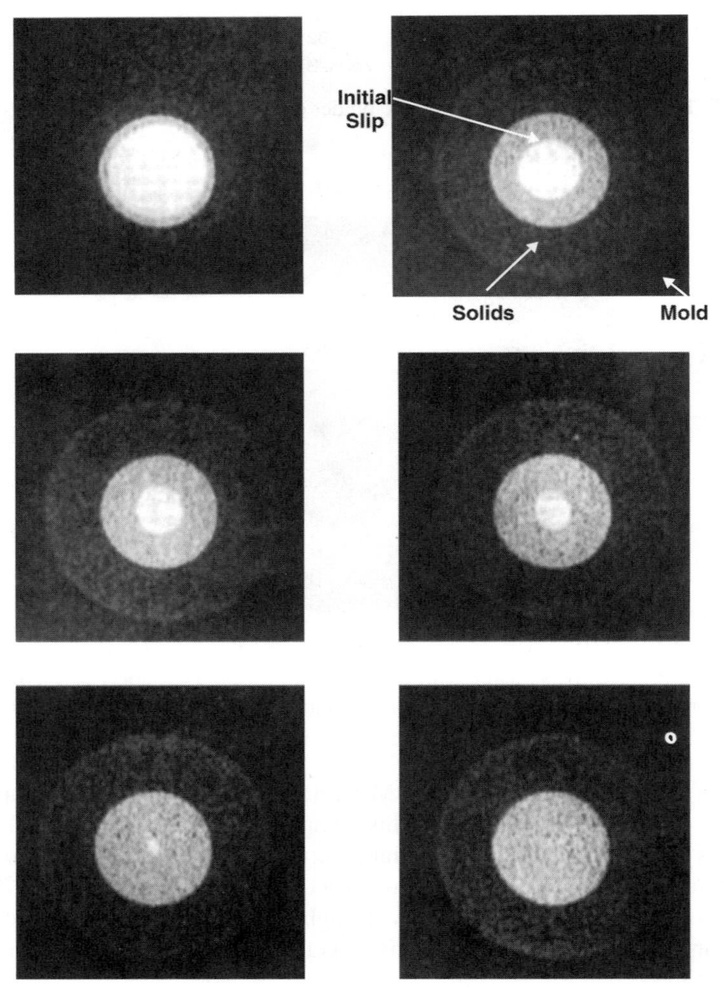

FIGURE 13.4 Sequence of time-dependent NMR images of slip-casting process of Al_2O_3 [27].

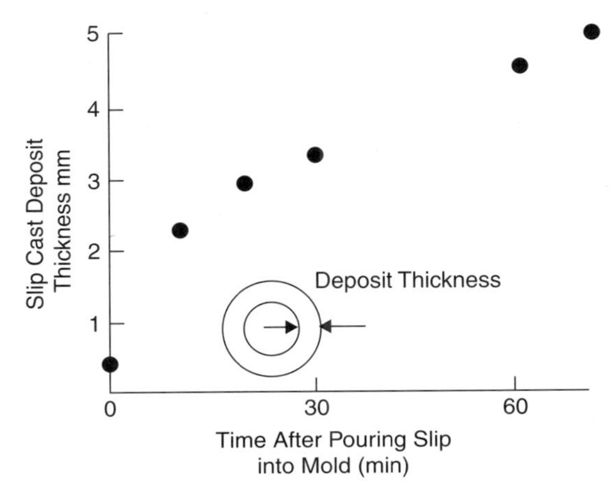

FIGURE 13.5 Rate of deposit buildup from Al_2O_3 slip in mold using NMR image data [27].

distinct rings of different "brightness" are apparent. The bright region in the center represents the slip, the next ring is the solids buildup, and the last ring is the mold itself. The mold is observable because water is migrating through it.

By measuring the thickness of the solids buildup as a function of time, one can obtain the rate of solids buildup. A typical result is illustrated in Fig. 13.5.

13.3.1.3 Porosity

Although mercury porosimetry has long been used to obtain porosity measurements [23], it has two limitations: (1) only small (<1 cm^3) specimens can be used, and (2) it does not yield spatial distribution information. Three-dimensional data for samples with connected porosity can be obtained with NMR imaging. This method provides information on specimens of any size and shape. To use NMR imaging methods special filler fluids with NMR-sensitive nuclei must be used, and adequate penetration of filler fluid must be obtained into all the internal volumes of interest. Ellingson et al. [25] have described one example of using MRI to measure bulk porosity on a set of partially sintered Al_2O_3 disks. The disks were each 25 mm in diameter and had densities of 1.640, 1.703, and 1.720 g/cm^3. The filler fluid used was benzene with a proton molarity of 67.3 *M* compared to water with 111 *M*. Figure 13.6 shows the relation between NMR image data for bulk porosity and density of an Al_2O_3 partially sintered compact. Discrimination of the porosity in the interior and at the edges of the disks was achieved. The higher porosity at the edges was expected based on knowledge of powder compaction.

13.3.2 Internal Defects

The most highly developed NDE method for detecting internal flaws such as voids, inclusions, or sharp density gradients is X-ray computed tomographic

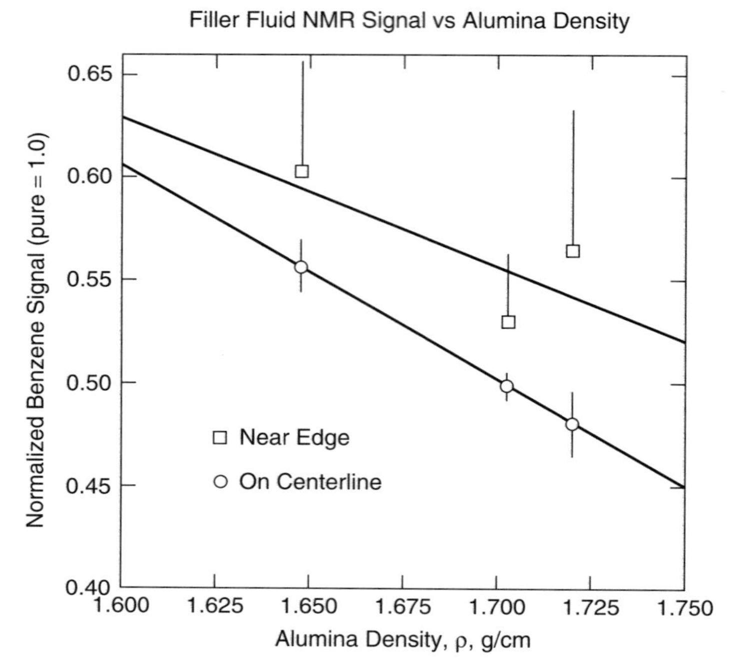

FIGURE 13.6 Correlation of NMR image data with porosity for Al_2O_3 cold-pressed specimen.

(XCT) imaging [16–18]. XCT imaging is a highly developed technology that has seen rapid developments over the past several years primarily through advances in more effective X-ray detectors and higher speed computers making image generation faster. Figure 13.7 is a schematic diagram of a typical XCT system. What is desirable about image data from XCT systems is the relative insensitivity to part shape. While conventional transmission X-ray imaging using film is a method in reasonably common practice [22, 28], reliable use of film radiography for detecting flaws in complex shaped components such as blades and vanes from a gas turbine is very difficult. Solid-state X-ray detectors [29] and fast computers have been shown to allow data to be acquired for an entire three-dimensional X-ray image in less than 6 min and images generated in less than 1 s. Figure 13.8 shows a 23-cm diameter radial flow monolithic ceramic rotor and one X-ray tomographic image from fast data acquisition with one slice reconstructed in less than 1 s. Features as small as 200 μm have been detected using such XCT systems.

13.3.3 Surface-Breaking Cracks and Machining Damage

There are two types of surface damage to be detected. Surface-breaking cracks and near-subsurface damage such as that induced from machining.

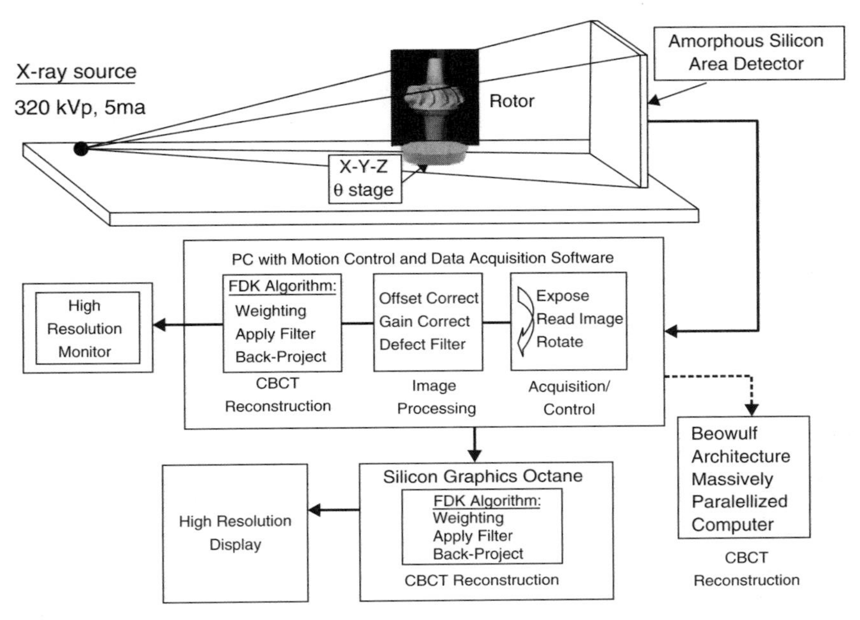

FIGURE 13.7 Schematic diagram of typical X-ray computed tomography imaging system used to characterize structural ceramics.

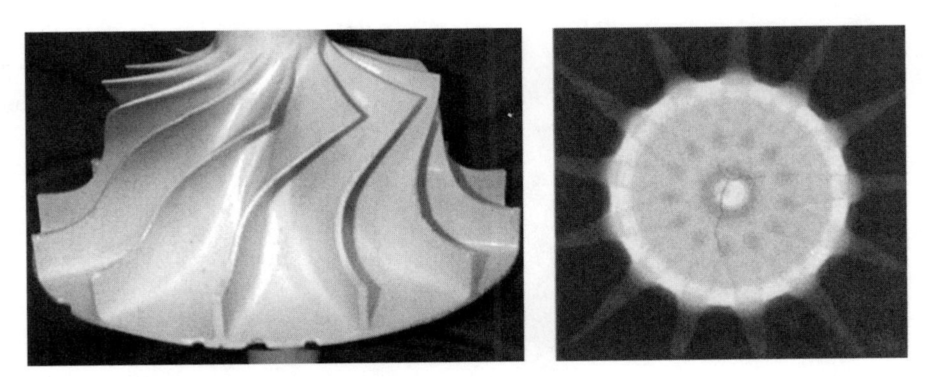

FIGURE 13.8 Large-diameter (23 cm) gelcast turbine rotor.

13.3.3.1 Surface-Breaking Cracks

Surface-breaking cracks have been shown to be detected by two NDE methods: fluorescent dye penetrants and elastic optical scatter based on a modified reflectometry setup. There are two penetrant methods: visible and fluorescent penetrants. For surface-breaking cracks, conventional visible penetrants have been shown to be unsatisfactory because the surface tension does not allow penetration into the very tight cracks associated with modern structural ceramics [30, 31]. Several investigators [30, 31] have demonstrated detection of tight surface-breaking cracks in Si_3N_4 structural ceramics using fluorescent methods. These results suggest that observation/detection of the cracks can be observed by using a 40–60X microscope equipped with an ultraviolet light source. It has been suggested that for best detection one should use an ultraviolet light intensity of about 1500 $\mu W/cm^2$ [31]. Crack detection sensitivity has been verified using Vickers indents with loads from 5 to 20 kg in several SiC and Si_3N_4 materials [30]. Fluorescent penetrant inspection was used to determine the length of the cracks across the Vickers indents. Figure 13.9 shows a comparison between what is seen in ordinary light compared to that seen with fluorescent for a typical 10-kg indent on a Si_3N_4 material. Examples of applying this penetrant method to a Si_3N_4 turbine blade are shown in Fig. 13.10.

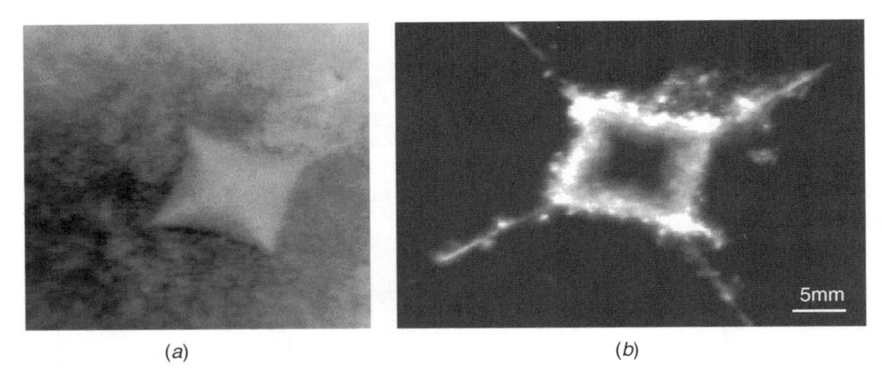

FIGURE 13.9 Example of use of fluorescent penetrant to detect cracks in Si_3N_4 from 10-kg Knoop indent: (*a*) no penetrant and (*b*) with penetrant.

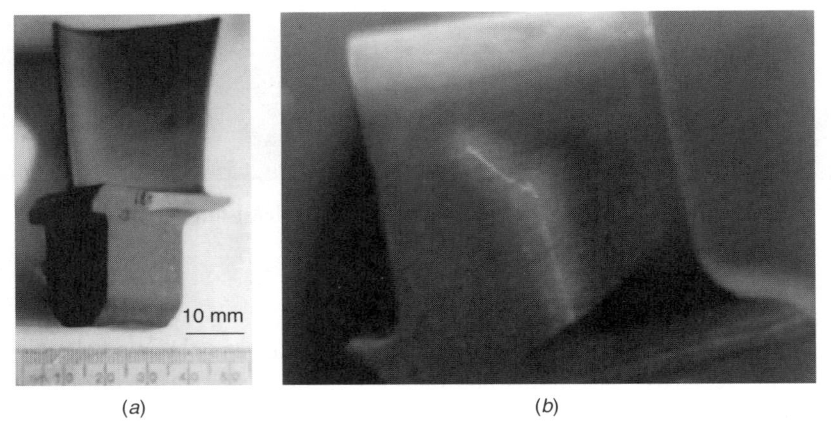

FIGURE 13.10 Example of fluorescent penetrant to crack detection on HIPped Si_3N_4: (*a*) photograph of turbine blade and (*b*) penetrant image showing crack.

13.3.3.2 Damage from Machining

Machining of ceramic materials is almost always necessary at some point in the manufacture of high-performance structural ceramics. It will not be the purpose of this brief section to review ceramic machining; rather the reader is referred to several excellent references on this topic [32–34]. While many methods are under development for machining of ceramics at various stages in the green state, significant machining in the sintered state is still being done. In sintered-state grinding, a grinding wheel is in contact with the workpiece, and the grit in the grinding wheel can introduce various damage levels in the material. This is shown schematically in Fig. 13.11. The grinding wheel contains embedded grit, usually diamond, and these grit particles can induce various types of damage. Of particular interest are: (a) the radial cracks that form normal to the tool-path direction, (b) lateral cracks that form below the surface and are parallel to the material surface, and (c) the median crack that forms to a greater depth and often follows the tool path mark.

From the point of view of NDE, it is the detection of the deep median crack and the subsurface lateral crack that becomes of great concern. One method that has shown promise is the polarized laser scatter method [35, 36]. In a recent series of tests, Si_3N_4 ceramic materials were aggressively machined using different types of grinding wheels, different grit sizes on the grinding wheels, different feed rates (or material removal rates) and other parameters [36]. These were studied by the laser scatter method. Figure 13.12 shows a diagram of the samples and typical back scatter laser image data. Difference in the back scatter image data can be seen by observing Fig. 13.12*c*. This figure shows two back scatter images from two different grinding parameters: material removal rate using all other parameters constant. Clearly the main characteristic difference between these scan "images" is the presence of the varying amounts of "black" dots. By examining

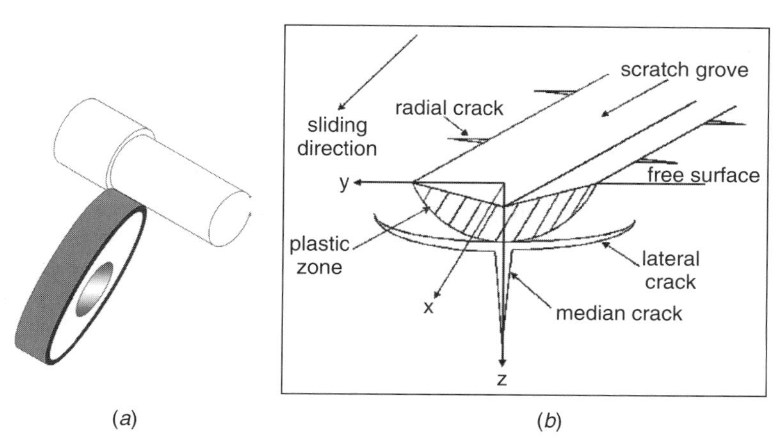

FIGURE 13.11 Schematic diagram of one grinding setup and descriptions of types of induced damage: (*a*) grinding wheel on test piece, and (*b*) diagram of typical damage types of brittle materials.

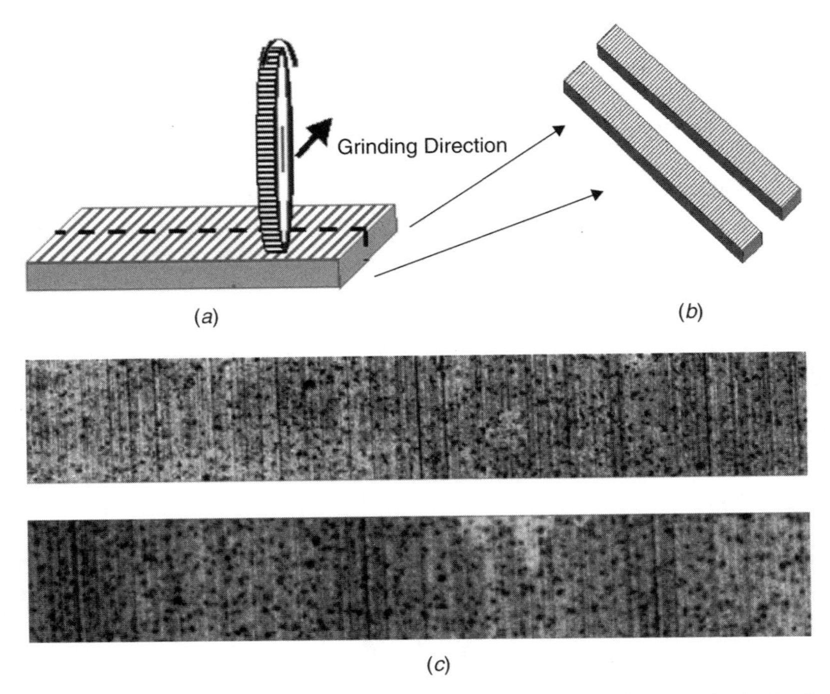

FIGURE 13.12 Machining samples used to study back scatter laser methods for detecting machining damage: (*a*) diagram of machining process, (*b*) samples sectioned, and (*c*) typical back scatter images of surfaces.

the laser data as a two-dimensional array of gray-scale values and calculating the coefficient of variation, C_ν, one can then plot C_ν versus the strength of the material. In this test four-point bend strength was used. Figure 13.13 shows the correlation between the laser scatter data, C_ν, and the four-point bend strength for two Si_3N_4 materials.

13.3.4 Creep

Creep is a concern in monolithic ceramics when used in high-temperature applications. Detection of creep is an active area of study for NDE. Over the past few years two NDE/C methods have been investigated to detect creep: One method uses backscattered laser light and one uses ultrasonic methods [35, 37].

13.3.4.1 Elastic Optical Scattering

Use of backscattered laser light for detection of creep is under development [35]. This method makes use of low-power laser-based elastic optical back scatter from the surface and near subsurface of structural ceramic materials. This technique is similar to classical ultrasonic technology, only rather than sound waves incident on the specimen, incident laser light probes the material's depth. As with C-scan

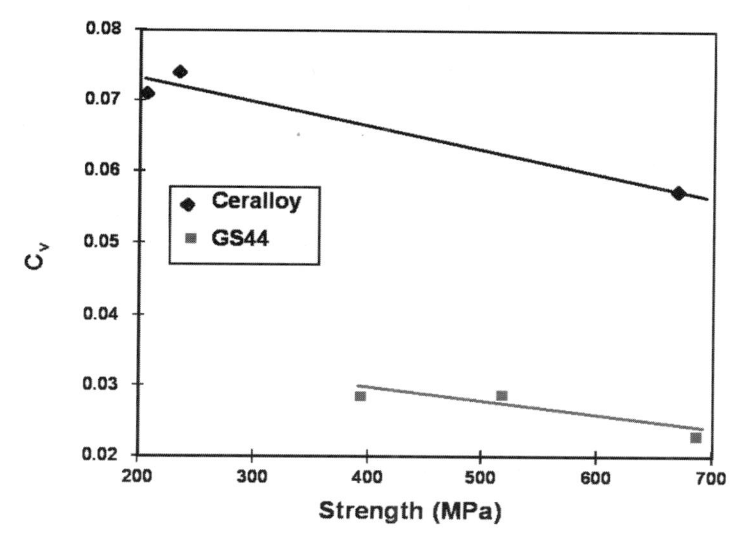

FIGURE 13.13 Correlation between laser back scatter data and four-point band strength for two Si_3N_4 materials.

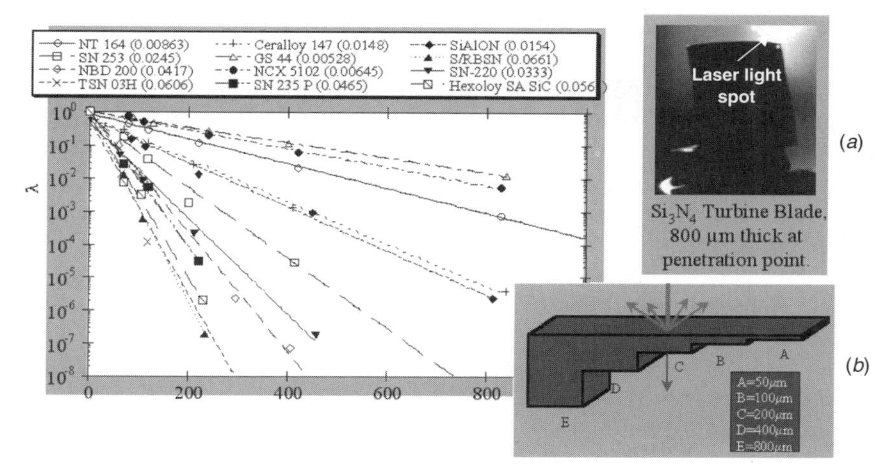

FIGURE 13.14 Optical transmission characteristics of several SiC and Si_3N_4 materials for $\lambda = 0.6328$ μm. Insets show: (*a*) visible observation of laser light on backside of Si_3N_4 turbine blade, (*b*) schematic of step wedge used to obtain through-transmission data.

ultrasonics, the data are presented in the form of a gray-scale image of the scanned area of the specimen. The elastic optical scatter method relies on the fact that many monolithic structural ceramic materials, including Si_3N_4, SiC, and zirconia, are optically translucent at various optical wavelengths as noted in Fig. 13.14.

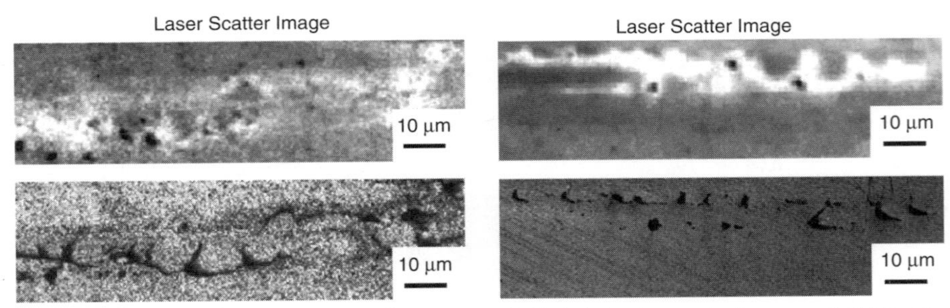

FIGURE 13.15 Comparison of optical photomicrographs and laser back scatter data on crept Si_3N_4.

To demonstrate polarized backscattered laser light detection of creep, an Si_3N_4 sample was crept at 1400°C under 41.5-MPa stress. Creep damage was verified by taking elastic optical back scatter data, polishing off 50–60 μm of material, and repeating these sequential steps through the thickness. Optical photomicrographs were taken of the surface before and after removing each layer. This sequence was repeated until the laser data suggested no more creep damage. The optical photomicrograph data and the corresponding laser scatter data consistently had one-to-one correspondence. Figure 13.15 shows two of these correlations.

The detection sensitivity of the laser back scatter method has been further tested using an Si_3N_4 polished surface flat specimen with intentionally induced subsurface Hertzian cone cracks. The Hertzian cracks were generated by loading small-diameter ceramic balls with different known loads. Figure 13.16 shows a schematic diagram of the specimen and resulting surface and subsurface laser scans. Note that in the surface scan, only the known surface breaking crack is detected, whereas in the subsurface scan, only the subsurface Hertzian cone cracks are detected. Figure 13.17 shows an enlarged view of the detection of the C-crack. Note that the subsurface detection suggest the larger diameter as would be expected of the "C"-crack.

13.3.4.2 Ultrasonics

Creep damage detection has also been under study by ultrasonic methods using precise ultrasonic velocity measurements [37]. Since creep changes the elastic property of the material, it also changes the acoustic velocity. Figure 13.18 shows typical profiles of ultrasonic longitudinal wave velocity in an Si_3N_4 crept at 1300°C for 200 h under various creep stress conditions. The profiles show the velocity of two parts of the creep specimens: the gage part and the grip part. Compared to the virgin specimen, the velocities in the gage part are shown to be lower than those in the grip parts. The velocity decreases with increasing creep stress. Another specimen crept under 200-MPa stress shows much larger reduction of the velocity in the gage part, which corresponds to larger creep strain and perhaps development of creep cavities. To obtain net change of velocities

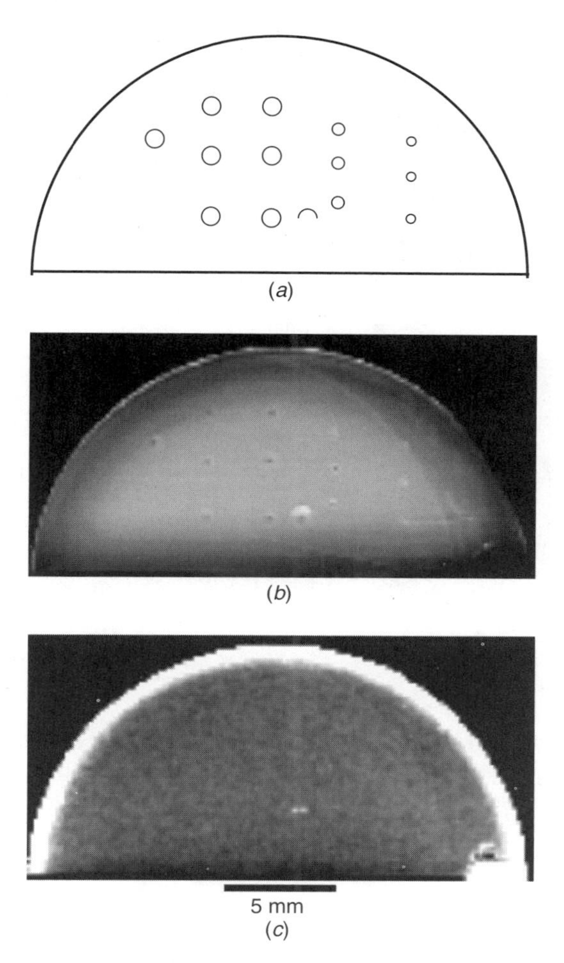

FIGURE 13.16 Hertzian crack test specimen of polished Si_3N_4 and resulting laser scatter data: (*a*) diagram, (*b*) surface scan, and (*c*) subsurface scan.

caused by creep cavities, each difference at the grip part was subtracted from the original total wave velocity. It was assumed that this provided the net effect of creep cavities.

13.3.5 Accumulated Damage

In some applications of monolithic materials, there is a need to assess the total accumulated damage as opposed to localization of individual or discrete fracture-causing flaws. Determination of accumulated damage is desired because mechanical properties are effected, and thus, if a correlation can be established between NDE data and a mechanical property, for example, strength, then reuse/replace

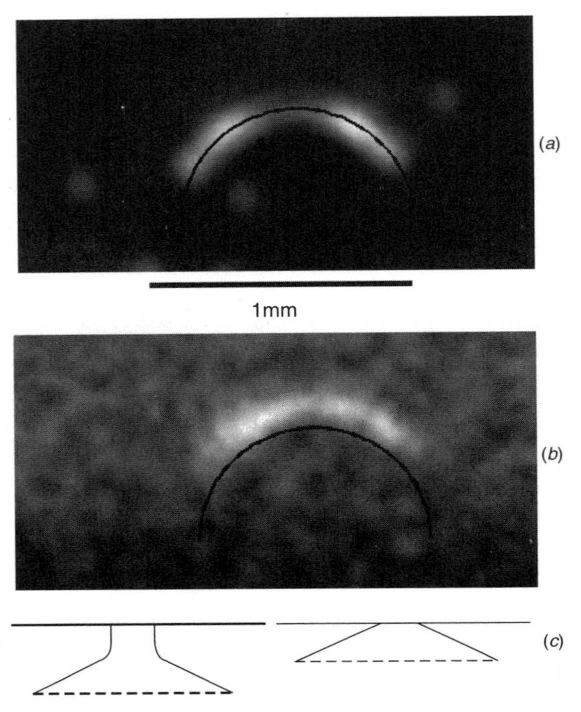

FIGURE 13.17 Enlarged view of "C" crack detection of test sample of Fig. 13.16: (*a*) surface, (*b*) subsurface. Line drawn in superposition shows position of surface-breaking crack. (*c*) Shows a diagram of subsurface crack types.

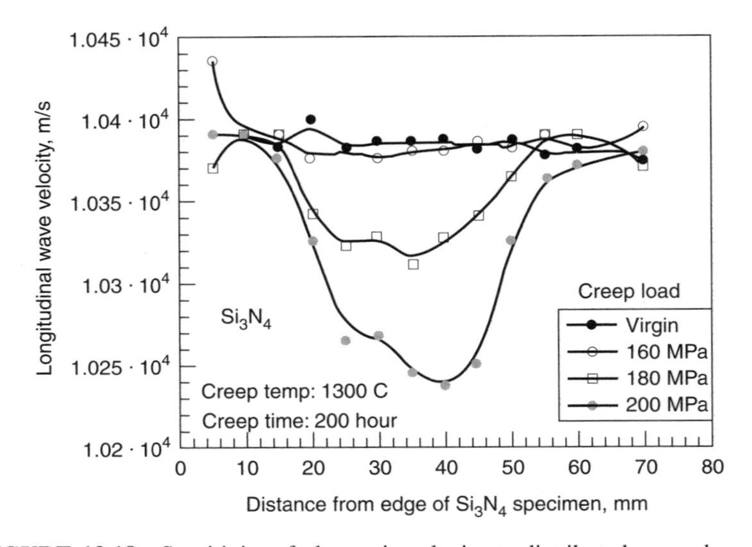

FIGURE 13.18 Sensitivity of ultrasonic velocity to distributed creep damage.

decisions can be made based on these data. One such example is in monolithic rigid hot-gas filters [38, 39]. Rigid ceramic hot-gas filters with about 15% porosity are being developed to clean particulate matter in hot-gas streams of various coal-fired plants. One principle application is in combined cycle power plants where the hot-gas is used for fuel in a gas turbine. In some plants, over 3000 such filters can be used and thus reuse/replace decisions are costly. In recent work [39], filters studied were made of clay-bonded SiC, recrystallized SiC, and alumina-mullite.

These filters were studied using an NDE technique referred to as acousto-ultrasound [40, 41]. An acousto-ultrasonic (AU) system (see Fig. 13.19) is a hybrid combination of classical ultrasonic techniques coupled with acoustic emission technology. A typical system consists of (a) two acoustic transducers placed in contact with the component under investigation, (b) supporting signal conditioning and detection electronics, (c) a computer for signal generation and detection, and (d) software packages for digital signal processing. When one transducer is pulsed, an acoustic wave travels along the wall of the specimen and is then detected by the second transducer. Previous research [41] has shown that the AU technique is applicable to a variety of materials. Plotting the stress-wave factor, obtained from the detected signal, versus strength of the materials yielded

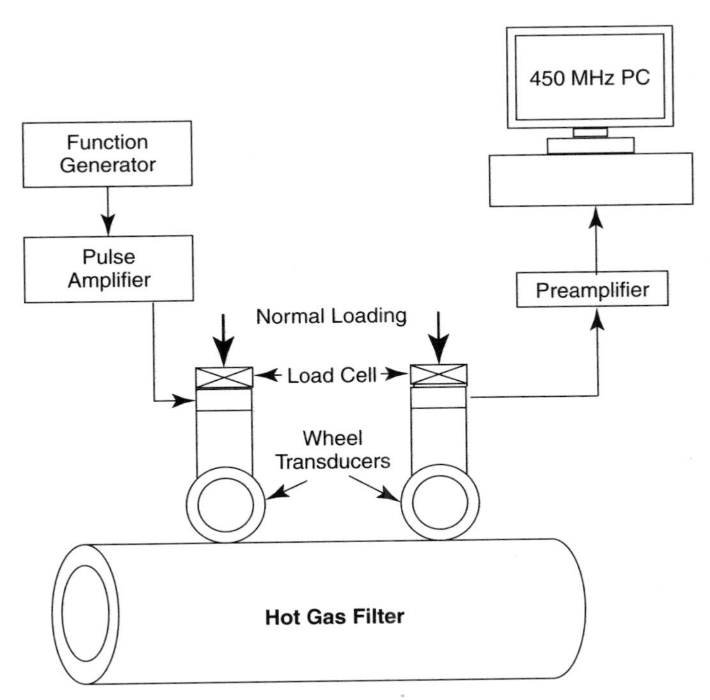

FIGURE 13.19 Schematic diagram of an automated acousto-ultrasonic method used to characterize accumulated damage in rigid-ceramic hot-gas filters.

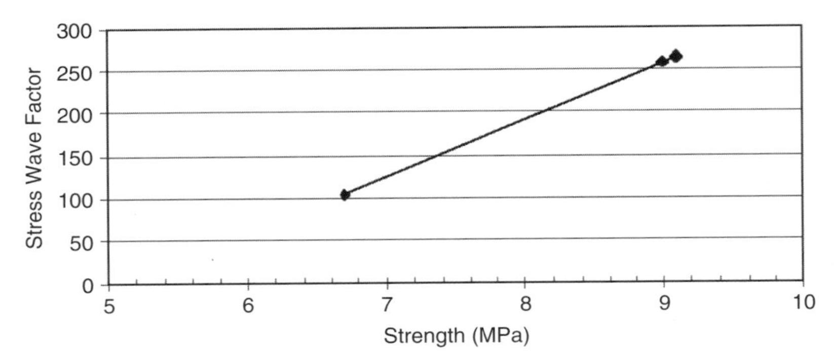

FIGURE 13.20 Correlation between stress-wave factor and retained strength of alumina-mullite and clad-bonded SiC hot-gas filters.

FIGURE 13.21 Photograph of automated acousto-ultrasound system for study of rigid-ceramic hot-gas candle filters.

the correlations shown in Fig. 13.20. An increasing stress-wave factor correlates with increasing strength. The acousto-ultrasound method has been automated; see Fig. 13.21. Two-wheel transducers are used as the send–receive transducers, and linear motion is controlled via a computer-controlled stepper motor.

13.3.6 Bioceramics

The general field of bioceramics has developed rapidly. Beginning in the 1970s it was established [16, 17] that several high-purity ceramics offered long-term biocompatibility. The desirable aspects of ceramics for biological applications stems from their high hardness, corrosion resistance, good lubricity, and low friction.

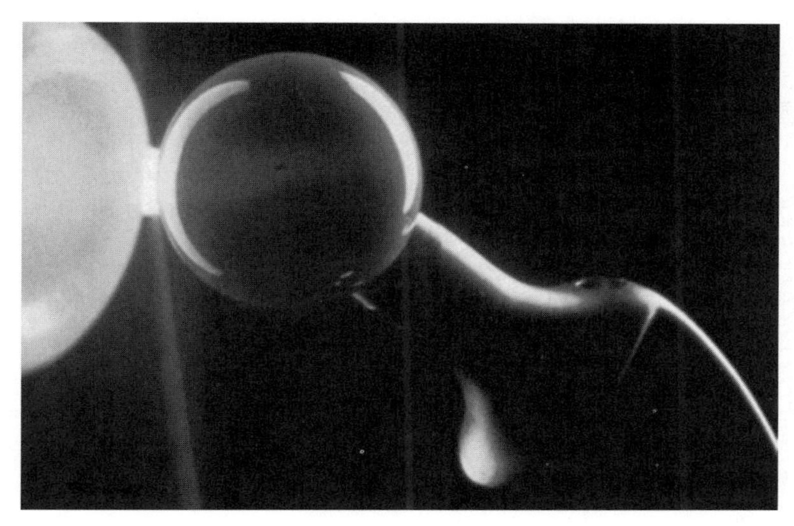

FIGURE 13.22 Photograph of ceramic head used in conjunction with metal shaft as part of a hip joint replacement.

Four ceramic materials are presently used. These are alumina, partially stabilized zirconia, β-tricalcium phosphate, and calcium hydroxyapatite. Of these, alumina and zirconia are used for prosthetic applications including hip joint replacements [6]. Figure 13.22 shows a photograph of a typical ceramic ball used as part of an artificial hip joint replacement.

From an NDE point of view, it is crucial to detect potential failure-causing internal defects as well as any machining damage done to the surface of the ball. The technologies discussed previously are applicable.

13.4 CHARACTERISTICS OF INTEREST TO COMPOSITE CERAMICS

Ceramic matrix composite materials [42–44] present different challenges to nondestructive evaluation technologies. The complex microstructure of these materials makes definition of a "flaw" rather uncertain. The significant differences in microstructure between monolithic ceramics and composite ceramics can be seen by looking at some of the various fiber architectures that are available for composites. Figure 13.23 shows diagrams of some of the various fiber architectures available. Ko [45] has presented the details of these fiber architectures in great detail, and the reader is referred to this excellent reference for information. As opposed to monolithic ceramic materials, composites exhibit what is commonly called "graceful failure" [46]. A typical stress–strain curve for an SiC/SiC ceramic composite is shown in Fig. 13.24. The significant differences in microstructure and mechanical properties between monolithic and composites changes the approaches used for NDE to detect "flaws." Characteristics of

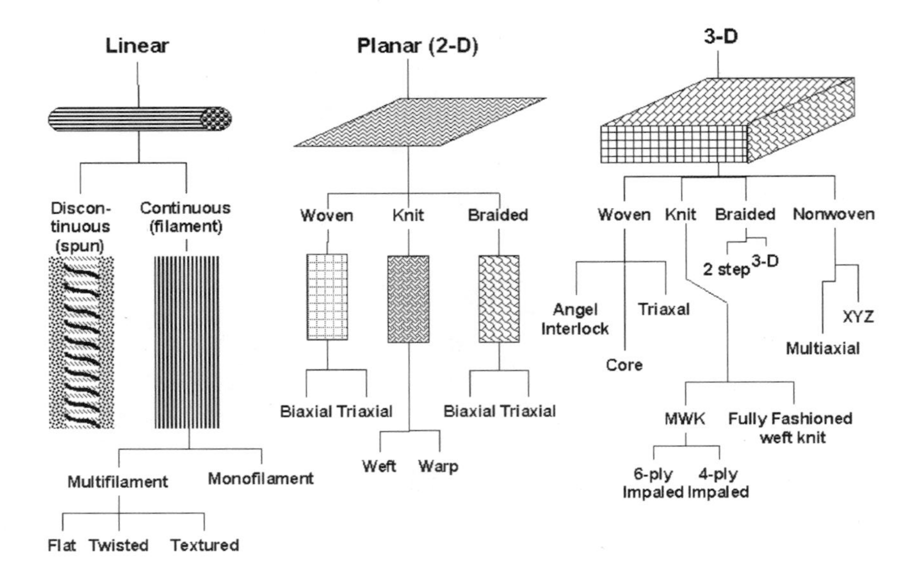

FIGURE 13.23 Schematic diagram of fiber architectures used for continuous fiber-reinforced ceramic matrix composites [63].

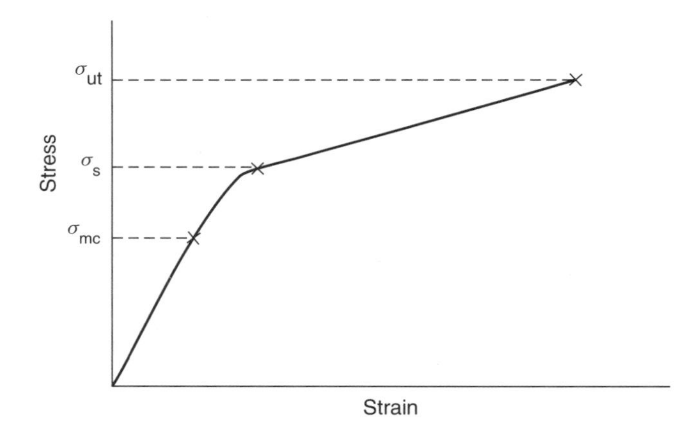

FIGURE 13.24 Typical stress–strain diagram for a two-dimensional cloth lay-up, melt-infiltrated SiC/SiC composite at room temperature.

composites that might be classified as a flaw include: delaminations, regions of open porosity, seams at plys or through-thickness cracks and damage zones caused by factors such as foreign object damage (FOD) or accelerated oxidation. Flaws in composites might better be described as "flaw regions." The development of nondestructive evaluation methods to detect these distributed flaw regions is described in the following subsections.

13.4.1 In-process Information

The processing stages used for most ceramic composites does not allow use of any contact or immersion in a liquid couplant as might be used for water-coupled ultrasonics. The reason is that any absorption of water might act as a contaminant in subsequent steps. For this reason, NDE technologies developed for composites primarily involve noncontact methods such as air-coupled ultrasonics [47–49], thermal imaging [50–52], X-ray imaging of various modalities, or very limited contact methods such as acousto-ultrasonics [40, 41] and resonance methods [53, 54].

13.4.1.1 Thermal Imaging

The thermal imaging method, recently advanced through new high-performance focal-plane array detectors, is based on the early work of Parker et al. [55]. This method assumes that the front surface of the sample is heated instantaneously. The rate of heat conduction through the sample is related to the thermal diffusivity of the material and is determined by measuring the rate of temperature rise at the back surface. Figure 13.25 shows a theoretically predicted back-surface temperature T as a function of time t and specimen thickness L, where T_M is the maximum back-surface temperature. One method to determine the thermal diffusivity is the "half-rise-time" ($T_{1/2}$) method [51]. When the back-surface temperature rise has reached half of its maximum, that is, $T/T_M = 0.5$, thermal diffusivity, α, can be determined as $\alpha = 1.37L^2/\pi^2 t_{1/2}$. The accuracy of the

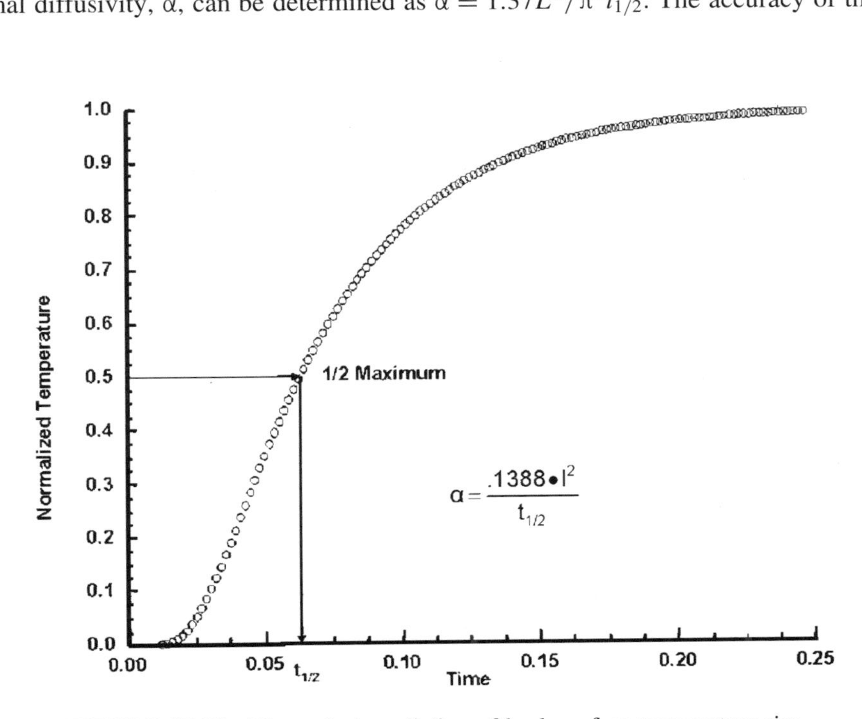

FIGURE 13.25 Theoretical prediction of back-surface temperature rise.

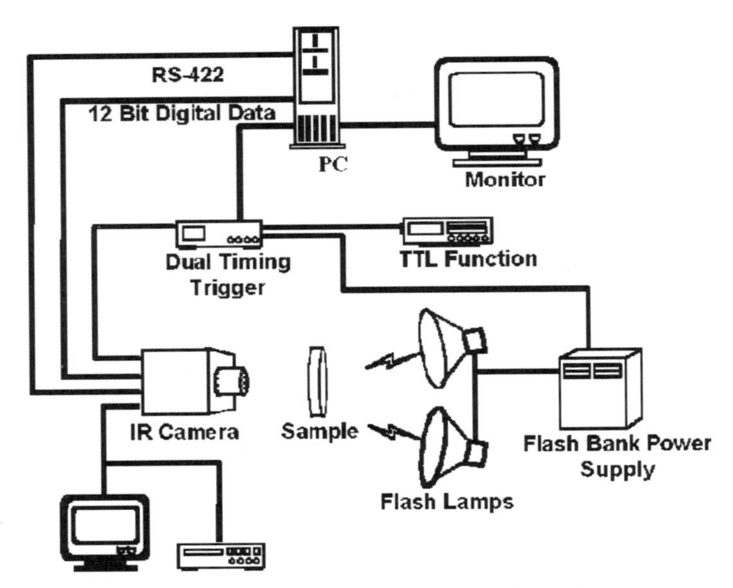

FIGURE 13.26 Schematic diagram of experimental thermal imaging apparatus.

thermal diffusivity measurement determined by this method can be calibrated with standards.

A typical experimental system used for obtaining diffusivity measurements over a predetermined area is shown in Fig. 13.26. The apparatus includes an infrared (IR) camera that consists of a focal-plane array of 256 × 256 InSb detectors, a 200-MHz Pentium-based personal computer (PC) equipped with a digital frame grabber, a flash lamp system for the thermal impulse, a function generator to operate the camera, and a dual-timing-trigger circuit for the camera and external trigger control. An analog video system is used to monitor the experiments. Processing time to measure a typical diffusivity image with 256 × 256 pixels ranges from 8 to 20 s, depending on the number of frames taken.

13.4.1.2 Air-Coupled Ultrasonics

The air-coupled ultrasonic method [47–49] is a relatively new method since 1995. The method has become feasible primarily through advances in piezoelectrics and digital signal processing. A typical air-coupled, through-transmission system consists of a traditional *x-y-z* positioning system with two matched air-coupled transducers in a coaxial transmission geometry (Fig. 13.27). The yoke assembly on which the transducers are mounted is connected to *x-y* scan stepper motors that are controlled by the host computer. The sample is mounted on an adjustable support so that the focal point of the transducers is within the thickness of the sample. A C-scan image of the sample is built up with a nominal 0.8-mm step size in both *x* and *y* directions. Tone bursts of acoustic energy at 0.4 MHz are incident on the sample from the transmitter side, and these ultrasonic

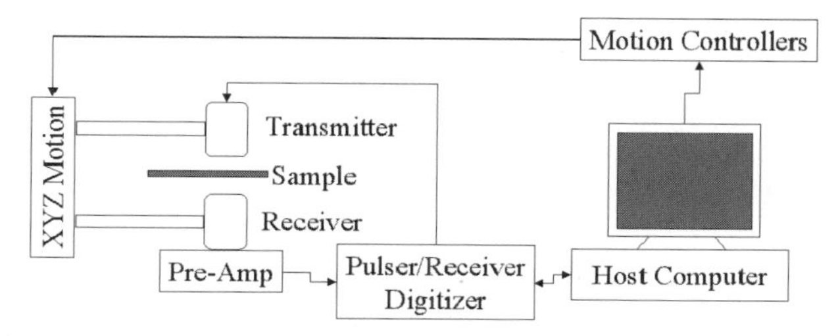

FIGURE 13.27 Schematic diagram of an air-coupled ultrasound system.

waves propagate through the sample and emerge to be detected by the receiving transducer. The detected signal is preamplified by a low-noise preamplified amp attached directly to the receiving transducer connected to a tuned amplifier. The digital value of the peak-transmitted signal is displayed and stored. The resulting image consists of a large number of pixels, whose gray level depends on the transmitted amplitude. As acoustic waves propagate through the material, they are scattered by defects. Areas with defects (such as pores and delaminations) appear with different gray-scale values on the image.

13.4.1.3 Impact Acoustic Resonance

In several applications of ceramic composites, the damping behavior and the resonant frequency is important. This is especially true for rotating components in turbomachinery. There are primarily two methods that can be used to determine resonant frequency of a component or subcomponent [53, 54]. These are continuous excitation with a swept frequency while monitoring the vibration of the component or by using a sharp impact excitation together with a method to detect the excited vibrations. There are advantages and disadvantages to both methods. Figure 13.28 shows a schematic diagram of one type of acoustic resonance

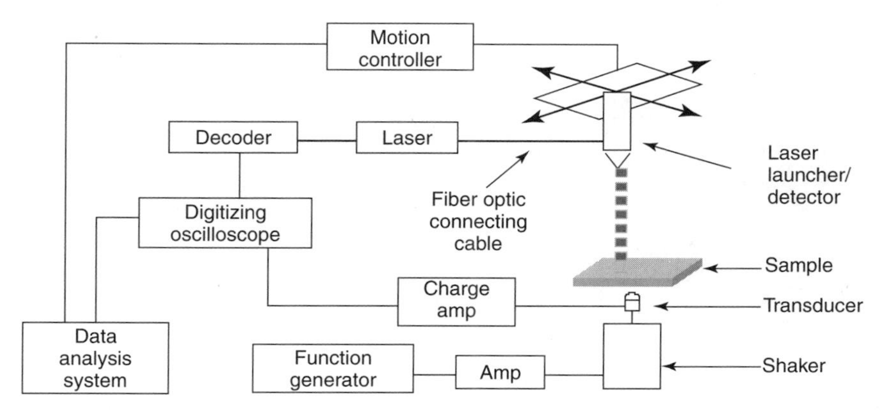

FIGURE 13.28 Schematic diagram of an impact acoustic resonance setup used to obtain resonant frequencies of ceramic composite components.

device. This particular diagram shows a noncontact laser vibrometer as the noncontact vibration detector. For impact excitation, the function generator delivers a single sharp spike pulse to the "stringer" on the shaker table, which impacts the specimen. For continuous excitation, the function generator sweeps through a broad range of frequencies and the laser vibrometer detects the displacement. Careful analysis of the resulting data allows determination of the fundamental frequencies as well as the damping behavior of the component.

13.4.2 Delamination Detection

The detection of delaminations in ceramic matrix composites can be achieved in several ways. The choice of NDE method depends to a great extent upon the properties of the material (e.g., oxides vs. nonoxides), the presence of any protective coating (e.g., an environmental barrier coating), and to a limited extent upon the part shape and size. Detection of delaminations in melt-infiltrated (MI) SiC/SiC, chemical vapor infiltration (CVI) SiC/SiC, polymer impregnation processes (PIP) SiC/SiC, CVI C/SiC, C/C, and sol–gel infiltrated oxides have been shown to be detectable with thermal imaging [50–52], as well as with air-coupled ultrasonic methods [48]. Three examples will be given to demonstrate detectability.

Example 1 is for a PIP SiC/SiC. Example 2 is for a sol–gel oxide/oxide, and Example 3 is for a MI SiC/SiC with an environmental barrier coating.

13.4.2.1 Polymer Impregnation Process, SiC/SiC

The test samples for these experiments were 8-ply and 16-ply cloth in a two-dimensional lay-up. Each specimen was 20.3 by 20.3 cm square. Since the PIP process allows sample removal between successive processing cycles, the same panels were studied by thermal imaging and air-coupled ultrasound after 1, 5, 10, and 15 cycles. Results of thermal diffusivity imaging of the 8-ply panel are shown in Fig. 13.29. What is noted after 1 PIP cycle are two large delaminated regions in the upper left and upper right corners. Visual inspection of the edge revealed ply separation. After 5 cycles, handling damage enlarged

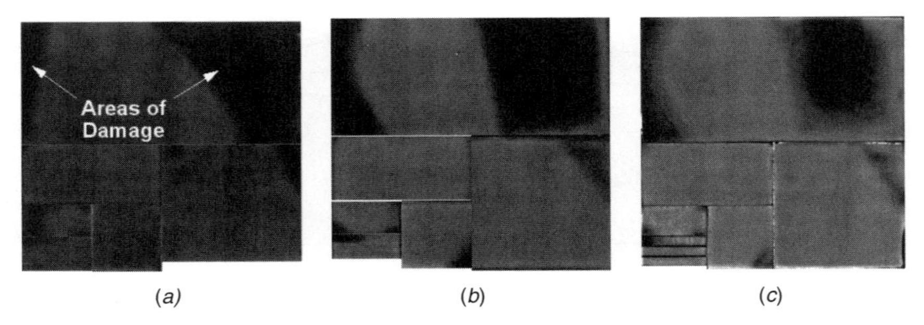

FIGURE 13.29 Thermal diffusivity image data of 20.3-cm square, 8-ply, two-dimensional cloth lay-up, PIP processes SiC/SiC flat plate: (*a*) after 1 PIP cycle, (*b*) after 5 PIP cycles, and (*c*) after 10 PIP cycles.

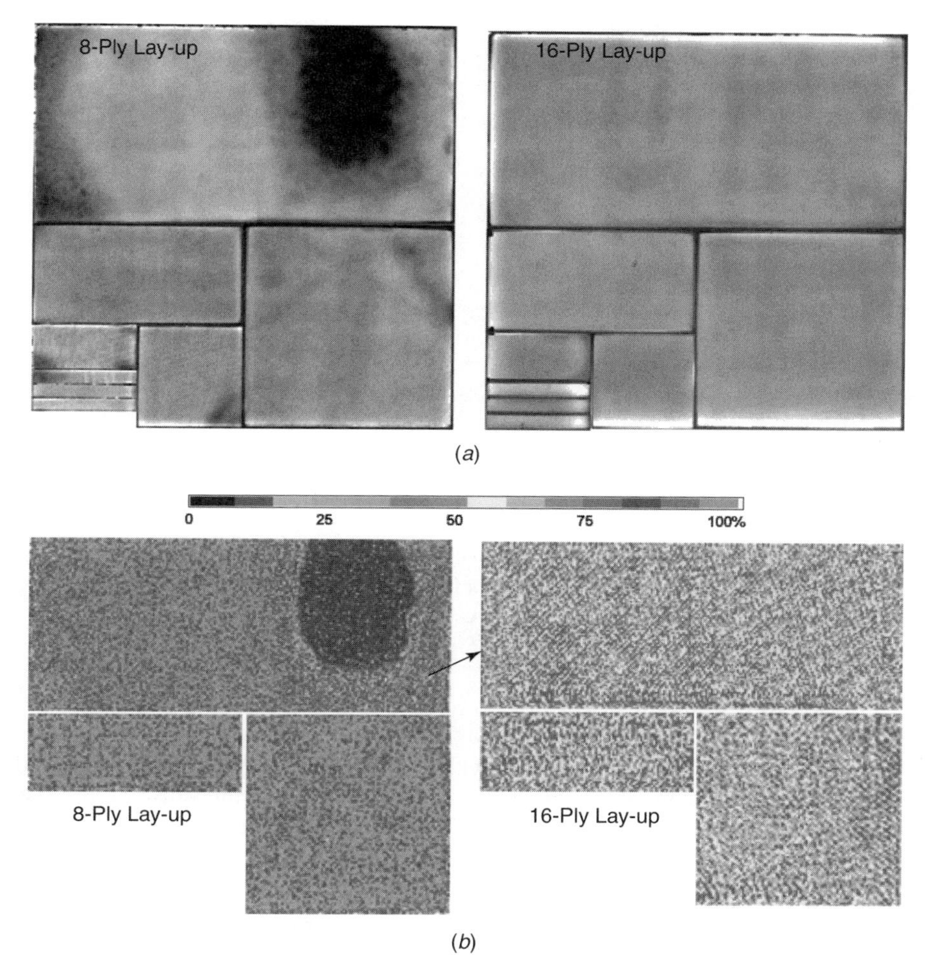

FIGURE 13.30 Thermal diffusivity and air-coupled through-transmission data for both 8-ply and 16-ply PIP SiC/SiC after 15 PIP cycles: (*a*) thermal diffusivity maps and (*b*) air-coupled ultrasound scans.

some of the delam regions, but also there was some infiltration into the delam region that acted as a form of repair. After 10 cycles, additional filling of the delam region is detected. Figure 13.30 shows both thermal diffusivity image data and air-coupled ultrasonic data for both the 8-ply and the 16-ply lay-up samples. While the 16-ply sample is unremarkable for features, the 8-ply specimen demonstrates the correlation between the thermal image data and the air-coupled ultrasonic data. Quantification of the density of this PIP SiC/SiC by the two NDE methods was also examined. A series of test samples were made, each 50 by 50 mm, of 8-ply two-dimensional lay-up. These were examined using both

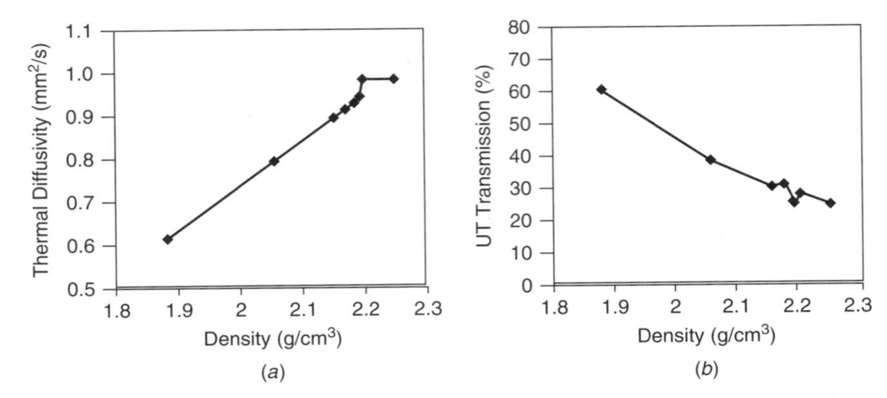

FIGURE 13.31 Relationship between PIP SiC/SiC, 8-ply two-dimensional cloth lay-up density and NDE data: (*a*) thermal diffusivity and (*b*) ultrasonic through-transmission signal strength.

thermal imaging and through transmission ultrasonic. The results are shown in Fig. 13.31. By calibrating the NDE methods for a particular fiber architecture and process method, quantification of certain physical parameters, for example, density, can be achieved.

13.4.2.2 Sol–gel Oxide/Oxide

The test sample used in this study was a 20.3-cm diameter cylinder, 20.3 cm long with approximate a 3-mm-thick wall. The fiber was 3M Corporation N720 cloth and was laid up layer by layer and infiltrated by the sol–gel process with an aluminosilicate matrix. Figure 13.32 shows the results of using both

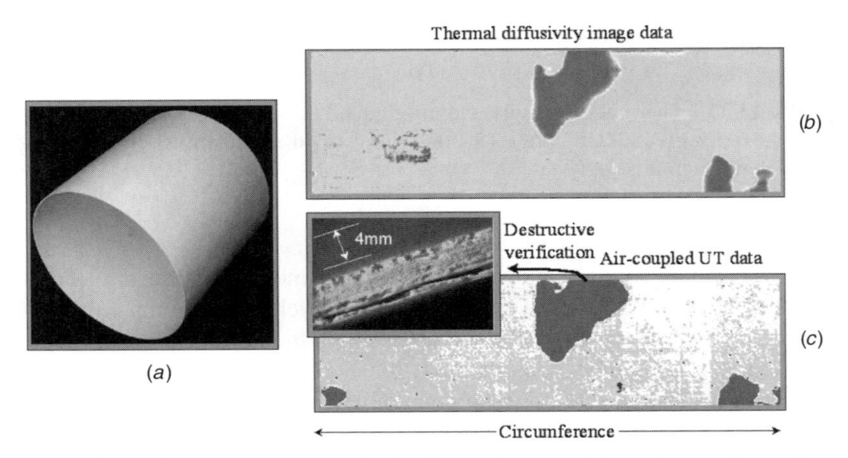

FIGURE 13.32 Delamination detection in 20-cm-diameter, 20-cm-long oxide/oxide combustor liner: (*a*) photograph of cylinder, (*b*) thermal diffusivity image map of surface, and (*c*) air-coupled through-transmission C scan.

through-transmission air-coupled ultrasound and through-thickness thermal diffusivity for detection of delams. The delamination was verified by destructive sectioning as shown in Fig. 13.32. One special feature about oxides is to be noted when conducting NDE using the thermal method. That is, in order to quantify the thermal diffusivity data, care must be exercised in the spectral output of the flash lamp. Recall that Parker's theory [55] requires that a step input in heat be applied at the surface of the sample under study. Oxide materials have a high optical translucency, and thus, if the optical band pass of the flash lamp is not correct, some heat of the flash lamp will penetrate into the volume of the test sample, thereby giving erroneous times for the back surface temperature to rise.

13.4.2.3 Melt-Infiltrated SiC/SiC with Environmental Barrier Coating

The third example of NDE data detecting a delamination concerns a MI SiC/SiC cylinder used as a combustor liner in a 4.5-MW gas turbine [56]. The annular combustor of the engine—see Fig. 13.33—consists of an outer liner 75 cm in diameter and an inner liner 33.5 cm in diameter. In this case, both liner components had been coated with a protective environmental barrier coating (EBC) [57]. Prior to insertion of the components into the engine, which was a full-size field test engine, both thermal diffusivity imaging and air-coupled ultrasonic imaging NDE data were obtained. Figure 13.34*a* shows the NDE thermal image data, which suggested a delaminated region. Figures 13.34*b* and *c* show the occurrence of a spallation of the coating. Figure 13.34*b* is after 357 h of operation

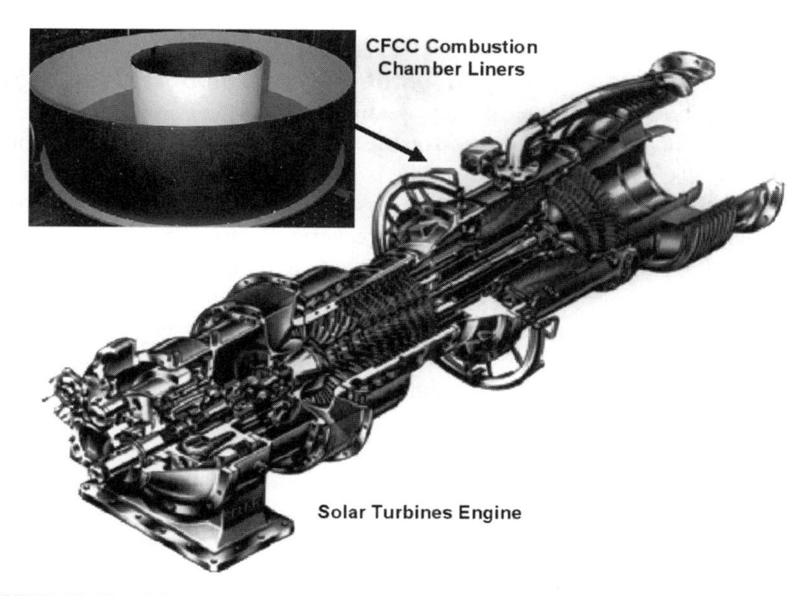

FIGURE 13.33 Diagram of gas turbine engine and photograph of annular combustor liner set made of SiC/SiC ceramic matrix composite.

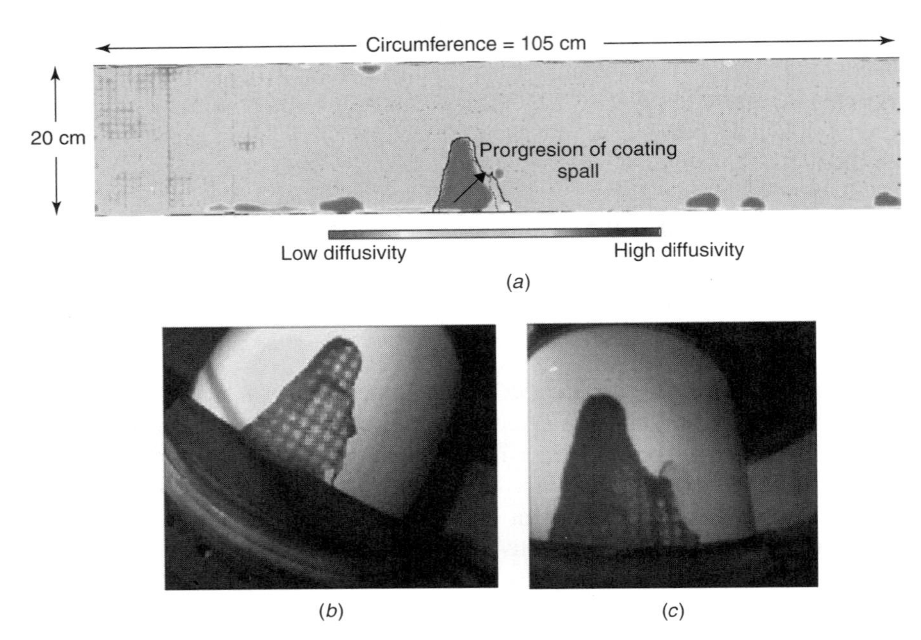

FIGURE 13.34 Use of thermal image NDE data to predict spall regions of EBC on MI SiC/SiC gas turbine combustor inner 33.5-cm-diameter liner as shown in Fig. 13.32.

and Fig. 13.34*c* is after 1573 h of operation. This demonstrates the ability of NDE data to accurately predict spall locations.

13.4.2.4 Delam Detection by Volumetric X-ray Tomographic Imaging

One NDE method that can be used for delam detection that is quite independent of the thermal or physical properties or the geometric shape is X-ray computed tomographic (XCT) imaging. An example is of a 75-cm-diameter combustor liner made of CVI SiC/SiC. Figure 13.35 shows how the initial thermal diffusivity data, which suggested a delamination, were verified through use of volumetric XCT. In this case, the entire 20-cm-long section of the liner was imaged by XCT. Displaying the volumetric X-ray CT image in a high-end workstation with advanced three-dimensional image display allows the user to "onion peel" off layer by layer. Figure 13.35*b* shows that after "peeling" off four layers, the delam appears. Thus the XCT method not only allows detection but size and depth location as well.

13.4.3 Porosity Detection

Most ceramic composites are designed for use in high-temperature environments. Therefore thermal properties and uniformity of the thermal properties over the entire component become important and regions of high porosity become an issue. Porosity detection can be accomplished by several NDE methods. However, the

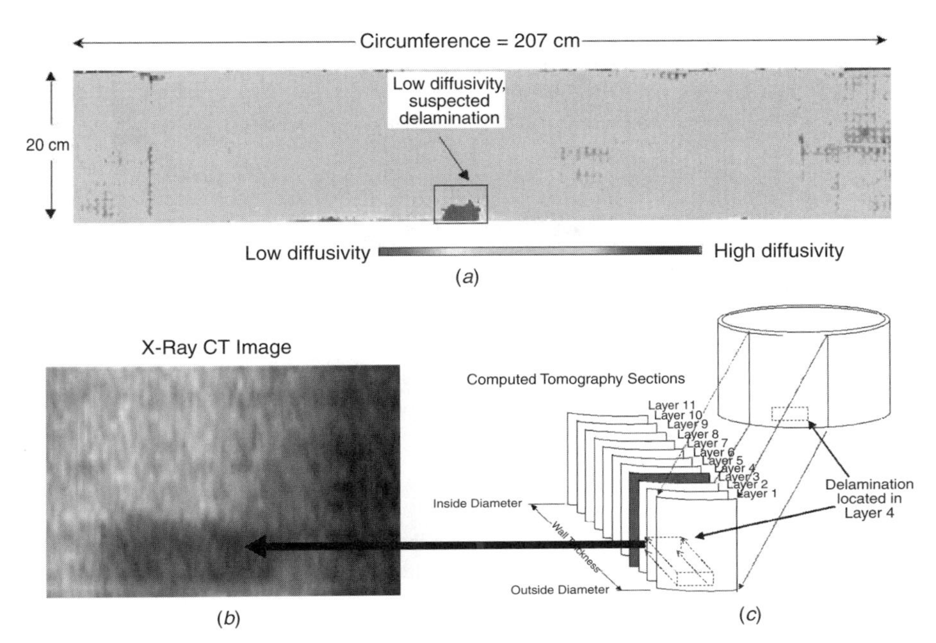

FIGURE 13.35 Detection of delamination in large 75-cm diameter, CVI SiC/SiC combustor liner: (*a*) initial thermal diffusivity, (*b*) X-ray CT volume image of one layer, and (*c*) diagram of (*b*).

primary methods used to date have been air-coupled ultrasonic methods, thermal imaging methods, and XCT. Figure 13.31 demonstrated correlations between density and thermal diffusivity and transmitted ultrasound signal strength for a PIP SiC/SiC. In a recent example [58], a melt-infiltrated SiC/SiC cylinder for a gas turbine combustor shell experienced an "upset" during processing. NDE data—see Fig. 13.36—suggested that there was a clear detectable difference at the process-upset position. Subsequent destructive analysis as noted in Fig. 13.36*c*, detected a difference in porosity.

13.4.4 Impact Damage Detection

Impact damage or foreign object damage (FOD) can occur on any component. To date, there has been a very limited effort to develop NDE methods to assess extent of FOD. Recently, detection of FOD has been approached by two NDE methods: thermal imaging and air-coupled ultrasound. In one recent set of experiments [47], five 76 × 152-mm 8-ply (3 mm thick) SYLRAMIC S200 CMC panels were studied for impact damage. Impact damage was introduced by using a 12.5-mm-diameter steel rod with a hemispherical head. Visual inspection revealed limited surface damage but could not assess the depth or extent of the damage. NDE data were used to determine extent of damage. In addition, since

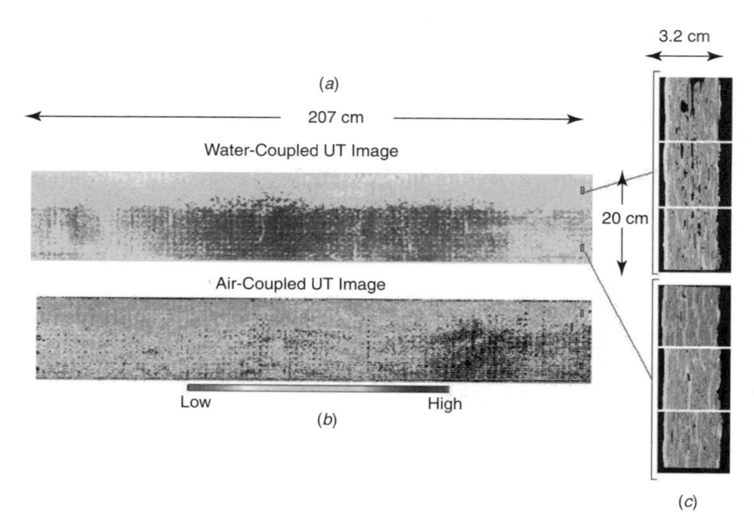

FIGURE 13.36 Detection of porosity in melt-infiltrated (MI) SiC/SiC: (*a*) air-coupled ultrasonic through-transmission data, (*b*) thermal diffusivity image data, and (*c*) destructive verification of porosity variation.

repair of CMC components is an area of interest, NDE/C methods are necessary to assess quality of the repair and assess condition of repaired regions at inspection intervals. As part of the effort on the impact-damaged samples, to assess repair, the samples were repaired and reexamined with NDE.

Conventional, through-transmission X-ray radiographs revealed few small, dark (low-density) lines near the impact, and it is well known that conventional through-transmission X-ray imaging does not detect planar delams. However, through-thickness air-coupled ultrasonics and through-thickness thermal imaging did reveal the impact damage. Figures 13.37 and 13.38 show through-thickness air-coupled ultrasonic and thermal diffusivity images, respectively, before and after repair processing. The correlation of the size and shape of the damaged region between diffusivity and ultrasound images is very good.

What is to be noted in the image data are the sizes, shapes, and the gray-scale levels. Recall, see Fig. 13.31, that as density increases, the through-transmitted air-coupled ultrasound signal decreases but the thermal diffusivity increases. Looking at Figs. 13.37 and 13.38, there is a significant increase in the density of the panel not impacted after repair by additional PIP cycles. Further, especially on panel A, the impact-damaged zone density has also increased. Observation of the repaired region increased suggesting an increase in density of the impact region after repair.

13.4.5 Through-Thickness Crack Detection

Through-thickness crack detection using NDE/C technology is important to CMC materials used as hot-gas confinement, such as a gas turbine combustor, because

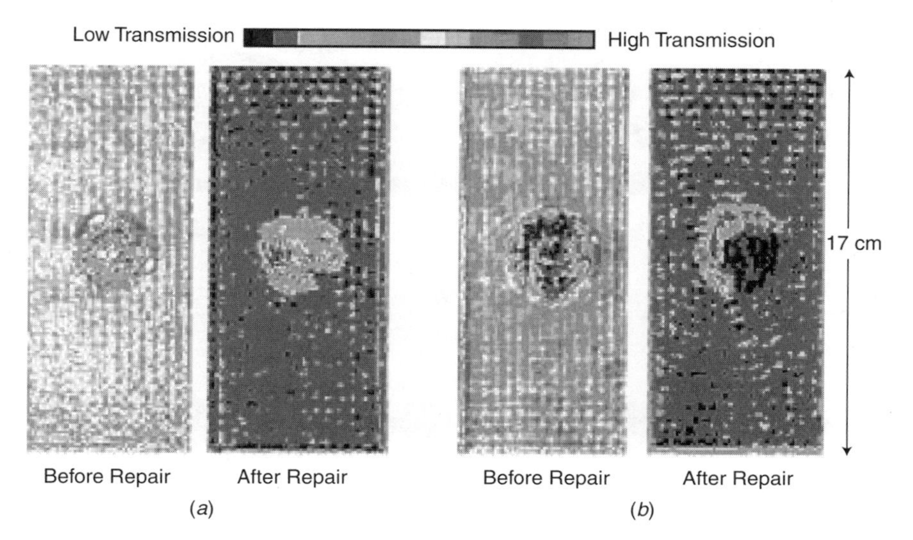

FIGURE 13.37 Through-transmission air-coupled ultrasonic C-scan image data of impact damaged PIP SiC/SiC 8-ply lay-up panels: (*a*) test panel E and (*b*) test panel A.

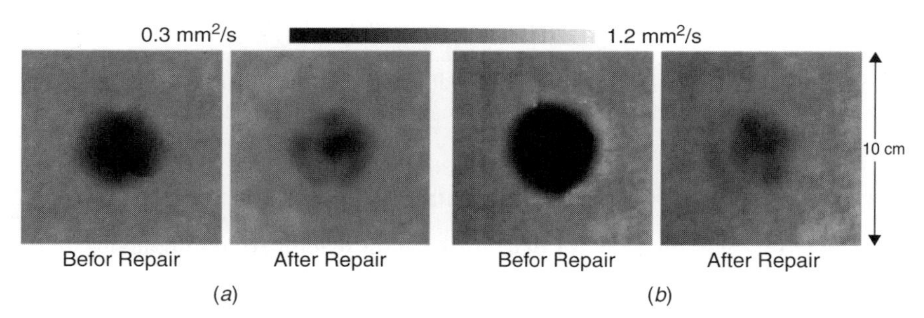

FIGURE 13.38 Through-thickness thermal diffusivity image data of impact-damaged PIP SiC/SiC, 8-ply lay-up: (*a*) test panel E and (*b*) test panel A.

cracks can provide a leakage path for hot gases to supporting structures. One recent example occurred in a large 75-cm-diameter CVI SiC/SiC combustor liner for a gas turbine. The NDE methods used for data acquisition were thermal imaging and air-coupled ultrasonics. All NDE data suggested an axial separation of unknown depth in the as-processed condition; see Fig. 13.39. The liner was installed and run for 5016 h and subsequently was removed for analysis. What the destructive analysis suggested (see Fig. 13.39*a*) is that at the time the liner was fabricated, the plys at the seam had been distorted such that the ply layers no longer were concentric but rather had an orientation toward the outer surface. Thus destructive analysis confirmed the early NDE data in that there was a seam with a weakness.

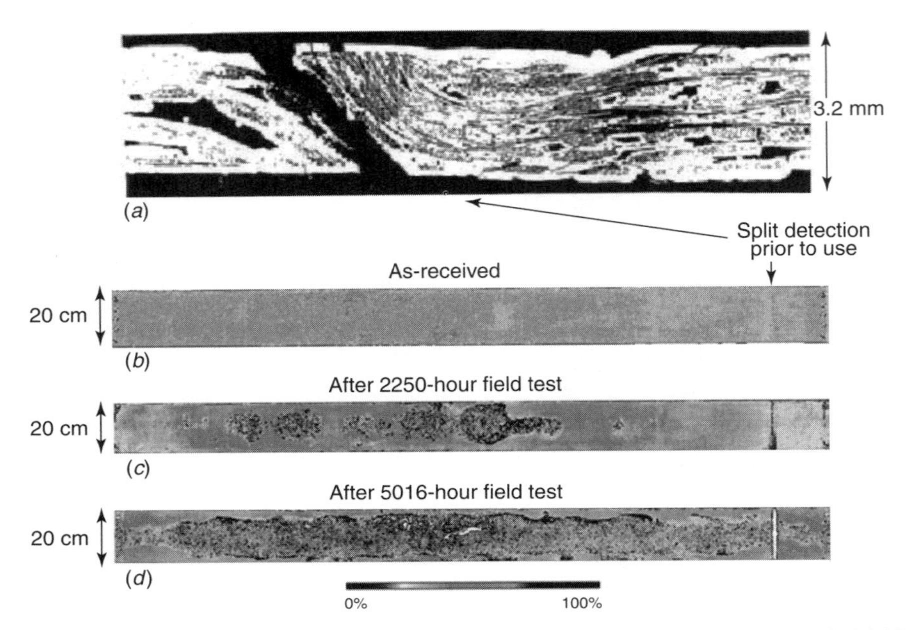

FIGURE 13.39 Detection of through-thickness crack in 3-mm-thick 8-ply, CVI SiC/SiC 75-cm-diameter combustor liner: (*a*) crack detected by thermal diffusivity image, (*b*) progression of crack as monitored by thermal diffusivity imaging. (*a*) Optical photo micrograph of cross section through liner wall after 5016 h, (*b*) initial thermal image, (*c*) thermal image after 2250 h, and (*d*) thermal image after 5016 h.

13.4.6 Resonant Frequency and Damping

It was noted in Section 13.4.1.3 that, for rotating machinery components such as bladed turbine wheels [59], it is important to know the resonant frequencies and damping characteristics. Knowledge of the uniformity of resonant frequencies helps to assure that the structure will not go into an unstable resonant condition thereby developing a potentially catastrophic condition. The NDE methods to measure resonant frequencies and damping capacity of ceramic composites were described in Section, 13.4.1.3. Spohnholtz [53] and Bemis et al. [54] have described the various details involved in making these measurements.

In a recent set of experiments, a blisk (an integrally bladed disk) made of $C_{(f)}$/SiC was examined [59]. Resonant frequency measurements were conducted using the setup described in Section 13.4.1.3. Figure 13.40 shows a photograph of the blisk understudy. Figure 13.41 shows a plot of normalized resonant frequency for each blade prior to spin testing. Note that blade number 21 was shown to have a significantly lower resonant frequency. Subsequent X-ray tomographic imaging revealed that a void about 1 mm in size was present.

FIGURE 13.40 Photograph of bladed disk (blisk) made of C(f)/SiC and studied by impact acoustic resonance and X-ray tomographic imaging.

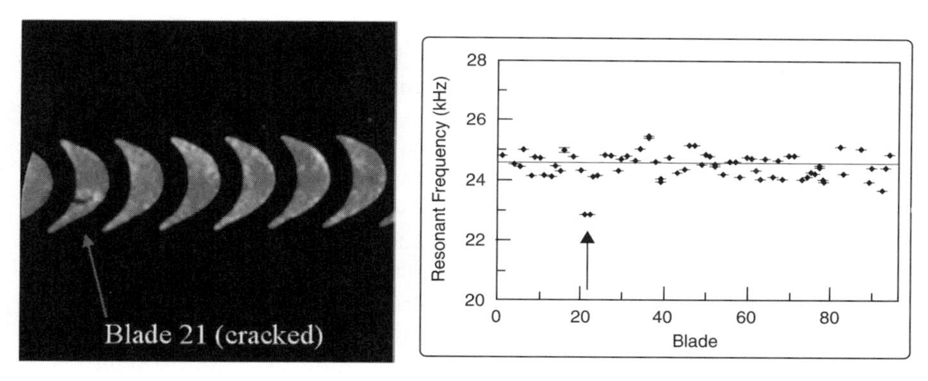

FIGURE 13.41 Resonant frequency of each blade of blisk shown in Fig. 13.40 and X-ray CT image showing the void in blade 21.

13.4.7 Remaining Useful Life

Many efforts have been put forth to develop analytical models for life prediction of ceramic composites [60, 61]. However, these models are based on micromechanics and require the constitutive properties of each component in the material. That is, properties of the individual fibers, the matrix, and the coating on the fibers and are necessary for each condition (e.g., temperature, time at load, etc.) that the material will be expected to encounter. An alternative approach is to try to utilize data from NDE methods that could be coupled to a more

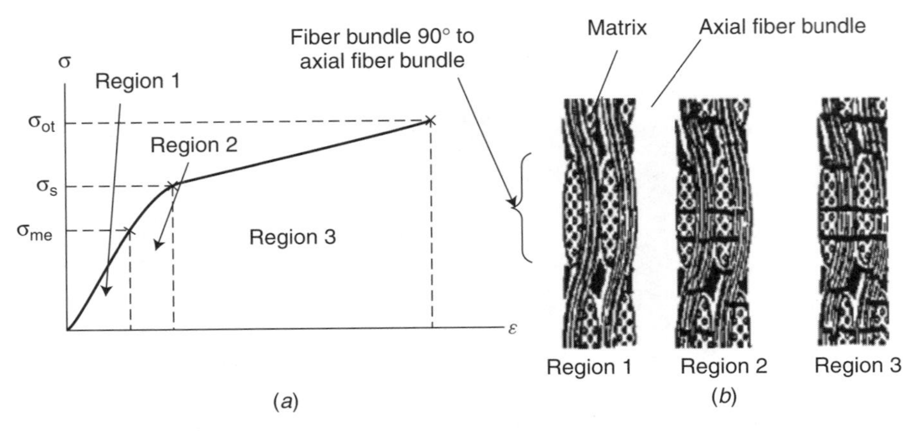

FIGURE 13.42 Diagram of stress–strain for typical SiC/SiC: (*a*) definition of regions and (*b*) schematics of damage expected in each region on stress–strain diagram.

macromechanic model to predict remaining useful life with the proviso that the component would not be subjected to conditions too far outside certain bounds. One such approach is to measure elastic modulus of the material. Figure 13.42 shows a typical stress–strain diagram for an SiC/SiC material as well as diagrams depicting damage in each of the "damage zones." Clearly there is a change in the elastic modulus with the detail of region 2 defining the upper limit of the elastic region of region 1 and the second elastic region of region 3.

Recent work has shown that guided plate waves [62] can be used to measure in-plane elastic modulus and that this can track damage states. The experimental setup used for initial measurements is shown in Fig. 13.43. Three acoustic-emission transducers are used each with a 400-kHz center frequency. One transducer is used to generate an acoustic pulse and two other transducers, with well-defined separation distance, are used to measure the time of flight (TOF) of the elastic wave. This TOF measurement can be used to determine the elastic modulus, which in turn can be related to a damage parameter [63] as

$$D = 1 - \frac{E}{E_0},$$

where D = damage parameter
E = measured elastic modulus after the material has been subjected to various road conditions
E_0 = elastic modulus of the new material prior to any exposure to loads

Two types of damage have been used for initial studies: impact and cyclic fatigue. In the first example, an $SiC_{(f)}$/barium–magnesium aluminosilicate matrix material was studied. Repeated pendulum-type impacts with 235 J per impact were induced in a 1-cm-wide specimen. The damage was assessed using the

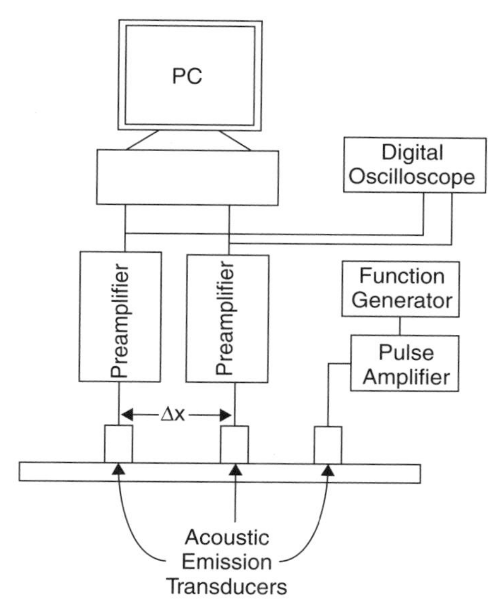

FIGURE 13.43 Schematic diagram of the guided plate wave data acquisition system.

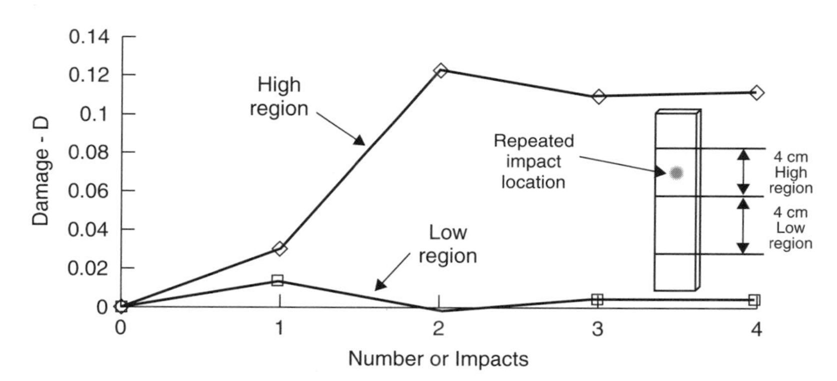

FIGURE 13.44 Measured accumulated damage as function of number of impacts for 8-ply SiC/SiC.

guided plate-wave method. The assessed damage after each impact is shown in Fig. 13.44. Two regions on the sample were studied. The high region, which included the impact-damaged region, and a low region away from the impact. Clearly the damage region is detected.

For the second set of tests, a specimen of the same material was used in tension–tension cyclic fatigue to induce damage. Again the experimental setup was used. However, data were obtained for regions under two load conditions: while under tensile load and with no load. Figure 13.45 shows the measured

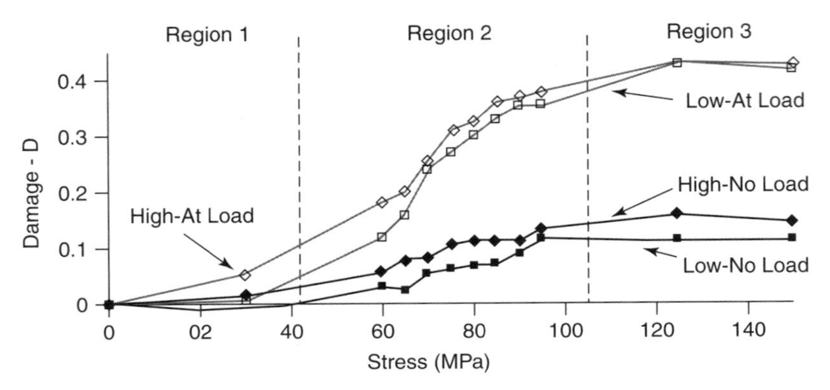

FIGURE 13.45 Accumulated damage induced through load/unload/reload tensile tests corresponding to three regions of the stress–strain curve shown in Fig. 13.41.

damage as a function of cyclic load stress. To be noted are two observations: (a) that the damage parameter clearly detects the three regions of the stress–strain curve and (b) that when the specimen is under load, the damage parameter is significantly higher. This is because the cracks are kept open at load and this impacts the elastic wave speed.

REFERENCES

1. D. S. Yan, X. R. Fu, and S. X. Shi (Eds.), *5th International Symposium on Ceramic Materials and Components for Engines*, World Scientific, Singapore, 1994.
2. V. J. Tennery, *Third International Symposium on Ceramic Materials and Components for Engines*, American Ceramic Society, Westerville, OH, 1988.
3. D. C. Larsen, J. W. Adams, L. R. Johnson, A. P. Teotia, and L. S. Hill, *Ceramic Materials for Advanced Heat Engines*, Noyes, Park Ridge, OH, 1985.
4. P. P. Lutton and B. Ben-Nissan, *Mat. Tech.*, **12**(2), 59–64 (1997).
5. P. P. Lutton and B. Ben-Nissan, *Mat. Tech.*, **12**(3/4), 107–111 (1997).
6. J. F. Shackelford, *Bioceramics: Applications of Ceramic and Glass Materials in Medicine*, Trans. Tech., Enfield, NH, 1999.
7. L. L. Hench, *J. Am. Cer. Soc.*, **74**(7), 1487–1510 (1991).
8. J. J. Schuldies and J. A. Branch, *Cer. Ind.*, **138**(5), 43–46 (1992).
9. W. A. Weibull, *J. Appl. Mech.*, **18**(3), 293–297 (1951).
10. P. N. Murthy, S. K. Mital, and A. R. Shah, Probabilistic Mircomechanics and Macromechanics for Ceramic Matrix Composites, A collection of technical papers: AIAA/ASME/ASCE/AHSASC 38th Structures, Structural Dynamics and Materials Conference, April 7–10, Kissimmee, FL, pp. 2532–2542.
11. Conference on Nondestructive Evaluation of Modern Ceramics, Am. Soc. Nondestructive Testing, Columbus, OH, 1990.
12. Conference on Nondestructive Testing of High-Performance Ceramics, American Ceramic Society, Westerville, OH, Ohio, 1987.

13. L. Goebbels and H. Reiter, in W. Bunk (Ed.), *Keramische Komponenten für Fahrzeug-Gasturbinen*, Vol. II, Springer, Berlin, 1983.
14. J. J. Schuldies and D. W. Richerson, in J. W. Fairbanks and R. W. Rice (Eds.), *Proc. 1977, DARPA/NAVSEA Ceramic Gas Turbine Demonstration Engine Program Review*, Metals and Ceramics Info. Center, Battelle Laboratories, Columbus, OH, 1978, pp. 381–402.
15. K. Nakajima, *J. Japanese Trade and Industry*, **5** (1983).
16. W. A. Ellingson and M. W. Vannier, X-ray Computed Tomography for Nondestructive Evaluation of Advanced Structural Ceramics, Argonne National Laboratory Report, ANL-87-52, 1987.
17. T. Rabe, J. Goebbels, A. Kunzmann, and H. Riesemeier, "Quantitative Determination of Density Gradients by Three-Dimensional Computer-Tomography," in *Proc. 5th European Ceramic Soc. Conf.*, **132–136**, 730–733 (1997).
18. J. Hsieh, O. E. Gurmen, and K. E. King, *IEEE Trans. Med. Imag.*, **19**(9), 930–940 (2000).
19. J. N. Panzarino, "NDE and Advanced Ceramics," in *Conf. on NDE of Modern Ceramics*, Am. Soc. for Nondestructive Testing, Columbus, OH, 1990, pp. 12–15.
20. S. Natenson, "Nondestructive Evaluation of Powders for Advanced Ceramics," in *Proc. Conf. on Nondestructive Evaluation of High-Performance Ceramics*, Am. Cer. Soc., Westerville, OH, 1987, pp. 73–87.
21. W. A. Ellingson, R. A. Roberts, M. W. Vannier, J. C. Ackerman, B. D. Sawicka, S. Gronemeyer, and R. J. Friz, Recent Developments in Nondestructive Evaluation for Structural Ceramics, Argonne National Laboratory Report ANL/FE-87-2, 1987.
22. K. E. Amin and T. P. Leo, "Radiographic Detectability Limits for Seeded Defects in Both Green and Densified Silicon Nitride," in *Proc. Conf. On Nondestructive Evaluation of High-Performance Ceramics*, Am. Cer. Soc., Westerville, OH, 1987, pp. 211–232.
23. D. W. Richerson, *Modern Engineering Ceramics: Properties, Processing and Use in Design*, 2nd ed., Marcel Dekker, New York, 1992.
24. H. C. Yeh, J. M. Wimmer, M. R. Huang, M. E. Rovabaugh, J. Schienle, and K. H. Styhr, Improved Silicon Nitride for Advanced Heat Engines, National Aeronautics and Space Administration Report, NASA-CR-175006, 1985.
25. W. A. Ellingson, J. L. Ackerman, L. Garrido, P. S. Wong, and S. Gronemeyer, Development of Nuclear Magnetic Resonance Imaging Technology for Advanced Ceramics, Argonne National Laboratory Report, ANL-87-53, 1987.
26. K. A. Carduner, G. R. Hatfield, W. A. Ellingson, and S. L. Dieckman, "Nuclear Magnetic Resonance Spectroscopy and Imaging of High-Performance Ceramics," in N. P. Cheremisinoff (Ed.), *Handbook of Ceramics and Composites, Vol. 2, Mechanical Properties and Specialty Applications*, Marcel Dekker, New York, 1992, pp. 465–502.
27. K. Hayashi and K. Kawashima, *J. Phys. D, App. Phys.*, **21**(6), 1037–1039 (1988).
28. P. McIntire (Ed.), *Nondestructive Testing Handbook*, Am. Soc. for Nondestructive Testing, Columbus, OH, 1990.
29. P. R. Grantors and R. Aufrichtig, "DQE(f) of an Amorphous-Silicon Flat Panel X-ray Detector: Detector Parameter Influences and Measurement Methodology," in J. T. Dobbins III and J. M. Boone (Eds.), *Proc. of 2000 Conf. On Medical Imaging*, SPIE, Vol. 3977, Bellingham, WA, 2000, pp. 2–13.

30. S. A. Horton, "Detection of Surface Defects in Ceramic Rolling Elements," Fourth Int. Symp. on Ceramic Materials and Components for Engines, Goteborg, Sweden, 1991, pp. 897–904.
31. H. Domon, K. Uemura, and K. Fujiwara, "Liquid Penetrant Test for Fine Ceramics," Proc. Eight Asia-Pacific Conference on Nondestructive Testing, Taipai, Taiwan, 1995, pp. 711–718.
32. S. Jahanmir, M. Ramulu, and P. Koshy, *Machining of Ceramics and Composites*, Marcel Dekker, New York, 1999.
33. S. Jahanmir, *Machining of Advanced Materials*, National Institute of Science and Technology (NIST), NIST Special Publication 847, 1993.
34. R. E. Barks, K. Subramanian, and K. E. Ball, *Intersociety Symposium on Machining of Advanced Ceramic Materials and Components*, Am. Cer. Soc., Westerville, OH, 1987.
35. W. A. Ellingson, J. A. Todd, and J. G. Sun, Optical Method and Apparatus for Detection of Defects and Microstructural Changes in Ceramics and Ceramic Coatings, US Patent 6,285,449, issued Sept. 2001.
36. J. G. Sun, M. H. Haselkorn, and W. A. Ellingson, "Laser Scattering Detection of Machining-Induced Damage in Si_3N_4 Components," in R. E. Green, Jr. (Ed.), *Nondestructive Characterization of Materials*, Vol. VIII, Plenum, New York, 1998, pp. 365–370.
37. I. Fujii and K. Kawashima, *Rev. Prog. in Quan. NDE*, **14**, 203–209 (1995).
38. W. A. Ellingson, E. R. Koehl, D. Sandberg, P. Pastila, A. P. Nikkila, and T. A. Mantyla, "Acousto-Ultrasonic Nondestructive Evaluation of Rigid Ceramic Hot-Gas Filters," in *Euro Ceramics*, Vol. VII, Key Eng. Mat. Vols. 206–213, Trans. Tech. Pubs, Switzerland, 2002, pp. 613–616.
39. P. Plastila, A. P. Nikkila, T. Mantyla, W. A. Ellingson, E. R. Koehl, and D. Sandberg, *Cer. Eng. Sci. Proc.* (2002).
40. A. Tiwari and E. G. Henneke, in R. E. Green, Jr., K. J. Kozacek and C. O. Rudd (Eds.), *Nondestructive Characterization of Materials*, Vol. VI, Plenum, New York, 1994, pp. 247–254.
41. H. Reis, "Acousto-ultrasonic Nondestructive Evaluation of Porosity in Graphite Fiber Reinforced PMR-15 Composite Panels," in W. J. McGonnagle (Ed.), *Intl. Advances in Nondestructive Testing*, Gordon and Breach, Longhorn, PA, 1994.
42. A. G. Evans and R. Naslain, "High-Temperature Ceramic-Matrix Composites I: Design, Durability and Performance," in *Cer. Trans.*, Vol. 57, Am. Cer. Soc., Westerville, OH, 1995.
43. A. G. Evans and R. Naslain, "High-Temperature Ceramic-Matrix Composites II: Manufacturing and Materials Development," in *Cer. Trans.*, Vol. 58, Am. Cer. Soc., Westerville, OH, 1995.
44. A. Russell, L. Mollinex, M. Holmquist, and O. Sudre, Oxide/oxide Ceramic Matrix Composites in Gas Turbine Combustors, ASME Paper 98-GT-30, 1998.
45. F. K. Ko, *Am. Cer. Bull.*, **68**(2), 401–414 1989.
46. R. A. Lowden, M. K. Ferber, J. R. Hellman, K. K. Chawla and S. G. Pietio, *Ceramic Matrix Composites—Advanced High Temperature Structural Materials*, Mat. Res. Soc, Vol. 365, 1995.

47. J. G. Sun, T. E. Easler, A. Szweda, T. A. K. Pillai, C. Deemer, and W. A. Ellingson, *Cer. Eng. Sci. Proc.*, **20**(3), 200–210 (1999).
48. W. A. Gradia and C. M. Fortunko, *Proc. IEEE Ultrasonic Symp.*, **1**, 697–709 (1995).
49. C. Wykes, *Nondestr. Test. Eval.*, **12**, 155–180 (1995).
50. W. P. Winfree and P. H. James, "Thermographic Detection of Disbonds," *Proc. 35th Int. Instru. Symp*, 1989, pp. 183–188.
51. J. G. Sun, C. Deemer, W. A. Ellingson, T. E. Easler, A. Szweda, and P. Craig, "Thermal Imaging Measurement and Correlation of Thermal Diffusivity in Continuous Fiber Ceramic Composites," in P. S. Gaal and D. E. Apostolescu (Eds)., *Thermal Conductivity V. 24*, Technomic Pub., Lancaster, PA, 1997, pp. 616–622.
52. P. Cielo, R. Lewak, X. Maldague, and M. Lamontagne, *J. Canadian Soc Nondestructive Testing*, **7**(2), 30–49 (1986).
53. T. W. Spohnholtz, Development of Impact Resonant Spectroscopy for Characterizing Structural Ceramic Components, MS Thesis, University of Illinois-Chicago, 1999.
54. R. A. Bemis, K. Shiloh, and W. A. Ellingson, *Trans. ASME, J. Eng. Gas Turbines Power*, **118**, 491–424 (1998).
55. W. J. Parker, R. J. Jenkins, C. P. Butler, and G. L. Abbott, *J. App. Phys.* **32**, 1679–1684 (1961).
56. J. Price, O. Jimenez, V. J. Parthasarthy, N. Miryala, and D. Levoux, Ceramic Stationary Gas Turbine Development Program, Seventh Annual Summary, ASME Paper 2000-GT-75, 2000.
57. H. E. Eaton, G. Linsey, E. Sun, K. L. More, J. Kimmel, J. Price, and N. Miryala, EBC Protection for SiC/SiC Composites in the Gas Turbine Combustion Environment: Continuing Evaluation and Refurbishment Consideration, ASME Paper 2001-GT-513, 2001.
58. W. A. Ellingson, J. G. Sun, K. L. More, and R. G. Hines, Characterization of Melt-Infiltrated SiC/SiC Composite Liners Using Meso and Macro NDE Techniques, ASME Paper 2000-GT-0067, 2000.
59. G. G. Genge and M. W. Marsh, Carbon Fiber Reinforced/Silicon Carbide Turbine Blisk Testing in the SIMPLEX Pump, Proc. of the 49th Joint Army, Navy, Air Force (JANNAF) Meeting, 1999, pp. 89–97.
60. R. Talreja, *J. of Strain Anal.*, **24**(4), 215–222 (1989).
61. G. N. Morscher, *Composite Sci. Tech.*, **59**, 687–697 (1999).
62. I. A. Viktorov, *Rayleigh and Lamb Waves, Physical Theory and Applications*, Plenum, New York, 1969.
63. D. Krajcinovic, *Appl. Mechanics Rev.*, **37**(1), 1–6 (1984).

CHAPTER 14

Advances in Rapid Prototyping and Manufacturing Using Laser-Based Solid Free-Form Fabrication

ERIC WHITNEY

Pennsylvania State University, State College, Pennsylvania 16804

14.1 OVERVIEW

It is well known that product cost is largely dependent on decisions made during its development stage [1]. As a product evolves through its development stage, it becomes difficult to alter design and manufacturing approaches without incurring additional cost. In fact, such design changes can be so costly as to affect adversely the financial health of a company. An example of this is the delay and

Handbook of Advanced Materials Edited by James K. Wessel
ISBN 0-471-45475-3

design changes that Rolls-Royce and Lockheed suffered during the development of the Rolls-Royce RB211 engine for the L-1011 aircraft [2]. The cost of changing the fan blade material from a composite material to a titanium alloy was so significant that Lockheed was forced from the commercial airliner business. It is not supposed here that if Lockheed had rapid manufacturing technology available to it during the 1970s that its problems could have been avoided. However, the example illustrates the degree in which early design and manufacturing decisions affect the profitability of a company.

From the observation that product costs are largely a result of poor conceptualization during the product development stage, the concepts of rapid prototyping and rapid manufacturing emerged. There are many rapid prototyping concepts, and these have been reviewed [3, 4] in detail. This chapter will address laser-based solid free-form fabrication techniques.

14.2 HISTORICAL REVIEW OF LASER-BASED SOLID FREE-FORM TECHNOLOGY

The development of laser-based solid free-form fabrication has its roots in two distinctly separate technological evolutions. The first evolution involved research organizations working on the development of laser materials processing technologies such as laser welding and cladding and the extension of these technologies to rapid manufacturing. The second evolution was the development of rapid prototyping processes based on the deposition of polymers.

14.3 RAPID PROTOTYPING DEVELOPMENT

There are three major rapid prototyping approaches that have influenced the development of laser-based rapid manufacturing in metals. These are stereolithography [5, 6], laminate object manufacturing [7], and selective laser sintering [8]. Each of these technologies use sophisticated computer codes to convert solid models into two-dimensional "slices" along a particular stacking axis. In the case of stereolithography, a photosensitive polymer is extruded through a nozzle and cured using an ultraviolet laser. In laminated object manufacturing, thin plastic sheets are laser cut and then stacked to form a three-dimensional structure. Selective laser sintering uses polymer powder as a feedstock and a laser to fuse them together. The resulting objects made from these three processes are prototypes used for the purposes of evaluating form and fit. Shapes made from these processes are not structural. Recently, work done on improving the strength and temperature capability of the resins and polymer feedstocks has resulted in pseudostructural parts (i.e., parts that have structural integrity over a very limited life span). The key element of these rapid prototyping processes are the computer codes used to control machine movement, and it is this technology that can be transitioned to rapid manufacturing of metal components. Of the three technologies mentioned above the selective laser sintering (SLS) process has been successively modified to be used with metal powders, as described later.

14.4 EARLY LASER STRUCTURAL SHAPE MAKING TECHNIQUES

Researchers developing laser materials processing technology also experimented with the fabrication of structural components by the successive buildup of layers completely melted and fused to the previous layer. One such process [9] involved using a laser to fuse added powder or wire on a rotating mandrel. Successive layers were built and found to be of high metallurgical quality. Graded compositions were also produced. In another process [10] a laser was used to melt and fuse layers of powder that was predeposited on the substrate or previous layer by means of a fluidized bed. The process was used to form crude titanium shapes. Neither of the previous two patents, however, discussed the method by which the shape would be generated from electronic data. In a third process [11] a laser was used to completely melt and fuse added powder to a substrate or previous layer using a computer-controlled machine with electronic data representing "slices" of the component being built. Examples of laser depositing of Ti-6Al-4V and alloy 718 were provided. However, all of these "early" efforts in making structural shapes using laser deposition technology did not demonstrate the complex shape-making capability that has been the signature of modern rapid prototyping technologies discussed above.

Only after the merger of these two independent technologies is it possible to claim that rapid manufacturing of structural metal components is possible.

14.5 DESCRIPTION OF VARIOUS LASER-BASED RAPID MANUFACTURING TECHNIQUES

All of the laser-based rapid manufacturing processes have the following three common machine elements:

1. A laser beam delivery system
2. A powder or wire feedstock delivery system
3. A means to manipulate the laser beam or the part or both in at least three linear dimensions

Other enhancements often included in laser-based solid free-form fabrication processes are some type of environmental control apparatus so that processing can be accomplished under an inert or reactive gas environment and a vision system. In some cases, it seems feasible to operate such systems under vacuum. In addition control features are often necessary to control or monitor the process. Each implementation of laser-based solid free-form fabrication claims a unique combination of the three common elements and enhancements that provides certain advantages. There are approximately seven laser-based solid free-form fabrication processes being actively developed and commercialized. Each one of these processes will be described briefly, and then a detailed description of one of these processes (the Penn State process) will be provided.

Los Alamos National Laboratory is developing a process called Directed Light Fabrication (DLF) [12]. In this process, a part is moved in X and Y and the laser-focusing head is moved in the Z direction as the structure is built. A fiber optically delivered Nd: YAG laser is used in this process. The process is conducted in an inert atmosphere. Layer thickness is between 0.075 and 0.25 mm. Materials reported to be processed include 410 stainless steel, P20 tool steel, Ni–Al, molybdenum disilicide, Ag–Cu alloys, and Fe-28%Ni. Deposition rates and other process particulars were not been reported, nor have basic mechanical properties of the deposits for many of these materials. More recently, DLF has been used to process rhenium and iridium with some success [13]. Porosity was reported in the deposits, but this was thought to be due to impure feedstock.

DTM Inc. has commercialized the SLS process using polymer-coated metal powder [14, 15]. In this process a laser is used to cure the polymer-coated metal powder such that the resulting structure has sufficient green strength to be handled subsequent to sintering in a furnace. The results show that dimensional control can be achieved by careful coating of the powder and selection of processing conditions. As expected sintered parts were porous as it is exceedingly difficult to get 100% dense material from a sintering operation. Nevertheless, the SLS process shows great promise in reducing the cost of press and sintered powder metal components by replacing traditional powder metal sintering processes. Materials reported to have been deposited by SLS are Ti-6Al-4V, 1018 steel, BNi-6 and a cobalt braze material [16], and intermetallics [17]. Unfortunately, mechanical properties for the material were not reported. Back infiltration of Cu in ferrous-based SLS structures has also been accomplished [18]. Infiltration of the structure with copper is reported to provide additional structural strength. Cu infiltration of ferrous powder metallurgy products is a common practice. Further, SLS has been used to form integral cans for subsequent hot isostatic pressing (HIP) of the SLS green part [19]. This is the so-called SLS/HIP process. Can removal of conventionally prepared hot isostatically pressed parts is expensive. If the can could be formed from the same material as the component then a near-net-shape part could be produced with little subsequent machining. In yet another adaptation of the process, SLS has been used to fabricate *in situ* cermets for turbine blade tip manufacture and restoration [20]. In this process a titanium-coated oxide particle, apparently based on a particle described by Cooper et al. [21], is used as the ceramic particle in a metal matrix. Titanium is chosen since it will promote wetting of the oxide particle by the metal matrix. Wear testing has shown this combination of metal matrix and titanium-coated oxide provides improved performance of the turbine.

The University of Liverpool has perfected a process called laser direct casting (LDC) [22]. In LDC a six-axis computer-controlled machine is capable of building three-dimensional objects without the limitation of the build angle or without the need for integral supports. In the LDC a 1.5-kW CO_2 laser was used to deposit stainless steel up to 9 mm^3/s. Travel speeds between 500 and 1000 mm/min were used, and it was found that fully dense, porosity-free deposits were readily obtained. The efficiency of powder usage was reported to be between 8 and 24% using a powder nozzle that dispersed powder over a wide area, much wider than the melt puddle. When an improved nozzle was used, powder utilization rose to as high as 85%. Microstructure of LDC deposited 316L stainless steel was that of a very fine grained casting with epitaxial growth between layers. Small pores were observed in the deposits of Ni, but its presence was discounted as insignificant, and no mechanical properties were reported.

The University of Michigan has developed a process called Direct Metal Deposition (DMD) [23]. In DMD a 5-kW CO_2 laser is used to completely melt and fuse powder feedstock. The DMD process is similar in nature to the previously described DLF processes. DMD has been successively used to deposit H13 tool steels in complex die patterns. It was found that the microstructure of the DMD deposited H13 was finer than that of conventionally processed material and that this may have beneficial effects in terms of reducing wear. Post-DMD heat treatments showed that the H13 material responds to conventional temper operations and that post-DMD tempering is desirable. Hardness, microstructure, and residual stress measurements were reported for DMD H13; however, no mechanical properties have thus far been published.

Sandia National Laboratory has pioneered the development of a process called laser engineered net shapes (LENS) [24]. In the LENS process a fiber optically delivered Nd:YAG laser is used to completely melt and fuse powder feedstock. The process is performed under an inert argon atmosphere where the oxygen content is reported to be less than 10 ppm. Currently, the LENS system utilizes a three-axis motion system where the part translates in X and Y and the laser beam is adjusted in the Z direction as the part builds. Many materials have been processed using LENS, including 316L stainless steel, alloy 625, Ti-6Al-4V, and H13 tool steels [25]. As will be shown later, tensile data has been reported for 316L stainless steel, alloy 625, and Ti-6Al-4V. A schematic diagram of the process, as implemented by Optomec Design Corporation, is shown in Fig. 14.1.

Figure 14.2 represents a complex, hollow shape that was fabricated using the LENS process. The shape encompasses features such as internal passages, angles, corners, and variation in deposit thickness.

Huffman Corporation (Clover, SC) has developed a unique laser cladding machine for the repair of turbine blade tips. It is a five-axis computer-controlled machine usually equipped with a 1- or 2-kW CO_2 laser. Although not explicitly used for laser-based free-form fabrication, it has been used to make some interesting shapes. Figure 14.3 shows golf tees fabricated from alloy 625. No metallurgical or mechanical tests were performed on the deposits.

Penn State University has developed a high-power laser-based solid free-form fabrication process under a DARPA/ONR contract (N00014-95-C0029), which was directed by the Applied Physics Laboratory, Johns Hopkins University, and for which MTS Inc. provided a preliminary design of a commercial unit. Material produced under the DARPA/ONR program was equivalent in tensile strength,

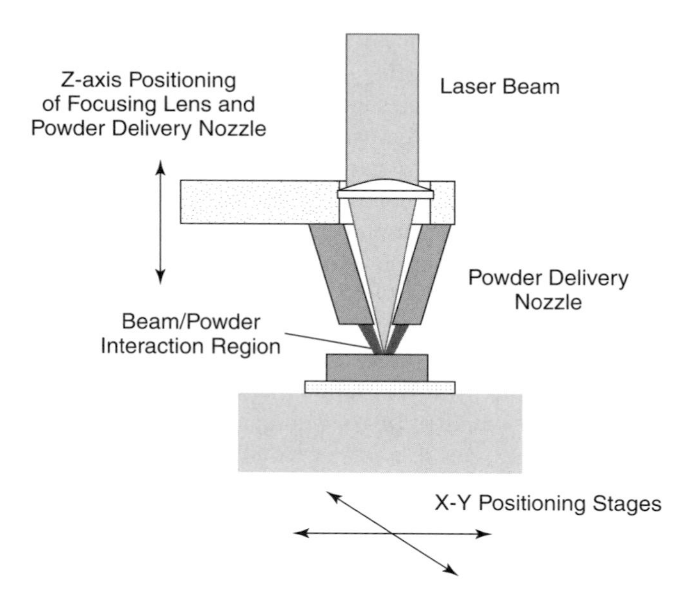

FIGURE 14.1 Schematic of LENS process.

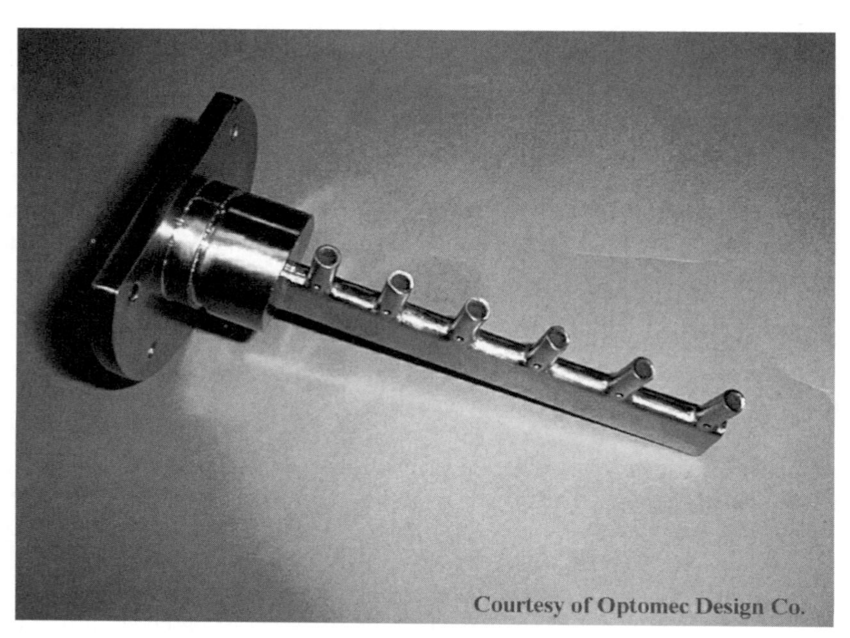

FIGURE 14.2 Intricate shape made by the LENS process (part is approximately 7 cm long).

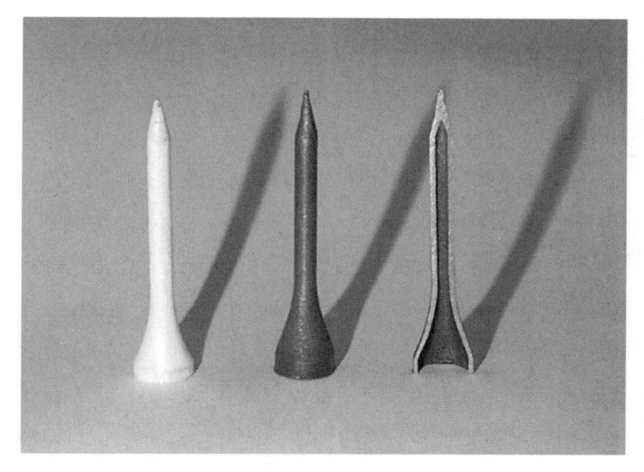

FIGURE 14.3 Intricate shapes made by Huffman Corp. (actual tee left, laser-fabricated tees center and right).

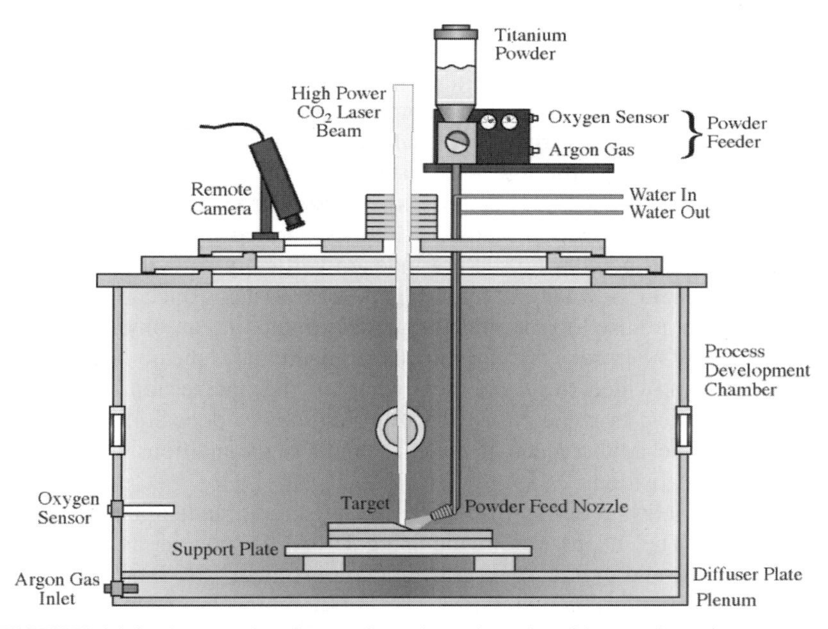

FIGURE 14.4 Schematic of Penn State laser-based rapid manufacturing process.

tensile ductility, charpy impact, and fatigue to conventionally cast and hot isostatically pressed material of similar composition [26, 27]. A schematic of the process is shown in Fig. 14.4.

In the Penn State process, a 14-kW CO_2 laser is introduced into a specially designed processing chamber via an opening in the lid. In this system the laser

FIGURE 14.5 Large laser-based rapidly manufactured Ti-6Al-4V shapes.

beam is translated in the X, Y, and Z directions; the part remains stationary or can be rotated in the X-Y plane. The laser beam manipulation device is attached to the center lid of the processing chamber so that as the laser beam is translated the lids move in unison, keeping the lid opening and the laser beam coincident. Argon or nitrogen gas enters the processing chamber near the bottom and is distributed throughout the chamber through a diffuser plate. The powder feeder is situated above the target and is fed into the processing chamber via a water-cooled copper feed tube. A nozzle directs the powder into the molten metal. After each layer the laser beam and powder nozzle are indexed vertically so that the relationship between laser beam, powder entry point, and target remains unchanged over time. Various sensors are used to monitor the process, including an oxygen sensor and vision system. Other sensors periodically measure laser beam attributes. The information is provided to an operator who can then make adjustments as the process progresses. In some cases brief interruptions in deposition can be made in order to make adjustments to machine and laser conditions that cannot be made during deposition.

Figure 14.5 shows a representative sample of shapes made under the DARPA/ONR contract. The shapes were made using Ti-6Al-4V powder as the feedstock. Corners, curves, and cylinders have been demonstrated.

Aeromet Corp. has commercialized the DARPA process under the name Lasform [28]. The process currently practiced by Aeromet is limited to titanium alloys and is capable of building complex components with a footprint of approximately $3.6 \times 1.2 \times 1.2$ m.

14.6 MATERIAL PROPERTIES

One of the discriminating factors between rapid prototyping and rapid manufacturing lies in the difference in material property and material quality requirements.

In rapid manufacturing the product must be functionally fit for use, and the manufacturing process must be capable of producing material of a consistent customer-defined quality level. As a result the demands are placed on rapid manufacturing processes that are not present for rapid prototyping. This difference is what separates rapid prototyping from rapid manufacturing.

14.7 COMMON DEFECT STRUCTURES

There are two common defect structures that can and have occurred in all laser-based rapid manufacturing processes. These are porosity and lack-of-fusion (LOF). In some cases porosity may result in a chemical reaction between different feedstocks. For example, if a high-carbon ferrous-based substrate is used in combination with a copper powder feedstock high in oxygen, then carbon may react with the oxygen and form CO and CO_2 gas, which can then be entrapped as porosity. Porosity can also be caused by out gassing of powder during processing [29]. Figure 14.6 shows an LOF defect in a Ti-6Al-4V deposit; such defects are hard to detect by nondestructive means. Process control must be such that the formation of LOF defects is completely avoided.

Of the common defects, LOF is the most damaging since it may act as a crack under certain loading conditions. A single LOF defect can lead to component failure. Excessive porosity (either size of individual pores or the number density of pores) can also be damaging particularly to tensile ductility.

Other defects that will be specific to the deposition process and material include various types of cracking such as hot shortness, strain age, and tearing during solidification. Cracking must be controlled by careful selection of deposition material and process parameters.

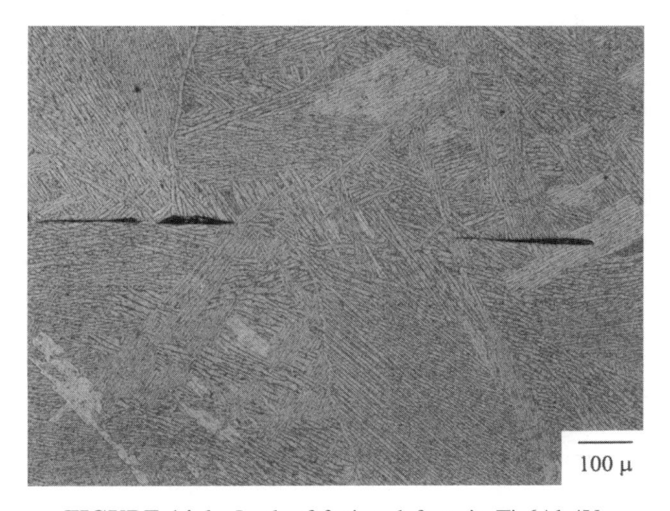

FIGURE 14.6 Lack-of-fusion defects in Ti-6Al-4V.

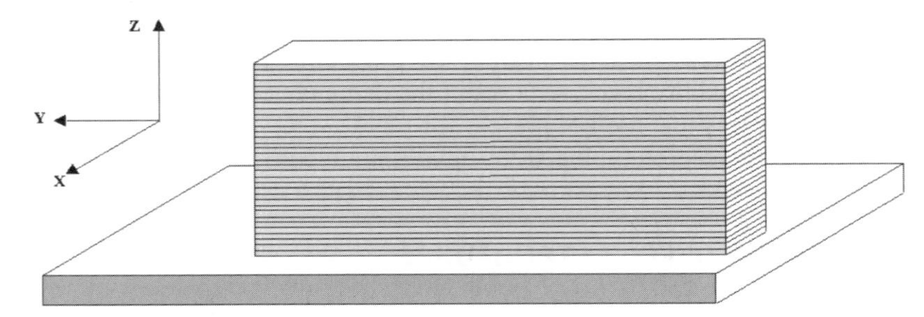

FIGURE 14.7 Normalized frame of reference for mechanical test data.

Anisotropic material properties are a real concern for those considering using laser-deposited material and is an area that requires additional investigation. Epitaxial growth between layers and a near unidirectional heat extraction result in a highly oriented microstructure. A possible exception to this observation would be the SLS/HIP process. SLS/HIP does not rely on the laser beam to completely melt powder and as a result may exhibit more isotropic material properties than other rapid manufacturing processes. The *Z* direction is the build direction. The laser beam axis is along a *Z* direction. The *Y* direction is the laser beam travel direction. Mechanical testing occurs in the *Z* or the *Y* direction since the samples are usually too thin to test in the *X* direction, as shown in Fig. 14.7

14.8 STAINLESS STEEL

Tensile properties for 316 stainless steel have been reported for the LENS [24] and for 316 for the directed light fabrication (DLF) process [29]. Table 14.1 provides the results. Note that the ultimate tensile yield strength reported for the LENS process is higher than typically observed for type 316 plate and far exceed the minimum expected values for bar. Ductility is excellent for the reported strength.

TABLE 14.1 Mechanical Properties of 316 Stainless Steel

Process	Reference	Condition	Test Direction	UTS (MPa)	YS (MPa)	% Elong.	% R of A
LENS	33	As deposited	*Z*	790	450	66	N/R
LENS	33	As deposited	*Y*	805	490	33	N/R
DLF	29	As deposited	N/R	579	296	41	N/R
316 Plate (typical)	34	Annealed	N/A	621	296	52	72
316 Bar (min)	35	Annealed	N/A	515	205	40	N/A

Note: UTS = ultimate tensile strength; YS = yield strength.

The increase in strength was attributed to classic Hall–Petch strengthening as the dislocation density of the LENS material was approximately the same as that for annealed bar. It should be noted that the LENS material was tested in the as-deposited condition while the handbook values are for annealed bar. There is some question as to whether the LENS material is thermally stable, that is, if the LENS material were annealed, there might be a reduction in strength. Chemical analysis of the powder and the deposit were not reported for the LENS or the DLF material.

14.9 ALLOY 625

Alloy 625 is nominally 21.5Cr-9Mo-3.5Nb and small amounts of Al and C. It is used widely in the aerospace, paper, and petrochemical industries because of its excellent balance of strength and corrosion and oxidation resistance. The results for alloy 625 processed by the LENS system are compared to annealed bar in Table 14.2.

Chemical analysis of the alloy 625 processed by Penn State is shown in Table 14.3, for the deposit, Unfortunately, the chemical composition of the powder was not available for comparison.

14.10 ALLOY 690

Alloy 690 is nominally 58Ni-29Cr-9Fe-0.15N. It is a nitrogen-strengthened austenitic material with excellent corrosion resistance. Alloy 690 has been successfully processed using the DLF process. Ultimate tensile strength (UTS)

TABLE 14.2 Mechanical Properties of Alloy 625

Process	References	Condition	Test Direction	UTS (MPa)	YS (MPa)	% Elong.	% R of A
LENS	33	As deposited	*Z*	930	635	38	N/R
LENS	33	As deposited	*Y*	930	515	37	N/R
Cast	Mil-Hdbk-5	Annealed	N/R	710	352	48	N/A
Annealed bar (typical)	35	Annealed	N/R	855	490	50	N/R

TABLE 14.3 Chemical Composition of Alloy 625

	Composition (wt %)						
Process	Ni	Cr	Mo	Nb	C	O	N
Penn State	63.14	21.34	8.53	3.66	0.02	0.0328	0.0954

TABLE 14.4 Mechanical Properties of Alloy 690

Process	Reference	Condition	Test Direction	UTS (MPa)	YS (MPa)	% Elong.	% R of A
DLF	29	As deposited	N/R	666	450	48.8	N/R
Bar	35	As rolled	N/R	765	434	40	N/R
Bar	35	Annealed	N/R	710	317	49	N/R
Plate	35	As rolled	N/R	765	483	36	N/R

and yield strength (YS) are shown in Table 14.4. It is interesting to note the high yield strength in combination with excellent ductility of the DLF compared to conventionally prepared material.

14.11 17-4PH STAINLESS STEEL

17-4PH stainless steel is a martensitic alloy that possesses a good balance of strength, toughness, and corrosion resistance for temperatures below about 350°C. It is susceptible to stress corrosion cracking when tempered to a hardness above about Rc40. 17-4PH contains about 4% copper. The copper precipitates as nearly pure copper during tempering. There are several commercial heat treatments for 17-4PH that yield a different combination of properties. The H1075 heat treatment is often used where the component may be exposed to corrosive conditions. Briefly, the H1075 heat treatment is an austenization at 1066°C and a temper at 634°C. It is this heat treatment that was given to the material deposited by the Penn State process; the results are shown in Table 14.5.

It is important to note that the data given for the material processed by Penn State are preliminary and were taken from the first batch of 17-4PH deposited; it should be assumed that this is a nonoptimized process. Improvement in the mechanical properties could be achieved with further process development.

The chemical analysis of the 17-4PH material is shown in Table 14.6. Note that the oxygen content in the powder is high. (Typically stainless steel powder

TABLE 14.5 Mechanical Properties of 17-4PH

Process	Source	Condition	Test Direction	UTS (MPa)	YS (MPa)	% Elong.	% R of A
Penn State	Penn State	H1075	*Y*	975	927	17.5	52.9
Cast min. (test bar)	Mil-Hdbk-5	H1000	—	1034	896	8	20
Cast (interpolated)	—	H1075	—	931	844	—	—
Cast min. (test bar)	Mil-Hdbk-5	H1100	—	896	827	8	15
Bar (typical)	TBD	H1075	N/R	1151	999	14	55

TABLE 14.6 Chemical Analysis of 17-4PH

	Composition (wt %)						
Process	Fe	Cr	Cu	Nb	Ni	O	N
Powder	Rem	15.58	2.95	0.095	4.02	0.212	0.0045
Penn State	75.5	15.6	3.24	0.16	4.4	0.073	0.0138

should have an oxygen content less than 0.03%). The oxygen content of the deposit is reported to be three times lower than that of the powder, while the nitrogen level showed about a three times increase. Clearly, additional testing and chemical analysis are needed to understand the reported results.

14.12 NICKEL–ALUMINUM–BRONZE

Nickel–aluminum–bronzes (NAB) are complex alloys consisting of Ni, Al, Fe, and Mn as major alloying elements. Cast NAB is used for propeller blades and hubs, valve bodies, and other components exposed to seawater. NAB alloys, as a class, have excellent corrosion and cavitation resistance in combination with relatively high strength and ductility. NAB alloys are used in both a cast and wrought form.

The microstructural development and resulting mechanical and environmental properties of cast NAB alloys are dependent on alloy chemistry and thermal history. In large castings local variation in chemistry may be outside the bulk chemical limits imposed by specification. In addition, local differences in cooling rates (large castings may take more than a week to cool) also affect the type, volume fraction, and morphological characteristics of matrix and precipitating phases.

One of the most widely used NAB alloys is UNS C95800, which has a nominal composition of Cu-9Al-5Ni-5Fe-1Mn. Laser-based rapid manufacturing was performed using the Penn State process and compared to conventionally cast NAB. Laser deposition was conducted under both argon and nitrogen atmospheres and no practical differences in chemistry or tensile properties were noted. Table 14.7 lists the results of tensile testing.

Table 14.8 shows the chemical analysis of the powder and laser-deposited material used for the mechanical testing shown above. Note that there is essentially no difference between the starting powder and the laser deposits using either argon or nitrogen as the inert cover gas. Obviously, nitrogen would be preferred since the cost of nitrogen is about five times lower than that of argon.

Microstructural analysis of laser deposits has also been made using powder feedstock with a slightly lower aluminum content and is shown in Table 14.9. The microstructures in Figs. 14.8– 14.10 show that the laser-deposited microstructure is much finer and more homogenous than a conventional casting. The laser-formed microstructure consists of a network of proeutectoid α surrounding prior β dispersed with various κ phases.

TABLE 14.7 Mechanical Properties of Nickel–Aluminum–Bronze

Process	Source	Condition	Test Direction	UTS (MPa)	YS (MPa)	% Elong.	% R of A
Penn State (argon)	ARL-Penn State	As deposited	*Z*	660	267	28	28.4
Penn State (argon)	ARL-Penn State	As deposited	*Y*	684	283	25	25.5
Penn State (nitrogen)	ARL-Penn State	As deposited	*Y*	656	256	24.5	24.9
Cast (min)	ASTM B148	As cast	N/R	585	240	15	N/A

TABLE 14.8 Chemical Analysis of NAB Made for Tensile Testing

	Composition (wt %)							
Process	Cu	Ni	Mn	Fe	Al	Si	O	N
Penn State (argon)	81.5	4.61	1.16	3.57	9.04	0.042	0.0011	<0.001
Penn State (nitrogen)	81.4	4.62	1.15	3.58	9.07	0.040	0.0015	<0.001
Powder	81.7	4.61	1.16	3.59	8.81	0.020	0.0022	0.00014

TABLE 14.9 Chemical Analysis of NAB Used for Microstructural Analysis

Element	UNS C98500 (wt %)	Starting Powder (wt %)	Laser-Formed Ingot (wt %)
Cu	Balance	83.1	83.4
Al	8.5–9.5	6.70	6.75
Ni	4.0–5.0 and (>%Fe)	4.65	4.64
Fe	3.5–4.5	3.72	3.52
Mn	0.8–1.5	1.14	1.14
Si	0.10 max.	0.43	0.36
Pb	0.03 max.	0.007	0.012
O	—	0.012	0.065
N	—	0.001	<0.001

14.13 Ti-6Al-4V

Ti-6Al-4V alloy has been investigated using the LENS, SLS/HIP, DLF, Aeromet, and the Penn State processes. Results of mechanical properties are shown in Table 14.10 and are compared to a variety of other forms of Ti-6Al-4V. The

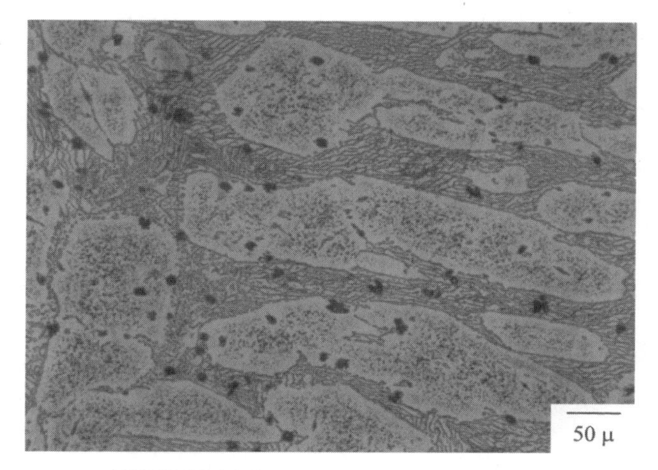

FIGURE 14.8 Conventionally cast NAB.

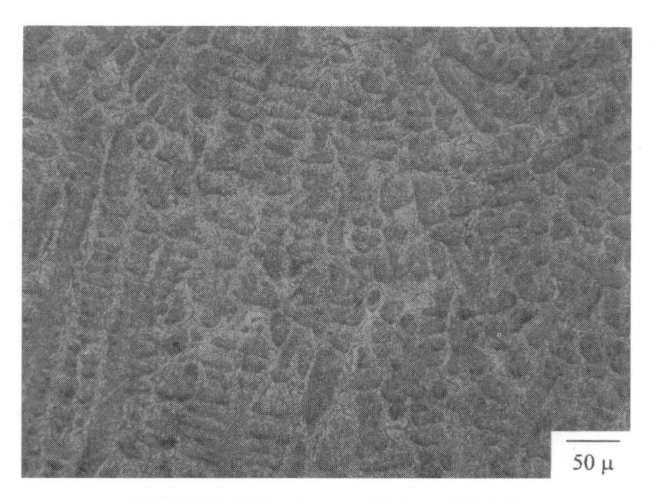

FIGURE 14.9 Laser-fabricated NAB.

LENS material has an excellent combination of strength and ductility, although its ductility is about what is typical for a casting and its strength is much higher. Again, it is important to consider the effect of heat treatments on mechanical property, and the LENS material was only tested in the as-deposited condition. The SLS/HIP material also has excellent strength, but its ductility is lower than all other processes and forms. The Penn State process has produced material properties between wrought and cast. The materials produced by the Penn

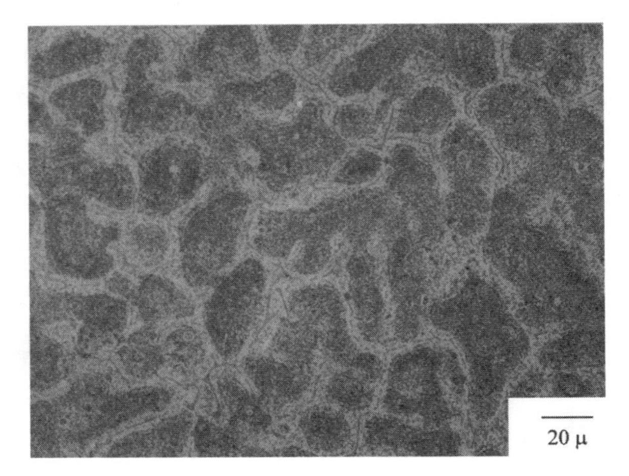

FIGURE 14.10 Laser-fabricated NAB.

TABLE 14.10 Mechanical Properties of Ti-6Al-4V

Process	Reference	Condition	Test Direction	UTS (MPa)	YS (MPa)	% Elong.	% R of A
LENS	24	As-deposited	*Z*	1172	1067	11	N/R
SLS/HIP	36	As-HIP'ed	N/R	1117	N/R	5	N/R
DLF	29	Mill annealed	N/R	1027	958	6.2	N/R
Aeromet	28	N/R	N/R	896–1000	827–896	9–12	18–22
Penn State	ARL-Penn State	Aged	*Y*	995	850	10.7	23.8
Penn State	ARL-Penn State	Mill annealed	*Y*	979	848	8.5	14.3
Press and Sinter	28	As-sintered	N/A	945	868	15	25
Cast (typical)	35	Annealed	N/A	1015	890	10	16
Cast (min)	ASTM B367 C5	Annealed	N/A	896	827	6	N/R
Wrought bar (typical)	35	Annealed	N/A	1000	925	16	34
Wrought (Armor, min)	Mil-A-40677	Annealed	N/A	896	827	12	30

State process exceed minimum values of both cast and wrought materials but were lower than typical values for the same material. The ductility of the Penn State material also exceeded minimum requirements but is less than typical for cast and wrought forms. It is interesting to note that the aging heat treatment for the Penn State material seemed to improve the material's ductility without much change in tensile and yield strength. Chemical analysis of SLS/HIP is shown in Table 14.11. Plasma rotating electrode powder (PREP) was used for the SLS/HIP investigation. The data are for powder that has been recycled six

TABLE 14.11 Chemical Analysis of SLS/HIP Ti-6Al-4V

	Composition (wt %)		
Process [36]	O	N	H
SLS/HIP	0.25	0.019	N/r
PREP powder	0.19	0.01	N/r

TABLE 14.12 Chemical Analysis of Penn State Processed Ti-6Al-4V

	Composition (wt %)						
Process [33]	Ti	Al	V	Fe	O	N	H
Penn State	Rem	5.83	3.75	N/c	0.23	0.048	0.0072
Powder	Rem	6.01	3.95	0.199	0.23	0.032	0.012

times; therefore, it is estimated that the average increase in oxygen per SLS/HIP cycle is 0.01%.

The chemical analysis of material produced by the Penn State process is given in Table 14.12. Results indicate that oxygen and nitrogen pick-up is low and manageable, similar to other processes.

14.14 Ti-48-2-2 (γ-TiAl)

Gamma titanium aluminides are a class of alloys undergoing significant development for use in a variety of aerospace applications [30]. The alloys are quite remarkable in that they have increasing yield strength with increasing temperature. At room temperature, however, the alloys are very brittle and are difficult to form. These alloys are difficult to cast, work, and weld, and there has been much work on the development of cost-effective forming techniques using powder metallurgy techniques [31]. Laser-based free-form techniques present interesting possibilities with respect to the cost-effective manufacture of γ-titanium aluminide alloys. There has been little work done on these alloys to date, although this will change in the near future.

Work performed at Penn State under the direction of Crucible Research was undertaken to obtain preliminary chemical and tensile data on laser forming of Ti-48Al-2Cr-2Nb alloy and has been reported by Moll et al. [32].

Figure 14.11 shows a cross section of the *X-Z* plane of a monolithic block of Ti-48Al-2Cr-2Nb produced at Penn State. The block was approximately 250 mm long by 75 mm high by 30 mm wide. Microstructures of the as-deposited and heat-treated material are shown in Figure 14.12. Note that pass lines are evident,

FIGURE 14.11 γ-TiAl (Ti-48Al-2Nb-2Cr) processed at Penn State.

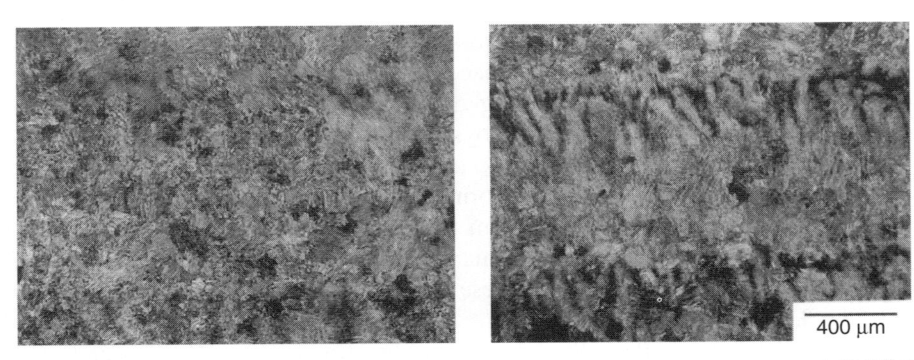

FIGURE 14.12 γ-TiAl (Ti-48Al-2Nb-2Cr) as-deposited (right) and heat treated 1338°C 0.5 h AC + 871°C 4 h AC.

these are probably due to solute banding similar to that observed in multipass weldments. The microstructure is fully lamellar.

Mechanical properties and chemical information for as-deposited material are shown Tables 14.13 and 14.14.

TABLE 14.13 Mechanical Properties of γ-TiAl (Ti-48Al-2Nb-2Cr)

Test Temp (°C)	Condition	Test Direction	UTS (MPa)	YS (MPa)	% Elong.	% R of A	Reference
21	As deposited	*Z*	586	550	1.5	2	32
21	Heat treated	*Z*	577	500	1.5	3	32
760	As deposited	*Z*	606	416	2	3	32
815	As deposited	*Z*	530	422	17.5	25.2	32
21	As deposited	*Y*	589	—	0.12	—	Crucible
815	As deposited	*Y*	589	386	2	5.5	Crucible
21	Heat treated	*Y*	468	454	1	< 1	Crucible
815	Heat treated	*Y*	451	344	2	1.6	Crucible
21	HIP	*Y*	386	—	0.34	1.8	Crucible
815	HIP	*Y*	451	332	5.5	7.8	Crucible

Heat treated: 1338°C 0.54 h AC + 781°C 4 h; HIP: 1200°C 103 MPa.

TABLE 14.14 Chemical Analysis of γ-TiAl (Ti-48Al-2Nb-2Cr)

	Composition (wt %)							
Process	Al	Nb	Cr	Fe	C	O	N	Ti
Penn State (sample 1)	32.26	4.74	2.68	0.06	0.018	0.061	0.007	Rem
Penn State (sample 2)	32.13	4.75	2.68	0.06	0.018	0.057	0.004	Rem
Powder	32.08	4.95	2.70	0.04	0.013	0.054	0.004	Rem

REFERENCES

1. S. C. Wheelwright and K. B. Clark, *Revolutionizing Product Development: Quantum Leaps in Speed, Efficiency, and Quality*, Free Press, New York, 1992, p. 364.
2. J. Newhouse, *The Sporty Game*, Knopf, New York, 1982, p. 242.
3. G. Winek and V. Sriraman, *J. Eng. Tech.*, **12**(2), 37–43 (1995).
4. X. Yan and P. Gu, *Computer Aided Design*, **28**(4), 307–318 (1996).
5. C. W. Hull, U.S. Patent 4,575,330, UVP, Inc., March 11, 1986.
6. C. W. Hull, S. T. Spence, D. J. Albert, D. R. Smalley, R. A. Harlow, P. Steinbaugh, H. Tarnoff, H. D. Nguyen, C. W. Lewis, T. J. Vorgitch, and D. Remba, U.S. Patent 5,059,359, 3D Systems, Inc., October 22, 1991.
7. M. Feygin, U.S. Patent 5,354,414, October 10, 1994.
8. C. R. Deckard, U.S. Patent 4,863,538, Board of Regents, University of Texas System, Austin, TX, September 5, 1988.
9. C. O. Brown, E. M. Breinan, and B. H. Kear, U.S. Patent 4,323,756, United Technologies Corporation, April 6, 1982.
10. F. G. Arcella and G. G. Lessmann, U.S. Patent 4,818,562, Westinghouse Electric Corporation, April 4, 1989.
11. V. Pratt, W. D. Scheidt and E. Whitney, U.S. Patent 5,038,014, General Electric Company, August 6, 1991.

12. G. K. Lewis and P. Lyons, *Mater. Technol.*, **10**(3–4), 51–54 (1995).
13. J. O. Milewski, D. J. Thoma, J. C. Fonesca, and G. K. Lewis, *Mater. Manu. Proc.*, **13**(5), 719–730 (1998).
14. D. L. Bourell, H. L. Marcus, J. W. Barlow, and J. J. Beaman, *Int. J. Powder Metall.*, **28**(4), 369–381 (1992).
15. B. Badrinarayan and J. W. Barlow, "Metal Parts from Selective Laser Sintering of Metal-Polymer Powders," in *Solid Freeform Fabrication Symposium 1992*, University of Texas at Austin, Austin, Tx, 1992.
16. D. E. Burrell, D. L. Bourell, and H. L. Marcus, "Solid Freeform Fabrication of Powders Using Laser Processing," in *Advances in Powder Metallurgy and Particulate Materials*, Metal Powder Industries Federation, Washington, DC, 1996.
17. D. L. Bourell and W. L. Weiss, *Metall. Trans. A*, **24A**(3), 757–759 (1993).
18. U. Lakshminarayan and K. P. McAlea, "Advances in Manufacturing Metal Objects by Selective Laser Sintering (SLS)," in *Advances in Powder Metallurgy and Particulate Materials*, Metal Powder Industries Federation, Washington DC, 1996.
19. S. Das, J. J. Beaman, D. L. Bourell, and M. Wohlert, "Direct Selective Laser Sintering of High Performance Metals for Containerless HIP," in *Advances in Powder Metallurgy and Particulate Materials*, Metal Powder Industries Federation, Chicago, 1997.
20. S. Das, N. Harlan, G. Lee, J. J. Beaman, D. L. Bourell, J. W. Barlow, T. Fuesting, L. Brown, and K. Sarget, *Mater. Manu. Proc.* **13**(2), 1–16 (1998).
21. E. B. Cooper, Jr., E. J. Whitney, and T. E. Mantkowski, U.S. Patent 5,134,032, General Electric Company, July 28, 1992.
22. M. A. McLean, G. J. Shannon, and W. M. Steen, "Laser Direct Casting High Nickel Alloy Components," in *International Conference on Powder Metallurgy and Particulate Materials*, Metal Powder Industries Federation, Princeton, NJ. 1997.
23. J. Mazumder, J. Choi, K. Nagarathnam, J. Koch, and D. Hetzner, *JOM*, **49**(5), 55–60 (1997).
24. D. M. Keicher, J. A. Romero, C. L. Atwood, M. L. Griffith, F. P. Jeantette, F. P. Harwell, D. L. Greene, and J. E. Smugeresky, "Free Form Fabrication Using the Laser Engineered Net Shaping LENS^{TM} Process," in *World Congress on Powder Metallurgy and Particulate Materials*, Metal Powder Industries Federation, Washington, DC, 1996.
25. J. Smugeresky and D. M. Keicher, *JOM*, **49**(5), 51–54 (1997).
26. J. T. Schriempf, E. J. Whitney, P. A. Blomquist, and F. G. Arcella, "Some Properties of Laser Clad and Laser Formed Materials," in *Advances in Powder Metallurgy and Particulate Materials*, Metal Powder Industries Federation, Princeton, NJ, 1997.
27. F. G. Arcella, E. J. Whitney, M. A. House, P. H. Cohen, and H. B. Bomberger, "Materials Characterization of Laser Cast^{TM} Titanium," in *Advances in Powder Metallurgy and Particulate Materials*, Metal Powder Industries Federation, Washington, DC, 1996.
28. D. H. Abbott and F. G. Arcella, *Adv. Mater. Proc. (USA)*, **153**(5), 29–30 (1998).
29. G. Lewis and E. Schlienger, "Practical Considerations and Capabilities for Laser Assisted Direct Metal Deposition," in *International Conference on Research and Development in Net Shape Manufacturing*, Birmingham, UK, 1999.

30. A. Partridge and M. R. Winstone, "Gamma–TiAl Alloys: Current Status and Future Potential," in *Advances in Turbine Materials, Design and Manufacturing, 4–6 Nov. 1997*, Newcastle upon Tyne, UK, Institute of Materials, London, 1997.
31. H. Clemens, P. Schretter, W. Glatz, C. F. Yolton, P. E. Jones, and D. Eylon, "Sheet Rolling of Ti-48Al-2Cr-2Nb Prealloyed Powder Compacts," in *Gamma Titanium Aluminides; Proceedings of the TMS Symposium*, 1995, Las Vegas, NV, Minerals, Metals & Materials Society, Warrendale, PA, 1995.
32. J. H. Moll, E. Whitney, C. F. Yolton, and U. Habel, "Laser Forming of Gamma Titanium Aluminide," in *Second International Symposium on Gamma Titanium Aluminides*, 1999. San Diego, TMS, Warrendale, PA, 1999.
33. P. W. Lee, Y. Trudel, R. Iacocca, R. M. German, B. L. Ferguson, W. B. Eisen, K. Moyer, D. Madan, and H. Sanderow, in S. R. Lampman (Ed.), *Powder Metal Technologies and Applications, ASM Handbook*, Vol. 7, ASM International, Materials Park, OH, 1998.
34. M. F. Rothman (Ed.), *High Temperature Property Data: Ferrous Alloys*, ASM International, Materials Park, OH, 1988.
35. H. E. Boyer and T. L. Gall (Eds.), *Metals Handbook(r), Desk Edition*, ASM International, Metals Park, OH, 1985.
36. S. Das, M. Wohlert, J. J. Beaman, and D. L. Bourell, *JOM*, **50**(12), 17–20 (1998).

INDEX

Handbook of Advanced Materials Edited by James K. Wessel
ISBN 0-471-45475-3